Power Converters

Power Converters

Editors

Diego Bellan
Jelena Loncarski

MDPI

Basel • Beijing • Wuhan • Barcelona • Belgrade • Novi Sad • Cluj • Manchester

Editors
Diego Bellan
Politecnico di Milano
Milan
Italy

Jelena Loncarski
University of Bologna
Bologna
Italy

Editorial Office
MDPI AG
Grosspeteranlage 5
4052 Basel, Switzerland

This is a reprint of articles from the Topic published online in the open access journals *Energies* (ISSN 1996-1073), *Electronics* (ISSN 2079-9292), *World Electric Vehicle Journal* (ISSN 2032-6653), and *Electricity* (ISSN 2673-4826) (available at: https://www.mdpi.com/topics/power).

For citation purposes, cite each article independently as indicated on the article page online and as indicated below:

Lastname, A.A.; Lastname, B.B. Article Title. *Journal Name* **Year**, *Volume Number*, Page Range.

ISBN 978-3-7258-1611-8 (Hbk)
ISBN 978-3-7258-1612-5 (PDF)
doi.org/10.3390/books978-3-7258-1612-5

Contents

About the Editors

Diego Bellan

Diego Bellan received his M.Sc. and the Ph.D. degrees in Electrical Engineering from the Politecnico di Milano, Milan, Italy, in 1994 and 1999, respectively. He is currently teaching Advanced Circuit Theory with the Department of Electronics, Information and Bioengineering, Politecnico di Milano.

His research interests are in the fields of power quality, power system analysis, electromagnetic compatibility (EMC), and statistical techniques. He was at part of the staff of several funded research projects, concerning EMC in the space sector (the European Space Agency) and in the railway sector (Trenitalia S.p.A.), and projects supported by the Italian Ministry of University and Research. He was also the principal investigator of several funded research activities concerning the electromagnetic modelling of electromechanical devices.

He is a TC member of several international conferences, and serves as an Associate Editor for the journal *IEEE Access*.

Jelena Loncarski

Jelena Loncarski (M'15) received her B.Sc. and M.Sc. degrees in Electrical Engineering from the Faculty of Electrical Engineering at the University of Belgrade, Serbia, in 2007 and 2010, respectively. She earned her Ph.D. in Electrical Engineering from the Department of Electrical Engineering at the University of Bologna, Italy, in 2014.

Dr. Loncarski has worked as a researcher in the Department of Engineering Sciences in the Division of Electricity at Uppsala University, Sweden, and in the Department of Electrical and Information Engineering at the Polytechnic University of Bari, Italy. She was an Assistant Professor of Power Electronics and Drives at the Department of Electrical, Electronic, and Information Engineering "Guglielmo Marconi," at the University of Bologna (until Jan. 2024) and is a core faculty member of Bologna Business School. She is currently constituting a startup oriented in the sustainability evaluation of companies working in various sectors. She is the author of more than 40 papers and the Associate Editor of the *IET PEL* journal, Topic Editor for MDPI's *Electronics*, and Guest Editor for various special sessions in both MDPI's *Energies* and *Electronics* and *IET PEL* journals.

Her research interests include power electronic circuits and converters for electric drives and renewable energy sources, Wide Band Gap (WBG) devices, electric vehicle charging systems, and sustainability.

electronics

rticle

Dual Voltage Forward Topology for High Efficiency at Universal Mains

Noam Ezra [1], Toine Werner [2] and Teng Long [1,*]

1 Department of Engineering, University of Cambridge, Cambridge CB3 0FA, UK; ne285@cam.ac.uk
2 Power Integrations, Cambridge CB4 1YG, UK; toine.werner@gmail.com
* Correspondence: tl322@cam.ac.uk

Abstract: This paper introduces a forward converter aimed at the universal mains voltages, i.e., 220–230 Vac and 115 Vac, named the 'dual voltage forward converter'. The suggested converter has a narrow dynamic range at the universal mains voltages, which results in lower stress on devices, optimal duty cycles, and better overall efficiency. The topology comprises two primary power loops reconfigurable by additional two-state switches and a passive diode, which allows the converter to run in parallel or in series modes and increase the performance over the full universal mains range of 90–265 Vac. The utilization of the devices is better, as they experience lower voltage and current stress by supporting two optimized working points. A converter operating at 100 kHz with an output power of 75 W and output voltage of 12 Vdc was designed and tested. The results were compared with a conventional forward converter with the identical specification. The results at the low mains were similar between the converters; however, at the high mains, the efficiency improvement was between 5% and 23%.

Keywords: dual voltage; forward converter; wide input range

Citation: Ezra, N.; Werner, T.; Long, T. Dual Voltage Forward Topology for High Efficiency at Universal Mains. *Electronics* **2022**, *11*, 1009. https://doi.org/10.3390/electronics11071009

Academic Editor: Diego Bellan

Received: 15 February 2022
Accepted: 20 March 2022
Published: 24 March 2022

1. Introduction

Electronic apparatus converting the mains Alternating Current (AC) voltage into a low Direct Current (DC) voltage usually use Switched Mode Power Supply (SMPS) for that purpose. Thus, SMPS can achieve high gain non-inverting output with low voltage stress. In some applications, such as renewable energy and fuel cells, it is possible to use non-isolated topology, such as Single Ended Primary Inductor Converter SEPIC [1]. However, other applications, such as chargers, severs and other systems with human interfaces, need isolation [2]. Flyback and forward topologies are popular due to their simplicity and are common in isolated DC–DC converters for the low power Power Supply Unit (PSU), less than 1 kW. The forward converter high-frequency transformer is not storing energy compared to the Flyback converter, which makes the forward converter more attractive for applications at high output current [2]. However, the main limitation of the forward converter is the limited input voltage range for optimal operation [3–5].

Figure 1 presents a conventional forward converter that supports the universal mains. The rectified input DC voltage from the mains AC voltage is crucial for converter design as it dictates the dynamic range and the voltage/current stress on devices. Below 75 W output power, there is no requirement for power factor correction and a simple passive diode bridge rectifier is usually used to reduce converter cost and complexity [3,4,6,7]. The mains voltage determined the rectified DC voltage, which cannot be controlled without an additional circuit. The world consists of two different ranges of mains voltages, 115 Vac Low Line (LL) used in the USA and Canada, and 220–230 Vac High Line (HL) used in Europe and China [6,7]. To support customer demands and simplify logistics, commercial PSUs supports the full range of mains voltages from 90 Vac to 265 Vac. The support of such a wide range compromises the performance and cost of the converter [3–7]. A universal mains forward converter need devices that withstand both high voltage stress when use at HL and high current stress when used at LL. That results in challenges in determinin suitable devices without compromise of performance and cost.

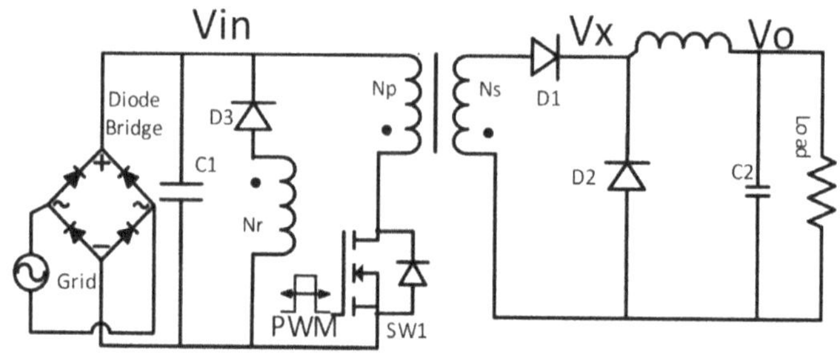

Figure 1. Conventional forward converter.

The requirement for better performance, especially efficiency, is growing due to stan dards and policies, such as Energy Star and 80 plus, which require a minimum efficiency of 86–92% [8]. Furthermore, enhanced efficiency supports a decrease in cooling arrangement which can reduce the volume and price of the PSU. There are many techniques to increase the efficiency of the conventional forward converter. For high output currents, synchronous rectification is a beneficial technique, which reduces the conduction losses of the output rectifier [9,10]. For high voltage stress, the active clamp [2,11], two-switch forward [4,12] or LC snubber circuit design [4,5,8] are beneficial methods to improve efficiency and, in some cases, even simplify the transformer construction. There are a few suggestions to support efficiency improvement of the forward converter at the universal mains wide range. However, most previous approaches only realize interleaving by mechanically combining several forward converters, connecting two forward converters' primary sides in series to reduce the voltage stress [3,5,13] or in parallel to reduce current stress [8,14]. The main limitations of these solutions are the wide dynamic range over the universal mains and the need for two transformers and two secondary side components. In this paper, the dual voltage forward solution is proposed to address these issues as it has a narrow dynamic range over the full universal mains range, a single transformer with single secondary-side components.

To achieve high efficiency, the converter duty cycle and the inductor ripple current ($K = \Delta I/2Io$) have a significant influence. These parameters are optimized through control and magnetic design to increase efficiency [3]. In most practical applications, the optimal performance (efficiency and size) is achieved at a duty cycle of 50% and a K of 0.1–0.2 [15].

The maximum duty cycle (Dmax) is set by the turns ratio between the primary and reset windings. However, those parameters also influence the voltage stress on the primary Metal Oxide Semiconductor Field Effect Transistor (MOSFET)—Vdsmax. Accordingly, setting low Dmax (Nr > Np) would result in low voltage stress on the primary switch, but a compressed current pulse with higher conduction losses [16]. Nevertheless, setting high Dmax would create the opposite aspects. Another benefit to designing the duty cycle to be about 50% is a better balance between the two rectifiers' (D1 and D2 in Figure 1) conduction losses due to better current stress division [12,15].

K factor is defined as the ripple on the output inductor divided by the load DC current. K linked to the duty cycle and a few other parameters, such as the transformer windings, input and output voltages, the switching frequency, and the inductor value. Optimizing K requires a balance between all the above parameters. Furthermore, a trade-off between the conduction losses of the forward devices (root mean square current) and the inductor size is inevitable.

This paper proposes a new forward converter topology named the Dual Voltage Forward (DVFW). This new topology endeavors to increase efficiency by operating a narrow dynamic range in both the HL and LL voltages, which enables optimizing parameters like duty cycle and K factor to achieve higher efficiency. Summarized comparisons of operation at the universal mains between the DVFW and the conventional forward converter are presented in Table 1. The operation principle of the DVFW converter is presented in Section 2. A comparison of simulation and hardware results is presented in Section 3, and conclusions of the article are in Section 4.

Table 1. Comparison between conventional forward converter and DVFW.

Parameter	Conventional Converter	DVFW
Duty Cycle (D)	D at 90 V, 0.25 D at 265 V	D at 90 V, 0.7 D at 265 V
Primary switch stress [1]	Vx at 90 V, 4 Vx at 265 V	Vx at 90 V, 1.45 Vx at 265 V
Secondary rectifier stress [2]	Vy at 90 V, 4 Vy at 265 V	Vy at 90 V, 1.45 Vy at 265 V

[1] Vx = 2 Vin assuming Nr = Np, at LL Vx ≈ 200 V and at HL Vx ≈ 750 V. [2] Vy = Vin*Ns/Np neglecting the diode voltage drop.

2. Proposed Topology and Operation

Figure 2 presents the proposed DVFW topology. This novel topology extends the efficiency at a broad range of input voltages to support the universal mains. The topology uses better device utilization, enabling the selection of more cost-effective devices. Note that different efficiency enhancement solutions, such as SR, snubbers, active clamp and two-switch forward are compatible with the DVFW and can further increase the overall efficiency. Another advantage of the DVFW is the ability to use an off-the-shelf PWM modulator controller, as the operation is similar to the conventional converter. Note that there is a need to control the delay of one sub-circuit for current sharing purposes, as will be explained later.

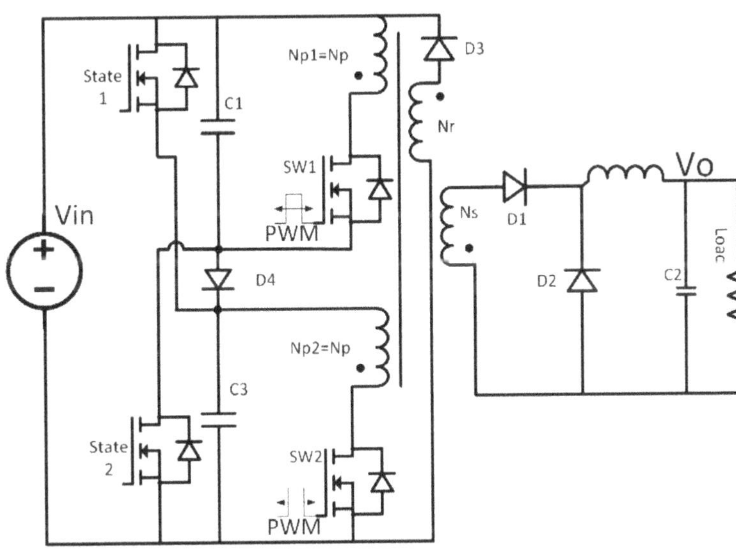

Figure 2. Proposed dual voltage forward converter (DVFW).

The DVFW structure is similar to the conventional converter built from two primary sides, which share one four winding transformer. The reset winding has a similar operation as the conventional forward converter reset winding and discharges the transformer magnetizing inductor. Similarly, the secondary winding has a similar role in both converters to transfer energy to the secondary side. The main benefit of the DVFW is the low voltage rating of the primary side MOSFETs, SW1 and SW2. At LL, they experience similar stress as the power switch of the conventional forward converter. However, at HL, they experience half the voltage stress of the conventional forward. The transformer is identical with a different pinout ($N_{p1} = N_{p2} = N_p$). The rectified input DC voltage divides between the input capacitors, C1 and C3. To select the operation mode, the state switches, State1 and State2, are employed. At the Low Line Mode (LLM), when the low voltage main is connected (90–130 Vac), the state switches are ON, forcing a parallel connection of the two sub-circuits while the input diode (D4) is in a blocking mode and experiencing LL. At the High Line Mode (HLM), when a high voltage main is connected (180–265 Vac), the state switches are OFF, allowing the two sub-circuits to be connected in series via a diode D4. It is possible to distinguish between the two modes by sensing the input rectified voltage and turning OFF the state switches below a certain threshold, for example, 160 Vac, which will be rectified into 220 Vdc. The state switches will be turned ON about one time at LL and zero at HL. Hence, switching losses can be ignored when selecting state switch MOSFETs, and only conduction resistance (Rds(on)) should be considered. Similarly, the input diode will not experience switching losses.

If the voltage is balanced between the input capacitors (C1 and C3), the state switches and the input diode (D4) experience only low voltage stress. At HLM, each state switch experiences half of the rectified DC input voltage, while at LLM, the input diode experiences the full low rectified DC input voltage. As for the currents, both the input diode and state switches conduct only the ripple current of the capacitors at HLM and LLM, respectively. The diode non-switching nature allows no switching loss, and for the same power rating, the HLM current is nearly half compared to the LLM. Similarly, at LLM, each state switch carries half the input current. Hence, selecting simple low conduction loss devices (input diode and state switches) regardless of the switching energy would keep the losses relatively low.

Figure 3 presents the operation of the conventional forward converter current loops during ON and OFF intervals. As explained in the design considerations, continuous conduction mode (CCM) is preferred; hence, idle mode is not discussed. The charge/discharge of the output capacitor and the reset winding are not presented in the figure because it has a similar operation in both converters.

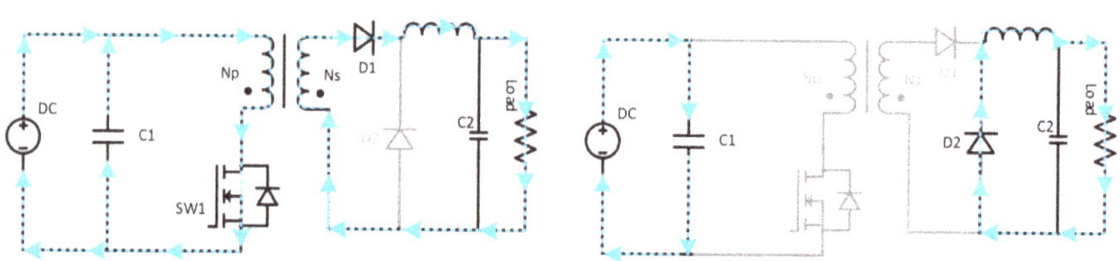

Figure 3. Conventional forward converter current loop during primary (**left**) ON time and (**right**) OFF time.

Figure 4 presents the operation of the DVFW at HLM (state switches are OFF). The primary side input DC loop is closed via D4. The input capacitors are connected in series by D4, which conducts the ripple current. The voltage stress on state switches is half of the input voltage, considering voltage balance. Figure 5 presents the operation of the DVFW at LLM (State switches are ON). The primary side input DC loop is closed via D4. The primary side input DC loop divides into two parallel loops, one via State1 and C3 and the second via State2 and C1. The voltage stress on D4 is the rectified DC voltage from the low voltage AC mains.

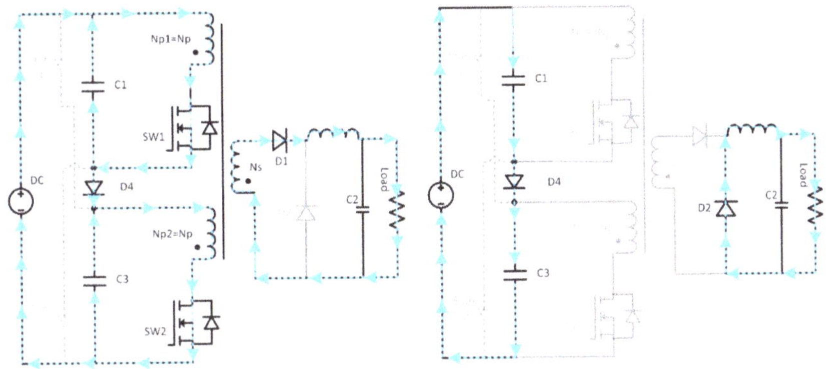

Figure 4. DVFW current loop at high line during (**left**) primary ON time and (**right**) OFF time.

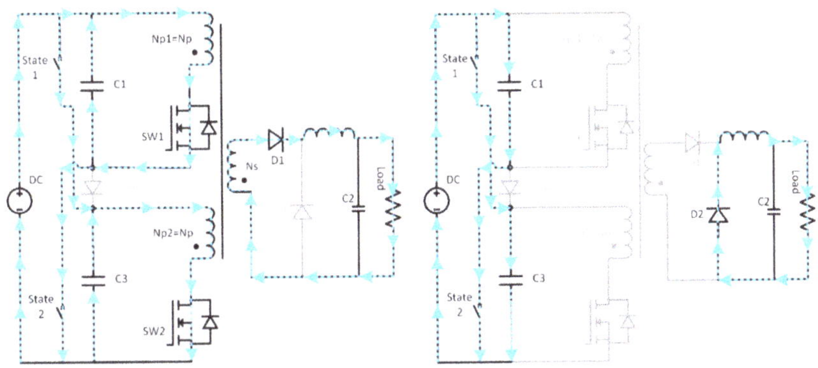

Figure 5. DVFW current loop at low line during (**left**) primary ON time and (**right**) OFF time.

The HLM series connection and the LLM parallel connection offers low-stress (voltage and current) requirements from the primary side devices. The secondary side current stress is identical to the conventional forward converter, but the voltage stress on the rectifier is lower due to the DVFW operation. Both modes resemble the conventional forward converter power loops, except that the DVFW use two primary loops instead of one.

3. Simulation and Results

3.1. Devices Stress Comparison

To compare the topologies performances, first it is necessary to make a device stress comparison, as shown in Table 2. The procedure used for the calculation is identical for the DVFW and the conventional forward converter; however, the DVFW converter experiences a voltage range of 90–132.5 Vac instead 90–265 Vac. That is because the DVFW have two forward converters that will experience 90–132.5 Vac each in both HLM and LLM. Both converters voltage and current stress have been verified by simulation and measurement, which will be presented later in this article. The result shows lower stress experienced by the DVFW devices than the conventional forward converters, which support better devices utilization. For example, in the worst case, the DVFW two primary MOSFET experience 375 V at 0.63 A, compared to the conventional forward converter single MOSFET that experiences 750 V, 1.26 A. Similarly, the DVFW rectifiers experience 67.5 V, 5.6 A and 67.5 V, 3.75 A, compared to 135 V, 5.9 A and 135 V, 3.75 A for the conventional solution. In both converters, the current stress of the D1 rectifier is higher than the D2 rectifier because the duty cycle is lower than 50%. The DRFW duty cycle values at LL is identical to the conventional forward converter. However, the DRFW duty cycle values at HL ar calculated according to the input voltage of Vin/2 because the voltage the series primarie experience is half.

Table 2. Parameters comparison between conventional forward converter and DVFW.

Parameter	Conventional Converter	DVFW
Duty Cycle	35% at 90 V [1], **9%** at 265 V [2]	35% at 90 V [1], **19%** at 265 V [2]
Primary switch stress	200 V@**1.26 A** at 90 V [1], **750 V**@0.65 A at 265 V [2]	200 V@**1.26 A/2** at 90 V [1], **375 V**@0.92 A/2 at 265 V [2]
Secondary rectifier stress sync/freewheel	36 V@3.75/5 A at 90 V [1], **135 V@2/5.9 A** at 265 V [2]	36 V@3.75/5 A at 90 V [1], **67.5 V@2.7/5.6 A** at 265 V [2]
Input capacitor	**375 V** at 265 V [2]	**188 V** at 265 V [2]
DVFW state switches [3]	NA	**0.42**–0.23 A at 90 V [1]–130 V 127–**187 V** at 180–265 V [2]
DVFW diodes [4]	NA	100–**185 V** at 90 V [1]–130 V **0.33**–0.22 A at 180–265 V [2]

[1] Assumed worst-case minimum voltage of 100 Vdc between diode bridge charge cycles (90 V*$\sqrt{2}$)-30 V. [2] Assumed worst-case maximum voltage of 375 Vdc (265 V*$\sqrt{2}$). [3] Voltage stress for state switches at HL only and current stress is LL only. [4] Voltage stress for diode at LL only and current stress is HL only.

3.2. *Price and Volume Comparison*

Although the DVFW requires five additional parts, the devices low-stress feature allows selection of lower-cost devices. Hence, the DVFW can compete with the conventional forward converter. The same product series devices were selected to make a fair comparison between the converters, as shown in Table 3. Note that by taking different product series, it is possible to get even lower-cost devices, which use other technologies.

Table 3. Price/size comparison between conventional forward converter and DVFW.

Parameter	Conventional Converter	DVFW
Primary switch	800 V@0.375 Ω STP15N80K5 $t_{d(on)}$ = 19 ns, t_r = 17.6 ns, $t_{d(off)}$ = 44 ns, t_f = 10 ns, TO220 package 1 × £1.69 [1]/0.4 cm^3	600 V@0.375 Ω STP15N60M2 $t_{d(on)}$ = 11 ns, t_r = 10 ns, $t_{d(off)}$ = 40 ns, t_f = 15 ns, TO220 package 2 X £0.83 [1]/0.4 cm^3
Secondary rectifier [2]	150 V@9.3 mΩ IRFB4115PBF $t_{d(on)}$ = 18 ns, t_r = 73 ns, $t_{d(off)}$ = 41 ns, t_f = 39 ns, TO220 package 2 × 2 × £1.23 [1]/0.4 cm^3	75@9.3 mΩ IRFB3607PBF $t_{d(on)}$ = 16 ns, t_r =110 ns, $t_{d(off)}$ = 43 ns, t_f = 96 ns, TO220 package 2 × 2 × £0.32 [1]/0.4 cm^3
Input capacitor	400 V@150 uF EKXG401ELL101MMN 1 × £1.63 [1]/40 cm^3	200 V@150 uF EKXG201ELL151ML25S 2 × £0.86 [1]/20 cm^3
Primary driver	2 A low side gate driver FAN3100 1 × £0.32 [1]/0.008 cm^3	2 A low side gate driver FAN3100 1 × £0.32 [1]/0.008 cm^3 and 4 A high side gate driver FAN37371 1 × £0.48 [1]/0.04 cm^3
DVFW current sense	NA	Optocoupler 140817142100 2 × £0.14 [1]/0.12 cm^3
DVFW state switches	NA	250 V@0.43Ω IPD5N25S3430 2 × £0.29 [1]/0.09 cm^3
DVFW diode	NA	200 V@1 A STPS4S200S 1 × £0.16 [1]/0.12 cm^3
Total	**£8.56/42 cm^3**	**£6.48/43 cm^3**

[1] All costs are in https://www.digikey.co.uk (accessed on 29 July 2021) for 1000 units. [2] Used the same MOSFET type for sync and freewheeling rectifier, used two parallel rectifiers due to high current stress.

The DVFW have additional parts, such as a high-side driver that drives the high-side primary MOSFET, an optocoupler that communicates the average current measurement for the current sharing control loop, state switches, and one input diode for the dual-voltage feature. However, the DVFW low-stress benefit, as presented in the previous section, supports a significant price reduction compared to the conventional forward converter. Typically, a low-stress requirement from a device enables the selection of lower-cost smaller parts. For example, MOSFET with lower breakdown voltage can achieve lower cost and/or better performance (low Rds(on) than a higher breakdown MOSFET; as presented in [17], 400 V MOSFET with lower Rds(on) than 800 V MOSFET is about half the price.

However, due to low-voltage stress, their size decreases dramatically [17], to the point in which the two low voltage input capacitors of the DVFW have a similar size to a single high voltage capacitor of the conventional forward converter. Note that the DVFW additional state switches and diode have very low stress. Therefore, the DVFW capacitors overall addition to the cost and size is small.

It is possible to reduce the DVFW volume and price further by better device selection and integration. In both converters, TO220 packages are used for the primary and secondary MOSFETs. However, due to better efficiency (lower losses at the devices) and spread thermal stress (two primary MOSFET instead of one), the DVFW converter could use smaller packages and less cooling assembly, which supports the further reduction of volume and price. The last opportunity for size and cost optimization is an integrated circuit solution.

The DVFW and the conventional forward converter can use the same transformer with the same amount of copper, i.e., the same length and type of winding wires because the magnetic stress is identical. The only difference is the pinout to accommodate the requirement of the DVFW and conventional forward converter. If the primary winding has an even number of windings connected in parallel, the conventional forward converter will have all these windings connected to the single primary MOSFET, while the DVFW divided them into the two sub-forwards. Therefore, neither the volume nor price of the transformer changed among DVFW and the conventional forward.

3.3. DVFW Simulation

The simulation verifies the DVFW features, such as optimal duty cycle, and better devices utilization at both LL and HL. The DVFW has been simulated using LTSpice, along with the conventional forward converter for comparison purposes. The simulation used the devices shown in Table 3 devised by STMicroelectronics, Infineon or LTspice library to achieve an accurate model. The waveforms of the simulation are presented in Figures 13 and 14. The results show that the DVFW can achieve an optimal duty cycle at both the HL and LL, while the conventional forward converter achieves it only at LL and a lower duty cycle at HL as expected.

As for the device stress, the simulation results verify the calculation shown in Table 2. For example, for the DVFW primary MOSFETs, the stress is 375 V at the worst case, while the single conventional forward converter MOSFET experiences 750 V (at HL) and 375 V (at LL). Similarly, the conventional forward converter output rectifiers experience about 136 V (at HL) compared to only 70 V (at HL and LL) in the DVFW topology. The DVFW's additional parts (state switches and input diode) low stress has been verified.

3.4. Voltage Balancing and Current Sharing

One of the benefits of the DVFW topology is the low-stress (current and voltage) devices. However, to achieve it, there is a prerequisite to accomplishing balanced voltage between the series-connected sub-forward circuits (MOSFETs and input capacitors) for the HLM and accomplishing balance current between the parallel-connected sub-forward for the LLM.

In LLM, the voltage is balanced because the state switches force parallel connections. In HLM, poor voltage balance would create uneven voltage stress on the devices. Therefore, one of the devices would experience higher voltage stress. The DVFW converter has been simulated with various parameter deviation between the two sub-forward circuits to discover the impact of device and control parameter tolerance on voltage imbalance. These parameters include gate drivers delay time, transformer primary inductances Lp, input capacitances Cin and MOSFETs resistance Rds(on). Figure 6 shows the simulation results for the voltage misbalance. The results show that the most dominant parameter in the voltage misbalance is the primary inductance, and for every 1% deviation of the primary inductance away from its nominal value $(L_{p1}\text{-}L_{p2})/L_{p\text{-}nom}$, a 0.25% imbalance of capacitor voltage occurs $(V_{cap1}\text{-}V_{cap2})/V_{nom}$. For example, at 10% inductance deviation (instead 2 mH for both primary inductances, one inductor of 2.1 mH and the other 1.9 mH), the capacitor imbalance would be 5% (instead 375 V divided into two, 196.875 V and 178.125 V obtained in each sub-forward).

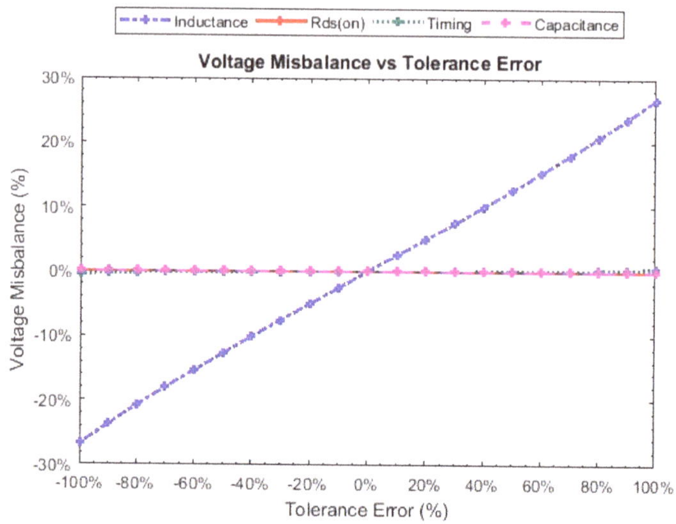

Figure 6. DVFW voltage balance vs. primary inductance tolerances at high line.

To explain the mechanism of the voltage imbalance caused by inductance deviation, two inductances, L_{p1} and L_{p2}, are assumed for Sub-Forward1 and Sub-Forward2, respectively. When both switches are ON, the reflected voltage that Sub-Forward1 induces on Sub-Forward2 is (1).

$$V_{2-ref} = V_{LP1}\frac{N_{P2}}{N_{P1}} = V_{LP1}\sqrt{\frac{L_{P2}}{L_{P1}}} \tag{1}$$

Assuming X% deviation between the inductances (L_{p1} is 0.5X% lower than nominal, while L_{p2} is 0.5X% higher) yields (2).

$$V_{LP1} = \frac{V_{2-ref}}{\sqrt{\frac{L_{nom}(1+0.5X\%)}{L_{nom}(1-0.5X\%)}}} \tag{2}$$

Similarly, analysing the other sub-forward circuit reflected voltage yields (3).

$$V_{LP2} = \frac{V_{1-ref}}{\sqrt{\frac{L_{nom}(1+0.5X\%)}{L_{nom}(1-0.5X\%)}}} \tag{3}$$

When the switches are ON, the voltage drop on the MOSFETs is negligible. Therefore, the capacitor voltage is approximately the reflected voltage. In consequence, neglecting the voltage drop on the DVFW diode D4, the reflected voltage sum is the input voltage as shown in (4).

$$V_{in} \approx V_{cap1} + V_{cap2} \approx V_{LP1} + V_{LP2} \quad (4)$$

Combining (2), (3) and (4) and assuming X = 10% yields (5) and (6).

$$V_{cap1} \approx V_{LP1} \approx Vin/2.05 \quad (5)$$

$$V_{cap2} \approx V_{LP2} \approx V_{in}/1.95 \quad (6)$$

Combining (5) and (6) yields (7).

$$V_{blance} = \frac{V_{cap1} - V_{cap2}}{V_{nom}} \approx \frac{\frac{V_{in}}{2.05} - \frac{V_{in}}{1.95}}{\frac{V_{in}}{2}} = 5\% \quad (7)$$

The calculation shows similar results to the simulation, with 5% misbalance when 10% inductance deviation. Therefore, other parameter differences between the sub-forward circuits, such as capacitance, resistance or gate drive delay time, would not significantly affect voltage balancing. The inductance mechanism forces voltage balance through the means of reflected voltage. The sub-forward circuit would act according to the reflected voltage (positive or negative current via the MOSFET) until the voltage is balanced. In real applications with the same number of turns, the difference between the primary inductances is insignificant. Therefore, the voltage imbalance should be negligible, as shown in Figure 14 of the experiment results.

As for current sharing, the critical parameter is the gate driver delay, which depends on the driver IC and the MOSFET source-to-gate capacitance. A simple but effective gate drive circuit for the DVFW can be applied to ensure good current sharing between MOSFETs. This method compares the average current in each sub-circuit and adjusts the delay accordingly, as shown in Figure 7. The additional parts in blue and pink are needed to support the control loop. The delay operation versus the differential amplifier results is explained in the timing diagram. The high side current sense and a fixed delay and the low side current sense (not necessarily isolated) together with the delay set the balancing mechanism. When the error is negative (higher current on the high side), the adjustable gate drive will reduce the low side gate delay and increase the duty cycle of the low side, resulting in a higher current in the low side to have better current sharing. A similar principle applies to the positive error (lower current on the high side). Because the system needs to support sharing in the steady-state, a relatively simple slow control loop is sufficient (much slower than the system control loop). A simple RC filter and an optocoupler or a current sensor followed by an RC filter is fast enough to measure the average current of the MOSFET.

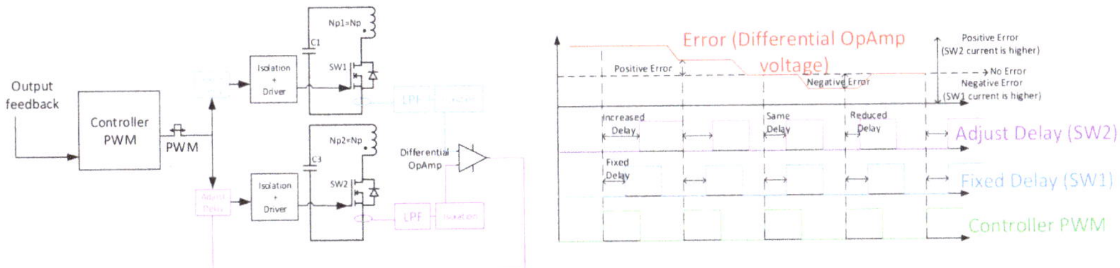

Figure 7. DVFW current sharing control loop schematic (**left**) and timing diagram (**right**).

3.5. Hardware Setup

The Evaluation Board (EVB) of the DVFW converter and the conventional forward converter has been planned and manufactured for comparison reasons. The models used the devices selected in the price and volume comparison section and presented in Table 3.

The DVFW two low-voltage MOSFETs and two low-voltage capacitors have a comparab volume to the conventional forward converter single high-voltage MOSFET and singl high-voltage capacitor, as shown in Figure 8. Furthermore, the DVFW three additiona devices (two state MOSFETs and input diode) rated at 200 V to 250 V were assembled on separate Printed Circuit Board (PCB).

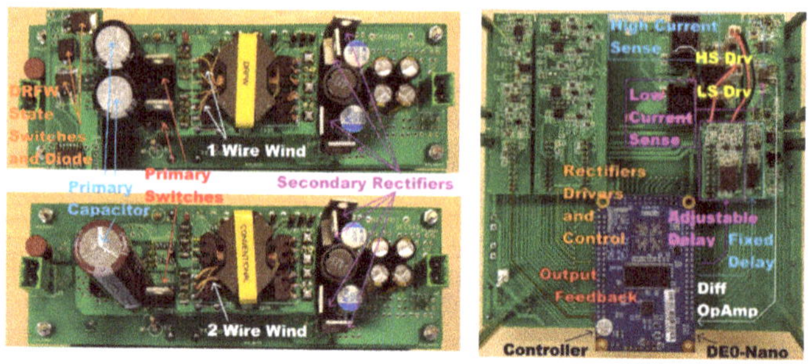

Figure 8. Power board of DVFW (**top left**), conventional forward (**bottom left**) and the curren sharing control (**right**).

The transformer is a dominant factor of the converter size and cost. The DVFW trans former might need to support more pins, but its overall stress is similar to the conventional transformer converter. Hence, it can have an identical volume and price. Both converters use the same magnetic cores (RM10) and the same amount of copper in the transformers The only difference is that the conventional forward converter transformer has two primary windings connected to pins 4 and 5, while the DVFW has the windings connected to pins 1 and 2 and pins 4 and 5, as shown in Figure 8. The figure also shows the hardware of the controller for the current sharing control. The current miss-sharing is measured and filtered with an RC circuit, then compared by using differential amplifiers. The comparison result is then sampled by the controller. Using this information, the controller sets the gate driver delay using the delay chips on the circuit.

3.6. Experimental Results

This section is divided into three sections: efficiency comparison, current sharing control, and operation comparison.

Efficiency results: The EVB efficiency measurement setup used AC PSU to supply the unit under test (conventional or DVFW converter), the load, and the power meter to measure the unit efficiency as shown in Figure 9. The measurement procedure was setting the target AC voltage and load, simultaneously measuring the input and output energy over a minute, followed by the efficiency calculation of the converter.

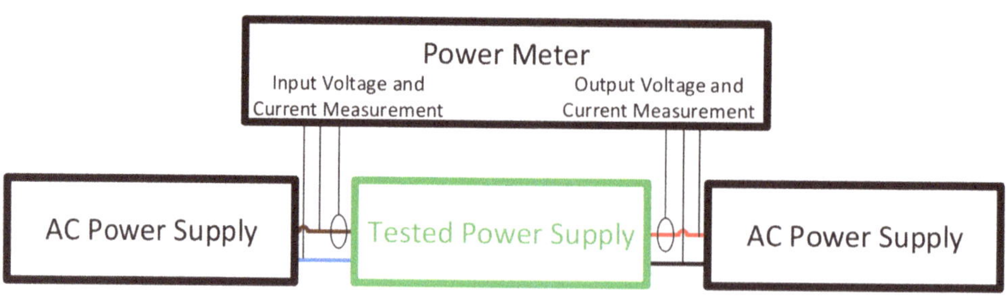

Figure 9. The efficiency measurement setup.

The EVB was measured for efficiency at loads from 20% to 100% at the universal mains, minimum voltage (90 Vac), nominal voltages (115 Vac and 230 Vac) and maximum voltage

(265 Vac), as shown in Figure 10. The DVFW achieves similar efficiency at HL and LL. That is due to a similar operating point which consists of parameters such as devices stress, duty cycle and K factor at both mains. Furthermore, at 230 Vac and 115 Vac, when the voltages are exactly double, the efficiency is almost identical because the operating point is alike, as expected. Due to a similar operating point at LL, the conventional forward converter efficiency is similar to the DVFW. However, at HL, the conventional forward converter efficiency is considerably lower at any given load because of the non-optimized nature of the conventional forward converter for universal mains, which cause low duty cycle and high voltage stress as explained above. The higher the voltage, the larger the difference between the converters, with a 5–17% difference at 230 V to 10–23% at 265 V. The results show that the DVFW overall performance at the universal mains is better than the conventional forward converter. The efficiency deviation between the converters demonstrates the DVFW efficiency gain over the full range of the universal mains. The graph takes the DVFW efficiency and substitutes the conventional forward converter efficiency. Therefore, in the sections that the graph is positive, the DVFW converter has better efficiency and vice-versa. Almost all the graph is positive. At HL, the graph is entirely positive and shows a significant efficiency benefit of the DVFW over the conventional forward converter due to the lower stress on devices and optimal duty cycle. At LL, on average, the efficiency of DVFW is only slightly higher than the conventional forward. The LL efficiency results are similar as both converters experience similar stress. However, the main benefit of the DVFW at the HL is due to its low voltage stress on the primary devices, which experience the same stress as DVFW LL operation.

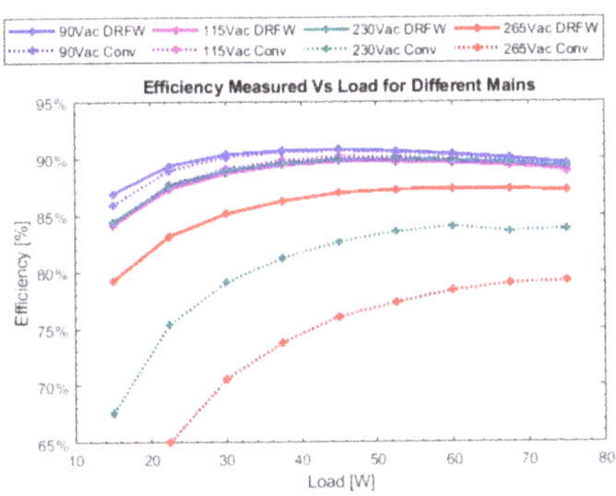

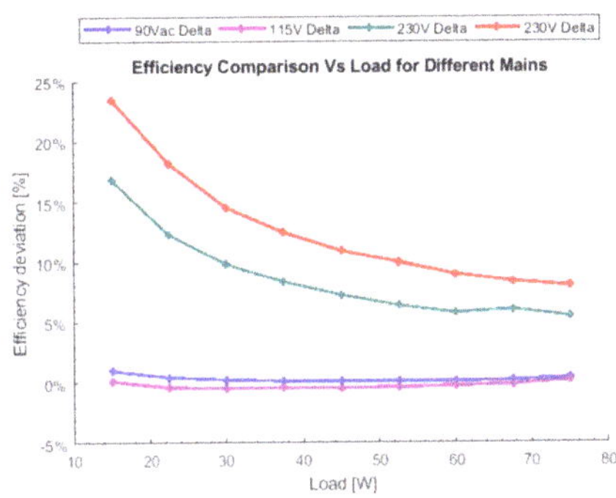

Figure 10. DVFW and conventional converter efficiency comparison (**left**) and efficiency deviation (**right**).

Current Sharing Control: To verify the miss-sharing control, the current waveforms were measured. The first step was to verify the current loop operation with two different MOSFETs (STP15N80K5 and STP15N60M2). That emulates extreme tolerance cases, far beyond the tolerances of using the same part numbers MOSFETs, which have similar gate capacitance. The following waveforms were captured in each sub-circuit: the gate signals (green and yellow), the current of the primary winding (purple and orange), the output inductor current (red) and the controller delay in hexadecimal value (blue). The current control loop managed to force these two different MOSFETs to share well, as shown in Figure 11, which shows the gate signals (yellow and green), the MOSFET currents (orange and blue) and the inductor current (red). To further test the control loop, an extreme condition was created by connecting a capacitor of about ten times larger than the gate capacitance in parallel with one of the MOSFETs between the gate and the source. That would change the MOSFET with the capacitor to have a much larger switching delay than the other MOSFET. The captured waveforms are similar to the previous experiment except; the differential amplifier, which measures the current difference, is now presented in pink, and the current of the primary winding is now red and orange (instead of purple and orange). The results show that the control changed the delay timing to address this extreme miss-sharing condition, and the current is now shared, as shown in Figure 11. The connection of the extra gate capacitance was manual with a through whole capacitor for a short interval to check the control dynamics. At the connection moment, the system had a high miss-sharing current and large current spikes, as shown in the red and orange waveform in the zoomed window 1, and the error (pink) was negative. The controller then adjusted the delay, from 72 to 3C Hex, until the error was small and the currents were balanced with much lower current spikes, as shown in the red and orange waveforms in the zoomed window 2. This extreme test indicates that the current sharing control loop can balance a current even when a large gate drive delay occurs between two sub-circuits.

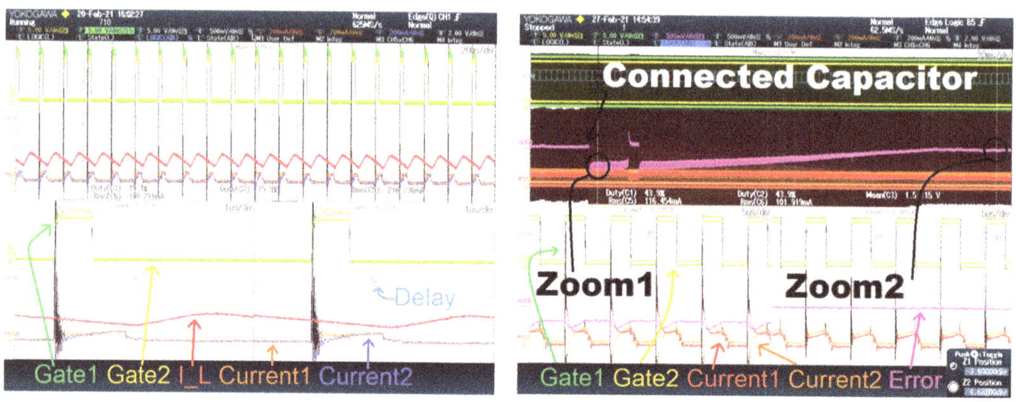

Figure 11. DVFW current sharing with different primary MOSFET (**left**), and responding to extreme change in one of the sub-forwards gate capacitance (**right**).

The last experiment illustrates how accurate the closed-loop control is by setting an open-loop with a ±62 ns fixed error. The same waveforms as the previous experiment were captured. Figure 12 shows positive and negative 62 ns timing error. In the positive delay test, the orange current had a smaller delay. Hence, it turned ON earlier and the error was positive. In the negative delay test, the orange current had a larger delay. Hence, it turned ON later and the error was negative. The currents have very large spikes due to the non-sharing effects of the DVFW and the voltage balancing mechanism. This experiment illustrates how sensitive the system is to a small timing error between the gates signals.

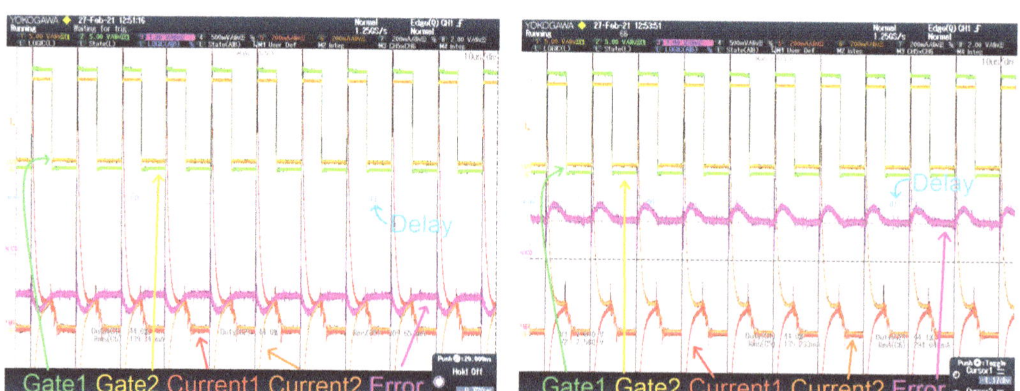

Figure 12. Fixed 62 ns timing error between the two sub-forwards: negative error (**left**) and positive error (**right**).

Steady-State results: The steady-state waveforms were measured on the EVB at the same working points as the efficiency measurements; hence, loads from 20% to 100% and mains from 90 Vac to 265 Vac, as presented in Figures 13 and 14. The waveforms measured on the EVB are presented alongside the waveform simulated in LTSpice. The waveforms presented are the full load results at different mains for both topologies. Table 4 contains information about the waveforms, such as the color and description. For simplicity, the waveform color of the simulation and the measurement is identical. The results show that the simulation model is accurate as the measurement results are comparable. For example, the stress on the MOSFETs, rectifiers and capacitors are about the same at all mains. When the operating point is LL, both topology performances are similar, as expected, due to similar stress on devices and working points. However, when the operating point is HL, there is a significant performance difference, as the DVFW devices experience similar stress as the LL operation, and the duty cycle is optimized, as expected. The measurements verified the DVFW balanced capacitor voltage, and the narrow dynamic range as the optimal duty cycle is kept at a wide range of input voltages.

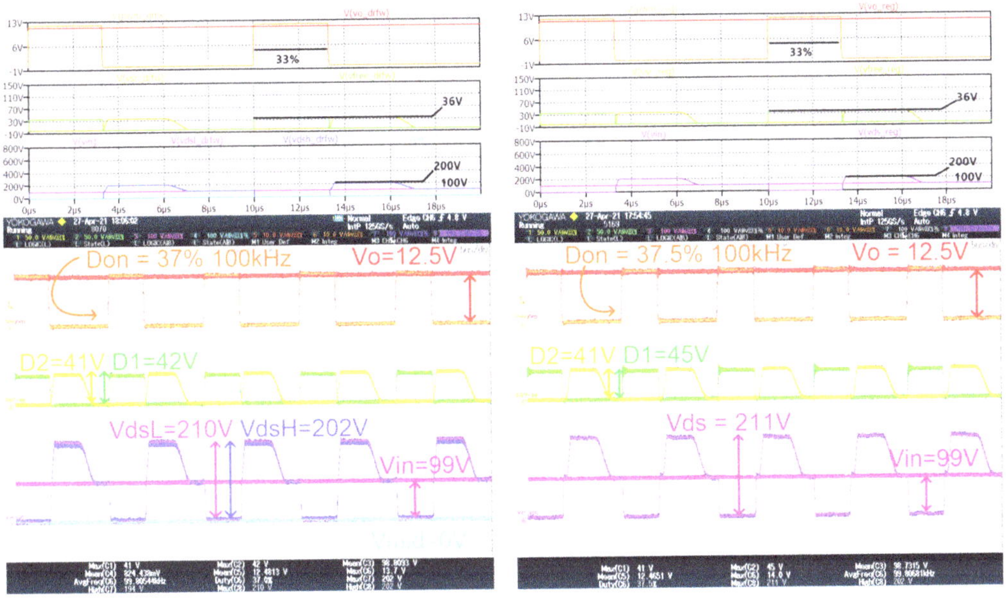

Figure 13. Low line minimum voltage (90 V) DVFW simulation (**top left**) and measurement (**bottom left**) vs. conventional forward converter simulation (**top right**) and measurement (**bottom right**).

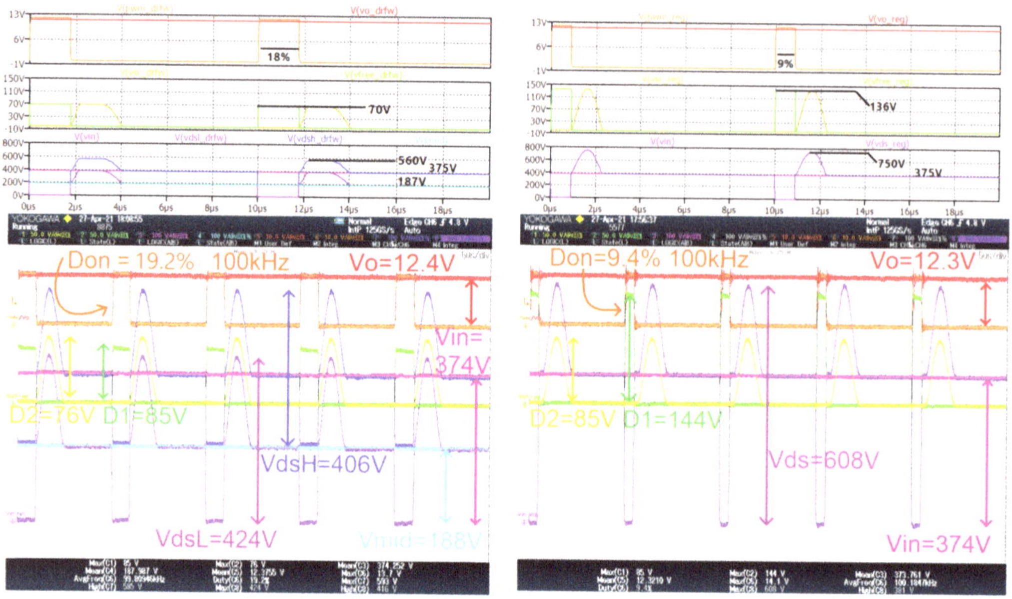

Figure 14. High line maximum voltage (265 V) DVFW simulation (**top left**) and measurement (**bottom left**) vs. conventional forward converter simulation (**top right**) and measurement (**bottom right**).

Table 4. Figures 13 and 14 waveforms description.

Signal	Scope Channel	Color	Description
SR Rectifier Voltage	1	Yellow	D1 [1]
Freewheeling Rectifier Voltage	2	Green	D2 [1]
Total Input Voltage	3	Pink	Vin [1]
Voltage on Low Sub-Forward Converter	4	Cyan	Voltage between C1 [1], SW1 [1] and D4 [1]
Output Voltage	5	Red	Vout [1]
Duty Cycle	6	Orange	Don
High Side MOSFET	7	Blue	SW1 [1]
Low Side MOSFET	8	Purple	SW2 [1]

[1] Symbols are indicated in Figures 1 and 2.

4. Conclusions

This paper introduces a new topology built on the conventional forward converter, the DVFW converter. The proposed topology main advantages are better device utilization and narrow dynamic range over the universal mains, which increase the overall efficiency at that range. Both converters have similar performance at the low lines (90 Vac and 115 Vac), and the DVFW performed considerably better at the high lines (230 Vac and 265 Vac), as it achieved better efficiency compared to the conventional forward converter. The DVFW converter can be optimized for both high and low mains, unlike the conventional forward converter that needs to select one of them. The proposed topology enables a narrower dynamic range optimal duty cycle for two different line voltages, such as 115 Vac and 230 Vac, while working with a fixed frequency and using low voltage devices. Although the DVFW part count is larger, its total dimension is comparable to the conventional forward converter, and its total price is lower due to lower rating devices that can be used.

The prototype experimental results verify the calculation and simulation. The comparison between the topologies shows expected benefits of the DVFW, such as lower losses and less stress on the parts than the conventional forward converter. The DVFW could substitute the conventional forward for universal mains, i.e., both the 90–130 Vac and 180–265 Vac. Better device selection and an integrated solution can further optimize the DVFW price and size. The integrated solution can also address the current sharing control loop challenge.

Author Contributions: Conceptualization, N.E., T.W., and T.L.; methodology, N.E.; software, N.E.; validation, N.E. and T.L.; formal analysis, N.E.; investigation, N.E.; resources, N.E. and T.L.; data curation, N.E.; writing—original draft preparation, N.E.; writing—review and editing, T.L.; visualization, N.E.; supervision, T.L.; project administration, N.E. All authors have read and agreed to the published version of the manuscript.

Funding: This research received no external funding.

Conflicts of Interest: The authors declare no conflict of interest.

Abbreviations

SMPS	Switched-Mode Power Supply
PSU	Power Supply Unit
LL and HL	Low Line (115 Vac) and High Line (220–230 Vac)
DRFW	Dual Voltage Forward
SR	Synchronous Rectifier
LLM and HLM	Low Line Mode and High Line Mode
EVB	Evaluation Board

References

1. Ansari, S.A.; Moghani, J.S. A Novel High Voltage Gain Noncoupled Inductor SEPIC Converter. *IEEE Trans. Ind. Electron.* **2019**, *66*, 7099–7108. [CrossRef]
2. Janga, R.; Malaji, S. Performance evaluation of active clamp forward converter with fuzzy logic controller. In Proceedings of the International Conference on Intelligent Computing and Control (I2C2), Coimbatore, India, 23–24 June 2017.
3. Wu, H.; Xu, P.; Liu, W.; Xing, Y. Series-Input Interleaved Forward Converter with a Shared Switching Leg for Wide Input Voltage Range Applications. *IEEE Trans. Ind. Electron.* **2013**, *60*, 5029–5039. [CrossRef]
4. Jang, P.; Cho, B.H. Two-Switch Forward Converter With Reset Winding and an Auxiliary Active-Clamp Circuit for a Wide Input Voltage Range. *IEEE Trans. Power Electron.* **2016**, *32*, 4491–4502. [CrossRef]
5. Wu, H.; Xing, Y. Families of Forward Converters Suitable for Wide Input Voltage Range Applications. *IEEE Trans. Power Electron.* **2014**, *29*, 6006–6017. [CrossRef]
6. Kim, C.E.; Baek, J.I.; Lee, J.B. High-Efficiency Single-Stage LLC Resonant Converter for Wide-Input-Voltage Range. *IEEE Trans. Power Electron.* **2017**, *33*, 7832–7840. [CrossRef]
7. Ezra, N.; Long, T. Dual Range Forward Topology for High Efficiency at Universal Mains. In Proceedings of the 2021 IEEE Energy Conversion Congress and Exposition (ECCE), Vancouver, BC, Canada, 10–14 October 2021.
8. Ke, C.M.; Liang, T.J.; Tseng, W.J.; Jiang, G.L. Design and Implementation of an Interleaved Forward Converter with Magnetizing Energy Recycled. In Proceedings of the IEEE Energy Conversion Congress and Exposition (ECCE), Baltimore, MD, USA, 29 September–3 October 2019.
9. Nagesh, L.; Venkatesha, L.; Nesh, J. Design and Implementation Of a Separate Control Driven High Current Synchronous Rectified Forward Converter with Non-Dissipative Snubber Reset. In Proceedings of the Global Conference for Advancement in Technology, Bangalore, India, 18–20 October 2019.
10. Patil, S.; Kumar Singh, B. Digital Controller for Active Clamp based Synchronous Forward Converter for High Power Applications. In Proceedings of the 6th International Conference for Convergence in Technology (I2CT), Maharashtra, India, 2–4 April 2021.
11. Ji, B.; Wang, J.; Zhao, J. Novel high step-up gain converters without forward and reverse recovery current issue of rectifying diodes. In Proceedings of the IEEE International Symposium on Industrial Electronics, Hangzhou, China, 28–31 May 2012.
12. Martis, J.; Vorel, P.; Knobloch, J. Two-switch forward converter efficiency optimization. In Proceedings of the IEEE International Conference on Environment and Electrical Engineering and Industrial and Commercial Power Systems Europe (EEEIC/I&CPS Europe), Milan, Italy, 6–9 June 2017.
13. Kroics, K.; Zakis, J.; Suzdalenko, A.; Gaigals, G. A simplified approach to input voltage balancing for series connected isolated DC-DC converters. In Proceedings of the 18th European Conference on Power Electronics and Applications, Karlsruhe, Germany, 5–9 September 2016.

14. Habumugisha, D.; Chowdhury, S.; Chowdhury, S. A DC-DC interleaved forward converter to step-up DC voltage for DC Micro grid applications. In Proceedings of the IEEE Power & Energy Society General Meeting, Vancouver, BC, Canada, 21–25 July 2013.
15. Anders Lind, Infineon Technologies Austria. *Design Note AN 2013-03*. 2013. Available online: www.mouser.com/pdfdocs/2-10.pdf (accessed on 21 March 2022).
16. Ezra, N.; Long, T. Dual Voltage Flyback Topology Operation With Efficiency Enhancers at Dual Voltage Mains. In Proceedings of the IEEE Energy Conversion Congress and Exposition (ECCE), Detroit, MI, USA, 11–15 October 2020.
17. Ezra, N.; Long, T. Dual range flyback topology for high efficiency at dual voltage mains. *IET Power Electron.* **2020**, *13*, 1565–1574. [CrossRef]

energies

MDPI

eview

A Review on Integrated Battery Chargers for Electric Vehicles

iago J. C. Sousa, Delfim Pedrosa, Vitor Monteiro * and Joao L. Afonso

Department of Industrial Electronics, School of Engineering, University of Minho, 4800-058 Guimaraes, Portugal; tsousa@dei.uminho.pt (T.J.C.S.); dpedrosa@dei.uminho.pt (D.P.); jla@dei.uminho.pt (J.L.A.)
* Correspondence: vmonteiro@dei.uminho.pt; Tel.: +351-253-510-190

Abstract: Electric vehicles (EVs) contain two main power electronics systems, namely, the traction system and the battery charging system, which are not used simultaneously since traction occurs when the EV is travelling and battery charging when the EV is parked. By taking advantage of this interchangeability, a single set of power converters that can perform the functions of both traction and battery charging can be assembled, classified in the literature as integrated battery chargers (IBCs). Several IBC topologies have been proposed in the literature, and the aim of this paper is to present a literature review of IBCs for EVs. In order to better organize the information presented in this paper, the analyzed topologies are divided into classical IBCs, IBCs for switched reluctance machines (SRMs), IBCs with galvanic isolation, IBCs based on multiple traction converters and IBCs based on multiphase machines. A comparison between all these IBCs is subsequently presented, based on both requirements and possible functionalities.

Keywords: integrated battery chargers; traction systems; battery charging systems; electric vehicles

Citation: Sousa, T.J.C.; Pedrosa, D.; Monteiro, V.; Afonso, J.L. A Review on Integrated Battery Chargers for Electric Vehicles. *Energies* **2022**, *15*, 2756. https://doi.org/10.3390/en15082756

Academic Editors: Diego Bellan and Jelena Loncarski

Received: 15 March 2022
Accepted: 7 April 2022
Published: 8 April 2022

1. Introduction

Electric vehicles (EVs) appear to be one of the most promising agents in the change towards a sustainable planet, where the aim is to decrease the levels of environmental pollution and the dependency on fossil fuels. Besides being an emissions-free system at the utilization level, EVs can also play a relevant role in terms of power grid support, since battery charging systems can allow bidirectional operation. Thus, from the power grid point of view, EVs are expected to be part of the solution rather than part of the problem, provided power grid congestion due to simultaneous charging of multiple EVs is avoided [1–3].

In terms of its constitution at the power electronics level, an EV contains a traction system, responsible for controlling one or more electric machines that guarantee the EV's propulsion, and a battery charging system, responsible for charging the battery from an external source, e.g., the power grid [4]. Regarding the traction system, in order to energize the electric machine windings from a battery power source, as well as to perform the opposite process during a braking operation, a dc–ac converter is needed [5]. In addition, in order to adapt the dc voltage of the dc–ac converter, as well as to regulate the energy generated by the electric machine and returned to the battery during a braking operation, a bidirectional dc–dc converter is used [6]. Regarding the battery charging system, since power is supplied by an ac power grid (either single-phase or three-phase), an ac–dc converter is needed. Moreover, in the traction system, a dc–dc converter is needed for controlling the battery current and voltage during the charging operation, though this converter can be the same as the one previously referred to. This constitution can be seen in Figure 1a, representing the conventional power electronics structure of an EV. Regarding conductive charging, considering that the EV must be stationary in order to charge its battery, it can clearly be seen that the traction and battery charging systems are never used simultaneously. Furthermore, it can also be seen that the front-end converters dedicated to each system are very similar, interfacing an ac side (electric machine or power grid) to a common dc link. Taking these two factors into consideration, integrated battery charger (IBCs) arose, which use the same power converter for traction and for battery chargin This approach can be seen in Figure 1b,c and is advantageous in the sense that it combine two functionalities in a single system, with the prospect of reducing the volume, weigh and cost of the power electronics comprising the EV [7].

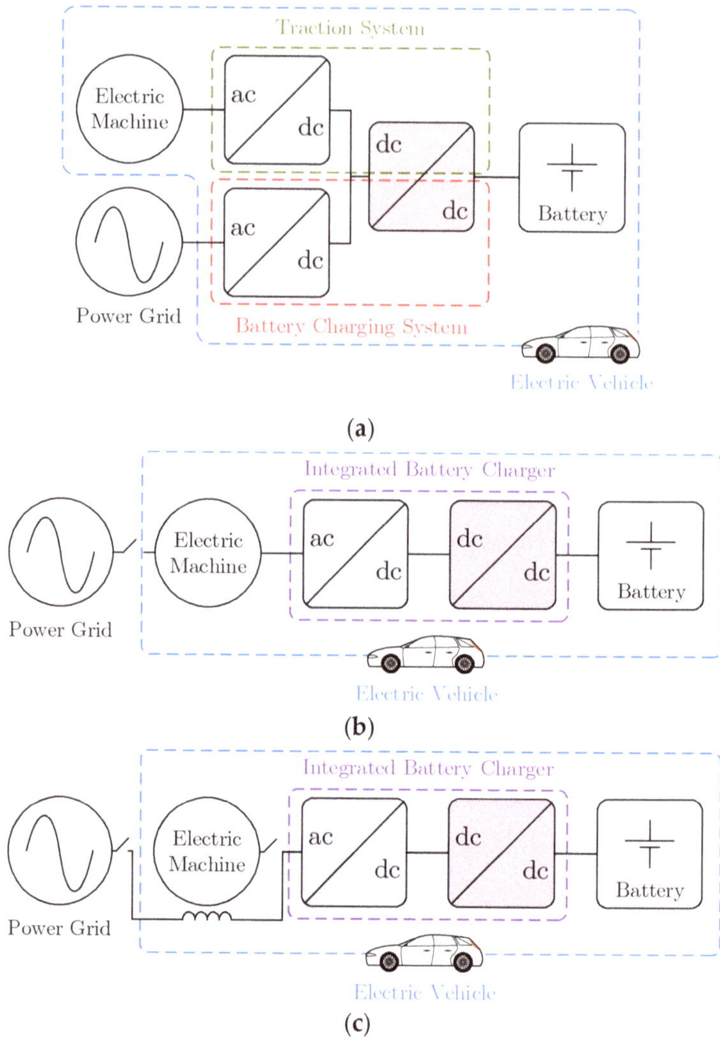

Figure 1. EV power electronics structure: (**a**) conventional; (**b**) integrated using the machine windings as coupling filters with the power grid; (**c**) integrated using external inductors as coupling filters with the power grid.

It can be seen that two types of IBCs are represented in the figure (Figure 1b,c). In Figure 1b, the electric machine windings are used as coupling filters with the power grid, while in Figure 1c external inductors are used. There are several ways of achieving an IBC, as can be seen by the different topologies published in the literature and patented [8]. Moreover, IBCs can present different specifications such as galvanic isolation, bidirectional operation with the power grid and the various possibilities for power grid interfacing (e.g., single-phase ac, three-phase ac or dc). However, they can also add complexity to the general system as additional inductors may be required to interface the converter with the power grid and/or contactors, to reconfigure the power converter. In [9] an analysis of IBCs is presented in terms of functionalities and implementation costs compared to dedicated systems that allow the same operation modes, with bidirectional operation also considered. Additionally, some proposed IBCs require access to the neutral point of the electric machine, which is not commonly necessary in conventional systems, and some IBCs even require access to all winding terminals in order to operate. Furthermore, there are IBCs that can be used with any electric machine type and others that are specifically designed for a given electric machine or for a number of given electric machines. In this context, this paper presents a review of the IBCs proposed in the literature and patented, as well as a comparison between them, both in terms of implementation requirements (e.g., the need for external inductors or contactors) and functionalities (e.g., galvanic isolation, bidirectional operation).

The rest of the paper is organized as follows. Section 2 presents the relevant IBCs that have been proposed in the literature or patented. Section 3 presents a comparison between all the IBCs presented in Section 2. Lastly, Section 4 presents the conclusions of this review paper.

2. Integrated Battery Chargers

This section presents relevant IBCs that have been proposed in the literature. In order to better organize the presented topologies, this section is divided into five subsections, grouping the IBCs into: (1) classical topologies; (2) IBCs especially developed for switched reluctance machines (SRMs); (3) IBCs with galvanic isolation; (4) IBCs based on multiple traction converters; (5) IBCs based on multiphase machines. It should be noted that the section on classical IBCs contains some topologies that are more recent than those of some of the IBCs presented in other sections. This is because these topologies do not match any of the other criteria, i.e., for use with SRMs, with galvanic isolation or based on multiple traction converters or multiphase machines.

2.1. Classical Integrated Battery Chargers

The first publication regarding EV IBCs dates back to 1983, a time when EVs were far from having the popularity they have nowadays. This publication consists of a United States Department of Energy report produced by NASA [10], and it was followed by a publication by the same author, D. Thimmesch, two years later [11]. The approach presented in both publications can be seen in Figure 2 and consists of a thyristor-based resonant three-phase dc–ac converter, sized for a mechanical peak power of 34 kW in the traction mode and a continuous power of 3.6 kW in the battery charging mode, which also uses a transformer comprising four windings (T_{1a}, T_{1b}, T_{2a}, T_{2b}). In traction mode, the system operates as a single converter, i.e., a resonant three-phase dc–ac converter (thyristors S_1–S_6), where the electric machine is of the three-phase type, while in battery charging mode the system operates with two separate converters (one ac–dc and the other dc–dc) connected to a single-phase ac power grid. The thyristor antiparallel diodes of the dc–ac converter (D_1–D_6) are employed as a diode full-bridge single-phase ac–dc converter, with the power grid connected in series in one of the converter phases (terminals a, n) and the stator windings of the electric machine (L_a, L_b, L_c) operating as coupling inductors with the power grid. The dc–dc converter, composed of thyristors S_7 and S_8 and diodes D_7 and D_8, operates as a resonant converter, similarly to the three-phase dc–ac converter used in traction mode. It should be noted that the battery charging operation is performed in an isolated way, using four contactors (K_1–K_4) that remain closed in traction mode and are opened in battery charging mode, isolating the batteries from the power grid.

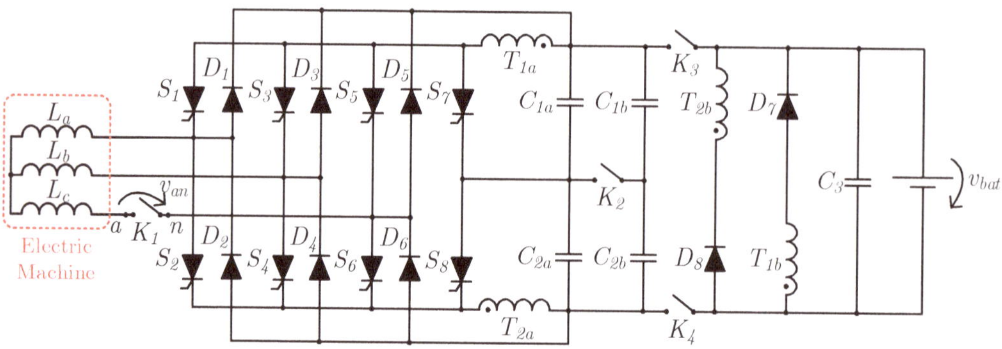

Figure 2. EV IBC proposed by D. Thimmesch in 1983 [10].

Several years later, in the early 1990s (1990 [12], 1992 [13] and 1994 [14]), three IBCs for EVs were patented by W. Ripple and A. Cocconi, as shown in Figure 3. The IBC patented in 1990 [12], shown in Figure 3a, consists of a bipolar-junction-transistor-based three-phase three-leg bidirectional ac–dc converter (S_1–S_6) and a diode full-bridge single-phase ac–dc converter (D_1–D_4). In traction mode, only the three-phase bidirectional ac–dc converter is used, and the electric machine is of the three-phase type. In battery charging mode, the diode full-bridge single-phase ac–dc converter is employed to interface the system with a single-phase ac power grid (terminals a, n), while the three-phase bidirectional ac–dc converter operates with only one leg, acting as a bidirectional buck–boost dc–dc converter. In battery charging mode, an additional inductor (L_1) is used to reduce the battery current ripple, although the authors note that it is possible to use the electric machine windings, with the trade-off of a higher current ripple. The IBC patented in 1992 [13] concerns the application of two induction machines or, alternatively, one electric machine with two sets of windings, composed of two three-phase three-leg bidirectional ac–dc converters, as shown in Figure 3b. In traction mode, each of these converters controls one electric machine independently. In battery charging mode, the two terminals of a single-phase ac power grid are connected to the neutral point of each of the electric machine stator windings, such that the three-phase bidirectional converters operate in a three-phase interleaved configuration of a full-bridge single-phase bidirectional ac–dc converter. In contrast to the solution patented in 1990, this IBC does not use external inductors. The IBC patented in 1994 [14] (Figure 3c) consists of an adaptation of the previously patented solutions with the purpose of operating either in single-phase or three-phase ac power grids, using eleven contactors (K_1–K_{11}). Contactors K_1–K_3 are responsible for connecting the system to a three-phase ac power grid (terminals a, b, c), whereas K_4 and K_5 have the role of connecting the electric machine stator windings in a star configuration. Contactors K_6–K_{11} are used to interface the system with a single-phase ac power grid (terminals a', n). In this case, K_6 and K_7 remain open and K_8–K_{11} are closed. As in the system patented in 1990 [12], an external inductor is used (L_1). Unlike the topology patented in 1992 [13], which only needs access to the neutral points of the electric machine stator windings, this topology needs access to all six terminals in order to perform the connection to a three-phase ac power grid, with the stator windings behaving as coupling inductors. The three IBCs patented by W. Rippel and A. Cocconi have a common characteristic, namely, the requirement for the power grid peak voltage to be lower than the EV battery voltage, as is the case for any voltage source ac–dc converter with a single dc link, i.e., without being split.

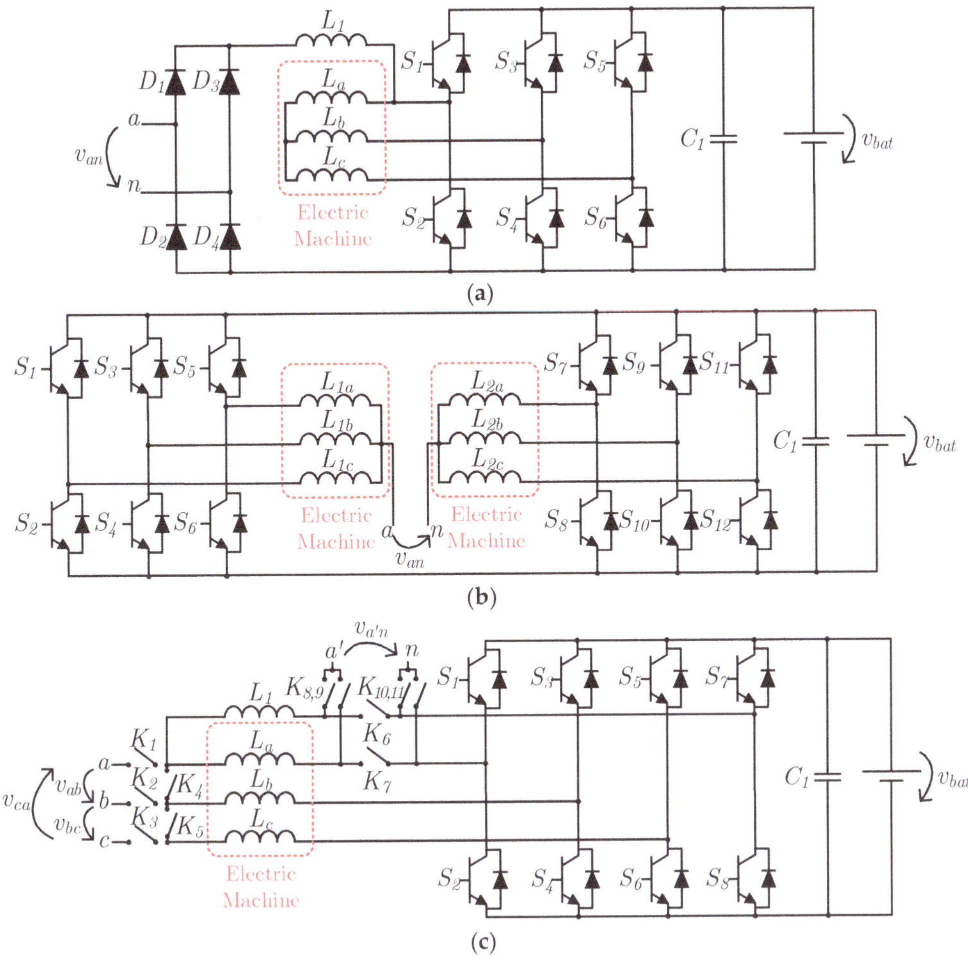

Figure 3. EV IBCs proposed by: (**a**) W. Rippel in 1990 [12]; (**b**) W. Rippel and A. Cocconi in 1992 [13]; (**c**) A. Cocconi in 1994 [14].

In 2001, L. Solero proposed an IBC for an electric scooter [15]. This system can be seen in Figure 4a. It is composed of a three-phase three-leg bidirectional ac–dc converter (insulated gate bipolar transistors (IGBTs) S_1–S_6) and a diode full-bridge single-phase ac–dc converter (diodes D_1–D_4). In traction mode, only the three-phase bidirectional ac–dc converter is used, controlling the electric machine, which is of the three-phase type. During battery charging mode, the diode full-bridge single-phase ac–dc converter is also used to interface the system with a single-phase ac power grid (terminals a, n), with the three-phase bidirectional ac–dc converter operating as a three-phase interleaved bidirectional buck–boost dc–dc converter. In this system, the stator windings behave as dc–dc converter inductors, and the neutral point is connected to the positive output terminal of the diode full-bridge single-phase ac–dc converter. In this case, the power grid peak voltage cannot be higher than the battery voltage due to the boost operation of the dc–dc converter from the power grid to the battery. In 2010, G. Pellegrino et al. proposed a similar system with an additional bidirectional dc–dc converter between the battery and the dc link of the three-phase bidirectional ac–dc converter, endowed with power factor correction (PFC) characteristics and allowing the battery charging operation from power grids with peak values that can be higher or lower than the battery voltage [16]. This topology can be seen in Figure 4b. It should be noted that the author did not specify the dc–dc converter topology.

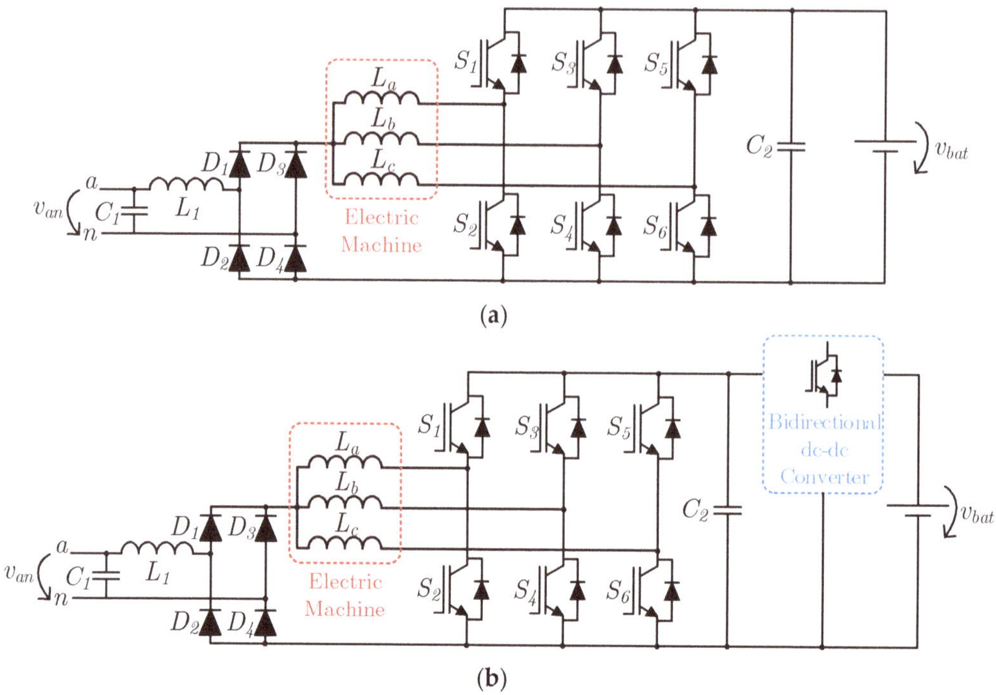

Figure 4. EV IBCs proposed by: (**a**) L. Solero in 2001 [15]; (**b**) G. Pellegrino et al. in 2010 [16].

In 2013, O. Hegazy et al. proposed the IBC presented in Figure 5, which is based on an eight-switch three-phase ac–dc converter (S_1–S_8) and a two-phase interleaved buck–boost dc–dc converter (S_9–S_{12}), both bidirectional [17]. The ac–dc converter is composed of three legs, one of them with two switches (S_7 and S_8) and the other two with three switches (S_1–S_6). This approach allows a three-phase three-leg bidirectional ac–dc converter (S_2, S_3, S_5, S_6, S_7, S_8) to be simultaneously established with a full-bridge single-phase bidirectional ac–dc converter (S_1, S_2, S_4, S_5 or S_1, S_3, S_4, S_6), connected with a three-phase electric machine and a single-phase ac power grid (terminals a, n), respectively. Since the electric machine and the power grid are connected to different terminals of the ac–dc converter, no system reconfiguration is necessary in order to change the operation mode between machine drive and battery charging, and hence no relays or contactors are used. The only additional element is an inductor (L_1), serving the purpose of interfacing the converter with the power grid. In order to guarantee the proper operation of the converter, IGBTs S_1 and S_4 are kept permanently closed in traction mode so that current flow in the three-phase three-leg bidirectional ac–dc converter (S_2, S_3, S_5, S_6, S_7, S_8) is allowed, while in battery charging mode IGBTs S_2 and S_5 (or S_3 and S_6) are kept permanently closed in order to allow current flow in the full-bridge single-phase bidirectional ac–dc converter (S_1, S_2, S_4, S_5 or S_1, S_3, S_4, S_6), while the third leg (S_7 and S_8) is kept permanently open.

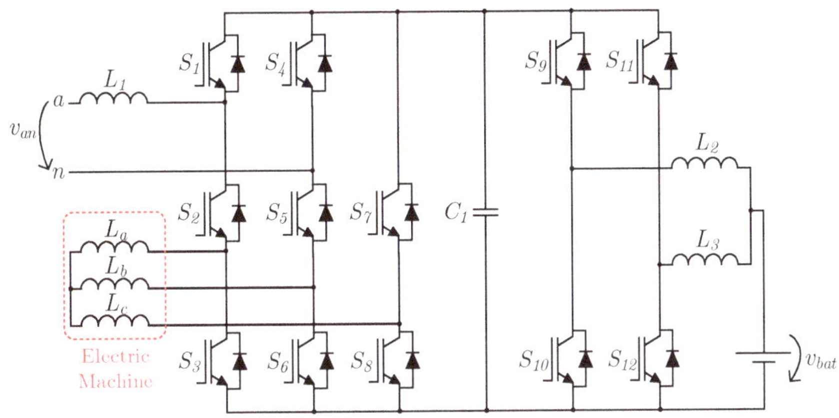

Figure 5. EV IBC proposed by O. Hegazy et al. in 2013 [17].

2.2. Integrated Battery Chargers for Switched Reluctance Machines

In 2000, C. Pollock and W. K. Thong proposed an IBC based on a two-phase SRM [18], as shown in Figure 6. As well as the converter that is typically used for traction systems in this type of machine (asymmetrical dc–ac converter), in this case of the two-phase type (comprising metal oxide semiconductor field effect transistors (MOSFETs) S_2, S_3 and diodes D_6, D_7), it also uses an energy recovery circuit (inductor L_2, capacitor C_4, MOSFET S_4 and diode D_8), an additional inductor (L_1) coupled to one of the machine phases (in this case, L_a) with the respective diode (D_5) and IGBT (S_1) and a diode full-bridge single-phase ac–dc converter (D_1–D_4) for interfacing with a single-phase ac power grid (terminals a, n). The additional inductor L_1 is only used in battery charging mode, making the system operate as a flyback isolated dc–dc converter. In addition, capacitor C_1 and resistor R_1 form an RC snubber and are also only used in battery charging mode.

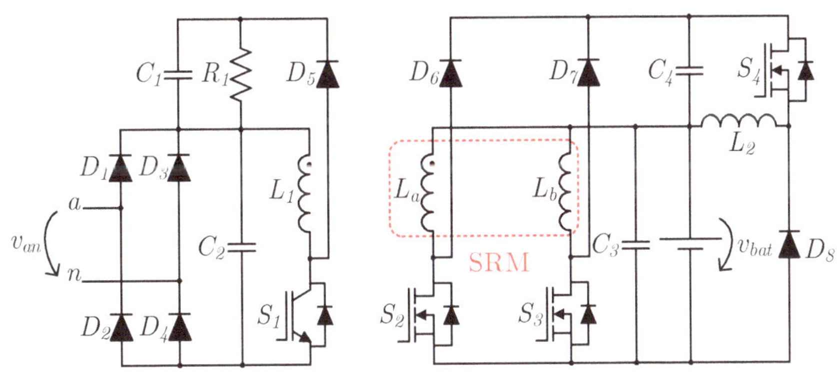

Figure 6. EV IBC based on SRMs proposed by C. Pollock and W. K. Thong in 2000 [18].

In 2009, H. C. Chang and C. M. Liaw proposed an IBC based on four-phase SRMs, as shown in Figure 7a, with the four-phase asymmetrical dc–ac converter (comprising IGBTs S_1–S_8 and diodes D_1–D_8) operating as a PFC converter in battery charging mode [19]. In this system, in traction mode, the unidirectional boost dc–dc converter formed by IGBT S_9, diode D_9 and inductor L_1 is used to step up the voltage from the battery to the dc link of the asymmetric dc–ac converter. On the other hand, in battery charging mode, the unidirectional boost dc–dc converter is not used, and two of the stator windings are used as input passive filters (in this case L_a, L_b) of a diode full-bridge single-phase ac–dc converter (D_1–D_4) used for interfacing with a single-phase ac power grid (terminals a, n). One of the remaining stator windings (in this case L_d) is used as an inductor for a unidirectional buck–boost dc–dc converter, which is composed of two IGBTs (S_7, S_8) and a diode (D_8) that belong to the traction converter. In battery charging mode, an extra diode (D_{10}) is also used to avoid short-circuiting the battery when S_8 is turned on. This diode exists on board the EV but its connection is performed through an off-board shunt; hence, it is represented with a dashed line. Moreover, two contactors are used, one with two positions (K_1) and the other with three (K_2). Contactor K_1 remains closed in traction mode and is opened in battery charging mode, connecting capacitor C_2 in parallel with the converter legs in traction mode. Contactor K_2 allows point x to be connected to one of two points, depending on the operation mode, namely, to point y in traction mode and point z in battery charging mode. In 2011, the same authors proposed modifications to this system, allowing the battery to be charged through a buck–boost or a buck configuration, with the latter being advantageous in the sense that it inserts an inductor in series with the battery, reducing its current ripple [20]. This topology is shown in Figure 7b, where it can be seen that the asymmetric dc–ac converter has a modified structure, using five IGBT legs (S_1–S_{10}) instead of four IGBT pairs and four diode pairs, reducing the semiconductor count from 16 to 10. In contrast to the system previously proposed by the authors, only one contactor (K_1) is used, which is responsible for connecting the capacitor C_2 in parallel with the converter legs in traction mode. Like diode D_{10} presented in the previous topology, diode D_1 serves the purpose of avoiding short-circuiting the battery during its charging process when IGBT S_{10} is conducting.

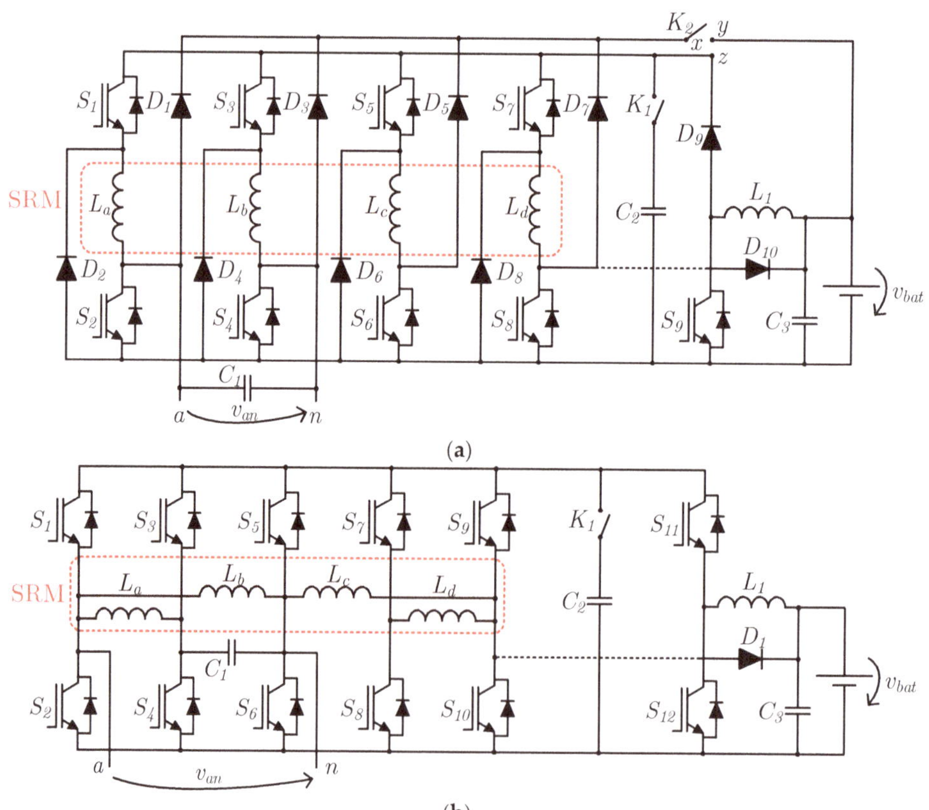

Figure 7. EV IBCs based on SRMs proposed by H. C. Chang and C. M. Liaw in: (**a**) 2009 [19]; (**b**) 2011 [20].

In 2014, Y. Hu et al. proposed an IBC for plug-in hybrid vehicles based on three-phase SRMs [21]. This topology is shown in Figure 8a. It is composed of an asymmetric dc–ac converter based on $n + 1$ switches (S_1–S_4) and $n + 1$ diodes (D_7–D_{10}), where n is the number of machine phases. This converter is also known as a Miller converter and is an alternative to the more commonly used version for SRM drives, in which $2n$ switches and $2n$ diodes are used. By adding a diode full-bridge three-phase ac–dc converter (D_1–D_6) and two contactors (K_3 and K_4), the asymmetric converter can operate as a dc–ac converter for driving the machine or as a buck dc–dc converter, where the machine stator windings behave as output inductors and the battery current is divided between the three windings. The input of the diode full-bridge three-phase ac–dc converter (terminals a, b, c) can be used for interfacing either with a power grid (single-phase ac, three-phase ac or dc) or a generator mechanically coupled with the internal combustion engine (ICE) of the vehicle via contactors K_1 and K_2. These contactors are closed whenever the ICE-coupled generator is used, as is the case for generator-only driving and hybrid driving in traction mode. Contactor K_3 is used in traction mode, either in generator-only driving, battery-only driving or hybrid driving, and should be kept open in battery charging mode in order to establish a buck dc–dc converter. Contactor K_4 is closed whenever the battery is used, either in traction mode (in battery-only driving and hybrid driving modes) and in battery charging mode.

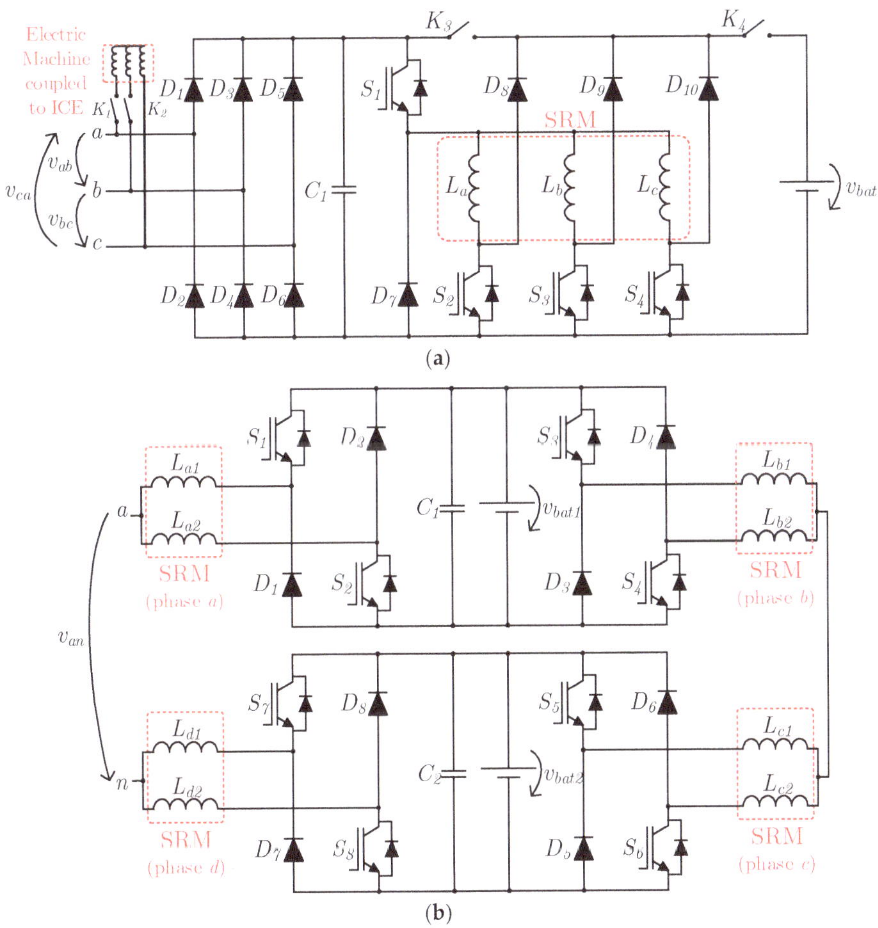

Figure 8. EV IBCs based on SRMs proposed by Y. Hu et al. in: (**a**) 2014 [21]; (**b**) 2015 [22].

In 2015, the same authors proposed the IBC presented in Figure 8b, for applicatio in EVs or plug-in hybrid vehicles, based on four-phase SRMs with two sets of statc windings [22]. Two asymmetrical converters with $2n$ switches and $2n$ diodes are employec with each one applied to two machine phases, and a battery set is connected to the dc lin of each converter. Hence, each of these converters can operate independently in tractio mode. Regarding the battery charging mode, this topology allows operation with dc an single-phase ac power grids, where both converters are connected to each other in tw of the phases (b, c) and the other two (a, d) operate as connection terminals to the powe grid (terminals a, n). In the case of a dc power grid, the converters operate as two cascad connected split-pi unidirectional buck–boost dc–dc converters, whereas in the case of single-phase ac power grid the converters operate as a five-level cascaded h-bridge ac–d converter (despite having more semiconductors than necessary). In both cases, due to th boost operation of the converter from the power grid to each set of batteries, the powe grid peak voltage should be lower than the battery voltage for proper operation in batter charging mode.

K.W. Hu et al. proposed, also in 2015, an IBC with additional operation modes, namel vehicle-to-grid (V2G) and vehicle-to-home (V2H) [23] modes. This topology can be seen ir Figure 9 and is based on a four-phase SRM and an asymmetric dc–ac converter similar to the one presented in 2011 [20] with five switch legs (S_1–S_6 and S_8–S_{11}) but also with an extra leg comprising a diode (D_1) and a switch (S_7). Moreover, a bidirectional buck–boost dc–dc converter is used (IGBTs S_{12}, S_{13} and inductor L_4), which may be reconfigured to operate in buck or boost mode in each direction, depending on the states of the three-position switches K_3 and K_4. In traction mode, these switches are configured in order to connect points x_1, x_2 to points y_1, y_2, respectively, linking the higher voltage side (leg formed by S_{12}, S_{13}) to the battery and linking the lower voltage side (inductor L_4) to the asymmetrical dc–ac converter dc link. In this regard, when the intention is to drive the machine, the bidirectional buck–boost dc–dc converter operates in buck mode during regular traction mode and in boost mode during regenerative braking, and hence the battery voltage must be higher than the dc link voltage of the asymmetrical dc–ac converter. On the other hand, in battery charging mode, the switches K_3, K_4 are configured in order to connect points x_1, x_2 to points z_1, z_2, respectively, linking the higher voltage side (leg formed by S_{12}, S_{13}) to the asymmetric dc–ac converter dc link and linking the lower voltage side (inductor L_4) to the battery. In battery charging mode, the bidirectional buck–boost dc–dc converter operates in buck mode during the conventional battery charging operation, i.e., in grid-to-vehicle (G2V) mode, and in boost mode during V2G or V2H operation modes, and therefore the dc link voltage must be higher than the battery voltage. The battery charging operation can be performed via a single-phase or a two-phase (using only one phase-to-phase voltage) ac power grid, where a single-phase three-wire bidirectional ac–dc converter is used, which is implicit in the asymmetric dc–ac converter (S_1–S_6). In order to connect the system to the power grid, two contactors (K_1, K_2), three inductors (L_1, L_2, L_3) and two capacitors (C_1, C_2) are used, where the three-wire interface (terminals a, n, b) allows two phase-to-neutral voltages (v_{an}, v_{bn}) and a phase-to-phase voltage (v_{ab}) to be obtained.

In 2018, M. Ma et al. proposed an IBC for plug-in hybrid vehicles based on SRMs, as shown in Figure 10 [24]. This topology consists of the asymmetric dc–ac converter typically used in SRMs (IGBTs S_2–S_7 and diodes D_9–D_{14}) with the addition of a boost PFC stage (diodes D_1–D_6, capacitor C_1, inductor L_1 and MOSFET S_1) to connect to a three-phase ac power grid (terminals a, b, c) or to a generator mechanically coupled to the ICE of the vehicle, using contactors K_1–K_3. Both stages are linked through a bridge formed by the diodes D_7 and D_8 and two battery sets (v_{bat1}, v_{bat2}), each with the respective capacitor connected in parallel (C_2, C_3) and a contactor connected in series (K_4, K_5), behaving as an energy recovery circuit. The contactors present in this topology have the purpose of enabling or preventing the use of the power provided by the generator (K_1–K_3) or by the batteries (K_4, K_5), making it possible to use the former, the latter or a combination of both, as required in a hybrid vehicle.

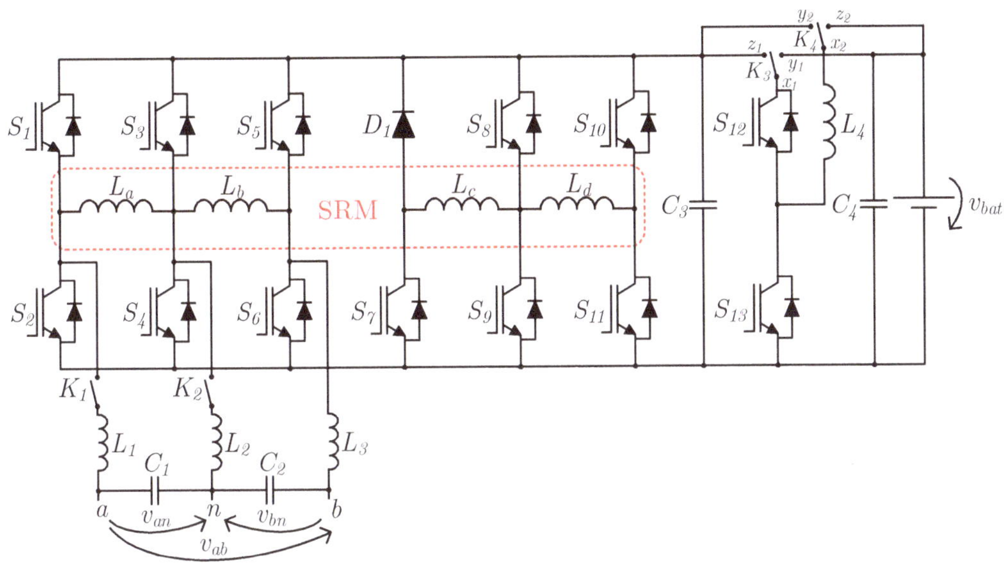

Figure 9. EV IBC based on SRMs proposed by K. W. Hu et al. in 2015 [23].

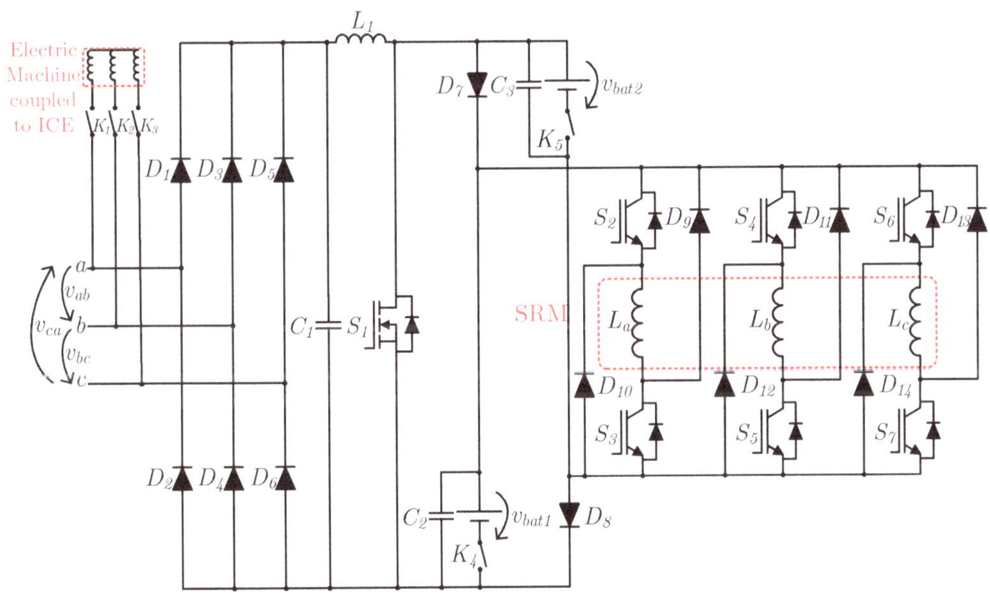

Figure 10. IBC for plug-in hybrid vehicles based on SRMs proposed by M. Ma et al. in 2018 [24].

In 2020, H. Cheng et al. proposed the IBC for plug-in hybrid vehicles based on SRMs presented in Figure 11a [25]. Like the previously presented IBC, this proposal comprises a connection to an electric machine coupled to the ICE of the vehicle or to a power grid, in this case not three-phase ac but single-phase ac. As well as allowing the connection to the traction battery, this topology also allows the connection to the auxiliary battery. The front-end converter employed in this case is a diode full-bridge ac–dc converter (diodes D_1–D_6), while the converter for driving the machine is formed by IGBTs S_1–S_4, diodes D_7–D_{10}, capacitor C_3 and inductor L_1. Like the IBC for SRMs first presented in this section, this

IBC uses an energy recovery circuit, which is based on IGBT S_4, diode D_{10}, capacitor C_3 and inductor L_1. The converter used to charge the auxiliary battery is a unidirectional isolated half-bridge dc–dc converter, which is formed by IGBTs S_5 and S_6, a high frequency transformer to step down the voltage, diodes D_{11} and D_{12}, inductor L_2 and capacitor C_4. The series connection of capacitors C_1 and C_2 establishes a midpoint available to this converter. In terms of operation modes, this IBC allows charging the traction battery from the power grid, charging the auxiliary battery from the electric machine coupled to the ICE or charging the auxiliary battery from the traction battery. In order to change the operation mode, five contactors are used (K_1–K_5), where the first three (K_1–K_3) have the purpose of disconnecting the electric machine coupled to the ICE (when the power grid is connected), K_4 is used when the vehicle is driving in hybrid mode, using the electric machine coupled to the ICE and K_5 is used when the traction battery is used in traction mode.

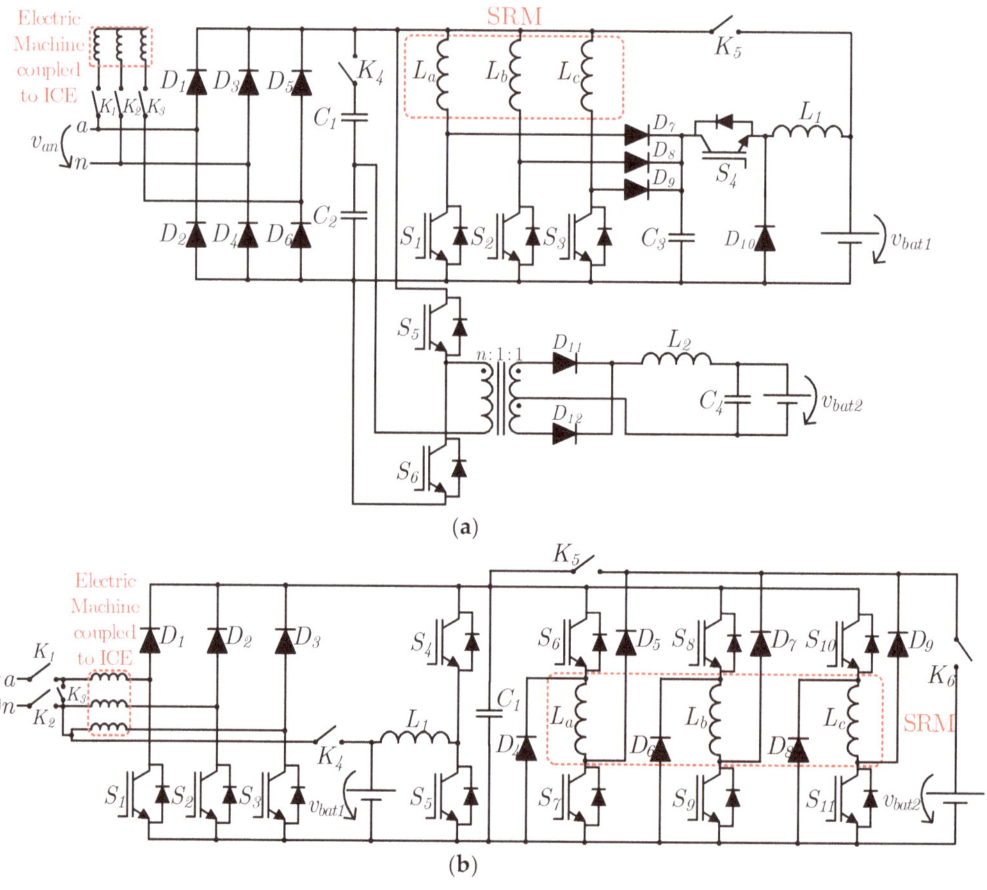

Figure 11. IBCs for plug-in hybrid vehicles based on SRMs proposed by H. Cheng et al. in 2020: (**a**) [25]; (**b**) [26].

In the same year, the same authors proposed a similar system in which the electric machine coupled to the ICE is also an SRM (Figure 11b) [26]. In this case, the converter used to drive the ICE-coupled machine is the same as the converter used to drive the main machine in the previous proposal, with the main difference that the windings can be reconfigured to be connected to a common neutral or isolated via contactor K_3. Similarly, interfacing with a single-phase power grid is feasible, and the topology also allows charging of both the traction and auxiliary batteries. As well as the converter for driving the ICE-coupled electric machine (IGBTs S_1–S_3 and diodes D_1–D_3), this system uses a typical asymmetrical dc–ac converter to drive the main SRM (IGBTs S_6–S_{11} and diodes D_4–D_9) and a bidirectional buck–boost dc–dc converter (IGBTs S_4 and S_5 and inductor L_1) interfacing with the traction battery (v_{bat1}). The system uses six contactors, where K_1 and K_2 are used to interface the converter with a single-phase ac power grid, K_3 is used to create the neutral point of the ICE-coupled SRM as noted, K_4 is used to charge the traction battery from the ICE-coupled SRM, K_5 is used to drive the main SRM and K_6 is used to charge the auxiliary battery (v_{bat2}). On this basis, there are five possible operation modes during traction operation, namely: (1) battery-only driving, where the vehicle behaves as an EV; (2) ICE-coupled SRM-only driving, where the power generated by the SRM is sufficient to drive the main SRM; (3) ICE-coupled SRM and battery hybrid driving, where the power to the main SRM comes from the two sources mentioned; (4) ICE-coupled SRM driving and battery charging, where the power generated by the SRM is more than the power required to drive the main SRM, and thus the surplus power is used to charge the traction battery; (5) regenerative braking, where the main SRM acts as a generator and charges the traction battery. Furthermore, during the battery charging operation (when the vehicle is stationary) there are three possible operation modes: (1) traction battery charged from the power grid; (2) auxiliary battery charged from the power grid; (3) auxiliary battery charged from the traction battery.

2.3. Integrated Battery Chargers with Galvanic Isolation

In 2005, F. Lacressonniere and B. Cassoret proposed two IBCs with galvanic isolation for industrial EVs [27]. One of these systems aims at application in pallet trucks based on 1.5 kW dc machines (Figure 12a), where galvanic isolation is achieved through a bidirectional isolated Cuk dc–dc converter (inductors L_1, L_2, capacitors C_1, C_2, IGBTs S_1, S_2 and a transformer with turns ratio N_1:N_2). As a dc traction drive system, this converter is responsible for both the battery interface and the machine drive. As well as this converter, it also uses a diode full-bridge single-phase ac–dc converter (D_1–D_4) to interface the system with a single-phase ac power grid (terminals a, n) in battery charging mode, with the dc–dc converter operating as a PFC converter. In order to switch between operation modes, two three-position switches are used (K_1, K_2) to change the dc–dc converter connections (points x_1, x_2), connecting it either to the machine (points y_1, y_2) or to the dc link of the ac–dc converter (points z_1, z_2). On the other hand, the second IBC proposed by the authors aims at application in forklifts based on a 6 kW wound rotor induction machine (Figure 12b). In this case, galvanic isolation is ensured by the machine itself in battery charging mode, where a three-phase ac power grid (terminals a, b, c) is connected to the rotor windings (L_{ra}, L_{rb}, L_{rc}), whose nominal voltage is 400 V, and a three-phase three-leg bidirectional ac–dc converter (S_1–S_6) is connected to the stator windings (L_{sa}, L_{sb}, L_{sc}), whose nominal voltage is 48 V. In order to switch between operation modes, three three-position contactors are used (K_1, K_2, K_3), which short-circuit the rotor windings in traction mode (points y_1, y_2, y_3), connecting them in a star configuration and leaving them independent in battery charging mode in order to connect to the power grid (points z_1, z_2, z_3). In this way, the bidirectional ac–dc converter operates as a dc–ac converter in traction mode and as an ac–dc converter in battery charging mode, controlling the currents absorbed from the power grid and the charging process of the battery. It should be noted that no additional inductors are used, with the machine stator windings behaving as inductive filters. It is also important to note that, in battery charging mode, the rotor must be mechanically braked since the machine is asynchronous and torque is created whenever its windings are energized.

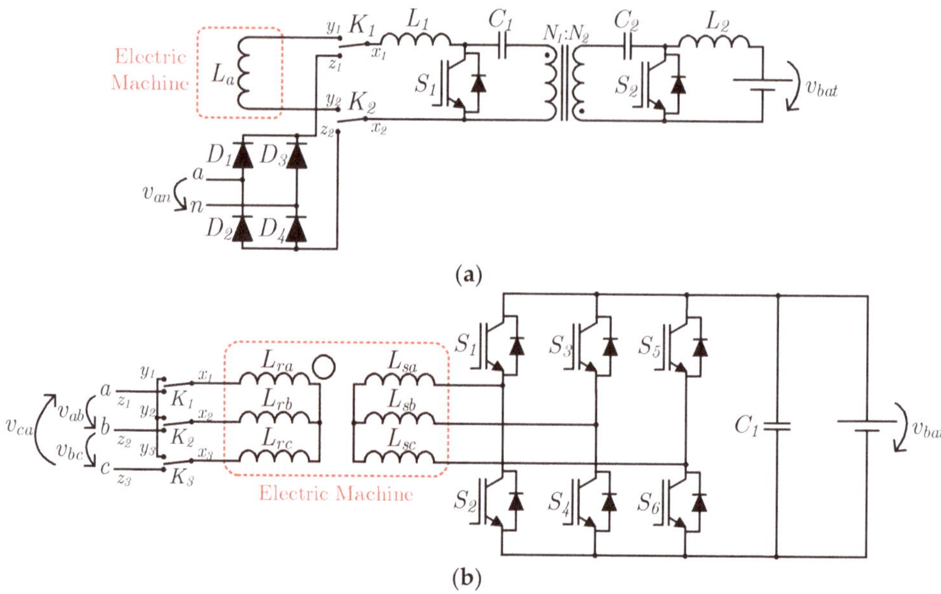

Figure 12. EV IBCs with galvanic isolation for industrial EVs proposed by F. Lacressonniere and B. Cassoret in 2005 [27] for application in: (**a**) pallet trucks; (**b**) forklifts.

In 2011, S. Haghbin et al. proposed an IBC with galvanic isolation based on stator winding reconfiguration of a three-phase electric machine [28]. The machine referred to is composed of four stator windings per phase, with all 24 winding terminals being externally accessible. As can be seen in Figure 13a, the only converter present in this topology is a three-phase three-leg bidirectional ac–dc converter (S_1–S_6), which is responsible for interfacing the battery and the machine. However, 12 relays and a three-phase contactor are used in order to reconfigure the stator windings. In Figure 13b, it is possible to see in detail the stator windings referring to phase a (L_{a1}, L_{a2}, L_{a3}, L_{a4}), as well as one of the phases of the three-phase contactor (K_1) and the four associated relays (K_2–K_5), with each relay being composed of a normally opened contact and a normally closed one. In traction mode, the windings of each phase are connected in a series–parallel arrangement (e.g., in phase a, L_{a1} and L_{a2} are connected in series, as well as L_{a3} and L_{a4}, with the two groups being connected in parallel with each other), and the machine is connected using a delta configuration. This configuration is respective to the normally closed state of the four relays. In the case of phase a, its windings are connected to phases a and b of the converter (terminals ca, cb, respectively). In battery charging mode, the windings of each phase are connected and isolated in pairs, where two of them are connected to the converter and the other two are connected to the respective phase of a three-phase ac power grid (terminals a, b, c), while the neutral point is connected to the power grid neutral (terminal n). For instance, in phase a, L_{a1} and L_{a3} are connected in parallel, with one of the resulting terminals connected to phase a of the converter (terminal ca) and the other to a point common to the other phases (terminal cn), i.e., the winding L_{ca} is formed by the parallel connection of L_{a1} and L_{a3}. On the other hand, L_{a2} and L_{a4} are connected in series, where one of the resulting terminals is connected to phase a of the power grid (terminal a) through contactor K_1 and the other is connected to a point common to the other phases, which in turn is connected to the power grid neutral (terminal n), i.e., the winding L_{ga} is formed by the series connection of L_{a2} and L_{a4}. Accordingly, it can be seen that the machine operates as a transformer, since there is no galvanic connection between the two sets of stator windings for the same phase. In 2013, the same authors analyzed this approach from a practical point of view [29], also giving attention to the electric machine design and to the phase shift used between both sets of stator windings [30].

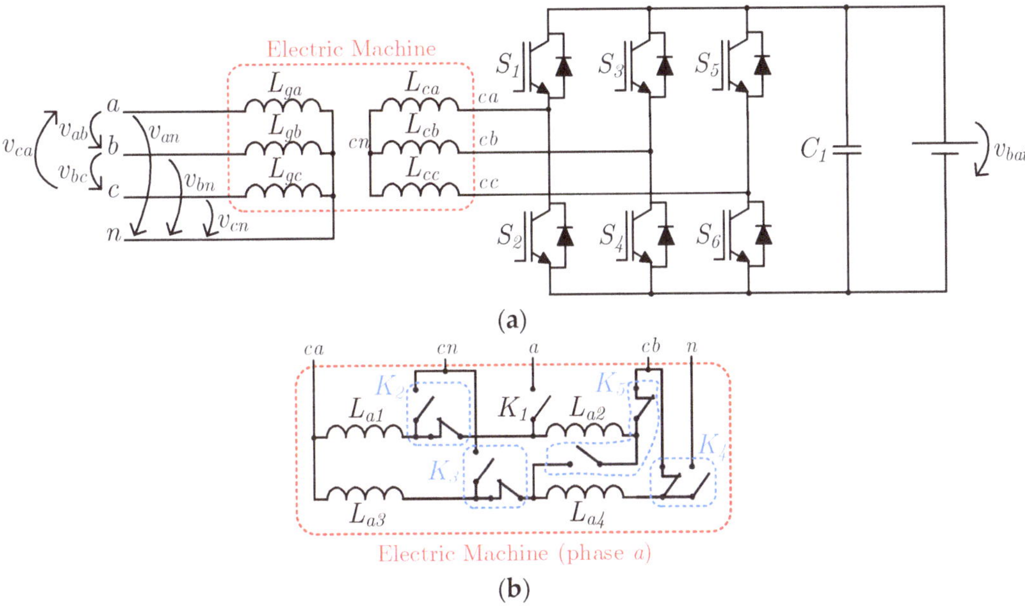

Figure 13. EV IBC with galvanic isolation proposed by S. Haghbin et al. in 2011 [28]: (**a**) topology; (**b**) stator winding arrangement for phase *a*.

In 2020, Z. Wang et al. proposed the galvanically isolated IBC presented in Figure 14 [31]. This topology is composed of a four-leg three-phase ac–dc converter (S_1–S_8), an active snubber in the dc link (S_9 and C_2) and a dual active bridge dc–dc converter (S_{10}–S_{17}), where the galvanic isolation is ensured. In traction mode, the ac–dc converter only operates with three legs during normal conditions; the fourth leg (S_7 and S_8) is only used in fault-tolerant operation, in which contactor K_1 is kept closed. In battery charging mode, the neutral point of the machine windings is used to connect to the phase terminal of a single-phase ac power grid, where the midpoint of the fourth leg is connected to neutral. This IBC allows bidirectional operation with the power grid and has a small capacitance value for the dc link capacitor C_2 due to the active snubber. In addition, owing to the voltage conversion ratio of the dual active bridge dc–dc converter, a low-voltage battery can be used.

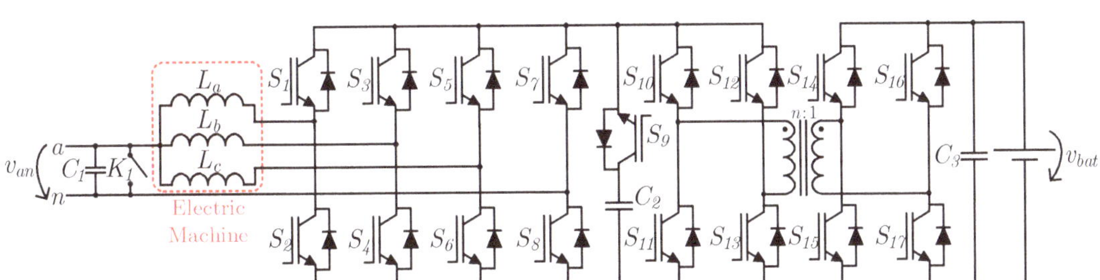

Figure 14. EV IBC with galvanic isolation proposed by Z. Wang et al. in 2020 [31].

2.4. Integrated Battery Chargers Based on Multiple Traction Converters

In 1995, S. K. Sul and S. J. Lee proposed an IBC for an all-wheel-drive EV, containing four electric machines and four three-phase three-leg bidirectional ac–dc converters, as well as a switch with two terminal pairs used for selecting the connection point of the positive battery terminal [32]. This topology can be seen in Figure 15. In traction mode, each ac–dc converter controls each electric machine independently, endowing the EV with all-wheel drive, with K_1 connecting the batteries to the dc link terminals of all converters (terminals a_1, a_2). In battery charging mode, each group formed by two ac–dc converters and two electric machines is used to employ two bidirectional interleaved converters, namely, a buck-boost dc–dc converter (S_1–S_{12}) and a full-bridge single-phase ac–dc converter (S_{13}–S_{24}) for interfacing with a single-phase ac power grid (terminals a, n). In this case, the neutral points of two of the electric machines are used for interfacing with a single-phase ac power grid, while the neutral points of the other two electric machines are connected to the positive battery terminal by means of the switch (terminals b_1, b_2). The utilization of two converters in battery charging mode allows the power grid peak voltage to be higher or lower than the battery voltage, provided it is lower than the dc link voltage due to the boost operation of the dc–dc converter from the battery to the dc link.

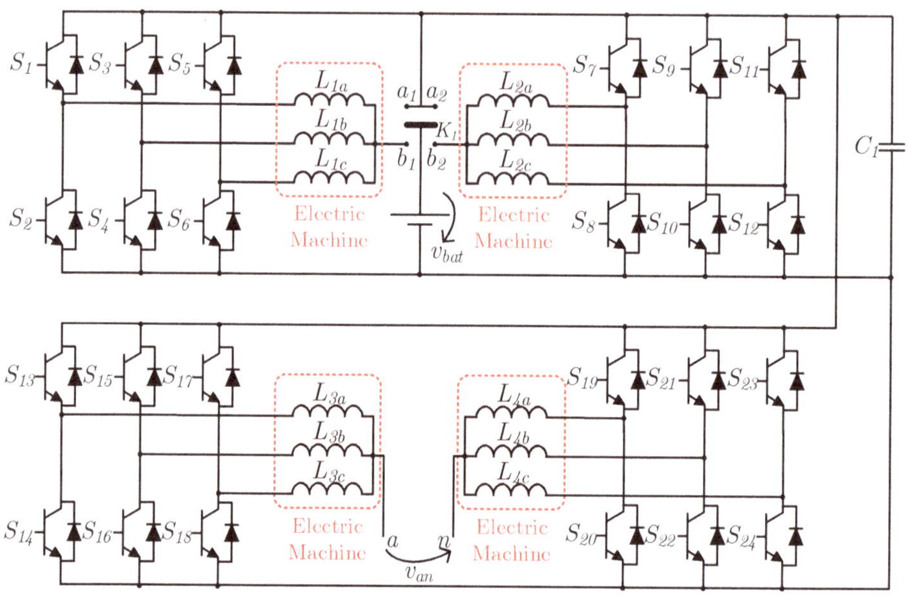

Figure 15. EV IBC proposed by S. K. Sul and S. J. Lee in 1995 [32].

In 2015, D. G. Woo et al. proposed an IBC based on two electric machines for interfacing with single-phase ac power grids, as shown in Figure 16a [33]. This topology is composed of two three-phase three-leg bidirectional ac–dc converters (S_1–S_6 and S_7–S_{12}), with each ac side connected to the stator windings of each electric machine, and the dc link is shared between these converters and connected to a bidirectional buck–boost dc–dc converter (IGBTs S_{13}, S_{14} and inductor L_3), which in turn interfaces with the battery. In battery charging mode, the three-phase ac–dc converters operate as boost dc–dc converters, with each converter operating in a power grid half cycle, and the stator windings of each machine are used as inductors in each converter. Furthermore, between the neutral point of each electric machine and the connection terminals to the power grid, two contactors (K_1 and K_2), two inductors (L_1 and L_2) and two diodes (D_1 and D_2) are used, where the diodes are used to reduce the common-mode noise, connecting the system ground to neutral (terminal n) during the power grid positive half cycle or to the phase (terminal a) during the power grid negative half cycle. The inductors are applied in order to compensate for the small common-mode inductance values of the stator windings.

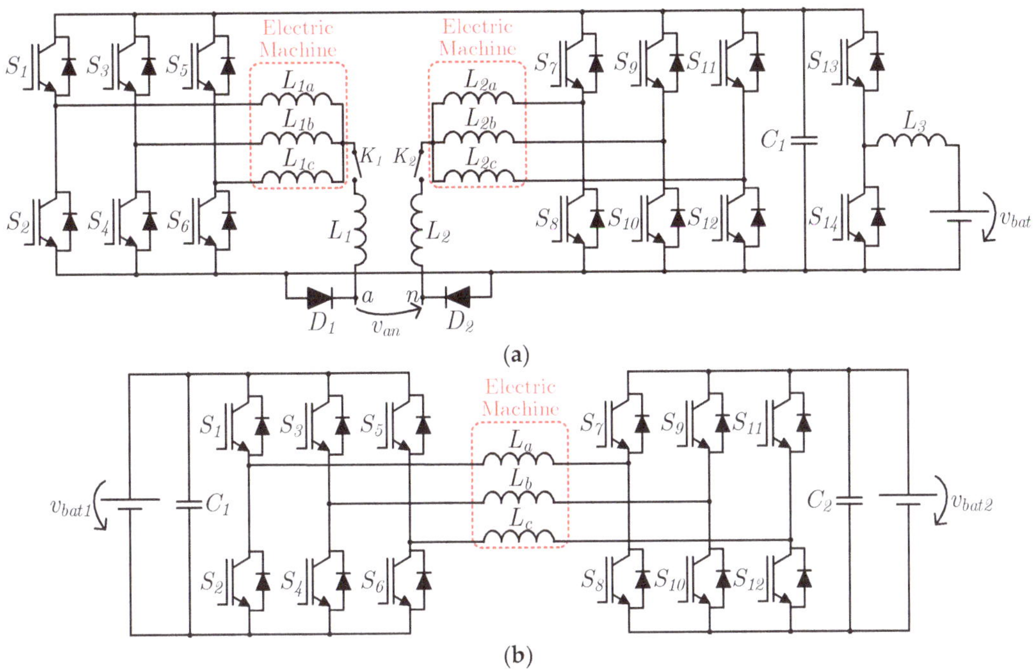

Figure 16. EV IBCs based on two traction converters proposed in 2015 by: (**a**) D. G. Woo et al. [33]; (**b**) J. Hong et al. [34].

In the same year, J. Hong et al. proposed an IBC based on an electric machine controlled by two converters for auxiliary battery charging, as shown in Figure 16b [34]. The power grid interface is not considered in this topology; instead, the interface between the traction and auxiliary batteries is considered. The dc links of both converters are isolated from each other, and one of the converters (S_1–S_6) is connected to the traction battery (v_{bat1}) and the other (S_7–S_{12}) to the auxiliary battery (v_{bat2}). The traction battery is used as the main power source for the electric machine, while the auxiliary battery can be used in high-speed operation, by adding its voltage to the main battery voltage and increasing the voltage supplied to the machine windings. The auxiliary battery can be bypassed, producing a zero-voltage vector in its connected converter (S_7–S_{12}), and it can also be charged from the traction battery while the electric machine is being powered, provided the converter connected to the traction battery (S_1–S_6) has enough voltage margin to produce the desired torque in the machine and to charge the auxiliary battery.

In 2018, S. Semsar et al. proposed a similar system, as shown in Figure 17a [35]. This system is capable of charging both batteries from a single-phase ac power grid (terminals a, n), using an additional diode full-bridge single-phase ac–dc converter (D_1–D_4) and a contactor (K_1) that should be closed in battery charging mode. In traction mode, K_1 is open, and the resulting power electronics system is the same as in the IBC previously shown. In battery charging mode, the voltage after the diode full-bridge single-phase ac–dc converter, i.e., the rectified v_{an} is the sum of the voltages produced by each shared leg of the two three-phase ac–dc converters, which, in this case, operate as PFC converters in conjunction with the diode bridge. The authors note that interleaved operation is possible in order to reduce the current ripple, but do not present details of this operation. Hence, when connected to a single-phase ac power grid, the currents in both batteries contain a high ac component, which is not desirable in these types of elements. In order to absorb sinusoidal current from the power grid, each converter is responsible for each half cycle, with the upper converter (S_1–S_6) synthesizing the positive current half cycle and the bottom converter (S_7–S_{12}) synthesizing the negative current half cycle.

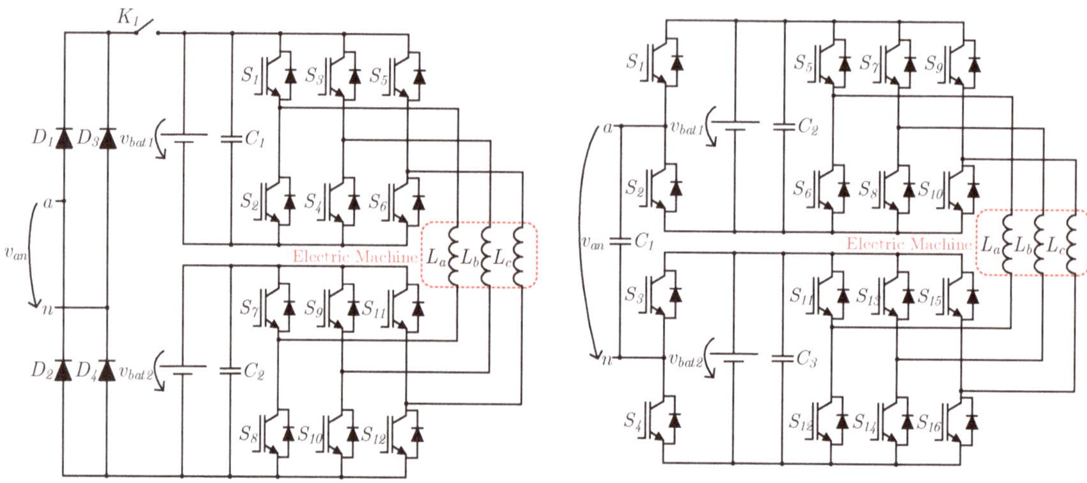

Figure 17. EV IBCs based on two traction converters proposed by S. Semsar et al. in: (**a**) 2018 [35], (**b**) 2020 [36].

Two years later, in 2020, the same authors proposed a similar system using an active front-end ac–dc converter, as shown in Figure 17b [36]. The diode full-bridge single-phase ac–dc converter and contactor were replaced by two half-bridge single-phase ac–dc converters (S_1–S_4) and a capacitor (C_1), allowing bidirectional operation with the power grid. The traction operation is the same as in the IBC previously described; only the power grid front-end converter is different. Regarding the battery charging operation, in order to achieve a sinusoidal current in the power grid side, the traction converters operate with a square-wave modulation, i.e., with the same frequency as the power grid, while the front-end converter operates with a modified sinusoidally modulated signal. This allows five voltage levels to be produced, contributing to a reduction in the harmonic distortion of the current absorbed or provided to the power grid. The machine winding currents must be in phase with each other in order to prevent the machine from producing torque and thus rotating, and if correctly balanced, each machine phase only carries one third of the amplitude of the power grid current. As happens with the topology previously presented by these authors, this topology has also the problem of a high ac component in the battery currents.

2.5. *Integrated Battery Chargers Based on Multiphase Machines*

In 2016, M. Diab et al. proposed an IBC based on a nine-switch ac–dc converter capable of controlling a six-phase machine [37]. This IBC is represented in Figure 18, where it can be seen that the nine-switch converter (S_1–S_9) is based on three legs with three semiconductors each, similar to two regular three-leg three-phase ac–dc converters on top of each other, where the bottom semiconductors of the upper converter are the same as the upper semiconductors of the bottom converter (S_2, S_5, S_8). Hence, six midpoints can be found, to which the windings of the six-phase machine are connected. This IBC has also a bidirectional buck–boost dc–dc converter (semiconductors S_{10} and S_{11} and inductor L_1) interfacing the dc link capacitor C_1 with the battery, as well as three sets of three-phase contactors to reconfigure the system. In traction mode, contactors K_1–K_6

are closed, in order to connect the neutral points of each group of machine windings, creating two neutral points in the six-phase machine. In battery charging mode, these contactors must be kept open, and K_7–K_9 are closed. The intermediate semiconductors of each leg (S_2, S_5, S_8) are kept closed, so that the nine-switch converter operates as a regular three-leg three-phase ac–dc converter, with each pair of machine windings being connected in parallel (L_a with L_d, L_b with L_e, L_c with L_f). This topology allows bidirectional operation with the power grid.

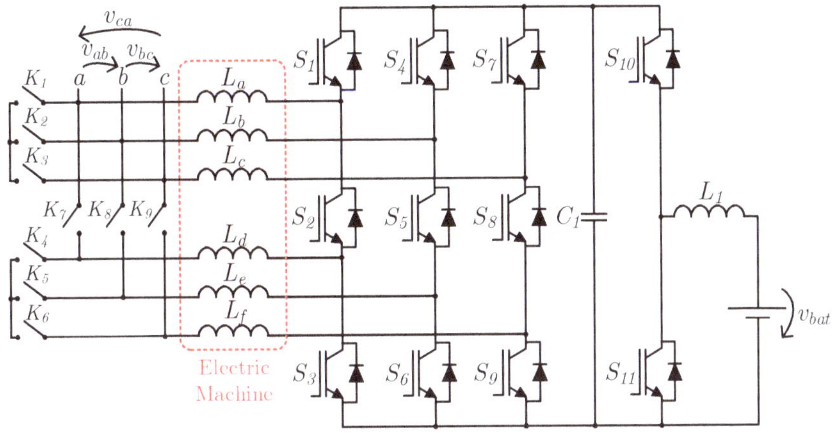

Figure 18. EV IBC proposed by M. Diab et al. in 2016 [37].

In 2016, I. Subotic et al. proposed an IBC based on a five-phase machine [38]. The same authors analyzed slow [39] and fast [40] battery charging operations using IBCs based on five-phase, six-phase and nine-phase electric machines. In the same year, the authors presented an IBC with galvanic isolation based on six-phase machines, where they simply added a three-phase transformer between the power grid and the converter [41]. It should be noted that these publications regarding multiphase machines only focus on the interface with the power grid (single-phase ac or three-phase ac) and do not consider a dc–dc converter for interfacing with the battery. In the following year, however, the same authors proposed two IBCs with direct interfacing with the power grid, including a bidirectional buck–boost dc–dc converter to interface with the battery. One of these systems is applied to six-phase machines and three-phase ac power grids [42], while the other is applied to nine-phase machines and can be connected to either single-phase or three-phase ac power grids [43]. Figure 19 shows the common structure between the proposed topologies from the power grid interface point of view, i.e., for five-phase (Figure 19a), six-phase (Figure 19b) and nine-phase (Figure 19c) electric machines. Regarding the five-phase topology, when it is connected to a single-phase ac power grid (terminals *a*, *n*), one of the terminals contains two stator windings (in this case *a*) and the other contains three stator windings (in this case *n*). On the other hand, when connected to a three-phase ac power grid (terminals *a*, *b*, *c*), two stator windings are used in two of the machine phases (in this case *a*, *c*) and only one is used in the remaining phase (in this case *b*). In the case of the six-phase topology, since the phase number is even and a multiple of three, the stator windings are equally divided in both types of interfaces and are assembled in pairs (in the case of a three-phase ac power grid) or in triplets (in the case of a single-phase ac power grid). Finally, for the nine-phase topology, the stator windings are grouped in triplets in both cases, using the three winding groups in the interface with a three-phase ac power grid and only two of them in the interface with a single-phase ac power grid.

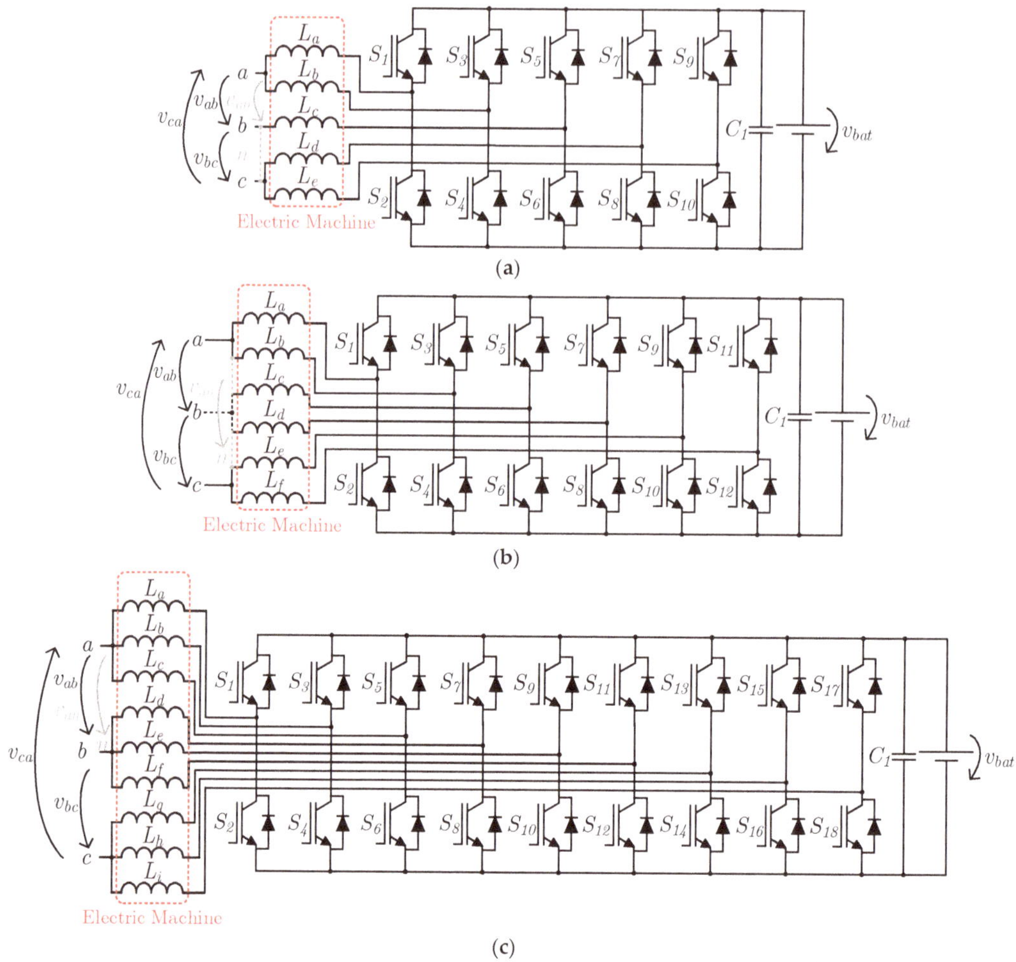

Figure 19. EV IBCs proposed by I. Subotic et al. and V. Katic et al. in 2016 [38–40] based on electric machines with phase counts equal to: (**a**) five; (**b**) six; (**c**) nine.

In 2020, M. Tong et al. proposed an IBC based on a five-phase hybrid excitation flux-switching machine, as shown in Figure 20 [44]. As well as the five converter legs (S_1–S_{10}) used to energize the machine armature windings (L_a–L_e), an extra leg (S_{11} and S_{12}) is used to energize a field winding (L_{fw}) for machine excitation. This topology also contains two contactors (K_1 and K_2) to switch the operation between traction and battery charging. In traction mode, the contactor K_1 connects terminals x_1, x_2 to terminal z, establishing the neutral point of the five-phase machine. Contactor K_2 is closed in this situation, connecting the battery and dc link capacitor C_1 in parallel. On the other hand, in battery charging mode, the contactor K_1 connects terminal x_1 to y_1 and x_2 to y_2, with this pair of terminals representing the connection points to a single-phase ac power grid. Hence, a two-phase interleaved bidirectional full-bridge ac–dc converter is assembled (semiconductors S_1–S_8 and inductors L_a–L_d), with the leg formed by semiconductors S_9 and S_{10} and machine winding L_e unutilized. In this case, the leg formed by semiconductors S_{11} and S_{12} and the machine field winding L_{fw}, form a bidirectional buck–boost dc–dc converter, where contactor K_2 must be kept open in order to separate the battery (low voltage side of the dc–dc converter) from the dc link capacitor C_1 (high voltage side of the dc–dc converter).

In 2021, the same authors proposed a modification to this system, changing contactor K_1 to accommodate the connection to a three-phase ac power grid and thus allowing fast battery charging [45].

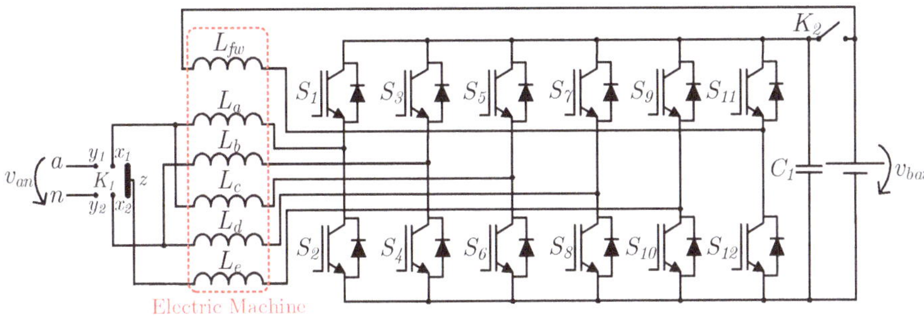

Figure 20. EV IBC proposed by M. Tong et al. in 2020 [44].

In 2020 also, H. Raherimihaja et al. proposed an IBC based on a six-phase open-end winding machine [46]. As illustrated in Figure 21, this topology is composed of a twelve-leg inverter (S_1–S_{24}) to drive the electric machine, as well as an additional leg (S_{25} and S_{26}) and inductor (L_1) that form a bidirectional buck–boost dc–dc converter to connect the inverter dc link to the battery. Instead of using contactors, this system contains a plug and socket to connect the machine terminals to a three-phase ac power grid. The machine windings are connected in pairs, and in this case (battery charging mode), two pairs of windings are connected to each other, where the midpoint of each pair is connected to a given phase of the power grid. With this configuration, it is possible to achieve a four-phase interleaved bidirectional three-phase ac–dc converter, since the twelve inverter legs can be used in an arrangement of four legs per power grid phase. Despite mentioning this feature, the authors do not employ this interleaved operation in the same paper. This system allows bidirectional operation with the power grid, and since it uses a six-phase machine, it allows a battery charging power twice that of the traction power, as well as having fault tolerance capability. In 2021, the same authors proposed a similar system based on a nine-leg inverter and a three-phase nine-winding machine [47], as well as giving a detailed analysis of the design of this type of machine for use in IBCs [48].

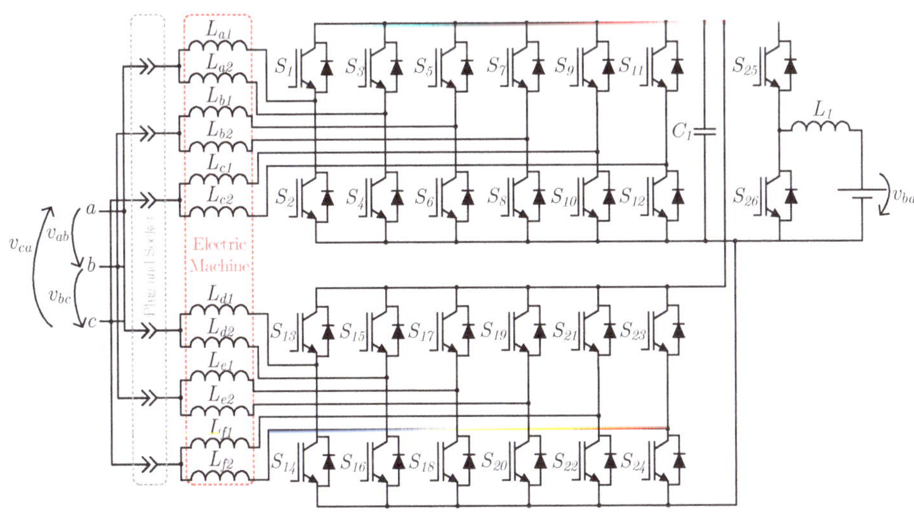

Figure 21. EV IBC proposed by H. Raherimihaja et al. in 2020 [46].

3. Comparison between Integrated Battery Chargers

The IBCs reviewed in this paper are compared in this section. The comparison i based on whether contactors are needed to reconfigure the system, whether additiona inductors are used to interface the system with the power grid, the type of machine tha the system supports and the winding access level that is required (where normal mean the conventional phase terminal access), the existence of galvanic isolation between th batteries and the power grid, the capability of bidirectional operation with the powe grid and the possibilities for interfacing that they allow in addition to interfacing witl the EV's main electric machine. Table 1 presents the comparison between the 29 IBC (31 IBCs were presented, but the features of the three IBCs proposed by I. Subotic et a and V. Katic et al. [38–40] are the same except for the number of electric machine phases The order used for the IBCs in the table is the same as the order in which they appear ir the paper.

Table 1. Qualitative comparison between the presented IBCs.

Proposed IBC	Contactors	Additional Inductors	Electric Machine Type	Machine Access	Galvanic Isolation	Bidirectional Operation	Interface Possibilities
D. Thimmesch (1983) [10]	Yes	No	3ph	Normal	Yes	No	1ph
W. Rippel (1990) [12]	No	Optional	3ph	Normal	No	No	1ph
W. Rippel, A. Cocconi (1992) [13]	No	No	2 × 3ph/ 1 × 3ph (w/2 stators)	Neutral point	No	Yes	1ph
A. Cocconi (1994) [14]	Yes	Optional	3ph	Total	No	Yes	1ph, 3ph
L. Solero (2001) [15]	No	No	3ph	Neutral point	No	No	1ph
G. Pellegrino (2010) [16]	No	No	3ph	Neutral point	No	No	1ph
O. Hegazy (2013) [17]	No	Yes	3ph	Normal	No	Yes	1ph
C. Pollock, W. K. Thong (2000) [18]	No	Yes	2ph SRM	Neutral point	Yes	No	1ph
H. C. Chang, C. M. Liaw (2009) [19]	Yes	No	4ph SRM	Total	No	No	1ph
H. C. Chang, C. M. Liaw (2011) [20]	Yes	No	4ph SRM	Total	No	Yes	1ph
Y. Hu et al. (2014) [21]	Yes	No	3ph SRM	Neutral point	No	No	1ph, 3ph, dc
Y. Hu et al. (2015) [22]	No	No	4ph SRM w/2 stators	Neutral point	No	No	1ph, 3ph, dc
K. W. Hu et al. (2015) [23]	Yes	Yes	4ph SRM	Total	No	Yes	1ph, 2ph
M. Ma et al. (2018) [24]	Yes	No	3ph SRM	Total	No	No	3ph
H. Cheng et al. (2020) [25]	Yes	No	3ph SRM	Neutral point	No	No	1ph
H. Cheng et al. (2020) [26]	Yes	No	3ph SRM	Total	No	No	1ph
F. Lacressonniere, B. Cassoret (2005) [27] (For pallet trucks)	Yes	No	Brushed dc	Total (just 2)	Yes	No	1ph
F. Lacressonniere, B. Cassoret (2005) [27] (For forklifts)	Yes	No	3ph wound rotor induction	Normal	Yes	Yes	3ph
S. Haghbin et al. (2011) [28]	Yes	No	3ph (4 windings per phase)	Total	Yes	Yes	1ph, 3ph
Z. Wang et al. (2020) [31]	Yes	No	3ph	Neutral point	Yes	Yes	1ph
S. K. Sul, S. J. Lee (1995) [32]	Yes	No	4 × 3ph	Neutral point	No	Yes	1ph
D. G. Woo et al. (2015) [33]	Yes	Optional	2 × 3ph	Neutral point	No	No	1ph

Table 1. *Cont.*

Proposed IBC	Contactors	Additional Inductors	Electric Machine Type	Machine Access	Galvanic Isolation	Bidirectional Operation	Interface Possibilities
J. Hong et al. (2015) [34] *	-	-	3ph	Total	-	-	-
S. Semsar et al. (2018) [35]	Yes	No	3ph	Total	No	No	1ph
S. Semsar et al. (2020) [36]	No	No	3ph	Total	No	Yes	1ph
M. Diab et al. (2016) [37]	Yes	No	6ph	Total	No	Yes	3ph
I. Subotic et al., V. Katic et al. (2016) [38–40]	No	No	5ph, 6ph, 9ph	Total	No	Yes	1ph, 3ph
M. Tong et al. (2020) [44]	Yes	No	5ph hybrid excitation flux switching	Total	No	Yes	1ph
H. Raherimihaja et al. (2020) [46]	No	No	6ph	Total	No	Yes	3ph

* An interface with the power grid is not considered.

It can be seen that most of the topologies do not require additional inductors to interface the system with the power grid; however, most of them do require contactors. Topologies that require neither contactors nor additional inductors require electric machine access to at least the neutral point, if not total access to the windings. Regarding bidirectional operation, it can be seen that when no contactors or additional inductors are used, total access to the machine windings is required, with the exception of the system patented by W. Rippel and A. Cocconi in 1992 [13], where only neutral-point access is required but with two three-phase electric machines or one machine with two sets of stator windings. It can also be verified that, when two or more power grid interface possibilities exist, at least neutral-point access is required. If two or more power grid interface possibilities are combined with bidirectional operation, total access to the electric machine windings is mandatory for the presented topologies. Therefore, it can be concluded that, in general, IBCs with higher functionalities have naturally higher requirements, either regarding the electric machine winding access or regarding external components, which is expected from a single set of power converters performing more than its two designed functions (traction and battery charging).

In order to establish a comparison between the IBCs in quantitative terms, Figure 22 presents five radar charts, where the number of switches, diodes, contactors, magnetics and capacitors is represented. Due to the high number of presented IBCs and in order to better analyze the provided information, five radar charts were used instead of a single chart, with each one representing a subsection of Section 2. Figure 22a represents the classical IBCs (Section 2.1), Figure 22b the IBCs for SRMs (Section 2.2), Figure 22c the IBCs with galvanic isolation (Section 2.3), Figure 22d the IBCs based on multiple traction converters (Section 2.4) and Figure 22e the IBCs based on multiphase machines (Section 2.5). The term "switches" refers to fully controlled semiconductors, either IGBTs, MOSFETs or bipolar junction transistors, with or without antiparallel diodes, while the term "diodes" refers to this single component, excluding the antiparallel diodes present in fully controlled semiconductors. "Magnetics" can refer to either inductors or transformer windings, excluding electric machine windings (e.g., one transformer with one primary and one secondary winding contributes two to the magnetics count). The term "capacitors" only counts each capacitor set (as in the capacitor numbering in the IBC figures). It should be noted that Figure 22e has only three comparative metrics, namely, switches, magnetics and contactors, since the diode count is nil and the capacitor count is one for all the topologies presented in this figure.

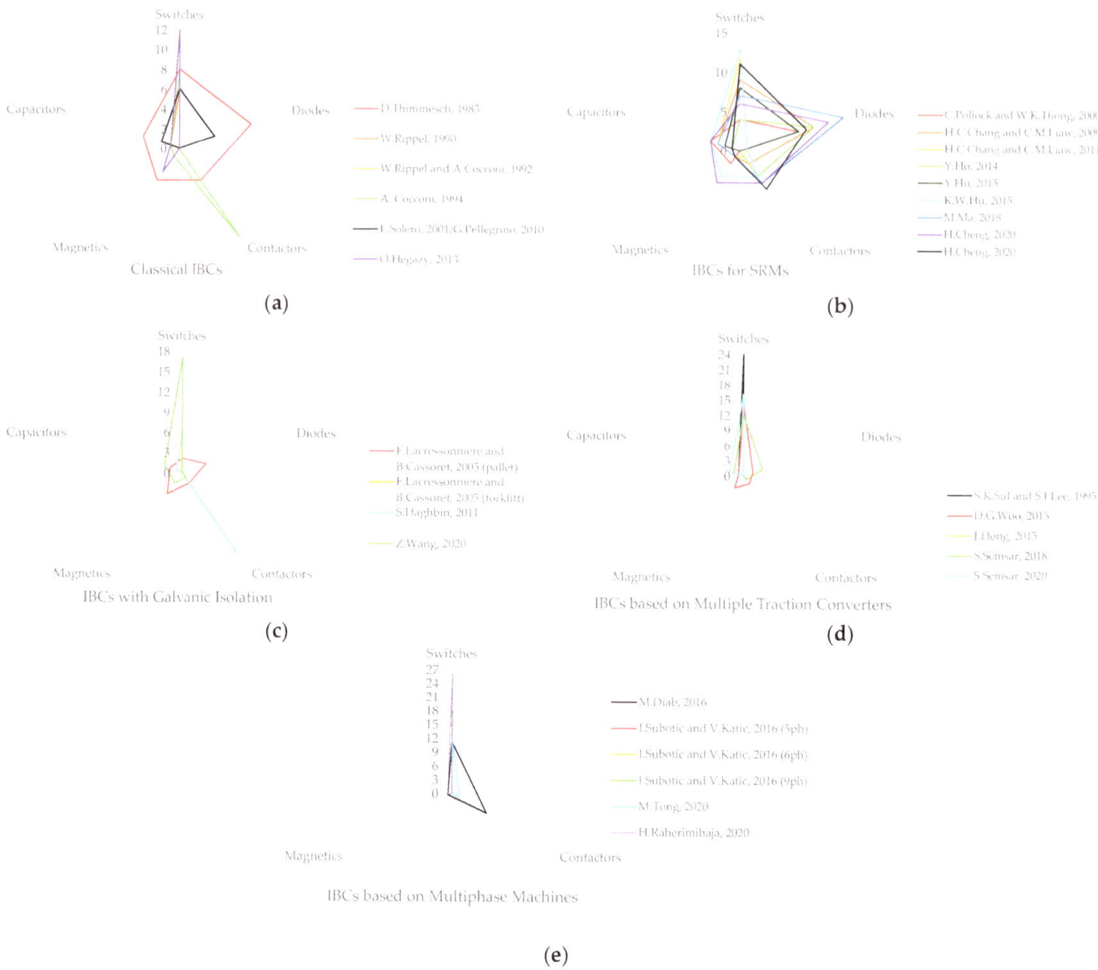

Figure 22. Radar charts illustrating the quantitative comparison between the presented: (**a**) classical IBCs; (**b**) IBCs for SRMs; (**c**) IBCs with galvanic isolation; (**d**) IBCs based on multiple traction converters; (**e**) IBCs based on multiphase machines.

4. Conclusions

Electric vehicles (EVs) have experienced tremendous growth in the past few years since they are an emissions-free means of transportation at the user level. The power electronics converters present in an EV are based on two main systems that operate only one at a time, namely, the traction system and the battery charging system. This operational nonconcurrence gave rise to the development of integrated battery chargers (IBCs) consisting of a single power electronics system with both functionalities. Several IBCs have been proposed in the literature and patented, and this paper aimed to provide a literature review of the most relevant IBCs. Due to their diversity, the paper divided the analyzed IBCs into five subsections, namely: (1) classical IBCs; (2) IBCs especially developed for switched reluctance machines (SRMs); (3) IBCs with galvanic isolation; (4) IBCs based on multiple traction converters; (5) IBCs based on multiphase machines. After presenting each type, the IBCs were compared both in terms of requirements (e.g., the need for additional inductors or contactors, the type of electric machine needed) and in terms of functionalities (e.g., galvanic isolation, bidirectional operation, power grid interface possibilities). It was verified that topologies that do not require external components such as inductors or contactors usually require special access to the electric machine as well as the conventional phase access. Additionally, some topologies allow interfacing to more than one type of power grid, and in most of these cases it is mandatory to have access to all electric machine windings. The same occurs with bidirectional operation with the power grid, where for topologies with no additional inductors or contactors, electric machine access to at least the neutral point is required. When combining bidirectional operation with more than two types of power grid interfaces, total winding access is a must. For the presented IBCs, it can be concluded that, in general, higher functionalities imply higher requirements, either from the electric machine winding access or from external components such as contactors to reconfigure the system. However, it should be kept in mind that this complexity refers to a single set of power converters performing more than its two designed functions (traction and battery charging), which still represents lower volume, weight and possibly cost than separate power electronics systems for performing the same functions.

Author Contributions: Conceptualization, T.J.C.S., D.P., V.M. and J.L.A.; investigation, T.J.C.S. and D.P.; resources, T.J.C.S. and D.P.; writing—original draft preparation, T.J.C.S.; writing—review and editing, D.P., V.M. and J.L.A.; supervision, V.M. and J.L.A.; funding acquisition, T.J.C.S., V.M. and J.L.A. All authors have read and agreed to the published version of the manuscript.

Funding: This work has been supported by FCT—Fundação para a Ciência e Tecnologia within the R&D Units Project Scope: UIDB/00319/2020. T.J.C.S. is supported by the FCT scholarships SFRH/BD/134353/2017 and COVID/BD/151993/2021.

Conflicts of Interest: The authors declare no conflict of interest.

References

1. Dudek, E. The Flexibility of Domestic Electric Vehicle Charging: The Electric Nation Project. *IEEE Power Energy Mag.* **2021**, *19*, 16–27. [CrossRef]
2. Lee, Z.J.; Sharma, S.; Low, S.H. Research Tools for Smart Electric Vehicle Charging: An introduction to the adaptive charging network research portal. *IEEE Electrif. Mag.* **2021**, *9*, 29–36. [CrossRef]
3. Sun, X.; Li, Z.; Wang, X.; Li, C. Technology Development of Electric Vehicles: A Review. *Energies* **2019**, *13*, 90. [CrossRef]
4. Afonso, J.L.; Cardoso, L.A.L.; Pedrosa, D.; Sousa, T.J.C.; Machado, L.; Tanta, M.; Monteiro, V. A Review on Power Electronics Technologies for Electric Mobility. *Energies* **2020**, *13*, 6343. [CrossRef]
5. Sayed, K.; Almutairi, A.; Albagami, N.; Alrumayh, O.; Abo-Khalil, A.G.; Saleeb, H. A Review of DC-AC Converters for Electric Vehicle Applications. *Energies* **2022**, *15*, 1241. [CrossRef]
6. Chakraborty, S.; Vu, H.-N.; Hasan, M.M.; Tran, D.-D.; Baghdadi, M.E.; Hegazy, O. DC-DC Converter Topologies for Electric Vehicles, Plug-in Hybrid Electric Vehicles and Fast Charging Stations: State of the Art and Future Trends. *Energies* **2019**, *12*, 1569. [CrossRef]
7. Pedrosa, D.; Monteiro, V.; Sousa, T.J.C.; Machado, L.; Afonso, J.L. Unified Power Converter Based on a Dual-Stator Permanent Magnet Synchronous Machine for Motor Drive and Battery Charging of Electric Vehicles. *Energies* **2021**, *14*, 3344. [CrossRef]
8. Yilmaz, M.; Krein, P.T. Review of Battery Charger Topologies, Charging Power Levels, and Infrastructure for Plug-In Electric and Hybrid Vehicles. *IEEE Trans. Power Electron.* **2013**, *28*, 2151–2169. [CrossRef]
9. Sousa, T.J.C.; Monteiro, V.; Pedrosa, D.; Machado, L.; Afonso, J.L. Unified Systems for Traction and Battery Charging of Electric Vehicles: A Sustainability Perspective. *EAI Endorsed Trans. Energy Web* **2021**, *8*, 170557. [CrossRef]
10. Thimmesch, D. *Integral Inverter/Battery Charger for Use in Electric Vehicles*; USA Department of Energy/NASA: Springfield, VA, USA, 1983.
11. Thimmesch, D. An SCR Inverter with an Integral Battery Charger for Electric Vehicles. *IEEE Trans. Ind. Appl.* **1985**, *IA-21*, 1023–1029. [CrossRef]
12. Rippel, W. Integrated Traction Inverter and Battery Charger Apparatus. U.S. Patent 1990.
13. Rippel, W.; Cocconi, A. Integrated Motor Drive and Recharge System. U.S. Patent 1992.
14. Cocconi, A. Combined Motor Drive and Battery Charger System. U.S. Patent 1994.
15. Solero, L. Nonconventional on-board charger for electric vehicle propulsion batteries. *IEEE Trans. Veh. Technol.* **2001**, *50*, 144–149. [CrossRef]
16. Pellegrino, G.; Armando, E.; Guglielmi, P. An Integral Battery Charger with Power Factor Correction for Electric Scooter. *IEEE Trans. Power Electron.* **2010**, *25*, 751–759. [CrossRef]
17. Hegazy, O.; Barrero, R.; Van Mierlo, J.; Lataire, P.; Omar, N.; Coosemans, T. An Advanced Power Electronics Interface for Electric Vehicles Applications. *IEEE Trans. Power Electron.* **2013**, *28*, 5508–5521. [CrossRef]

18. Pollock, C.; Thong, W.K. Low-cost battery-powered switched reluctance drives with integral battery-charging capability. *IEEE Trans. Ind. Appl.* **2000**, *36*, 1676–1681. [CrossRef]
19. Chang, H.-C.; Liaw, C.-M. Development of a Compact Switched-Reluctance Motor Drive for EV Propulsion with Voltage-Boosting and PFC Charging Capabilities. *IEEE Trans. Veh. Technol.* **2009**, *58*, 3198–3215. [CrossRef]
20. Chang, H.; Liaw, C. An Integrated Driving/Charging Switched Reluctance Motor Drive Using Three-Phase Power Module. *IEEE Trans. Ind. Electron.* **2011**, *58*, 1763–1775. [CrossRef]
21. Hu, Y.H.; Song, X.G.; Cao, W.P.; Ji, B. New SR Drive With Integrated Charging Capacity for Plug-In Hybrid Electric Vehicles (PHEVs). *IEEE Trans. Ind. Electron.* **2014**, *61*, 5722–5731. [CrossRef]
22. Hu, Y.; Gan, C.; Cao, W.; Li, C.; Finney, S. Split Converter-Fed SRM Drive for Flexible Charging in EV/HEV Applications. *IEEE Trans. Ind. Electron.* **2015**, *62*, 6085–6095. [CrossRef]
23. Hu, K.W.; Yi, P.H.; Liaw, C.M. An EV SRM Drive Powered by Battery/Supercapacitor with G2V and V2H/V2G Capabilities. *IEEE Trans. Ind. Electron.* **2015**, *62*, 4714–4727. [CrossRef]
24. Ma, M.; Chang, Z.; Hu, Y.; Li, F.; Gan, C.; Cao, W. An Integrated Switched Reluctance Motor Drive Topology with Voltage-Boosting and On-Board Charging Capabilities for Plug-In Hybrid Electric Vehicles (PHEVs). *IEEE Access* **2018**, *6*, 1550–1559. [CrossRef]
25. Cheng, H.; Wang, Z.; Yang, S.; Huang, J.; Ge, X. An Integrated SRM Powertrain Topology for Plug-In Hybrid Electric Vehicles with Multiple Driving and Onboard Charging Capabilities. *IEEE Trans. Transp. Electrif.* **2020**, *6*, 578–591. [CrossRef]
26. Cheng, H.; Wang, L.; Xu, L.; Ge, X.; Yang, S. An Integrated Electrified Powertrain Topology with SRG and SRM for Plug-In Hybrid Electrical Vehicle. *IEEE Trans. Ind. Electron.* **2020**, *67*, 8231–8241. [CrossRef]
27. Lacressonniere, F.; Cassoret, B. Converter used as a battery charger and a motor speed controller in an industrial truck. In Proceedings of the 2005 European Conference on Power Electronics and Applications, Dresden, Germany, 11–14 September 2005; Volume 9, p. 7.
28. Haghbin, S.; Lundmark, S.; Alakula, M.; Carlson, O. An Isolated High-Power Integrated Charger in Electrified-Vehicle Applications. *IEEE Trans. Veh. Technol.* **2011**, *60*, 4115–4126. [CrossRef]
29. Haghbin, S.; Lundmark, S.; Alakula, M.; Carlson, O. Grid-Connected Integrated Battery Chargers in Vehicle Applications: Review and New Solution. *IEEE Trans. Ind. Electron.* **2013**, *60*, 459–473. [CrossRef]
30. Haghbin, S.; Khan, K.; Zhao, S.; Alakula, M.; Lundmark, S.; Carlson, O. An Integrated 20-kW Motor Drive and Isolated Battery Charger for Plug-In Vehicles. *IEEE Trans. Power Electron.* **2013**, *28*, 4013–4029. [CrossRef]
31. Wang, Z.; Zhang, Y.; You, S.; Xiao, H.; Cheng, M. An Integrated Power Conversion System for Electric Traction and V2G Operation in Electric Vehicles with a Small Film Capacitor. *IEEE Trans. Power Electron.* **2020**, *35*, 5066–5077. [CrossRef]
32. Sul, S.-K.; Lee, S.-J. An integral battery charger for four-wheel drive electric vehicle. *IEEE Trans. Ind. Appl.* **1995**, *31*, 1096–1099. [CrossRef]
33. Woo, D.-G.; Joo, D.-M.; Lee, B.-K. On the Feasibility of Integrated Battery Charger Utilizing Traction Motor and Inverter in Plug-In Hybrid Electric Vehicles. *IEEE Trans. Power Electron.* **2015**, *30*, 7270–7281. [CrossRef]
34. Hong, J.; Lee, H.; Nam, K. Charging method for the secondary battery in dual-inverter drive systems for electric vehicles. *IEEE Trans. Power Electron.* **2015**, *30*, 909–921. [CrossRef]
35. Semsar, S.; Soong, T.; Lehn, P.W. Integrated Single-Phase Electric Vehicle Charging Using a Dual-Inverter Drive. In Proceedings of the 2018 IEEE Transportation Electrification Conference and Expo (ITEC), Long Beach, CA, USA, 13–15 June 2018; pp. 320–325.
36. Semsar, S.; Soong, T.; Lehn, P.W. On-Board Single-Phase Integrated Electric Vehicle Charger With V2G Functionality. *IEEE Trans. Power Electron.* **2020**, *35*, 12072–12084. [CrossRef]
37. Diab, M.S.; Elserougi, A.A.; Abdel-Khalik, A.S.; Massoud, A.M.; Ahmed, S. A Nine-Switch-Converter-Based Integrated Motor Drive and Battery Charger System for EVs Using Symmetrical Six-Phase Machines. *IEEE Trans. Ind. Electron.* **2016**, *63*, 5326–5335. [CrossRef]
38. Subotic, I.; Bodo, N.; Levi, E. An EV Drive-Train with Integrated Fast Charging Capability. *IEEE Trans. Power Electron.* **2016**, *31*, 1461–1471. [CrossRef]
39. Subotic, I.; Bodo, N.; Levi, E. Single-phase on-board integrated battery chargers for EVs based on multiphase machines. *IEEE Trans. Power Electron.* **2016**, *31*, 6511–6523. [CrossRef]
40. Katic, V.; Subotic, I.; Bodo, N.; Levi, E.; Dumnic, B.; Milicevic, D. Overview of fast on-board integrated battery chargers for electric vehicles based on multiphase machines and power electronics. *IET Electr. Power Appl.* **2016**, *10*, 217–229. [CrossRef]
41. Subotic, I.; Bodo, N.; Levi, E.; Jones, M.; Levi, V. Isolated Chargers for EVs Incorporating Six-Phase Machines. *IEEE Trans. Ind. Electron.* **2016**, *63*, 653–664. [CrossRef]
42. Subotic, I.; Bodo, N.; Levi, E. Integration of Six-Phase EV Drivetrains into Battery Charging Process with Direct Grid Connection. *IEEE Trans. Energy Convers.* **2017**, *32*, 1012–1022. [CrossRef]
43. Bodo, N.; Levi, E.; Subotic, I.; Espina, J.; Empringham, L.; Johnson, C.M. Efficiency Evaluation of Fully Integrated On-Board EV Battery Chargers with Nine-Phase Machines. *IEEE Trans. Energy Convers.* **2017**, *32*, 257–266. [CrossRef]
44. Tong, M.; Cheng, M.; Hua, W.; Ding, S. A Single-Phase On-Board Two-Stage Integrated Battery Charger for EVs Based on a Five-Phase Hybrid-Excitation Flux-Switching Machine. *IEEE Trans. Veh. Technol.* **2020**, *69*, 3793–3804. [CrossRef]
45. Tong, M.; Cheng, M.; Wang, S.; Hua, W. An On-Board Two-Stage Integrated Fast Battery Charger for EVs Based on a Five-Phase Hybrid-Excitation Flux-Switching Machine. *IEEE Trans. Ind. Electron.* **2021**, *68*, 1780–1790. [CrossRef]

46. Raherimihaja, H.J.; Zhang, Q.; Na, T.; Shao, M.; Wang, J. A Three-Phase Integrated Battery Charger for EVs Based on Six-Phase Open-End Winding Machine. *IEEE Trans. Power Electron.* **2020**, *35*, 12122–12132. [CrossRef]
47. Raherimihaja, H.J.; Zhang, Q.; Xu, G.; Zhang, X. Integration of Battery Charging Process for EVs into Segmented Three-Phase Motor Drive with V2G-Mode Capability. *IEEE Trans. Ind. Electron.* **2021**, *68*, 2834–2844. [CrossRef]
48. Zhang, Q.; Raherimihaja, H.J.; Xu, G.; Zhang, X. Design and Performance Analysis of Segmented Three-Phase IPMSM for EVs Integrated Battery Charger. *IEEE Trans. Ind. Electron.* **2021**, *68*, 9114–9124. [CrossRef]

energies

MDPI

Review

Technical Review of Traction Drive Systems for Light Railways

Nursaid Polater * and Pietro Tricoli

Department of Electronic, Electrical and Computer Engineering, University of Birmingham, Birmingham B15 2TT, UK; p.tricoli@bham.ac.uk
* Correspondence: nxp857@bham.ac.uk

Abstract: Due to environmental concerns, governments around the world are taking measures to decarbonise railway transport by replacing diesel traction systems with cleaner alternatives. While the electrification of railway systems is spreading rapidly, it is unlikely that all routes will be electrified as the volume of passengers will not justify the high infrastructure costs. Therefore, it is expected that, for several lines, a combination of hydrogen and electric traction will be used, with the latter partly provided by fixed infrastructure and partly by batteries. Railway traction drives will then need to change to accommodate these new types of power supply. A detailed review of the available traction motors and drives is provided with this review, given application to the new hybrid-electric systems. In particular, permanent magnet synchronous motors with multiphase windings are evaluated in comparison with traditional three-phase machines. Additionally, low and medium-voltage multisource power converters have been reviewed, taking into account the introduction of wide band-gap semiconductor devices.

Keywords: traction drive systems; traction motors; power converter; fuel cell propulsion; wide band-gap devices

Citation: Polater, N.; Tricoli, P. Technical Review of Traction Drive Systems for Light Railways. *Energies* **2022**, *15*, 3187. https://doi.org/10.3390/en15093187

Academic Editors: Jelena Loncarski and Diego Bellan

Received: 28 March 2022
Accepted: 25 April 2022
Published: 27 April 2022

1. Introduction

The electrification of transportation is crucial for transport stakeholders to meet existing policies of decarbonisation that have been issued by many governments around the world [1–4]. In this regard, the International Railway Association (with 240 members worldwide) has two main objectives: reducing emissions by 2030 and switching to more sustainable traction modes by 2050. The first objective will be addressed by improving the energy management systems, while the second objective will be achieved by a modal shift scenario that involves shifting the transport of goods and passengers from road to rail, which may lead to greenhouse gas (GHG) reduction by duplicating railway passengers within the transport sector [5].

Diesel trains have higher emission levels and are noisier than electric trains. However, the electrification of railway lines has a high capital cost, and it is unlikely that all routes will be electrified. These considerations suggest that hybrid trains, i.e., those capable of travelling both under the wire and independently, will see a substantial development over the next few years. For these reasons, batteries and hydrogen fuel cells are currently considered the most viable alternatives to railway electrification [6].

Electric and non-electric traction systems are combined in hybrid traction systems [7]. Diesel–electric is a popular type of hybrid propulsion because it does not need a gearbox for power transmission [8]. Electric motors offer better torque performance in the whole speed range and only need a fixed-ratio gearbox [8]. When tank-to-wheel analysis is taken into consideration for the traction system, Battery–electric, battery–fuel cell, battery–fuel cell and electrified systems are the greener options and are being investigated extensively nowadays. The main target of the combination of battery and hydrogen propulsion is to achieve a low-carbon-emission alternative to diesel. When using batteries with fuel cells, they provide direct electrical power and operate at higher efficiencies than internal combustion engines.

For traction motors, power density and efficiency are the key properties for railway applications. The most widely used traction motors are induction motors, but there are still a large number of DC motors and synchronous motors in legacy systems, and now, permanent magnet synchronous motors have emerged for several new applications [9,10].

Traction converters are classified as DC/AC-fed, current and voltage source inverters for AC motors, which will be investigated in this article. In addition, power converters can be classified for vehicular applications as medium-voltage converters or low-voltage converters, with unidirectional, bidirectional and multilevel topologies [11]. Additionally, references [12,13] investigated a cascaded resonant DC/DC converter and a transformerless conversion unit for a light DC railway, respectively.

This paper reviews in detail the main traction drive systems used in independently powered trains, with a focus on the comparison between permanent magnet synchronous traction motors and induction traction motors. Additionally, the number of motors with dedicated converter arrangements is surveyed to understand the overall volume and weight of the traction drive system. Then, conventional power converter units are reviewed considering the recent introduction of wide-bandgap (WBG) equipped converters that promise a significant increase in the energy efficiency of the traction system.

2. Traction Drive Systems

2.1. Diesel Traction Systems

Diesel-only traction systems have been considered mainly for single railcars [14], as diesel engines are not sufficiently reliable and powerful for railway traction [15]. Dependency on diesel fuel, the level of CO_2 emissions and the overall efficiency of the diesel engine are considerable negative aspects of the traction system. Recent studies have focused on reducing emissions, especially those of hydrocarbon waste, NOx and CO [16]. These pollutants can be reduced by using greener fuel types, such as oxygenated fuels, biofuels, biodiesel, alcohols, vegetable oil, acetone–butanol–ethanol blends, dieselhols (blends of types of diesel, biodiesels and alcohols), water–diesel emulsions, etc. Some blended fuels release more NOx and have a lower energy content, making them more costly [17].

Amongst the blended fuels, biofuels are interesting for the railway transport sector due to the reduction of greenhouse gases, unburnt hydrocarbons, CO and particulate matter with the reduced increase of NOx [17]. Biofuels are compatible with ordinary diesel engines, with only minor modifications [18]. They have also been proposed in combination with fossil diesel without engine modification, which requires water-blended fuels, such as water taker apparatus, wear minimisation system and oil contamination structure [17].

2.2. Electric Traction System

Independent powered electric traction has been introduced to increase train performance by removing the power generation unit from the train. The main advantages of catenary-free electric traction systems can be listed as:

- ✓ There are no emissions at the point of use;
- ✓ Lower maintenance costs;
- ✓ Regenerative braking to reduce energy consumption;
- ✓ Lower electrified-track capital cost investment compared to electric-only traction.

The disadvantages are:

- ✗ They require a larger onboard energy storage system;
- ✗ Complex energy management;
- ✗ The cost of capital for an energy storage unit is high;
- ✗ Power interruptions affect services;
- ✗ Interference with command, control and communication systems.

Figure 1 shows the battery with an electrified multiple unit configuration, whic uses the catenary system to charge the battery and can reach an area where an overhea installation does not exist due to the battery unit. In [19], its efficiency and regenerativ braking were reviewed and compared with diesel multiple units.

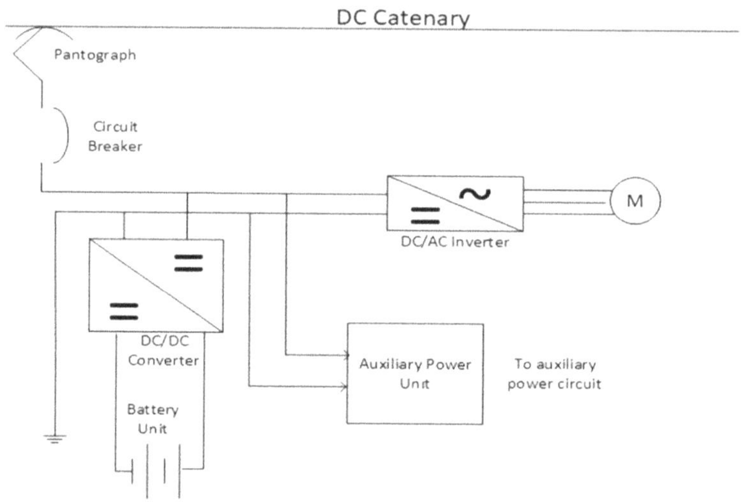

Figure 1. Electric with battery traction system [20].

2.3. Hybrid Traction System

Hybrid systems can be classified as fuel cell hybrid systems and diesel–electric systems.

A diesel–electric traction system uses a diesel engine to drive an electric generator, which then sends electricity to the traction motors, as illustrated in Figure 2.

By replacing the fixed gear ratio with mechanical transmission, diesel–electric traction has been used successfully since the 1920s [21] to overcome the problems of mechanical transmissions, which allows the gear ratio between the engine and the drive wheels to change as the vehicle speeds up and slows down [22]. On the other hand, diesel–electric traction combines the advantages of electric machines and diesel engines. Diesel–electric traction has lower emissions than diesel-only traction, better working performance and higher reliability [23], since this configuration increases the diesel engine's lifespan and provides better starting torque with electric transmission.

Existing studies have investigated the application of energy storage systems (ESSs) to increase efficiency and reduce CO_2 levels. ESSs can be used to either reduce the peak power of a diesel engine or reduce fuel consumption.

Nickel–cadmium (NiCd) batteries have been used for peak-shaving operations for shunting locomotives [24] by providing up to 25% of the demanded power. The Li-ion batteries proposed by [25] have shown a reduction of 16.5% in fuel costs. In [26], flywheels showed a similar reduction of 16.65% of fuel consumption. Both batteries and flywheels are pre-charged by either the electric power network or regenerative energy. The main benefits derived from this configuration are the reduced use of fuel and NOx emissions, which are cut by 30% [27].

Recent proposals focused on how to improve the recuperation of the braking energy in the ESS [28], as shown in Figures 3 and 4. As stated earlier, this excess energy can be stored most efficiently in flywheels, supercapacitors and batteries for railway applications. Ref. [29] addressed the solution overcharging or discharging problem of a $LiFePO_4$ battery pack by implementing the state-of-charge estimation algorithm based on the lookup table approach. In [30], the authors proposed high-speed flywheels to mitigate high-self-discharge problems depending on the load profile. The effect of the distance between the supercapacitor and the traction substation was elaborated in [31], then [32] included a control strategy with a variable threshold for trackside energy storage systems. Additionally, Ref. [33] provided a case study that used trackside flywheel energy storage and proposed a modified nodal analysis method based on train movement and electrical-network load-flow simulation. The results showed that energy recuperation increased at the overhead line.

In [26], the flywheel storage mechanism was employed to replace one of the diesel–electric locomotives as the intermediate energy power plant for three diesel–electric locomotives. In [34], the authors evaluated onboard and wayside energy storage systems to optimise energy consumption and increase performance.

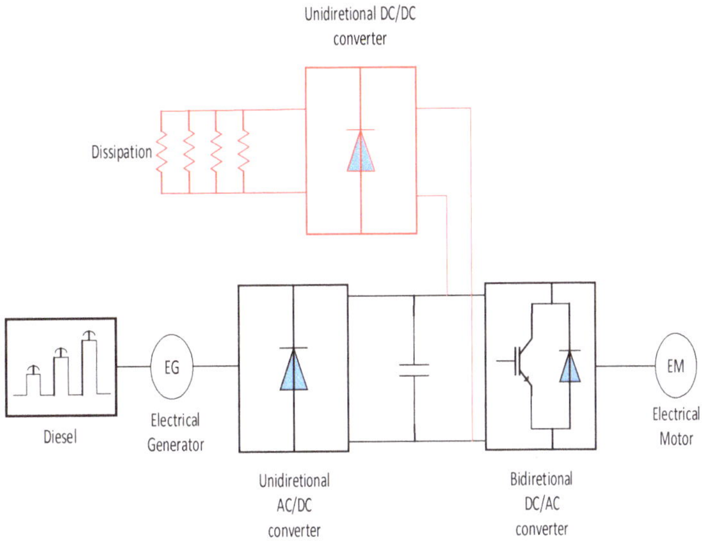

Figure 2. Schematic of diesel–electric power conversion chain with dissipation resistors [35].

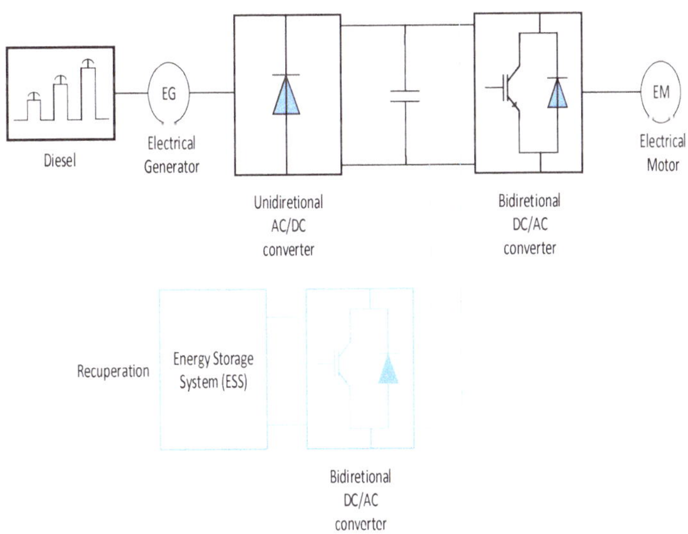

Figure 3. Block diagram of diesel–electric traction with an energy recuperation system [35].

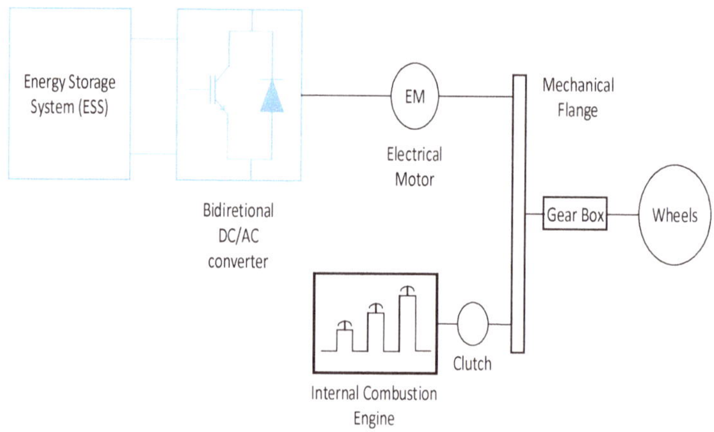

Figure 4. Parallel hybrid diesel–electric traction system [36].

Diesel with batteries is used in JR East Ki-Ha E200 and JR Hokkaido [37,38] as well as [36] as hybrid diesel cars. The main targets are the regeneration of energy and reducing the emission level by using a series hybrid system that stores the braking energy efficiently within two sets of batteries. Therefore, the diesel engine is operated and controlled by the secondary battery at its most efficient point.

Hybrid diesel traction is shown in Figure 5 and consists of a diesel generator and a catenary connection [39]. It is becoming less costly and has been discussed in literature in terms of fuel consumption [25] and architecture [40]. The existence of a battery storage unit enables one to scale down the engine size, which then requires less fuel and becomes more efficient with variable SOC references by using the algorithm given in [25]. Additionally, mild, parallel hybrid and series hybrid architectures were analysed in [39] to reduce both powertrain cost and fuel consumption. Mild powertrain architecture enabled us to save 14% on fuel in the case used for comparison.

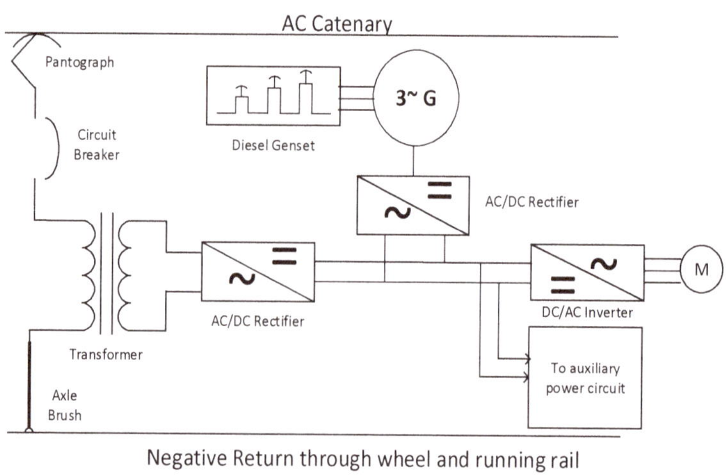

Figure 5. Hybrid diesel–electric multiple-unit system [39].

In another aspect, the fuel cell with battery system in Figure 6 is utilised more effectively as a hybrid traction system. For the example configuration, the battery unit can be connected in parallel to the fuel cell and supply energy during start-up and acceleration, as well as store energy during regenerative braking. In most circumstances, a fuel cell is designed for the train's average power demand, while a battery is designed for peak power and regenerative braking [19].

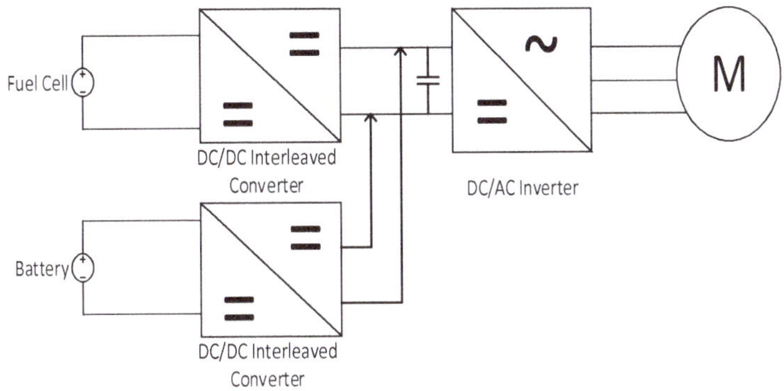

Figure 6. Fuel cell–battery hybrid traction [19].

By comparing the fuel cell with an internal combustion engine (ICE), the following statements can be derived according to the efficiency power maps given in Figure 7.

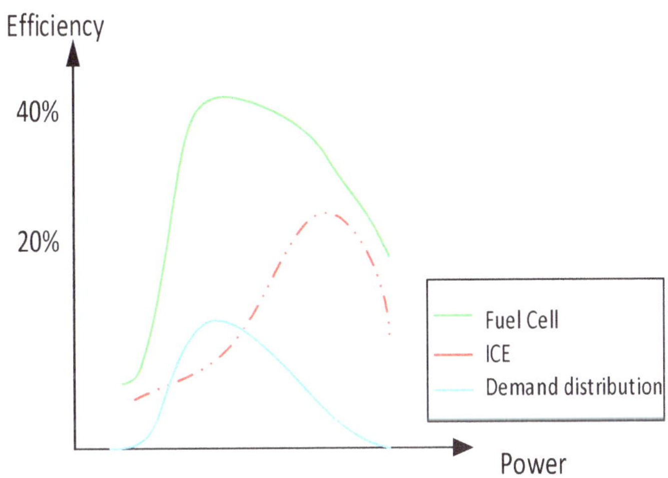

Figure 7. Fuel Cell & ICE Power Efficiency Curves [41].

- The fuel-to-wheel efficiency of the fuel cell is 40–50%;
- The fuel cell operates at its maximum efficiency of around 20–30% of its rated power, but can be increased up to 60%;
- The diesel ICE efficiency is 20–25% and a maximum occurs at 70–80% power;
- Fuel cells have better efficiency maps for driving power demands [41].

3. Traction Motors and Arrangements for Railway Applications

Although DC motors and wound-rotor synchronous motors (WRSMs) are still available on several trains, they are mostly legacy systems and will not be covered in this review, as the vast majority of recent research has focused on induction motors, permanent magnet synchronous motors and improved WRSMs, such as brushless WRSMs in several studies.

Although PMSMs' magnetic material costs and the demagnetising issue of the magnets are clear disadvantages, some studies have addressed the demagnetising issue of PMSMs and enabled them to have an equally distributed demagnetised field to achieve a smooth torque transition [42–45]. Most of the researchers adopted particular software to ove come the problems, despite the large amount of research having been carried out on thi topic [46–48]. However, the cost of magnetic materials is still a problem, since one countr is holding the market due to having all the rare earth element reserves. Therefore, brushles WRSMs have also been studied and used for modern railways recently to overcome th deficiencies of the WRSMs and provide a wider speed spectrum. However, this moto uses the third harmonic current to create a voltage across the rotor windings, which lead to high harmonic distortions and then develops ripples in the machine's electromagneti torque [49]. In [50], the authors proposed a configuration that employs stator harmoni windings, a transformer and a diode rectifier circuit to reduce the torque ripple, but i became bulky. For this reason, thyristor-controlled stator harmonic windings have emergec by offering smaller sizes and cheaper solutions, as explained in [51]. On the other hand having rotor windings will result in rotor losses, and the speed of the motor might be restricted by rotor windings, as well as the rotor integrity.

There is also another important feature of electric motors called the constant powe speed range (CPSR). The speed capability of an electric motor can be extended by using the field-weakening technique, but there are still restrictions on the limits of the motor. Fo instance, the CPSR is four or five times the rated speed of an induction motor [52], while this is five times for PMSMs [53]. This speed range can be extended by the optimisation o the reverse-salient permanent magnet synchronous motor up to 7.25 times that of the rated value [53]. Additionally, the constant power region was monitored for brushless wound rotor synchronous machines (BWRSM) in [54] and the speed was extended from 900 rpm the rated value, to 3500 rpm by using the injection of a negative id current.

Eventually, this is a trade-off to utilise traction motors that possess different advantages and disadvantages. In this regard, the paralleling of IM and PMSMs utilising fewer inverters is reviewed.

3.1. Permanent Magnet Synchronous Motors

The current traction motor trend is to use PMSMs and advanced control algorithms to improve system reliability [55].

Although the main requirement for traction motors is to require less maintenance, recent studies have focused on increasing the efficiency of traction motors. Therefore, induction motors were adopted by designers for the purpose of reducing maintenance costs, but PMSMs are more efficient. For the efficiency target of PMSMs, entirely enclosed self-cooled PMSMs for narrow gauge trains were explored and tested in [56].

The main benefits of PMSMs are their energy-saving features and their high level of power density. For instance, 200 kW PMSMs have 97% efficiency, while their induction machine counterparts have 92% [57].

There are few PMSM manufacturers for railway applications, as shown in Table 1.

Table 1. Selected light railways using permanent magnet motors [8].

Operator	Trains	Manufacturer
SNCF	31xCitadis-Dualis tram-train vehicles	Alstom
Praha	15T For City low-floor tram	Skoda
Prototypes and Other Test Trains		
München U-Bahn	C19 metro trainset with Syntegra bogies	Siemens
China	Fuel cell loco prototype	CNR Yongji
Turkey	Citadis X04 low-floor tram	Alstom
Japan	Gauge-Changing Train 2	n/a

The stator windings of permanent magnet machines are quite sensitive to failures, accounting for 50% to 90% of all failures of the machines [55–58]. There are several reasons for the winding failure of permanent magnet motors, causing the deformation of electrical insulation, which is affected by the resistive and capacitive features of the machine.

3.2. Paralleling the Motors

For railway applications, induction motors can be paralleled and fed by one inverter. The volume and cost of the traction system are reduced by paralleling multiple motors with one inverter. However, this case is not possible for synchronous motors.

It is worth mentioning that [59] proposed double-layered winding induction motors for railway traction to enable the motor parameters to be nearly the same for the paralleling purpose. A multi-machine drive system for both induction machines and permanent magnet synchronous machines used in electric multiple units was investigated in [60] in terms of a highly reliable traction system. Then, open-end-winding PMSMs were considered an alternative for the paralleling of PMSMs. Additionally, Ref. [61] indicated a load sharing problem with paralleling motors, which happened because of the fact that they do not have identical wheel sizes powered by parallel machines.

Some studies also referred to the paralleling of motors for electric vehicle applications, which might be useful to consider for railway applications. For example, Ref. [62] provided a control technique for hybrid electric vehicles that utilised IM and [63] addressed how employing two inverters can upgrade the efficiency of IM by extending the constant power region. Furthermore, a six-phase IM with a pole-changing technique was adopted in [63,64] to keep the volume within the acceptable margins while broadening the constant power region.

Paralleling two PMSMs is quite difficult because of the non-coincidence of the pole positions of multiple motors, which are unable to control the current of multiple motors simultaneously. Hence, losing the synchronism results in torque pulsations [65,66]. Most of the research has addressed how to eliminate this torque ripple for paralleled permanent magnet motors, such as damping control studied in [67], the five-leg inverter utilised in [68,69] and the nine switch inverter proposed in [70,71]. Every single one of these solutions has drawbacks, namely limited inverter output voltage, requiring voltage balancing technique, low voltage ratio due to sharing the DC-link voltage between the motors, etc. Additionally, to limit torque vibration, Refs. [72,73] drawing attention, a driver uses an extra inverter and motor windings.

To be able to understand the difference between the utilisation of PMSMs, IMs and the effect of paralleling the motors on the overall volume and weight of the traction system, a lightweight, four-carriage with eight-motor railway case study [74] was taken into consideration. Two similar datasheets [75–77] were used for 200 kW motor dimensions for the sake of a fair comparison of IMs and PMSMs. The volume and weight of the traction drive system are compared in the following charts. It is quite obvious that the inverter volume and weight are reduced in the paralleled IM case due to feeding two motors from one inverter, but PMSMs are more power-dense than IMs based on the analysis given in Figure 8. The overall comparison of the traction drive size is illustrated in Figure 9. Hence, PMSMs offer fewer volume and weight options, which means more efficient paralleling of two motors.

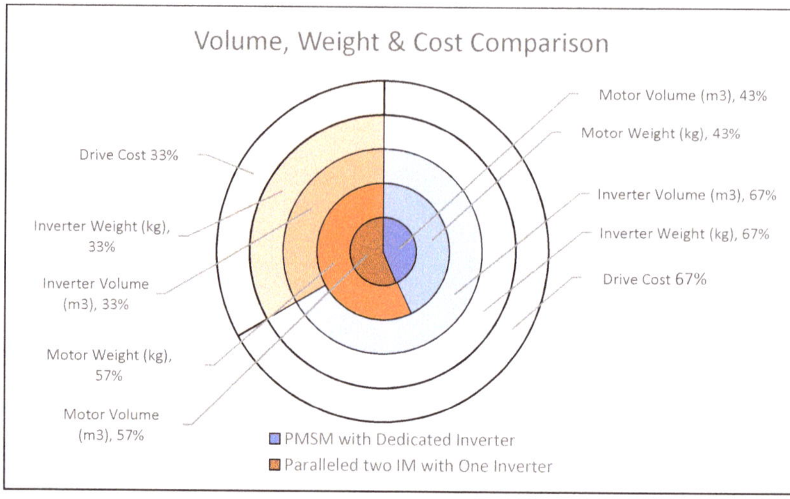

Figure 8. Volume and weight comparison charts for two configurations.

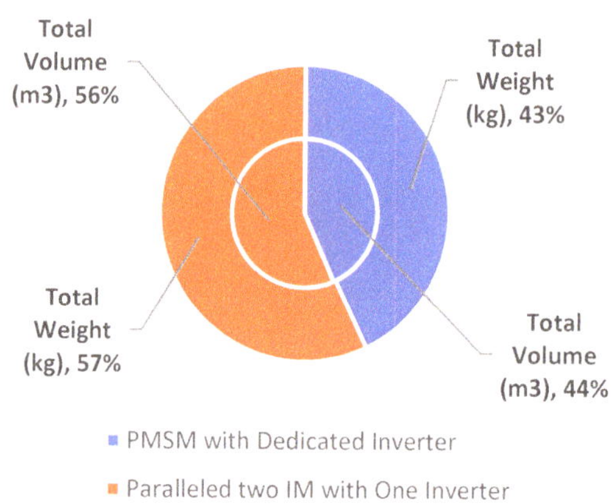

Figure 9. Overall volume and weight comparison.

3.3. Multiphase/Multi-Three Phase Structure

Despite the wide usage of three-phase machines, multiphase machines have been gaining popularity recently. The main reason for the interest in multiphase machines is their reliability, which has crucial importance, particularly for aerospace and traction applications. These machines also have other benefits, such as the lower rating of their drive circuits' components, lower harmonic distortion and lower total losses when compared to three-phase machines [78].

The other purpose of the multiphase/multi-three phase system is to make the machine windings electrically and magnetically independent to eliminate the interaction of the two or more three-phases with one another in failure. This means that one-phase breakdown does not affect the system [78]. The potential options for six-phase PMSM configurations are summarised here.

Two design ideas are common for six-phase machines according to the connection of the neutral point and phase displacement angle. In [79], dual-three-phase (0°), asymmetrical six-phases (30°) and symmetrical six-phase (0°) machines were investigated in terms of their phase displacement angle. It should be noted that dual three-phase machines are known by different names, such as asymmetrical six-phase, split-phase, or double-star machines, in many studies [80]. Therefore, the name of the asymmetrical machine is accepted as a dual-three-phase machine in this article.

As for open-end winding configurations, the most important benefit is fault tolerance when compared to their one-side powered counterparts. This topology pursues operation during the breakdown of one of the inverters or DC suppliers [81]. Additionally, the voltage is halved by using this configuration, since each winding group is energised by individual inverters located on both sides. Additionally, an extra semiconductor and passive components, such as capacitors, diodes, etc. are needed in multilevel converters to enable voltage balancing techniques. Nevertheless, the open-end topology makes the system simpler by reducing the number of passive components due to having two individual energy suppliers [79,82]. Furthermore, the extenuation of the odd-numbered harmonic content owing to the 30° phase displacement angle of the input voltage is another benefit of the open-ended configuration [83].

It is well-known that the current back-EMF of an electrical machine can remove the harmonic content in the associated currents. As a result of the lack of these back-EMF currents, massive deteriorations show on the current waveforms. Henceforth, both the availability of $6n \pm$ first harmonic components and the inexistence of current back-EMFs will be a handicap of the six-phase machines [84]. As previously stated, an open-ended topology can alleviate this problem by a 30° phase displacement angle.

Reference [85] studied asymmetrical machines, which are famous for their torque smoothness and steadiness. Fortunately, new methods have also been proposed to overcome torque ripple and fluctuation issues for symmetrical machines. For example, a high-frequency PWM was adopted by [86] to achieve a stable torque output. Moreover, Ref. [87] claimed that asymmetrical six-phase machines are capable of being fault-tolerant by implementing one neutral connection, rather than two.

Finally, there exists a study on dual-three-phase machines in [80], which stated that shifting the machine's coils 30° apart from each other will result in the mitigation of the fifth and seventh harmonic contents. Since there is no zero sequence component in this winding structure, it enables for easier control and efficient voltage utilisation [87].

The multiplication of phases can reduce the torque ripple and harmonic components in the machine's current waveform, which results in a more efficient system. Additionally, electrical power can be shared efficiently between each phase and provide much more tolerance against faults [88].

These machines can remain in operation even if one or two phases fail [89], and once torque ripple arises due to magnetic saturation, the torque density can be improved by injecting a third harmonic [90]. Dual-three-phase PMSMs achieve dual redundancy against faults with two inverter circuits [91]. Additionally, for multiphase structures, the extra switching leg provides additional redundancy [92].

However, there are some issues with multiphase machines, such as being unable to create an air-gap flux at some of the stator current harmonics [88]. All of these harmonics are limited by the stator resistance (R_S) and leakage inductance (L_S). Additionally, these harmonics can occur in the case of any voltage excitation due to a low-impedance current path, where the current flows with small resistance, regardless of whether it is normal current or fault current [93].

In terms of permanent magnet synchronous machines (PMSMs), it should be noted that back-EMF also contributes to current harmonics [91,92,94] and that the torque ripple becomes relatively larger as the amplitude of the harmonic current gets larger.

In [78], there is a comparison of three and six-phase permanent magnet machines. As inverter circuits exist, fifth and seventh harmonics are induced on the stator windings, potentially causing extra losses in multiphase machines.

Figure 10, produced based on the literature reviewed in Section 3, shows a comparison chart of traction motors in terms of efficiency, torque density, control complexity, maintenance cost and durability. It can be seen that there is no best option for every aspect, but an optimum one, multiphase PMSM, does exist. In light of the given advantages, such a smooth torque transition, torque density, control simplicity and fault-tolerant properties these machines are great candidates for railway traction systems.

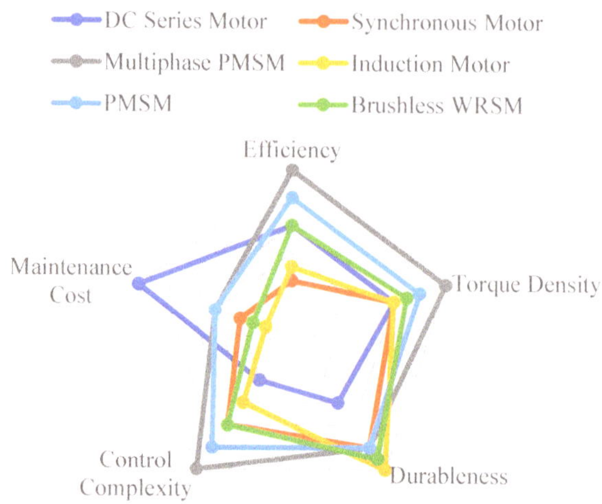

Figure 10. Spider web comparison chart of traction motors.

4. Power Converter Unit of Railway Applications with Wide Band Gap Devices

Power converter units have recently been extensively developed for railway applications. As explained for the hybrid traction drive system, inserting ESSs is one of the most popular topics, particularly for DC-powered railway systems [95]. Then, the regenerative inverter is a surrogate option for gaining excess energy through the AC grid. Additionally, different old methods were adopted to compensate for voltage drop/fluctuations, power factor correction and three-phase imbalances within a given time period, such as static VAR compensators, railway static power converters, frequency converters, etc. [95]. The most recent solutions address power electronics and switches to increase the efficiency of railway traction systems.

4.1. DC Motor Drives

Despite the fact that DC traction motor drives are practically obsolete, a summary of their capabilities is given here.

DC traction motors and drives were reviewed in [96] and are not popular nowadays because of the disadvantages of DC motors, such as the requirements for maintenance, efficiency, robustness and reliability. As such, a DC–DC chopper converter, a two-quadrant traction chopper, an AC/DC rectifier and a semi-controlled phase converter are employed as DC motor drives for railway applications.

4.2. AC Motor Drives

After the development of semiconductor switches, GTO thyristors became popular for AC motor drive circuits, and these motors have played an important role in railway traction systems so far. Traction drive systems for AC induction motors were reviewed in [97]. The main drives for AC motors, used for either induction or permanent magnet motors, are DC-fed current source inverters, DC-fed voltage source inverters and AC-fed voltage source inverters.

4.2.1. DC-Fed Current Source Inverters

Figure 11 shows how current source inverters (CSI) were utilised in electric vehicles and hybrid electric vehicles to overcome the shortcomings of voltage source inverters (VSI) [98]. They have fewer switches and a smaller dv/dt rate, as well as improved reliability in terms of excess current protection. As might be expected, the presence of large inductance restricts dynamic performance [99]. It should have a chopper circuit before the inverter to supply a constant current through the inverter. Having a DC chopper increases reliability in terms of the current defect and reduces power losses, thereby enabling a high starting torque [97].

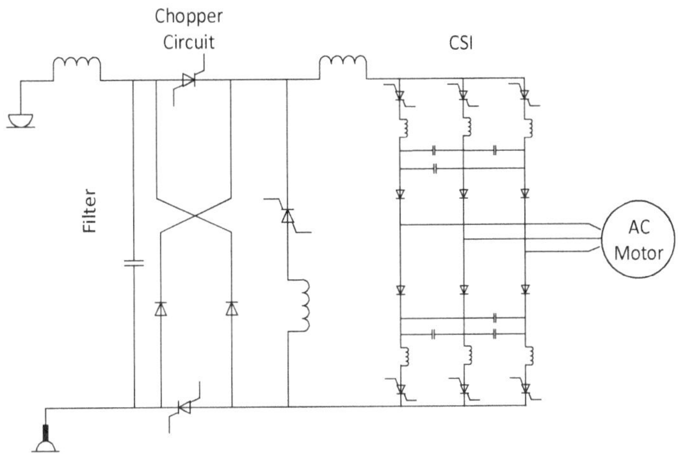

Figure 11. DC-fed CSI.

4.2.2. DC-Fed Voltage Source Inverter

Two-level and three-level voltage source inverters (VSI) are the most common types of inverters used in railway applications, and DC-fed VSIs have been fed from a DC source as shown in Figure 12. Due to its energy-saving features, regenerative braking is going to be essential for modern trains. Hence, recent development is centred around the compatibility of inverters with regenerative braking. In the literature, there are IGBT based, cost-effective, PV-supported reversible converters [100–102]. Additionally, the working characteristics of inverters need to be considered and were discussed in [103–105] in terms of voltage control techniques.

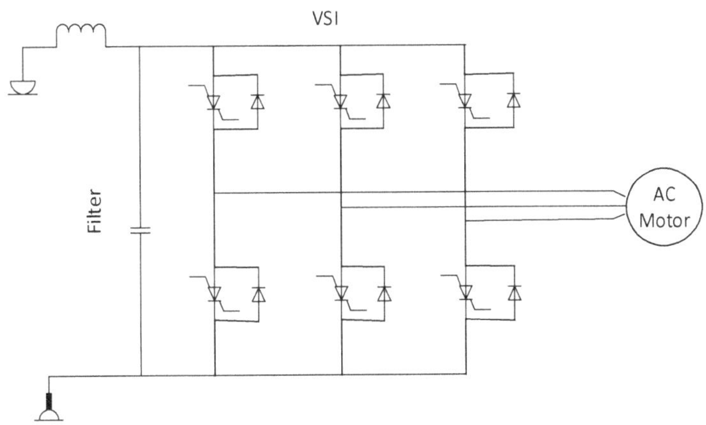

Figure 12. DC-fed VSI.

Furthermore, the requirement of a large capacitor for the filter is a disadvantage of DC railway power converter units. Therefore, PWM converters [106] and double converters [107] are employed to convert power from DC to AC systems, and [108] proposed a dual-winding isolated CUK converter that requires smaller capacitors and delivers a continuous current.

Two-level converters are most suitable for low-voltage applications (<1000 V AC RMS and <1500 DC) by utilising wide bandgap semiconductors, which offer higher efficiency with high power density [11]. On the other hand, multilevel converters provide lower switching losses and semiconductor stress with better power quality compared to two-level converters in medium voltage applications. Therefore, the comparison is based on two-level and three-level inverters.

A two-level converter generates $+V_{dc}$, $-V_{dc}$ and 0 at the output, either voltage or current. It has some restrictions in terms of frequency level that are implicitly affected by switching losses and semiconductor ratings in high-power applications [109]. However, three-level converters produce $+V_{dc}/2$, $-V_{dc}/2$ and 0 states, and have three types, namely diode-clamped, capacitor-clamped and cascaded inverters [97,109]. The main advantages of three-level converters are that they have lower DC components and a higher switching frequency that can be double that of the real value [109]. In addition, the voltage waveform became more sinusoidal with NPCs due to the fact that the rate of frequency could be four times that of the real frequency. With this contribution, the level of current ripples, torque pulsations and overall losses might be reduced [110].

Utilising fault-tolerant T-type NPCs provides fault-free operation due to having a greater degree of freedom for switching scenarios. Three-level T-NPCs also have outstanding harmonic performance and efficiency, small common-mode voltage and EMI, as well as high reliability [111]. In addition to T-NPC converters, D-NPC, two-level and three-level converters [112,113] can be considered fault-tolerant power-conditioning units. A brief comparison is given in Table 2 and [111].

Table 2. Performance comparison [111].

	Two-Level	**D-NPC Three-Level**	**T-NPC Three-Level**
IGBT Voltage	U_{DC}	$U_{DC}/2$	$U_{DC}/2$, U_{DC}
Harmonics	High	Low	Low
Loss (5–30 kHz)	Large	Middle	Small
EMI	Large	Small	Small
Fault-Tolerant Capability	Low	Middle	High

4.2.3. AC-Fed Voltage Source Inverter

AC fed VSIs are fed from DC, which is derived from the AC source, as visualised in Figure 13. It consists of an intermediate rectifier circuit. Diode clamped, fly-capacitor clamped, active neutral point clamped and multilevel converters have been investigated [114]. Most research has focused on the following issues: power quality, negative sequence current, harmonics and reactive power. For example, static VAR compensators [115], active power filters [116] and static synchronous compensators [117] were studied to mitigate the aforementioned issues. Additionally, ref. [118] provided a solution for power quality and, in particular, the neutral section.

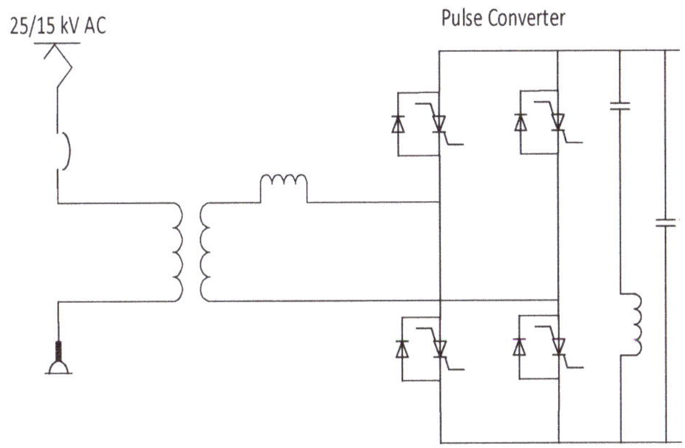

Figure 13. AC-fed VSI.

4.3. Recent Trend

Power converter units have been updated regarding the energy efficiency of the traction drive systems. Herewith, the use of energy storage units in traction drive systems has become a hot topic. The usual DC traction drive system includes a traction transformer and either a 6-pulse or 12-pulse uncontrolled rectifier, which was covered in [119]. Total harmonic distortion is the main disadvantage of this traction system. Therefore, Ref. [120] proposed adding a passive filter and [121] proposed a filter with a static var compensator. For the sake of power quality, a shunt active power filter is used to remove the harmonics in the [122,123] configuration given in Figure 14.

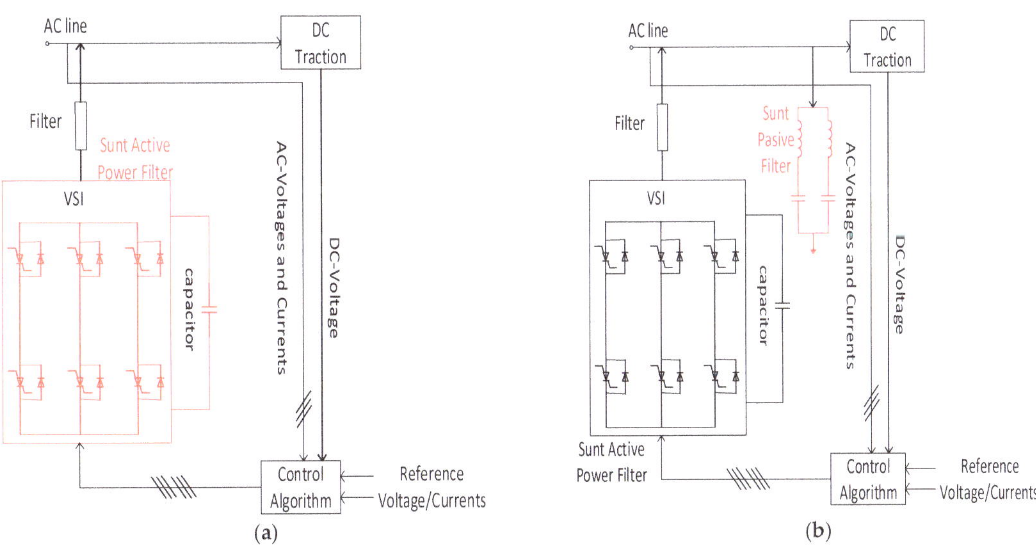

Figure 14. Shunt active power filter: (**a**) simplified model [122], (**b**) hybrid version [123].

Additionally, there exist some studies on reversible traction substations that increase the flexibility and receptivity of DC lines. For example, bidirectional reversible traction substations have been considered to improve power quality. Therefore, Siemens developed a commercial power converter unit called Sitras-TCI (thyristor line commutated inverter), as shown in Figure 15. It allows storing the braking energy at the medium voltage grid side, enabling energy transfer between the vehicles and reducing current ratings by half compared to the topology used in the forward rectifier.

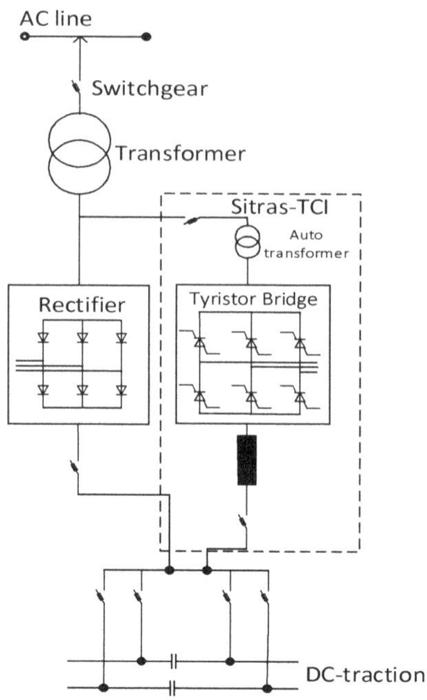

Figure 15. Siemens Sitras-TCI substation.

A traction company called INGETEAM commercialised the INGEBER system illustrated in Figure 16. The topology has an energy-recovery system (ERS) integrated between the secondary side of the transformer and the catenary line. An ERS includes an inverter, a chopper, a coupled inductor and a DC side connection. For the grid side, this configuration produces high-quality AC [124].

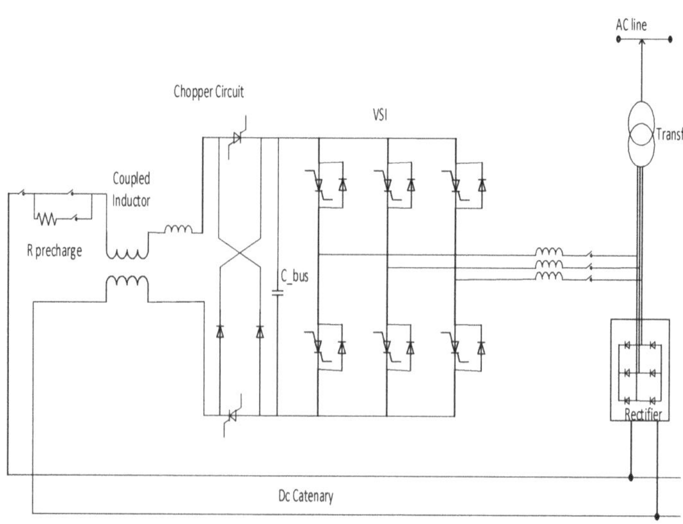

Figure 16. INGEBER configuration.

Alstom designed a converter that reduces harmonics and saves energy by paralleling a thyristor rectifier with an inverter and inserting it between the transformer and the railway network, as given in Figure 17. With this configuration, the main target is to increase the regenerative braking performance by over 99% [125].

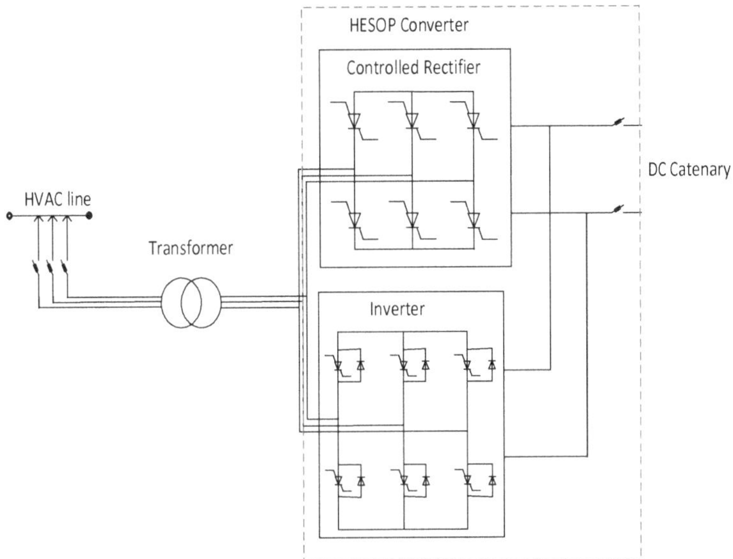

Figure 17. Harmonic and energy-saving optimizer (HESOP) converter [101].

On the other hand, HESOP is improved by the French railway company (SCNF) in Figure 18 to ameliorate the energy performance by reusing the braking energy. The control system of this substation schematic allows harmonic and short-circuit current reduction with voltage regulation [126]; additionally, it is used actively to transmit energy to the AC grid from the catenary directly.

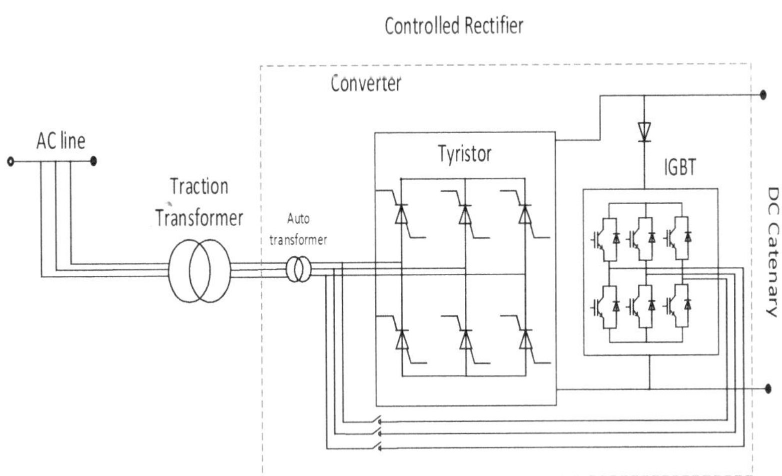

Figure 18. Improved HESOP converter by SNCF [126].

Then, Alstom proposed another HESOP configuration to combine the rectifier and the inverter shown in Figure 19. In this substation model, the converter works as a bidirectional for both rectification and inversion. It was employed on the London Underground as a trial, as illustrated in Figure 20. Therefore, it was connected separately to the substation rather than paralleling with the rectifier or using it as a bi-directional converter.

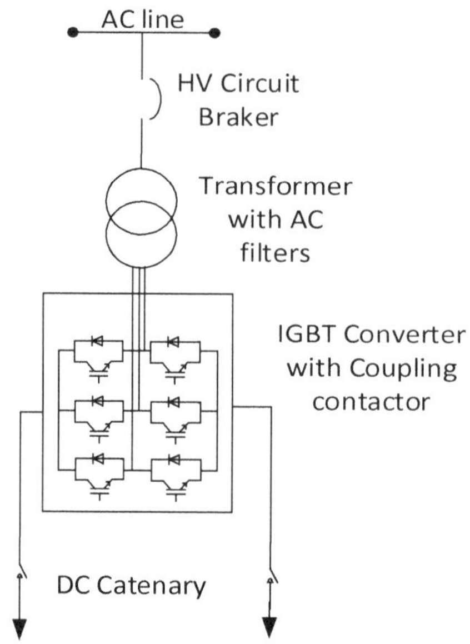

Figure 19. Updated HESOP Configuration [127].

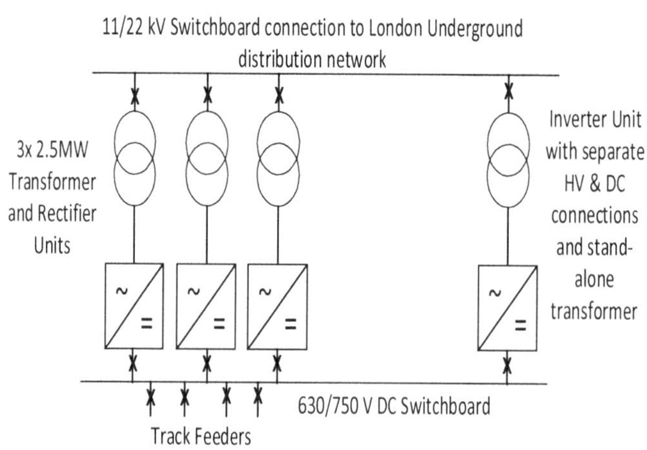

Figure 20. Upgraded version of the traction system of the London Underground [128].

4.4. Multi-Source Converters

Despite the fact that multisource converters were proposed and implemented for automobile applications, light railways should be mentioned as a potential future application.

Typically, electric vehicle power converter units use a DC link to connect the energy sources, traction inverter and the motor [126,128]. Even though the multisource concept reduces cost and control complexity, it decreases efficiency during light loads due to the utilisation of a constant DC-link voltage [129]. Recently, manufacturers have inserted a DC to DC boost converter between the DC source and DC link to overcome this issue for electric vehicle applications [128,130–132].

On the other hand, the purpose of multi-source converters is to connect independent supply sources, which might be DC sources, to the same AC output to reduce the conversion stage level. Without the use of an additional power converter, the multi-source inverter can drive a traction motor from varying DC voltages [129].

One topology has the converters being fed from two isolated DC sources [82]. The other configuration is powered by one DC supply to two inverters in a five-phase motor drive and enables the drive to have doubled output voltage as it connects each phase to an H bridge. Additionally, the concept design was given in [82], which consisted of four inverters, two sources and a common capacitor located between the coupled inverters. It is obvious that an extra capacitor is required for this topology, and a three-level output can be derived with this configuration. Technical details, such as the output current and converter performance, were discussed in [133,134].

The production of individual voltages from a few DC sources is targeted by a multi-source power converter and uses a single conversion stage. The output voltage is derived from distinct and independent input sources in a multi-source inverter as compared with a multilevel inverter that produces many output voltages from a fixed ratio of the shared DC input source [135].

A couple of multi-source converters were proposed in the literature, but some of them suffer from modularity. The others can only include two DC input sources and lack enough operation modes. However, the most important factor is modularity, which enables us to raise the number of DC sources, and that makes the system much more reliable during the breakdown of one or more sources [136]. For instance, the topology in [136] enabled an extensive number of DC sources, including fuel cells, battery packs and supercapacitors, to be employed in such a structure.

It can be clearly stated that the need for a DC/DC step-up converter is eliminated by using the multi-source inverter. The motor can be operated directly from energy storage units without boosting its voltage with a DC/DC step-up converter. Hence, the overall conversion efficiency is upgraded due to the disuse of the second boost converter in the topology given in [132].

Moreover, ref. [137] proposed two-input two-output converters that could control power flow between each source and load. These converters are then interfaced with a multilevel inverter. The outputs of the constructed converter are fed as inputs to the multilevel inverter. This proposed converter uses just one inductor, and the converter can be employed for transferring energy between different energy resources such as photovoltaic (PV), fuel cells and energy storage systems, such as batteries and supercapacitors [137].

4.5. Impact of Wide-Band-Gap Devices

The use of insulated gate bipolar transistor (IGBT) was deemed appropriate in vehicular applications, because the high switching frequency and low voltage are key parameters for the converter; therefore, owing to the lower cost and higher performance of the IGBT devices, they are the best choice for the main switches [138]. On the other hand, the newest Wide-Band-Gap (WBG) semiconductors, such as Gallium Nitride (GaN) and Silicon Carbide (SiC), offer better switching performance.

The SiC power converter unit was first introduced by Odakyu Electric Railways for a 1.5 kV DC traction inverter, which was adequate to drive four 180 kW induction motors [139]. The volume of the power converter unit was reduced by 80% compared to the traditional 4.5 kV GTO inverter [139]. A 3.3 kV SiC MOSFET is employed on Series N700S Shinkansen electric multiple units to reduce the mass of the train, and it drives four 300 kW six-pole induction motors. Ultimately, the traction power converter unit occupies 30% less space compared to the traditional IGBT converter [140]. Mitsubishi launched a 140 kVA, 600 Vdc SiC commercial product in 2013 for the auxiliary power supply system of the Tokyo Metro. It reduces power loss by up to 30% and occupies 20% less space with 15% less weight and increased regenerative energy by 51%, while it was 22.7% for the Si-based module [141]. Additionally, Alstom implemented a 1.2 kV/100 A SiC device in Line 3 o the Metro Milan for charging energy storage units and auxiliary systems [142].

Additionally, Toshiba's European branch introduced a 1.7 kV/1.2 kA Injection-Enhance Gate Transistor (IEGT), which includes SiC and a fast recovery diode for the propulsion converter [143]. Mitsubishi integrated and implemented a 3.3 kV full SiC module for a 1.5 kV DC traction inverter application that consisted of four 190 kW traction motors. By doing this, the volume and switching loss of the converter unit were decreased by 65% in size, 30% in weight and 55% in switching loss, respectively [144].

However, there are some technical challenges to the implementation of SiC power semiconductors in high-power applications. One of them is the design difficulty of parasitic inductance and the value of inductance itself, which were addressed in [145]. It is known that high voltage and current ratings, which are typically >1.7 kV and >500 A, of semiconductor modules are desired for railway applications [146]. A 3.3 kV/1500 A all-SiC traction inverter was developed with 55% fewer switching losses [147] and a 3.3 kV/500 A all-SiC module was recently designed for railway traction applications [146]. Furthermore a 1.7 kV/1.6 kA hybrid SiC, which is utilised by Si IGBT and assisted by Schottky Barrier Diode (SBD), was proposed to decrease the design complexity [148].

The deterioration of gate oxide, which is a material degraded because of the electric field, is more critical for full SiC. Since it can affect the reliability and quality control of the module [149], due to the fast switching feature of WBG devices, the gate driver needs to be designed carefully to keep the performance of the module high. Therefore, signal isolators [150], integrated circuits of gate drivers [151], thermal capability [152] of the module and the switching transients [153] of semiconductors have been redesigned intelligently to achieve the best performance. Additionally, the filtering of EMI is becoming harder with high frequency. Some solutions have been proposed to find the optimum size and value of filters [154] and eliminate coupling effects [155] on the component.

The effects of dead time on modulation and voltage/current detection are particular challenges for these devices in terms of control. Adaptive dead-time adjustment was used in [156] to meet the desired efficiency and reliability balance. The quality of power is reduced at current zero-crossing due to the high switching rate. Although this causes voltage loss and harmonic distortion, ref. [157] regulated this by duty cycle, refs. [158,159] a proposed compensation of modulation and [160–163] adjusted the feedback control to mitigate this issue.

The transition to WBG devices will increase and be more efficient with the proposed remedies in railway applications.

5. Conclusions

An extensive review of railway traction systems and their requirements is given in this article. First, the efficiency of the traction drive system and traction motor designs was investigated, followed by a detailed discussion of the power converter and power semiconductors using various topologies.

Most light railway networks are electrified within urban areas all around the world. However, there are still unelectrified and diesel-propelled railways. Therefore, the superiority of electric and hybrid electric traction against non-electrified systems is revealed, and different diesel–electric configurations are displayed.

Furthermore, two major flaws in contemporary railway systems have prompted stakeholders to examine fuel cell-powered rail rolling stock. One of them is the demand for significant investment in the construction of electric infrastructure alongside railroad tracks, despite the fact that electrification reduces carbon emissions. The other is that, despite the lack of a catenary, diesel propulsion emits the same amount of pollution as ICE-equipped vehicles on the road. As a result, for the decarbonization of vehicular applications', fuel cell battery-powered rolling stock and hybrid autonomous systems come to mind. They are both zero-emission and low-cost to install.

Afterwards, three-phase and multiphase PMSM traction motors were analysed, and the prodigiousness of PMSMs was underpinned by successfully implemented applications.

The main benefits of a multiphase machine structure are the increased torque per unit volume and reduced torque pulsations in the motor. For example, typical light railway systems have four, six or eight motors that, depending on their ratings, take up a lot of room onboard. When multiphase machines with the same ratings are compared to three-phase machines with the same ratings, the multiphase motor power is unavoidably denser. As a result, the traction motors take up less room in the car. Furthermore, torque pulsation plays a vital role in motor vibration that could be reduced with multiphase machine construction.

Multi-phase machines, as previously stated, have a higher torque density than their three-phase counterparts. Therefore, adopting multi-three-phase machines could reduce the number of motors. However, the adhesion issue also arises in this instance. The adhesion force is particularly significant for high-speed and high-power freight trains. As a consequence, there are two possibilities for railway traction drive systems: either the use of a medium-voltage converter with concentrated (fewer motors) motors or a low-voltage converter with the same number of motors.

Higher efficiency is achieved by using medium voltage and a concentrated motor converter architecture. Additionally, because a medium-level DC connection voltage is necessary, a boost converter should be employed. The number of motors will remain the same as before, but a low-voltage power converter will be used. There is no need to boost the converter in this situation, and the power element ratings are much lower, resulting in significantly lower power and switching losses. The power conditioning unit will be less expensive and more reliable than the medium-voltage topology.

Then, a more in-depth investigation of the traction power converter units was given with a brief introduction to multisource converters and WBG power semiconductor technologies. Multisource power converter units are offered for the multiphase structure to increase the reliability of the system. In addition, a few proposed structures support these converter topologies to allow fault-tolerant functioning.

Finally, WBG devices were analysed, and possible obstacles were inspected to make them more efficient for implementing railway applications. The thermal durability and packaging materials have a key role in the purpose of efficiency.

Author Contributions: Conceptualization, methodology, investigation, software, visualization, writing—original draft N.P. Supervision, revision, editing, proofreading, analysis and interpretation of data P.T. All authors have read and agreed to the published version of the manuscript.

Funding: This research was funded by the Ministry of National Education, Republic of Turkey, for doctoral studies of the first authors at the University of Birmingham under the scholarship of Higher Education, grant number 64243970/150.02/8603043.

Institutional Review Board Statement: Not applicable.

Informed Consent Statement: Not applicable.

Data Availability Statement: Not applicable.

Conflicts of Interest: The authors declare no conflict of interest.

References

1. Cuff, M. *More Trains, More Buses: How to Send UK Transport Emissions into Reverse Gear*; Business Green: London, UK, 2019.
2. Henley, W. Is it Greener to Travel by Rail or Car. The Guardian, 24 July 2013. Available online: https://www.theguardian.com/sustainable-business/greener-travel-car-or-rail (accessed on 24 April 2022).
3. Owen-Jones, J. New Study Will Assess Potential of Fuel Cell and Hydrogen for the Rail Sector. Gasworld, 21 March 2018. Available online: https://www.gasworld.com/potential-of-fuel-cells-and-hydrogen-for-the-rail-sector-/2014391.article (accessed on 24 April 2022).
4. Wiseman, E. Hydrogen Fuel Cell Trains to Run on British Railways from 2022. The Telegraph, 7 January 2019. Available online: https://www.telegraph.co.uk/cars/news/hydrogen-fuel-cell-trains-run-british-railways-2022/ (accessed on 24 April 2022).

5. Hoen, A.; van Grinsven, A.; Kampman, B.; Faber, J.; van Essen, H.; Skinner, I. *Research for TRAN Committee—Decarbonisation of EU Transport, European Parliament, Policy Department for Structural and Cohesion Policies;* European Union: Brussels, Belgium, 2017.
6. Brenna, M.; Foiadelli, F.; Zaninelli, D. *Electrical Railway Transportation Systems;* IEEE Press Wiley: Piscataway, NJ, USA, 2018.
7. Lu, S. *Optimising Power Management Strategies for Railway Traction Systems;* The University of Birmingham: Birmingham, UK, 2011
8. Hillmansen, S.; Schmid, F.; Schmid, T. The rise of the permanent-magnet traction motor. *Railw. Gaz. Int.* **2011**, *167*, 30–34.
9. Chan, C.C.; Liu, C.; Chau, K.T. Overview of permanent-magnet brushless drives for electric and hybrid electric vehicles. *IEEE Trans. Ind. Electron.* **2008**, *55*, 2246–2257.
10. Kuşdoğan, Ş. Elektrikli otomobillerde enerji depolama sistemlerindeki gelişmeler. *Mühendis Ve Makine* **2009**, *50*, 2–11.
11. Szczesniak, P. *Challenges and Design Requirements for Industrial Applications of AC/AC Power Converters without DC-Link;* University of Zielona Góra: Zielona Góra, Poland, 2019.
12. Lin, B.-R. Investigation of a resonant dc-dc converter for light rail transportation applications. *Energies* **2018**, *11*, 1078. [CrossRef]
13. Chua, T.Z.Y.; Ong, Y.T.; Toh, C.L. Transformerless DC traction power conversion system design for light-rail-transit (LRT). In Proceedings of the IEEE Energy Conversion Conference (CENCON), Kuala Lumpur, Malaysia, 30–31 October 2017.
14. Schafer, M.W. *The American Passenger Train;* MBI Publishing Company: Saint Paul, MN, USA, 2001.
15. Duffy, M.C. *Electric Railways: 1880–1990;* The Institution of Electrical Engineers: London, UK, 2003.
16. Saylam, A. Methodology to reduce diesel engine pollutant emissions. *Int. J. Petrochem. Nat. Gas* **2022**, *1*, 4–8.
17. Mwangi, J.K.; Lee, W.J.; Chang, Y.C.; Chen, C.Y.; Wang, L.C. An overview: Energy saving and pollution reduction by using green fuel blends in diesel engines. *Appl. Energy* **2015**, *159*, 214–236. [CrossRef]
18. Misra, R.D.; Murthy, M.S. Straight vegetable oils usage in a compression ignition engine—A review. *Renew. Sustain. Energy Rev.* **2010**, *14*, 3005–3013. [CrossRef]
19. Pagenkopf, J.; Kaimer, S. Potentials of Alternative Propulsion Systems for Railway Vehicles—A Techno-Economic Evaluation. In Proceedings of the Ninth International Conference on Ecological Vehicles and Renewable Energies (EVER), Monte-Carlo, Monaco, 25–27 March 2014.
20. Iwase, T.; Kawamura, J.; Tokai, K.; Kageyama, M. Development of battery system for railway vehicle. In Proceedings of the International Conference on Electrical Systems for Aircraft, Railway, Ship Propulsion and Road Vehicles (ESARS), Aachen, Germany, 3–5 March 2015.
21. Welsh, J. *Union Pacific's Streamliners;* Voyageur Press: Minneapolis, MN, USA, 2008.
22. Zimmermann, K. *Burlington's Zephyrs;* MBI Publishing Company: Saint Paul, MN, USA, 2004.
23. Hollingsworth, B.; Cook, A. *Das Handbuch der Lokomotiven;* Bechtermuenz Verlag im Weltbildverlag GmbH: Augsburg, Germany, 1996.
24. Letrouve, T.; Lhomme, W.; Pouget, J.; Bouscayrol, A. Different hybridization rate of a diesel-electric locomotive. In Proceedings of the IEEE Vehicle Power and Propulsion Conference (VPPC), Coimbra, Portugal, 27–30 October 2014.
25. Cipek, M.; Pavkovic, D.; Kljaic, Z.; Mlinaric, T.J. Assessment of battery-hybrid diesel-electric locomotive fuel savings and emission reduction potentials based on a realistic mountainous rail route. *Energy* **2019**, *173*, 1154–1171. [CrossRef]
26. Spiryagin, M.; Wolfs, P.; Szanto, F.; Sun, Y.Q.; Cole, C.; Nielsen, D. Application of flywheel energy storage for heavy haul locomotives. *Appl. Energy* **2015**, *157*, 607–618. [CrossRef]
27. Moraes, C.G.S.; Junior, S.L.B.; Heldwein, M.L.; Franca, A.S.; Vaccari, A.S.; Waltrich, G. Power Conversion Technologies for a Hybrid Energy Storage System in Diesel-Electric Locomotives. *IEEE Trans. Ind. Electron.* **2021**, *68*, 9081–9091. [CrossRef]
28. Mayrink, S.; Oliveira, J.G.; Dias, B.H.; Oliveira, L.W.; Ochoa, J.S.; Rosseti, G.S. Regenerative Braking for Energy Recovering in Diesel-Electric Freight Trains: A Technical and Economic Evaluation. *Energies* **2020**, *13*, 963. [CrossRef]
29. Gao, Z.; Chin, C.; Woo, W.; Jia, J. Integrated Equivalent Circuit and Thermal Model for Simulation of Temperature-dependent $LiFePO_4$ Battery in Actual Embedded Application. *Energies* **2017**, *10*, 85. [CrossRef]
30. Irena, A. *Electricity Storage and Renewables: Costs and Markets to 2030;* International Renewable Energy Agency: Abu Dhabi, United Arab Emirates, 2017.
31. Killer, A.; Armstorfer, A.; Diez, A.E.; Biechl, H. Ultracapacitor Assisted Regenerative Braking in Metropolitan Railway Systems. In Proceedings of the 2012 IEEE Colombian Intelligent Transportation Systems Symposium (CITSS), Bogota, Colombia, 30 August 2012.
32. Lin, F.; Li, X.; Zhao, Y.; Yang, Z. Control Strategies with Dynamic Threshold Adjustment for Supercapacitor Energy Storage System Considering the Train and Substation Characteristics in Urban Rail Transit. *Energies* **2016**, *9*, 257. [CrossRef]
33. Gee, A.M.; Dunn, R.W. Analysis of Trackside Flywheel Energy Storage in Light Rail Systems. *IEEE Trans. Veh. Technol.* **2015**, *64*, 3858–3869. [CrossRef]
34. García-Garre, A.; Gabaldón, A. Analysis, Evaluation and Simulation of Railway Diesel-Electric and Hybrid Units as Distributed Energy Resources. *Appl. Sci.* **2019**, *9*, 3605. [CrossRef]
35. Destraz, B.; Barrade, P.; Rufer, A. *A Solution for Increasing the Efficiency of Diesel—Electric Locomotives with Supercapacitive Energy Storage;* Laboratory of Industrial Electronics Swiss Federal Institute of Technology Lausanne: Lausanne, Switzerland, 2004.
36. HybridFLEX Battery-Diesel Train, UK. *Railway Technology;* 2021. Available online: https://www.electricandhybridrail.com/content/news/hybridflex-battery-diesel-train-enters-passenger-service-in-london-uk#:~{}:text=British%20train%20operating%20company%20Chiltern,the%20UK\T1\textquoterights%20national%20rail%20network (accessed on 24 April 2022).

37. Ihara, H.; Kakinuma, H.; Sato, I.; Ihaba, T.; Anada, K.; Morimoto, M.; Oda, T.; Kobayashi, S.; Ono, T.; Karasawa, R. Development of Motor-Assisted Hybrid Traction System. In Proceedings of the World Congress on Railway Research (WCRR) 2008, Seoul, Korea, 18–20 May 2008.
38. Furuta, R.; Kawasaki, J.; Kondo, K. Hybrid Traction Technologies with Energy Storage Devices for Nonelectrified Railway Lines. *IEEJ Trans. Electr. Electron. Eng.* **2010**, *5*, 291–297. [CrossRef]
39. Olmos, J.; Gandiaga, I.; Lopez, D.; Larrea, X.; Nieva, T.; Aizpuru, I. Impact of Li-ion Battery Technology and Energy Management on the Life Cycle Costs of Hybrid Diesel-Electric Multiple Units. In Proceedings of the Fifteenth International Conference on Ecological Vehicles and Renewable Energies (EVER), Monte-Carlo, Monaco, 10–12 September 2020.
40. Giglioli, R.; Lutzemberger, G.; Poli, D.; Sani, L. Hybridisation of Railcars for Usage in Non-Electrified Lines. In Proceedings of the 6th International Conference on Clean Electrical Power (ICCEP), Santa Margherita Ligure, Italy, 27–29 June 2017.
41. Walker, G. *Technologies for the Hydrogen Transport Economy-Lecture Notes*; University of Nottingham: Nottingham, UK, 2016.
42. Arkkio, S.R.A. Partial demagnetization of permanent magnets in electrical machines caused by an inclined field. *IEEE Trans. Magn.* **2008**, *44*, 1773–1778.
43. Ruoho, S.; Haavisto, M.; Takala, E.; Santa-Nokki, T.; Paju, M. Temperature dependence of resistivity of sintered rare-earth permanent-magnet materials. *IEEE Trans. Magn.* **2010**, *46*, 15–20. [CrossRef]
44. Katter, M. Angular dependence of the demagnetization stability of sintered Nd-Fe-B magnets. *IEEE Trans. Magn.* **2005**, *41*, 3853–3855. [CrossRef]
45. Goldemberg, C.; Lebensztajn, L.; Lobosco, O.S. Analysis of short-circuit transients of a PM machine. In Proceedings of the IEEE International 41 Electric Machines and Drives Conference Record, Milwaukee, WI, USA, 18–21 May 1997.
46. Ruoho, S.; Dlala, E.; Arkkio, A. Comparison of Demagnetization Models for Finite-Element Analysis of Permanent Magnet Synchronous Machines. *IEEE Trans. Magn.* **2007**, *43*, 3964–3968. [CrossRef]
47. Ruschetti, C.; Verucchi, C.; Bossio, G.; Angelo, C.D.; García, G. Rotor demagnetization effects on permanent magnet synchronous machines. *Energy Convers. Manag.* **2013**, *74*, 1–8. [CrossRef]
48. Kim, K.-C.; Lim, S.-B.; Koo, D.-H.; Lee, J. The shape design of permanent magnet for permanent magnet synchronous motor considering partial demagnetization. *IEEE Trans. Magn.* **2006**, *42*, 3485–3487. [CrossRef]
49. Yao, F.; An, Q.; Sun, L.; Lipo, T.A. Performance investigation of a brushless synchronous machine with additional harmonic _eld windings. *IEEE Trans. Ind. Electron.* **2016**, *63*, 6756–6766. [CrossRef]
50. Jawad, G.; Ali, Q.; Lipo, T.A.; Kwon, B.-I. Novel brushless wound rotor synchronous machine with zero-sequence third-harmonic field excitation. *IEEE Trans. Magn.* **2016**, *52*, 1–4. [CrossRef]
51. Sirewal, G.J.; Ayub, M.; Kwon, S.A.B.-I. Analysis of a Brushless Wound Rotor Synchronous Machine Employing a Stator Harmonic Winding. *IEEE Access* **2020**, *8*, 151392–151402. [CrossRef]
52. Patil, N.A.; Lawler, J.S.; McKeever, J.W. Determining Constant Power Speed Ratio of the Induction Motor from Equivalent Circuit Parameters. In Proceedings of the IEEE SoutheastCon, Huntsville, AL, USA, 3–6 April 2008.
53. Zhao, X.; Kou, B.; Huang, C.; Zhang, L. Optimization Design and Performance Analysis of a Reverse-Salient Permanent Magnet Synchronous Motor. *Machines* **2022**, *10*, 204. [CrossRef]
54. Ayub, M.; Atiq, S.; Ali, Q.; Kwon, B. Dual-Mode Wound Rotor Synchronous Machine for Variable Speed Applications. *IEEE Access* **2020**, *8*, 115812–115822. [CrossRef]
55. Sun, X.; Hu, C.; Zhu, J.; Wang, S.; Zhou, W.; Yang, Z.; Lei, G.; Li, K.; Zhu, B.; Guo, Y. MPTC for PMSMs of EVs with multi-motor driven system considering optimal energy allocation. *IEEE Trans. Magn.* **2019**, *55*, 1–6. [CrossRef]
56. Matsuoka, K.; Kondo, M. Energy Saving Technologies for Railway Traction Motors. *IEEJ Trans. Electr. Electron. Eng.* **2010**, *5*, 278–284. [CrossRef]
57. Kondo, K. Recent Energy Saving Technologies on Railway Traction Systems. *IEEJ Trans. Electr. Electron. Eng.* **2010**, *5*, 298–303. [CrossRef]
58. Ryabinin, I.A.; Parfenov, Y.M. *Reliability, Survivability and Safety of Ships Electric Power Systems*; Kuzentsov Naval Academy: St. Petersburg, Russia, 1997.
59. Bernatt, J.; Gawron, S.; Glinka, T.; Polak, A. Traction induction motor. In Proceedings of the 13th International Conference Modern Electrified Transport (MET), Warsaw, Poland, 5–7 October 2017.
60. Mohammadi Pirouz, H.; Hajizadeh, A. A Highly Reliable Propulsion System with Onboard Uninterruptible Power Supply for Train Application: Topology and Control. *Sustainability* **2020**, *12*, 3943. [CrossRef]
61. Nategh, S.; Boglietti, A.; Liu, Y.; Barber, D.; Brammer, R.; Lindberg, D.; Aglen, O. A Review on Different Aspects of Traction Motor Design for Railway Applications. *IEEE Trans. Ind. Appl.* **2020**, *56*, 2148–2157. [CrossRef]
62. Kou, B.; Li, L.; Cheng, S.; Meng, F. Operating control of efficiently generating induction motor for driving hybrid electric vehicle. *IEEE Trans. Magn.* **2005**, *41*, 488–491.
63. Kelly, J.W.; Strangas, E.G.; Miller, J.M. Control of a continuously operated pole-changing induction machine. *IEEE Int. Conf. Electr. Mach. Drives* **2003**, *1*, 211–217.
64. Jiang, S.; Chau, K.; Chan, C. Spectral analysis of a new six-phase pole-changing induction motor drive for electric vehicles. *IEEE Trans. Ind. Electron.* **2003**, *50*, 123–131. [CrossRef]
65. Bouscayrol, A.; Pietrzak-David, M.; Delarue, P.; Pefta-Eguiluz, R.; Kestelyn, P.V. Weighted Control of Traction Drives with Parallel-Connected AC Machine. *IEEE Trans. Ind. Electron.* **2006**, *53*, 1799–1806. [CrossRef]

66. Nagano, T.; Itoh, J. Parallel Connected Multiple Drive System Using Small Auxiliary Inverter for Numbers of PMSM. In Proceedings of the International Power Electronics Conference (IPEC-Hiroshima 2014-ECCE ASIA), Hiroshima, Japan, 18–21 May 201
67. Itoh, J.-I.; Nomura, N.; Ohsawa, H. A Comparison between V/f Control and Position-Sensorless Vector Control for the Permanent Magnet Synchronous Motor. In Proceedings of the Power Conversion Conference (PCC), Osaka, Japan, 2–5 April 2002.
68. Nozawa, Y.; Oka, K.; Omata, R.; Matsuse, K. Performance for Position Control of Two Permanent Magnet Synchronous Motor with the Five-Leg Inverter. In Proceedings of the IECON 2006-32nd Annual Conference on IEEE Industrial Electronics, Paris, France, 6–10 November 2006.
69. Ibrahim, Z.; Lazi, I.M.; Sulaiman, M. Independent Speed Sensorless Control of Dual Parallel PMSM Based on Five-Leg Inverter. In Proceedings of the 9th International Multi-Conference on Systems, Signals and Devices, Chemnitz, Germany, 20–23 March 2012.
70. Dehnavi, S.M.D.; Mohamadian, M.; Yazdian, A.; Ashrafzadeh, F. Space Vectors Modulation for Nine-Switch Converters. *IEEE Trans. Power Electron.* **2010**, *25*, 1488–1496. [CrossRef]
71. Gao, F.; Zhang, L.; Li, D.; Loh, P.C.; Tang, Y.; Gao, H. Optimal Pulsewidth Modulation of Nine-Switch Converter. *IEEE Trans. Power Electron.* **2010**, *25*, 2331–2342. [CrossRef]
72. Nagano, T.; Itoh, J. Design of multi-parallel drive technique for system with numbers of Permanent Magnet Synchronous Motors. In Proceedings of the IEEE 10th International Conference on Power Electronics and Drive Systems (PEDS), Kitakyushu, Japan, 22–25 April 2013.
73. Nagano, T.; Chiang, G.T.; Itoh, J.; Kato, K. Verification of parallel connected multiple motor drive system with numbers of permanent magnet synchronous motors. In Proceedings of the 15th European Conference on Power Electronics and Applications (EPE), Lille, France, 2–6 September 2013.
74. Ratniyomchai, T. *Optimal Design and Control of Stationary Electrochemical Double-Layer Capacitors for Light Railways*; The University of Birmingham: Birmingham, UK, 2016.
75. Vem-Group: VEM Motors GmbH. Available online: https://www.vem-group.com/fileadmin/content/pdf/Download/Kataloge/Kataloge/pm_en.pdf (accessed on 21 April 2022).
76. Vem-Group: VEM Holding GmbH. 2017. Available online: https://www.vem-group.com/fileadmin/content/pdf/Download/Kataloge/Kataloge/Hauptkatalog_2017_KAP12_en.pdf (accessed on 21 April 2022).
77. WEG S/A, Inverterdrive. Industrial Automation Distributor. Available online: https://inverterdrive.com/group/Motors-AC/ac-Motor-WEG-12862505/ (accessed on 10 December 2021).
78. Khan, K.S. *Comparative Analysis of Multiphase Machines*; Chalmers University: Göteborg, Sweden, 2008.
79. Che, H.S.; Hew, W.P. Dual Three-phase Operation of Single Neutral Symmetrical Six-phase Machine for Improved Performance. In Proceedings of the IECON 2015-41st Annual Conference of the IEEE Industrial Electronics Society Yokohama, Yokohama, Japan, 9–12 November 2015.
80. Khan, K.S.; Arshad, W.M.; Kanerva, S. On performance figures of multiphase machines. In Proceedings of the 2008 18th International Conference on Electrical Machines, Vilamoura, Portugal, 6–9 September 2008.
81. Wang, Y.; Lipo, T.A.; Pan, D. Robust Operation of Double-Output AC Machine Drive. In Proceedings of the Conference on Power Electronics-ECCE Asia (ICPE & ECCE), Jeju, Korea, 30 May–3 June 2011.
82. Salem, A.; Narimani, M. A Review on Multiphase Drives for Automotive Traction Applications. *IEEE Trans. Transp. Electrif.* **2019**, *5*, 1329–1348. [CrossRef]
83. Mohapatra, K.K.; Gopakumar, K. A novel split phase induction motor drive without harmonic filters and with linear voltage control for the full modulation range. *EPE J.* **2006**, *16*, 20–28. [CrossRef]
84. Azeez, N.A.; Mathew, J.; Gopakumar, K.; Cecati, C. A 5th and 7th Order Harmonic Suppression Scheme for Open-end Winding Asymmetrical Six-phase IM Drive Using Capacitor-fed Inverter. *IECON Proc.* **2013**, *1*, 5118–5123.
85. Nelson, R.H.; Krause, P.C. Induction machine analysis for arbitrary displacement between multiple winding sets. *IEEE Trans. Power Appar. Syst.* **1974**, *PAS-93*, 841–848.
86. Dujic, D.; Iqbal, A.; Levi, E. A space vector PWM technique for symmetrical six-phase voltage source inverters. *EPE J.* **2007**, *17*, 24–32. [CrossRef]
87. Munim, W.N.W.A.; Duran, M.J.; Che, H.S.; Bermúdez, M.; González-Prieto, I.; Rahim, N.A. A unified analysis of the fault tolerance capability in six-phase induction motor drives. *IEEE Trans. Power Electron.* **2017**, *32*, 7824–7836. [CrossRef]
88. Levi, E. Multiphase electrical machines for variable-speed applications. *IEEE Trans. Ind. Electron.* **2008**, *55*, 1893–1909. [CrossRef]
89. Valente, G.; Papini, L.; Formentini, A.; Gerada, C.; Zanchetta, P. Open-Circuit Fault Tolerant Study of Bearingless Multi-Sector Permanent Magnet Machines. In Proceedings of the 2018 International Power Electronics Conference, Niigata, Japan, 20–24 May 2018.
90. Hu, Y.; Zhu, Z.Q.; Odavic, M. Torque Capability Enhancement of Dual Three-Phase PMSM Drive with Fifth and Seventh Current Harmonics Injection. *IEEE Trans. Ind. Appl.* **2017**, *53*, 4526–4535. [CrossRef]
91. Kuang, X.; Guo, H.; Xu, J.; Zhou, T. Research on a six-phase permanent magnet synchronous motor system at dual-redundant and fault tolerant modes in aviation application. *Chin. J. Aeronaut.* **2017**, *30*, 1548–1560. [CrossRef]
92. Nobel, I.O. Fault Tolerant Control and Reliability Assessment of Modular PMSM Drive Systems. Master's Thesis, Delft University of Technology, Delft, The Netherlands, 2018.
93. Karttunen, J.; Kallio, S.; Honkanen, J.; Peltoniemi, P.; Silventoinen, P. Stability and performance of current harmonic controllers for multiphase PMSMs. *Control Eng. Pract.* **2017**, *65*, 59–69. [CrossRef]

94. Hu, Y.; Zhu, Z.; Liu, K. Current Control for Dual Three-Phase Permanent Magnet Synchronous Motors Accounting for Current Unbalance and Harmonics. *IEEE J. Emerg. Sel. Top. Power Electron.* **2014**, *2*, 272–284.
95. Hayashiya, H.; Kondo, K. Recent Trends in Power Electronics Applications as Solutions in Electric Railways. *IEEJ Trans. Electr. Electron. Eng.* **2020**, *15*, 632–645. [CrossRef]
96. Hill, R. Electric railway traction: Part 1 Electric traction and DC traction motor drives. *Power Eng. J.* **1994**, *8*, 47–56. [CrossRef]
97. Hill, R.J. Electric railway traction: Part 2 traction drives with three-phase induction motors. *Power Eng. J.* **1994**, *8*, 143–152. [CrossRef]
98. Su, G.J.; Tang, L. Current source inverter based traction drive for EV battery charging applications. In Proceedings of the 2011 IEEE Vehicle Power and Propulsion Conference, Chicago, IL, USA, 6–9 September 2011.
99. Khosravi, M.; Larijani, S.N.; Fazel, S.S. Analytical Review of Voltage Source, Current Source and Impedance Source Inverters for Electrical Drive Applications. In Proceedings of the 31th Power System Conference, Tehran, Iran, 24–26 October 2016.
100. Zhang, G.; Tian, Z.; Du, H.; Liu, Z. A Novel Hybrid DC Traction Power Supply System Integrating PV and Reversible Converters. *Energies* **2018**, *11*, 1661. [CrossRef]
101. Cornic, D. Efficient recovery of braking energy through a reversible dc substation. In Proceedings of the Electrical Systems for Aircraft, Railway and Ship Propulsion (ESARS), Bologna, Italy, 19–21 October 2010.
102. Nomura, J.; Inagaki, A.K.K. Development of a Hybrid Inverter and a Hybrid Converter for an electric railway. In Proceedings of the 2007 Power Conversion Conference, Nagoya, Japan, 2–5 April 2007.
103. Szeląg, W.J.A. The multi-criteria optimization method for implementation of a regenerative inverter in a 3 kV DC traction system. *Electr. Power Syst. Res.* **2018**, *161*, 61–73.
104. Bae, C.H. A simulation study of installation locations and capacity of regenerative absorption inverters in DC 1500 V electric railways system. *Simul. Model. Pract. Theory* **2009**, *17*, 829–838. [CrossRef]
105. Bae, C.H.; Han, M.S.; Kim, Y.K.; Choi, C.Y.; Jang, S.J. Simulation Study of Regenerative Inverter for DC Traction Substation. In Proceedings of the International Conference on Electrical Machines and Systems, Nanjing, China, 27–29 September 2005.
106. Sonetaka, M.; Kaneko, T. Development and installation of PWM converter for Tsukuba express. *Railw. Electr. Eng.* **2005**, *16*, 12–16.
107. Harada, T. Regenerative energy utilization in abroad and examples of installations of regenerative energy utilization devices. *Railw. Electr. Eng.* **2015**, *26*, 44–48.
108. Alotaibi, S.; Darwish, A.; Ma, X.; Williams, B.W. A New Four-Quadrant Inverter Based on Dual-Winding Isolated Cuk Converters for Railway and Renewable Energy Applications. In Proceedings of the 10th International Conference on Power Electronics, Machines and Drives, Online, 15–17 December 2020.
109. Honade, T.; Udapure, S.; Timande, S.; Rodge, S.; Burde, V.; Gudadhe, S. Comparative Study Between Two And Three Level Converter For Electric Application. *Int. J. Adv. Eng. Technol.* **2016**, *2*, 529–540.
110. Wang, X.; Wang, Z.; Cheng, M.; Hu, Y. Remedial Strategies of T-NPC Three-Level Asymmetric Six-phase PMSM Drives Based on SVM-DTC. *IEEE Trans. Ind. Electron.* **2017**, *64*, 6841–6853. [CrossRef]
111. Wang, Z.; Zhang, B.; Cheng, Y.W.M. Analysis and Control of Active Neutral-Point-Clamping Three-Level Inverters Under Fault Tolerant Operation Modes. In Proceedings of the International Conference on Electrical, Pattaya, Thailand, 25–28 October 2015; pp. 2140–2146.
112. Correa, M.B.R.; Jacobina, C.B.; Silva, E.R.C.; Lima, A.M.N. A general PWM strategy for four-switch three-phase inverters. *IEEE Trans. Power Electron.* **2006**, *21*, 1618–1627. [CrossRef]
113. Ye, Z.; Chen, A.; Mao, S.; Wang, T.; Yu, D.; Deng, X. A Novel Three-Level Voltage Source Converter for AC–DC–AC Conversion. *Energies* **2018**, *11*, 1147. [CrossRef]
114. Lokhande, S.; Patil, S.; Shende, K.; Patil, D.; Mulla, A. Introduction to FC-TBSR based SVC for voltage regulation and reactive power compensation. In Proceedings of the IEEE International Conference on Power Electronics, Drives and Energy Systems (PEDES), Trivandrum, India, 14–17 December 2016.
115. Kwon, K.-M.; Song, Y.-S.; Choi, J. 6MVA single-phase APF for high speed train line in Korea. In Proceedings of the IEEE Conference on Power Engineering and Renewable Energy (ICPERE), Bali, Indonesia, 9–11 December 2014.
116. Ahmadinia, M.; Ghazi, R. Coordinated control of STATCOM and ULTC to reduce capacity of STATCOM. In Proceedings of the Iranian Conference Electrical Engineering (ICEE), Mashhad, Iran, 8–10 May 2018.
117. Li, L.; Wu, M.; Wu, S.; Li, J.; Song, K. A Three-Phase to Single-Phase AC-DC-AC Topology Based on Multi-Converter in AC Electric Railway Application. *IEEE Access* **2019**, *7*, 111539–111558. [CrossRef]
118. Popescu, M.; Bitoleanu, A. A Review of the Energy Efficiency Improvement in DC Railway Systems. *Energies* **2019**, *12*, 1092. [CrossRef]
119. Djeghader, Y.; Zellouma, L.; Labar, U.; Toufouti, R.; Chelli, Z. Study and filtering of harmonics in a dc electrified railway system. In Proceedings of the 7th International Conference on Modelling, Identification, Sousse, Tunisia, 18–20 December 2015.
120. Martinez, J.; Ramos, G. Reactive power and harmonic distortion control in electric traction systems. In Proceedings of the IEEE/PES Transmission and Distribution Conference and Exposition: Latin America, Sao Paulo, Brazil, 8–10 November 2010.
121. Ramos, G.; Cantor, E.; Rios, M.A.; Roa, L.F. Instantaneous p-q Theory for Harmonic Compensation with Active Power Filter in DC Traction Systems. In Proceedings of the International Conference on Power Engineering, Energy and Electrical Drives, Malaga, Spain, 11–13 May 2011.

122. Hosseini, S.H.; Shahnia, F.; Sarhangzadeh, M.; Babaei, E. Power Quality Improvement of DC electrified Railway Distribution Systems Using Hybrid Filters. In Proceedings of the International Conference on Electrical Machines and Systems, Nanjing, China, 27–29 September 2005.
123. Devaux, F.O.; Tackoen, X. WP2B Energy Recovery. Overview of Braking Energy Recovery Technologies in the Public Transport Field. March 2011. Available online: https://www.stib-mivb.be/irj/go/km/docs/resource/tickettokyoto/sites/default/files/downloads/T2K_ER_overview_technologies_web.pdf (accessed on 21 April 2022).
124. González-Gil, A.; Palacin, R.; Batty, P. Sustainable urban rail systems: Strategies and technologies for optimal management of regenerative braking energy. *Energy Convers. Manag.* **2013**, *75*, 374–388. [CrossRef]
125. Warin, Y.; Lanselle, R.; Thiounn, M. Active substation. In Proceedings of the 9th World Congress on Railway Research (WCRR 2011), Lille, France, 22–26 May 2011.
126. ALSTOM, "RE-USE Layman's Report: September 2012–May 2018. Available online: https://www.alstom.com/sites/alstom.com/files/2018/10/30/re-use_laymans_report_en.pdf (accessed on 21 April 2022).
127. Gray, A.J.; Stinton, D. Designing reversible substations using inverters. In Proceedings of the 7th IET Professional Development Course on Railway Electrification Infrastructure and Systems (REIS 2015), London, UK, 8–11 June 2015.
128. Burress, T.; Campbell, S. Benchmarking EV and HEV power electronics and electric machines. In Proceedings of the IEEE Transportation Electrification Conference and Expo, Detroit, MI, USA, 16–19 June 2013.
129. Dorn-Gomba, L.; Magne, P.; Danen, B.; Emadi, A. On the Concept of the Multi-Source Inverter for Hybrid Electric Vehicle Powertrain. *IEEE Trans. Power Electron.* **2018**, *33*, 7376–7386. [CrossRef]
130. Burress, T.A.; Campbell, S.L.; Coomer, C.; Ayers, C.W.; Wereszczak, A.A.; Cunningham, J.P.; Marlino, L.D.; Seiber, L.E.; Lin, H.T. *Evaluation of the 2010 Toyota Prius Hybrid Synergy Drive System*; Oak Ridge National Lab: Oak Ridge, TN, USA, 2008.
131. Staunton, R.; Burress, T.A.; Marlino, L.D. *Evaluation of the 2005 Honda Accord Hybrid Synergy Drive System*; Oak Ridge National Lab: Oak Ridge, TN, USA, 2006.
132. Dorn-Gomba, L.; Chemali, E.; Emadi, A. Novel Hybrid Energy Storage System Using the Multi-Source Inverter. In Proceedings of the EEE Applied Power Electronics Conference and Exposition (APEC), Antonio, TX, USA, 4–8 March 2018.
133. Loncarski, J.; Leijon, M.; Rossi, C.; Srndovic, M.; Grandi, G. Current ripple evaluation in dual three-phase inverters for open-end winding EV drives. In Proceedings of the International Conference Connected Vehicles Expo (ICCVE), Vienna, Austria, 3–7 November 2014; pp. 507–513.
134. Patkar, F.; Jidin, A.; Levi, E.; Jones, M. Performance comparison of symmetrical and asymmetrical six-phase open-end winding drive with carrier-based PWM. In Proceedings of the 6th International Conference on Electrical Engineering and Informatics (ICEEI), Langkawi, Malaysia, 25–27 November 2017; pp. 1–6.
135. Abu-Rub, H.; Holtz, J.; Rodriguez, J.; Baoming, G. Medium-voltage multilevel converters state of the art, challenges, and requirements in industrial applications. *IEEE Trans. Ind. Electron.* **2010**, *57*, 2581–2596. [CrossRef]
136. Salari, O.; Nouri, M.; Zaad, K.H.; Bakhshai, A.; Jain, P. A Multi-Source Inverter for Electric Drive Vehicles. In Proceedings of the IEEE Energy Conversion Congress and Exposition (ECCE), Portland, OR, USA, 23–27 September 2018.
137. Umamaheswari, N.; Priya, M.B.; Sengolan, M. Multi-input multi-output DC-DC converter with multilevel inverter application. In Proceedings of the IEEE National Conference on Emerging Trends In New & Renewable Energy Sources And Energy Management (NCET NRES EM), Chennai, India, 16–17 December 2014.
138. Andreiciks, A.; Steiks, I.; Krievs, O.; Blaabjerg, F. Design of Current Source DC/DC Converter and Inverter for 2kW Fuel Cell Application. In Proceedings of the 9th IEEE International Symposium on Diagnostics for Electric Machines, Power Electronics and Drives (SDEMPED), Valencia, Spain, 27–30 August 2013.
139. Huang, X.; Zhao, J.; Li, F.; Lin, F. Loss Characteristics of Input-Series Output-Parallel SiC DC-DC Converter Used in Auxiliary Power Systems. *IEEJ J. Ind. Appl.* **2019**, *8*, 677–684. [CrossRef]
140. Sato, K.; Kato, H.; Fukushima, T. Development of SiC Applied Traction System for Shinkansen High-speed Train. In Proceedings of the 2018 International Power Electronics Conference (IPEC-Niigata 2018-ECCE Asia), Niigata, Japan, 20–24 May 2018.
141. Wang, F.; Zhang, Z. Overview of Silicon Carbide Technology: Device, Converter, System, and Application. *CPSS Trans. Power Electron. Appl.* **2016**, *1*, 13–32. [CrossRef]
142. Alkayal, F.; Saada, J.B. Compact three phase inverter in Silicon Carbide technology for auxiliary converter used in railway applications. In Proceedings of the 2013 15th European Conference on Power Electronics and Applications (EPE), Lille, France, 2–6 September 2013.
143. Release, T.P. *Toshiba Launches SiC Hybrid Power Module for High Power Switching, Inverter and Motor Control Applications*; PCIM Europe: Nuremberg, Germany, 2013.
144. Mitsubishi Electric Corporation Public Relations. *Division, Mitsubishi Electric to Launch Railcar Traction Inverter with All-SiC Power Module*; Mitsubishi Electric: Tokyo, Japan, 2013.
145. Wada, K.; Hino, A.; Ando, M. High-Speed Analysis of Bus Bar Inductance for a Laminated Structure. *IEEJ J. Ind. Appl.* **2013**, *2*, 189–194. [CrossRef]
146. Ke, H.; Chang, G.; Zhou, W.; Li, C.; Peng, Y.; Dai, X. 3.3 kV/500 A SiC Power Module for Railway Traction Application. In Proceedings of the PCIM Asia 2018; International Exhibition and Conference for Power Electronics, Intelligent Motion, Renewable Energy and Energy Management, Shanghai, China, 26–28 June 2018.

147. Hamada, K.; Hino, S.; Miura, N.; Watanabe, H.; Nakata, S.; Suekawa, E.; Ebiike, Y.; Imaizumi, M.; Umezaki, I.; Yamakawa, S. 3.3 kV/1500 A Power Modules for The World's First All-SiC Traction Inverter. *Jpn. J. Appl. Phys.* **2015**, *54*, 04DP07. [CrossRef]
148. Liu, G.; Wu, Y.; Li, K.; Wang, Y.; Li, C.Z. Development of high power SiC devices for rail traction power systems. *J. Cryst. Growth* **2018**, *507*, 442–452. [CrossRef]
149. Karki, U.; Peng, F.Z. Precursors of Gate-Oxide Degradation in Silicon Carbide MOSFETs. In Proceedings of the 2018 IEEE Energy Conversion Congress and Exposition (ECCE), Portland, OR, USA, 23–27 September 2018.
150. Wei, Z.; Xiucheng, H.; Lee, F.C.; Qiang, L. Gate Drive Design Considerations for High Voltage Cascade GaN HEMT. In Proceedings of the Applied Power Electronics Conference and Exposition (APEC), Fort Worth, TX, USA, 16–20 March 2014.
151. Zhang, Z.; Wang, F.; Tolbert, L.M.; Blalock, B.J.; Costinett, D.J. Active gate driver for fast switching and cross-talk suppression of SiC devices in a phase-leg configuration. In Proceedings of the 2015 IEEE Applied Power Electronics Conference and Exposition (APEC), Charlotte, NC, USA, 15–19 March 2015.
152. Greenwell, R.L.; McCue, B.M.; Tolbert, L.M.; Blalock, B.J.; Islam, S.K. High-temperature SOI-based gate driver IC for WBG power switches. In Proceedings of the Applied Power Electronics Conference and Exposition (APEC), Long Beach, CA, USA, 17–21 March 2013.
153. Colmenares, J.; Peftitsis, D.; Rabkowski, J.; Sadik, D.P.; Nee, H.P. Dual-Function Gate Driver for a Power Module with SiC Junction Field-Effect Transistors. *Power Electron. IEEE Trans.* **2014**, *29*, 2367–2379. [CrossRef]
154. Wang, S.; Maillet, Y.Y.; Wang, F.; Boroyevich, D.; Burgos, R. Investigation of Hybrid EMI Filters for Common-Mode EMI Suppression in a Motor Drive System. *IEEE Trans. Power Electron.* **2010**, *25*, 1034–1045. [CrossRef]
155. Wang, S.; Lee, F.C.; Wyk, J.D.V. Study of Integration of Parasitic Cancellation Techniques for EMI Filter Design With Discrete Components. *IEEE Trans. Power Electron.* **2008**, *23*, 3094–3102. [CrossRef]
156. Zhang, Z.; Wang, F.; Costinett, D.J.; Tolbert, L.M.; Blalock, B.J.; Haifeng, L. Dead-time optimization of SiC devices for voltage source converter. In Proceedings of the 2015 IEEE Applied Power Electronics Conference and Exposition (APEC), Charlotte, NC, USA, 15–19 March 2015; pp. 1145–1152.
157. Zhang, Z.; Xu, L. Dead-Time Compensation of Inverters Considering Snubber and Parasitic Capacitance. *IEEE Trans. Power Electron.* **2014**, *29*, 179–3187. [CrossRef]
158. Liu, B.; Ren, R.; Jones, E.A.; Wang, F.; Costinett, D.; Zhang, Z. A compensation scheme to reduce input current distortion in GaN based 450 kHz three-phase Vienna type PFC. In Proceedings of the IEEE Energy Conversion Congress and Exposition, Milwaukee, WI, USA, 18–22 September 2016.
159. Hartmann, M.; Ertl, H.; Kolar, J.W. On the Tradeoff Between Input Current Quality and Efficiency of High Switching Frequency PWM Rectifiers. *IEEE Trans. Power Electron.* **2012**, *27*, 3137–3149. [CrossRef]
160. Bedetti, N.; Calligaro, S.; Petrella, R. Self-Commissioning of Inverter Dead-Time Compensation by Multiple Linear Regression Based on a Physical Model. *IEEE Trans. Ind. Appl.* **2015**, *51*, 3954–3964. [CrossRef]
161. Li, C.; Gu, Y.; Li, W.; He, X.; Dong, Z.; Chen, G.; Ma, C.; Zhang, L. Analysis and compensation of dead-time effect considering parasitic capacitance and ripple current. In Proceedings of the 2015 IEEE Applied Power Electronics Conference and Exposition (APEC), Charlotte, NC, USA, 15–19 March 2015; pp. 501–1506.
162. Yuan, J.; Zhao, Z.; Chen, B.; Li, C.; Wang, J.; Tian, C.; Chen, Y. An Immune-Algorithm-Based Dead-Time Elimination PWM Control Strategy in a Single-Phase Inverter. *IEEE Trans. Power Electron.* **2015**, *30*, 3964–3975. [CrossRef]
163. Guo, B.; Xu, F.; Zhang, Z.; Xu, Z.; Wang, F.; Tolbert, L.M.; Blalock, B.J. Compensation of input current distortion in three-phase buck rectifiers. In Proceedings of the 2013 Twenty-Eighth Annual IEEE Applied Power Electronics Conference and Exposition (APEC), Long Beach, CA, USA, 17–21 March 2013; pp. 930–938.

energies

MDPI

Review

Power Converter Solutions for Industrial PV Applications—A Review

Ievgen Verbytskyi [1], Mykola Lukianov [1], Kawsar Nassereddine [2,*], Bohdan Pakhaliuk [3], Oleksandr Husev [4,*] and Ryszard Michał Strzelecki [3]

[1] Igor Sikorsky Kyiv Polytechnic Institute, National Technical University of Ukraine, 03056 Kyiv, Ukraine; verbytskyi.ievgen@gmail.com (I.V.); zebrahead097@gmail.com (M.L.)
[2] Faculty of Engineering, Lebanese University, Beirut 1002, Lebanon
[3] Department of Power Electronics and Electrical Machines, Faculty of Electrical and Control Engineering, Gdańsk University of Technology, 80-233 Gdansk, Poland; bohdan.pakhaliuk@pg.edu.pl (B.P.); ryszard.strzelecki@pg.edu.pl (R.M.S.)
[4] Department of Electrical Power Engineering and Mechatronics, Tallinn University of Technology, 19086 Tallinn, Estonia
* Correspondence: nassereddinejana@gmail.com (K.N.); oleksandr.husev@taltech.ee (O.H.)

Abstract: As the use of photovoltaics becomes more widespread, new technologies for more efficient energy generation, transmission, and distribution based on power electronics converters are being developed. The most common applications are grid-on, energy storage, hybrid, and high voltage gain applications. These applications impose several additional requirements in the design of power converters associated with the solar battery's maximum power tracking and operation in a wide range of input currents and voltages. The practical realization of such solutions can be implemented on the basis of various topologies, which requires a preliminary application of criteria for assessing their effectiveness. The paper conducts a comparison of different topologies on power converters based on two parameters that describe their cost and power loss for various PV applications. For a straightforward study, these parameters are represented using the gain factor, which allows for an accurate comparison of the efficiency of various types of converters.

Keywords: power converters; PV applications review; cost factor; power loss

Citation: Verbytskyi, I.; Lukianov, M.; Nassereddine, K.; Pakhaliuk, B.; Husev, O.; Strzelecki, R.M. Power Converter Solutions for Industrial PV Applications—A Review. *Energies* **2022**, *15*, 3295. https://doi.org/10.3390/en15093295

Academic Editors: Diego Bellan and Jelena Loncarski

Received: 4 April 2022
Accepted: 28 April 2022
Published: 30 April 2022

1. Introduction

The modern development of the energy industry implies a gradual abandonment of fossil energy sources and a transition to renewable energy, i.e., solar, wind, geothermal, hydro energy, and biofuels [1]. Wind power plants currently produce 52% of total electricity generated from renewable sources (excluding hydropower), while solar power plants produce 26% [2]. Solar power plants, on the other hand, rank first in terms of installed capacity for 2020, with 125 GW compared to 110 GW for wind power plants. There are cheaper materials for solar panels, such as silicon [3], and it costs less to keep them up and running [4].

The main economic constraint on the use of solar power plants, namely the higher cost of electricity than for fossil energy sources, is gradually being eliminated. However, solar energy loses in terms of technical indicators: specific capacity, generation stability, cost of maintenance, and infrastructure, which is the main deterrent to its rapid spread [5,6]. So, to overcome this gap, it is necessary to modernize power plant infrastructure. Among these actions, here are the ones that should be highlighted [7]:

- Maximum power point tracking;
- Energy storage and balancing;
- Electric parameter transformation and stabilization.

The generation of solar battery power is characterized by a family of power curves for different solar insolation powers: P_{SI1}, P_{SI2}, and P_{SI3} as functions of voltage U, $P = f(U)$, or current I, $P = f(I)$, as shown in Figure 1a,b. A high level of efficiency is reached when the solar battery is near the maximum power point (MPP) [8], where $P_{MPP} = f(U_{MPP})$ or $P_{MPP} = f(I_{MPP})$ with power from the solar panel 1, 2, and 3.

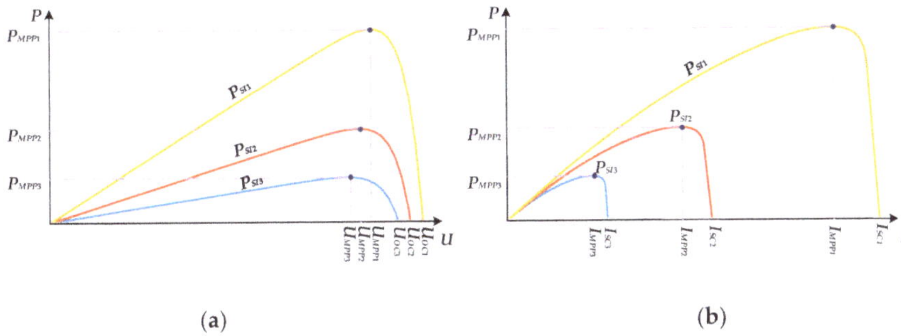

Figure 1. A family of power curves for different solar insolation powers: (**a**) as functions of voltage U, $P = f(U)$; (**b**) as functions of current I, $P = f(I)$.

For voltage curves, MPP coordinates U_{MPP} are defined by the open circuit solar battery voltage U_{OC} which is almost constant and belongs to the narrow range $U_{MPP} \in [0.72 \cdot U_{OC}; 0.76 \cdot U_{OC}]$ that slightly depends on solar insolation power P_{SI}. The current MPP co-ordinate I_{MPP} is influenced by the short-circuit current I_{SC}, $I_{MPP} \in [0.9 \cdot I_{SC}; 0.93 \cdot I_{SC}]$. However, the short-circuit current, I_{SC}, in contrast to the open-circuit voltage, varies widely and is proportional to P_{SI} [9].

If a load with resistance R_L is connected directly to a solar battery, maximum power P_{MPP} is delivered to the load when the equivalent resistance of the solar battery R_{SB} is equal to the load resistance R_L in MPT:

$$R_{SB} = \frac{U_{MPP}}{I_{MPP}} = R_L, \tag{1}$$

in another case, when $R_{SB} \neq R_L$, power converters as maximum power point trackers (MPPT) are used [10]. The power converter via output voltage (current) U_L (I_L) regulation tunes equivalent load resistance $R_{L(eq)}$ on the converter primary side according to Equation (2) as shown in Figure 2:

$$R_{SB} = R_{L(eq)} = \frac{U_{MPP}}{I_{MPP}} = \frac{U_L}{G} \cdot \frac{1}{I_L G} = \frac{R_L}{G^2}, \tag{2}$$

where G is a converter voltage gain.

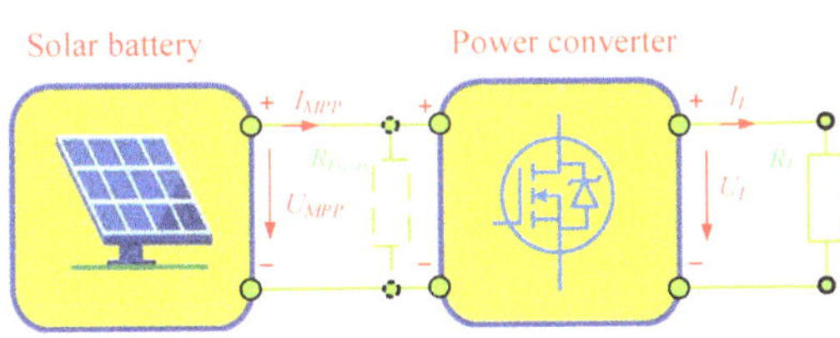

Figure 2. Equivalent load resistance $R_{L(eq)}$ tuning with a power converter.

However, even a power converter such as MPPT does not always provide an efficient mode of solar battery operation. As shown in Figure 3a, only constant solar battery current I_{MPT} that fits MPP allows for generating of maximum power for defined solar insolation P_{SI}. If the solar battery current has ripple ΔI, output power pulsation ΔP appears and decreases average power P proportionally to ripple amplitude ΔI as shown in Figure 3b [11]. If the input current is discontinuous, then solar battery energy is used inefficiently as shown in Figure 3c. Therefore, the aforementioned features of the solar batteries impose additional requirements on the power converter as an MPPT, specifically:

- Continuous low ripple input current I;
- High efficiency in a wide power range.

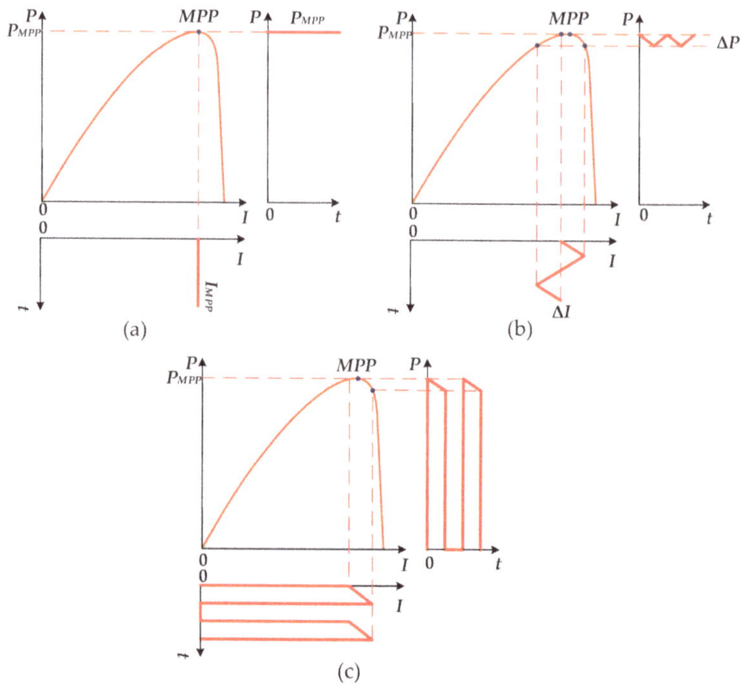

Figure 3. The Impact of MPPT input current shape on the value of solar battery power: (**a**) continuous current; (**b**) current with ripple ΔI; (**c**) discontinuous current.

There is an input inductor L in the boost converter [12]. This inductor produces continuous current and can be used directly as an MPPT (Figure 4a). Because it has a series-connected transistor, buck and buck/boost converters have an extra capacitor C_F that filters the input current, as shown in Figure 4b.

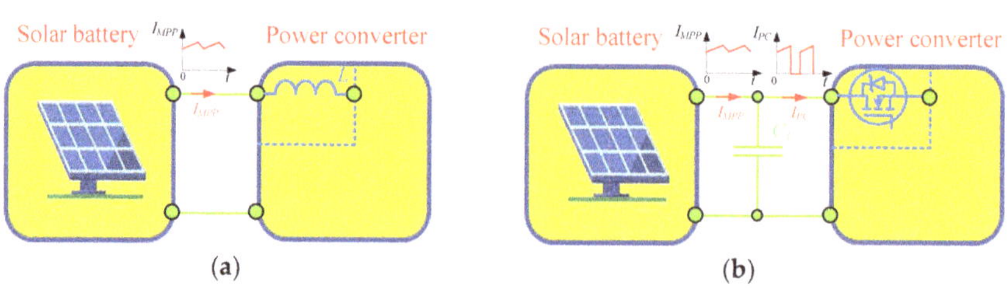

Figure 4. Providing continuous current: (**a**) natural for power converters with the series-connected inductor L; (**b**) with an additional capacitor C_F for a converter with a series transistor.

The equivalent resistance of solar battery, R_{SB}, varies as a function of solar insolation power and is greater than the minimum value $R_{SB(min)}$ attained for maximum insolation:

$$R_{SB(\min)} = \left. \frac{U_{MPP}}{I_{MPP}} \right| P_{SB} \rightarrow \max \quad . \tag{3}$$

When solar insolation is zero, R_{SB} tends to infinity. In practice, MPPT works effectively with some range of solar battery resistance $R_{SB} \in [R_{SB(\min)}; R_{SB(\max)}]$, where $R_{SB(\max)}$ defines the minimum solar insolation power $P_{SB(\min)}$ where MPPT may operate [13,14]. Therefore, MPPT has to effectively shift the operating point of the solar battery to MPP in the defined range of resistance R_{SB}. MPPT converters with different voltage gains G are used depending on the relationship between the resistances of a solar battery R_{SB} and the load R_L:

- Boost converters ($G > 1$), if $R_{SB(\max)} < R_L$;
- Buck/boost converters ($G = G_{(\min)} \ldots 1 \ldots G_{(\max)}$), if $R_{SB(\min)} < R_L < R_{SB(\max)}$;
- Buck converters ($G < 1$), if $R_{SB(\min)} > R_L$.

The common foregoing conditions of effective PV system design are complemented by the technical aspects of connecting the solar batteries to the power grid or its standalone operation. In Table 1, the topologies of common PV applications and their modifications are listed.

Table 1. Common PV application descriptions and requirements.

PV Application	Modifications
1. Standalone PV applications [15,16] 2. On-grid PV applications [17,18] 3. Hybrid PV applications [19,20]	- DC or AC; - with DC-link or without it (only for AC systems) - low, middle, or high voltage; - with electric battery or without it; - DC or AC battery connected; - isolated or transformerless; - one, three, or multiphase.

For the efficient design of PV systems, the key task is the correct choice of power converter topology for providing effective energy transfer and conversion. Therefore, it is necessary to pay attention to the following features of power converter design for PV systems:

1. The multifunctional purpose of the power converters is:
    - MPPT in a standalone DC or AC system simultaneously provides maximum power and charges an electric accumulator, as shown in Figure 5a,b [21];
    - In AC on-grid applications that directly deliver energy to the AC power grid, MPPT maximizes energy, converts it from DC to AC form, and provides power factor correction and reactive power generation as needed, as shown in Figure 5c [22].
2. Advanced schematic and control algorithms:
    - For energy balancing and stable generation, often for AC grid-connected applications, electric batteries with bidirectional chargers are connected [23], as shown in Figure 5c. To make sure that electricity has the right dynamics and quality, you need to use advanced control algorithms with multi-loop and predictive control [24].
    - Co-ordination of parallel operations of several energy sources and determination of their priority in a hybrid system [25], as seen in Figure 5d.
3. Special issues in converter design:
    - For high voltage DC and AC on-grid systems, specialized converters adapted for PV applications are being developed [26];

- Hybrid systems provide parallel operation of several alternative power supplie connected to the power grid and load that in general may be considered a mul tiport power system [27]. As a result, a reduced-component multiport powe system can be made instead of having a lot of separate power converters that dc the same thing [28].

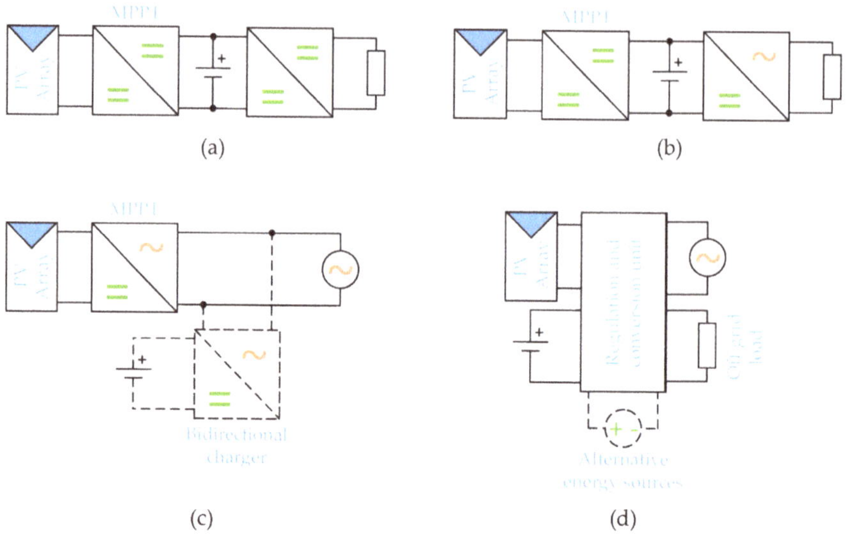

Figure 5. Common PV applications: (**a**) standalone DC systems; (**b**) standalone AC systems; (**c**) on-grid AC systems with energy storage; and (**d**) hybrid systems.

Implementation of the aforementioned PV applications is covered by a set of converter topologies with their own advantages and disadvantages. The choice of the most appropriate solution depends on many factors [29]:

- System type (standalone or grid-on, DC or AC);
- Voltage and power level (low, middle, or high);
- The relationship between the voltage of the solar battery and the grid;
- Additional requirements on power factor value and power stability.

For instance, an MPPT unit may be effectively realized with either a DC–AC converter or a two-stage power system with DC–DC and DC–AC converter; a storage system may be installed on the DC–DC side of the PV system or directly on the DC or AC system; a PV system may be connected to the grid with one or few high-power converters, or realized as a distributed system with a set of low or middle power converters, etc.

This paper looks at different topologies that could be used to solve the discussed problems It also tries to show the good and bad points of the solutions based on proper generalization parameters that allow comparing their effectiveness.

In general, power converter comparative analysis is performed based on their complexity, i.e., number of inductors, capacitors, and semiconductor components [30], also taking in to account voltage gain characteristics, control strategy, efficiency, etc. However, the most fruitful results are obtained by comparing converters designed to solve a particular problem that put forward a number of specific requirements to the power converter. For instance, for designing a dynamic voltage restorer [31], these parameters are high dynamics, wide reactive and active power regulation range, and easy energy storage integration, whereas for AC/DC Microgrids [32] parameters are the value of output impedance, type of power flow control, voltage modulation strategy, and voltage and frequency regulation. For solar applications, the main requirement, as mentioned before, is high efficiency in a wide power range and, as for industrial solutions, low cost per unit of power.

As a numerical criteria of the converter cost per unit of output power, P_{out} may be measured by the cost factor k_C which is inverse to the transistor utilization factor k_U [33]:

$$k_C = \frac{1}{k_U} = \frac{S}{P_{out}}, \quad (4)$$

where S is total switch stress and:

$$S = \sum_{k=1}^{N} U_{k\max} \cdot I_{k\max}, \quad (5)$$

where N is the switch number of the power converter, and $U_{k\max}$ and $I_{k\max}$ are the maximum voltage and current stress of the switch k, respectively. Normalized power loss P^* [34] of the switch can be used to estimate power converter efficiency.

Detailed and comprehensive comparative PV converter analysis in space of cost factor and normalized power loss is carried out in the paper in such a sequence.

Section 2 discloses a methodology of deriving cost factor and normalized power loss for DC–DC and DC–AC converters with hard and soft-switching transistor commutation.

In Sections 3–6, classification, cost, and efficiency comparisons of power converters for AC grid, energy storage, high voltage gain, and hybrid PV applications are given.

Discussion (Section 7) deals with the overall analysis of PV application design and recommendations.

2. Materials and Methods

According to standards for solar batteries and inverters [35–38], the maximum operating voltage of PV equipment currently does not exceed $U_{\max}$ = 1500 V. In this case, the converter is connected to 1–3 solar battery strings with a short-circuit current of 10 A per string. Application design with such operating voltage and current, as usual, is performed based on Si IGBT or SiC MOSFET transistors [39]. Due to less static and dynamic power loss, and a permanent price reduction in SiC transistors, Cree C2M0080170P is used for loss calculation [40]. Static loss of the transistor P_{st}* with a constant current I_{DC} that provides output power P_{out} at voltage $U_{\max}$ is normalized and defined as P_{Tst}* = 1 p.u. It is well known that the resistance $R_{DS(on)}$ of SiC and MOSFET transistors quadratically increases with voltage. Additionally, power loss depends on transistor current shape, which in a general case for DC-DC converters is shown in Figure 6. There are shown peak and minimum transistor values, I_{peak} and $I_{\min}$, correspondingly. Average value of transistor I_{Tav} on interval DT is calculated as $I_{Tav} = (I_{peak} + I_{\min})/2$.

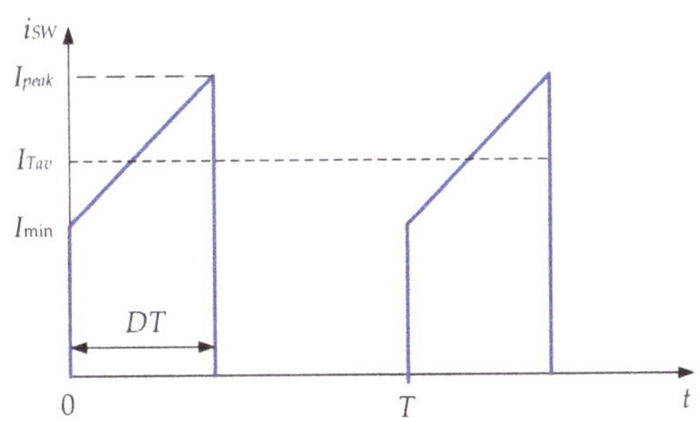

Figure 6. The current shape of a transistor in a DC–DC hard-switching converter.

Thus, the RMS value of $I_{SW(RMS)}$ compared with I_{DC} is calculated as follows:

$$\frac{I_{SW(RMS)}}{I_{DC}} = \frac{\sqrt{D(I_{peak}{}^2 + I_{peak}I_{\min} + I_{\min}^2)}}{\sqrt{3}I_{DC}} = k_{Tst}\sqrt{\frac{D}{3}}, \quad (6)$$

where k_{Tst} is the factor that depends on converter topology.

If there is no pulsation in the current, $I_{peak} = I_{\min} = I_{Tav}$, and the RMS value is calculated using the average transistor current value I_{Tav} on the interval DT:

$$\frac{I_{SW(RMS)}}{I_{DC}} = \frac{I_{Tav}}{I_{DC}}\sqrt{D}.$$

Therefore, k switch power loss $P_{Tst(k)ht}{}^*$ for hard-switching converters with overvoltage $U_{k\max}$ is defined as:

$$P_{Tst(k)ht}{}^* = P_{Tst}{}^*\left(\frac{U_{k\max}}{U_{\max}}\right)^2\frac{D\left(I_{peak}{}^2 + I_{peak}I_{\min} + I_{\min}^2\right)}{3I_{DC}{}^2} = \frac{D}{3}k_{Tst(k)}{}^2\left(\frac{U_{k\max}}{U_{\max}}\right)^2, \quad (7)$$

Whereas in resonant converters with sine wave current shape, the power loss $P_{Tst(k)s}{}^*$ is multiplied by shape factor $k_{sh} = \pi^2/8$ [41]:

$$P_{Tst(k)s}{}^* = \frac{\pi^2}{8}Dk_{Tst(k)}{}^2\left(\frac{U_{k\max}}{U_{\max}}\right)^2. \quad (8)$$

Due to the nearly rectangular shape of the current in soft-switching converters, the static power loss is considered the same as for constant current $P_{Tst}{}^*$.

The dynamic power loss $P_{Td}*$ for hard-switching mode is calculated using a comparison of transistor static and dynamic losses at maximum power $I_{DC} = 30$ A, $U_{\max} = 1500$ V at switching frequency $f_{sw} = 40$ kHz, $P_{Td}* = 0.5$ p.u. The dynamic loss $P_{T(k)\,d}{}^*$ for defined transistor current I_{Tk} and voltage U_{Tk} is:

$$P_{Td(k)}{}^* = P_{Td}{}^*\left(\frac{U_{Tk}}{U_{\max}}\right)\left(\frac{I_{Tk}}{I_{DC}}\right) = 0.5\left(\frac{U_{Tk}}{U_{\max}}\right)\left(\frac{I_{Tk}}{I_{DC}}\right). \quad (9)$$

It is considered that in resonant and soft-switching converters, dynamical loss is absent.

A diode's static power loss $P_{Dst}{}^*$ is calculated based on Cree C5D25170H [42] and expressed in units of $P_{Tst}{}^*$, $P_{Dst}{}^* = 0.8$ p.u for constant current I_{DC}. If the diode's forward V_D is assumed to be constant and proportional to reverse voltage $U_{k\max}$, its static loss for the hard-switching converter is calculated as follows:

$$P_{Dst(k)hd}{}^* = P_{Dst}{}^*k_{Dst(k)}\left(\frac{U_{k\max}}{U_{\max}}\right)(1-D) = \frac{0.8k_{Dst(k)}U_{k\max}(1-D)}{U_{\max}}, \quad (10)$$

where $k_{Dst(k)}$ is the relationship between the average value of the diode current I_{Dk} on the interval $(1-D)$ to current I_{DC}:

$$k_{Dst(k)} = \frac{I_{Dk}}{I_{DC}}, \quad (11)$$

whereas in the case of soft-switching:

$$P_{Dst(k)s}{}^* = P_{Dst}{}^*k_{Dst(k)}\frac{\pi}{2\sqrt{2}}\left(\frac{U_{k\max}}{U_{\max}}\right)(1-D) = \frac{0.8k_{Dst(k)}\pi U_{k\max}(1-D)}{2\sqrt{2}U_{\max}}. \quad (12)$$

Dynamic loss of the diode for current I_{DC} is $P_{Dd}* = 1$ p.u. For defined diode current I_{Dk} and voltage U_{Dk}, dynamic loss $P_{D(k)d^*}$ is calculated similarly to the Formula (9):

$$P_{Dd(k)}{}^* = P_{Dd}{}^*\left(\frac{U_{Dk}}{U_{\max}}\right)\left(\frac{I_{Dk}}{I_{DC}}\right) = \left(\frac{U_{Dk}}{U_{\max}}\right)\left(\frac{I_{Dk}}{I_{DC}}\right). \quad (13)$$

The final values of static and dynamic power losses for DC–DC converters are given in Table 2.

Table 2. Normalized power loss of the transistor and diode for DC–DC converters.

Power Loss Type	Converter Type		
	Hard-Switching	**Soft-Switching**	**Resonant**
Transistor static power loss, $P_{Tst}{}^*$	$\frac{D}{3}k_{Tst(k)}{}^2\left(\frac{U_{k\max}}{U_{\max}}\right)^2$	$\frac{D}{3}k_{Tst(k)}{}^2\left(\frac{U_{k\max}}{U_{\max}}\right)^2$	$\frac{\pi^2 D}{8}k_{Tst(k)}{}^2\left(\frac{U_{k\max}}{U_{\max}}\right)^2$
Transistor dynamic power loss, $P_{Td}{}^*$	$\frac{0.5U_{Tk}I_{Tk}}{U_{\max}I_{DC}}$	0	0
Diode static power loss, $P_{Dst}{}^*$	$\frac{0.8k_{Dst(k)}U_{k\max}(1-D)}{U_{\max}}$	$\frac{0.8k_{Dst(k)}U_{k\max}(1-D)}{U_{\max}}$	$\frac{0.8k_{Dst(k)}\pi U_{k\max}(1-D)}{2\sqrt{2}U_{\max}}$
Diode dynamic power loss, $P_{Dd}{}^*$	$\frac{0.5U_{Dk}I_{Dk}}{U_{\max}I_{DC}}$	0	0

For DC–AC and AC–DC converter output, AC side voltage and current are sinusoidal; therefore, diodes and transistor current, as well as power loss, depend on grid voltage phase $\varphi = 2\pi \cdot t/T_f$, where T_f is the fundamental period of the grid voltage that is shown in Figure 7.

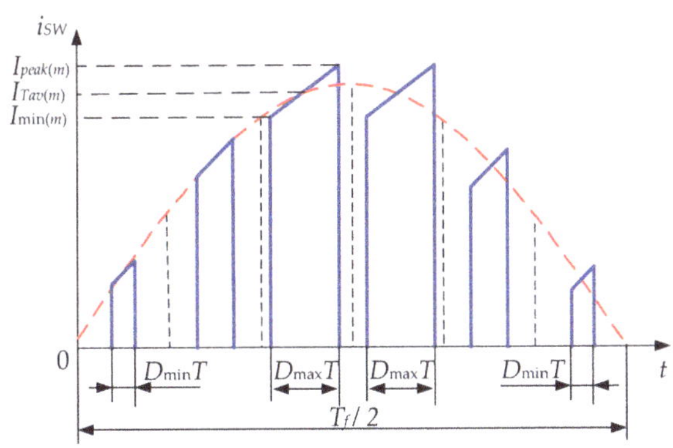

Figure 7. Transistor current shape for AC–DC and DC–AC hard-switching converters.

The amplitude peak current values $I_{peak(m)}$, $I_{Tav(m)}$, and $I_{min(m)}$ are used to calculate the AC shaped current cost factor kc, whereas the power loss $P_{con(AC)}*$ analysis is performed based on the integration of instantaneous power loss $P_{con(AC)}*$ from Table 2, where duty cycle D is a function of power grid phase φ, D [0; D_{max}], and normalized with current i_{norm}:

$$P_{con(AC)}{}^* = \frac{1}{2\pi}\int_0^{2\pi} P_{con}{}^*(\varphi) i_{norm} d\varphi, \tag{14}$$

where $i_{norm} = \pi \cdot \sin(\varphi)/2$ has a half-period average value of unity:

$$\frac{1}{\pi}\int_0^{\pi} i_{norm} d\varphi = 1. \tag{15}$$

Thus, equations have been derived that can be used to calculate the total converter power loss $P_{con}*$ and the cost factor kc for both AC and DC applications. These numbers are used to compare different PV applications with different converters for each type of PV.

3. AC Grid on PV Applications

AC grid-connected renewable applications aim to deliver the maximum possible energy to the grid as efficiently as possible. This will be accomplished using single-stage or two-stage systems. The single-stage application is an inverter that simultaneously performs the functions of MPPT, boosting, and voltage conversion from DC to AC [43]. In the two-stage converter, these functions are split between DC–DC and DC–AC inverters [44]. Depending on the safety conditions or voltage gain requirements, applications may be equipped with transformers or be transformerless. Typical structures of on-grid applications are shown in Figure 8.

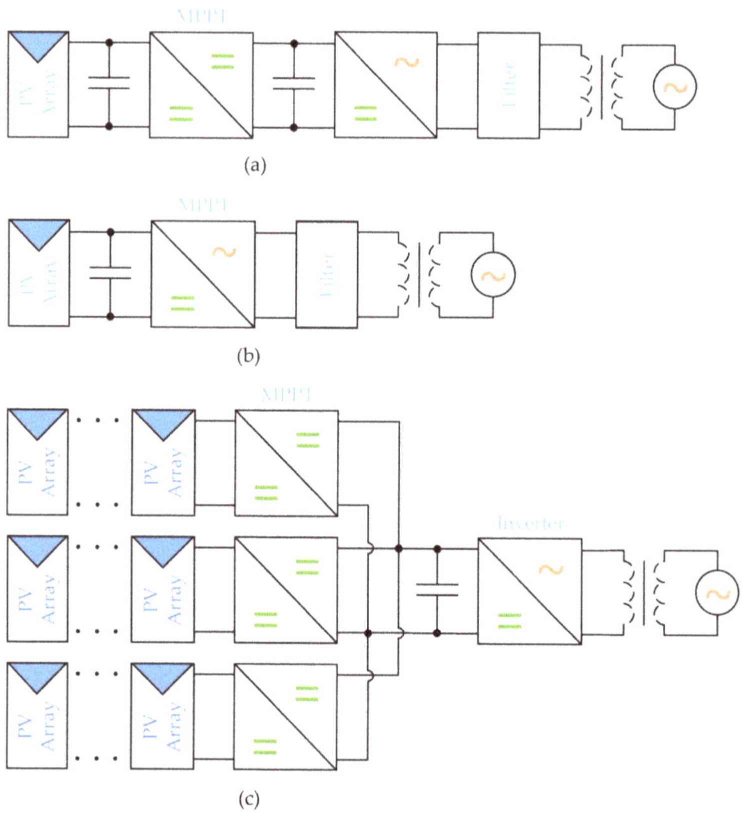

Figure 8. On-grid application structures that are commonly used include: (**a**) two-stage on-grid application; (**b**) single-stage on-grid application; and (**c**) distributed two-stage applications.

In terms of improving efficiency and reducing the cost of the system, single-stage systems have clear advantages over two-stage. Nevertheless, two-stage systems are actively used due to simple system control and the opportunity to connect energy storage to DC-link or design multi-string PV systems [45]. In two-stage systems, any non-isolated DC–DC converter can be used as MPPT [12], but mostly boost, buck, and buck-boost converters are used due to less transistor stress [46]. For reducing power loss, soft-switching [47] or interleaved converters [48,49] are used.

As usual, single-phase or three-phase bridge topologies are implemented for the design of DC–AC converters in single and two-stage topologies [50]. Soft-switching commutation and reduced power loss are possible with advanced control methods [51,52].

Non-isolated one-phase inverters with common full or half-bridge topology suffer from leakage current that flows between AC grid ground and PV through parasitic capacitors [53] and brings about safety and interference issues [54]. Additional transistors implemented in H5, H6, HERIC, and hybrid-bridge topologies [55] cut off the leakage current path and eliminate it. For transistor overvoltage limitation, multilevel inverters, for instance, 3L-NPC [56] and 3L-SC [57], are used. The aforementioned DC–AC PV application schematics are illustrated in Figure 9 and listed in Table 3 for one stage DC–DC and DC–AC converters, respectively, with cost factor k_C and normalized power loss of the converter $P_{con}{}^{*}$.

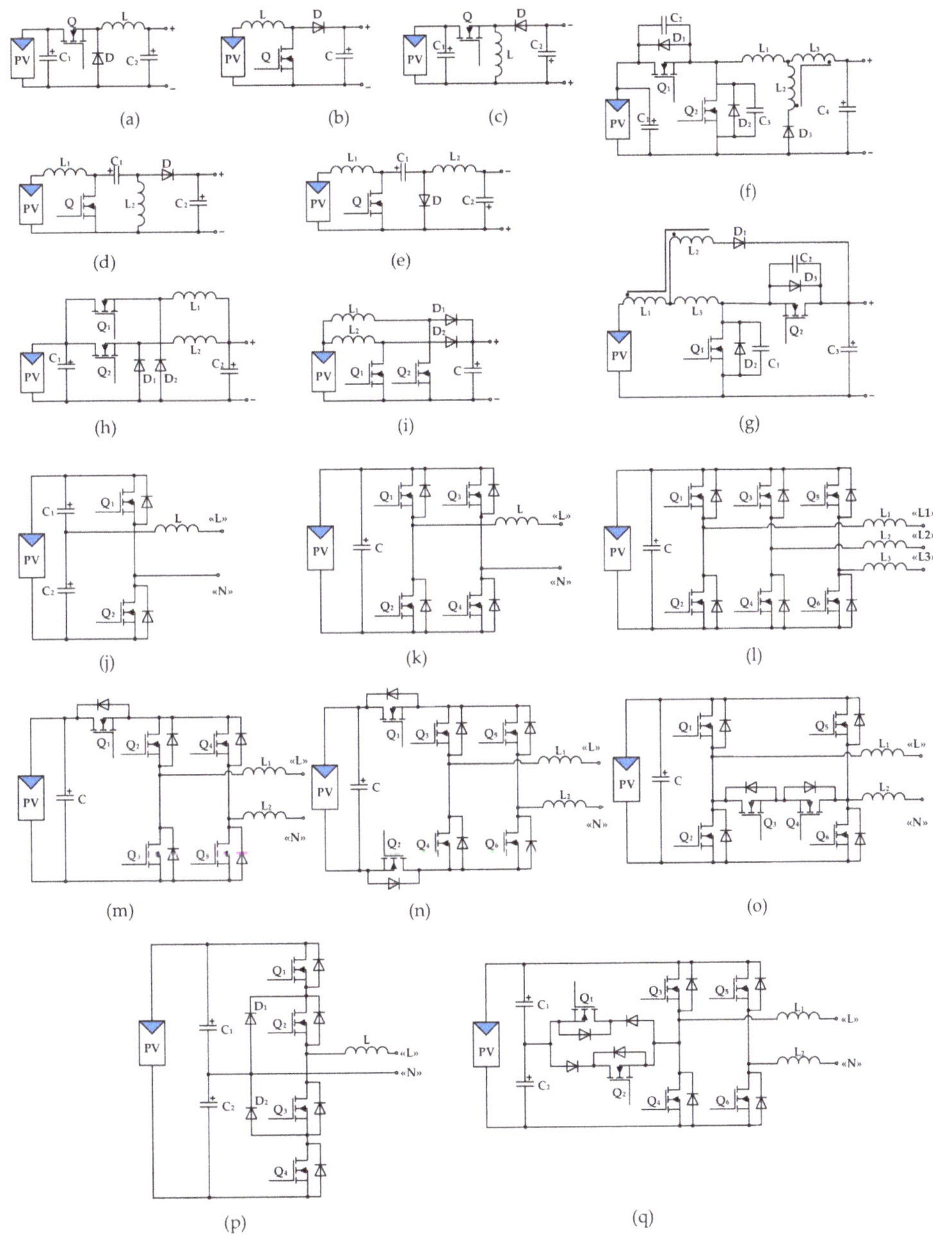

Figure 9. Converter topologies for grid-based applications: (**a**) buck; (**b**) boost; (**c**) buck-boost; (**d**) SEPIC; (**e**) Cuk′; (**f**) soft-switching buck; (**g**) soft-switching boost; (**h**) interleaved buck; (**i**) interleaved boost; (**j**) half-bridge; (**k**) full-bridge; (**l**) three phase full-bridge; (**m**) H5; (**n**) H6; (**o**) HERIC; (**p**) 3L-NPC; (**q**) 3L-SC.

Table 3. The cost factor k_C and the normalized power loss for common DC–DC and DC–AC converters.

Converter Topology	k_C	P_{con} *
	DC–DC converters	
Buck [12], Figure 9a	$\frac{I_{peak}}{DI_{Tav}}$	$\frac{4}{3D}\left(1-\frac{I_{peak}I_{\min}}{(I_{peak}+I_{\min})^2}\right)+\frac{0.8(1-D)}{D}+\frac{I_{peak}}{DI_{Tav}}$
Boost [12], Figure 9b	$\frac{I_{peak}}{D(1-D)I_{Tav}}$	$\frac{4D}{3(1-D)^2}\left(1-\frac{I_{peak}I_{\min}}{(I_{peak}+I_{\min})^2}\right)+0.8+\frac{I_{peak}}{(1-D)I_{Tav}}$
Buck-boost, SEPIC, Cuk' [12], Figure 9c–e	$\frac{I_{peak}}{D(1-D)I_{Tav}}$	$\frac{4}{3D(1-D)^2}\left(1-\frac{I_{peak}I_{\min}}{(I_{peak}+I_{\min})^2}\right)+\frac{0.8}{D}+\frac{I_{peak}}{(1-D)DI_{Tav}}$
Soft-switching buck [47], Figure 9f	$\frac{2I_{peak}}{DI_{Tav}}$	$\frac{4}{3D}\left(1-\frac{I_{peak}I_{\min}}{(I_{peak}+I_{\min})^2}\right)+\frac{0.8(1-D)}{D}$
Soft-switching boost [47], Figure 9g	$\frac{2I_{peak}}{(1-D)I_{Tav}}$	$\frac{4D}{3(1-D)^2}\left(1-\frac{I_{peak}I_{\min}}{(I_{peak}+I_{\min})^2}\right)+0.8$
Interleaved buck [48], Figure 9h, m cells	$\frac{I_{peak}}{DI_{Tav}}$	$\frac{4}{3Dm}\left(1-\frac{I_{peak}I_{\min}}{(I_{peak}+I_{\min})^2}\right)+\frac{0.8(1-D)}{D}+\frac{I_{peak}}{DmI_{Tav}}$
Interleaved boost [49], Figure 9i, m cells	$\frac{I_{peak}}{(1-D)I_{Tav}}$	$\frac{4D}{3(1-D)^2m}\left(1-\frac{I_{peak}I_{\min}}{(I_{peak}+I_{\min})^2}\right)+0.8+\frac{I_{peak}}{(1-D)mI_{Tav}}$
	DC–AC converters	
Half-bridge [50],Figure 9j	$\frac{8I_{peak(m)}}{D_{\max}I_{Tav(m)}}$	$\frac{4\pi^2}{3D_{\max}^2}\left(1-\frac{I_{peak(m)}I_{\min(m)}}{(I_{peak(m)}+I_{\min(m)})^2}\right)+\frac{1.6}{D_{\max}}\left(\frac{\pi}{2}+1\right)+\frac{2I_{peak(m)}}{D_{\max}I_{Tav(m)}}$
Full-bridge [50], Figure 9k	$\frac{8I_{peak(m)}}{D_{\max}I_{Tav(m)}}$	$\left(\frac{2\pi}{3D_{\max}}+\frac{\pi^2}{3D_{\max}^2}\right)\left(1-\frac{I_{peak(m)}I_{\min(m)}}{(I_{peak(m)}+I_{\min(m)})^2}\right)+\frac{0.8}{D_{\max}}\left(\frac{\pi}{2}+1\right)+\frac{I_{peak(m)}}{D_{\max}I_{Tav(m)}}$
Triple-bridge [50], Figure 9l	$\frac{12I_{peak(m)}}{\sqrt{3}D_{\max}I_{Tav(m)}}$	$\frac{2\pi}{9D_{\max}}\left(1-\frac{I_{peak(m)}I_{\min(m)}}{(I_{peak(m)}+I_{\min(m)})^2}\right)+\frac{0.8}{D_{\max}}\left(\frac{\pi}{2}+1\right)+\frac{I_{peak(m)}}{D_{\max}I_{Tav(m)}}$
H5 [55], Figure 9m	$\frac{10I_{peak(m)}}{D_{\max}I_{Tav(m)}}$	$\left(\frac{2\pi}{3D_{\max}}+\frac{\pi^2}{3D_{\max}^2}\right)\left(1-\frac{I_{peak(m)}I_{\min(m)}}{(I_{peak(m)}+I_{\min(m)})^2}\right)+\frac{0.8}{D_{\max}}\left(\frac{\pi}{2}+1\right)+\frac{1.5I_{peak(m)}}{D_{\max}I_{Tav(m)}}$
H6 [55], Figure 9n	$\frac{12I_{peak(m)}}{D_{\max}I_{Tav(m)}}$	$\left(\frac{2\pi}{D_{\max}}+\frac{\pi^2}{3D_{\max}^2}\right)\left(1-\frac{I_{peak(m)}I_{\min(m)}}{(I_{peak(m)}+I_{\min(m)})^2}\right)+\frac{0.8}{D_{\max}}\left(\frac{\pi}{2}+1\right)+\frac{1.5I_{peak(m)}}{D_{\max}I_{Tav(m)}}$
HERIC [55],Figure 9o	$\frac{12I_{peak(m)}}{D_{\max}I_{Tav(m)}}$	$\left(\frac{2\pi}{3D_{\max}}+\frac{\pi^2}{3D_{\max}^2}\right)\left(1-\frac{I_{peak(m)}I_{\min(m)}}{(I_{peak(m)}+I_{\min(m)})^2}\right)+\frac{0.8}{D_{\max}}\left(\frac{\pi}{2}+1\right)+\frac{1.5I_{peak(m)}}{D_{\max}I_{Tav(m)}}$
3L-NPC [56], Figure 9p	$\frac{8I_{peak(m)}}{D_1I_{Tav(m)}}$	$\left(\frac{2\pi}{3D_{\max}}+\frac{\pi^2}{3D_{\max}^2}\right)\left(1-\frac{I_{peak(m)}I_{\min(m)}}{(I_{peak(m)}+I_{\min(m)})^2}\right)+\frac{0.8}{D_{\max}}\left(\frac{\pi}{2}+1\right)+\frac{I_{peak(m)}}{D_{\max}I_{Tav(m)}}$
3L-SC [57], Figure 9q	$\frac{10I_{peak(m)}}{D_1I_{Tav(m)}}$	$\left(\frac{\pi}{2D_{\max}}+\frac{5\pi^2}{12D_{\max}^2}\right)\left(1-\frac{I_{peak(m)}I_{\min(m)}}{(I_{peak(m)}+I_{\min(m)})^2}\right)+\frac{0.8}{D_{\max}}\left(\frac{\pi}{2}+1\right)+\frac{I_{peak(m)}}{D_{\max}I_{Tav(m)}}$

Because low current ripple ΔI is a key requirement of the grid on applications for sufficient power factor, the majority of converters use $I_{\min}$ = 0.9 $I_{Tav(m)}$, I_{peak} = 1.1 I_{Tav}. However, interleaved converters operate in boundary mode, when $I_{\min}$ = 0 and I_{peak} = 2 I_{Tav}. Based on the aforementioned assumptions, the cost factor k_C and normalized power loss P_{con}* of DC–DC converters are shown in Figure 10a,b, respectively. The same curves for DC–AC converters are shown in Figure 11a,b.

Because of a larger number of transistors, DC–AC converters are weaker in terms of parameter values k_C and P_{con}* than their DC–DC counterparts. Among DC–DC converters, the most effective solutions are based on buck and boost converters. Whereas buck-boost solutions (buck-boost, SEPIC, Cuk' converters) increase costs and losses. interleaved converters allow for reduced losses at the expense of an increase in price and dimensions.

Among DC–AC converters, three-phase is the most efficient converter because of the minimum number of transistors per phase. For single-phase applications, full-bridge, H5, H6, and HERIC converters have very similar k_C and P_{con}* values.

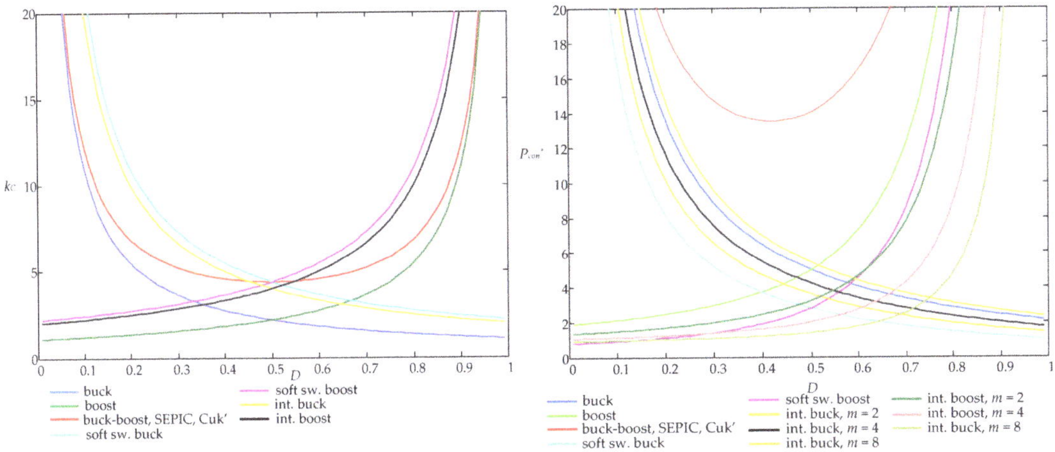

(a) (b)

Figure 10. DC–DC converters: (**a**) cost factor k_C; (**b**) normalized power loss P_{con}*.

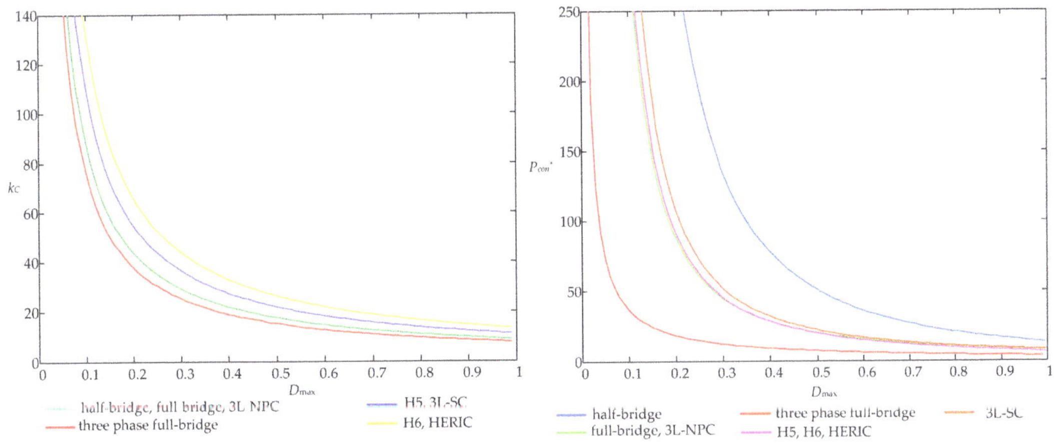

(a) (b)

Figure 11. DC–AC converters: (**a**) cost factor k_C; (**b**) normalized power loss P_{con}*.

Two-stage DC–AC applications may be used for PV applications in the following cases:

- Boosting DC voltage;
- Decreasing k_C and P_{con}* to allow for a wider range of input/output voltage operations;
- Inconsistency of power grid and solar battery voltages.

In a two-stage system, voltage regulation is performed at the DC–DC stage, whereas the DC–AC stage operates with the maximum duty cycle, $D_{\max} \rightarrow 1$, that provides minimum power loss. For instance, Figure 12a compares the cost factor k_C of single-stage applications based on a full-bridge converter and a two-stage application designed on buck and full-bridge converters. Figure 12b compares power loss for the same configurations with $D_{\max} = 1$ for the DC–AC converter of the two-stage system. Formulas for the considered cases are shown in Table 4.

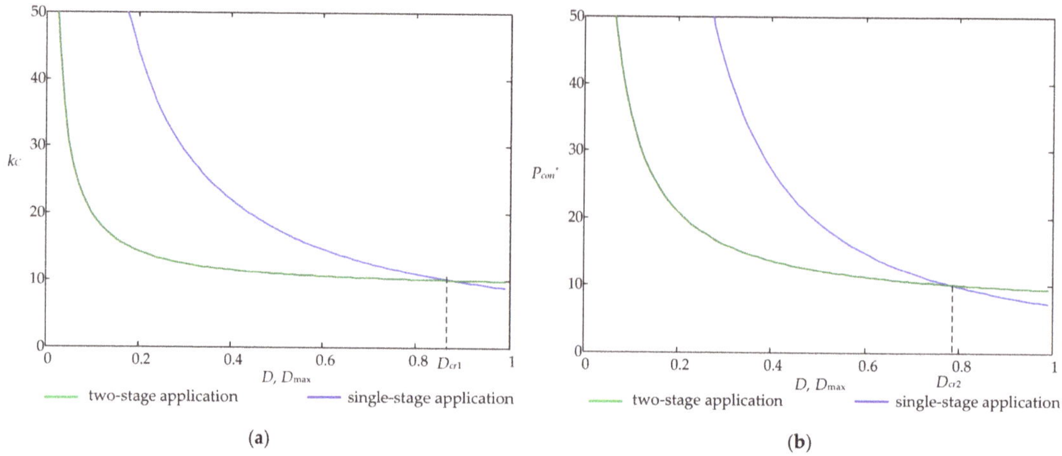

Figure 12. DC–AC converters: (**a**) the cost factor k_C; (**b**) the normalized power loss P_{con}*.

Table 4. The full-bridge converter is used to calculate the cost factor k_C and the normalized power loss for single- and two-stage applications.

	k_C	P_{con} *
Single-stage application with a full-bridge converter	$\frac{8I_{peak(m)}}{D_{max}I_{Tav(m)}}$	$\left(\frac{2\pi}{3D_{max}}+\frac{\pi^2}{3D_{max}^2}\right)\left(1-\frac{I_{peak(m)}I_{\min(m)}}{(I_{peak(m)}+I_{\min(m)})^2}\right)+\frac{0.8}{D_{max}}\left(\frac{\pi}{2}+1\right)+\frac{I_{peak(m)}}{D_{max}I_{Tav(m)}}$
Two-stage application with buck and full-bridge converters	$\frac{I_{peak(m)}}{I_{Tav(m)}}\left(8+\frac{1}{D}\right)$	$\left(\frac{2\pi}{3}+\frac{\pi^2}{3}+\frac{4}{3D}\right)\left(1-\frac{I_{peak(m)}I_{\min(m)}}{(I_{peak(m)}+I_{\min(m)})^2}\right)+0.8\left(\frac{\pi}{2}+1\right)+$ $\frac{I_{peak(m)}}{I_{Tav(m)}}\left(1+\frac{1}{D}\right)+\frac{0.8(1-D)}{D}$

Figure 12 shows that for $D_{max} < D_{cr1} = 0.87$ and $D_{max} < D_{cr2} = 0.78$, the two-stage system has a lower cost factor k_C and a lower power loss P_{con*} than the one-stage system. This means that the two-stage system is more efficient than the one-stage system.

Aside from energy-delivery, common DC–AC applications have additional functionality [58]:

- Voltage conversion and power grid synchronization;
- Disconnection and anti-islanding protection when power grid fault appears; Correction of the power factor of the input current.

The rapid evolution of on-grid PV applications with unstable generation complicates the grid's stable operation. Due to the common DC–AC application's disconnection from the power grid fault, the grid operation only worsened and became unbalanced. As a result, using converters with advanced functions that remain connected to the grid during faults and attempt to maintain a stable operation not only mitigates the impact of renewable energy instability, but also creates additional opportunities for power grid control and improves reliability [59].

The list of additional converter functions that improve power grid operation and converter control approaches, as well as topologies that may perform them, are given in Table 5.

Table 5. List of additional converter functions.

Opportunities for Power Grid Control	Converter Function	Control Strategy
Reactive power control [58]	Independent reactive and active power generation	Sinusoidal pulse width modulation (SPWM), space vector modulation (SVM) or hysteresis modulation in dq or $\alpha\beta$ space
Voltage stabilization [60–65]	Voltage and frequency ride through, voltage sag detection, reactive power generation	For frequency synchronization: zero-crossing method and the phase-locked-loop. For sag detection: RMS value estimator, synchronous rotating reference frame, wavelet, and Fourier transform. For power generation: SPWM, SVM or hysteresis modulation in dq or $\alpha\beta$ space
Grid power quality control [62]	Controlled injected current	SPWM or hysteresis modulation, frequency synchronization

As usual, the inverter's advanced functions are realized with basic inverter topologies with improved control strategies and require no extra elements except for sensors.

4. Energy Storage Applications

An energy storage system is an integral part of renewable power supply systems. The main purpose of the storage system is to provide a powerful balancing of unstable renewable sources that operate at MPP and variable load. Typical converter topologies of energy storage applications are shown in Figure 13.

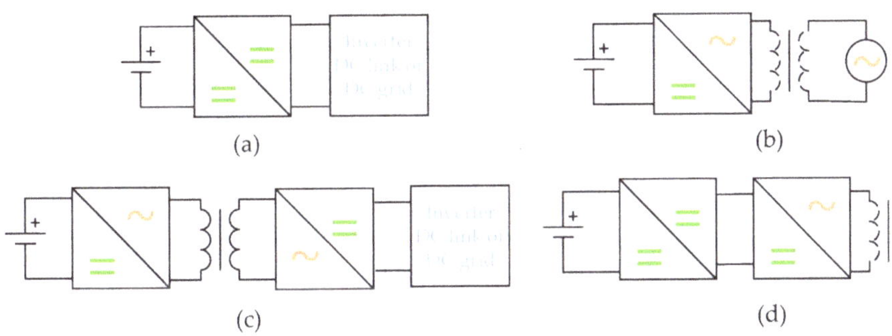

Figure 13. The following are typical energy storage converter topologies: (**a**) single-stage DC–DC application; (**b**) single-stage DC–AC application; (**c**) two-stage isolated DC–DC application; (**d**) two-stage DC–AC application.

Single-stage DC–DC and DC–AC topologies in Figure 13a and 13b respectively have simple structures and functionality. DC–DC converters for built-in energy storage connect to two-stage on-grid PV inverters or DC power grids, whereas DC-AC topologies are connected directly to the AC power grid. Figure 13c,d are used for energy storage isolation in DC power grids or AC grids and provide much more functionality, i.e., multi-mode charging, a wide range of battery charging voltages and currents, advanced control, improved performance, and energy quality control.

With increasing renewable energy penetration, energy storage systems are becoming a necessary element to maintain stable grid operation. In particular, in the concepts of intelligent transmission and control of distributed systems FACTS [66], Smart Grid [67], and Vehicle to Grid (V2G) [68], energy storage is considered as a system forming unit or unit that strongly improves the system operation, specifically:

- Load shifting occurs when renewable energy mostly charges the energy storage during the day, and the energy storage is discharged in the late hours of peak power demand [69];

- Shutdown protection in smart distributed power grids that allows supplying end-users when loss of power arises [70];
- Energy quality control (voltage, frequency, reactive power compensation, high harmonic reduction) [71].

Therefore, distributed systems usually use a single common storage system which improves the functioning of the entire system.

Because of the unstable generation of renewable sources, energy storage is permanently switched between charge and discharge modes with the unstable current. Therefore, energy storage control is much more intelligent than for a common charging device [72,73].

For DC–DC in stage energy storage applications, some of the converters discussed in the section about grid applications with modified control laws, such as full or half-bridge [74] and 3L-NPC [75] converters, require minor modifications and an increase in the number of transistors, for instance, SEPIC, Cuk' [76], or cascaded [77] and interleaved [78] half-bridge converters.

Electric isolation in DC applications is provided with a high-frequency transformer intermediate AC link. An AC link may be organized with dual full- or half-bridge converters and their soft-switching modifications [79–81].

Multi-level [82,83] or high-frequency link AC–AC converters [84] are used for single-stage AC applications, whereas full- and half-bridge converter modifications [85,86] are used for two-stage solutions.

Figure 14 depicts converters for single-stage storage applications, and Table 6 compares them based on the parameters k_C and P_{con}*, whereas two-stage solutions are shown in Figure 15 and Table 7, respectively.

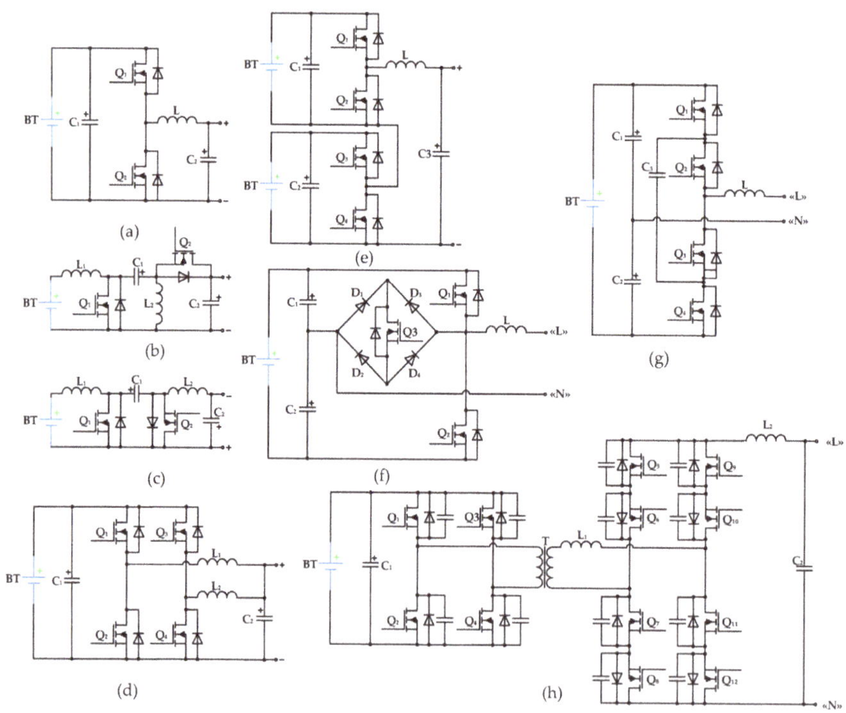

Figure 14. Converters for single-stage storage applications: (**a**) half-bridge; (**b**) Cuk'; (**c**) SEPIC/Luo; (**d**) interleaved half-bridge; (**e**) cascaded half-bridge; (**f**) half-bridge rectifier with neutral point switch clamped scheme; (**g**) capacitor clamped three-level; (**h**) high-frequency-link inverter.

Table 6. A Comparative analysis of single-stage energy storage converter effectiveness.

Converter Topology	k_C	P_{con} *
DC–DC converters		
Half-bridge [74], Figure 14a	$\frac{4I_{peak}}{D_1 I_{Tav}}$	$\frac{4}{3D_1^2}\left(1-\frac{I_{peak}I_{\min}}{(I_{peak}+I_{\min})^2}\right)+\frac{0.8}{D_1}+\frac{I_{peak}}{D_1 I_{Tav}}$
3L-NPC [75], Figure 9q	$\frac{4I_{peak}}{D_1 I_{Tav}}$	$\left(\frac{4}{3D_1}+\frac{4}{3D_1^2}\right)\left(1-\frac{I_{peak}I_{\min}}{(I_{peak}+I_{\min})^2}\right)+\frac{0.8(1-D_1)}{D_1}+\frac{I_{peak}}{D_1 I_{Tav}}$
Cuk' [76], Figure 14b, SEPIC/Luo [76], Figure 14c	$\frac{2I_{peak}}{D_1(1-D_1)I_{Tav}}$	$\frac{4}{3D_1(1-D_1)^2}\left(1-\frac{I_{peak}I_{\min}}{(I_{peak}+I_{\min})^2}\right)+\frac{0.4}{D_1(1-D_1)}+\frac{I_{peak}}{D_1(1-D_1)I_{Tav}}$
Interleaved half-bridge [77], Figure 14d	$\frac{2I_{peak}}{D_1 I_{Tav}}$	$\frac{2}{3mD_1^2}\left(1-\frac{I_{peak}I_{\min}}{(I_{peak}+I_{\min})^2}\right)+\frac{0.8}{D_1}+\frac{I_{peak}}{D_1 m I_{Tav}}$
If $U_{BT} > U_g \rightarrow$ Cascaded half-bridge [78], Figure 14e If $U_{BT} < U_g \rightarrow$	$\frac{2I_{peak}(1+D_1)}{D_1 I_{Tav}}$ $\frac{2I_{peak}(2-D_1)}{(1-D_1)I_{Tav}}$	$\frac{4}{3D_1^2}\left(1-\frac{I_{peak}I_{\min}}{(I_{peak}+I_{\min})^2}\right)+\frac{0.8}{D_1}+$ $\frac{I_{peak}}{D_1(1-D_1)I_{Tav}}$ $\frac{4}{3(1-D_1)^2}\left(1-\frac{I_{peak}I_{\min}}{(I_{peak}+I_{\min})^2}\right)+\frac{0.8}{1-D_1}+\frac{I_{peak}}{D_1(1-D_1)I_{Tav}}$
DC–AC converters		
Half-bridge, Figure 9j	$\frac{8I_{peak(m)}}{D_{\max 1} I_{Tav(m)}}$	$\frac{4\pi^2}{3D_{\max 1}^2}\left(1-\frac{I_{peak(m)}I_{\min(m)}}{(I_{peak(m)}+I_{\min(m)})^2}\right)+\frac{1.6}{D_{\max 1}}\left(\frac{\pi}{2}+1\right)+\frac{2I_{peak(m)}}{D_{\max 1} I_{Tav(m)}}$
Full-bridge, Figure 9k	$\frac{8I_{peak(m)}}{D_{\max 1} I_{Tav(m)}}$	$\left(\frac{2\pi}{3D_{\max 1}}+\frac{\pi^2}{3D_{\max 1}^2}\right)\left(1-\frac{I_{peak(m)}I_{\min(m)}}{(I_{peak(m)}+I_{\min(m)})^2}\right)+$ $\frac{0.8}{D_{\max 1}}\left(\frac{\pi}{2}+1\right)+\frac{I_{peak(m)}}{D_{\max 1} I_{Tav(m)}}$
Half-bridge rectifier with neutral point switch clamped scheme [82], Figure 14f	$\frac{5I_{peak(m)}}{D_{\max 1} I_{Tav(m)}}$	$\frac{4\pi^2}{3D_{\max 1}^2}\left(1-\frac{I_{peak(m)}I_{\min(m)}}{(I_{peak(m)}+I_{\min(m)})^2}\right)+\frac{3.2}{D_{\max 1}}\left(\frac{\pi}{2}+1\right)+\frac{2I_{peak(m)}}{D_{\max 1} I_{Tav(m)}}$
Capacitor clamped three-level PWM converter [83], Figure 14g	$\frac{8I_{peak(m)}}{D_{\max 1} I_{Tav(m)}}$	$\frac{8\pi^2}{3D_{\max 1}^2}\left(1-\frac{I_{peak(m)}I_{\min(m)}}{(I_{peak(m)}+I_{\min(m)})^2}\right)+\frac{0.5I_{peak(m)}}{D_{\max 1} I_{Tav(m)}}$
High-frequency link inverter [84], Figure 14h	$\frac{24I_{peak(m)}}{D_{\max 1} I_{Tav(m)}}$	$\left(\frac{2\pi}{D_{\max}}+\frac{\pi^2}{3D_{\max}^2}\right)\left(1-\frac{I_{peak(m)}I_{\min(m)}}{(I_{peak(m)}+I_{\min(m)})^2}\right)+\frac{0.8}{D_{\max}}\left(\frac{\pi}{2}+1\right)+\frac{3.2}{\pi}$

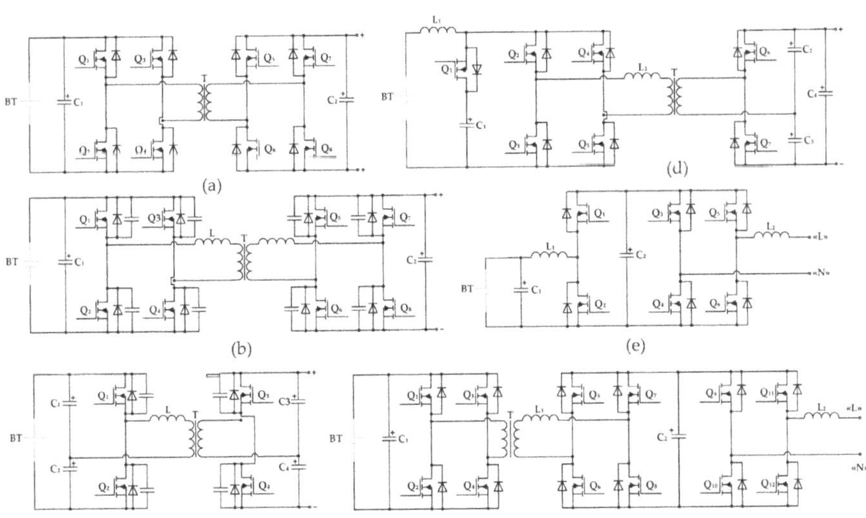

Figure 15. Converters for two-stage storage applications: (**a**) dual active bridge; (**b**) dual active bridge soft-switching; (**c**) with two voltage-fed half-bridges; (**d**) combined voltage- half-bridge and current-fed full-bridge; (**e**) half-bridge and full-bridge; (**f**) full-bridge DC–AC and dual active bridge DC–DC.

Table 7. A Comparative analysis of the two-stage energy storage converters' effectiveness.

Converter Topology	k_C	P_{con} *
	DC–DC converters	
Dual active bridge, Figure 15a	$\frac{8I_{peak}}{D_1 I_{Tav}}$	$\frac{4}{3D_1^2}(1+D_1)\left(1-\frac{I_{peak}I_{\min}}{(I_{peak}+I_{\min})^2}\right)+\frac{0.8}{D_1}+0.8+\frac{3I_{peak}}{D_1 I_{Tav}}$
Dual active bridge soft-switching [79], Figure 15b	$\frac{8I_{peak}}{D_1 I_{Tav}}$	$\frac{4}{3D_1^2}(1+D_1)\left(1-\frac{I_{peak}I_{\min}}{(I_{peak}+I_{\min})^2}\right)+\frac{0.8}{D_1}+0.8$
Converter with two voltage-fed half-bridges [80], Figure 15c	$\frac{8I_{peak}}{D_1 I_{Tav}}$	8
Combined-voltage half-bridge and current-fed full-bridge [81], Figure 15d	$\frac{11I_{peak}}{(1-D_1)I_{Tav}}$	$\left(\frac{5+4D_1}{3(1-D_1)^2}+\frac{8(1-D_1)}{3D_1{}^2}\right)\left(1-\frac{I_{peak}I_{\min}}{(I_{peak}+I_{\min})^2}\right)+\frac{0.8}{1-D_1}$
	DC–AC converters	
Half-bridge and full-bridge [85], Figure 15e	$\frac{I_{peak(m)}}{I_{Tav(m)}}\left(\frac{8}{D_{\max 1}}+\frac{2}{(1-D_3)}\right)$	$\left(\frac{2\pi}{3D_{\max 1}}+\frac{\pi^2}{3D_{\max 1}^2}+\frac{4D_3}{3(1-D_3)^2}\right)\left(1-\frac{I_{peak(m)}I_{\min(m)}}{(I_{peak}+I_{\min})^2}\right)+\frac{0.8}{D_{\max 1}}\left(\frac{\pi}{2}+1\right)+0.8+\frac{I_{peak(m)}}{I_{Tav(m)}}\left(\frac{1}{D_{\max 1}}+\frac{1}{(1-D_3)}\right)$
Full-bridge DC–AC and dual active bridge DC–DC [86], Figure 15f	$\frac{8I_{peak(m)}}{I_{Tav(m)}}\times\left(\frac{1}{D_{\max 1}}+\frac{1}{D_3}\right)$	$\left(\frac{2\pi}{3D_{\max 1}}+\frac{\pi^2}{3D_{\max 1}^2}+\frac{4}{3D_3}+\frac{4}{3D_3{}^2}\right)\left(1-\frac{I_{peak(m)}I_{\min(m)}}{(I_{peak}+I_{\min})^2}\right)+\frac{0.8}{D_{\max 1}}\left(\frac{\pi}{2}+1\right)+0.8+\frac{0.8}{D_3}+\frac{I_{peak(m)}}{D_{\max 1}I_{Tav(m)}}$

As usual, the voltage gains of the power converters used in energy storage applications are different. This means that the duty cycle values and power loss values (P_{con}*) are different depending on which way the power converters are used. For instance, a full-bridge converter, shown in Figure 9k, operates as a boost converter with duty cycle D_1 when energy is stored and as a buck converter with duty cycle D_2 when delivering energy to the power grid. The relationship between the parameters D_1 and D_2 is equal to:

$$D_2 = 1 - D_1. \tag{16}$$

Thus, power loss P_{con}* estimation depends on the direction of energy transfer. As long as the same amount of energy is sent in both directions, the power loss weight formula is used with weight factor w = 0.5:

$$P_{con}{}^* = 0.5P_{con1}{}^*(D_1) + 0.5P_{con2}{}^*(D_2), \tag{17}$$

where P_{con1}* is the power loss value for operation in energy storage mode with duty cycle D_1, and P_{con2}* is the power loss value for operation in delivering energy to the power grid with duty cycle D_2.

In two-stage DC–AC energy storage applications, the first and second stages operate independently, so their operation is defined by four duty cycles, i.e., D_1, D_2 for the first stage and D_3, D_4 for the second stage. However, DC–DC two-stage applications always contain AC links that are used for electric isolation and soft-switching. Therefore, one of the stages always works in passive mode as a rectifier with the same duty cycle.

Derived results in Figures 16 and 17 make it clear that more effective applications are designed with basic half- and full-bridge topologies with improved dynamic loss. For DC–DC applications, there are cascaded and interleaved half-bridge topologies and 3L-NPC converters, whereas effective DC–AC applications are based on full-bridge, and full-bridge and half-bridge topologies.

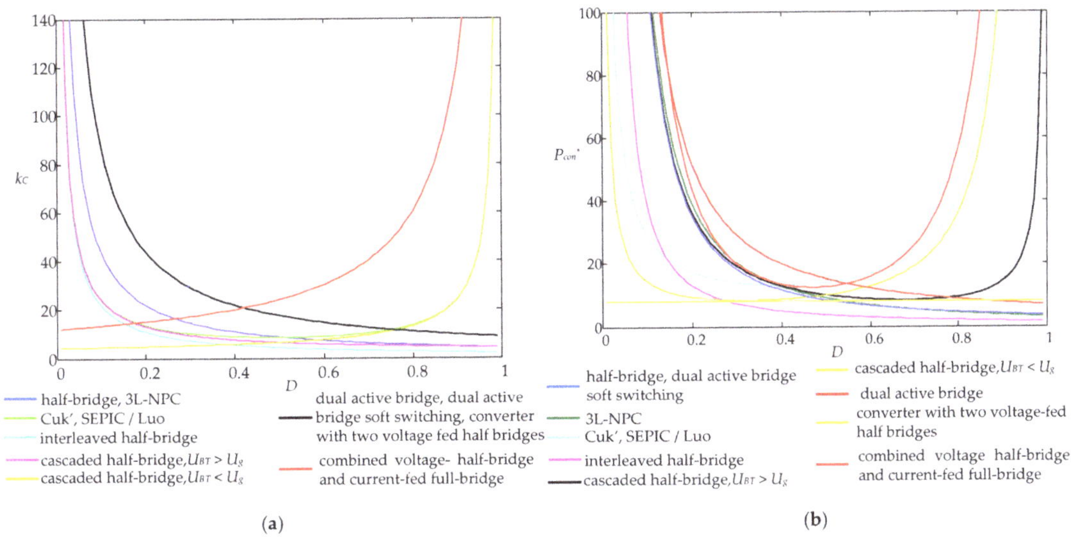

Figure 16. The cost factor k_C and normalized power loss $P_{con}*$ of DC–DC energy storage (**a**) Cost factor k_C; (**b**) normalized power loss $P_{con}*$ in DC–DC energy storage converters.

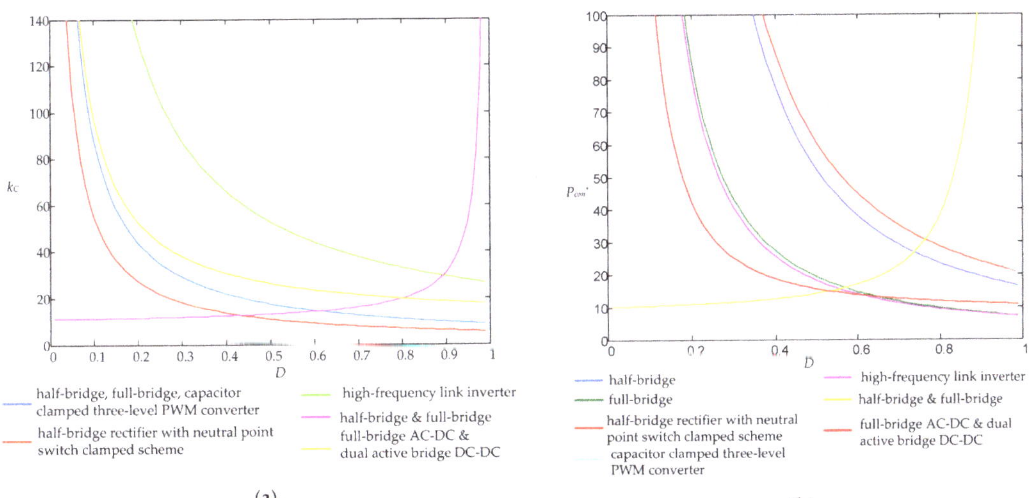

Figure 17. The parameters for DC–AC applications (**a**) Cost factor k_C; (**b**) normalized power loss $P_{con}*$ in DC–AC energy storage converters.

Figure 16a,b show the cost factor k_C and normalized power loss $P_{con}*$ of DC–DC energy storage, while Figure 17a,b show the parameters for DC–AC applications.

5. High Voltage Gain Converters

Solar panels have relatively low voltage and current and are usually combined with batteries in parallel and series connections. The variation of power in solar radiation and in the parameters of the panels in the battery leads to a decrease in overall efficiency [87] and may cause system instability due to partial shadowing effects [88]. For mitigation of the aforegoing drawbacks, high voltage gain converters are used. The common voltage boosting method, with a high-frequency transformer and high turns ratio n, increases component stress on the secondary size and causes leakage inductance voltage spikes, thus non-isolated converters are often used in practice. A high gain value G for the conventional boost converter is achieved to the detriment of power loss with operation in modes with a close to unity duty cycle [89]. A cascading technique involves a series connection of several converters and expands the total voltage gain value G several times [90]. It increases the cost and complicates the converter design and control. For efficiency improvement the interleaving technique is used [91].

Other DC–DC converters, such as SEPIC and Flyback, can provide galvanic isolation via additional winding on the inductor core. These converters are used for high voltage gain but require additional snubber circuits to mitigate issues caused by leakage inductance [92] and usually suffer from increased transistor overvoltage with factors of 1.5–2.0. Passive dissipative RCD snubbers are commonly used [93], but for high-efficiency solutions more sophisticated active [94,95] or passive [96,97] regenerative snubbers deliver energy leakage inductance on the load or primary energy source. In addition, an isolated solution with leakage inductance allows the implementation of transistor soft-switching as well as voltage doubling, for instance, in an LC parallel current source converter with a voltage doubler [98]. The idea of capacitor voltage doubling is used for a set of multi-cell flying capacitor converters: super lift voltage converter [99], modified voltage lift converter [100], Cockcroft Walton multiplier based boost converter [101], Dickson multiplier based boost converter [101], boost derived MIESC SC-cell converter [102], and buck-boost derived MIESC SC-cell converter [102]. Capacitor voltage doubling is also realized in converters based on a three-state switching cell (3SSC) that is a combination of two switching PWM cells (2SSC) [103]. Such converters have reduced the size, weight, and volume of magnetics and reduced the current stress of switches.

Figure 18 shows the reviewed topologies of high gain converters, whereas Table 8 analyzes the main features of high gain converters, i.e., voltage gain G, switch voltage stress, and basic parameters k_C and $P_{con}{}^*$.

As shown in Figure 19a, an essential problem for high gain applications is significant transistor overvoltage. As a result, power converters with lower transistor stress recalculated on the gain unit, such as Cockcroft–Walton and Dickson multiplier-based boost converters, boost derived MIESC SC-cell converters, and boost 3SSC cell converters, are more appealing for high voltage application design.

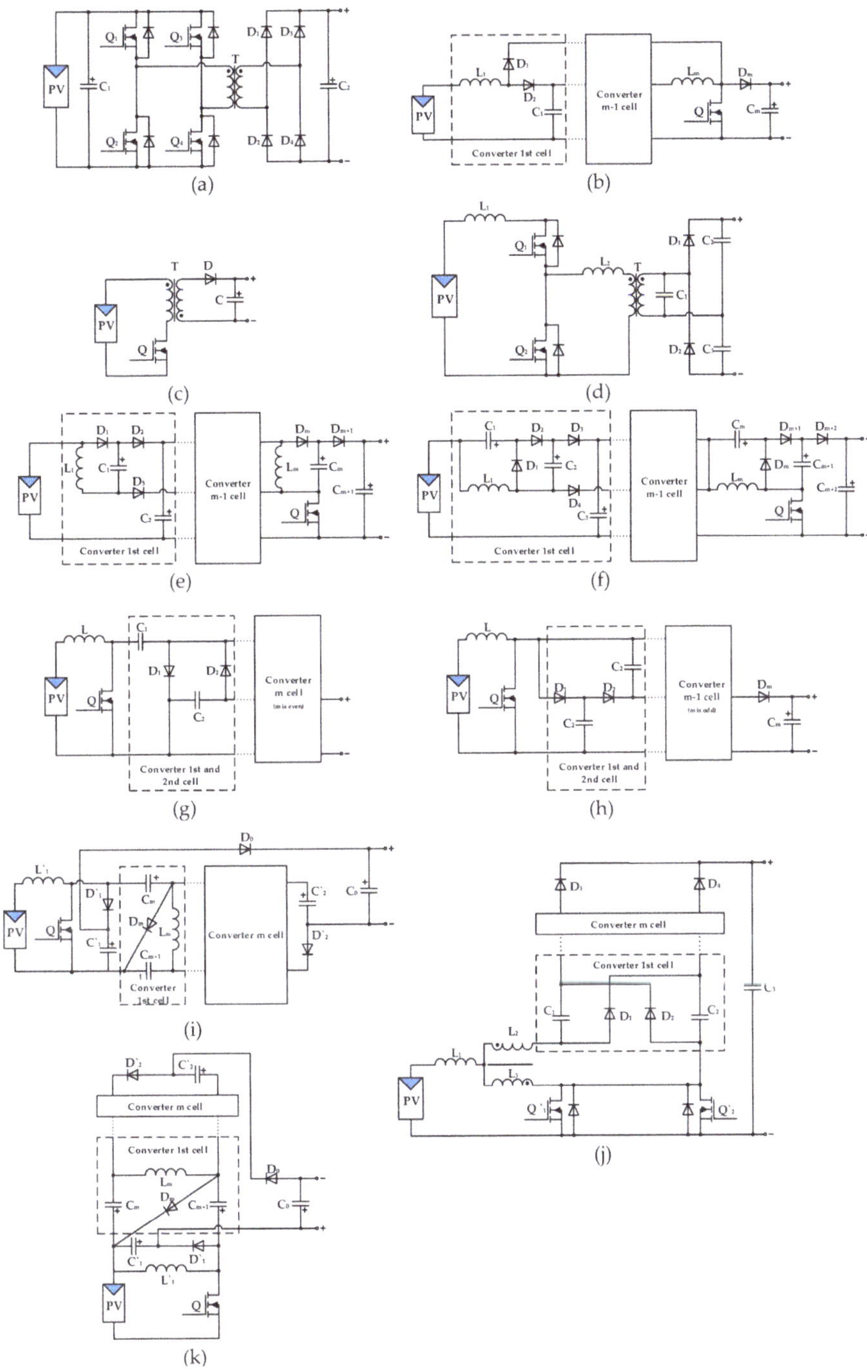

Figure 18. High voltage gain converters: (**a**) isolated full-bridge; (**b**) cascaded boost converter; (**c**) flyback; (**d**) LC parallel resonant converter with voltage doubler; (**e**) super lift voltage converter; (**f**) modified voltage lift converter; (**g**) Cockcroft–Walton multiplier based boost converter; (**h**) Dickson multiplier based boost converter; (**i**) boost derived MIESC SC-cell converter; (**j**) boost 3SSC cell converter; (**k**) buck-boost derived MIESC SC-cell converter.

Table 8. High voltage gain converters: comparative analysis.

Converter Topology	G	k_C	P_{con}*
Isolated full-bridge, Figure 18a	nD	$\frac{8I_{peak}}{DI_{Tav}}$	$\frac{32}{3D}\left(1-\frac{I_{peak}I_{\min}}{(I_{peak}+I_{\min})^2}\right)+\frac{1.6}{D}+\frac{2I_{peak}}{DI_{Tav}}$
Conventional boost converter, Figure 9b	$\frac{1}{1-D}$	$\frac{I_{peak}}{(1-D)I_{Tav}}$	$\frac{4D}{3(1-D)^2}\left(1-\frac{I_{peak}I_{\min}}{(I_{peak}+I_{\min})^2}\right)+0.8+\frac{I_{peak}}{(1-D)I_{Tav}}$
Cascaded boost converter [90], Figure 18b	$\frac{1}{(1-D)^2}$	$\frac{I_{peak}}{(1-D)^2 I_{Tav}}$	$\frac{4D}{3(1-D)^2}\left(1-\frac{I_{peak}I_{\min}}{(I_{peak}+I_{\min})^2}\right)+\frac{0.8D^2}{(1-D)^2}+0.8+\frac{0.8}{1-D}+\frac{2I_{peak}}{(1-D)I_{Tav}}$
SEPIC [92], Figure 9d, Flyback [92], Figure 18c	$\frac{nD}{1-D}$	$\frac{1.5I_{peak}}{(1-D)DI_{Tav}}$	$\frac{3}{D(1-D)^2}\left(1-\frac{I_{peak}I_{\min}}{(I_{peak}+I_{\min})^2}\right)+\frac{0.8}{D}+\frac{I_{peak}}{(1-D)DI_{Tav}}$
LC parallel current source converter with voltage doubler [98], Figure 18d	$\frac{2n}{1-D}$	$\frac{1.1I_{peak}}{(1-D)I_{Tav}}$	$\left(\frac{2.42D}{3(1-D)^4}+\frac{8}{3(1-D)}\right)\left(1-\frac{I_{peak}I_{\min}}{(I_{peak}+I_{\min})^2}\right)+0.8+\frac{I_{peak}}{(1-D)I_{Tav}}$
Super lift voltage converter [99], m cells, Figure 18e	$\left(\frac{2-D}{1-D}\right)^m$	$\left(\left(\frac{2-D}{1-D}\right)^m-1\right)\frac{I_{peak}}{I_{Tav}}$	$\frac{4D}{3(1-D)^2}\left(1-\frac{I_{peak}I_{\min}}{(I_{peak}+I_{\min})^2}\right)+\frac{(1.6(m-1)+0.8)D}{(1-D)}+0.8m+\frac{1.5mI_{peak}}{(1-D)I_{Tav}}$
Modified voltage lift converter [100], m cells, Figure 18f	$\left(\frac{2}{1-D}\right)^m$	$\left(\left(\frac{2}{1-D}\right)^m-1\right)\frac{I_{peak}}{I_{Tav}}$	$\frac{4D}{3(1-D)^2}\left(1-\frac{I_{peak}I_{\min}}{(I_{peak}+I_{\min})^2}\right)+\frac{(1.6(m-1)+0.8)D}{(1-D)}+1.6m+\frac{2mI_{peak}}{(1-D)I_{Tav}}$
Cockcroft–Walton and Dickson multiplier based boost converter [101], m cells, Figure 18g,h	$\frac{m+D}{1-D}$	$\frac{I_{peak}}{(1-D)I_{Tav}}$	$\frac{4D}{3(1-D)^2}\left(1-\frac{I_{peak}I_{\min}}{(I_{peak}+I_{\min})^2}\right)+\frac{0.8mD}{(1-D)}+0.8(m+1)+\frac{(m+0.5)I_{peak}}{(1-D)I_{Tav}}$
Boost derived MIESC SC-cell converter [102], m cells, Figure 18i	$\left(\frac{2}{1-D}\right)^m$	$\frac{I_{peak}}{(1-D)I_{Tav}}$	$\frac{4D}{3(1-D)^2}\left(1-\frac{I_{peak}I_{\min}}{(I_{peak}+I_{\min})^2}\right)+\frac{0.8D}{(1-D)}+0.8(m+2)+\frac{0.5(m+4)I_{peak}}{(1-D)I_{Tav}}$
Buck-boost derived MIESC SC-cell converter [102], m cells, Figure 18j	$\left(\frac{1+D}{1-D}\right)^m$	$\frac{I_{peak}}{D(1-D)I_{Tav}}$	$\frac{4}{3D(1-D)^2}\left(1-\frac{I_{peak}I_{\min}}{(I_{peak}+I_{\min})^2}\right)+\frac{0.8}{1-D}+\frac{0.8(m+2)}{D}+\frac{0.5(m+4)I_{peak}}{(1-D)DI_{Tav}}$
Boost 3SSC cell converter [103], m cells, Figure 18k	$\frac{m+1}{1-D}$	$\frac{2I_{peak}}{(1-D)I_{Tav}}$	$\left(\frac{2D}{3(1-D)^2}+\frac{4}{3(1-D)}\right)\left(1-\frac{I_{peak}I_{\min}}{(I_{peak}+I_{\min})^2}\right)+1.6+\frac{(m+0.5)I_{peak}}{(1-D)I_{Tav}}$

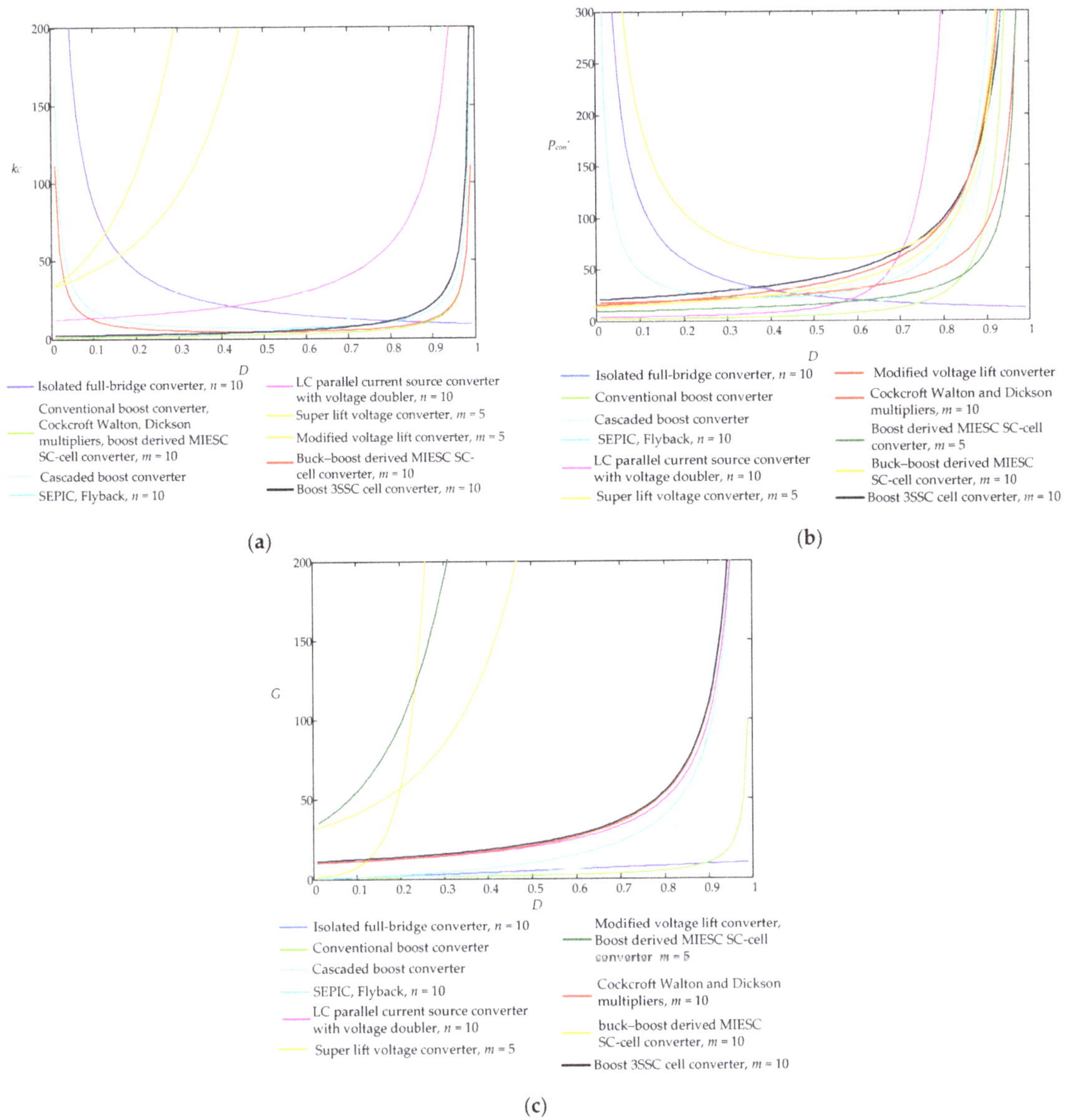

Figure 19. In high voltage gain converters: (**a**) cost factor k_C; (**b**) normalized power loss P_{con}*; and (**c**) voltage gain G.

6. Hybrid PV Applications

Hybrid PV systems are complex solutions that include multiple energy sources, such as a diesel generator [104], a wind turbine or a fuel cell [105]; integrated energy storage with an interruptible power supply function; and multiple AC and/or DC outputs. The system is often used for standalone applications for reliable power supply [106] or as an advanced uninterruptable power system that allows generating or consuming energy from the grid for imbalance elimination of PV systems and load [107]. The hybrid system consists of several stages which, in the common case, are realized with individual power converters, which is the redundant solution. Combining identical converter links makes it possible to reduce the number of elements in the system and simplify control [108]. The simplest case is the use of multi-winding transformers [109]. In non-isolated applications, combinations of cells with boost, buck, and buck-boost converters with unidirectional [110] or bidirectional [111–113] energy flow are implemented. For distributed systems, interleaved multi-input solutions are also used [114].

As usual, multi-port solutions are designed for DC–DC [115] or DC–AC [116–119] applications with renewable sources and energy storage suitable for PV and wind applications as well. Table 9 lists the analytical expressions of the multiport converter parameters shown in Figure 20.

Table 9. A comparative analysis of the effectiveness of the hybrid converter.

Converter Topology	k_C	P_{con} *
Boost three-port converter [110], Figure 20a	$\frac{3}{1-D}\frac{I_{peak}}{I_{Tav}}$	$\frac{16D}{9(1-D)^2}\left(1-\frac{I_{peak}I_{\min}}{(I_{peak}+I_{\min})^2}\right)+0.8+\frac{I_{peak}}{(1-D)I_{Tav}}$
Buck three-port converter [110], Figure 20b	$\left(\frac{2}{D}+1\right)\frac{I_{peak}}{I_{Tav}}$	$\frac{4}{3D}\left(1-\frac{I_{peak}I_{\min}}{(I_{peak}+I_{\min})^2}\right)\left(1+\frac{1}{3D}\right)+\frac{0.8(1.33-D)}{D}+\frac{I_{peak}}{DI_{Tav}}$
Buck-boost three-port converter [110], Figure 20c	$\frac{2}{D(1-D)}\frac{I_{peak}}{I_{Tav}}$	$\frac{4}{3}\left(\frac{1}{D(1-D)^2}+\frac{1}{3(1-D)}\right)\left(1-\frac{I_{peak}I_{\min}}{(I_{peak}+I_{\min})^2}\right)+$ $\frac{0.8}{D}-\frac{0.8}{3}+\frac{I_{peak}}{3I_{Tav}}\left(\frac{2}{D(1-D)}+\frac{1}{D}\right)$
Bidirectional buck-boost converter [111], Figure 20d	$\frac{8I_{peak}}{DI_{Tav}}$	$\frac{4D}{(1-D)^2}\left(1-\frac{I_{peak}I_{\min}}{(I_{peak}+I_{\min})^2}\right)+\frac{0.8\cdot(1-D)}{3D}+\frac{3.2}{3}+\frac{0.5I_{peak}}{3DI_{Tav}}$
Switched capacitor multi-port converter [112], Figure 20e	$\frac{2I_{peak}}{DI_{Tav}}$	$\frac{1}{3D}\left(1-\frac{I_{peak}I_{\min}}{(I_{peak}+I_{\min})^2}\right)+\frac{5\cdot 0.8(1-D)}{3D}$
Dual active bridge multi-port converter [112], Figure 20f	$\frac{4I_{peak}}{DI_{Tav}}$	$\left(\frac{4}{9D}+\frac{8D}{9(1-D)^2}\left(1+\frac{1}{2D}\right)\right)\left(1-\frac{I_{peak}I_{\min}}{(I_{peak}+I_{\min})^2}\right)+$ $\frac{0.8}{3D}+\frac{1.6}{3}+\frac{0.8D}{3(1-D)}+\frac{I_{peak}}{3DI_{Tav}}$
Double-stage with battery boost converter [116], Figure 20g	$\frac{I_{peak}}{I_{Tav}}\left(\frac{3}{1-D}+8\right)$	$\left(\frac{4D}{3(1-D)^2}+\frac{\pi}{3}+\frac{\pi^2}{6}\right)\left(1-\frac{I_{peak}I_{\min}}{(I_{peak}+I_{\min})^2}\right)+$ $0.8+0.4\left(\frac{\pi}{2}+1\right)+\frac{0.5I_{peak}}{I_{Tav}(1-D)}+\frac{0.5I_{peak}}{I_{Tav}}$
Fully soft-switched multi-port DC-DC converter [113], Figure 20h	$\frac{5I_{peak}}{DI_{Tav}}$	$\frac{4}{9}\left(\frac{1}{D^2}+\frac{D+1}{(1-D)^2}\right)\left(1-\frac{I_{peak}I_{\min}}{(I_{peak}+I_{\min})^2}\right)+\frac{0.8(1-D)}{3D}$
Multiple-input SEPIC converter, *m* cells [114], Figure 20i	$\frac{I_{peak}}{D(1-D)I_{Tav}}$	$\frac{4}{3Dm(1-D)^2}\left(1-\frac{I_{peak}I_{\min}}{(I_{peak}+I_{\min})^2}\right)+\frac{0.8}{D}+\frac{0.8}{1-D}+\frac{1.5I_{peak}}{I_{Tav}mD(1-D)}$
NPC multiport converter [110], Figure 20j	$\frac{8I_{peak}}{D_{\max}I_{Tav}}$	$\left(\frac{2\pi}{9D_{\max}}+\frac{\pi^2}{9D_{\max}^2}\right)\left(1-\frac{I_{peak}I_{\min}}{(I_{peak}+I_{\min})^2}\right)+\frac{0.8}{D_{\max}}\left(\frac{\pi}{2}+1\right)+\frac{I_{peak}}{D_{\max}I_{Tav}}$

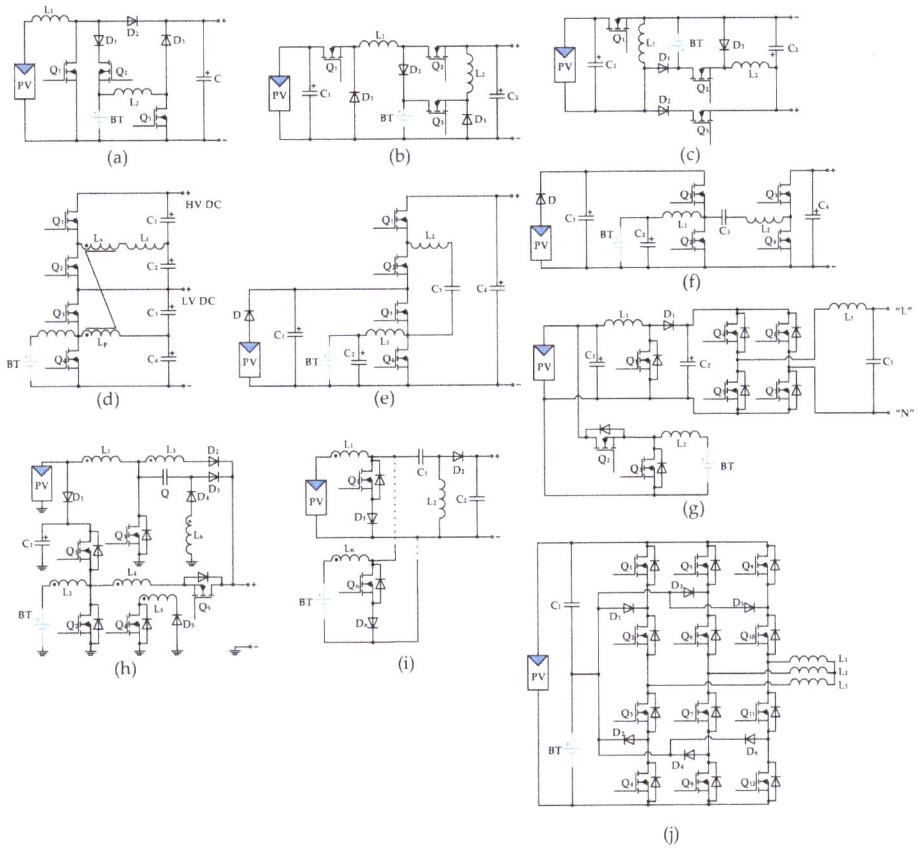

Figure 20. Hybrid converters: (**a**) boost three-port converter; (**b**) buck three-port converter; (**c**) buck-boost three-port converter; (**d**) bidirectional buck-boost converter; (**e**) switched capacitor multi-port converter; (**f**) dual active bridge multi-port converter; (**g**) double-stage with battery boost converter; (**h**) fully soft-switched multi-port DC–DC converter; (**i**) multiple-input SEPIC converter; (**j**) NPC multiport converter.

Hybrid converter parameters k_C and P_{con}* are shown in Figure 21a,b, respectively.

As shown in Figure 21, the lowest values of k_C and P_{con}* parameters are achieved for converters with basic buck and boost topologies, as well as for more complex switched-capacitor multi-port converters and dual active bridge multi-port converters with soft-switching transistor commutation. However, the analyzed hybrid converters generally have the same number of transistors as common power converters with the same features. The hybrid converters, on the other hand, have about the same power loss and cost as their counterparts with common topologies.

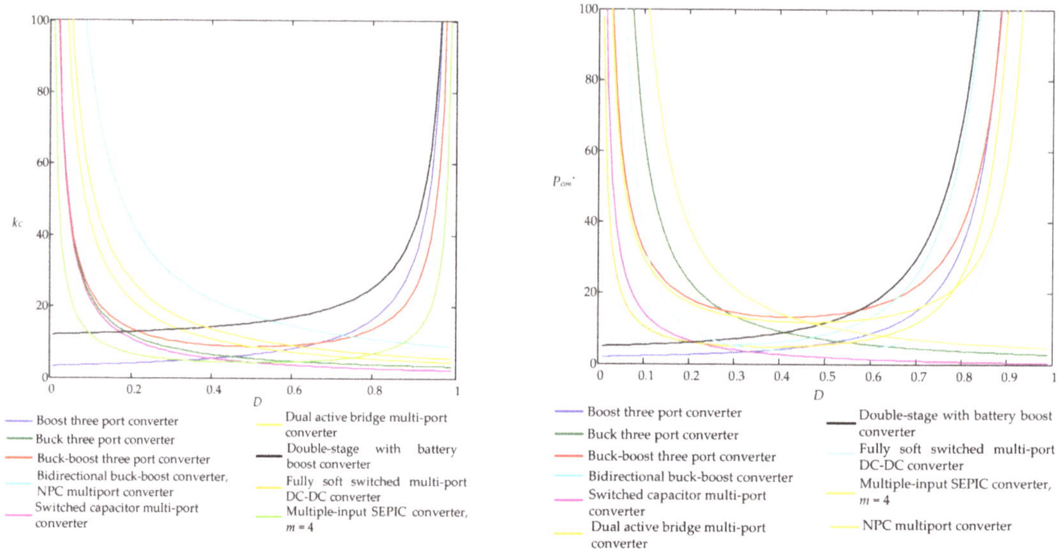

(a) (b)

Figure 21. Hybrid converter parameters: (**a**) cost factor k_C; (**b**) normalized power loss P_{con}*.

7. Discussion

Derived analytical expressions for cost k_c and power loss P_{con}* factors are evaluation indicators for selecting the converter topology for PV application design based on duty cycle D. Such parameter representations are strictly related to the converter operation mode and help to define the duty cycle D range of the PV application. However, the comparative analysis of the converter based on k_C and P_{con}* factors as functions of D, $k_c = f(D)$, and P_{con}* $= f(D)$, is inconvenient due to their different voltage gains. This is especially true when comparing different types of converters, i.e., buck, boost, and buck-boost converters. It is more informative to analyze parameters k_C and P_{con}* in some load voltage ranges ($U_{L(\min)}$, $U_{L(\max)}$) where the PV application operates for comparative analysis unification. If voltage $U_{L(\min)}$ is associated with voltage gain $G_{\min}$ and voltage $U_{L(\max)}$ with voltage gain $G_{\max}$, we can define the maximum gain factor $G_{\max}$*:

$$G_{\max}{}^* = \frac{G_{\max}}{G_{\min}}. \tag{18}$$

The real gain factor G* of a converter is always varied in the range [1, $G_{\max}$*] and may be used as a universal parameter for comparing different types of power converters.

Due to the opportunity to achieve gain factor $G_{\max}$* on different duty cycle D ranges [$D_{\min}$; $D_{\max}$], the range is defined according to providing the lowest cost and/or power loss. According to the specified conditions and obtained results:

- In buck converters, the output voltage maximum value $U_{L(\max)}$ is fixed to the input voltage U_{in}, $U_{L(\max)} = U_{in}$ that corresponds to $D_{\max} = 1$;
- In boost converters, the output voltage minimum value $U_{L(\min)}$ is fixed to the input voltage U_{in}, $U_{L(\min)} = U_{in}$ that corresponds to $D_{\min} = 0$;
- In buck-boost converters, the lowest values of k_c and P_{con}* are achieved in the middle of the duty cycle range, $D = 0.5$. Therefore, for the proper definition of $D_{\min} < 0.5$ and $D_{\max} > 0.5$, one of the following equations is solved:

$$k_C(D_{\min}) = k_C(D_{\max});\\ \text{or}\\ P_{con}{}^*(D_{\min}) = P_{con}{}^*(D_{\max}), \tag{19}$$

The values of $U_{L(\min)} = f(D_{\min})$ and $U_{L(\max)} = f(D_{\max})$ are defined depending on the choice of minimization parameter.

After the proposed transformation, parameters k_C and $P_{con}{}^*$ may be defined based on a function of the parameter G^* that varies in the range [1, $G_{\max}{}^*$]. Due to this representation, a comprehensive comparison analysis of converters for the defined value of parameter G^* may be conducted. However, in real PV applications, because of variations in solar insolation, temperature, battery state of charge, grid voltage, etc., the gain parameter G^* varies in range of 1 ... $G^*_{\max}$, and the parameters k_c and $P_{con}{}^*$ change, respectively. The cost factor kc of the PV application is obviously defined by the largest value of $G^*_{\max}$, whereas the power loss parameter $P_{con}{}^*$ is defined as an intermediate mean value $P_{con(av)}{}^*$ between $P_{con}{}^*_{(\min)} = f(G^* = 1)$ and $P_{con}{}^*_{(\max)} = f(G^* = G_{\max}{}^*)$. The precise $P_{con(av)}{}^*$ value is determined by a weighted function that defines a probability distribution law of G^*, $p(G^*)$, with values ranging from 1 to $G_{\max^*}$ [120]:

$$P_{con(av)}{}^* = \int_1^{G_{\max}{}^*} p(G^*) \cdot P_{con}{}^*(G^*) dG^*. \tag{20}$$

For example, in the case of a continuous uniform distribution, $p = 1/(G_{\max}{}^* - 1)$ [121] Formula (20) yields:

$$P_{con(av)}{}^* = \frac{1}{G_{\max}{}^* - 1} \int_1^{G_{\max}{}^*} P_{con}{}^*(G^*) dG^*. \tag{21}$$

Additionally, analytical expressions of converter efficiency parameters k_C and $P_{con(av)}{}^*$ for DC–DC converters listed in Table 3 in space of the variable G^* are shown in Table 10 and illustrated in Figure 22a,b, respectively.

A representation of the parameters k_C and $P_{con(av)}{}^*$ with the gain factor G^* aids in the estimation of converter features within a given voltage range, as well as the comparison of different converter types (buck, boost and buck-boost). For example, a buck converter has the lowest cost factor k_C value for $G^* \in (1; 4.5)$, whereas a buck-boost converter is preferable for $G^* \in (4.5; \infty)$ applications. A similar decision may be made for mean power loss $P_{con(av)}{}^*$, which has the lowest value for buck converter modifications in all G^* ranges.

Obviously, for different models of semiconductor devices and other probability distribution laws, the results would be different. This allows you to compare power converters based on the PV application's start-up conditions.

Table 10. Comparative analysis of the DC–DC converter's effectiveness in space of the variable G^*.

Converter Topology	k_C	$P_{con(av)}{}^*$
Buck	$\frac{G_{max}{}^* I_{peak}}{I_{Tav}}$	$\frac{1}{G_{max}{}^*-1}\left(\frac{G_{max}{}^{*2}-1}{2}\left(\frac{4}{3}\left(1-\frac{I_{peak}I_{min}}{(I_{peak}+I_{min})^2}\right)+0.8+\frac{I_{peak}}{I_{Tav}}\right)-0.8(G_{max}{}^*-1)\right)$
Boost	$\frac{G_{max}{}^* I_{peak}}{I_{Tav}}$	$\frac{1}{G_{max}{}^*-1}\left(\left(\frac{4(G_{max}{}^{*3}-1)}{9}-\frac{2(G_{max}{}^{*2}-1)}{3}\right)\left(1-\frac{I_{peak}I_{min}}{(I_{peak}+I_{min})^2}\right)+ +0.8(G_{max}{}^*-1)+\frac{(G_{max}{}^{*2}-1)I_{peak}}{2I_{Tav}}\right)$
Buck-boost, SEPIC, Cuk'	$\frac{\left(1+\sqrt{G_{max}{}^*}\right)^2 I_{peak}}{\sqrt{G_{max}{}^*}I_{Tav}}$	$\frac{1}{G_{max}{}^*-1}\left(\left(\frac{4}{3}\left(1-\frac{I_{peak}I_{min}}{(I_{peak}+I_{min})^2}\right)+\frac{I_{peak}}{I_{Tav}}\right)\times \left(\frac{2}{3}(G_{max}{}^*)^{3/4}+3(G_{max}{}^*)^{1/2}+6(G_{max}{}^*)^{1/4}+\frac{1}{2}\ln(G_{max}{}^*)-\frac{29}{3}\right)+ +0.8\left((G_{max}{}^*)^{1/2}-1\right)+0.53\left((G_{max}{}^*)^{3/4}-1\right)+ 1.6\left((G_{max}{}^*)^{1/2}-(G_{max}{}^*)^{1/4}\right)+ 0.8\left(G_{max}{}^*-(G_{max}{}^*)^{1/2}\right)+\left(\frac{4}{3}\left(1-\frac{I_{peak}I_{min}}{(I_{peak}+I_{min})^2}\right)+\frac{I_{peak}}{I_{Tav}}\right)\times \left(2\left((G_{max}{}^*)^{3/2}-(G_{max}{}^*)^{3/4}\right)+\frac{1}{2}\left(G_{max}{}^{*2}-G_{max}{}^*\right) +2\left((G_{max}{}^*)^{1/2}-(G_{max}{}^*)^{1/4}\right)+3\left(G_{max}{}^*-(G_{max}{}^*)^{1/2}\right)\right)$
Soft-switching buck	$\frac{2G_{max}{}^* I_{peak}}{I_{Tav}}$	$\frac{1}{G_{max}{}^*-1}\left(\frac{G_{max}{}^{*2}-1}{2}\left(\frac{4}{3}\left(1-\frac{I_{peak}I_{min}}{(I_{peak}+I_{min})^2}\right)+0.8\right)-0.8(G_{max}{}^*-1)\right)$
Soft-switching boost	$\frac{2G_{max}{}^* I_{peak}}{I_{Tav}}$	$\frac{1}{G_{max}{}^*-1}\left(\left(\frac{4(G_{max}{}^{*3}-1)}{9}-\frac{2(G_{max}{}^{*2}-1)}{3}\right)\left(1-\frac{I_{peak}I_{min}}{(I_{peak}+I_{min})^2}\right)+ +0.8(G_{max}{}^*-1)\right)$
Interleaved buck	$\frac{G_{max}{}^* I_{peak}}{I_{Tav}}$	$\frac{1}{G_{max}{}^*-1}\times \left(\frac{G_{max}{}^{*2}-1}{2}\left(\frac{4}{3m}\left(1-\frac{I_{peak}I_{min}}{(I_{peak}+I_{min})^2}\right)+0.8+\frac{I_{peak}}{mI_{Tav}}\right)-0.8(G_{max}{}^*-1)\right)$
Interleaved boost	$\frac{G_{max}{}^* I_{peak}}{I_{Tav}}$	$\frac{1}{G_{max}{}^*-1}\left(\frac{1}{m}\left(\frac{4(G_{max}{}^{*3}-1)}{9}-\frac{2(G_{max}{}^{*2}-1)}{3}\right)\left(1-\frac{I_{peak}I_{min}}{(I_{peak}+I_{min})^2}\right)+ +0.8(G_{max}{}^*-1)+\frac{(G_{max}{}^{*2}-1)I_{peak}}{2mI_{Tav}}\right)$

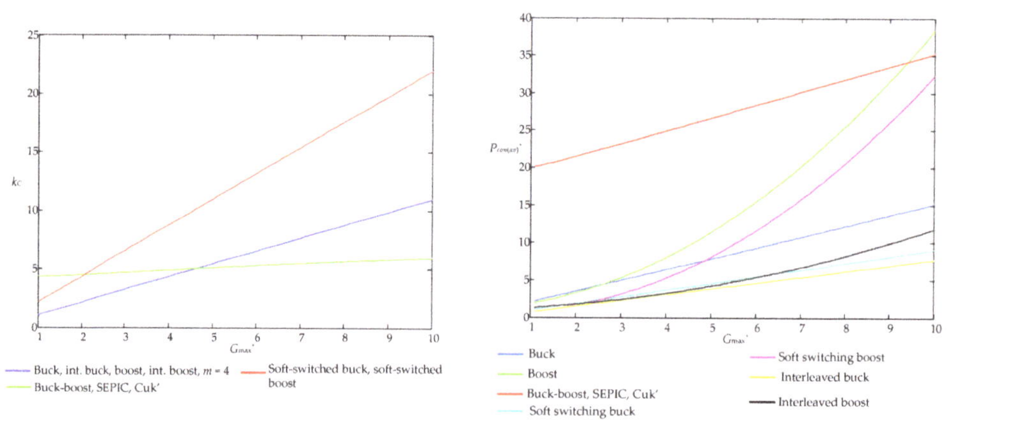

(**a**) (**b**)

Figure 22. Comparative analysis of the DC–DC converter's effectiveness in space of the variable G^*: (**a**) cost factor k_C; (**b**) normalized mean power loss $P_{con(av)}{}^*$.

8. Conclusions

Industrial PV solutions contain different kinds of applications designed with power converters as electric energy transformers and interconnectors. Thus, the specific converter requirements vary depending on the application which complicates the selection of the appropriate type of converter for a specific task. The paper highlights generalized criteria that have a critical impact on the efficiency and cost of converter topologies for PV applications that allows performing a comparative analysis in space of two numerical parameters, namely cost and power loss factors, of common PV applications, i.e., grid-on, energy storage, hybrid, and high voltage gain, based on cost and power loss factors.

According to the results of the analysis, the following conclusions are made:

- Basic DC–DC buck and boost topologies, as well as full-bridge topologies for DC-AC applications, have lower cost and power loss factors, whereas more complex interleaved or soft-switching topologies may decrease power loss by increasing the converter total cost;
- For DC–AC applications with a low or medium voltage range, it is advisable to use single-stage DC–AC converters, whereas for wide voltage range applications, two-stage converters have better cost and power loss factors;
- High voltage gain applications suffer from high transistor voltage stress. Therefore, specialized power converter topologies with reduced voltage stress, such as Cockcroft–Walton and Dickson multiplier-based boost converters, boost derived MIESC SC-cell converters, and boost 3SSC cell converters, have the advantage over their counterparts;
- Hybrid converters have approximately the same power loss and cost as power converters with common topologies because of the same number of power transistors;
- For clear analysis of different types of converters, it is better to represent cost and power loss factors in the space of gain factor and analyze the impact of the environment on gain factor probability distribution during operation.

Author Contributions: Conceptualization, I.V.; Formal analysis, I.V. and M.L.; Validation, I.V. and M.L.; Methodology, I.V.; Visualization, M.L. and K.N.; Investigation, M.L., K.N., and B.P.; Writing — original draft, K.N. and B.P.; Resources, O.H.; Writing—review and editing, B.P. and O.H.; Supervision, R.M.S. and O.H; Project administration, R.M.S.; Funding acquisition, R.M.S. All authors have read and agreed to the published version of the manuscript.

Funding: This research was funded by the EU Horizon 2020 Framework Programme Project: 955614—SMARTGYsum—H2020-MSCA-ITN-2020.

Institutional Review Board Statement: Not applicable.

Informed Consent Statement: Not applicable.

Data Availability Statement: Not applicable.

Conflicts of Interest: The authors declare no conflict of interest.

References

1. Qazi, A.; Hussain, F.; Rahim, N.A.; Hardaker, G.; Alghazzawi, D.; Shaban, K.; Haruna, K. Towards Sustainable Energy: A Systematic Review of Renewable Energy Sources, Technologies, and Public Opinions. *IEEE Access* **2019**, *7*, 63837–63851. [CrossRef]
2. Statistical Review of World Energy 2021, 70th ed. Available online: https://www.bp.com/content/dam/bp/business-sites/en/global/corporate/pdfs/energy-economics/statistical-review/bp-stats-review-2021-full-report.pdf (accessed on 30 August 2021).
3. Shubbak, M.H. Advances in solar photovoltaics: Technology review and patent trends. *Renew. Sustain. Energy Rev.* **2019**, *115*, 109383. [CrossRef]
4. Mehra, V.; Amatya, R.; Ram, R.J. Estimating the value of demand-side management in low-cost, solar micro-grids. *Energy* **2018**, *163*, 74–87. [CrossRef]
5. Romashko, V.J.; Verbitsky, I.V.; Kyrychik, I.I. Energy losses analyze in solar battery maximum power picking system. *Tech. Electrodyn.* **2014**, *4*, 55–57.
6. Clauser, C.; Ewert, M. The renewables cost challenge: Levelized cost of geothermal electric energy compared to other sources of primary energy–Review and case study. *Renew. Sustain. Energy Rev.* **2018**, *82*, 3683–3693. [CrossRef]
7. Tummuru, N.R.; Mishra, M.K.; Srinivas, S. Integration of PV/battery hybrid energy conversion system to the grid with power quality improvement features. In Proceedings of the IEEE International Conference on Industrial Technology (ICIT), Cape Town, South Africa, 25–28 February 2013. [CrossRef]
8. Hua, C.-C.; Fang, Y.-H.; Wong, C.-J. Improved solar system with maximum power point tracking. *IET Renew. Power Gener.* **2018**, *12*, 806–814. [CrossRef]

9. Darvishzadeh, P.; Redzwan, G.; Ahmadi, R.; Gorji, N.E. Modeling the degradation/recovery of short-circuit current density in perovskite and thin film photovoltaics. *Org. Electron.* **2017**, *43*, 247–252. [CrossRef]
10. Hussaian Basha, C.H.; Rani, C. Performance Analysis of MPPT Techniques for Dynamic Irradiation Condition of Solar PV. *Int. J. Fuzzy Syst.* **2020**, *22*, 2577–2598. [CrossRef]
11. Zhuo, S.; Gaillard, A.; Li, Q.; MA, R.; Paire, D.; Gao, F. Current Ripple Optimization of Four-Phase Floating Interleaved DC-DC Boost Converter under Switch Fault. *IEEE Trans. Ind. Appl.* **2020**, *56*, 4214–4224. [CrossRef]
12. Raghavendra, K.V.G.; Zeb, K.; Muthusamy, A.; Krishna, T.N.V.; Kumar, S.V.S.V.P.; Kim, D.-H.; Kim, M.-S.; Cho, H.-G.; Kim, H.-J. A Comprehensive Review of DC–DC Converter Topologies and Modulation Strategies with Recent Advances in Solar Photovoltaic Systems. *Electronics* **2020**, *9*, 31. [CrossRef]
13. Vinnikov, D.; Chub, A.; Korkh, O.; Liivik, E.; Blaabjerg, F.; Kouro, S. MPPT performance enhancement of low-cost PV microconverters. *Sol. Energy* **2019**, *187*, 156–166. [CrossRef]
14. Kolsi, S.; Samet, H.; Amar, M.B. Design Analysis of DC-DC Converters Connected to a Photovoltaic Generator and Controlled by MPPT for Optimal Energy Transfer throughout a Clear Day. *J. Power Energy Eng.* **2014**, *2*, 27–34. [CrossRef]
15. Bukar, A.L.; Tan, C.W.A. Review on Stand-alone Photovoltaic-Wind Energy System with Fuel Cell: System Optimization and Energy Management Strategy. *J. Clean. Prod.* **2019**, *221*, 73–88. [CrossRef]
16. Bonkile, M.P.; Ramadesigan, V. Power management control strategy using physics-based battery models in standalone PV-battery hybrid systems. *J. Energy Storage* **2019**, *23*, 258–268. [CrossRef]
17. Kumar, A.; Gupta, N.; Gupta, V. A Comprehensive Review on Grid-Tied Solar Photovoltaic System. *J. Green Eng.* **2017**, *7*, 213–254 [CrossRef]
18. Ali Khan, M.Y.; Liu, H.; Yang, Z.; Yuan, X. A Comprehensive Review on Grid Connected Photovoltaic Inverters, Their Modulation Techniques, and Control Strategies. *Energies* **2020**, *13*, 4185. [CrossRef]
19. Mahmood, H.; Michaelson, D.; Jiang, J. A Power Management Strategy for PV/Battery Hybrid Systems in Islanded Microgrids. *IEEE J. Emerg. Sel. Top. Power Electron.* **2014**, *2*, 870–882. [CrossRef]
20. Mira, M.C.; Zhang, Z.; Knott, A.; Andersen, M.A.E. Analysis, Design, Modeling, and Control of an Interleaved-Boost Full-Bridge Three-Port Converter for Hybrid Renewable Energy Systems. *IEEE Trans. Power Electron.* **2017**, *32*, 1138–1155. [CrossRef]
21. Bhattacharjee, A.; Samanta, H.; Banerjee, N.; Saha, H. Development and validation of a real-time flow control integrated MPPT charger for solar PV applications of vanadium redox flow battery. *Energy Convers. Manag.* **2018**, *171*, 1449–1462. [CrossRef]
22. El Aroudi, A.; Haroun, R.; Al-Numay, M.; Huang, M. Multiple-Loop Control Design for a Single-Stage PV-Fed Grid-Tied Differential Boost Inverter. *Appl. Sci.* **2020**, *10*, 4808. [CrossRef]
23. Nassary, M.; Orabi, M.; Ghoneima, M.; El-Nemr, M.K. Single-Phase Isolated Bidirectional AC-DC Battery Charger for Electric Vehicle–Review. In Proceedings of the International Conference on Innovative Trends in Computer Engineering (ITCE), Aswan, Egypt, 2–4 February 2019. [CrossRef]
24. Sultana, W.R.; Sahoo, S.K.; Sukchai, S.; Yamuna, S.; Venkatesh, D. A review on state of art development of model predictive control for renewable energy applications. *Renew. Sustain. Energy Rev.* **2017**, *76*, 391–406. [CrossRef]
25. Deng, W.; Pei, W.; Li, N.; Zhang, G.; Ding, L.; Kong, L. AC/DC Hybrid Renewable Energy System Coordinated Control and Test Platform. In Proceedings of the 2021 3rd Asia Energy and Electrical Engineering Symposium (AEEES), Chengdu, China, 26–29 March 2021. [CrossRef]
26. Amir, A.; Amir, A.; Che, H.S.; El Khateb, A.; Rahim, N.A. Comparative Analysis of High Voltage Gain DC-DC Converter Topologies for Photovoltaic Systems. *Renew. Energy* **2018**, *136*, 1147–1163. [CrossRef]
27. Rahman, S.; Khan, I.; Rahman, K.; Al Otaibi, S.; Alkhammash, H.I.; Iqbal, A. Scalable Multiport Converter Structure for Easy Grid Integration of Alternate Energy Sources for Generation of Isolated Voltage Sources for MMC. *Electronics* **2021**, *10*, 1779. [CrossRef]
28. Chen, G.; Liu, Y.; Qing, X.; Wang, F. Synthesis of Integrated Multi-Port DC-DC Converters with Reduced Switches. *IEEE Trans. Ind. Electron.* **2019**, *67*, 4536–4546. [CrossRef]
29. Başoğlu, M.E.; Çakır, B. Comparisons of MPPT performances of isolated and non-isolated DC–DC converters by using a new approach. *Renew. Sustain. Energy Rev.* **2016**, *60*, 1100–1113. [CrossRef]
30. Gorji, S.A.; Sahebi, H.G.; Ektesabi, M.; Rad, A.B. Topologies and Control Schemes of Bidirectional DC–DC Power Converters: An Overview. *IEEE Access* **2019**, *7*, 117997–118019. [CrossRef]
31. Moghassemi, A.; Padmanaban, S. Dynamic Voltage Restorer (DVR): A Comprehensive Review of Topologies, Power Converters, Control Methods, and Modified Configurations. *Energies* **2020**, *13*, 4152. [CrossRef]
32. Ansari, S.; Chandel, A.; Tariq, M. A Comprehensive Review on Power Converters Control and Control Strategies of AC/DC Microgrid. *IEEE Access* **2020**, *9*, 17998–18015. [CrossRef]
33. Undeland, M.N.; Robbins, W.P.; Mohan, N. Power Electronics. In *Converters, Applications, and Design*, 3rd ed.; John Whiley & Sons: Hoboken, NJ, USA, 1995.
34. Ha, K.; Lee, C.; Kim, J.; Krishnan, R.; Oh, S.G. Design and development of brushless variable speed motor drive for low cost and high efficiency. *Proc. IAS Conf.* **2006**, *4*, 1649–1656. [CrossRef]
35. *IEC 61730-1*; Photovoltaic (PV) Module Safety Qualification-Part 1: Requirements for Construction. CENELEC: Brussels, Belgium, 2018.
36. *IEC 61730-2*; Photovoltaic (PV) Module safety Qualification-Part 2: Requirements for Testing. CENELEC: Brussels, Belgium, 2018.
37. *IEC 62109-1*; Safety of Power Converters for Use in Photovoltaic Power Systems–Part 1: General Requirements. CENELEC: Brussels, Belgium, 2010.

38. *IEC 62109-2*; Safety of power Converters for Use in Photovoltaic Power Systems-Part 2: Particular Requirements for Inverters. CENELEC: Brussels, Belgium, 2011.
39. Wang, G.; Wang, F.; Magai, G.; Lei, Y.; Huang, A.; Das, M. Performance comparison of 1200V 100A SiC MOSFET and 1200V 100A silicon IGBT. In Proceedings of the 2013 IEEE Energy Conversion Congress and Exposition, Denver, CO, USA, 15–19 September 2013. [CrossRef]
40. Datasheet of the Transistor C2M0080170P. Available online: https://cms.wolfspeed.com/app/uploads/2020/12/C2M0080170P.pdf (accessed on 9 October 2021).
41. Erickson, R.; Maksimovic, D. *Fundamentals of Power Electronics, Ser. Power Electronics*; Springer: Berlin/Heidelberg, Germany, 2001.
42. Datasheet of the diode C5D25170H. Available online: https://cms.wolfspeed.com/app/uploads/2020/12/C5D25170H.pdf (accessed on 9 October 2021).
43. Jain, S.; Agarwal, V. A Single-Stage Grid-Connected Inverter Topology for Solar PV Systems with Maximum Power Point Tracking. *IEEE Trans. Power Electron.* **2007**, *22*, 1928–1940. [CrossRef]
44. Altin, N.; Ozdemir, S.; Komurcugil, H.; Sefa, I.; Biricik, S. Two-stage grid-connected inverter for PV systems. In Proceedings of the 2018 IEEE 12th International Conference on Compatibility, Power Electronics and Power Engineering, Doha, Qatar, 10–12 April 2018. [CrossRef]
45. Zebarjadi, M.; Askarzadeh, A. Optimization of a reliable grid-connected PV-based power plant with/without energy storage system by a heuristic approach. *Sol. Energy* **2016**, *125*, 12–21. [CrossRef]
46. Xiao, W.; El Moursi, M.S.; Khan, O.; Infield, D. Review of grid-tied converter topologies used in photovoltaic systems. *IET Renew. Power Gener.* **2016**, *10*, 1543–1551. [CrossRef]
47. Zhang, Y..; Sen, P.C. A new soft-switching technique for buck, boost, and buck~boost converters. *IEEE Trans. Ind. Appl.* **2003**, *39*, 1775–1782. [CrossRef]
48. Verbytskyi, I.; Bondarenko, O.; Liivik, E. Control features of multi cell-type current regulator for resistance welding. In Proceedings of the 2017 IEEE 58th International Scientific Conference on Power and Electrical Engineering of Riga Technical University (RTUCON), Doha, Qatar, 10–12 April 2018. [CrossRef]
49. Jyotheeswara Reddy, K.; Sudhakar, N. High Voltage Gain Interleaved Boost Converter With Neural Network Based MPPT Controller for Fuel Cell Based Electric Vehicle Applications. *IEEE Access* **2018**, *6*, 3899–3908. [CrossRef]
50. Zeb, K.; Uddin, W.; Khan, M.A.; Ali, Z.; Ali, M.U.; Christofides, N.; Kim, H.J. A comprehensive review on inverter topologies and control strategies for grid connected photovoltaic system. *Renew. Sustain. Energy Rev.* **2018**, *94*, 1120–1141. [CrossRef]
51. Keyhani, H.; Toliyat, H.A. Single-Stage Multi string PV Inverter With an Isolated High-Frequency Link and Soft-Switching Operation. *IEEE Trans. Power Electron.* **2014**, *29*, 3919–3929. [CrossRef]
52. Khodabandeh, M.; Afshari, E.; Amirabadi, M. A Single-Stage Soft-Switching High-Frequency AC-Link PV Inverter: Design, Analysis, and Evaluation of Si-based and SiC-based Prototypes. *IEEE Trans. Power Electron.* **2018**, *34*, 2312–2326. [CrossRef]
53. Lopez, O.; Freijedo, F.D.; Yepes, A.G.; Fernandez-Comesana, P.; Malvar, J.; Teodorescu, R.; Doval-Gandoy, J. Eliminating ground current in a transformerless photovoltaic application. *IEEE Trans. Energy Convers.* **2010**, *25*, 140–147. [CrossRef]
54. Xiao, H.; Xie, S. Leakage current analytical model and application in single-phase transformerless photovoltaic grid-connected inverter. *IEEE Trans. Electromagn Compat.* **2010**, *52*, 902–913. [CrossRef]
55. Zhang, L.; Sun, K.; Xing, Y.; Xing, M. H6 Transformerless Full-Bridge PV Grid-Tied Inverters. *IEEE Trans. Power Electron.* **2014**, *29*, 1229–1238. [CrossRef]
56. Ahmed, M.H.; Wang, M.; Hassan, M.A.S.; Ullah, I. Power Loss Model and Efficiency Analysis of Three phase Inverter Based on SiC MOSFETs for PV Applications. *IEEE Access* **2019**, *7*, 75768–75781. [CrossRef]
57. Escobar, G.; Martinez-Rodriguez, P.R.; Pool, E.I.; Pena-Quintal, A.E.; Vazquez, G.; Sosa, J.M. A model-based controller of a three-level stacked-cell grid connected converter. In Proceedings of the IECON 2015-41st Annual Conference of the IEEE Industrial Electronics Society, Yokohama, Japan, 9–12 November 2015. [CrossRef]
58. Mamadaminov, U.M. *Advanced Inverters and Their Functionalities for Distributed Solar Generation*; Energy Engineering III, Spring; Oregon Institute of Technology: Wilsonville, OR, USA, 2014.
59. Hidalgo, R.; Abbey, C. Integrating distributed generation with Smart Grid enabling technologies. In Proceedings of the IEEE PES Conference on Innovative Smart Grid Technologies Latin America SGT LA, Medellin, Colombia, 19–21 October 2011. [CrossRef]
60. Talha, M.; Amir, A.; Raihan, S.R.S.; Abd Rahim, N. Grid-connected photovoltaic inverters with low-voltage ride through for a residential-scale system: A review. *Int. Trans. Electr. Energy Syst.* **2020**, *31*, e12630. [CrossRef]
61. Taul, M.G.; Wang, X.; Davari, P.; Blaabjerg, F. An Overview of Assessment Methods for Synchronization Stability of Grid-Connected Converters Under Severe Symmetrical Grid Faults. *IEEE Trans. Power Electron.* **2019**, *34*, 9655–9670. [CrossRef]
62. Geddada, N.; Karanki, S.B.; Mishra, M.K. Synchronous reference frame based current controller with SPWM switching strategy for DSTATCOM applications. In Proceedings of the 2012 IEEE International Conference on Power Electronics, Drives and Energy Systems (PEDES), Bengaluru, India, 16–19 December 2012. [CrossRef]
63. Chappa, H.; Thakur, T. Voltage instability detection using synchrophasor measurements: A review. *Int. Trans. Electr. Energy Syst.* **2020**, *30*, 12343. [CrossRef]
64. Calleja, H.; Jimenez, H. Performance of a grid connected PV system used as active filter. *Energy Convers. Manag.* **2004**, *45*, 2417–2428. [CrossRef]

65. Verbytskyi, Y.V. Double Fourier series using for calculating modulating signals spectrum. *Tekhnichna Elektrodynamika* **2014**, *4*, 96–98.
66. Adewuyi, O.B.; Shigenobu, R.; Ooya, K.; Senjyu, T.; Howlader, A.M. Static voltage stability improvement with battery energy storage considering optimal control of active and reactive power injection. *Electr. Power Syst. Res.* **2019**, *172*, 303–312. [CrossRef]
67. Pinnangudi, B.; Kuykendal, M.; Bhadra, S. Smart Grid Energy Storage. *Power Grid* **2017**, 93–135. [CrossRef]
68. Liu, S.; Xie, X.; Yang, L. Analysis, Modeling and Implementation of a Switching Bi-directional Buck-Boost Converter Based on Electric Vehicle Hybrid Energy Storage for V2G System. *IEEE Access* **2020**, *8*, 65868–65879. [CrossRef]
69. Pakkiraiah, B.; Sukumar, G.D. Research Survey on Various MPPT Performance Issues to Improve the Solar PV System Efficiency. *J. Sol. Energy* **2016**, *2016*, 8012432. [CrossRef]
70. Yang, J.-H.; Jeong, K.-I.; Kwon, J.-M. Energy storage system with PV generation and online UPS functions. In Proceedings of the 2014 IEEE 36th International Telecommunications Energy Conference (INTELEC), Vancouver, BC, Canada, 28 September–2 October 2014. [CrossRef]
71. Li, K.; Zhao, J.; Ma, Q.; Xu, H. Hierarchy control of power quality for wind–battery energy storage system. *IET Power Electron.* **2014**, *7*, 2123–2132. [CrossRef]
72. Li, X.; Wang, S. A review on energy management, operation control and application methods for grid battery energy storage systems. *CSEE J. Power Energy Syst.* **2019**, *7*, 1026–1040. [CrossRef]
73. Hannan, M.A.; Hoque, M.M.; Hussain, A.; Yusof, Y.; Ker, P.J. State-of-the-Art and Energy Management System of Lithium-Ion Batteries in Electric Vehicle Applications: Issues and Recommendations. *IEEE Access* **2018**, *6*, 19362–19378. [CrossRef]
74. Ortúzar, M.; Dixon, J.; Moreno, J. Ultracapacitor-Based Auxiliary Energy System for an Electric Vehicle: Implementation and Evaluation. *IEEE Trans. Ind. Electron.* **2007**, *54*, 2147–2156. [CrossRef]
75. Fernandez, G. A Bidirectional Buffered Charging Unit for EV's (BBCU). In Proceedings of the International Power Electronics Conference (IPEC-Niigata 2018-ECCE Asia), Niigata, Japan, 20–24 May 2018. [CrossRef]
76. Schupbachj, R.M.; Bald, C. Comparing DC-DC Converters for Power Management in Hybrid Electric Vehicles. In Proceedings of the IEEE International Electric Machines and Drives Conference, Madison, WI, USA, 1–4 June 2003; Volume 3, pp. 1369–1374. [CrossRef]
77. Czogalla, J.; Li, J.; Sullivan, C.R. Automotive Application of Multi-Phase Coupled-Inductor DC-DC Converter. In Proceedings of the IEEE Industry Applications Conference, Salt Lake City, UT, USA, 12–16 October 2003; Volume 3, pp. 1524–1529. [CrossRef]
78. Farhangi, B.; Toliyat, H.A. Modeling and Analyzing Multiport Isolation Transformer Capacitive Components for Onboard Vehicular Power Conditioners. *IEEE Trans. Ind. Electron.* **2015**, *62*, 3134–3142. [CrossRef]
79. Inoue, S.; Akagi, H. A Bi-Directional DC/DC Converter for an Energy Storage System. In Proceedings of the APEC 07-Twenty-Second Annual IEEE Applied Power Electronics Conference and Exposition, Anaheim, CA, USA, 25 February–1 March 2007. [CrossRef]
80. Xu, X.; Khambadkone, A.M.; Oruganti, R. A Soft-Switched Back-to-Back Bi-directional DC/DC Converter with an FPGA based Digital Control for Automotive applications. In Proceedings of the IECON 2007-33rd Annual Conference of the IEEE Industrial Electronics Society, Taipei, Taiwan, 5–8 November 2007. [CrossRef]
81. Jang, S.-J.; Lee, T.W.; Lee, W.C.; Won, C.Y. Bidirectional DC to DC Converters for Fuel Cell Generation System. *Power Electron. Spec. Conf.* **2004**, *6*, 4722–4728. [CrossRef]
82. Lin, B.R.; Hung, T.L. Single-phase half-bridge converter topology for power quality compensation. *Electr. Power Appl. IEE Proc.* **2002**, *149*, 351–359. [CrossRef]
83. Lin, B.R.; Hung, T.L.; Huang, C.H. Bi-directional single-phase half-bridge rectifier for power quality compensation. *Electr. Power Appl. IEE Proc.* **2003**, *150*, 397–406. [CrossRef]
84. Korkh, O.; Blinov, A.; Vinnikov, D. Analysis of Oscillation Suppression Methods in the AC-AC Stage of High Frequency Link Converters. In Proceedings of the IEEE 60th International Scientific Conference on Power and Electrical Engineering of Riga Technical University (RTUCON), Riga, Latvia, 7–9 October 2019. [CrossRef]
85. Bilgin, B.; Emadi, A.; Krishnamurthy, M. Universal input battery charger circuit for PHEV applications with the simplified controller. In Proceedings of the Applied Power Electronics Conference and Exposition (APEC), Fort Worth, TX, USA, 6–11 March 2011; pp. 815–820. [CrossRef]
86. Segaran, D.; Holmes, D.G.; McGrath, B.P. High performance bi-directional AC-DC converters for PHEV with minimized DC bus capacitance. In Proceedings of the IECON 37th Annual Conference on IEEE Industrial Electronics Society, Melbourne, Australia, 7–10 November 2011; pp. 3620–3625. [CrossRef]
87. Odamov, U.O.; Kamilov, M.M.; Niyazov, S.K.; Song, K. Research of the efficiency of the solar battery operations in real exploitation conditions. *Sci. Rep. Bukhara State Univ.* **2021**, *5*, 2–17. [CrossRef]
88. Saidi, A.S.; Slimene, M.B.; Khlifi, M.A. Transient stability analysis of photovoltaic system with experimental shading effects. *Eng. Technol. Appl. Sci. Res.* **2018**, *8*, 3592–3597. [CrossRef]
89. Araujo, S.V.; Torrico-Bascope, R.P.; Torrico-Bascope, G.V.; Menezes, L. Step-up converter with high voltage gain employing three-state switching cell and voltage multiplier. In Proceedings of the 2008 IEEE Power Electronics Specialists Conference, Rhodes, Greece, 15–19 June 2008; pp. 2271–2277. [CrossRef]

90. Ahmad, J.; Pervez, I.; Sarwar, A.; Tariq, M.; Fahad, M.; Chakrabortty, R.K.; Ryan, M.J. Performance Analysis and Hardware-in-the-Loop (HIL) Validation of Single Switch High Voltage Gain DC-DC Converters for MPP Tracking in Solar PV System. *IEEE Access* **2021**, *9*, 48811–48830. [CrossRef]
91. Arunkumari, T.; Indragandhi, V. An overview of high voltage conversion ratio DC-DC converter configurations used in DC micro-grid architectures. *Renew. Sustain. Energy Rev.* **2017**, *77*, 670–687. [CrossRef]
92. Blinov, A.; Verbytskyi, I.; Zinchenko, D.; Vinnikov, D.; Galkin, I. Modular Battery Charger for Light Electric Vehicles. *Energies* **2020**, *13*, 774. [CrossRef]
93. Norrga, S. Experimental Study of a Soft-Switched Isolated Bidirectional AC DC Converter Without Auxiliary Circuit. *IEEE Trans. Power Electron.* **2006**, *21*, 1580–1587. [CrossRef]
94. Kummari, N.; Chakraborty, S.; Chattopadhyay, S. An Isolated High-Frequency Link Microinverter Operated with Secondary-Side Modulation for Efficiency Improvement. *IEEE Trans. Power Electron.* **2018**, *33*, 2187–2200. [CrossRef]
95. Nayak, P.; Rajashekara, K.; Pramanick, S. Soft-Switched Modulation Technique for a Single-Stage Matrix-Type Isolated DC–AC Converter. *IEEE Trans. Ind. Appl.* **2019**, *55*, 7642–7656. [CrossRef]
96. Tibola, G.; Lemmen, E.; Duarte, J.L.; Barbi, I. Passive Regenerative and Dissipative Snubber Cells for Isolated SEPIC Converters: Analysis, Design, and Comparison. *IEEE Trans. Power Electron.* **2017**, *32*, 9210–9222. [CrossRef]
97. Blinov, A.; Verbytskyi, I.; Peftitsis, D.; Vinnikov, D. Regenerative Passive Snubber Circuit for High-Frequency Link Converters. *IEEE J. Emerg. Sel. Top. Ind. Electron.* **2021**, *3*, 252–257. [CrossRef]
98. Blinov, A.; Kosenko, R.; Chub, A.; Vinnikov, D. Snubberless boost full-bridge converters: Analysis of soft-switching performance and limitations. *Int. J. Circ. Theor. Appl.* **2019**, *47*, 884–908. [CrossRef]
99. Luo, F.L.; Ye, H. Positive output super-lift converters. *IEEE Trans. Power Electron.* **2003**, *18*, 105–113. [CrossRef]
100. Zhang, S.; Xu, J.; Yang, P. A single-switch high gain quadratic boost converter based on voltage-lift technique. In Proceedings of the 10th International Power & Energy Conference (IPEC), Ho Chi Minh City, Vietnam, 12–14 December 2012; pp. 71–75. [CrossRef]
101. Axelrod, B.; Berkovich, Y.; Shenkman, A.; Golan, G. Diode-capacitor voltage multipliers combined with boost-converters: Topologies and characteristics. *IET Power Electron.* **2012**, *5*, 873–884. [CrossRef]
102. Wu, G.; Ruan, X.; Ye, Z. Nonisolated High Step-Up DC-DC Converters Adopting Switched-Capacitor Cell. *IEEE Trans. Ind. Electron.* **2015**, *62*, 383–393. [CrossRef]
103. Tofoli, F.L.; de Souza Oliveira, D.; Torrico-Bascopé, R.P.; Alcazar, Y.J.A. Novel Nonisolated High-Voltage Gain DC–DC Converters Based on 3SSC and VMC. *IEEE Trans. Power Electron.* **2012**, *27*, 3897–3907. [CrossRef]
104. Zhang, J.; Huang, L.; Shu, J.; Wang, H.; Ding, J. Energy Management of PV-diesel-battery Hybrid Power System for Island Stand-alone Micro-grid. *Energy Procedia* **2017**, *105*, 2201–2206. [CrossRef]
105. Cano, A.; Jurado, F.; Sánchez, H.; Fernández, L.M.; Castañeda, M. Optimal sizing of stand-alone hybrid systems based on PV/WT/FC by using several methodologies. *J. Energy Inst.* **2014**, *87*, 330–340. [CrossRef]
106. Ghaffari, A.; Askarzadeh, A. Design optimization of a hybrid system subject to reliability level and renewable energy penetration. *Energy* **2020**, *193*, 116754. [CrossRef]
107. Noshahr, J.B.; Mohamadi, B.; Kermani, M.; Kermani, M. Operational Planning of Inverter Control in a grid-connected Microgrid with hybrid PV and BESS. In Proceedings of the 2020 IEEE International Conference on Environment and Electrical Engineering and 2020 IEEE Industrial and Commercial Power Systems Europe (EEEIC/I & CPS Europe), Madrid, Spain, 9–12 June 2020. [CrossRef]
108. Vahid, S.; EL-Refaie, A. A Novel Topology for an Extendable Isolated DC-DC Multi-port Power Converter with a Multipurpose Hybrid Energy Storage System. In Proceedings of the 2020 IEEE Energy Conversion Congress and Exposition (ECCE), Detroit, MI, USA, 11–15 October 2020. [CrossRef]
109. Pereira, T.; Hoffmann, F.; Zhu, R.; Liserre, M. A Comprehensive Assessment of Multiwinding Transformer-Based DC-DC Converters. *IEEE Trans. Power Electron.* **2021**, *36*, 10020–10036. [CrossRef]
110. Wu, H.; Xing, Y.; Xia, Y.; Sun, K. A family of non-isolated three-port converters for stand-alone renewable power systems. In Proceedings of the IECON 2011-37th Annual Conference of the IEEE Industrial Electronics Society, Melbourne, Australia, 7–10 November 2011. [CrossRef]
111. Zhang, M.; Xing, Y.; Wu, H.; Lu, Y.; Sun, K. Performance evaluation of a non-isolated bidirectional three-port power converter for energy storage applications. In Proceedings of the 2016 IEEE 8th International Power Electronics and Motion Control Conference (IPEMC-ECCE Asia), Hefei, China, 22–26 May 2016; pp. 2703–2708. [CrossRef]
112. Sato, Y.; Uno, M.; Nagata, H. Nonisolated Multiport Converters Based on Integration of PWM Converter and Phase-Shift Switched-Capacitor Converter. *IEEE Trans. Power Electron.* **2019**, *35*, 455–470. [CrossRef]
113. Faraji, R.; Farzanehfard, H. Fully Soft Switched Multi-Port DC-DC Converter with High Integration. *IEEE Trans. Power Electron.* **2020**, *36*, 1901–1908. [CrossRef]
114. Kwasinski, A. Identification of Feasible Topologies for Multiple-Input DC-DC Converters. *IEEE Trans. Power Electron.* **2009**, *24*, 856–861. [CrossRef]
115. Balaji, C.; Chellammal, N.; Sanjeevikumar, P.; Bhaskar, M.S.; Holm-Nielsen, J.B.; Leonowicz, Z.; Masebinu, S.O. Non-Isolated High-Gain Triple Port DC-DC Buck-Boost Converter with Positive Output Voltage for Photovoltaic Application. *IEEE Access* **2020**, *2020*, 3003192. [CrossRef]

116. Thang, T.V.; Ahmed, A.; Kim, C.I.; Park, J.H. The flexible system architecture of stand-alone PV power generation with the energy storage device. *IEEE Trans. Energy Convers.* **2015**, *30*, 1386–1396. [CrossRef]
117. Amirabadi, M.; Toliyat, H.A.; Alexander, W.C. A multiport ac link PV inverter with reduced size and weight for a stand-alone application. *IEEE Trans. Ind. Appl.* **2013**, *49*, 2217–2228. [CrossRef]
118. Hassan, S.Z.; Mumtaz, S.; Kamal, T.; Khan, L. Performance of grid-integrated photovoltaic/fuel cell/ electrolyzer/battery hybrid power system. In Proceedings of the 2015 Power Generation Systems and Renewable Energy Technologies, PGSRET, Islamabad, Pakistan, 10–11 June 2015. [CrossRef]
119. Jayasinghe, S.D.; Vilathgamuwa, D.M.; Madawala, U.K. Diodeclamped three-level inverter-based battery/supercapacitor direct integration scheme for renewable energy systems. *IEEE Trans. Power Electron.* **2011**, *26*, 3720–3729. [CrossRef]
120. Rangu, S.K.; Lolla, P.R.; Dhenuvakonda, K.R.; Singh, A.R. Recent trends in power management strategies for optimal operation of distributed energy resources in microgrids: A comprehensive review. *Int. J. Energy Res.* **2020**, *44*, 9889–9911. [CrossRef]
121. David, F. *Probability and Statistics for Computer Science*; Springer International Publishing: Berlin/Heidelberg, Germany, 2018; p. 367. [CrossRef]

energies

Article

High-Frequency Oscillation of the Active-Bridge-Transformer-Based DC/DC Converter

Shusheng Wei [1] and Wusong Wen [1,2,*]

1 Department of Electrical Engineering, Tsinghua University, Beijing 100084, China; weishusheng13@163.com
2 Key Laboratory of Military Special Power Supply, Army Engineering University of PLA, Chongqing 400035, China
* Correspondence: wenwusong@163.com

Abstract: The dual-active-bridge converter (DAB) has attracted tremendous attention in recent years. However, its EMI issues, especially the high-frequency oscillation (HFO) induced by the dv/dt and parasitic elements of the transformer, are significant challenges. The multi-active-bridge converter (MAB) based on the multi-winding transformer also faces similar problems, which are even more complicated. This article investigates the HFO of active-bridge-transformer-based DC/DC converters including DAB and MAB. Firstly, the general HFO model is studied using the analysis of the AC equivalent circuit considering the asymmetrical parameters. Ignoring the AC resistance in the circuit, the high-order model of the voltage oscillation could be reduced to a second-order system. Based on the simplified model, the oscillation voltage generated by an active bridge is analyzed in the time domain. Then, a universal active voltage-oscillation-suppression method-selected harmonic-elimination phase-shift (SHE PS) modulation method is proposed. The impacts of the system parameters on the method are also revealed. The experimental results show the excellent performance of the proposed active suppression method, with voltage spike amplitude (VSA) reductions of 92.1% and 77.8% for the DAB and MAB prototypes, respectively.

Keywords: high-frequency oscillation (HFO); dual-active bridge (DAB); selected harmonic elimination (SHE)

Citation: Wei, S.; Wen, W. High-Frequency Oscillation of the Active-Bridge-Transformer-Based DC/DC Converter. *Energies* **2022**, *15*, 3311. https://doi.org/10.3390/en15093311

Academic Editors: Diego Bellan and Jelena Loncarski

Received: 5 April 2022
Accepted: 27 April 2022
Published: 2 May 2022

1. Introduction

The dual-active bridge (DAB) converter has continued to be a hot topic in both academia and industry in recent years, due to its properties of bidirectional power flow, high power density, soft switching, and inherent galvanic isolation [1,2]. The multi-active bridge converter (MAB), regarded as the natural expansion of DAB, could provide multi-ports to integrate multiple DC voltage domains, making it a competitive candidate in many applications such as power electronic transformers (PETs), electric aircrafts and all-electric ships, and energy routers for smart homes [3,4]. To make the MAB more flexible and scalable, the modular multi-active bridge (MMAB) converter has been proposed, where one multi-winding transformer is replaced by several dependent two-winding transformers.

With the increasing switching speed, the high-frequency oscillation (HFO), which is excited by the output voltage of the active bridge and the transformer stray parameters in the DAB, has raised concern. Recently, wideband gap (WBG) devices have been widely used in DAB converters and other H-bridge-based converters [5–9], which could increase the system's efficiency. With the WBG switching devices and the nanocrystalline alloys utilized, the HFO at the ac side of the DAB converter represents an urgent problem that needs to be solved [8]. The HFO in the DAB could increase the voltage stress of the transformer, induce EMI and common mode noise, and distort the waveforms [8]. Additionally, it may increase the high-frequency losses, decreasing the system's efficiency. The HFO of the MAB converter has not been reported in the existing literature, but can be naturally predicted. Figure 1 shows a three-winding MAB, which is also called a triple-active bridge (TAB) converter, and the waveforms. The dv/dt is 3 kV/us and obvious voltage HFO exists with the high frequency near 1 MHz.

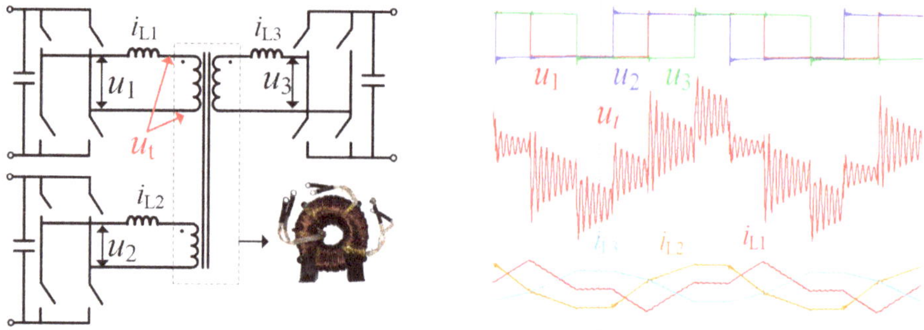

(**a**) (**b**)

Figure 1. HFO in a MAB converter prototype: (**a**) TAB circuit; (**b**) waveforms.

To modify the dv/dt [8,9] to suppress the voltage spike amplitude (VSA) using the extra parallel capacitor involves deceasing the switching speed and adding extra hardware components. The optimization of the inductor and transformer stray capacitors [10,11] will increase the complexity of the design procedure. The split inductance tunning between the HV and LV side [12] is often a special case without generality, which may lose effectiveness for multi-port converters. The active selective harmonic-elimination general phase-shift PWM method is introduced to suppress the HFO of the AC-link in MMAB [13], which has the potential to suppress the HFO of the transformer of the DAB and MAB, but has not been derived or verified. On the other hand, the branches of the DAB [10] and MMAB converters [13] are symmetrical with 1:1 transformer turn ratios and the same phase-shift inductors.

The HFO mechanism of active-bridge-transformer-based DC/DC converters, including DABs and MABs, is investigated in this article. The converters with asymmetrical inductances are considered for the general case study. On top of that, the universal SHE PS modulation method to suppress the voltage oscillation is proposed and the impacts of the system parameters on the method are studied. Finally, the effectiveness of the proposed voltage suppression method is verified by the experimental results of a DAB and MAB prototype.

2. High-Frequency Oscillation of the Active-Bridge-Based DC/DC Converter

2.1. Equivalent Circuit of the Transformers

The active bridge with the transformer is the core unit in active-bridge-transformer-based converters. Stray capacitances are regarded as the major factor causing the HFO phenomenon. A classical π-shaped network with three capacitors [6,7,9] is used to model the HFT. The high-frequency (HF) equivalent circuits of the DAB and MAB converters considering the stray parameters of the transformers are shown in Figure 2. The parameters are referred to the primary side. The symbols are defined as follows: u_i is the AC voltage of the active bridge i, L_i and R_i are the inductance and the equivalent resistance of the phase-shift inductor i. L_s and R_s are the leakage inductor and the equivalent winding resistance of the transformer, L_m is magnetizing inductance, and R_m is the magnetizing resistor. C_1 and C_2 are self-capacitances of the primary side and the secondary side, which is referred to the primary side, respectively. C_{12} is the mutual capacitance between the two windings. Similarly, the extended π-shaped network could be utilized to model the circuit for the multi-winding transformer [14]. C_i is the self-capacitance of the windings and C_{ij} is the mutual capacitance between the two windings i and j.

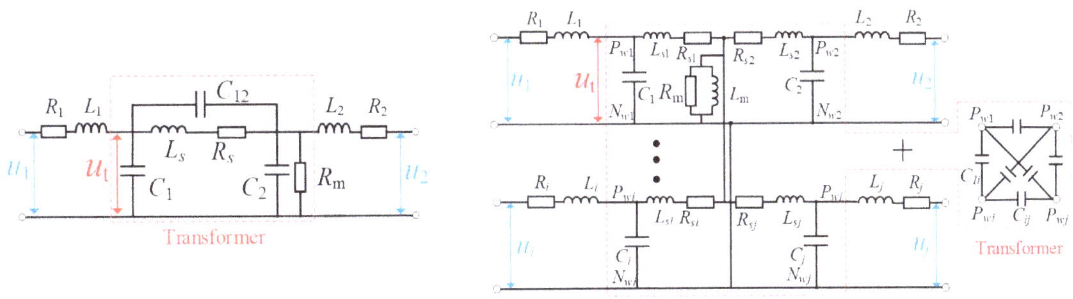

(a) (b)

Figure 2. Equivalent HF circuit considering the stray parameters of the transformer. (**a**) DAB converter; (**b**) MAB converter.

As shown in Figure 2, the HF voltages of the transformer u_t of the DAB and MAB are the superposition of the responses of each excitation voltage, which can be regarded as the multi-input single-output (MISO) impedance model.

2.2. General HFO Model with the Asymmetrical Branch Parameters

Assuming $L_s << L_m$ and L_s is small compared with phase-shift inductance L, the AC equivalent of the DAB can be simplified to be an equivalent model [8]. Considering the general case for the DAB and MAB converters, one excitation voltage u_1 was taken into consideration while the other active bridges short-circuited. Then, we could obtain the single-input single-output (SISO) impedance model shown in Figure 3, where u_{e1} is the equivalent excitation voltage, Z_o is the internal resistance, and C_t is the equivalent capacitance.

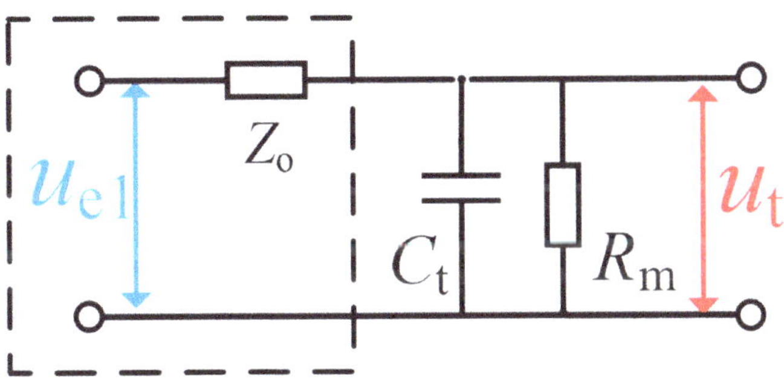

Figure 3. Thevenin equivalent circuit.

The variables in the two-port network in Figure 3 can be derived using Thevenin's theorem:

$$u_{e1}(s) = \frac{Z_2||Z_3|||\ldots||Z_n}{Z_1 + Z_2||Z_3|||\ldots||Z_n} u_1(s),\ Z_o = Z_1||Z_2||Z_3|||\ldots||Z_N \tag{1}$$

$$Z_i = R_i + L_i s,\ C_t = \sum_{i=1}^{N} C_{si} \tag{2}$$

The voltage transfer function of $G_h(s) = u_t(s)/u_1(s)$ for the DAB circuit in Figure 2b was derived by:

$$G_h(s) = \frac{u_t(s)}{u_1(s)} = \frac{R_m(L_2 s + R_2)}{L_1 L_2 R_m C_t s^3 + [(L_2 R_1 + L_1 R_2) R_m C_t + L_1 L_2] s^2 + [R_m C_t R_1 R_2 + L_1 R_2 + L_2 R_1 + R_m(L_1 + L_2)] s + R_1 R_2 + R_m R_1 + R_m R_2} \tag{3}$$

The transfer function (4) is a third control system. For MAB, with the increasing branches, the order of the transfer would be increased, making it difficult to calculate the analytical solution. Thus, it was necessary to simplify the model.

The impedance of the inductor was much larger compared with the AC resistance including the wire, inductor and transformer resistance at the oscillation frequency. It was reasonable to reduce the system order by ignoring the resistance. The external resistors R_{ext} were inserted into the AC loop of the DAB converter to investigate the effect of AC resistance on the HFO, as shown in Figure 4.

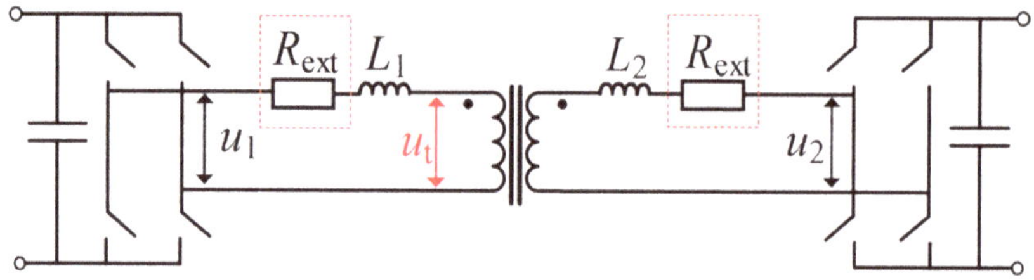

Figure 4. DAB circuit with external AC resistors.

The magnitude and phase of the voltage transfer function (3) are illustrated in Figure 5. $R = 0$ means $R_1 = R_2 = 0$ and no external resistors were inserted to the AC loop. It can be seen that the AC resistance only affected the magnitude at low frequency, while the magnitude and phase were irrelevant to the AC resistance. Figure 6 shows the experimental waveforms of a DAB converter using different external AC resistors. Both the oscillation frequency and amplitude were the same with different AC resistances.

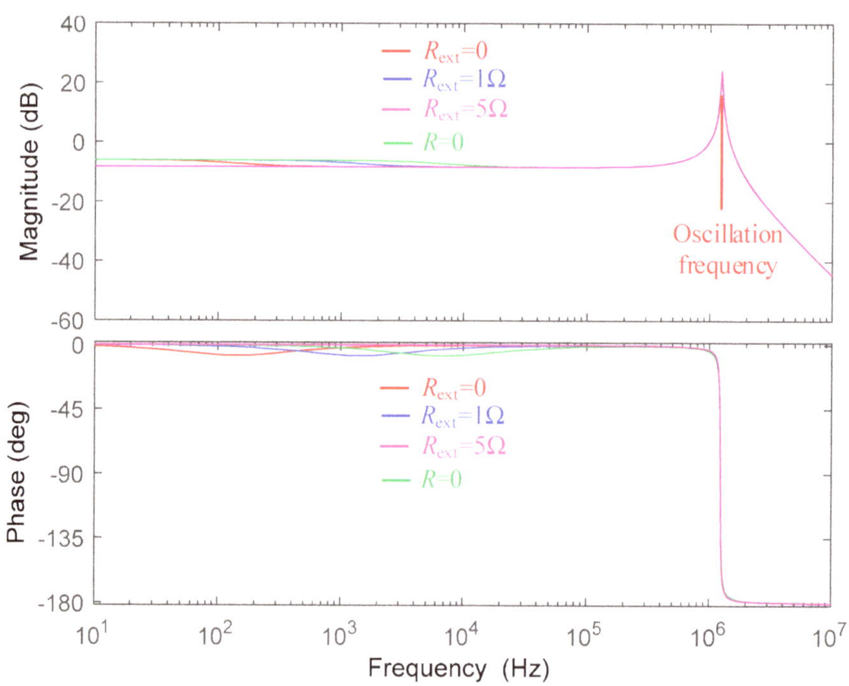

Figure 5. Bode plot of the voltage transfer function with different AC resistances.

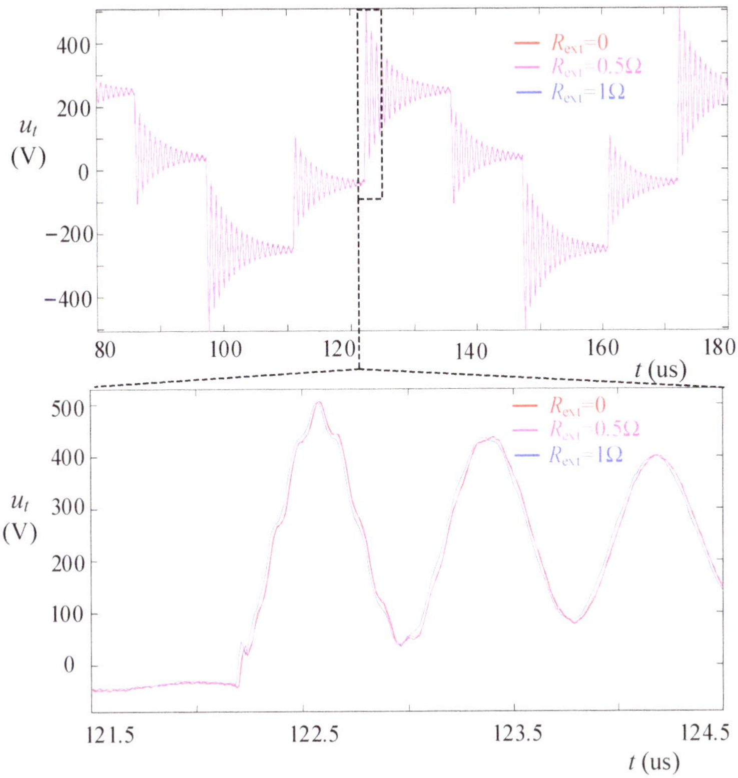

Figure 6. Experimental voltage waveform with different AC resistances.

Then, ignoring the AC resistance of the loop, the transfer function could be simplified from the third-order control system to a standard second-order control system, as shown in Equation (4). The steady state value of the u_t considering both u_1 and u_2 was calculated as follows in (5).

$$G_{hS}(s) = \frac{u_t(s)}{u_1(s)} = \frac{1}{L_1C_t}\frac{1}{s^2 + \frac{1}{R_mC_t}s + \frac{L_1+L_2}{L_1L_2C_t}} \tag{4}$$

$$u_{t,S} = \frac{L_2}{L_1+L_2}u_1 + \frac{L_1}{L_1+L_2}u_2 \tag{5}$$

Thus, for MAB, the transfer function could be derived as (6):

$$G_{hS}(s) = \frac{u_t(s)}{u_1(s)} = \frac{1}{L_1C_t}\frac{1}{s^2 + \frac{1}{R_mC_t}s + \frac{L_1+L_{2e}}{L_1L_{2e}C_t}} \tag{6}$$

where L_{2e} is the equivalent inductance as $L_{2e} = L_2 \,||\, \ldots \,||\, L_N$. The simplified transfer function (6) can be utilized to analyze the HFO of the DAB and MAB converters, generally.

3. Model Analysis

The transfer functions (4) and (6) were standard underdamped second-order control systems when the damping ratio met the inequality:

$$\xi = \frac{1}{2R_mC_t}\sqrt{\frac{L_1L_2C_t}{L_1+L_2}} < 1 \tag{7}$$

For a highly efficient HFT with little iron loss, the inequality (7) was easily met. When the excitation (i.e., the step change voltage) appeared, there was voltage oscillation, which gradually decayed. The natural frequency ω_n and oscillation frequency ω_{OSC} were calculated as follows:

$$\omega_n = \sqrt{\frac{L_1 + L_2}{L_1 L_2 C_t}}, \omega_{OSC} = \sqrt{1 - \xi^2}\omega_n \tag{8}$$

Furthermore, ignoring R_m, which yielded $\xi = 0$ and $\omega_{osc} = \omega_n$, ω_{osc} could be derived as follows:

$$\omega_{OSC} = \omega_n = 1/\sqrt{\left[1/(\sum_{i=1}^{N} 1/L_i)\sum_{k=1}^{N} C_{sk}\right]} \tag{9}$$

The result was symmetrical, which means the period of the oscillation induced by the excitation source at any port was the same. According to the rated transferred power of DAB, the splitting inductors should meet:

$$L_1 + L_2 = 2L_{ph} \tag{10}$$

Combing (9), the oscillation of DAB is written as:

$$\begin{cases} \omega_{OSC} = \sqrt{2L_{ph}/(C_1 + C_2)}/\sqrt{L_1/(2L_{ph} - L_1)} \\ \omega_{OSC,min} = 1/\sqrt{(C_1 + C_2)L_{ph}/2} \end{cases} \tag{11}$$

The minimum oscillation frequency $\omega_{osc,min}$ is achieved when $L_1 = L_2 = L_{ph}$. Tuning the splitting inductances to move the oscillation frequency, which represents the pole point in (4), to the first zero cross point (ZCP) of the excitation voltage is the basic principle in [12].

The DAB and TAB prototypes were setup to test the model. The parameters are shown in Table 1.

Table 1. Parameters of the DAB and MAB prototypes.

Parameters	DAB Converter	TAB Converter
Swithing devices	C3M0075120K(SiC MOSFET)	C3M0075120K (SiC MOSFET)
Swithing frequency	20 kHz	20 kHz
Transformer turn ratio	N_1:N_2 = 1:1	N_1:N_2:N_3 = 4:5:6
Transformer capacitances	$C_1 = C_2 = 130$ pF, $C_{12} = 55$ pF	$C_1 = 85$ pF, $C_2 = 100$ pF, $C_3 = 140$ pF,$C_{12} = 70$ pF, $C_{23} = 70$ pF, $C_{13} = 70$ pF
Transformer leakage inductance	$L_{S1} = 3$ μH, $L_{S2} = 3$ μH	$L_{S1} = 3$ μH, $L_{S2} = 3.5$ μH, $L_{S2} = 4$ μH
Transformer winding resistance	$R_S = 0.016$ Ω	$R_{S1} = 0.03$ Ω, $R_S = 0.035$ Ω, $R_S = 0.05$ Ω
Transformer magnetizing inductance	$L_m = 63$ mH	$L_m = 24$ mH
Transformer magnetizing resistor	$R_m = 10$ kΩ	$R_m = 25$ kΩ
Phase-shift inductance	$L_1 = 160$ μH, $L_2 = 100$ μH	$L_1 = 140$ μH, $L_2 = 160$ μH, $L_3 = 100$ μH
Equivalent resistance	$R_1 = 0.016$ Ω, $R_2 = 0.016$ Ω	$R_1 = 0.016$ Ω, $R_2 = 0.016$ Ω, $R_3 = 0.016$ Ω

Square wave u_1 at the primary AC side was used as the excitation source while the secondary AC side short-circuited. It was easy to run the test with a small excitation amplitude of 50 V. Obvious oscillation appeared after the u_1 step changed and decayed to zero with several cycles, as shown in Figure 7c. To substitute the parameters of the transformer and inductor into (5), we obtained $\xi = 0.039$, $\omega_{osc} = 0.99\omega_n$, and $t_{osc} = 810$ ns. The theoretical oscillation period matched well with the test results.

The test circuit could be utilized to estimate the damping characteristics and to pre-obtain the oscillation frequency. For multiport converters, all the other branches should be short-circuited except the excitation port. Figure 7b,d show the test short circuit and the waveforms of the TAB converter, where branch 2 and 3 were short-circuited.

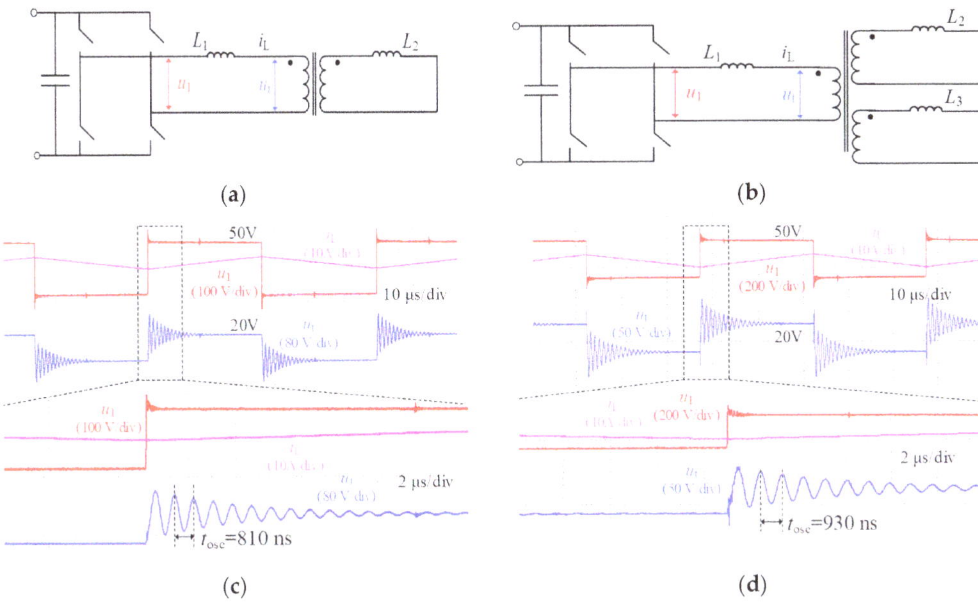

Figure 7. Short-circuit test. (**a**,**c**) Test circuit and waveforms of the DAB converter; (**b**,**d**) test circuit and waveforms of the TAB converter.

4. Active Suppression Method for the HFO

4.1. General Description of the Voltage Excitation

The excitation source of the active-bridge-transformer-based converter, generated by the active bridge, was the superposition of the two half bridges [15]. As shown in Figure 8, the phase voltages u_A, u_B and the output voltage u_{AB} were modeled by segmental linear functions considering the inner phase shift and transient time as:

$$\begin{cases} u_A(t) = V_P/(t_r)[te(t) + (t - t_r)e(t - t_r)]\ , u_B(t) = -u_A(t - t_{Iph}) \\ u_{AB}(t) = u_A(t) - u_B(t) = u_A(t) + u_A(t - t_{Iph}) \end{cases} \tag{12}$$

where V_P is the DC bus voltage, u_B lags behind u_A with time t_{Iph}, and t_r is the transient time.

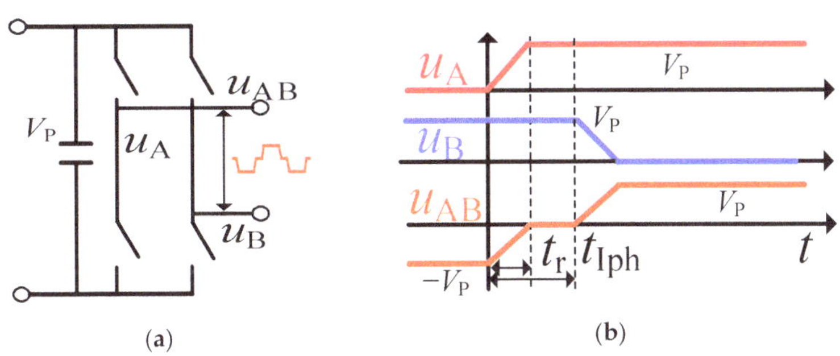

Figure 8. Output voltage of the active bridge. (**a**) H bridge; (**b**) output voltage.

4.2. Active SHE PS Suppression Method

To modify the inner phase-shift t_{Iph}, reducing the harmonics of the oscillation frequency is an effective way to suppress the oscillation, which is called the selective harmonic elimination phase-shift (SHE PS) modulation method. Combing (12) with the transfer function (4), $u_{t,A}$, the high-frequency voltage which represents the zero-state response of $u_{A,}$ was derived as:

$$\begin{cases} u_{t,A}(t) = 0.5k_e k_r(t_r/(\alpha^2+\beta^2) + Me^{\alpha t}\cos(\beta t + \theta_1 + \theta_2)) \\ \theta_1 = \arctan(\alpha^2-\beta^2)/(2\alpha\beta),\ \theta_2 = -\arctan(\sin(\beta t_r)/(1-e^{-\alpha t_r}\cos(\beta t_r)) \\ k_e = 1/(L_1 C_t),\ k_r = 2V_P/t_r \end{cases} \quad (13)$$

where α and β are written as:

$$\alpha = -\frac{1}{2R_m C_t},\ \beta = \frac{1}{2}\sqrt{\frac{4(L_1+L_2)}{L_1 L_2 C_t} - (\frac{1}{R_m C_t})^2} \quad (14)$$

Thus, the zero-state response of $u_{t,AB}$ could be derived, combining (12) and (13) as:

$$\begin{cases} u_{t,AB}(t) = V_1 L_1/(L_1+L_2) + Me^{\alpha t} f(t_{Iph})\cos(\beta t + \theta_1 + \theta_2 + \theta_3) \\ \theta_3 = -\arctan(e^{-\alpha t_{Iph}}\sin(\beta t_{Iph})/(1+e^{-\alpha t_{Iph}}\cos(\beta t_{Iph})) \\ M = V_P\sqrt{(1-e^{-\alpha t_r}\cos(\beta t_r))^2 + \sin^2(\beta t_r)}/(L_1 L_2 C_t t_r(\alpha^2+\beta^2)\beta/(L_1+L_2)) \end{cases} \quad (15)$$

M is constant. The amplitude $f(t_{Iph})$, which is a function of inner phase-shift t_{Iph}, was derived as:

$$f(t_{Iph}) = \sqrt{(1+e^{-\alpha t_{Iph}}\cos(\beta t_{Iph}))^2 + e^{-2\alpha t_{Iph}}\sin^2(\beta t_{Iph})} \quad (16)$$

The $f(t_{Iph})$ in Equation (16) could be derived as:

$$-\alpha t_{Iph} \approx 0 \Rightarrow e^{\alpha t} \approx 1 \Rightarrow f(t_{Iph}) \approx \sqrt{2+2\cos(\beta t_{Iph})} \quad (17)$$

Thus, the minimum of the VSA could nearly be achieved at the following condition:

$$\beta t_{Iph} = \pi,\ t_{Iph} = \pi/\beta = 0.5T_{osc} \quad (18)$$

To substitute (18) into (16), the minimum $f_{min}(t_{Iph})$ and the corresponding VSA A_{min} were derived as:

$$f_{min}(t_{Iph}) = 1 - e^{-\frac{\alpha\pi}{\beta}},\ A_{min}(t_{Iph}) = \frac{1}{2}k_e k_r M(1 - e^{-\frac{\alpha\pi}{\beta}}) \quad (19)$$

4.3. *Impact of the System Parameters on the SHE PS Modulation Suppression Method*

4.3.1. Inner Phase-Shift Accuracy

As seen in Equation (18), the SHE PS modulation utilizes the inner phase-shift, which equals half of the oscillation period to eliminate the HFO. When the frequency was high, the phase-shift accuracy should be considered due to the time resolution of the controller and the non-ideal factors of the circuit, i.e., different time delay of the driver signals.

Considering the $f(t_{Iph})$ with different inner phase-shifts, we drew the curve of oscillation amplitude vs. the inner phase-shift, as shown in Figure 9. The amplitude value decreased when the inner phase-shift was applied, and it reached its optimal point when $t_{Iph} = 0.5T_{osc}$. The VSA still fell to the half of the unsuppressed value when applying the inner phase-shift even with $\pm$30% error, which verifies the robustness of the method in terms of inner-phase accuracy.

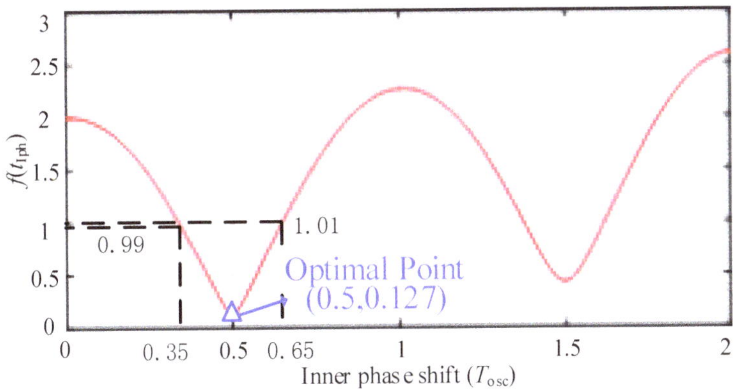

Figure 9. Oscillation amplitude vs. inner phase-shift.

4.3.2. Asymmetrical Splitting Inductances

The oscillation frequency changed with the varying splitting inductances according to (11), which could be verified in the simulation results in Figure 10. Two cases of splitting inductances were considered: the symmetrical case with $L_{ph1} = L_{ph2} = 160$ µH, and the asymmetrical case with $L_{ph1} = 50$ µH, $L_{ph2} = 274$ µH. The simulated waveforms show that the voltage and current oscillations were immigrated to a negligible level for both the symmetrical and asymmetrical situations when the SHE PS modulation method was adopted.

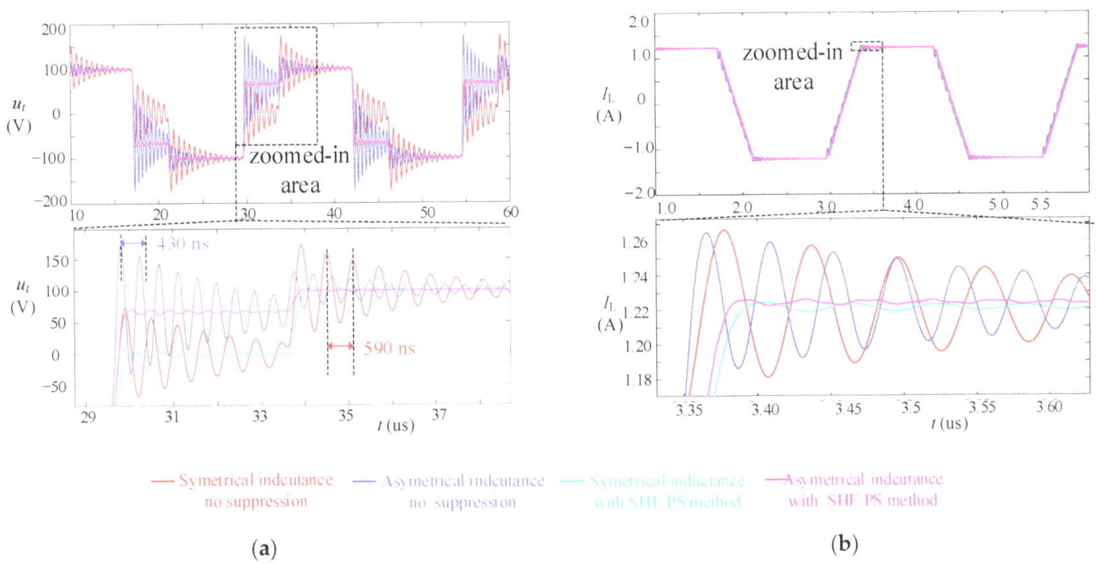

Figure 10. Simulated waveforms with different splitting inductances. (**a**) Transformer voltage; (**b**) transformer current.

4.4. Comparison with Other Suppression Methods

The proposed active method was from the perspective of the excitation source. The comparisons between the proposed method and the existing methods in the literature [8–13] are listed in Table 2.

Table 2. Overall comparison with other methods.

Methods	Mode	Applicable Topology	Branch Parameters	Implementation Complexity
[8,9]	Passive	DAB	Only symmetrical parameters are considered.	Moderate
[10]	Passive	DAB	Only asymmetrical parameters are considered with $N_1 : N_2 \neq 1$	Moderate
[11]	Passive	DAB	Only asymmetrical parameters are considered.	Difficult
[12]	Passive	DAB	Only asymmetrical parameters are considered with $N_1 : N_2 \neq 1$	Moderate
[13]	Active	Modular MAB	Only symmetrical parameters are considered.	Easy
Proposed method	Active	DAB, MAB	Effective with both the symmetrical and asymmetrical parameters	Easy

The existing methods in [8–12] are passive methods. For [8,9], external capacitors were connected in parallel with the drain and source of the MOSFET to modify the dv/dt of the AC voltage of the bridges, which limited the switching speed of the devices. Additionally, the analytical model in [8] was based on the symmetrical parameters with the same two phase-shift inductors on the primary and secondary sides, respectively. On the contrary, the model in this article considered the asymmetrical parameters. In practice, the parameters for the MAB were not always symmetrical for the different branches.

The method in [11] optimizes the stray capacitance of the transformer by the optimal design of the transformer in the DAB converter, which is difficult. For the MAB with more windings, the design complexity was increased.

The different effects of the phase-shift inductors of the primary and secondary side on the impedance of the resonant tank were analyzed. The asymmetrical splitting inductances with the transformer turn ratio N_1:$N_2 \neq 1$ were utilized to suppress the oscillation by modifying the ZCP of the AC impedance of the circuit in [10,12]. For MAB, the optimization of the splitting inductances for all the branches was quite difficult.

The method in [13] utilized a similar method to eliminate the high-frequency oscillation of the AC bus in MMAB. However, only the symmetrical parameters were considered. Compared with [13], the more general situations with the asymmetrical inductances were taken into consideration when establishing the HF voltage oscillation model. The proposed active method showed the effectiveness with both the symmetrical and asymmetrical parameters, as shown in Figure 10.

5. Experimental Validation

The TAB prototype described in the article was set up to verify the proposed method, which is shown in Figure 11. The SiC MOSFET from Cree C3M0075120K (1200 V 23 A) was utilized as the switching device in the H bridge. Nanocrystalline alloy magnetic core VAC W342 was used to make the transformer. It was convenient to set up the DAB converter by replacing the three-windings transformer with a two-windings transformer.

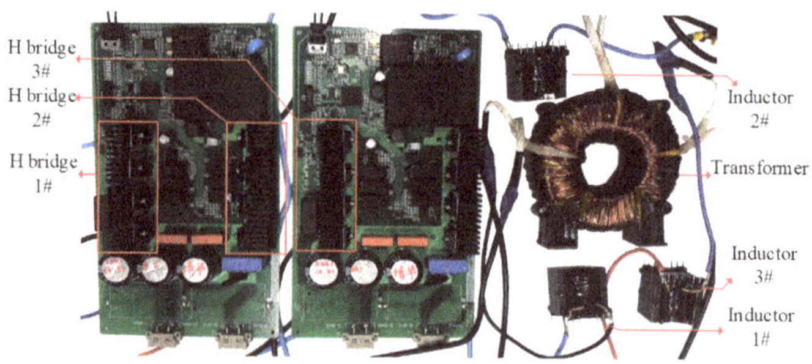

Figure 11. Experimental prototype.

Figure 12 shows the waveforms of the DAB prototype for the forward power flow: $V_{P1} = V_{P2} = 250$ V. The phase shift duty between the primary side and second side was 0.4. The VSA was 238 V. Figure 8b shows the waveforms after the SHE PS modulation method was utilized. The inner phase-shift time was 400 ns, which was half of the oscillation period. The VSA was reduced by 94.1% from 238 V to 14 V.

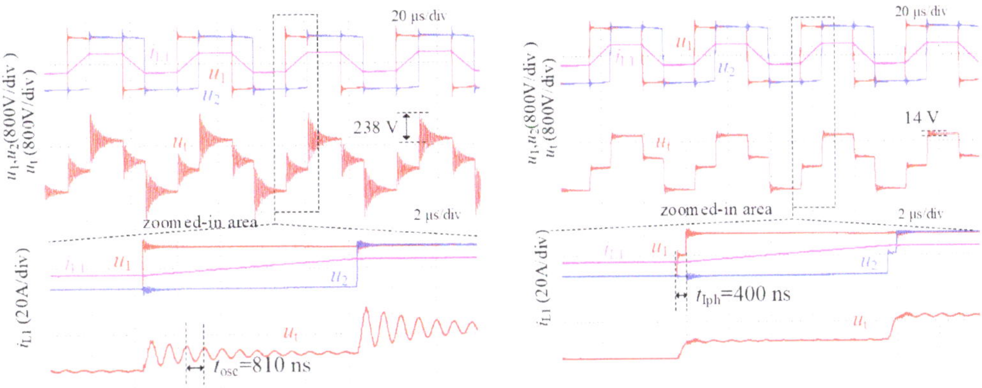

(a) (b)

Figure 12. Waveforms of the DAB converter with forward power flow ($D = 0.4$): (**a**) no suppression; (**b**) with the SHE PS method.

Figure 13 shows the waveforms for the reverse power flow. The phase-shift duty between the primary side and second side was −0.4. The VSA was 173 V. Figure 13b shows the waveforms after the SHE PS modulation method was utilized. The inner phase-shift time was 400 ns, which was half of the oscillation period. The VSA was reduced by 93.1% from 173 V to 12 V.

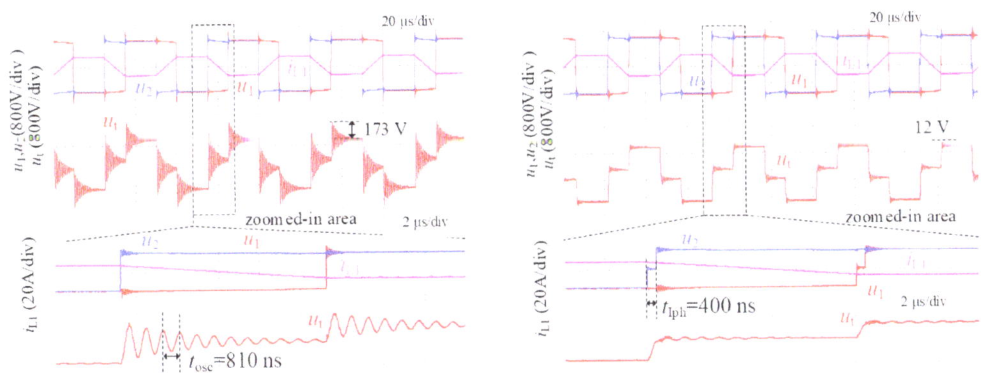

(a) (b)

Figure 13. Waveforms of the DAB converter with reverse power flow ($D = -0.4$): (**a**) no suppression; (**b**) with SHE PS method.

Figure 14 shows the waveforms of the TAB prototype: $V_{P1} = V_{P2} = V_{P3} = 250$ V. The phase shift duties D_1, D_2 and D_3 were 0, 0.2 and 0.3, respectively. The oscillation period was 0.92 μs. Figure 14b shows the waveforms after the SHE PS modulation method was utilized. The inner phase-shift time t_{Iph} was 460 ns, which was half of the oscillation period. The VSA was reduced by 92.1% from 151 to 18 V.

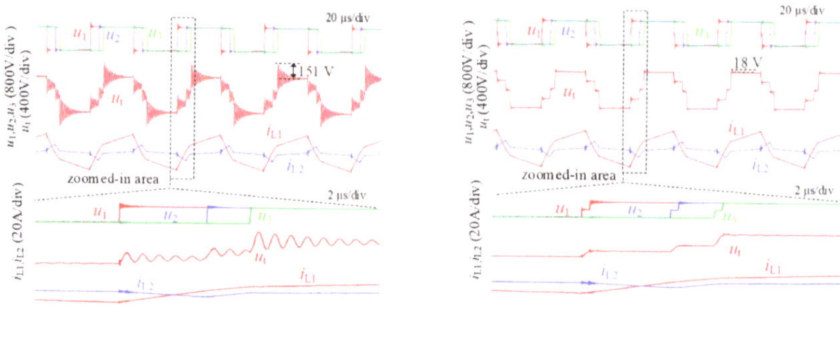

(a) (b)

Figure 14. Waveforms of the TAB converter with forward power flow (D_1 = 0, D_2 = 0.2, D_3 = 0.3) (**a**) no suppression; (**b**) with the SHE PS method.

Figure 15 shows that the waveforms with the phase-shift duties D_1, D_2 and D_3 were 0, −0.2 and −0.3, respectively. Figure 15b shows the waveforms after SHE PS modulation method was utilized. The VSA was reduced by 86.0% from 107 to 15 V. Figure 16 shows that the waveforms with the phase-shift duties D_1, D_2 and D_3 were 0, 0.2 and −0.3, respectively. The VSA was reduced by 77.8% from 63 to 14 V.

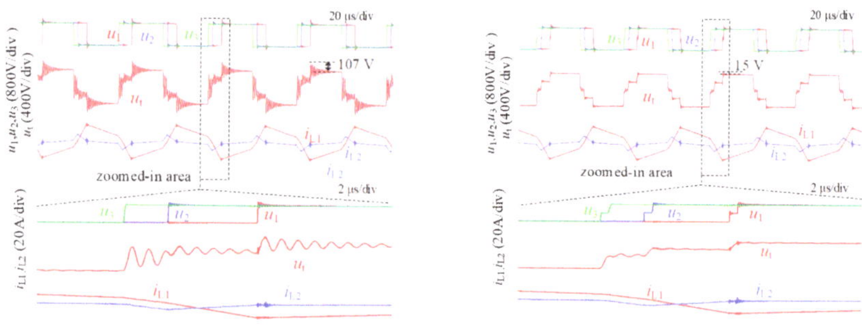

(a) (b)

Figure 15. Waveforms of the TAB converter with reverse power flow (D_1 = 0, D_2 = −0.2, D_3 = −0.3): (**a**) no suppression; (**b**) with SHE PS method.

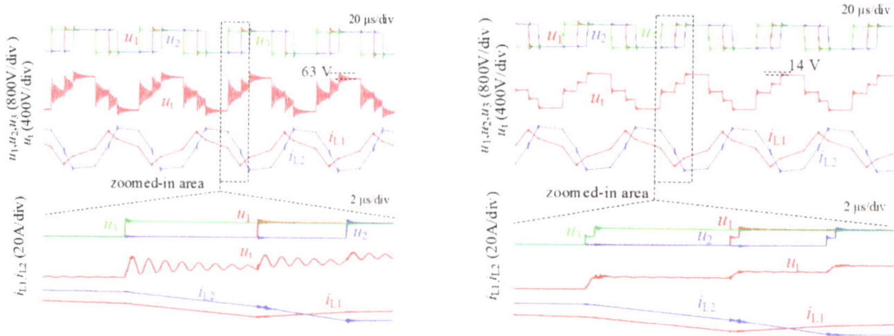

(a) (b)

Figure 16. Waveforms of the TAB converter with reverse power flow (D_1 = 0, D_2 = 0.2, D_3 = −0.3): (**a**) no suppression; (**b**) with the SHE PS method.

Figure 17 shows the harmonics spectrum of the transformer voltage u_t in the TAB in Figure 14. The harmonics around the oscillation frequency were obvious mitigated where the amplitude of the harmonics with the oscillation frequency was decreased by 77.1% from 35 V to 8 V when the SHE PS method was utilized.

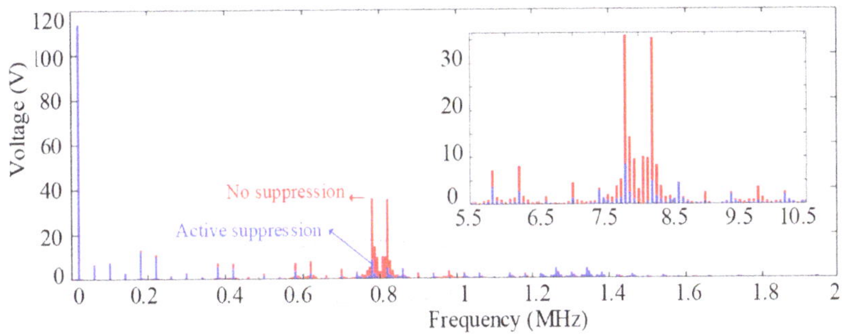

Figure 17. Harmonics spectrum of the transformer voltage in the TAB converter.

The efficiency curve of the DAB prototype is shown in Figure 18, where $V_{P1} = V_{P2} = 220$ V. When the HFO was suppressed, the efficiencies were increased by 0.3~0.4% after the SHE PS modulation method was utilized.

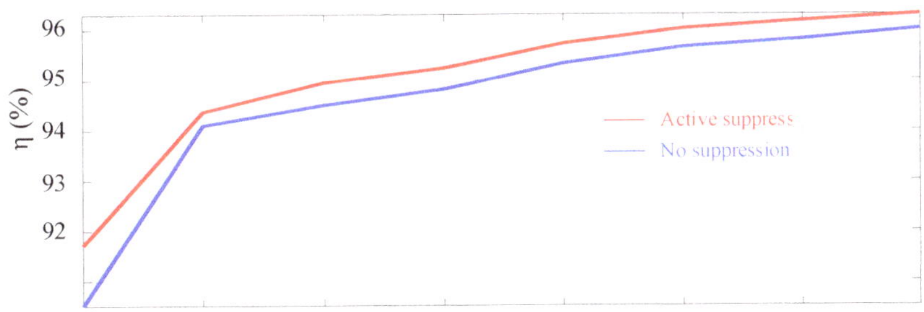

Figure 18. Efficiency curve of the DAB prototype.

6. Conclusions

This article studied the HFO of active-bridge-transformer-based DC/DC converters including DAB and MAB converters. Ignoring the AC resistance in the circuit, the high-order model of the voltage oscillation could be reduced to a second-order system. On top of that, a simplified and general HFO model for DABs and MABs with asymmetrical inductances was proposed. Then, the HFO was analyzed from a time domain prospective. Finally, a universal active SHE PS modulation method to suppress oscillation was investigated. Finally, the effectiveness and generality of the method were verified by DAB and MAB experimental results. To summarize, the SHE PS modulation suppression was applicable for DAB and MAB. No additional circuits were required to mitigate the HFO problem for the existing equipment, so redesigning or changing the transformers and inductors was not needed. It was easy to implement. This simple short-circuit test could be used to pre-obtain the inner phase-shift as well as for the accurate measurement of the stray parameters.

Author Contributions: Conceptualization, S.W. and W.W.; methodology, S.W.; software, S.W.; validation, W.W.; formal analysis, W.W.; investigation, W.W.; resources, S.W.; data curation, S.W.; writing—original draft preparation, S.W.; writing—review and editing, S.W.; visualization, S.W.; supervision, W.W.; project administration, W.W.; funding acquisition, W.W. All authors have read and agreed to the published version of the manuscript.

Funding: This research was supported by the key project of Science and Technology Innovation Program of Army Engineering University of PLA under Grant KYCQJQZL2119.

Data Availability Statement: Not applicable.

Conflicts of Interest: The authors declare no conflict of interest.

References

1. Hou, N.; Li, Y. Overview and comparison of modulation and control strategies for non-resonant single-phase dual-active-bridge dc-dc converter. *IEEE Trans. Power Electron.* **2020**, *35*, 3148–3172. [CrossRef]
2. Michal, G.; Lech, M.G.; Kaszewski, A. A Dual Rising Edge Shift Algorithm for Eliminating the Transient DC-Bias Current in Transformer for a Dual Active Bridge Converter. *Energies* **2021**, *14*, 4264.
3. Bandyopadhyay, S.; Purgat, P.; Qin, Z. A Multi-active Bridge Converter with Inherently Decoupled Power Flows. *IEEE Trans. Power Electron.* **2021**, *36*, 2231–2245. [CrossRef]
4. Falcones, S.; Ayyanar, R.; Mao, X. A DC–DC Multiport-Converter-Based Solid-State Transformer Integrating Distributed Generation and Storage. *IEEE Trans. Power Electron.* **2013**, *28*, 2192–2203. [CrossRef]
5. Tehrani, K.; Yunbi, L.I.; Rousseau, L.; Normand, A.; Vurpillot, F. Design of A Multistage Marx Generator Topology based on SiC-MOSFET Device for Atomic Probe Tomography Applications. In Proceedings of the 2020 IEEE 15th International Conference of System of Systems Engineering SoSE, Budapest, Hungary, 2–4 June 2020; pp. 423–428.
6. Tuluhong, A.; Wang, W.; Li, Y.; Wang, H.; Xu, L. Research on Modelling and Stability Characteristics of Electric Traffic Energy System Based on ZVS-DAB Converter. *J. Electr. Comput. Eng.* **2020**, *2020*, 5450628. [CrossRef]
7. Tehrani, K.; Weber, M.; Rasoanarivo, I. Design of High Voltage Pulse Generator with Back-to-Back Multilevel Boost Buck Converters Using SiC-MOSFET Switches. In Proceedings of the 2020 IEEE 15th International Conference of System of Systems Engineering SoSE, Budapest, Hungary, 2–4 June 2020; pp. 507–512.
8. Cui, B.; Shi, H.; Sun, Q. A Novel Analysis Design and Optimal Methodology of High-Frequency Oscillation for Dual Active Bridge Converters with WBG Switching Devices and Nanocrystalline Transformer Cores. *IEEE Trans. Power Electron.* **2021**, *36*, 7665–7678. [CrossRef]
9. Cui, B.; Xue, P.; Jiang, X. Elimination of High Frequency Oscillation in Dual Active Bridge Converters by dv/dt Optimization. *IEEE Access* **2019**, *7*, 55554–55564. [CrossRef]
10. Qin, Z.; Shen, Z.; Blaabjerg, F. Transformer Current Ringing in Dual Active Bridge Converters. *IEEE Trans. Ind. Electron.* **2021**, *68*, 12130–12140. [CrossRef]
11. Hassan, I.; Keshmiri, N.; Dorneles, A. Design Optimization Methodology for Planar Transformers for More Electric Aircraft. *IEEE Open J. Ind. Electron. Soc.* **2021**, *2*, 568–583. [CrossRef]
12. Wang, C.; Zsurzsan, G.T.; Zhang, Z. Genetic Algorithm Assisted Parametric Design of Splitting Inductance in High Frequency GaN-based Dual Active Bridge Converter. *IEEE Trans. Ind. Electron.* **2021**, 1. [CrossRef]
13. Wei, S.; Zhao, Z.; Yuan, L. Voltage Oscillation Suppression for the High-Frequency Bus in Modular-Multiactive-Bridge Converter. *IEEE Trans. Power Electron.* **2021**, *36*, 9737–9742. [CrossRef]
14. Farhangi, B.; Toliyat, H.A. Modeling and Analyzing Multiport Isolation Transformer Capacitive Components for Onboard Vehicular Power Conditioners. *IEEE Trans. Ind. Electron.* **2015**, *62*, 3134–3142. [CrossRef]
15. Alawieh, H.; Riachy, L.; Tehrani, K.A.; Azzouz, Y.; Dakyo, B. A new dead-time effect elimination method for H-bridge inverters. In Proceedings of the IECON 2016—42nd Annual Conference of the IEEE Industrial Electronics Society, Florence, Italy, 23–26 October 2016; pp. 3153–3159.

energies

Article

Rapid Evaluation Method for Modular Converter Topologies

Damiano Lanzarotto *, Florent Morel *, Pierre-Baptiste Steckler and Konstantin Vershinin

Supergrid Institute, 69100 Villeurbanne, France; pierre-baptiste.steckler@supergrid-institute.com (P.-B.S.); konstantin.vershinin@supergrid-institute.com (K.V.)
* Correspondence: damiano.lanzarotto@gmail.com (D.L.); florent.morel@supergrid-institute.com (F.M.)

Abstract: The success of modular multilevel converters (MMCs) in high-voltage direct current (HVDC) applications has fueled the research on modular converter topologies. New modular converter topologies are often proposed, discussed, and sometimes applied in HVDC, as well as other industrial application such as STATCOMs, DC/DC HVDC, medium-voltage direct current (MVDC), etc. The performance evaluation of new modular converter topologies is a complex and time-consuming process that typically involves dynamic simulations and the design of a control system for the new converter topology. Sadly, many topologies do not progress to the implementation stage. This paper proposes a set of key performance indicators (KPIs) related to the cost and footprint of the converter and a procedure designed to rapidly evaluate these indicators for new converter topologies. The proposed methodology eliminates the need for dynamic simulations and control-system design, and is capable of identifying whether a particular converter is worth considering or not for further studies of a specific application, depending on the operating requirements. Thanks to the method outlined in this work and via the key parameters quantifying the "relevance" of the analyzed converters, promising topologies were easily identified, while the others could be rapidly discarded, resulting in saving valuable time in the study of the solutions that have a real potential. The proposed method is first described from a general point of view and then applied to a case study of the new converter topology—Open-Delta CLSC—and its application in two use cases.

Keywords: HVDC; converter topologies; sizing; Open-Delta; key performance indicators

Citation: Lanzarotto, D.; Morel, F.; Steckler, P.-B.; Vershinin, K. Rapid Evaluation Method for Modular Converter Topologies. *Energies* **2022**, *15*, 3492. https://doi.org/10.3390/en15103492

Academic Editors: Jelena Loncarski and Diego Bellan

Received: 30 March 2022
Accepted: 7 May 2022
Published: 10 May 2022

1. Introduction

High-voltage direct current (HVDC) technology nowadays represents the most advantageous technical solution to problems such as long-distance energy transmission, asynchronous AC system interconnection, interconnection of different regions requiring submarine and underground cables, and transmission of offshore wind power to shore [1,2]. The ability to efficiently connect large renewable energy sources located far away from the main loads is rapidly expanding the installation of HVDC lines in areas such as northern Europe and across China [3–6].

The crucial elements of HVDC transmission, from both a technological and ultimately a cost point of view, are the power electronic converters, which allow the AC/DC energy conversion and vice versa. Thus, a great amount of research has been and still is currently directed toward the investigation of new and more advantageous HVDC converter topologies. The modular multilevel converter (MMC) [7] currently represents the most accepted solution in new installations, in spite of the many variations that have been proposed [8–11] and the many more new topologies that have attempted to challenge it [12–15] (to cite only a few of them).

The advantages of the MMC, such as independent active and reactive power controls, modularity, and reduced filtering requirements, have made this modular technology interesting even in other applications such as STATCOMs [16] and HV DC-DC converters, [17,18] and in medium-voltage (MV) applications [19].

To understand whether a new topology is beneficial and can potentially compete in the marketplace with established designs, one should identify the key strengths of the proposed solution and any potential weaknesses. This process should be done in the most efficient manner possible in order to discard topologies that do not bring significant benefits early in the development process. A typical approach is to define a set of key performance indicators (KPIs), but it is difficult to find a general agreement on what those KPIs should be, as the works comparing different topologies often make use of different ones [17,18]. A significant attempt to harmonize them can be traced to [20]. Moreover, a typical approach to calculation of KPIs adopted up to now involves an extensive use of dynamic simulations even for the calculation of steady-state parameters and the acquisition of steady-state waveforms. The main drawbacks of this approach can be summarized in the following two points: (1) running simulations makes it necessary to design the control system, which is not a straightforward and quick task, especially for new topologies; and (2) for a complete evaluation of the converter performance, multiple simulations must be run in order to analyze the behavior in different operating conditions, which takes valuable time. Such an approach is especially wasteful when it is found that benefits offered by the new topology are not enough to justify commercial interest compared to an already-established solution in the market.

In this paper, we formalize a methodology that can be applied to the assessment of a new converter topology in much more efficient manner and that does not require a substantial simulation effort to assess whether further development of the converter topology is worth pursuing for a particular application. This methodology rapidly provides the data necessary for a manufacturer to choose whether the converter under scrutiny has no interest or represents a valid solution, thereby justifying additional studies on it. To the best of the authors' knowledge, there are no papers that describe and discuss similar general procedures. In addition, we identify and propose a set of general KPIs that can be used universally in the assessment of topologies with different weightings applied, depending on the application. A practical application of the proposed methodology is also presented in order to demonstrate the efficiency of the method.

The paper is structured as follows. In Section 2, the methodology is presented, and particular aspects of its application are discussed. In Section 3, generic KPIs are defined and a calculation method is presented. In Sections 4 and 5, the proposed methodology is applied in a case study of a new topology called the "Open-Delta Capacitor Link Series Converter (CLSC)" [21], which is compared to the half-bridge modular multilevel converter (MMC) used as a reference topology. Finally, in Section 6, the KPIs are presented and compared for the topologies under study. The conclusions drawn in given in Section 7.

2. Topology Assessment Methodology

The goal of the proposed methodology was to minimize the development effort required to identify whether a new topology can be a serious contender to replace an existing design in practical applications. Such a methodology must avoid the necessity of the control-system design and the repetition of a significant number of simulations, so that weak topologies can be immediately identified and discarded, while time can be saved and better invested in analyzing converter architectures that have a real potential. To address this goal, we proposed to focus our analysis on the steady-state operation of the topology; if no significant benefits are shown, it is unlikely the topology would be of interest in industrial applications, and there would be no need to develop it further.

A simplified block diagram of the proposed procedure is shown in Figure 1. It begins with the definition of the requirements of the application and identification of the topology optimization approached, followed by the derivation of the converter equations that shall consider the energy balance within the converter. Theoretical investigation are continued with the implementation of the derived equations in a mathematical analysis tool such as MATLAB. Once implemented, the full operation domain can be swept quickly to identify the best and worst operation conditions and compute the associated KPIs. These steps are explained in detail in the next section.

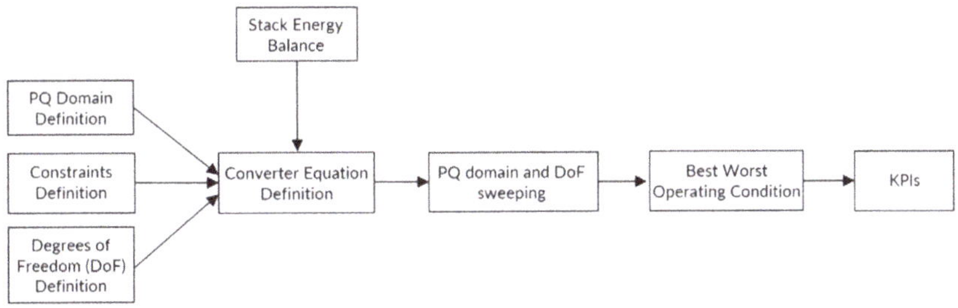

Figure 1. Topology assessment methodology.

2.1. Definition of the PQ Operating Domain

When assessing the benefits of a topology, it is very important to define the application and the requirements that are expected by the converter in the target application. It might be tempting to define the widest requirements possible to cover all possible applications; this approach is interesting academically, but it can miss some of the topologies that can bring advantages to one application but cannot be applied universally. Moreover, this approach can lead to oversizing of the converter. In a grid-connected converter, the key features defining the operating domain are the requirements for the active and reactive power values that the converter must provide. These determine the "PQ-domain" for which the converter must be sized.

Four different PQ-domains can be identified for applications within electrical grids. These profiles are shown in Figure 2. The rectangular PQ domain [22] shown in Figure 2a is common in HVDC applications, and represents the maximum requirements. On the other hand, for converters used in medium-voltage direct current (MVDC) applications, a "butterfly PQ domain" [23] is more common; this domain is characterized by the fixed minimum $\cos(\varphi)$ as shown in Figure 2b. Other typical PQ domains are represented by the STATCOM shown in Figure 2c and the MVDC load converter shown in Figure 2d. In the case of STATCOM, the domain can be asymmetrical; i.e., the max inductive power is greater than the max capacitive power (or vice versa). For MV loads, the converter only absorbs the active power, and the sign of the active power never changes.

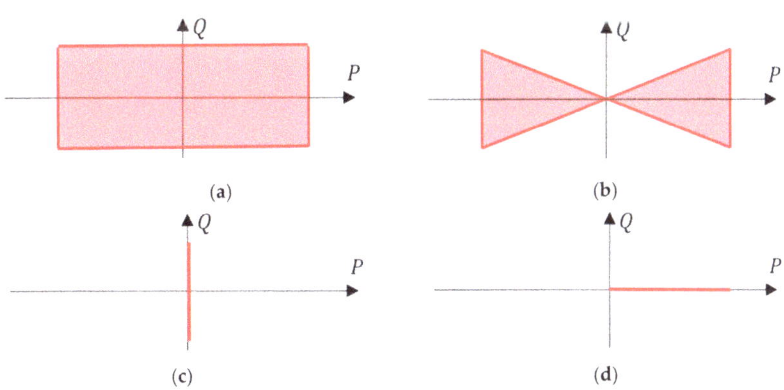

Figure 2. Typical PQ domains for grid-connected converters: (**a**) rectangular profile; (**b**) butterfly profile; (**c**) reactive power only; (**d**) zero-Q load.

2.2. Definition of Constraints and Degrees of Freedom

It is usually possible to optimize/design a topology in different ways that somehow depend on the application. Therefore, it is important to decide which parameters of the topology shall be fixed and which ones can be changed to achieve the optimum design. This leads to the definition of the:

- Degrees of freedom—parameters used to optimize the overall sizing.
- Constraints or constant parameters—parameters that must be the same for all the converter topologies in order to allow for a fair comparison between them.

These parameters change depending on the topology and the application. To provide some guidance, typical options are presented below.

Typical degrees of freedom:

1. Transformer secondary side voltage.

Typical constant parameters:

1. Submodule rated voltage;
2. Maximum stack voltage ripple;
3. Maximum DC voltage ripple.

2.3. Definition of the Steady-State Characteristic Equations Guaranteeing the Converter Energy Balance

The next step is to write the characteristic equations of the converters in a steady state based on the parameters reflecting constraints and degrees of freedom, as well as the PQ operating domain. This means analytically defining the current and voltage across each element of the topology. At this stage, only the ideal voltage and current waveforms during steady-state operation are considered. This implies that the voltage of the stacks does not present the staircase shape, which is characteristic of multilevel converters, but is rather assumed to be smooth and as close as possible to its ideal waveform. For most applications, the waveforms are considered to have just DC and one AC components.

An important point to consider during this stage is the energy balance inside the converter, and each single stack must be satisfied; i.e., $e_{stack}(t) = e_{stack}(t+T)$, where e_{stack} is the instantaneous energy stored in the stack and T is the fundamental period. Moreover, for sizing purposes, the components are considered to be ideal; i.e., all the power losses are neglected. This means that the waveforms are calculated while assuming that the component voltage drops have a negligible impact on them. The waveforms computed with this assumption are later used to calculate losses.

2.4. PQ Domain Sweeping, Sizing Work Point Identification, and KPI Calculations

Once the characteristic equations have been defined, the entire PQ operating domain can be swept in order to identify the most critical operating condition, on which the converter sizing has to be based. At the same time, the degrees of freedom must be selected so that the most advantageous sizing in the most critical working conditions is obtained.

Finally, all the KPIs are computed and are ready to be compared between the topologies under investigation.

3. Converter KPI Identification

As described in [24], the interest in a particular converter configuration depends mainly on the converter's cost, size, and efficiency. The cost data is normally confidential, and only manufactures can accurately assess it. Regarding converter size, at least a preliminary design must be established to estimate the converter's footprint. For that, the size of each element should be roughly assessed, as well as the needed clearance distance between them. This is not generally done during the early stages of development of a new promising topology, as it requires a good understanding of the considered topology. Therefore, it is important to identify the parameters or KPIs able to suitably represent those aspects without entering into a time-consuming detailed design.

Before presenting the KPIs, it is important to define the basic element of a modular converter, commonly known as the "stack". The term "stack" is usually adopted to describe a certain number of submodules, which can be both of the half-bridge (HB) and/or the full-bridge (FB) type, which are connected in series as depicted in Figure 3. The voltage created by the stack v_{stack}^{real} is given instant-by-instant by the number of capacitors that are inserted in the circuit, and it is characterized by the typical "staircase" profile. Whereas the v_{stack}^{ideal} considered in this study is the first harmonic approximation of the instantaneous voltage v_{stack}^{real}. A "stack" is further characterized by N_{SM}, the total number of SMs inside it; and V_{SM}, the rated SM voltage.

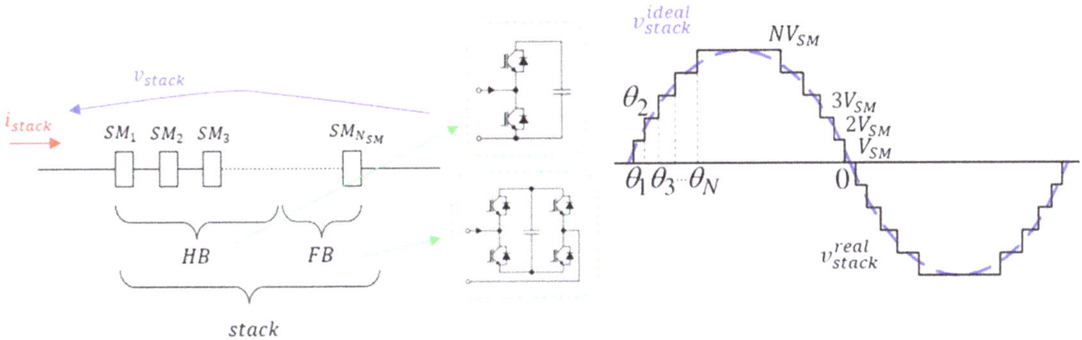

Figure 3. General submodule stack consisting of both HB and FB SMs.

This structure is used in MMCs, as well as in STATCOMs, cascaded H-bridges (CHB), and other converters adopted for the AC/DC or DC/DC conversion in HVDC or MVDC applications.

In our view, the KPIs presented below provide a solid basis on which the converter costs and footprint can be evaluated in the early stages of development.

1. **Transformer number and their sizing power (N_T,S_T).** these parameters are related to the cost and volume of the transformers. In HVDC applications, single-phase transformers are usually preferred over three-phase transformers (due to transportation constraints) and, as long bushings and clearance distances are needed, the number of transformers has a significant impact on the station footprint. This KPI is highlighted here; even if the standard number of transformers is three for AC/DC HVDC converters, it can be different for other topologies.
2. **Submodule (SM) number (N_{SM}).** In the set of the KPIs proposed, this represents the footprint of the converter to the highest degree. It is related to the number of interconnections between the submodules and mechanical assemblies, the number of capacitor voltages to measure, and the number of discharge circuits (as well as the number of bypass circuits, depending on the manufacturer's technical choices). As the submodules correspond to the major cost of the converter, they also relate to the cost, but other KPIs provide a better representation of the cost.
3. **Semiconductor switch total sizing power (S_{SW}).** This is related to the "quantity of silicon" (voltage the semiconductors must withstand and the current passing through them), and therefore is related to the converter cost. In simple terms, it represents the sum of the sizing power of all the switches.
4. **DC voltage ripple (R_{02pk}).** This is related to the converter's cost and volume, as it indicates whether an additional filter is needed on the DC side.
5. **Submodule cell capacitance (C_{SM}).** This is mainly related to the SM size (which is important regarding its ability to handle it during the construction phase and during replacement operations for faulty ones). It is also related to the energy stored in an individual submodule, which is a constraint for the devices in the fault current path in the case of an SM internal short-circuit.

6. **Stored energy (W_{stored}).** This parameter quantifies the energy stored in the converter, which is mainly due to the SM capacitors (but it also takes into account the energy stored in the inductors), which represent the major part of the SM volume. Therefore this parameter is linked to the converter volume.
7. **Switch number (N_{SW}).** This has a main influence on cost.
8. **Power loss (P_L).** This is related to the converter's efficiency and then the operation costs, but also to the constraints on the thermal-management system (impacts on cost and footprint).

Most of the KPIs defined above are straightforward; however, some others must be clearly defined mathematically. More specifically, a closer focus must be placed on the calculations of the SM capacitance C_{SM} and the total switch sizing power S_{SW}.

3.1. Per Unit System

In this paper, the calculations were carried out on a per unit (PU) basis in order to generalize the results and the comparison between topologies. The bases used for the PU calculations were:

- The DC voltage V_{DC};
- The maximum DC current $I_{DC,max}$;
- The power base, which is given by: $P_{DC,max} = V_{DC} I_{DC,\ max}$.

3.2. Submodule Capacitance Calculation

The sizing of the submodule capacitors was carried out by following the approach described in [25]. For a given submodule stack, the submodule capacitors can be found using the following equation:

$$C_{SM} = \frac{\Delta e_{stack}^{\max}}{2N_{SM} V_{SM,N}^{pu}{}^{2} \delta V_{stack}^{\max}} \frac{I_{DC,\max}}{V_{DC}} \quad (1)$$

where N_{SM} is the number of submodules in the stack, $V_{SM,n}^{pu}$ is the submodule rated voltage on a PU basis, and $\delta V_{stack}^{\max}$ is the maximum stack voltage variation allowed, defined as:

$$V_{SM,N}^{pu} = \frac{V_{SM,N}}{V_{DC}} \quad (2)$$

$$\delta V_{stack}^{\max} = \frac{\max\left\{\left|v_{stack}^{real}(t) - N_{SM}^{on}(t) V_{SM,N}\right|\right\}}{V_{SM,N}} \quad (3)$$

where N_{SM}^{on} is the instantaneous number of SMs inserted in the circuit, v_{real}^{stack} is the istanteneus stack voltage in V, and $V_{SM,N}$ is the rated SM voltage in V, $\Delta e_{stack}^{\max}$ is the maximum energy variation over the period, and it can be found by using the following equation:

$$\Delta e_{stack} = e_{stack}^{\max} - e_{stack}^{\min} \quad (4)$$

where:

$$e_{stack}^{\max} = \max\{e_{stack}(t)\} \quad , \quad e_{stack}^{\min} = \min\{e_{stack}(t)\} \quad (5)$$

In which:

$$e_{stack}(t) = \int_0^t v_{stack}^{pu}(\tau) i_{stack}^{pu}(\tau) d\tau \quad (6)$$

Please note that the unit of e_{stack} is seconds; namely: $e_{stack}[s] = E_{stack}[J] / P_{DC,max}[W]$.

3.3. Total Semiconductor Switch Sizing Power

The total semiconductor switch sizing power for a single stack depends on the maximum current and voltage ratings of the stack and the type and the number of the submodules in the stack, and is defined by Equation (7):

$$S_{SW} = \frac{2N_{HB} + 4N_{FB}}{N_{HB} + N_{FB}} V_{\max}^{pu} I_{\max}^{pu} = N_{SW} V_{SM,N}^{pu} I_{\max}^{pu} \quad (7)$$

where N_{HB} and N_{FB} are the numbers of HB and FB SMs in the stack, respectively. Knowing that:

$$N_{SW} = 2N_{HB} + 4N_{FB} \quad (8)$$

and

$$N_{SM} = N_{HB} + N_{FB} \quad (9)$$

then:

$$S_{SW} = N_{SW} V_{SM,N}^{pu} I_{\max}^{pu} \quad (10)$$

in which:

$$I_{\max}^{pu} = \max\{i_{stack}(t)\}/I_{DC,\max} \quad , \quad V_{\max}^{pu} = \max\left\{v_{stack(t)}\right\}/V_{DC} \quad (11)$$

In other words, Equation (7) is the sum of the power ratings of each switch of the converters, and therefore is linked to the "quantity of silicon" necessary.

3.4. Power Loss

To calculate power losses, a choice of the semiconductor device must be made. Once this choice has been made, then power losses can be evaluated. To estimate the power loss, both conduction and switching losses must be evaluated. For high-power MMC converters, the device's switching frequency is close to the line frequency, and switching losses contribute less than 25% of the total losses; these are highly dependent on the capacitor voltage-balancing algorithm (VBA) and cannot be analytically calculated. Moreover, conduction losses and switching losses are related to the number of switches and the current passing through them. Therefore, we proposed to concentrate on the conduction losses only as the indication of the converter's efficiency.

Without a loss of generality, when considering an IGBT, its conduction power losses are quantified by the following relation:

$$P_L = \frac{1}{T}\int_0^T |v_{CE}(t) i_C(t)| dt \quad (12)$$

where v_{CE} is the IGBT collector–emitter voltage and i_C is the collector current. It can be easily proven that:

$$P_L = V_{CE0}\overline{|i_C|} + R_{CE0} I_{C,rms}^2 \quad (13)$$

where V_{CE0} and R_{CE0} can be extracted from the manufacturer's datasheet, as shown in Figure 4; $\overline{|i_c|}$ is the mean value of the absolute value of the collector current; and $I_{C,rms}$ is the rms collector current. The same procedure can be applied to the free-wheeling diode.

Once the conduction power loss is defined for the single switch, the total conduction power loss calculation can be extended to the entire converter, as shown in the following sections.

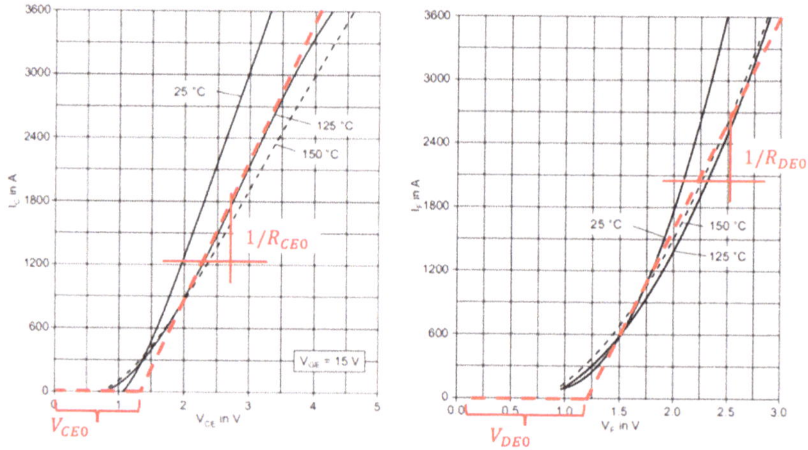

Figure 4. Collector current–collector emitter voltage curves for the IGBT (**left**); forward current–forward voltage curves for the associated freewheeling diode (**right**).

4. Definition of Case Studies

To demonstrate the proposed methodology, it was applied to a new topology called the "Open-Delta Capacitor Link Series Converter (CLSC)" [21], which was compared to the half-bridge modular multilevel converter (MMC) that was used as a reference topology. In order to demonstrate the importance of the defining target application when analyzing new topologies, two scenarios were considered. These two scenarios were defined by different PQ domains, as they corresponded to different real applications (HVDC and load converter). In the first scenario, a rectangular PQ domain was used; for the second scenario, a zero-Q load was used.

The characteristic equations are reported as a function of the PQ work point, the degrees of freedom, and the constraints.

For the analysis of the topologies, a PU system based on DC side values was adopted as described in Section 3.1. For both topologies, the secondary transformer voltage (peak value, phase to phase) Rv was considered as a degree of freedom. It was defined as:

$$R_v = \sqrt{2}V''/V_{DC} \tag{14}$$

4.1. MMC

The well-known structure of the MMC is shown in Figure 5.

To derive steady-state equations for the converter, the following assumptions were made:

- The voltage drop determined by the transformer was negligible.
- The voltage drop on the arm inductance was negligible.
- The DC current source was ideal.

With reference to Figure 5, and considering the upper arm connected to the a phase, one has in PU:

$$\begin{cases} v_{a,arm}^{U,pu} = v_{a,arm,AC}^{U,pu} + 1/2 \\ i_{a,arm}^{U,pu} = i_{a,arm,AC}^{U,pu} + 1/3 \end{cases}, \tag{15}$$

where:

$$v_{a,arm,AC}^{U} = \sqrt{2}E_{Ua}^{pu}\sin(\omega t) + \sqrt{2}E_{Ua}^{pu}\left(1 - k^{III}\right)\sin(3\omega t) \tag{16}$$

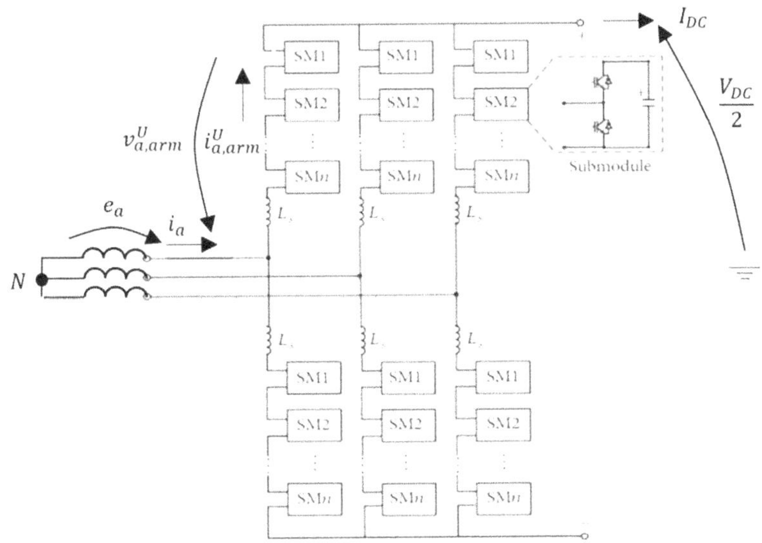

Figure 5. Modular multilevel converter.

The second term in the formula describes a third harmonic injection with the following amplitude [26]:

$$E_{Ua}^{pu} = -\frac{R_v}{\sqrt{6}k^{III}}, \tag{17}$$

where $k^{III} = \sqrt{3}/2$.

On the other hand, the AC components of the current are defined as:

$$i_{a,arm,AC}^{U} = \sqrt{2} I_{AC}^{pu*} \sin(\omega t - \varphi) \tag{18}$$

where:

$$I_{AC}^{pu*} = \frac{I_{AC}^{pu}}{2k^{III}} \tag{19}$$

in which:

$$I_{AC}^{pu} = \sqrt{\frac{2}{3}} \frac{\sqrt{p^2+q^2}}{R_v} \tag{20}$$

$$\varphi = \text{atan2}(q, p) \tag{21}$$

where p and q are the active and reactive power in PU, respectively:

$$p\,[pu] = P\,[W] / P_{DC.\max}\,[W] \tag{22}$$

$$q\,[pu] = Q\,[Var] / P_{DC,\max}[W] \tag{23}$$

It could be easily verified for all p, q, and R_v that the arm energy balance was satisfied; in other words, the condition $e_{arm,a}^{U}(t_0) = e_{arm,a}^{U}(t_0 + T)$, where $T = 20$ ms (for a 50 Hz system) and $e_{arm,a}^{U}(t)$ is defined as follows:

$$e_{arm,a}^{U}(t) = \int_0^t v_{a,arm}^{U,pu}(\tau) i_{a,arm}^{U,pu}(\tau) d\tau \tag{24}$$

The MMC equations as a function of p, q, and R_v that were able to satisfy the arm energy balance in every operating condition were obtained. These equations were implemented in a MATLAB script able to sweep a large number of working points in the expected PQ domain and return the sizing of the converter based on the most critical condition. Figure 6 shows the energy variation over the period for the following p and q values: $p = [0\ 0.25\ 0.5\ 0.75\ 1]$, $q = [-0.3 - 0.15\ 0\ 0.15\ 0.3]$; and for $R_v = 0.866$.

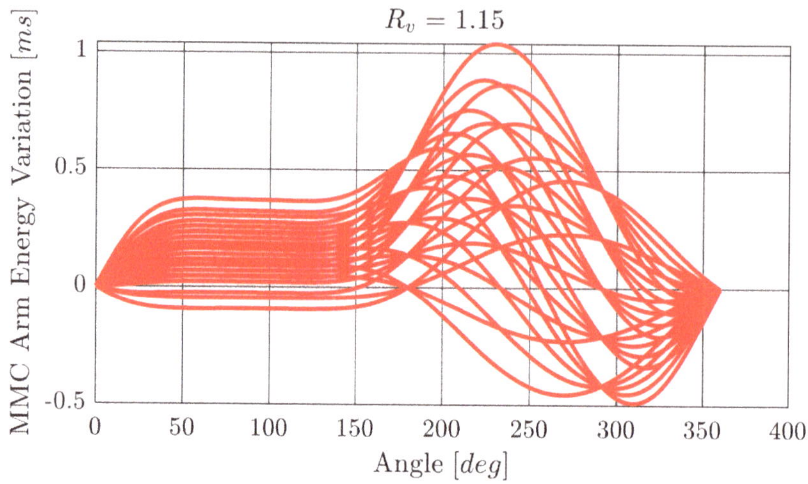

Figure 6. MMC arm energy variation over the period in 25 different PQ work points.

Considering that at any given time, only half of the switches in the submodule conducted the current, then conduction current losses could be calculated according to (25):

$$P_{L,cond}^{MMC} = 3N_{SW}^{arm}\left[V_{eq}\,\overline{|i_{arm}|}_{\max} + R_{eq}{I_{arm,\max}^{rms}}^2\right] \tag{25}$$

where: $V_{eq} = (V_{DE0} + V_{CE0})/2$ and $R_{eq} = (R_{CE0} + R_{DE0})/2$, and N_{SW}^{arm} represents the number of installed arm switches.

4.2. Open-Delta CLSC

The open-delta CLSC converter topology is shown in Figure 7. It consisted of two transformers, two phase elements (PEs) comprising a SM stack and the series capacitor C_s, an active filter (AF) (which was an SM stack), and the DC-link capacitor C_{DC}. The only nonphysical components in the schematic are the reactors, which represent the transformer leakage inductance.

The open-delta CLSC converter adopted the same PEs of the three-phase CLSC [27]; however, the transformers were connected following the open-delta scheme [21]. This converter belongs to family of the converters whose PEs are connected in series, such as the SBC [22,28] or the converter presented in [15], but only two single-phase transformers were employed. In order to create a symmetric and balanced load/generator from the grid standpoint, the current and voltage had to be properly controlled by the two phase elements.

This topology had the advantage of using only two transformers and no arm inductors but it was not possible to verify at first glance how it compared with a state-of-the-art MMC regarding other KPIs: even if there were only two phase elements, the number of submodules and switches could not be immediately found; in addition, it was not obvious how the submodules' ratings compared with those of the MMC. Those considerations paved the way for the analysis of this new topology via the procedure outlined in this paper. The resulting KPIs were then compared with the ones from the MMC.

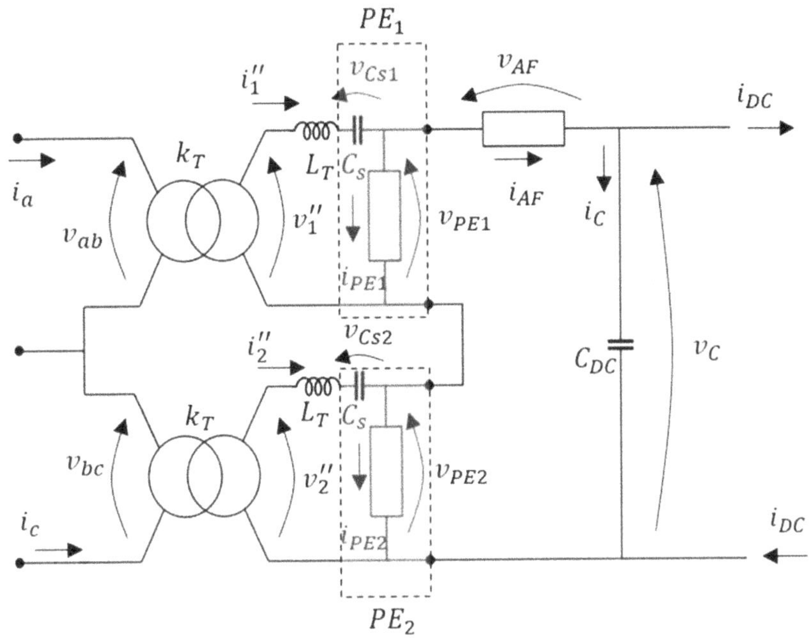

Figure 7. Open-delta CLSC converter.

In order to form a symmetric and balanced three-phase system, the open-delta current and voltage vector diagram must be the one shown in Figure 8. In particular, the converter control must always ensure that:

$$\begin{cases} v_1'' = v_{ab} \\ v_2'' = v_{bc} \end{cases} \quad \begin{cases} i_1'' = i_a \\ i_2'' = -i_c \end{cases} \tag{26}$$

It appeared that the phases carried the same amount of active power only when the reactive power was zero (i.e., when $\varphi = 0$).

It was verified that:

$$\begin{cases} p_{AC1} = \frac{3p-\sqrt{3}q}{6} \\ q_{AC1} = \frac{\sqrt{3}p+3q}{6} \end{cases}, \quad \begin{cases} p_{AC2} = \frac{3p+\sqrt{3}q}{6} \\ q_{AC2} = \frac{-\sqrt{3}p+3q}{6} \end{cases} \tag{27}$$

where p_{AC1}, q_{AC1}, p_{AC2}, and q_{AC2} are the active and reactive power flowing through phases 1 and 2, respectively, as shown in Figure 9. In general, since $\varphi \neq 0$, then $p_{AC1} \neq p_{AC2}$; and since the average values of v_{PE1} and v_{PE2} were equal to $V_{DC}/2$, then an additional flow transferring power from PE1 to PE2 (or vice-versa) had to arise in order to maintain the energy balance in the phase elements. Such additional power flow, which can be called "circulating power flow" (p_C in Figure 9), was controlled by an appropriate voltage injection from the active filter (AF). Finally, the sizing of the DC-link capacitor was uniquely determined by the max acceptable voltage ripple on the DC side.

Equation Definition

The following assumptions were made:

- The series connection of the capacitor C_s and the transformer inductance L_T determined a perfect series-resonance.
- An ideal DC current source.

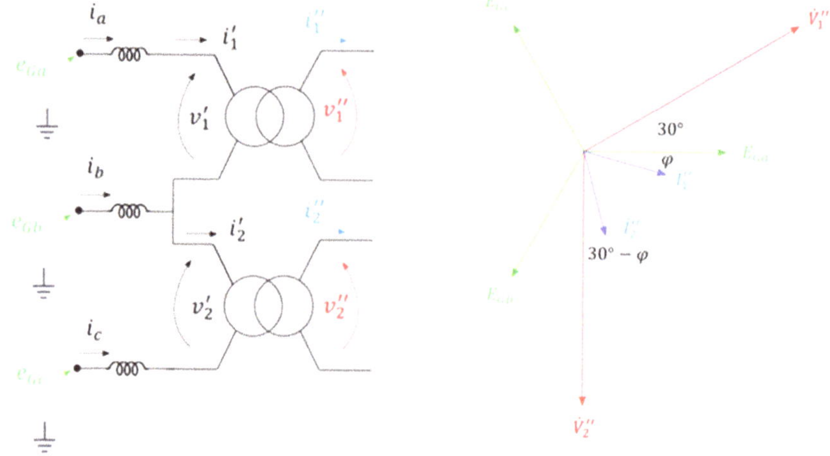

Figure 8. Open-delta phasor diagram.

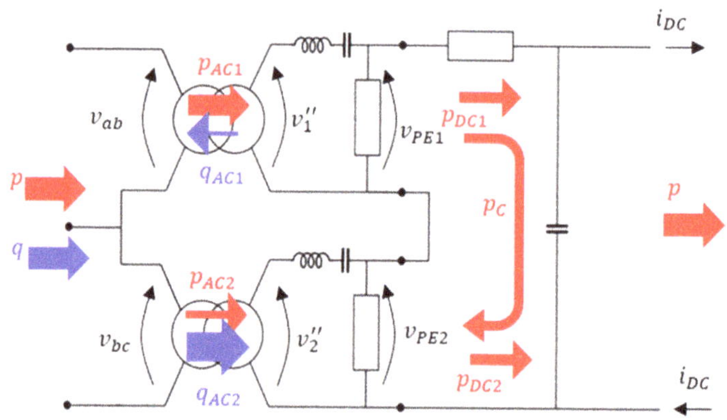

Figure 9. Series open-delta CLSC power flows.

With reference to Figure 7, the characteristic equations of the open-delta CLSC converter are the following:

$$\begin{cases} v_{ab} = R_v \sin(\omega t) \\ v_{bc} = R_v \sin(\omega t - 2\pi/3) \end{cases} \quad \begin{cases} i_a = \sqrt{2} I_{AC}^{pi} \sin(\omega t - \varphi - \pi/6) \\ i_c = \sqrt{2} I_{AC}^{pu} \sin(\omega t - \varphi - 3\pi/2) \end{cases} \tag{28}$$

where I_{AC}^{pu} is defined in (20), and φ in (21).

$$\begin{cases} v_{PE1} = v_{ab}/k_T + 1/2 \\ v_{PE2} = v_{bc}/k_T + 1/2 \end{cases} \quad \begin{cases} i''_1 = k_T i_a \\ i''_2 = -k_T i_c \end{cases} \tag{29}$$

where k_T is the transformation ratio. It could be easily verified that the voltage that had to be injected by the AF in order to be in an energy-balanced operating condition and that guaranteed the Pes' energy balance at the same time is:

$$v_{AF} = \left(R_v + r_{02pk}^{\max} \frac{q}{q_{\max}} \right) \sin(\omega t + 2\pi/3) \tag{30}$$

where r_{02pk}^{max} is the maximum zero to peak voltage ripple on the DC capacitor and $q_{\max}$ is the maximum reactive power in PU. Therefore, the DC side voltage can now be written as:

$$v_C = v_{PE1} + v_{PE2} + v_{AF} = 1 + r_{02pk}^{\max}\frac{q}{q_{\max}}\sin(\omega t + 2\pi/3) \tag{31}$$

It is interesting to note when using (31) that no voltage ripple appears when $q = 0$, which is also in accordance with what is stated by Equation (27). In other words, when the reactive power was zero, there was no power circulation between the Pes, and thus no voltage ripple on the DC side. Indeed, the current flowing through C_{DC} is:

$$i_C = r_{02pk}^{\max}\omega C_{DC}^{pu}\frac{q}{q_{\max}}\sin\left[\omega t + \frac{\pi}{6} + \frac{\pi}{2}(1 - sign(q))\right] \tag{32}$$

where:

$$C_{DC}^{pu} = \frac{2R_v q_{\max}}{3\omega r_{02pk}^{\max}} \tag{33}$$

being:

$$C_{DC} = C_{DC}^{pu}\frac{I_{DC,\max}}{V_{DC}} \tag{34}$$

Again, using (32), it is possible to observe that no current flows in the DC-link capacitor when $q = 0$. Figure 10 depicts the current and voltage phasors involved in the power circulation between the PEs.

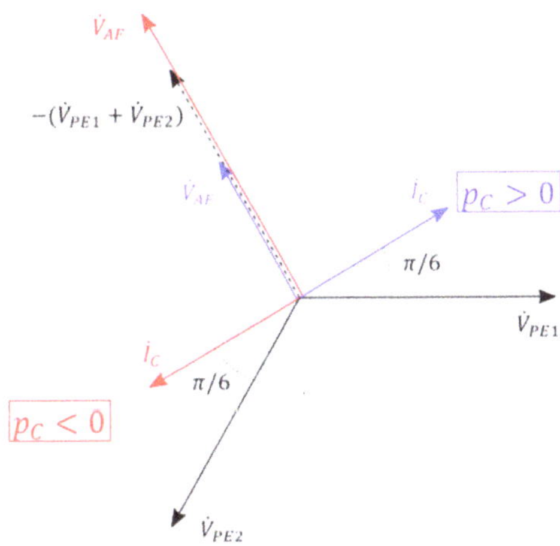

Figure 10. Open-delta CLSC PE and AF phasor diagram.

All the remaining equations can be found straightforwardly. The voltage on the series capacitors is defined in (35):

$$\begin{cases} v_{Cs1} = -V_{DC}/2 + \frac{1}{C}\int_0^t i_1''(\tau)d\tau \\ v_{Cs2} = -V_{DC}/2 + \frac{1}{C}\int_0^t i_2''(\tau)d\tau \end{cases} \tag{35}$$

The current flowing through the *PE*s and the *AF* is defined in (36):

$$\begin{cases} i_{PE1} = i_1'' - i_{DC} - i_C \\ i_{PE2} = i_2'' - i_{DC} - i_C \\ i_{AF} = i_{DC} + i_C \end{cases} \tag{36}$$

Finally, remembering that only half of the installed switches conducted the current at the same time, then the conduction current losses can be calculated using (37):

$$P_{L,cond}^{O\Delta-CLSC} = V_{eq}\left[N_{SW}^{PE}\overline{|i_{PE}|} + \frac{N_{SW}^{AF}}{2}\overline{|i_{AF}|}\right] + R_{eq}\left[N_{SW}^{PE}I_{PE}^{rms\ 2} + \frac{N_{SW}^{AF}}{2}I_{AF}^{rms\ 2}\right] \quad (37)$$

where N_{SW}^{PE} and N_{SW}^{AF} represent the number of PE and AF installed switches, respectively. The same switch chosen for the MMC was utilized here for the power-loss computation [29].

5. Sizing Results

The sizing results obtained following the outlined procedure are shown in this section as a function of R_v. Table 1 reports the numerical values of the main parameters: the extreme p and q values were chosen to be ± 1 and ± 0.3, respectively.

For calculation of the conduction power losses, the 5SNA 1800G330400 HiPak IGBT module was adopted; its datasheet can be found in [29].

Table 1. Parameter numerical values for both of the converters.

Name	Symbol	Value
DC voltage	V_{DC}	640 kV
Rated power	$P_{DC,max}$	1 GW
Max reactive power	Q_{max}	$\pm 0.3 P_{DC,max} = \pm 330$ MVar
Transformer impedance	$X_{cc\%}$	10%
IGBT collector–emitter forward voltage	V_{CE0}	1.3 V
Freewheeling diode forward voltage	V_{DE0}	1.3 V
IGBT on resistance	R_{CE0}	0.79 mΩ
Freewheeling diode forward voltage	R_{DE0}	0.46 mΩ

5.1. MMC

Figure 11a shows the voltage and the current waveforms for the upper arm connected to the phase *a* in one particular working condition. In the voltage waveform, it is possible to notice the third harmonic injection. Figure 11b, on the other hand, shows the maximum arm voltage and current as a function of the transformer secondary side voltage. It can be observed that the max arm voltage increased linearly with R_v, as this parameter was proportional to the secondary side voltage of the transformer; consequently, the maximum current was proportional to $1/R_v$.

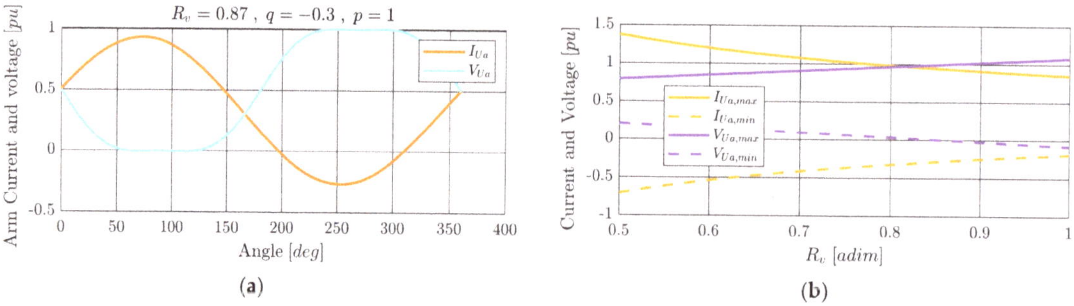

Figure 11. Upper *a* arm current and voltage waveforms at $p = 1$, $q = -0.3$, $R_v = 0.866$ (**a**); max and min arm voltage and current (**b**).

The total number of submodules and the total switch sizing power can be observed in Figure 12a,b, respectively. Again, it can be observed that the SM number increased linearly with R_v as expected, while the sizing power slowly decreased. Assuming that only half-bridge (HB) SMs were present, then the total switch number could be obtained by simply multiplying the SM number by 2.

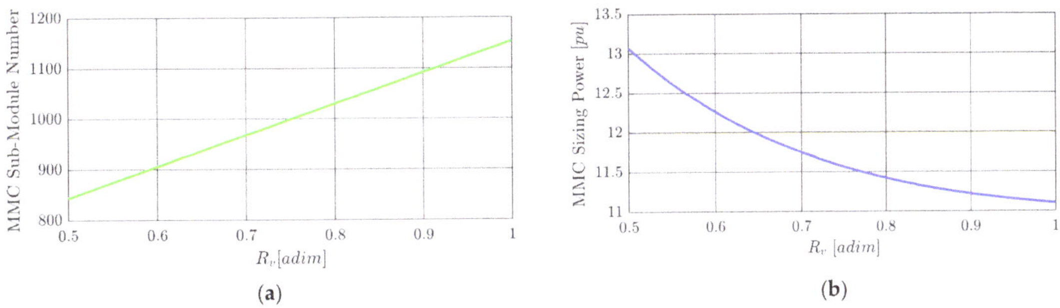

Figure 12. MMC submodule number (**a**); MMC total switch sizing power (**b**).

Figure 13a shows the dependence of the SM capacitance on R_v, while the energy stored in the MMC while also taking into account the arm inductors is shown in Figure 13b.

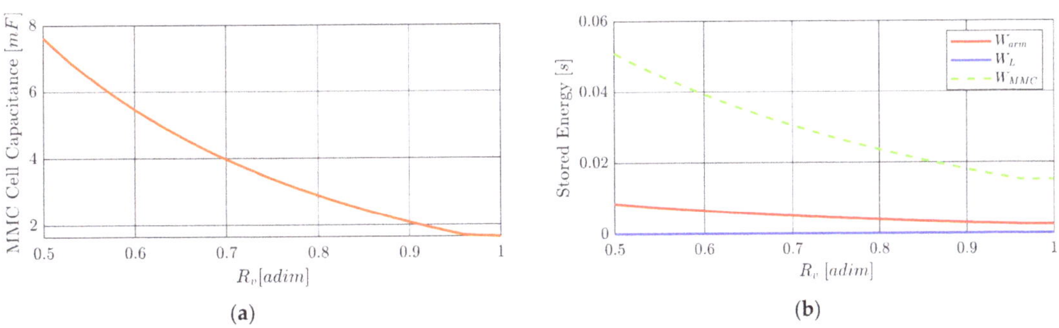

Figure 13. MMC submodule capacitor (**a**); MMC total stored energy (**b**).

It can be seen that for the MMC, there was no obvious optimum value for the R_v. Therefore, to minimize the losses of the converter, it was chosen to be the maximum possible value. For an HB MMC, the peak of the secondary phase to ground voltage had to be below Vdc/2 to ensure the controllability of the converter. Therefore, $R_{vN} \cong 0.866$ was chosen, where the subscript "N" means "nominal".

Once the nominal R_v was chosen ($R_{vN}^{MMC} = 0.866$ in this case), then all the KPIs could be computed. The KPI values are presented in the final section, where they are also compared to those of the open-delta CLSC converter.

5.2. Open-Delta CLSC

The sizing results obtained by following the outlined procedure are shown in this section as a function of R_v. The max zero to peak DC voltage ripple r_{02pk}^{max} was set to be equal to 0.1 PU. The current and voltage waveforms are shown in Figure 14 for a specific working condition. With reference to Figure 14, since $q \neq 0$, the sum $v_{PE1} + v_{PE2} + v_{AF}$ determined a voltage waveform characterized by a nonzero mean value and a certain ripple that can be seen in v_c. Therefore, the AF was responsible for the power exchange between PEs, as shown by Equations (30) and (31).

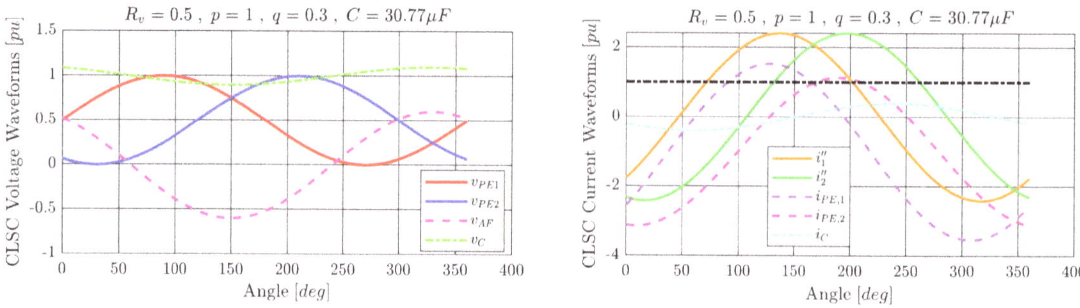

(**a**) (**b**)

Figure 14. Open-delta CLSC voltages (**a**); and currents (**b**) for $p = 1$, $q = 0.3$, and $R_v = 0.5$.

The maximum and minimum currents flowing through the converter stacks are shown in Figure 15. Again, it can be noticed that the max and min voltages increased linearly with R_v while the maximum and minimum currents were proportional to $1/R_v$. Additionally, a DC component was noticeable in the PE voltages, PE currents, and AF current.

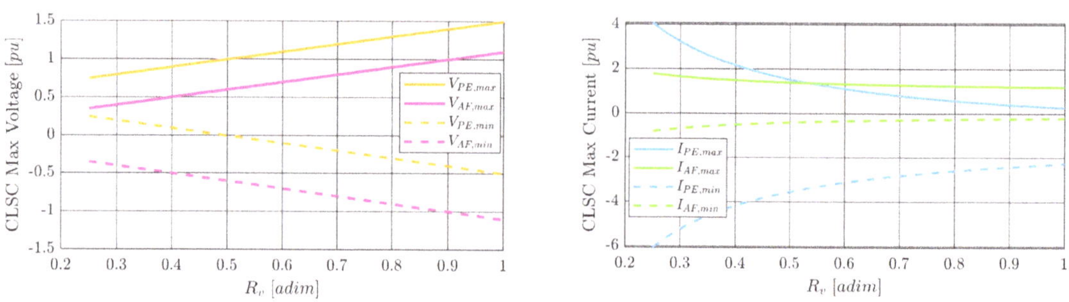

(**a**) (**b**)

Figure 15. Open-delta CLSC max and min voltages (**a**), and max and min currents (**b**).

The total SM number and the SM number divided by type and stack are shown in Figure 16.

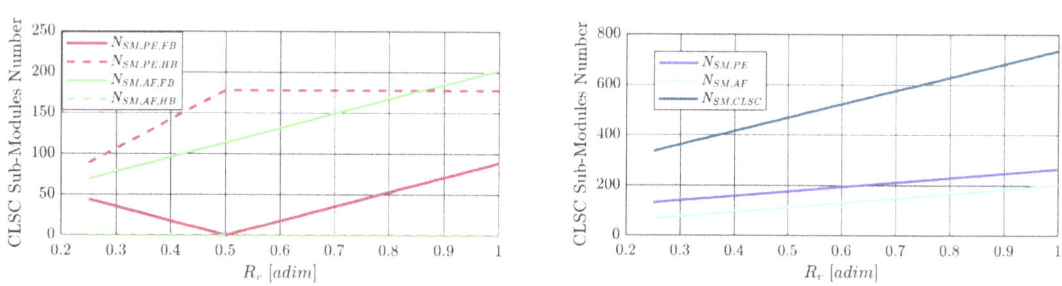

(**a**) (**b**)

Figure 16. Open-delta CLSC SM number divided by type (**a**), and divided by stack (**b**).

The switch number and the total switch sizing power are reported in Figure 17. Firstly, we immediately noticed the presence of a minimum point at $R_v = 0.5$. This was explained by the fact that, as R_v increased above 0.5, the number of FB SMs increased, resulting in an increase in the number of switches, and therefore in the sizing power at the same time. Secondly, the sizing power curve was quite "steep" as a function of R_v, and therefore was strongly dependent on the transformer secondary side voltage (which was not the case for the MMC, see Figure 12b). This underlined the importance of exploring the converter sizing as a function of the available degrees of freedom (R_v in this case).

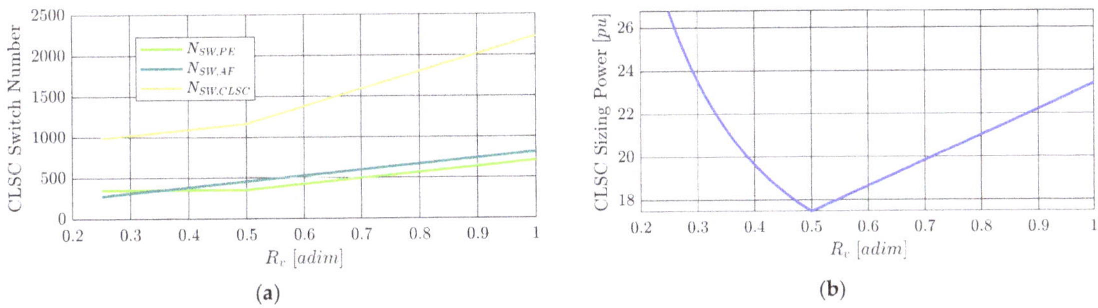

Figure 17. Open-delta CLSC switch number (a), and total switch sizing power (b).

The SM capacitance and the converter's stored energy are shown in Figure 18.

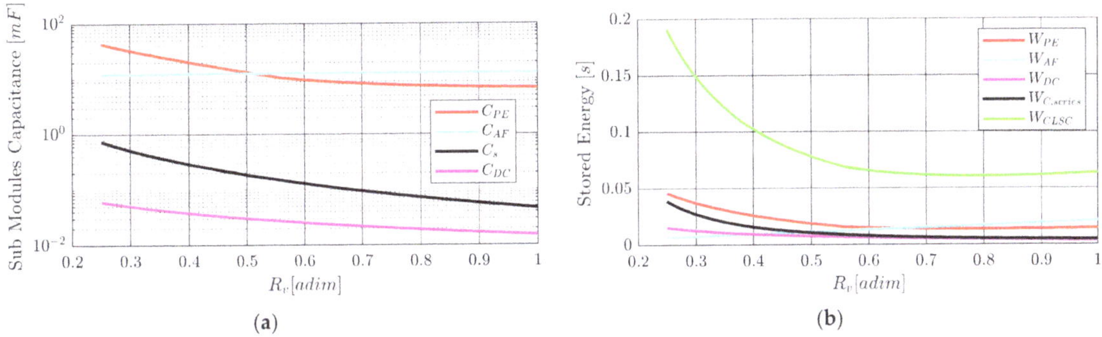

Figure 18. Open-delta CLSC SM capacitors (a), and stored energy (b).

6. KPI Comparison

This section presents a comparison between the KPIs of the two converters identified thanks to the procedure outlined above. Two cases were studied in order to show the importance of the required operating domain in the converter sizing. More specifically, the KPIs of the two converters derived from a rectangular PQ domain and the $\cos(\varphi) = 1$ line domain were compared.

In order to compute the KPIs, the degree of freedom R_v must be fixed. It was already pointed out in Section 5.1 that for the MMC, the choice was $R_{vN} = 0.866$; however, since the open-delta CLSC is a new topology, the choice is up to the designer. In Figures 16–18, it can be noticed that $R_{vN} = 0.5$ was a good operating point, not only because it corresponded to the minimum sizing power, but also because it was associated with a low SM and switch number, and it was close to the minimum stored energy point. The KPIs could then be obtained simply by selecting the values in the plots shown in Sections 5.1 and 5.2 that corresponded to the chosen R_{vN}.

Those KPIs were collected in the spider plot shown in Figure 19 in order to compare the two converters. For a better representation, each KPI belonging to the same type was normalized with respect to their maximum, so that the external perimeter was always equal to 1, and their difference was relative.

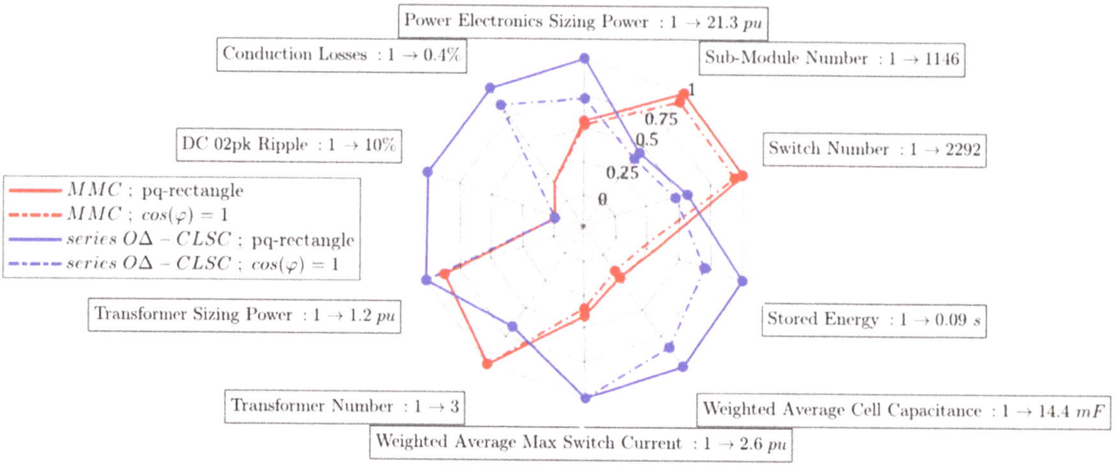

Figure 19. KPI spider plot. The notation "$1 \rightarrow x$" means that 1 in the spider plot corresponds to the value x.

It can be easily observed that although the MMC had a larger SM and switch number, it had a much lower sizing power, stored energy, and SM capacitance. This was mainly because the SMs in the open-delta CLSC were significantly larger than in the MMC. This can be seen in both the weighted average capacitance and the weighted average current. The necessity of creating these weighted average quantities resulted from the fact that the open-delta CLSC had two different types of SM stacks (in PEs and in AF) that were of different ratings, while the MMC was composed of six identical stacks. Therefore, in order to be able to compare the SM max current and capacitance with the MMC, it was useful to define the following quantities for the open-delta CLSC:

$$C_{SM}^{av} = \frac{N_{SM,PE}C_{PE} + N_{SM,AF}C_{AF}}{N_{SM} + N_{SM,AF}} \tag{38}$$

$$I_{\max}^{av} = \frac{N_{SM,PE}I_{PE,\max} + N_{SM,AF}I_{AF,\max}}{N_{SM} + N_{SM,AF}} \tag{39}$$

where C_{SM}^{av} and I_{SM}^{av} are the weigthed average capacitance and max switch current, respectively.

In Figure 19, it can be seen that the MMC was far less sensitive to the choice of the operating domain than the open-delta CLSC. As a matter of fact, the blue perimeter shrank significantly when passing from the rectangular domain to the $\cos(\varphi) = 1$ line, while the red polygon associated with the MMC remained almost unchanged. This fact underlined the importance of clearly defining the application's PQ domain. As an example, if a DC fault-blocking capability is required from the converter, the difference in terms of sizing parameters becomes smaller, especially the switch sizing power. The DC fault-blocking capability consists of the ability of the converter to block faults on the DC side by applying, thanks to the SM capacitors, a sufficiently high voltage of the opposite polarity [30]. The max voltage available depends on the number of FB SMs in the short-circuit current path; HB SMs are not able to counteract the fault, as they cannot change the voltage polarity at their terminals. While the open-delta CLSC had a sufficient number of FB SMs to block the max DC short-circuit current, at least 50% of the HB SMs had to be replaced by FB SMs in the MMC. This SM replacement increased the switch number and therefore the MMC switch sizing power.

7. Conclusions

A new methodology to assess the benefits of a modular converter was described in this paper. The methodology allows quick informed decisions to be made regarding whether research into a new topology should be pursued further in the early stages of development. The particular strengths of this methodology are that setting up dynamic simulations and development of the converter's control-system design are not required, resulting in significant time savings. A rapid analysis of all the possible work points inside a prescribed PQ operating domain allows assessment of the converter sizing and evaluation of critical KPIs. Unless a significant improvement in one of the KPIs is shown, there is very little probability that the new topology will find success in practical applications, and further research into it should be stopped.

To demonstrate an application of the proposed methodology, it was applied to two HVDC converter topologies in order to investigate the potential of a candidate challenger (the open-delta CLSC) as compared to a reference in the VSC HVDC domain (the half-bridge MMC). For each topology, the steady-state equations, the PQ operating domain, and the degrees of freedom were defined; and the converter sizing, together with the critical KPIs, were calculated. The KPIs extracted from the two converters showed their differences in terms of their "sizing performance", and highlighted their dependencies on the prescribed PQ operating domain (the MMC sizing was slightly affected by the PQ diagram, contrary to the open-delta CLSC). This not only enabled a quick quantification of the advantages related to the open-delta CLSC (reduced numbers for submodules and switches), but also the identification of its drawbacks (sizing power, switch current ratings, and stored energy). Overall, this case study indicated that the new open-delta topology did not represents a valid alternative to the classic MMC solution.

The procedure developed in this paper is valid for modular multilevel converters, as they produce voltage waveforms with a low harmonic content and can be accurately modeled analytically by ideal waveforms. Future works will investigate the generalization of the proposed procedure to converters that are not of the modular multilevel type.

Author Contributions: Conceptualization, D.L., F.M. and P.-B.S.; methodology, D.L., F.M., P.-B.S. and K.V.; software, D.L.; validation, D.L. and F.M.; formal analysis, D.L., F.M., P.-B.S. and K.V.; investigation, D.L., F.M., P.-B.S. and K.V.; resources, F.M. and P.-B.S.; writing—original draft prepa ration, D.L.; writing—review and editing, F.M., P.-B.S. and K.V.; visualization, F.M., P.-B.S. and K.V.; supervision, F.M., P.-B.S. and K.V. All authors have read and agreed to the published version of the manuscript.

Funding: This work was supported by a grant overseen by the French National Research Agency (ANR) as part of the "Investissements d'Avenir" Program ANE-ITE-002-01.

Institutional Review Board Statement: Not applicable.

Informed Consent Statement: Not applicable.

Data Availability Statement: The study did not report any data.

Conflicts of Interest: The authors declare no conflict of interest.

References

1. Bahrman, M.P.; Johnson, B.K. The ABCs of HVDC transmission technologies. *IEEE Power Energy Mag.* **2007**, *5*, 32–44. [CrossRef]
2. Bresesti, P.; Kling, W.L.; Hendriks, R.L.; Vailati, R. HVDC connection of offshore wind farms to the transmission system. *IEEE Trans. Energy Convers.* **2007**, *22*, 37–43. [CrossRef]
3. Qin, X.; Zeng, P.; Zhou, Q.; Dai, Q.; Chen, J. Study on the development and reliability of HVDC transmission systems in China. In Proceedings of the 2016 IEEE International Conference on Power System Technology (POWERCON), Wollongong, Australia, 28 September–1 October 2016; pp. 1–6.
4. Xie, H.; Bie, Z.; Li, G. Reliability-oriented networking planning for meshed VSC-HVDC grids. *IEEE Trans. Power Syst.* **2018**, *34*, 1342–1351. [CrossRef]
5. Orths, A.; Hiorns, A.; van Houtert, R.; Fisher, L.; Fourment, C. The European North seas countries' offshore grid initiative—The way forward. In Proceedings of the 2012 IEEE Power and Energy Society General Meeting, San Diego, CA, USA, 22–26 July 2012; pp. 1–8.
6. Gomez, D.; Páez, J.D.; Cheah-Mane, M.; Maneiro, J.; Dworakowski, P.; Gomis-Bellmunt, O.; Morel, F. Requirements for interconnection of HVDC links with DC-DC converters. In Proceedings of the IECON 2019-45th Annual Conference of the IEEE Industrial Electronics Society, Lisbon, Portugal, 14–17 October 2019; pp. 4854–4860.

7. Lesnicar, A.; Marquardt, R. An innovative modular multilevel converter topology suitable for a wide power range. In Proceedings of the 2003 IEEE Bologna Power Tech Conference Proceedings, Bologna, Italy, 23–26 June 2003; Volume 3, p. 6.
8. Merlin, M.; Green, T.; Mitcheson, P.D.; Trainer, D.R.; Critchley, R.; Crookes, W.; Hassan, F. The alternate arm converter: A new hybrid multilevel converter with DC-fault blocking capability. *IEEE Trans. Power Deliv.* **2013**, *29*, 310–317. [CrossRef]
9. Hao, Q.; Li, G.; Ooi, B. Approximate model and low-order harmonic reduction for high-voltage direct current tap based on series single-phase modular multilevel converter. *IET Gener. Transm. Distrib.* **2013**, *7*, 1046–1054. [CrossRef]
10. Hao, Q.; Ooi, B.-T.; Gao, F.; Wang, C.; Li, N. Three-phase series-connected modular multilevel converter for HVDC application. *IEEE Trans. Power Deliv.* **2015**, *31*, 50–58. [CrossRef]
11. Feldman, R.; Tomasini, M.; Amankwah, E.; Clare, J.; Wheeler, P.; Trainer, D.R.; Whitehouse, R.S. A hybrid modular multilevel voltage source converter for HVDC power transmission. *IEEE Trans. Ind. Appl.* **2013**, *49*, 1577–1588. [CrossRef]
12. Amankwah, E.; Costabeber, A.; Watson, A.; Trainer, D.; Jasim, O.; Chivite-Zabalza, J.; Clare, J. The series bridge converter (SBC): A hybrid modular multilevel converter for HVDC applications. In Proceedings of the 2016 18th European Conference on Power Electronics and Applications (EPE'16 ECCE Europe), Karlsruhe, Germany, 5–9 September 2016; pp. 1–9.
13. Patro, S.K.; Shukla, A.; Ghat, M.B. Hybrid series converter: A DC fault-tolerant HVDC converter with wide operating range. *IEEE J. Emerg. Sel. Top. Power Electron.* **2019**, *9*, 765–779. [CrossRef]
14. Adam, G.P.; Abdelsalam, I.A.; Ahmed, K.H.; Williams, B.W. Hybrid multilevel converter with cascaded H-bridge cells for HVDC applications: Operating principle and scalability. *IEEE Trans. Power Electron.* **2014**, *30*, 65–77. [CrossRef]
15. Kaya, M.; Costabeber, A.; Watson, A.J.; Tardelli, F.; Clare, J.C. A Push–Pull Series Connected Modular Multilevel Converter for HVdc Applications. *IEEE Trans. Power Electron.* **2021**, *37*, 3111–3129. [CrossRef]
16. HITACHI. Static Compensator (STATCOM). Available online: https://www.hitachienergy.com/offering/product-and-system/facts/statcom (accessed on 15 June 2021).
17. Schön, A.; Bakran, M.-M. Comparison of modular multilevel converter based HV DC-DC-converters. In Proceedings of the 2016 18th European Conference on Power Electronics and Applications (EPE'16 ECCE Europe), Karlsruhe, Germany, 5–9 September 2016; pp. 1–10.
18. Kung, S.H.; Kish, G.J. Multiport modular multilevel converter for DC systems. *IEEE Trans. Power Deliv.* **2018**, *34*, 73–83. [CrossRef]
19. Camurca, L.; Langwasser, M.; Zhu, R.; Liserre, M. Future MVDC Applications Using Modular Multilevel Converter. In Proceedings of the 2020 6th IEEE International Energy Conference (ENERGYCon), Gammarth, Tunisia, 28 September–1 October 2020; pp. 1024–1029.
20. Alvarez, J.P. DC-DC Converter for the Interconnection of HVDC Grids. Ph.D. Dissertation, Université Grenoble Alpes, Greenoble, France, 2019.
21. Steckler, P.-B. Convertisseur de Tension AC/DC Triphasé Comprenant Uniquement deux Modules de Conversion Electrique. EU Patent FR3112042A1, 28 April 2020.
22. Farr, E.M.; Trainer, D.R.; Idehen, O.E.; Vershinin, K. The series bridge converter (SBC): AC faults. *IEEE Trans. Power Electron.* **2019**, *35*, 4467–4471. [CrossRef]
23. ANGLE-DC. 2015 Electricity Network Innovation Competion. Available online: https://www.ofgem.gov.uk/2015 (accessed on 15 June 2021).
24. Evans, N.; Dworakowski, P.; Al-Kharaz, M.; Hegde, S.; Perez, E.; Morel, F. Cost-performance framework for the assessment of Modular Multilevel Converter in HVDC transmission applications. In Proceedings of the IECON 2019-45th Annual Conference of the IEEE Industrial Electronics Society, Lisbon, Portugal, 14–17 October 2019; pp. 4793–4798.
25. Merlin, M.; Green, T.; Mitcheson, P.; Moreno, F.; Dyke, K.; Trainer, D. Cell capacitor sizing in modular multilevel converters and hybrid topologies. In Proceedings of the 2014 16th European Conference on Power Electronics and Applications, Lappeenranta, Finland, 26–28 August 2014; pp. 1–10.
26. Li, R.; Fletcher, J.E.; Williams, B.W. Influence of third harmonic injection on modular multilevel converter-based high-voltage direct current transmission systems. *IET Gener. Transm. Distrib.* **2016**, *10*, 2764–2770. [CrossRef]
27. ABB Research Ltd Switzerland, ABB Research Ltd Sweden. AC/DC Converter with Series Connected Multicell Phase Modules Each in Parallel with a Series Connection of DC Blocking Capacitor and AC Terminals. EU Patent EP2569858B1, 1 May 2010.
28. Amankwah, E.; Costabeber, A.; Watson, A.; Trainer, D.; Jasim, O.; Chivite-Zabalza, J.; Clare, J. The series bridge converter (SBC): Design of a compact modular multilevel converter for grid applications. In Proceedings of the IECON 2016-42nd Annual Conference of the IEEE Industrial Electronics Society, Florence, Italy, 23–26 October 2016; pp. 2588–2593.
29. ABB. 5SNA 1800G330400 HiPak IGBT Module, Data Sheet, Doc. No. 5SYA 1471-00 11-2019. Available online: https://search.abb.com/library/Download.aspx?DocumentID=5SYA1471&LanguageCode=en&DocumentPartId=&Action=Launch (accessed on 6 November 2021).
30. Judge, P.D.; Chaffey, G.; Wang, M.; Dejene, F.Z.; Beerten, J.; Green, T.C.; Van Hertem, D.; Leterme, W. Power-system level classification of voltage-source HVDC converter stations based upon DC fault handling capabilities. *IET Renew. Power Gener.* **2019**, *13*, 2899–2912. [CrossRef]

World Electric Vehicle Journal

MDPI

Article

Study of the Effects of Current Imbalance in a Multiphase Buck Converter for Electric Vehicles

Iván Alfonso Reyes-Portillo [1,*], Abraham Claudio-Sanchéz [2], Jorge Alberto Morales-Saldaña [1], Jesús Darío Mina-Antonio [2], Edgardo Marvel Netzahuatl-Huerta [1], Luisana Claudio-Pachecano [3], Mario Ponce-Silva [2] and Ericka Reyes-Sánchez [1]

1 Engineering Electronic Department, Universidad Autonoma de San Luis Potosi—CIEP, San Luis Potosi 78290, Mexico; jmorales@uaslp.mx (J.A.M.-S.); A325084@alumnos.uaslp.mx (E.M.N.-H.); ericka.sanchez@uaslp.mx (E.S.-R.)
2 Engineering Electronic Department, Tecnologico Nacional de Mexico—CENIDET, Cuernavaca 62490, Mexico; abraham.cs@cenidet.tecnm.mx (A.C.-S.); jmina@cenidet.edu.mx (J.D.M.-A.); mario.ps@cenidet.tecnm.mx (M.P.-S.)
3 Engineering Mechanical Department, Tecnologico Nacional de Mexico—CENIDET, Cuernavaca 62490, Mexico; luisana.cp@cenidet.tecnm.mx
* Correspondence: A318057@alumnos.uaslp.mx; Tel.: +52-294-121-45-77

Abstract: The excessive use of fossil fuels has caused great concern due to modern environmental problems, particularly air pollution. The above situation demands that different areas of research aim at a sustainable movement to reduce CO_2 emissions caused by non-renewable organic fuels. A solution to this problem is the use of Electric Vehicles (EV) for mass transportation of people. However, these systems require high-power DC/DC converters capable of handling high current levels and should feature high efficiencies to charge their batteries. For this application, a single-stage converter is not viable for these applications due to the high current stress in a switch, the low power density, and its low efficiency due to higher switching losses. One solution to this problem is Multiphase Converters, which offer high efficiency, high power density, and low current ripple on the battery side. However, these characteristics are affected by the current imbalance in the phases. This paper is focused on the study of the effects of the current imbalance in a Multiphase Buck Converter, used as an intermediate cover between a power supply and the battery of an EV. Analyzing the efficiency and thermal stress parameters in different scenarios of current balance and current imbalance in each phase.

Keywords: battery charger; DC/DC; Multiphase Buck Converter; current balance

Citation: Reyes-Portillo, I.A.; Claudio-Sanchéz, A.; Morales-Saldaña, J.A.; Mina-Antonio, J.D.; Netzahuatl-Huerta, E.M.; Claudio-Pachecano, L.; Ponce-Silva, M.; Reyes-Sánchez, E. Study of the Effects of Current Imbalance in a Multiphase Buck Converter for Electric Vehicles. *World Electr. Veh. J.* **2022**, *13*, 88. https://doi.org/10.3390/wevj13050088

Academic Editor: Diego Bellan

Received: 25 February 2022
Accepted: 3 May 2022
Published: 12 May 2022

1. Introduction

Global warming, climate change, and the destruction of the ozone layer are factors that are forcing governments to look for feasible solutions to reduce the use of fossil fuels [1,2]. Automotive transportation is a major source of pollution, due to a large number of vehicles on the road every day around the world. For this reason, the progressive change from vehicles powered by fossil fuels to electric vehicles, as sustainable mobility, is essential for helping to solve the environmental problems mentioned above [3]. Although the majority of electrical energy is indeed generated by non-renewable energy sources, the trends are being directed towards the generation of energy with renewable energy sources (photovoltaic panels, wind generators, fuel cells) [4]. Electric vehicles are a solution that promises to reduce the damage to the environment, due to their sustainable characteristics described in [5,6]. Another aspect is the reduction of noise pollution because the electrical engine noise is almost imperceptible [7,8]. In this way, emissions and noise pollution levels are reduced in the places where these vehicles circulate, thus reducing respiratory diseases. Other advantages of the electric vehicle are related to the efficient use of the energy consumed by the electric motor, which is 90% [9]. Unlike the internal combustion engine whose energy efficiency percentage is between 20% and 25%, that is, only that percentage of thermal energy is transformed into mechanical energy [10]. The absence of a gearbox contributes to better acceleration response and excellent kinematic behavior. Moreover, electric vehicles implement regenerative braking systems recovering energy during this process, a characteristic that conventional vehicles do not have, which produce losses in the form of heat dissipated due to friction [11,12]. However, there are still many challenges to be addressed in these complex electrical systems [13]. A clear example is the charging process of electric vehicle batteries, whose voltage is 48V, and requires high current levels to achieve a fast and complete state of charge [14–17]. The characteristics of rechargeable batteries have always been critical in the electric vehicle development [18–21]. To carry out a fast and adequate process of battery charge, a high power density DC/DC converter is required [2,22,23]. The converters for EV´s battery charging applications must have high power density with low ripple of current and voltage, especially on the battery side. In addition, the converter has to meet basic industry requirements, such as high efficiency, low cost, and compact component size [24]. Several circuit designs for high power applications have been published [25,26]. Most of these designs require large coupled inductors and high voltage, high current devices [27]. The size of the components are large, which decreases the power density [28]. A multiphase converter is a good solution for high power, high current, and low ripple of voltage and current output applications [29]. The advantages of multiphase techniques are: reducing the stress on the devices, reducing the filter size, and decreasing the voltage and current ripple at the converter output [30,31]. However, most published papers require current sensors in the control loops of each phase to achieve balanced phase currents and improve the dynamic response of the converter [32–34]. The current imbalance depends mainly on the duty cycle, the inductance, and the parasitic resistances in each phase that integrate the converter [34,35]. The current imbalance is one of the most important problems for multiphase converters in practical applications. Several approaches have been proposed to solve this problem, including current sensor-based [35–37] and non-sensor-based methods [38,39]. In [38], the current imbalance is caused by variations in the parasitic resistances of each phase. Most papers are concerned with correcting the current balance in multiphase converters but do not focus on the study of the physical effects of this current imbalance on the converter. This paper presents a study of the effects of the current imbalance in the Multiphase Buck Converter (MBC). To analyze the disadvantages of the operating converter with current imbalance, an average modelwith parasitic resistances is proposed to obtain the parameters involved in the current imbalance phenomenon in each phase. In addition, the physical effects of the converter in current balance and imbalance are experimentally compared. Figure 1 shows the schematic of the four-phase MBC for electric vehicle battery charging. The EV battery charger considers an MBC, which has been designed with four phases, where (V_G) is the input voltage and V_o is the output voltage. The output current is formed by the four phases of the converter, making this structure ideal for fast battery charging. Finally, this paper is organized as follows: Section 2 describes the converter operation and shows the equations for sizing of the passive elements. In Section 3, the average model and the effects of parasitic resistors on the DC current balance are presented. Section 4 presents the effects of parasitic resistances in the converter. Section 5 presents the experimental results and, finally, Section 6 presents the conclusions and recommendations.

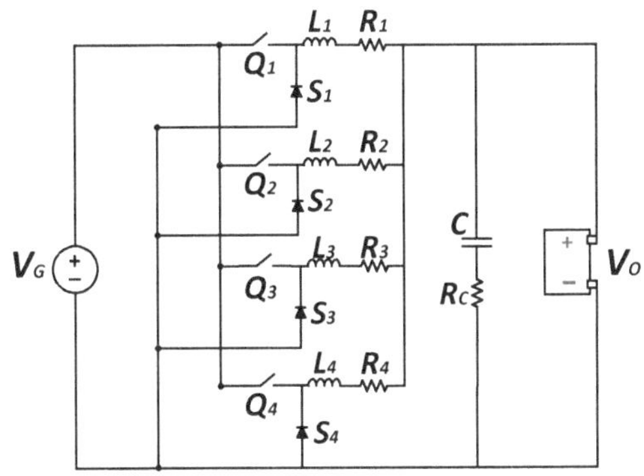

Figure 1. MBC consisting of four phases for battery charging in EVs.

2. Operation of the Multiphase Buck Converter

The converter presents two modes of conduction: Continuous Conduction Mode (CCM) where the inductor current never reaches zero, and Discontinuous Conduction Mode (DCM) where the inductor current is zero during an interval of time in this paper; only the CCM is analyzed. The MBC is proposed for EV battery charging applications operating with overlapping control signals. The equivalent circuits proposed for the "ON and OFF" operating modes are only valid for the following duty cycle operating range $0.25 < D < 0.5$. The CCM has different operating states during a switching period (T_s), four in "ON mode" which correspond to turning ON switches Q_1, Q_2, Q_3 and Q_4 with 90° delay between each phase, while the diodes remain ON except for the diode where switch Q is active. Figure 2 presents the operating states in "ON mode" considering the duty cycle signals (D) of each phase in a switching period.

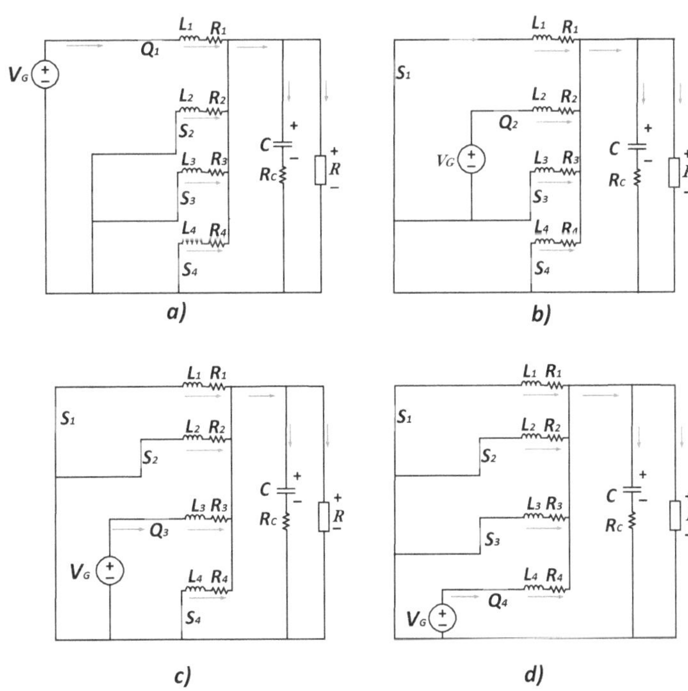

Figure 2. During the battery charging process, the equivalent circuits in "ON mode" for each phase elements, (**a**) D_1t_2, (**b**) D_2t_4, (**c**) D_3t_6, (**d**) D_4t_8.

During OFF mode, there are overlaps in the control signals, during a lapse of the switching period. The switch Q_1 overlaps with the next phase that is 90° away; this phenomenon occurs for each phase in turn. Phase four overlaps with the next switching period. With the overlap of the control signals, the source (V_G) is always present in the eight operating states of the converter. Figure 3 shows the four equivalent circuits of the OFF mode.

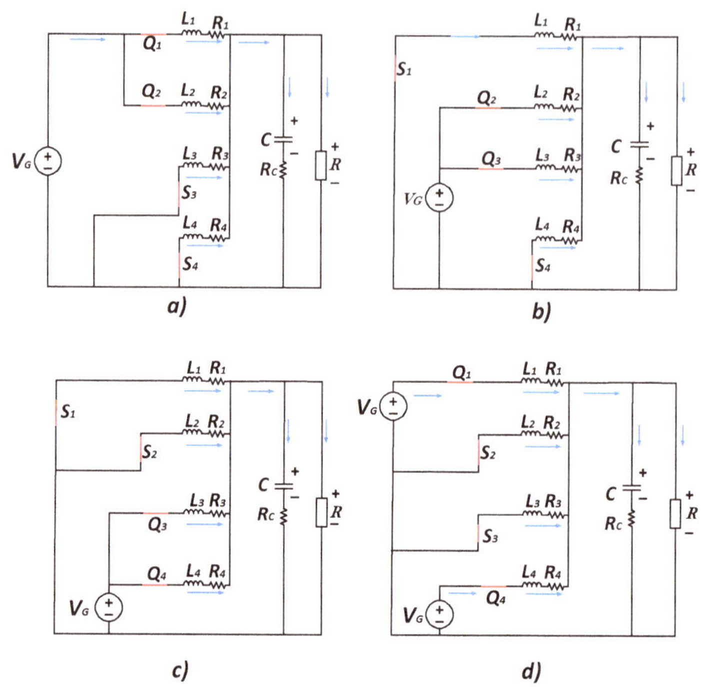

Figure 3. During the battery charging process, the equivalent circuits in "OFF mode" for each phase elements, (**a**) $D_1D_2t_3$; (**b**) $D_2D_3t_5$; (**c**) $D_3D_4t_7$; (**d**) $D_4D_1t_1$.

Figure 4 shows the waveforms, during one switching period, for the steady-state analysis of the MBC.

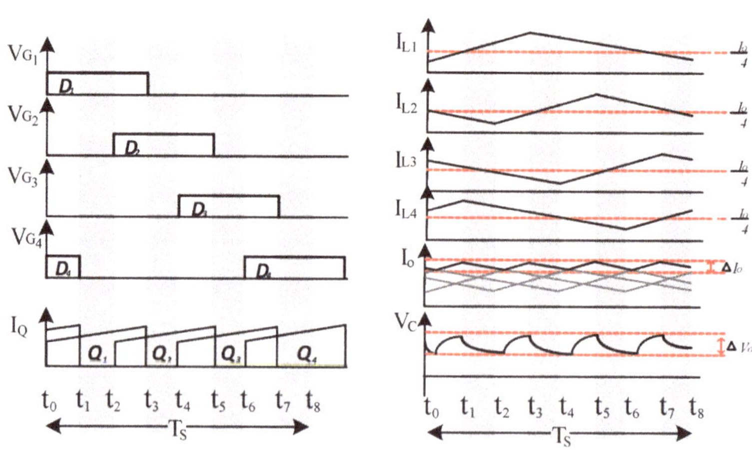

Figure 4. Waveforms during a commutation period.

Figure 4 shows that the inductor currents at the output of the MBC are interleaved, due to the effect of the phase difference between the control signals. The same improves the behavior of the converter by reducing the output current ripple and the converter output voltage. It also increases the frequency of the voltage and current at the output. It is important to note that reducing the current ripple size at the converter output also helps in decreasing the capacitance value at the output [40–42]. Another advantage of the MBC is that the pulsed current demanded from the source (V_G) is four times lower compared to a conventional buck converter, considering the same power requirements. The advantage of operating the converter with overlapping control signals is that the discontinuity of the input current demanded from the V_G source is eliminated. By analyzing the steady-state signals (Figure 4), the steady-state (1)–(6) of the MBC was determined:

$$I_{L_1} = \frac{D_1 V_G}{4R} \quad (1)$$

$$I_{L_2} = \frac{D_2 V_G}{4R} \quad (2)$$

$$I_{L_3} = \frac{D_3 V_G}{4R} \quad (3)$$

$$I_{L_4} = \frac{D_4 V_G}{4R} \quad (4)$$

$$V_C = D_1 V_G \quad (5)$$

To obtain an effective voltage transformation ratio for CCM, the volt–second balance of the voltage signal in one of the inductors is applied, considering that the duty cycle (D) is the same in each phase. The voltage transformation ratio, also known as the gain, is given by:

$$M = \frac{Vo}{V_G} = D \quad (6)$$

The following expressions establish the value of the inductance that the converter needs in each phase to operate in its steady state. It is important to consider the percentage of the current ripples (Δ_{I_L}) according to the load:

$$L_1 = \frac{V_o(1 - D_1)}{\Delta I_{L_1} f_s} \quad (7)$$

$$L_2 = \frac{V_o(1 - D_2)}{\Delta I_{L_2} f_s} \quad (8)$$

$$L_3 = \frac{V_o(1 - D_3)}{\Delta I_{L_3} f_s} \quad (9)$$

$$L_4 = \frac{V_o(1 - D_4)}{\Delta I_{L_4} f_s} \quad (10)$$

To determine the value of the capacitor, it is necessary to consider the converter output frequency ($f_{S_{out}}$) given by:

$$f_{S_{out}} = 4fs \quad (11)$$

The capacitor value is determined with the next expression:

$$C = \frac{V_o(1D_1)D_1}{8\Delta V_C L_1 f_{S_{out}}^2} \quad (12)$$

In (12), the output frequency is four times higher than the switching frequency (f_s) as a result of the ripple cancellation effect [30]. Consequently, the output capacitor reduces its capacitance value considerably.

3. Modeling of the Converter

The modeling of the converter was carried out through the state space technique, analyzing the different configurations defined by the operation modes presented in Figures 2 and 3. This modeling considers the parasitic resistances in the electronic elements. To simplify the analysis, the following considerations were made:

$$R_1 = R_{Q_1} + R_{L_1} + R_{track} \tag{13}$$

$$R_2 = R_{Q_2} + R_{L_2} + R_{track} \tag{14}$$

$$R_3 = R_{Q_3} + R_{L_3} + R_{track} \tag{15}$$

$$R_4 = R_{Q_4} + R_{L_4} + R_{track} \tag{16}$$

$$R_C = R_{ESR} + R_{track} \tag{17}$$

where $R_{Q_{1,2,3,4}}$ are the resistances $R_{DS_{on}}$ present in the switches, $R_{L_{1,2,3,4}}$ are the resistances present in the inductors and R_{track} is the resistance of the copper track, present in each phase. State variables are chosen as: $x_1 = I_{L_1}$, $x_2 = I_{L_2}$, $x_3 = I_{L_3}$, $x_4 = I_{L_4}$ and $x_5 = V_C$. The state variables correspond to the currents in the inductors and the voltage of the output capacitor. Using the technique of averaged states, it is possible to obtain a unified average model, given by:

$$\dot{x}_1 = -\frac{R_1 + R_C}{L_1}x_1 - \frac{R_C}{L_1}x_2 - \frac{R_C}{L_1}x_3 - \frac{R_C}{L_1}x_4 - \frac{1}{L_1}x_5 + \frac{D_1}{L_1}V_G + \frac{R_C}{L_1}I_o \tag{18}$$

$$\dot{x}_2 = -\frac{R_C}{L_2}x_1 - \frac{R_2 + R_C}{L_2}x_2 - \frac{R_C}{L_2}x_3 - \frac{R_C}{L_2}x_4 - \frac{1}{L_2}x_5 + \frac{D_2}{L_2}V_G + \frac{R_C}{L_2}I_o \tag{19}$$

$$\dot{x}_3 = -\frac{R_C}{L_3}x_1 - \frac{R_C}{L_2}x_2 - \frac{R_3 + R_C}{L_3}x_3 - \frac{R_C}{L_3}x_4 - \frac{1}{L_3}x_5 + \frac{D_3}{L_3}V_G + \frac{R_C}{L_3}I_o \tag{20}$$

$$\dot{x}_4 = -\frac{R_C}{L_4}x_1 - \frac{R_C}{L_3}x_2 - \frac{R_C}{L_4}x_3 - \frac{R_4 + R_C}{L_4}x_4 - \frac{1}{L_4}x_5 + \frac{D_4}{L_4}V_G + \frac{R_C}{L_4}I_o \tag{21}$$

$$\dot{x}_5 = -\frac{1}{C}x_1 + \frac{1}{C}x_2 + \frac{1}{C}x_3 + \frac{1}{C}x_4 - \frac{1}{C}I_o \tag{22}$$

and in the form $\dot{x} = Ax + Bu$, we obtain the average model matrix

$$\begin{bmatrix} \dot{x}_1 \\ \dot{x}_2 \\ \dot{x}_3 \\ \dot{x}_4 \\ \dot{x}_5 \end{bmatrix} = \begin{bmatrix} -\frac{R_1+R_C}{L_1} & -\frac{R_C}{L_1} & -\frac{R_C}{L_1} & -\frac{R_C}{L_1} & -\frac{1}{L_1} \\ -\frac{R_C}{L_2} & -\frac{R_2+R_C}{L_2} & -\frac{R_C}{L_2} & -\frac{R_C}{L_2} & -\frac{1}{L_2} \\ -\frac{R_C}{L_3} & -\frac{R_C}{L_3} & -\frac{R_3+R_C}{L_3} & -\frac{R_C}{L_3} & -\frac{1}{L_3} \\ -\frac{R_C}{L_4} & -\frac{R_C}{L_4} & -\frac{R_C}{L_4} & -\frac{R_4+R_C}{L_4} & -\frac{1}{L_4} \\ \frac{1}{C} & \frac{1}{C} & \frac{1}{C} & \frac{1}{C} & 0 \end{bmatrix} \begin{bmatrix} x_1 \\ x_2 \\ x_3 \\ x_4 \\ x_5 \end{bmatrix} + \begin{bmatrix} \frac{D_1}{L_1} & \frac{R_C}{L_1} \\ \frac{D_2}{L_2} & \frac{R_C}{L_2} \\ \frac{D_3}{L_3} & \frac{R_C}{L_3} \\ \frac{D_4}{L_4} & \frac{R_C}{L_4} \\ 0 & -\frac{1}{C} \end{bmatrix} \begin{bmatrix} V_G \\ I_o \end{bmatrix}$$

Solving for the DC values of the state variables, we employ expression (23)

$$X = -A^{-1}BU \tag{23}$$

where U is the input voltage (V_G) and the output current (I_o). The equations of the currents in each phase of the converter and the output voltage capacitor are given by:

$$I_{L_1} = \frac{V_G[R_3R_4(D_1D_2) + R_2R_4(D_1D_3) + R_2R_3(D_1D_4)] + I_o(R_2R_3R_4)}{R_1R_2R_3 + R_1R_2R_4 + R_1R_3R_4 + R_2R_3R_4} \tag{24}$$

$$I_{L_2} = \frac{V_G[R_3R_4(D_2D_1) + R_1R_4(D_2D_3) + R_1R_3(D_2D_4)] + I_o(R_1R_3R_4)}{R_1R_2R_3 + R_1R_2R_4 + R_1R_3R_4 + R_2R_3R_4} \tag{25}$$

$$I_{L_3} = \frac{V_G[R_2R_4(D_3D_1) + R_1R_4(D_3D_2) + R_1R_2(D_3D_4)] + I_o(R_1R_2R_4)}{R_1R_2R_3 + R_1R_2R_4 + R_1R_3R_4 + R_2R_3R_4} \tag{26}$$

$$I_{L_4} = \frac{V_G[R_2R_3(D_4 - D_1) + R_1R_4(D_4 - D_2) + R_1R_2(D_4 - D_3)] + I_o(R_1R_2R_3)}{R_1R_2R_3 + R_1R_2R_4 + R_1R_3R_4 + R_2R_3R_4} \quad (27)$$

$$V_C = \frac{V_G[D_1R_2R_3R_4 + D_2R_1R_3R_4 + D_3R_1R_2R_4 + D_4R_1R_2R_3] - I_o(R_1R_2R_3R_4)}{R_1R_2R_3 + R_1R_2R_4 + R_1R_3R_4 + R_2R_3R_4} \quad (28)$$

Equations (24)–(28) contain the parameters involved in the effect of current imbalance, describing that the MBC converter is susceptible to current imbalance due to factors such as small variations of the duty cycle (D) between each phase and variations of the values of the parasitic resistances present in the converter.

Current Balance Control Strategy

The paper presents a hysteresis controller that integrates the functions of voltage regulation, current balance, and the 90° phase shift of the control signals for the MBC. The implementation of these functions is done employing analog circuits taking into account simplicity, low cost, and functionality as main objectives. The controller architecture is shown in Figure 5 in the form of a block diagram. The scheme was implemented in a commercial controller [43]. The voltage regulation function only requires sensing of the output voltage V_o and, for the current balance function, it requires sensing of the voltage at the chopper output between the switches Q and S. It is not necessary to sense the current directly in each phase, which reduces the controller complexity and implementation cost. The input voltage (V_G) of the MBC is 120 V, the reference voltage (V_{ref}) is 3 V, and the output voltage (V_o) was considered to be 48 V. With these considerations, the current balance scheme was implemented in the commercial controller [43]. This paper focuses on the effects of current imbalance on the MBC converter and not on the controller performance.

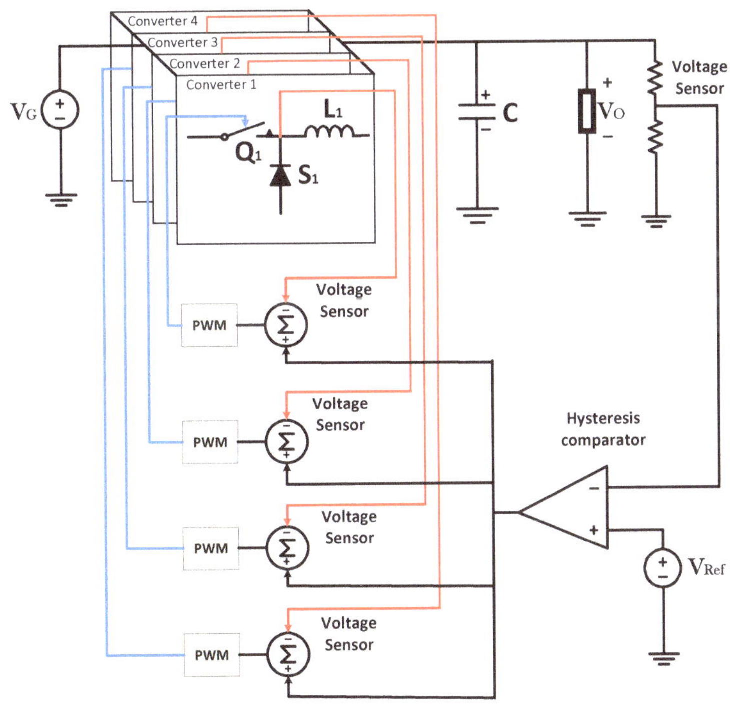

Figure 5. General control diagram for current balance by a phase based on [43].

4. Parasitic Resistance Effect

One of the main causes of current imbalance in the different phases is the variation of the parasitic resistances present in power semiconductor devices and other elements (L and C) [35,44]. This paper only considers the effect of parasitic resistance variations, as this parameter is more susceptible to variations due to external effects, such as temperature, copper track length, and component manufacturing. The typical variations in the values of parasitic resistances in inductors and capacitors reported by suppliers are ±2%. These variations increase in value, due to temperature effects as shown in Figure 6. The temperatures recommended by the manufacturers at which the switches should operate are between 20∼80 °C. Figure 6 shows the behavior of $R_{DS_{on}}$ concerning the temperature present in the switches, where the red frame indicates the temperatures recommended by the manufacturers and their interaction with the variation of the resistance $R_{DS_{on}}$.

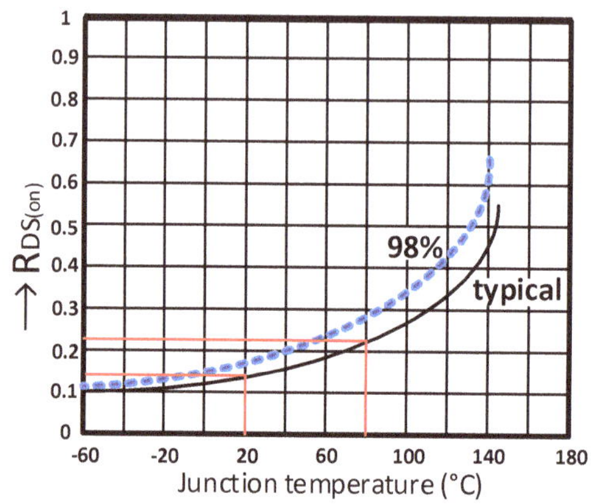

Figure 6. Variation of the $R_{DS_{ON}}$ with respect to temperature variation.

The typical variation of $R_{DS_{on}}$ at the same temperature, considering the fabrication linked to each particular switch is 2%. These variations increase in value when the power switches operate at different temperatures. This effect is observed in the curves described in Figure 6. The data of these curves are provided by the manufacturer of the switches. The variations of the parasitic resistances in the MBC directly affect the current balance [31]. This effect is described by expressions (24)–(27). To analyze the current imbalance due to the effect of parasitic elements, an MBC was designed considering an input voltage (V_G) of 120 V and an output current (I_o) of 40 A. Table 1 shows the list of parameters and devices used to perform the current imbalance analysis.

Table 1. Design parameters.

Symbol	Parameter	Value
V_G	Input Voltage	120 V
P_o	Output Power	1920 W
V_o	Output voltage	48 V
I_o	Output current	40 A
f_s	Frequency	103 kHz
L_1, L_2, L_3, L_4	Inductors	150 uH
C	Capacitor	2 uF
D	Duty cycle per phase	40%
Q_1, Q_2, Q_3, Q_4	MOSFETs	C20N60CFD
S_1, S_2, S_3, S_4	Diodes	TO-220-L
R_1, R_2, R_3, R_4	Parasitic resistance	130 mΩ
R_C	Capacitor ESR	30 mΩ

It should be noted that an output voltage of 48 V was selected based on current trends for battery voltage in electric vehicles presented in [14–17]. Current imbalance analysis was simulated in PSpice. To emulate these effects, parasitic resistance variations and an open-loop operation were considered to establish critical scenarios. Table 2 shows the values of the variations of the parasitic resistors due to the effect of temperature.

Table 2. Percentage of parasitic variations per phase.

Case	Phase 1-R_1 %	mΩ	Phase 2-R_2 %	mΩ	Phase 3-R_3 %	mΩ	Phase 4-R_4 %	mΩ
I	+10	143	+10	143	+10	143	0	130
II	+10	143	+10	143	0	130	0	130
III	+10	143	+10	143	0	130	−10	117
IV	0	130	+10	143	−10	117	−10	117

Figure 7 presents the current waveforms in the MBC, considering the different cases established in Table 2. In the MBC simulation, all duty cycles (D) of the same value were considered.

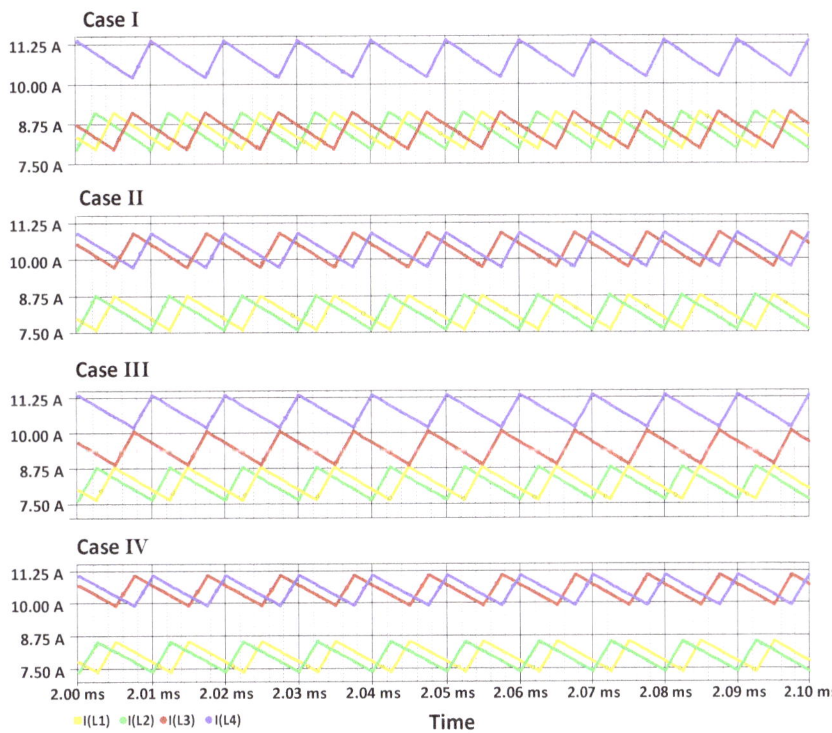

Figure 7. Four cases of current imbalance with respect to Table 2.

- Case I: In phase four, a current of 10.7 A flows, this is due to the lower value of parasitic resistance. Having the highest current in this phase, the switching losses and thermal stress will be higher, while the other phases maintain a balanced current flow with a value of 8.7 A.
- Case II: In phases three and four, the current flow is 10.7 A; this is because the parasitic resistors have less opposition to the passage of current in these phases, while in phases one and two, the current value is 8.2 A.

- Case III: In phase four, it demands a current of 10.7 A, phase three demands 9.5 A while phases one and two only demand 8.2 A.
- Case IV: In phases three and four, a current of 10.7 A is demanded, in phase one a current of 9.5 A is demanded, and, in phase two, the current is 8.2 A.

5. Experimental Analysis

A functional prototype of the MBC was built (Figure 8), to analyze the different current imbalance scenarios according to the variables established in Table 2. The current imbalance tests were carried out in an open-loop, where the converter operates in the scenarios used for the simulation.

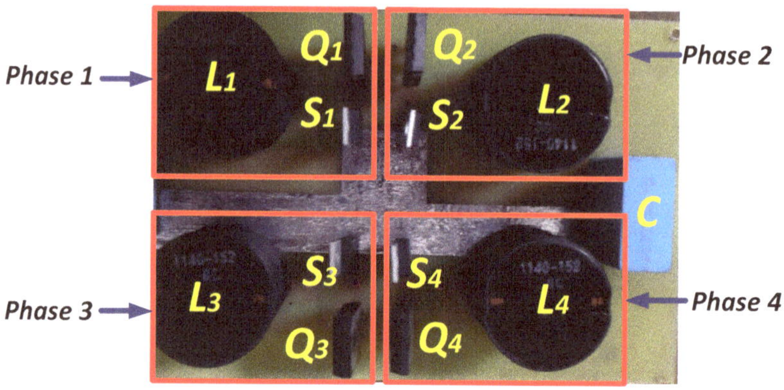

Figure 8. Functional prototype of the MBC.

Figure 9 shows the control signals used in the switches. The control signals have an offset of 90° between each phase, a frequency of 103 kHz, and a duty cycle (D) of 40%.

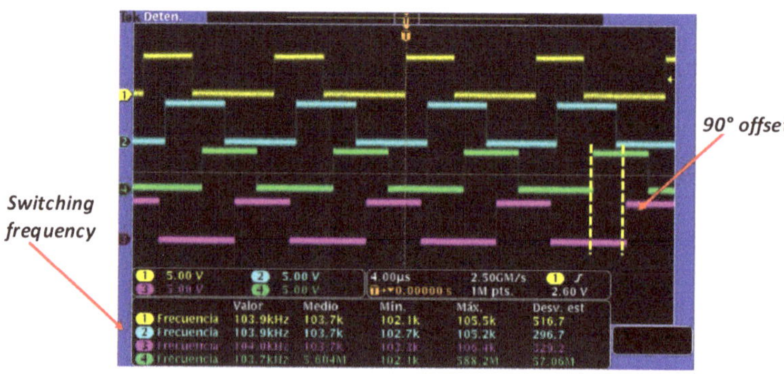

Figure 9. Commutation signals for phases with the corresponding 90° phase shift and duty cycle (40%).

The functional prototype was tested with different current demands at the output, maintaining a constant duty cycle. Table 3 shows the current values per phase in the MBC when operating in an open-loop and with current imbalance. Figure 10 presents the current imbalance cases described in Table 3. Due to the complexity of varying the parasitic resistances of the components, the external parameter R_{track} of the copper tracks was used to consider the different current imbalance scenarios.

Table 3. Experimental cases of current imbalance per phase.

Case	Phase 1 (A)	Phase 2 (A)	Phase 3 (A)	Phase 4 (A)
I	7.02	6.61	6.37	2.99
II	8.83	8.39	3.47	3.86
III	7.06	6.68	5.13	3.21
IV	13.1	3.78	2.58	8.29

The cases presented in Table 3 are described below:

- Case I: In Phase one, it transfers 7.02 A, being the phase with the highest current. Phases two and three present similar current values, with a value of 6.61 A and 6.37 A, while phase three is the one with the lowest current value transferred to the output. In this case, the output current (I_0) was 23 A.
- Case II: Phases one and two present more switching losses and high levels of thermal stress; this is due to the fact that they are the branches that transfer more current to the output, while phases three and four present values close to 3.6 A. In this case, the output current (I_0) was 25.55 A.
- Case III: Phase one transfers 7.02 A, in phase two, a current of 6.68 A circulates and phase three sends 5.13 A to the converter output, while phase four is where less current circulates. In this case, the output current (I_0) was 22.08 A.
- Case IV: This case presents the most critical case of current imbalance, phase one transfers 13.1 A to the output of the converter, phase four 8.29 A and phase two 3.78 A, while only 2.58 A circulates through phase three. In this case, the output current (I_0) was 27.75 A.

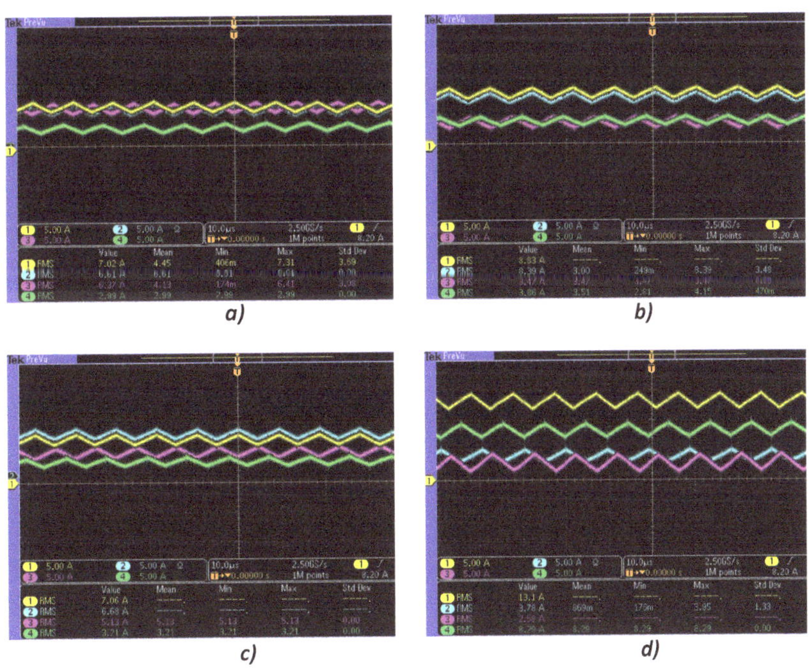

Figure 10. Current imbalance in the MBC prototype. (**a**) Case I, (**b**) Case II, (**c**) Case III, (**d**) Case IV.

The experimental tests of the previous cases were performed in a controlled environment. The initial ambient temperature for each test was set at 22 °C. In each test, it was corroborated that all semiconductor devices started at the aforementioned ambient temperature. To carry out the current imbalance tests, the semiconductor devices were oversized for the protection of the experimental prototype. In order to reduce the effects of current imbalance, it is of utmost importance to have a scheme that guarantees the current balance in each phase during the battery charging process in an electric vehicle.

Figure 11 shows the operation of the control scheme based on Figure 5 to balance the current flow per phase. It is observed that the system does not manage to equalize the currents in all phases; however, there is a great improvement with respect to the critical cases presented in Figure 10. Figure 11a shows the balanced currents, where phase one transfers 8.30 A, being the branch where the highest current flows, while, in phase three, 6.52 A circulates, being the phase that transfers less current to the output. The other two phases operate with current values close to 7.4 A. In this test, the value of the output current is 30 A.

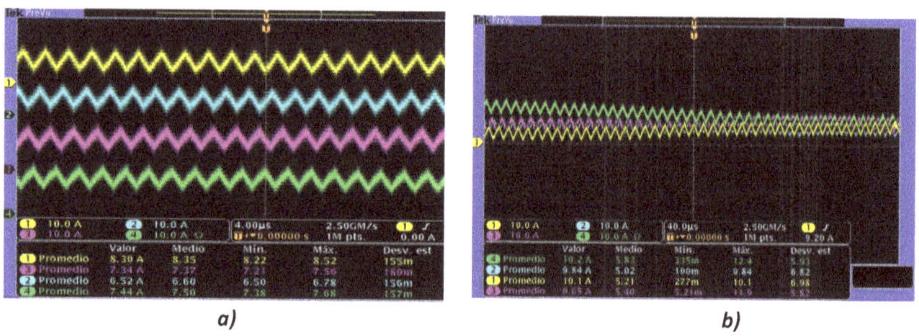

Figure 11. Current signals with the current balance scheme (**a**) balanced currents, (**b**) operation of the current balance scheme ($\approx 240\ \mu S$ stabilization response).

The theoretical balanced current in each phase for the case of Figure 11a is 7.5 A per phase. The current balance for phase one varies by +15%, phase two by −2%, phase three has a variation of −13%, and phase four has an imbalance of −1%. In Figure 11b, the converter was subjected to a current demand of $10A$ per phase, considering a theoretical balance. In phase one, the variation between the theoretical balance and the experimental balance was +2%, in phase two, the variation was −1.6%, in phase three, the variation was +1% and for phase four the variation was −3.5%. Figure 11b shows how the control scheme forces the currents to balance in each phase. This test was subjected to the nominal output current of 40 A. It is important to note that the study focuses on the physical effects suffered by the MBC converter during critical cases of current imbalance. However, to have comparative data of the results obtained with balanced currents scenarios, it was necessary to have a control scheme that offers these characteristics. However, the objective of this work is not to evaluate the performance of the control scheme.

Thermal Stress Analysis of the Converter

The current imbalance directly influences the temperatures of the switches. The phase in which the higher current flows suffers higher thermal stress, which damages the switching devices and causes more frequent system failures. Figure 12 shows the temperature spectrum of the switches in a current imbalance scenario. In this test, it is observed that the highest thermal stress is found in the phase two and phase four transistor.

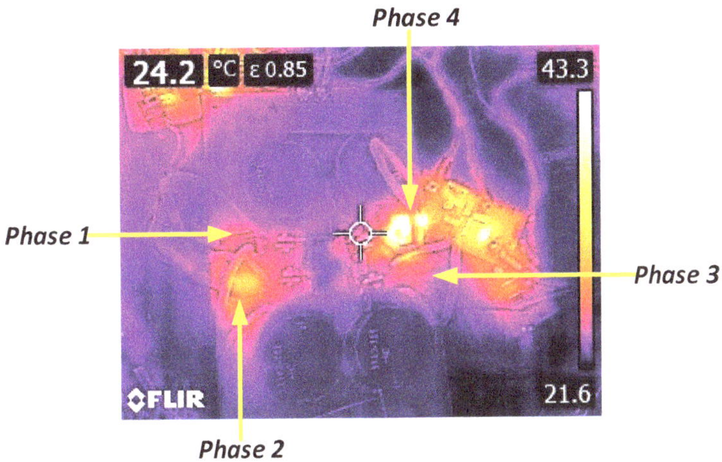

Figure 12. Temperature spectrum with imbalanced phases of the converter.

The current balance in the phases benefits from an equal distribution of thermal stress in the switches. Figure 13 presents the temperature spectrum during a test of the MBC operating with the current balance scheme.

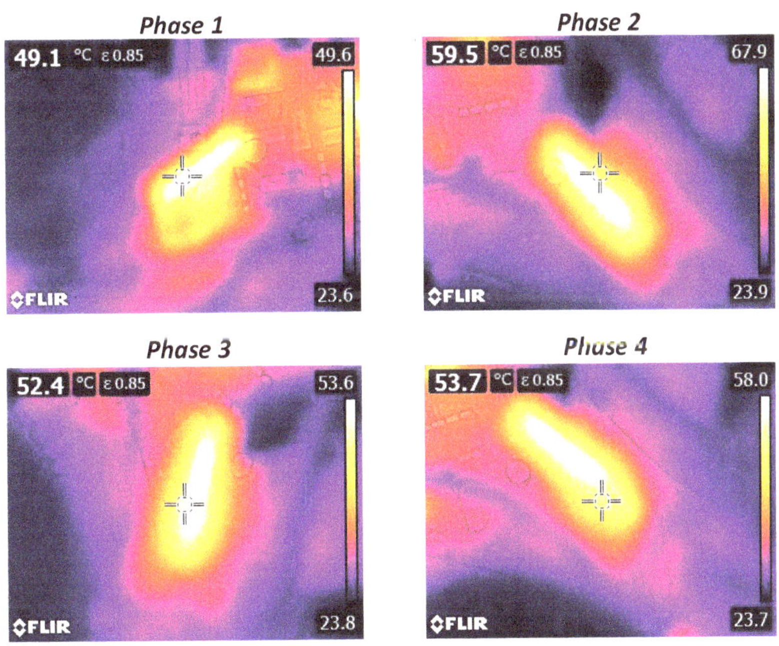

Figure 13. Temperature spectrum with balanced phases of the converter.

The temperature variations shown by the switches in Figure 13 are between 49.1 °C and 59.5 °C, being a suitable temperature to operate the manufacturer recommendations. Figure 14 shows the temperature curves for different power scenarios. The data obtained are with and without the current balance scheme.

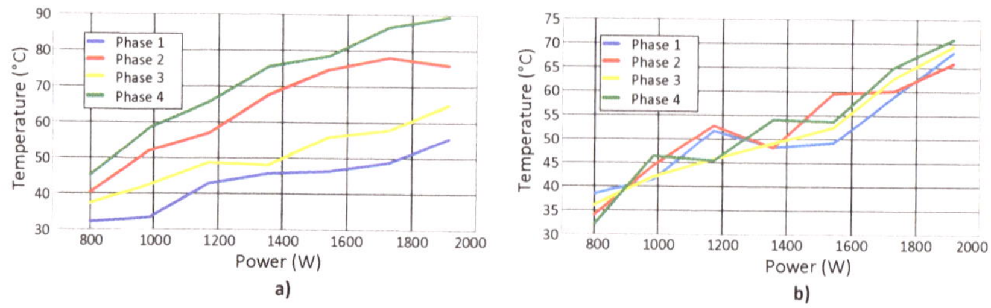

Figure 14. Behavior of temperature in switches (**a**) imbalanced currents, (**b**) balanced currents.

Figure 14a shows the behavior of the temperature of the switches during the current imbalance, after 60 s of operation in a controlled environment of 22 °C as the initial temperature. An indirect way to measure the current imbalance per phase is by employing the temperature of the semiconductors. On the other hand, Figure 14b shows the temperature behavior of the switches in each phase with the balanced current. The temperatures of the transistors in each phase present insignificant variations compared to the temperatures presented without the current compensation scheme per phase. The efficiency is directly affected by the balance of currents in each phase, due to thermal stress and higher dissipated power losses in switches. Figure 15 shows the comparison of the efficiency curves in the MBC with balanced and imbalanced currents. The nominal power of the converter is 1.9 kW. The converter achieved an efficiency of 91.87% with the current balance scheme operating at its nominal power, while, with imbalanced currents, the efficiency was 89.41%.

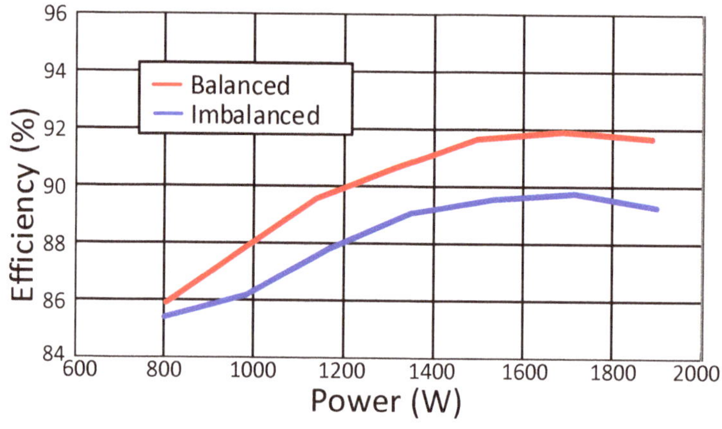

Figure 15. Converter efficiency as a function of nominal power.

6. Conclusions

This paper presented the study of the effects of current imbalance in the MBC during the battery charging process. It focused mainly on the comparative study of the physical phenomena suffered by the converter, such as thermal stress and efficiency. It was proved that the effect of current imbalance directly affects the favorable characteristics of the MBC, such as power density and efficiency. To corroborate the theoretical results, an experimental prototype of the converter with a nominal power of 1.9 kW was built and tests were carried out under different power scenarios. The behavior of the switches temperatures in different power scenarios was presented experimentally, considering the balanced and imbalanced currents. The efficiency curves of the converter with and without the current balance scheme were presented, showing that current imbalance has a negative influence on efficiency. The efficiency at nominal power of the MBC shows a difference of 2.46% between the balanced and imbalanced currents. This converter has interesting and favorable characteristics for battery charging in electric vehicles due to its high power density and simple dynamic characteristics, but it is important to consider a good current balancing scheme in order not to lose these favorable characteristics. Future work includes the study of the effects of current imbalance on the battery life cycle and the analysis of the different current balancing strategies considering aspects such as cost and performance in the MBC.

Author Contributions: Conceptualization, I.A.R.-P., A.C.-S.; validation, J.D.M.-A.; formal analysis; L.C.-P., E.R.-S.; investigation, J.A.M.-S., M.P.-S., E.M.N.-H. All authors have read and agreed to the published version of the manuscript.

Funding: This research received no external funding.

Institutional Review Board Statement: Not applicable.

Informed Consent Statement: Not applicable.

Data Availability Statement: Study did not report any data.

Acknowledgments: The authors would like to thank the Tecnologico Nacional de Mexico—CENIDET and the Universidad Autonoma de San Luis—CIEP for the material resources provided.

Conflicts of Interest: The authors declare no conflict of interest.

References

1. Ju, F.; Zhang, Q.; Deng,W.; Li, J. Review of structures and control of battery-supercapacitor hybrid energy storage system for electric vehicles. In Proceedings of the 2014 IEEE International Conference on Automation Science and Engineering (CASE), Taipei, Taiwan, 18–22 August 2014; pp. 143–148.
2. Liu, Z.; Du, J.; Yu, B. Design Method of Double-Boost DC/DC Converter with High Voltage Gain for Electric Vehicles. *World Electr. Veh. J.* **2020**, *11*, 64. [CrossRef]
3. Yujie, W.; Zhendong, S.; Zonghai, C. Energy management strategy for battery/supercapacitor/fuel cell hybrid source vehicles based on finite state machine. *Appl. Energy* **2019**, *254*, 113707.
4. Portillo, I.A.R.; Saldaña, J.A.M.; Huerta, E.M.N.; Hernández, E.R.P.; Elizondo, S.R.M.; Sánchez, E.R. Design of a DC Transformer-less Step-Down Regulator. In Proceedings of the 2021 IEEE International Autumn Meeting on Power, Electronics and Computing (ROPEC), Ixtapa, Mexico, 10–12 November 2021; pp. 1–6.
5. Chen, T.; Fu, P.; Chen, X.; Dou, S.; Huang, L.; He, S.; Wang, Z. An Optimal Control Algorithm with Reduced DC-Bus Current Fluctuation for Multiple Charging Modes of Electric Vehicle Charging Station. *World Electr. Veh. J.* **2021**, *12*, 107. [CrossRef]
6. Milad, B.; Jean-Philippe, M.; Gael, M.; Serge, P.; Mathieu, W.; Farid, M.; Majid, Z. Design and modeling of anequalizer for fuel cell energy management systems. *IEEE Trans. Power Electron.* **2019**, *34*, 10925–10935.
7. Le Comte, A.; Reynier, Y.; Vincens, C.; Leys, C.; Azaïs, P. First prototypes of hybrid potassium-ion capacitor (KIC): An innovative, cost-effective energy storage technology for transportation applications. *J. Power Sources* **2017**, *363*, 34–43. [CrossRef]
8. Jing, W.; Lai, C.H.; Wong, W.S.H.; Wong, M.L.D. A comprehensive study of battery-supercapacitor hybrid energy storage system for standalone PV power system in rural electrification. *Appl. Energy* **2018**, *224*, 340–356 [CrossRef]
9. Shirsath, R.B.; Shaikh, S.L. Power optimisation Scheme for Electric Vehicle using Induction Motor. In Proceedings of the 2021 International Conference on Smart Generation Computing, Communication and Networking (SMART GENCON), Pune, Maharashtra, India, 29–30 October 2021; pp. 1–9.
10. Feng, L.; Mears, L.; Schulte, J. Key variable analysis and identification on energy consumption of automotive manufacturing plant. In Proceedings of the 2016 IEEE Conference on Technologies for Sustainability (SusTech), Phoenix, AZ, USA, 9–11 October 2016; pp. 162–168.
11. Khan, M.A.; Zeb, K.; Sathishkumar, P.; Ali, M.U.; Uddin, W.; Hussain, S.; Ishfaq, M.; Khan, I.; Cho, H.G.; Kim, H.J. A novel supercapacitor/lithium-ion hybrid energy system with a fuzzy logic-controlled fast charging and intelligent energy management system. *Electronics* **2018**, *7*, 63. [CrossRef]
12. Trashorras Montecelos, J. Motor Eléctrico. In *Libro Vehículos Eléctricos*; Ediciones Paraninfo, SA: Madrid, Spain, 2019; pp. 14–19.
13. Strano, G.; Hao, L.; Evans, K.E.; Everson, R.M. Optimisation of quality and energy consumption for additive layer manufacturing processes. In 5th International Conference on Responsive Manufacturing-Green Manufacturing (ICRM 2010), Ningbo, China, 11–13 January 2010; pp. 364–369.
14. Chiang, G.T.; Shuji, T.; Takahide, S.; Hand, Y.; Fukada, M.; Kitamura, Y. Coupled Magnetic-based Integrated Isolated On-board Battery Charger and Boost Motor Drive Unit for Electric Vehicles. *IEEE Trans. Transp. Electrif.* **2021**, *8*, 135–148. [CrossRef]
15. Kushwaha, R.; Singh, B.; Khadkikar, V. An Isolated Bridgeless Cuk-SEPIC Converter Fed Electric Vehicle Charger. *IEEE Trans. Ind. Appl.* **2021**, *58*, 2512–2526. [CrossRef]
16. He, Y.; Kwak, K.H.; Kim, Y.; Jung, D.; Lee, J.; Ha, J. Real-time Torque-split Strategy for P0+P4 Mild Hybrid Vehicles with eAWD Capability. *IEEE Trans. Transp. Electrif.* **2021**, *8*, 1401–1413. [CrossRef]

17. Liu, Z.; Onori, S.; Ivanco, A. Synthesis and Experimental Validation of Battery Aging Test Profiles Based on Real-World Duty Cycles for 48-V Mild Hybrid Vehicles. *IEEE Trans. Veh. Technol.* **2021**, *66*, 8702–8709. [CrossRef]
18. García-Vite, P.M.; Rosas-Caro, J.C.; Martínez-Salazar, A.L.; de Jesus Chavez, J.; Valderrábano-González, A.; Sánchez-Huerta, V.M. Quadratic buck-boost converter with reduced input current ripple and wide conversion range. *IET Power Electron.* **2019**, *12*, 3977–3986. [CrossRef]
19. Martinez-Heredia, J.M.; Colodro, F.; Mora-Jiménez, J.L.; Remujo, A.; Soriano, J.; Esteban, S. Development of GaN Technology-Based DC/DC Converter for Hybrid UAV. *IEEE Access* **2020**, *8*, 88014–88025. [CrossRef]
20. Matsushita, Y.; Noguchi, T.; Taguchi, N.; Ishii, M. 2 kW Dual-Output Isolated DC/DC Converter Based on Current Doubler and Step-Down Chopper. *World Electr. Veh. J.* **2020**, *11*, 78. [CrossRef]
21. Dao, N.D.; Lee, D.; Phan, Q.D. High-Effciency SiC-Based Isolated Three-Port DC/DC Converters for Hybrid Charging Stations. *IEEE Trans. Power Electron.* **2020**, *35*, 10455–10465. [CrossRef]
22. Al Attar, H.; Hamida, M.A.; Ghanes, M.; Taleb, M. LLC DC-DC Converter Performances Improvement for Bidirectional Electric Vehicle Charger Application. *World Electr. Veh. J.* **2022**, *13*, 2. [CrossRef]
23. Feng, L.; Ulutan, D.; Mears, L. Energy consumption modeling and analyses in automotive manufacturing final assembly process. In Proceedings of the 2015 IEEE Conference on Technologies for Sustainability (SusTech), Ogden, UT, USA, 30 July–1 August 2015; pp. 224–228.
24. Alexandru, I.C.; Mircea, B. Analysis and design of a current mode buck converter with digitally controlled output voltage. In Proceedings of the 2019 International Semiconductor Conference (CAS), Sinaia, Romania, 9–11 October 2019; pp. 309–312.
25. Zhu, Y.; Ye, Z.; Ge, T.; Abramson, R.; Pilawa-Podgurski, R.C.N. A Multi-Phase Cascaded Series-Parallel (CaSP) Hybrid Converter for Direct 48 V to Point-of-Load Applications. In Proceedings of the 2021 IEEE Energy Conversion Congress and Exposition (ECCE), Vancouver, BC, Canada, 10–14 October 2021; pp. 1973–1980.
26. Qian, W.; Peng, F.Z.; Tolbert, L.M. Development of a 55 kW 3X dc-dc converter for HEV systems. In Proceedings of the 2009 IEEE Vehicle Power and Propulsion Conference, Dearborn, MI, USA, 7–10 September 2009; pp. 433–439.
27. Kwan, A.K.; Younes, M.; Darraji, R.; Ghannouchi, F.M. On Track for Efficiency: Concurrent Multiband Envelope-Tracking Power Amplifiers. *IEEE Microw. Mag.* **2016**, *17*, 46–59. [CrossRef]
28. Aggeler, D.; Biela, J.; Kolar, J.W. A compact, high voltage 25 kW, 50 kHz dc-dc converter based on SiC JFETs. In Proceedings of the 2008 23rd Annual IEEE Applied Power Electronics Conference and Exposition, (APEC), Austin, TX, USA, 24–28 February 2008; pp. 801–807.
29. Ni, L.; Patterson, D.J.; Hudgins, J.L. High Power Current Sensorless Bidirectional 16-Phase Interleaved DC-DC Converter for Hybrid Vehicle Application. *IEEE Trans. Power Electron.* **2011**, *27*, 1141–1151. [CrossRef]
30. Reyes, I.; Claudio, A.; Flores, E.; Lopez, M. Analysis of kinetic energy recovery system based on inertial flywheel. In Proceedings of the 2018 IEEE 14th International Conference on Power Electronics (CIEP), Cholula, Mexico, 24–26 October 2018; pp. 130–136.
31. Gordillo, J.; Aguilar, C. A Simple Sensorless Current Sharing Technique for Multiphase DC–DC Buck Converters. *IEEE Trans. Power Electron.* **2017**, *32*, 3480–3489. [CrossRef]
32. Miaja, P.F.; Rodríguez, A.; Sebastián, J. Buck-Derived Converters Based on Gallium Nitride Devices for Envelope Tracking Applications. *IEEE Trans. Power Electron.* **2015**, *30*, 2084–2095. [CrossRef]
33. Vasić, M.; Cheng, P.; García, O.; Oliver, J.A.; Alou, P.; Cobos, J.A.; Tena, D.; Ortega-González, F.J. The Design of a Multilevel Envelope Tracking Amplifier Based on a Multiphase Buck Converter. *IEEE Trans. Power Electron.* **2016**, *31*, 4611–4627. [CrossRef]
34. Choi, H.-W.; Kim, S.-M.; Kim, J.; Cho, Y.; Lee, K.-B. Current-balancing strategy for multileg interleaved DC/DC converters of electric-vehicle chargers. *J. Power Electron.* **2021**, *21*, 94–102. [CrossRef]
35. Han, J.; Song, J. Phase Current-Balance Control Using DC-Link Current Sensor for Multiphase Converters With Discontinuous Current Mode Considered. *IEEE Trans. Ind. Electron.* **2016**, *63*, 4020–4030. [CrossRef]
36. Chen, H.; Lu, C.; Rout, U.S. Decoupled Master-Slave Current Balancing Control for Three-Phase Interleaved Boost Converters. *IEEE Trans. Power Electron.* **2018**, *33*, 3683–3687. [CrossRef]
37. Kim, H.; Falahi, M.; Jahns, T.M.; Degner, M.W. Inductor Current Measurement and Regulation Using a Single DC Link Current Sensor for Interleaved DC–DC Converters. *IEEE Trans. Power Electron.* **2011**, *26*, 1503–1510. [CrossRef]
38. Foley, R.F.; Kavanagh, R.C.; Egan, M.G. Sensorless Current Estimation and Sharing in Multiphase Buck Converters. *IEEE Trans. Power Electron.* **2012**, *27*, 2936–2946. [CrossRef]
39. Baek, J.; Elasser, Y.; Radhakrishnan, K.; Gan, H.; Douglas, J.P.; Krishnamurthy, H.K.; Li, X.; Jiang, S.; Sullivan, C.R.; Chen, M. Vertical Stacked LEGO-PoL CPU Voltage Regulator. *IEEE Trans. Power Electron.* **2022**, *37*, 6305–6322. [CrossRef]
40. Kumar, S.S.; Panda, A.K.; Ramesh, T. A ZVT-PWM multiphase synchronous buck converter with an active auxiliary circuit for portable applications. In Proceedings of the 2014 IEEE 6th India International Conference on Power Electronics (IICPE), Kurukshetra, India, 8–10 December 2014; pp. 1–6.
41. Valdez-Reséndiz, J.E.; Claudio-Sánchez, A.; Guerrero-Ramírez, G.V.; Tapia-Hernández, A.; Juárez, A.N.H.; Núñez, A.R.L. Voltage balancing in an interleaved high gain boost converter. In Proceedings of the 2014 IEEE Energy Conversion Congress and Exposition (ECCE), Pittsburgh, PA, USA, 14–18 September 2014; pp. 988–992.
42. Costabeber, A.; Mattavelli, P.; Saggini, S. Digital Time- Optimal Phase Shedding in Multiphase Buck Converters. *IEEE Trans. Power Electron.* **2010**, *25*, 2242–2247. [CrossRef]

43. ISL658 Datasheet. Multi-Purpose Precision Multi-Phase PWM Controller With Optional Active Voltage Positioning. 2015. Available online: https://www.renesas.com/us/en/document/dst/isl6558-datasheet (accessed on 11 March 2022).
44. Hu, K.; Chen, Y.; Tsai, C. A Digital Multiphase Converter with Sensor-less Current and Thermal Balance Mechanism. In Proceedings of the 2018 IEEE Asian Solid-State Circuits Conference (A-SSCC), Tainan, Taiwan, 5–7 November 2018; pp. 175–178.

energies

Article

Pulse-Amplitude-Modulation Full-Bridge Diode-Clamped Multilevel *LLC* Resonant Converter Using Multi-Neighboring Reference Vector Discontinuous PWM

Min-Sup Song [1] and Jae-Bum Lee [2,*]

[1] Railroad Safety Division, Korea Railroad Research Institute, Uiwang-si 16105, Korea; mssong@krri.re.kr
[2] Department of Railroad Electrical and Electronics Engineering, Korea National University of Transportation, Uiwang-si 16106, Korea
* Correspondence: leejb@ut.ac.kr; Tel.: +82-31-460-0573

Abstract: A full-bridge diode-clamped multilevel *LLC* resonant converter suitable for power conversion systems that use high input voltage, such as railway vehicles, is proposed in this paper. In order to eliminate the voltage deviations of the capacitors connected in series to the high voltage input DC link, a novel modulation strategy referred to as multi-neighboring reference vector discontinuous pulse-width modulation (MNRV DPWM) is proposed. Unlike the existing two-level resonant converter that varies the operating frequency to hold the output voltage constant, the proposed multilevel resonant converter modulates the amplitude of the fundamental wave input to a resonance tank while fixing the operating frequency at the resonance point. Therefore, the design of passive elements becomes easier, and stable operation is possible over a wide operating range with only one power conversion stage. In this paper, the control algorithm and operation characteristics of the newly proposed full-bridge diode-clamped four-level *LLC* resonant converter are analyzed in detail and design guidelines are presented. The feasibility of the proposed converter is verified through a simulation and an experiment with a prototype converter.

Keywords: amplitude modulating; full-bridge; diode-clamped converter; multilevel; *LLC*; DPWM

Citation: Song, M.-S.; Lee, J.-B. Pulse-Amplitude-Modulation Full-Bridge Diode-Clamped Multilevel *LLC* Resonant Converter Using Multi-Neighboring Reference Vector Discontinuous PWM. *Energies* **2022**, *15*, 4045. https://doi.org/10.3390/en15114045

Academic Editors: Jelena Loncarski and Diego Bellan

Received: 15 May 2022
Accepted: 30 May 2022
Published: 31 May 2022

1. Introduction

Due to excellent soft switching characteristics such as primary-side zero-voltage switching (ZVS) and secondary-side zero-current switching (ZCS), *LLC* resonant converters are widely used in many areas such as electric vehicles, servers, uninterruptible power supplies, and TV power adapters. Furthermore, with regard to an *LLC* resonant converter operating at a high switching frequency, the power density can be greatly improved due to the reduction in the size of the passive element [1,2]. However, because the output voltage is controlled through frequency variations, a wide range of frequency sweeps is required in a wide input/load variation system, making it difficult to design optimal passive devices [3]. In addition, because most resonant converters developed thus far are based on a frequency-controlled two-level topology, they are not widely used in high-power systems with very high input voltages, such as railroad vehicles, due to the limited endurance voltage of the switching devices. Currently, as part of the effort to replace the existing low-frequency transformers, resonance topology-based power converters are widely applied to the auxiliary power systems of railway vehicles, with most being composed of two stages. One type is an *LLC* resonant converter controlled in an open loop with a fixed operating frequency, and the other is a constant-voltage regulator connected before or after the *LLC* converter, which also serves to provide proper voltage for the normal operation of the *LLC* resonant converter. This complicates the overall system configuration [4,5].

Meanwhile, multilevel converters have been studied extensively in an effort to overcome the limitations of the two-level topology [6–10]. The multilevel converter, as exemplified by the diode-clamped method, the flying capacitor method, and the cascaded H-bridge, can be applied to a high-voltage, high-power system with devices with low withstand voltages. It also has the advantage of a low harmonic content and low dv/dt generation. Among these three, in a system that receives power from a single external power supply line such as a railroad vehicle, the diode-clamped topology with a simple power stage configuration is suitable. However, with diode-clamped topologies, it is not easy to balance the DC link capacitor voltages to ensure equal voltage sharing and good performance, especially for level four and higher. Over the past three decades, numerous researchers have attempted to find appropriate modulation strategies to solve the voltage balance problem in a three-phase diode-clamped multilevel converter [11–14]. The addition of an external circuit to compensate for voltage deviations and a hybrid method that utilizes a diode-clamped circuit and a flying capacitor together both complicate the system and are not cost effective [15–17]. Recently, the voltage imbalance problem of a three-phase diode-clamped DC/AC inverter with three or more levels was resolved using a virtual vector scheme [18–20]. The voltage imbalance was eliminated by combining reference vectors whose associated current sum is zero with the aid of a virtual reference vector. However, in a three-phase system, the amount of calculation is increased because the phase angle as well as the magnitude of the reference voltage must be considered. In addition, it is not easy to find an optimal switching pattern due to the considerable redundancy of the switching pairs. The increased switching loss compared to the general space vector pulse-width modulation (SVPWM) scheme is another issue. Recently, carrier-overlapped PWM (COPWM) applicable to diode-clamped multilevel inverters was proposed [21]. With multiple carrier-based modulation, it can control capacitor voltage deviation while maintaining volt-second balance in a diode-clamped multilevel inverter with four or more levels. However, this method is effective under ideal conditions and requires a little complex closed-loop control under non-ideal conditions and increases switching losses compared to the conventional single-carrier method [22]. As previously discussed, most of the multilevel topologies known to date have been applied to three-phase inverter systems. Therefore, in order to apply the diode-clamped multilevel topology to the single-phase DC/DC converter, a new switching modulation technique is required that can effectively eliminate voltage deviations in series-connected DC link capacitors while simultaneously controlling the output voltage simply in a closed-loop control manner.

A multi-phase, multi-level *LLC* converter with a two-level topology-based modular structure has been proposed for high-voltage and high-power systems [23]. Because it is based on the existing frequency modulation method, it is difficult to optimally design passive components under wide input/load variation conditions. Additionally, a separate resonance tank is required for each sub-module, increasing the required number of passive elements. A suitable solution for voltage deviation between series-connected DC link capacitors has not been proposed. Input-series-output-parallel (ISOP) topology based on a two-level DAB converter can handle high voltage and high power [24]. However, the ISOP topology requires an additional passive element for each sub-module, which complicates the overall system. An additional control algorithm to guarantee a stable operation and an adequate input voltage distribution is also mandatory.

On the other hand, a fixed-frequency three-level full-bridge *LLC* resonant converter has been proposed [25]. The output voltage can be controlled by modulating the magnitude of the fundamental wave component of the voltage input to the resonance tank. However, since it does not operate as a typical diode-clamped topology, it is difficult to control because each switch must be duty-controlled separately. In addition, there is no mention of a DC link voltage deviation problem. Although various modulation techniques have been proposed for the *LLC* resonant converter, a suitable method for a diode-clamped multilevel converter with four levels or more has not yet been reported [26].

Meanwhile, single-phase diode-clamped multilevel AC/DC and DC/AC converters based on MNRV DPWM were proposed recently [27,28]. This MNRV DPWM is characterized by the presence of several adjacent reference vectors with different capacitor charging/discharging characteristics depending on the position of the command voltage in order to offset the voltage deviation of series-connected DC link capacitors and to match the magnitude of the command voltage on average. However, because AC/DC and DC/AC converters use a relatively low fundamental frequency to utilize a commercial grid or drive a motor, they use a relatively high frequency modulation index (m_f) to reduce capacitor voltage fluctuations due to the limited capacitance. To apply this MNRV DPWM to a DC/DC converter, some modifications of the modulation strategy are required, as in [29].

Based on [29], this paper presents a novel switching modulation scheme suitable for single-phase full-bridge diode-clamped multilevel *LLC* resonant converters with an addition of detailed circuit analysis, various types according to the carrier deformation, loss analysis, and experimental results. By using the linear amplitude modulation characteristics of the proposed diode-clamped multilevel converter, an *LLC* resonant converter based on voltage amplitude modulation with a fixed switching frequency at the resonant point is proposed, thus enabling an optimal passive device design and ensuring stable operation over a wide operating range. In addition, capacitor voltage deviation compensation is implemented using only $m_f = 2$ owing to the high fundamental switching frequency of the DC/DC converter. The proposed method utilizes only the clamped switching pair based on DPWM to facilitate the design of the switching pattern and reduce the switching losses. This paper briefly introduces a modified MNRV DPWM suitable for DC/DC converters, analyzes the operating characteristics of the proposed voltage-magnitude-modulation-based diode-clamped multilevel *LLC* resonant converter, and presents design guidelines. In addition, various types of *LLC* resonant converters are examined according to the deformation of the carrier. The performance and feasibility of the proposed converter are verified through simulations and experiments.

2. General Approach: Multi-Neighboring Reference Vector Discontinuous PWM (MNRV DPWM)

Figure 1 shows the circuit diagram of the single-phase diode-clamped four-level PWM converter. The AC voltage source V_S supplies input power to the full-bridge switching stack through boost inductor L_B with an equivalent series resistance of R_B. The full-bridge switching stage is composed of the switches Q_{A1}~Q_{A6} and Q_{B1}~Q_{B6} and the clamping diodes D_{A1}~D_{A6} and D_{B1}~D_{B6}. The output DC link stage consists of three series-connected capacitors, C_{dc1}, C_{dc2}, and C_{dc3}. R_L implies output load resistance.

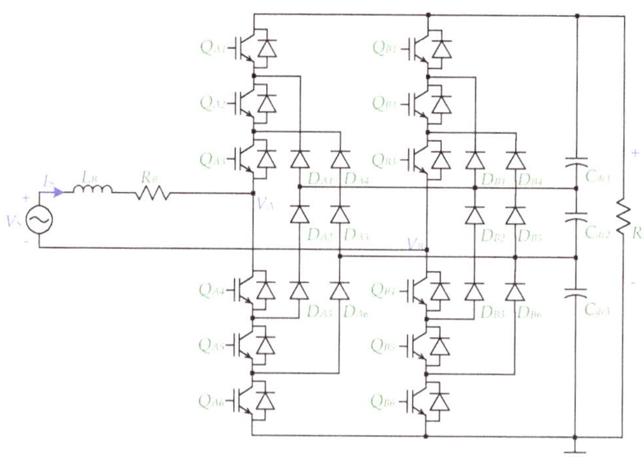

Figure 1. Single-phase diode-clamped four-level PWM converter.

Figure 2a shows the switching pairs of A and B legs and the capacitor charging/discharging status according to the position of the command voltage V_C^* of the single-phase diode-clamped four-level PWM converter. In the full-bridge four-level topology, *0*, *E*, *2E*, and

3*E* imply reference vectors representing each step voltage that can be output as a leg-to-leg voltage. The numbers in parentheses refer to switching pairs (AB) that can express the reference vectors, and *C* and *D* positioned to the right of them are the charging and discharging states of the capacitor, respectively. For example, if V_C^* is located at ②, in carrier-based sinusoidal pulse-width modulation (CB-SPWM), adjacent reference vectors are *E* and 2*E*, and (AB) expressing *E* has redundancy via (10), (21), and (32). With regard to (10), the charging/discharging state of the three capacitors connected in series as the top capacitor (C_{dc1}), intermediate capacitor (C_{dc2}), and bottom capacitor (C_{dc3}) is *DDC*, which means that that C_{dc1} and C_{dc2} are discharging and C_{dc3} is charging. For CB-PWM, which is widely used in single-phase AC/DC converters, two reference vectors adjacent to V_C^* are selected and their duration is adjusted to satisfy the magnitude of V_C^* and generate a symmetrical PWM pattern to minimize the ripple, as shown in Figure 2b. However, from level 4 or higher, a voltage imbalance inevitably occurs due to the limited selectable reference vectors capable of actively controlling the charging/discharging state of the capacitors. In particular, when the unity power factor is controlled as in the PWM converter, the voltage of the middle capacitor among the capacitors constituting the DC link stage is excessively charged compared to the other capacitors. This occurs because the charging/discharging behavior cannot be actively controlled with only two adjacent reference vectors.

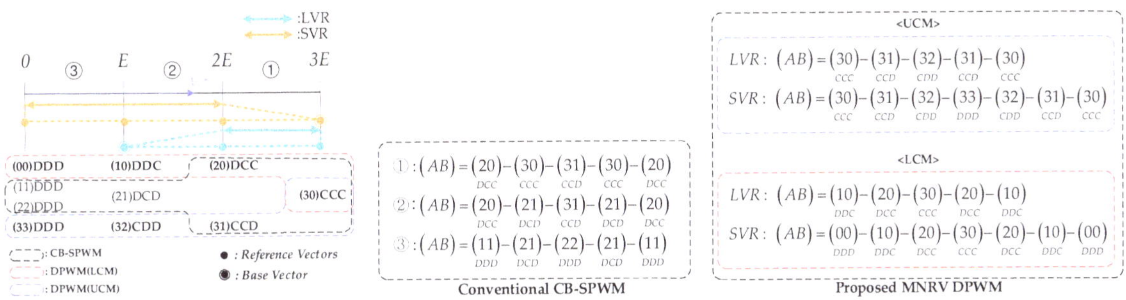

(a) (b) (c)

Figure 2. Implementation example of MNRV DPWM for a single-phase diode-clamped four-level PWM converter: (**a**) selection of MNRVs according to the V_C^* position, (**b**) switching patterns of the conventional CB-SPWM, and (**c**) those of the proposed MNRV DPWM.

When (AB) is (20), (30), and (31), the capacitor charging/discharging states are *DCC*, *CCC*, and *CCD*, respectively, and it can be confirmed that C_{dc2} is always in the charged state. In particular, for a PWM converter in which the phase difference between V_C^* and the input current is small, the capacitor charging current is largest when V_C^* is located in ① (large vector region (LRV)); moreover, even if V_C^* is located in different regions such as ② and ③ (the small vector region (SVR)), this overcharged state is not overcome, and the voltage deviation intensifies. Due to the characteristics of the diode-clamped multilevel converter, if the switching state is implemented by selecting two adjacent vectors as in the conventional CB-SPWM, the voltage deviation of the capacitor cannot be eliminated. Figure 3 shows the voltage imbalance problem of the PWM converter when applying CB-SPWM. V_S is the input voltage source; I_S is the input current; and V_{dc1}, V_{dc2}, and V_{dc3} are the terminal voltages of C_{dc1}, C_{dc2}, and C_{dc3}, respectively. V_{AB} is the switching leg voltage. As expected, the middle capacitor voltage V_{dc2} increases continually after applying CB-SPWM.

Meanwhile, total harmonic distortion (THD) of the switching leg voltage V_{AB} was analyzed under the following conditions of $V_S = 220V_{rms}$(60 Hz ac), output voltage $V_{dc} = 500$ V, output power $P_O = 30$ kW, source frequency $f_S = 60$ Hz, $m_f = 55$, $L_B = 4$ mH, and $R_B = 1$ mΩ. THD results were 0.23 and 0.43 for the CB-SPWM and the proposed MNRV DPWM, respectively, which reveals the much-increased high-frequency harmonic components of V_{AB} in the proposed method. However, the CB-SPWM-based multilevel converter corresponds to an ideal case that is difficult to implement in practice due to a voltage imbalance problem. When compared with the existing two-level converter, the THD value of MNRV DPWM was reduced by 0.15. On the other hand, the multilevel topology enables the use of the low withstand voltage devices with lower switching losses than higher withstand voltage devices. Therefore, in the proposed method, switching frequency can be further increased, thereby reducing the size of filter for harmonic reduction and facilitating its design.

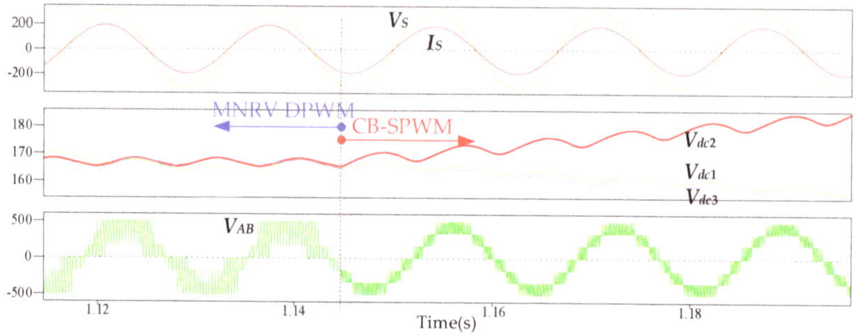

Figure 3. Voltage imbalance problem of a diode-clamped four-level PWM converter using CB-SPWM.

In order to solve the voltage deviation problem of the single-phase diode-clamped multilevel converter, MNRV DPWM was proposed [27]. Figure 2c shows the reference vectors selected according to the location of V_C^* and the switching pairs that can represent them, also showing the capacitor charging and discharging states according to the clamping mode (CM). When V_C^* is located in LVR, E, $2E$, and $3E$ are selected as reference vectors. If E is selected, which was not chosen in CB-SPWM, (10) and (32) can be utilized, and these switching pairs can serve to discharge C_2 (*DDC*, *CDD*). Therefore, if E is used as a reference vector along with $2E$ and $3E$, the voltage increase of C_2 can be suppressed, and if the duration times of reference vectors E, $2E$, and $3E$ are properly adjusted, the magnitude of V_C^* can be tracked on average. In this paper, to reduce the switching redundancy and minimize the switching loss, the clamped switching states in $\pm 180°$ DPWM are employed [30]. This means that only (3×), (×3), (0×), and (×0) are used among all switching pairs that can represent the reference vectors. (3×) and (×3) indicate that legs A or B, respectively, are clamped with a positive DC rail, called the upper clamping mode (UCM), where CM = 1. Similarly, (0×) and (×0) indicate the clamping of legs A or B, respectively, to the negative DC rail, called the lower clamping mode (LCM), where CM = −1. The generation of multiple references to eliminate voltage deviations among DC link capacitors in [21] is similar to the proposed method. However, MNRV DPWM is used, since V_C^* is clamped to the positive or negative DC rail during every half cycle, resulting in the advantage of reducing the switching redundancy and minimizing the switching loss. Meanwhile, in the UCM and LCM, the symmetry of the charging and discharging patterns for the same reference vector are horizontally opposite to each other. For example, (31) and (20) are switching pairs expressing reference vector $2E$ in the UCM and LCM, respectively, but the capacitor charging/discharging patterns are symmetric in those opposite to each other as *CCD* and *DCC*. As a symmetrical characteristic, the charging/discharging characteristics when V_C^* is negative are identical to the case when V_C^* is positive.

The procedure for calculating the duration time of the MNRV based on the reduced capacitor voltage deviation is as follows. First, V_{dc1}, V_{dc2}, and V_{dc3} are input. Between V_{dc1} and V_{dc3}, which are the top and bottom capacitor voltages, respectively, we select a voltage with a large absolute difference from the reference value, $V_{dc\_ref}/3$, which is the target voltage of the unit capacitor. If the selected voltage is V_{dc1} and $V_{dc\_ref}/3 - V_{dc1}$ is greater than zero, the UCM is selected. On the other hand, if the difference is less than zero, the LCM is selected. If the selected voltage is V_{dc3} and $V_{dc\_ref}/3 - V_{dc3}$ is greater than zero, we select the LCM, and if the difference is less than zero, we select the UCM, because when the charging/discharging behavior of the capacitor according to CM is analyzed, for the UCM, V_{dc1} tends to increase and V_{dc3} decrease, and vice versa in the LCM. When CM is selected, the compensation controller calculates the duty compensation values for the reference vectors proportional to the voltage deviations among the capacitors. In a four-level converter, three capacitors and two voltage-deviation compensators are needed. There are several methods for configuring the controllers implementing the two voltage deviation compensators. Here, two controllers, one compensating for the difference between $(V_{dc1} + V_{dc2})/2$ and V_{dc3} and the other compensating for that between V_{dc1} and $(V_{dc2} + V_{dc3})/2$, are used. These design methods are natural charging/discharging characteristics of reference voltage vectors and easy to expand to a general high-dimensional multilevel converter.

As shown in Equation (1), the two compensation controllers output d_{comp12_3} and d_{comp1_23}, which are used to compensate for the voltage difference between $(V_{dc1} + V_{dc2})/2$ and V_{dc3} and the voltage difference between V_{dc1} and $(V_{dc2} + V_{dc3})/2$, respectively. In Equation (1), K_P and K_I mean the proportional and integral gains of the PI controller, respectively. According to the position of V_C^* and CM, the duties of the each MNRV are calculated as follows. If V_C^* belongs to the LVR, we select E, $2E$, and $3E$ as the MNRV. Here, with E as a basic vector for the duty calculation, the duties of the E, $2E$, and $3E$ vectors are calculated as Equation (2). Here, d_E, d_{2E}, and d_{3E} are the duties of E, $2E$, and $3E$, respectively. V_E, V_{2E}, and V_{3E} are the voltage levels of E, $2E$, and $3E$, respectively, meaning $V_{dc\_ref}/3$, $V_{dc\_ref}\cdot 2/3$, and V_{dc_ref}. Equation (2) is applied differently depending on CM, and its meaning is as follows. The duties of the MNRVs must basically satisfy the magnitude of V_C^*. At the same time, the voltage deviations among all capacitors are reduced by applying two controller outputs (d_{comp12_3} and d_{comp1_23}) differently according to CM. For example, if $(V_{dc1} + V_{dc2})/2 > V_{dc3}$ in the UCM, d_{comp12_3} increases and d_{2E} decreases. In the UCM, d_{2E} is the duty of the switching pair (31), which serves to increase V_{dc1} and V_{dc2} and decrease V_{dc3}. Therefore, V_{dc1} and V_{dc2} decrease and V_{dc3} increases due to the reduced d_{2E}. According to the negative feedback, d_{comp12_3} is stabilized and the capacitor voltage deviations are also reduced. The same can be explained in the LCM. Using the previously calculated duties of each reference vector, the PWM command values of each of the switches in the LRV are determined via Equation (3). Here, N_{max} refers to the maximum value at the period of a single carrier considering the DSP implementation. For a positive V_C^*, leg A is clamped to the positive DC rail in the UCM and leg B is clamped to the negative DC rail in the LCM.

$$\begin{aligned} d_{comp12\_3} &= K_P\left(\frac{V_{dc1}+V_{dc2}}{2} - V_{dc3}\right) + K_I\int_0^t \left(\frac{V_{dc1}+V_{dc2}}{2} - V_{dc3}\right)dt, \\ d_{comp1\_23} &= K_P\left(V_{dc1} - \frac{V_{dc2}+V_{dc3}}{2}\right) + K_I\int_0^t \left(V_{dc1} - \frac{V_{dc2}+V_{dc3}}{2}\right)dt. \end{aligned} \tag{1}$$

$$\begin{array}{ll} LVR@UCM & LVR@LCM \\ d_{2E} = d_E - d_{comp12\_3}, & d_{2E} = d_E + d_{comp1\_23}, \\ d_{3E} = 1 - d_E - d_{2E}, & d_{3E} = 1 - d_E - d_{2E}, \\ V_C^* = V_E d_E + V_{2E} d_{2E} + V_{3E} d_{3E}, & V_C^* = V_E d_E + V_{2E} d_{2E} + V_{3E} d_{3E}, \\ d_E = \frac{V_C^* + d_{comp12\_3}(V_{2E}-V_{3E}) - V_{3E}}{V_E + V_{2E} - 2V_{3E}}, & d_E = \frac{V_C^* - d_{comp1\_23}(V_{2E}-V_{3E}) - V_{3E}}{V_E + V_{2E} - 2V_{3E}}. \end{array} \tag{2}$$

$$\begin{array}{l} LVR@UCM \\ \begin{pmatrix} PWM\_CMD\_A1 \\ PWM\_CMD\_A2 \\ PWM\_CMD\_A3 \end{pmatrix} = N_{max}\begin{pmatrix} 1 \\ 1 \\ 1 \end{pmatrix}, \quad \begin{pmatrix} PWM\_CMD\_B1 \\ PWM\_CMD\_B2 \\ PWM\_CMD\_B3 \end{pmatrix} = N_{max}\begin{pmatrix} 0 \\ d_E \\ d_E + d_{2E} \end{pmatrix}, \\ LVR@LCM \\ \begin{pmatrix} PWM\_CMD\_A1 \\ PWM\_CMD\_A2 \\ PWM\_CMD\_A3 \end{pmatrix} = N_{max}\begin{pmatrix} d_{3E} \\ d_{2E} + d_{3E} \\ 1 \end{pmatrix}, \quad \begin{pmatrix} PWM\_CMD\_B1 \\ PWM\_CMD\_B2 \\ PWM\_CMD\_B3 \end{pmatrix} = \begin{pmatrix} 0 \\ 0 \\ 0 \end{pmatrix}. \end{array} \tag{3}$$

On the other hand, when $V_C{}^*$ is located in the SVR, 0, E, $2E$, and $3E$ are selected as reference vectors for a similar principle. To achieve a smooth transition at the moment of a region change, the range of the reference vectors in SVR must be wider than that of the LVR to increase the common reference vectors between the LVR and SVR. Each duty is calculated as in Equation (4). Here, d_0 is the duty of the 0 vector. Using the calculated duties of each reference vector, the PWM command values of each of the switches in the SRV are determined as in Equation (5).

Using the symmetric characteristic of the full-bridge topology, the duties of the MNRVs when $V_C{}^*$ is negative are determined to be identical to those in the positive $V_C{}^*$ case except for the interchange of legs A and B.

$$\begin{array}{ll} SVR@UCM & SVR@LCM \\ d_{2E} = d_{3E} - d_{comp12\_3}, & d_{2E} = d_{3E} + d_{compl\_23}, \\ d_E = d_{3E} - d_{compl\_23}, & d_E = d_{3E} + d_{comp12\_3}, \\ d_0 = 1 - d_E - d_{2E} - d_{3E}, & d_0 = 1 - d_E - d_{2E} - d_{3E}, \\ V_C{}^* = V_0 d_0 + V_E d_E + V_{2E} d_{2E} + V_{3E} d_{3E}, & V_C{}^* = V_0 d_0 + V_E d_E + V_{2E} d_{2E} + V_{3E} d_{3E}, \\ d_{3E} = \frac{V_C{}^* + d_{compl\_23} V_E + d_{comp12\_3} V_{2E}}{V_E + V_{2E} + V_{3E}}, & d_{3E} = \frac{V_C{}^* - d_{comp12\_3} V_E - d_{compl\_23} V_{2E}}{V_E + V_{2E} + V_{3E}}. \end{array} \tag{4}$$

$$\begin{array}{l} SVR@UCM \\ \begin{pmatrix} PWM\_CMD\_A1 \\ PWM\_CMD\_A2 \\ PWM\_CMD\_A3 \end{pmatrix} = N_{max} \begin{pmatrix} 1 \\ 1 \\ 1 \end{pmatrix}, \quad \begin{pmatrix} PWM\_CMD\_B1 \\ PWM\_CMD\_B2 \\ PWM\_CMD\_B3 \end{pmatrix} = N_{max} \begin{pmatrix} d_0 \\ d_0 + d_E \\ d_0 + d_E + d_{2E} \end{pmatrix}, \\ SVR@LCM \\ \begin{pmatrix} PWM\_CMD\_A1 \\ PWM\_CMD\_A2 \\ PWM\_CMD\_A3 \end{pmatrix} = N_{max} \begin{pmatrix} d_{3E} \\ d_{2E} + d_{3E} \\ d_E + d_{2E} + d_{3E} \end{pmatrix}, \quad \begin{pmatrix} PWM\_CMD\_B1 \\ PWM\_CMD\_B2 \\ PWM\_CMD\_B3 \end{pmatrix} = \begin{pmatrix} 0 \\ 0 \\ 0 \end{pmatrix}. \end{array} \tag{5}$$

The MNRV DPWM applied to a single-phase diode-clamped four-level converter can be expanded to a general diode-clamped N-level converter as follows. Based on the analysis of the voltage fluctuation characteristics of the DC link capacitors of the reference step voltage vectors, appropriate MNRVs are selected according to the position of $V_C{}^*$ and the capacitor voltage deviation compensation parameters are designed. MNRV selections applied to the five-level, six-level, and general N-level diode-clamped topologies are shown in Figure 4.

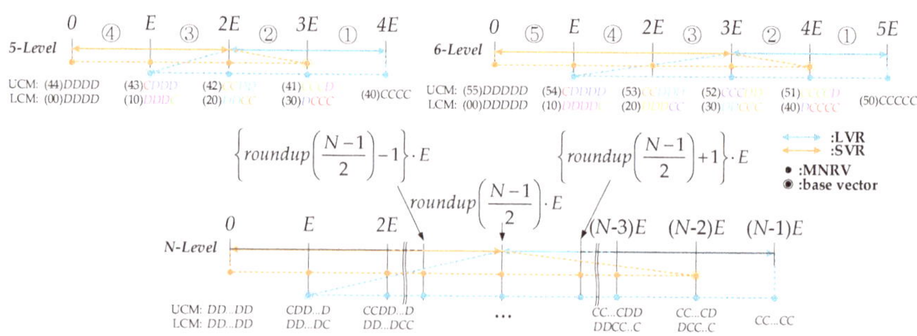

Figure 4. MNRV selection rule according to the $V_C{}^*$ position in single-phase full-bridge diode-clamped multilevel topologies.

The general rule for selecting MNRVs according to the $V_C{}^*$ position is as follows. Compensation parameters for controlling the voltage deviations of capacitors are $N-2$ in the $N-$ level case. First, the charging/discharging states of the capacitors according to the reference step voltage are analyzed as follows. In the N-level case, the maximum step voltage is $(N-1)\cdot E$, and the capacitor charging/discharging state is $CC \ldots CC$ regardless of CM. Because all capacitors are charged with the same amount of current, there is no voltage deviation. The step voltage one step lower is $(N-2)\cdot E$ and the capacitor charging/discharging states are $CC \ldots CD$ and $DC \ldots CC$ in the UCM and LCM, respectively. The outermost capacitor is changed to the discharge mode. In the UCM, the capacitor located at the bottom of the DC link stage, and in the LCM, the capacitor located at the top of the DC link stage, change from charging to discharging mode. The next step voltage is $(N-3)\cdot E$, and the capacitor charging/discharging states are $CC \ldots CDD$ and $DDC \ldots CC$ in the UCM and LCM, respectively. That is, whenever the step voltage decreases by one step, it can be seen that the capacitor charging/discharging states change from charging to discharging from the lower end of the DC link stage in the UCM and from the upper end of the DC link stage in the LCM. Applying this to all step voltages, the following rule can be found. The total number of different charging/discharging states of capacitors is $2(N-2)$. At this time, there are step voltage pairs in which the sum of two step voltages is $(N-1)\cdot E$, and the capacitor charging/discharging states in the UCM and LCM of these two step voltages are related to the same capacitor voltage compensation parameters with opposite signs. The two step voltages in this case are cross oppositely coupled to each other; therefore, there are $N-2$ independent capacitor charging/discharging states in the overall step voltage. Thus, $N-2$ independent d_{comp} parameters can be generated, i.e., $d_{comp1\_2\ldots(N-1)}$, $d_{comp12\_3\ldots(N-1)}, \ldots, d_{comp1\ldots(N-2)\_(N-1)}$. In order to control the capacitor voltage deviation completely regardless of the position of V_C^*, the MNRV should be selected so that all independent compensation parameters are included. For every V_C^* location, it is sufficient to set the MNRV such that $N-2$ independent duty compensation parameters are included in order to eliminate the voltage deviations of all capacitors. However, large fluctuations in the duty compensation value may adversely affect normal operation; i.e., the duty of a particular reference vector can be negative or greater than one due to the addition of duty-compensation parameters. Accordingly, it is better to include all different charging/discharging states of the capacitors for more reliable operation by limiting the duty-compensation effort. Including all capacitor charging and discharging states is good for stable operation but increases the switching losses. However, as will be described later, by designing an appropriate carrier, a voltage-deviation-compensation operation may occur in the ZVS region, thereby minimizing an increase in the switching loss. Because the outermost vectors $(0, (N-1)\cdot E)$ do not affect the capacitor voltage deviation, the capacitor-voltage-deviation-compensation parameters are designed from independent capacitor charging/discharging states of the remaining intermediate reference vectors.

In the five-level case, for example, there are three independent charging/discharging states: *CDDD*(*DCCC*), *CCDD*(*DDCC*), and *CCCD*(*DDDC*). The capacitor charging/discharging states of the reference voltages E in the UCM and $3E$ in the LCM are *CDDD* and *DCCC*, respectively. They both can be used to control the voltage deviation between V_{dc1} and $(V_{dc2} + V_{dc3} + V_{dc4})/3$. It can be seen that the charging/discharging states of $2E$ in the UCM and LCM are *CCDD* and *DDCC*, respectively, and are related to the voltage deviations between $(V_{dc1} + V_{dc2})/2$ and $(V_{dc3} + V_{dc4})/2$. Similarly, the charge/discharge states of $3E$ in the UCM and E in the LCM are *CCCD* and *DDDC*, respectively, and they can be utilized to control the voltage deviation between $(V_{dc1} + V_{dc2} + V_{dc3})/3$ and V_{dc4} with opposite signs. Therefore, in the five-level case, three independent compensation parameters are required to control the capacitor voltage deviation: d_{comp1_234}, d_{comp12_34}, and $d_{comp123\_4}$. They are related to the voltage differences between V_{dc1} and $(V_{dc2} + V_{dc3} + V_{dc4})/3$, $(V_{dc1} + V_{dc2})/2$ and $(V_{dc3} + V_{dc4})/2$, and $(V_{dc1} + V_{dc2} + V_{dc3})/3$ and V_{dc4}, respectively. Therefore, if V_C^* belongs to the LVR, the range of the MNRV should be $E \sim 4E$ in order to utilize all six different charging/discharging states of the capacitors. On the other hand, if V_C^* is located in the SVR, the MNRV range should be $0 \sim 3E$ for the same reason. In addition, when V_C^* moves between the LVR and SVR, the duties of the reference voltage vectors may suddenly change, which is a factor that degrades the linearity of the output voltage. Therefore, in order to minimize duty changes of the reference voltage vectors and thus ensure smooth transitions between the LRV and SRV, the minimum reference voltage in the LVR and the maximum reference voltage in the SVR should be selected as the basic vector for the MNRV.

In this way, by selecting the MNRV according to the positions of V_C^* and designing compensation parameters for the deviations of the capacitor voltages, extension to a high-dimensional *N*-level diode-clamped converter is possible. For a general *N*-level case, the LRV and SRV range from roundup$\{(N-1)/2\}\cdot E$ to $(N-1)\cdot E$ and from *0* to roundup$\{(N-1)/2\}\cdot E$, respectively. The corresponding MNRVs are from *E* to $(N-1)\cdot E$ and from *0* to $(N-2)\cdot E$, as shown in Figure 4. It should be noted that thus far MNRV DPWM has been described based on the incoming current to the switching legs, as in a PWM converter. Therefore, the charging/discharging states and CM selection should be reversed in DC/DC owing to the reversed reference current direction.

3. Proposed Full-Bridge Diode-Clamped Four-Level *LLC* Resonant Converter

Circuit and Control Algorithm for the Proposed Diode-Clamped 4-Level LLC Resonant Converter

Figure 5 shows the circuit of the proposed full-bridge diode-clamped four-level *LLC* resonant converter. The voltage source V_{dc} supplies input power through the DC link stage composed of three series-connected capacitors, C_{dc1}, C_{dc2}, and C_{dc3}. The full-bridge switching stage is composed of the switches Q_{A1}~Q_{A6} and Q_{B1}~Q_{B6}, and the clamping diodes D_{A1}~D_{A6} and D_{B1}~D_{B6} that connect the unit step capacitor voltages of the DC link stage to each switching node. The resonance tank stage composed of resonant inductor L_r, resonant capacitor C_r, and magnetizing inductor L_m receives voltage amplitude modulated sag-type voltage with a fixed frequency from switching leg voltage $V_{leg}(=V_{AB}=V_A-V_B)$. The resonant current I_{Lr} charges the output capacitor C_O through an *n*:1:1 center-tapped transformer and the output rectification stage composed of diodes D_{O1} and D_{O2} and supplies power to the load R_L.

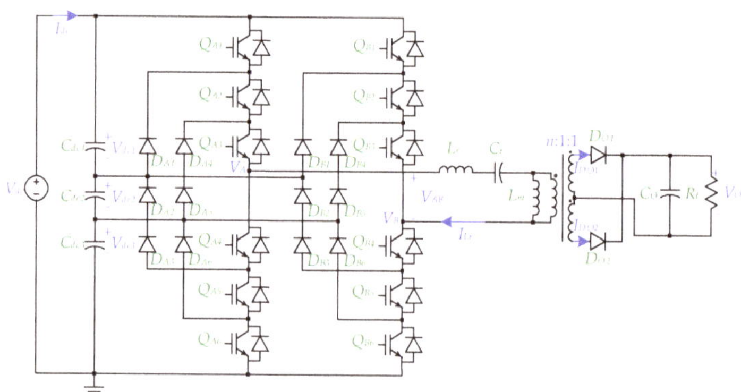

Figure 5. Circuit diagram of the proposed full-bridge diode-clamped four-level *LLC* converter.

AC/DC or DC/AC converters have a relatively high m_f to reduce the harmonic components and to control the fundamental component on average during one fundamental period. However, in a DC/DC converter, it is desirable to increase the frequency of the AC switching pulse until it is within an acceptable switching loss range to reduce the size of the passive element. Given that such a high fundamental frequency increases the number of compensations for capacitor voltage deviation during the unit time, MNRV DPWM for a DC/DC converter can be implemented with small m_f, i.e., $m_f = 2$. A DC/DC converter to which MNRV DPWM is applied can also linearly modulate the magnitude of the reference voltage akin to a multilevel inverter. Therefore, when applied to a resonant converter, the output voltage can be controlled by modulating the magnitude of V_C^* while fixing the frequency to the resonance point. This simplifies the design of the passive components compared to those in conventional frequency-swept resonant converters.

The control block diagram of the proposed MNRV DPWM-based diode-clamped four-level *LLC* resonant converter is shown in Figure 6. The output V_{ampl} of the constant-voltage controller is multiplied by a pulse that swings between 1 and −1 at a rate of 50%

with switching frequency f_{sw} to obtain pulse-shaped command voltage V_C^*. Here, f_{sw} is designed as the resonance frequency ($f_r = 1/[2\pi \cdot (L_r C_r)^{0.5}]$) for maximum efficiency. CM is set to be between 1 and −1, referring to the UCM and LCM, respectively, according to V_{dc1} and V_{dc3}. In order to remove the voltage deviations among the capacitors, V_{dc1}, V_{dc2}, and V_{dc3} are input to the PI controller to calculate the duty compensation parameters (d_{comp1_23}, d_{comp12_3}). According to the position of V_C^*, appropriate MNRVs are selected and the duration times of each reference vectors are calculated, with the PWM CMDs of the A and B legs finally output. These values are compared to the triangular carrier to determine the on/off status of every upper switch in the A and B legs. According to the complementary rule of the diode-clamped topology, the on/off statuses of the lower switches are determined as opposite to the corresponding upper switches' on/off states in the same leg. Q_{A1} and Q_{A4}, Q_{A2} and Q_{A5}, and Q_{A3} and Q_{A6} are the complementary switch pairs in leg A. The same applies to leg B.

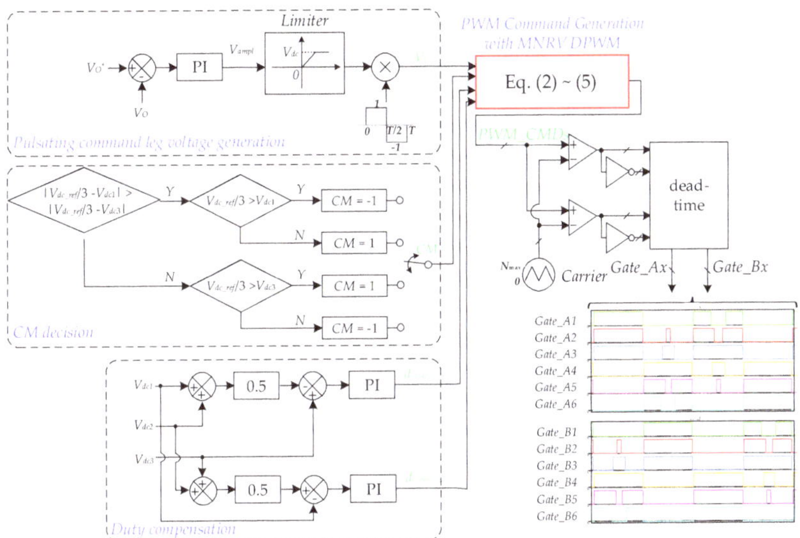

Figure 6. Control block diagram for the proposed converter.

4. Analysis

4.1. Overall Operation Characteristics of the Proposed Converter

In this section, we analyze the operation mode of the proposed diode-clamped four-level resonant *LLC* converter. The operation state of the circuit is largely classified according to CM. Figure 7 shows the circuit operation status in the UCM and LCM, and (AB) on the right side of each circuit represents the switching status of the A and B phases. The shaded red line in the figure represents the current path flowing through the channel or body diode of the conducting switches. Except for (30), (03), (33), and (00), the magnitude and direction of the current flowing through the series-connected capacitors C_{dc1}, C_{dc2}, and C_{dc3} differ depending on the switching pattern as follows. In the switching pairs of (31) and (13), the charging/discharging state is *DDC*, and in (32) and (23) the charging/discharging state is *DCC*. In (20) and (02) the charging/discharging state is *CDD*, and in (10) and (01) the charging/discharging state is *CCD*. Therefore, it is expected that a voltage deviation will occur because the charging/discharging patterns of each capacitor differ depending on the duration of the specific switching pair. Here, the amount of charging/discharging current flowing through each capacitor is related to I_{Lr}. If there are two capacitors connected in series, the charging/discharging current becomes $1/3 \times I_{Lr}$, and if there is one capacitor, it becomes $2/3 \times I_{Lr}$. For example, for (32), C_{dc1} is discharged with a current of $2/3 \times I_{Lr}$

and C_{dc2} and C_{dc3} are charged with $1/3 \times I_{Lr}$. The core principle of MNRV DPWM is to set several reference vectors, including all independent charging/discharging characteristics of DC link capacitors depending on the location of V_C^*, and adjust their duration time to reduce the voltage deviations of the capacitors and to satisfy the magnitude of V_C^* on average at the same time.

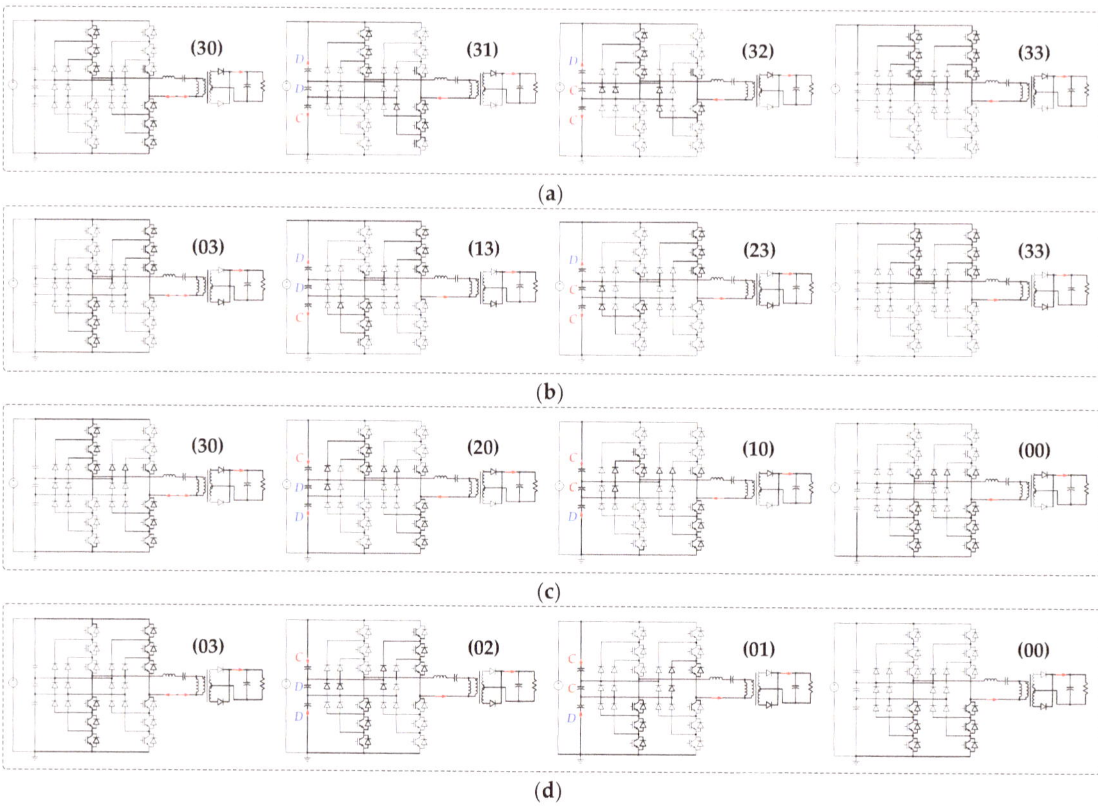

Figure 7. Operating circuit states when (**a**) $V_C^* > 0$ and (**b**) $V_C^* < 0$ in the UCM and when (**c**) $V_C^* > 0$ and (**d**) $V_C^* < 0$ in the LCM.

Figure 8 depicts the equivalent circuit of the resonance tank at the resonance point. Regardless of CM, when $V_C^* > 0$, one from among V_{dc}, $2/3 \times V_{dc}$, and $1/3 \times V_{dc}$ is applied to V_{leg}, and nV_O is applied to the magnetizing inductor L_m. When the voltage of $V_{leg} - nV_O$ is supplied to L_r-C_r, I_{Lr} and I_{Lm} increase in the positive direction. Meanwhile, when $V_C^* < 0$, one from among $-V_{dc}$, $-2/3 \times V_{dc}$, and $-1/3 \times V_{dc}$ is applied to V_{leg} and $-nV_O$ is applied to L_m. I_{Lr} and I_{Lm} decrease by $V_{leg} + nV_O$ supplied to L_r-C_r.

Based on this circuit operation state, Figure 9 shows the main operating waveforms of the proposed full-bridge diode-clamped four-level *LLC* resonant converter. CM is determined as the UCM or LCM according to the characteristics of the capacitor voltage deviation at the start of one switching period. Here, it is assumed that the converter is operated in the UCM for one cycle and operated in the LCM during the next cycle in a steady state. The operation during one cycle is a series of symmetrical half-cycle operations, and the operation during the half-cycle is divided into six modes.

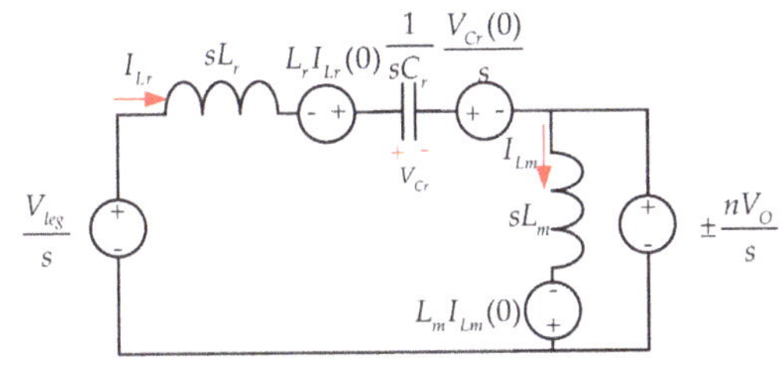

Figure 8. Equivalent circuit of the resonant tank at the resonance point.

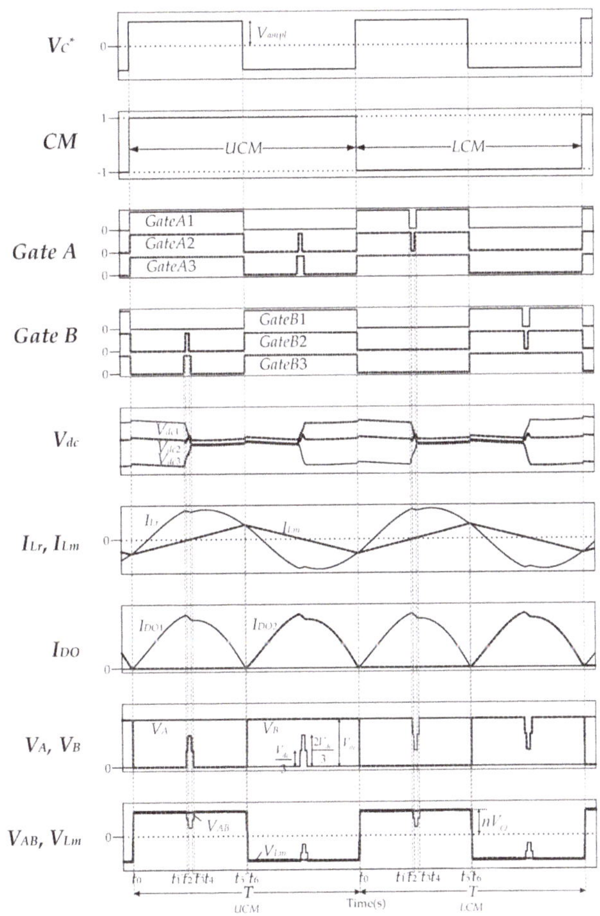

Figure 9. Operating mode analysis of the proposed converter.

4.1.1. Operation under the UCM

Mode 1 $[t_0,t_1]$: At t_0, CM is changed from the LCM to the UCM; the A leg upper switches Q_{A1}, Q_{A2}, and Q_{A3} are all turned on with the ZVS condition owing to the negative current of I_{Lr}, and the B leg upper switches Q_{B1}, Q_{B2}, and Q_{B3} are all turned off. V_{leg} with V_{dc} is input to the resonance tank stage, and the resonance current I_{Lr} formed by L_r and C_r is transferred to the secondary side of the transformer; D_{O1} then conducts, and the output voltage V_O multiplied by the turn ratio of n is applied to L_m. At this time, the switching state is (30), and the currents flowing through C_{dc1}, C_{dc2}, and C_{dc3} are all identical to $I_{dc}-I_{Lr}$ here, I_{dc} is the outgoing current of V_{dc} such that voltage deviations do not occur in the capacitor. This mode is terminated at t_1 when Q_{B6} turns off (when Q_{B3} turns on). In mode 1, resonant current I_{Lr} and resonant voltage V_{Cr} are determined by Equation (6). Here, V_{leg} is V_{dc} and $I_{Lr}(0)$ and $V_{Cr}(0)$ refer to the initial values of I_{Lr} and V_{Cr} at t_0, respectively. The characteristic impedance Z equals $(L_r/C_r)^{0.5}$, and the angular frequency ω is $2\pi f_r$.

$$\begin{aligned} I_{Lr}(t) &= I_{Lr}(0)\cos\omega t + \tfrac{V_{leg}-nV_o-V_{Cr}(0)}{Z}\sin\omega t, \\ V_{Cr}(t) &= ZI_{Lr}(0)\sin\omega t + [V_{leg}-nV_o-V_{Cr}(0)]\cdot(1-\cos\omega t) + V_{Cr}(0). \end{aligned} \tag{6}$$

Mode 2 [t_1,t_2]: When Q_{B6} is off at t_1, the switching state becomes (31), and C_{dc1} and C_{dc2} are discharged with a current of $1/3 \times I_{Lr}$ and C_{dc3} is charged with a current of $2/3 \times I_{Lr}$. V_{leg} becomes $2/3 \times V_{dc}$, and the voltage applied to L_r-C_r is changed from V_{dc} to $2/3 \times V_{dc}$; thus, the slope of I_{Lr} changes. L_m is still charged with nV_O due to the conduction of D_{O1}. This mode continues until Q_{B5} turns off at t_2. In mode 2, I_{Lr} and V_{Cr} are also determined by Equation (6), except for $V_{leg} = 2/3 \times V_{dc}$, and $I_{Lr}(0)$ and $V_{Cr}(0)$ are changed to $I_{Lr}(t_1)$ and $V_{Cr}(t_1)$, respectively.

Mode 3 [t_2,t_3]: When Q_{B5} is turned off at t_2, the switching state is (32); accordingly, C_{dc1} is discharged with a current of $2/3 \times I_{Lr}$, and C_{dc2} and C_{dc3} are charged with a current of $1/3 \times I_{Lr}$. Because V_{leg} is changed from $2/3 \times V_{dc}$ to $1/3 \times$ Vdc, the slope of the I_{Lr} also changes. L_m is still charged with nV_O. This mode ends when Q_{B5} is turned on at t_3. In mode 3, I_{Lr} and V_{Cr} are also determined by Equation (6), except for $V_{leg} = 1/3 \times V_{dc}$, and $I_{Lr}(0)$ and $V_{Cr}(0)$ are changed to $I_{Lr}(t_2)$ and $V_{Cr}(t_2)$, respectively.

Mode 4 [t_3,t_4]: When Q_{B5} is turned on at t_3, the switching state is (31) again, and operation in this mode is identical to that in mode 2. This mode ends when Q_{B6} turns on at t_4.

Mode 5 [t_4,t_5]: When Q_{B6} is turned on at t_4, the switching state is (30), and operation in this mode is identical to that in mode 1. This mode is terminated when I_{Lr} equals I_{Lm} at t_5. At the end of this mode, D_{O1} is turned off with the ZCS condition.

Mode 6 [t_5,t_6]: At t_5, $I_{Lr} = I_{Lm}$, the primary and the secondary sides of the transformer are separated, and the resonance period becomes longer due to the large inductance of L_m contributing to the resonance. This mode continues until the polarity of V_C^* becomes negative. In mode 6, I_{Lr} and V_{Cr} are determined by Equation (7). Here, V_{leg} is V_{dc} and $I_{Lr}(0)$ and $V_{Cr}(0)$ refer to the initial values of I_{Lr} and V_{Cr} at t_5, respectively. The characteristic impedance Z' equals $[(L_r + L_m)/C_r]^{0.5}$, and the angular frequency ω' is $1/[(L_r + L_m)\cdot C_r]^{0.5}$.

$$\begin{aligned} I_{Lr}(t) &= I_{Lr}(0)\cos\omega' t + \tfrac{V_{leg}-V_{Cr}(0)}{Z'}\sin\omega' t \\ V_{Cr}(t) &= Z'I_{Lr}(0)\sin\omega' t + [V_{leg}-V_{Cr}(0)]\cdot(1-\cos\omega' t) + V_{Cr}(0) \end{aligned} \tag{7}$$

Due to the symmetry of the DC/DC converter, operation during the remaining half cycle where V_C^* is negative can easily be inferred from the previous positive half cycle of V_C^*; accordingly, a description thereof will be omitted. Regarding the capacitor charging and discharging status, in the UCM, V_{dc1} continues to discharge and V_{dc3} continues to charge regardless of the polarity of V_C^*. Therefore, under actual operating conditions, the UCM is selected when $V_{dc1} > V_{dc3}$.

4.1.2. Operation under the LCM

When V_C^* changes from negative to positive again, one cycle of LCM operation begins. Operation in the LCM has cross-symmetric duality with that of the UCM as follows. When $Q_{A(m)}/Q_{B(m)}$ is turned on/off in the UCM, $Q_{B(7\text{-}m)}/Q_{A(7\text{-}m)}$ is turned on/off in the LCM at the same timing, where $m = 1, 2, \ldots, 6$. That is, the turn on/off timing of switch pairs Q_{A1} and Q_{B6}, Q_{A2} and Q_{B5}, Q_{A3} and Q_{B4}, Q_{A4} and Q_{B3}, Q_{A5} and Q_{B2}, and Q_{A6} and Q_{B1}, located in cross-opposite directions, are correspondingly matched. Thus, when the A/B leg is clamped to the positive DC rail in the UCM, the B/A leg is clamped to the negative DC rail in the LCM at the same timing. This duality can be confirmed by examining Figures 7 and 9, and thus a detailed description of operation in the LCM is skipped here to save space.

In the LCM, V_{dc3} continues to discharge and V_{dc1} continues to charge regardless of the polarity of V_C^*. Therefore, under actual operating conditions, the LCM is selected when $V_{dc1} < V_{dc3}$.

4.2. Voltage Transfer Gain

In this section, the input–output voltage transfer gain M is derived through a fundamental harmonic analysis (FHA) [31]. Figure 10 shows the V_{leg} waveform when $1/3 \times V_{dc} \leq |V_C^*| < 2/3 \times V_{dc}$. For convenience of the calculation, the original sag-type V_{leg} is divided into a square-wave swinging at $\pm V_{dc}$ with full duty and three square waves with the opposite phase, swinging at $\pm V_{dc}/3$ during angular sections $[\pi/2 \pm \alpha]$, $[\pi/2 \pm \beta]$, and $[\pi/2 \pm \gamma]$, respectively. The fundamental component is analyzed as Equation (8).

$$V_{leg}^F = \frac{4}{\pi} V_{dc} - \frac{4}{\pi} \cdot \frac{V_{dc}}{3} (\sin\alpha + \sin\beta + \sin\gamma) \tag{8}$$

Here, α, β, and γ are given by Equation (9). On the other hand, if $2/3 \times V_{dc} \leq |V_C^*|$, γ becomes 0, as shown in Figure 9.

$$\alpha = 0.5\pi(d_0 + d_E + d_{2E}),\ \beta = 0.5\pi(d_0 + d_E),\ \gamma = 0.5\pi d_0 \tag{9}$$

The fundamental component is reduced compared to the full duty square wave due to the voltage sag section, as shown in Equation (10). Here, $k = L_m/L_r$, $Q = Z/R_{ac}$, and $R_{ac} = 8n^2 R_L/\pi^2$. The length of the sag section is related to the duty of the MNRV, which is also related to V_C^*. This means that even if the frequency is fixed, the output voltage can be adjusted by changing V_C^*. This is a distinguishing feature of the voltage modulation method used in MNRV DPWM compared to the conventional two-level resonant converter using the frequency modulation method.

$$|M| = \frac{nV_O}{V_{dc}} = \frac{1 - \frac{\sin\alpha + \sin\beta + \sin\gamma}{3}}{\sqrt{\left[1 + \frac{1}{k}\{1 - \left(\frac{f_r}{f}\right)^2\}\right]^2 + \left[Q\left(\frac{f}{f_r} - \frac{f_r}{f}\right)\right]^2}} \tag{10}$$

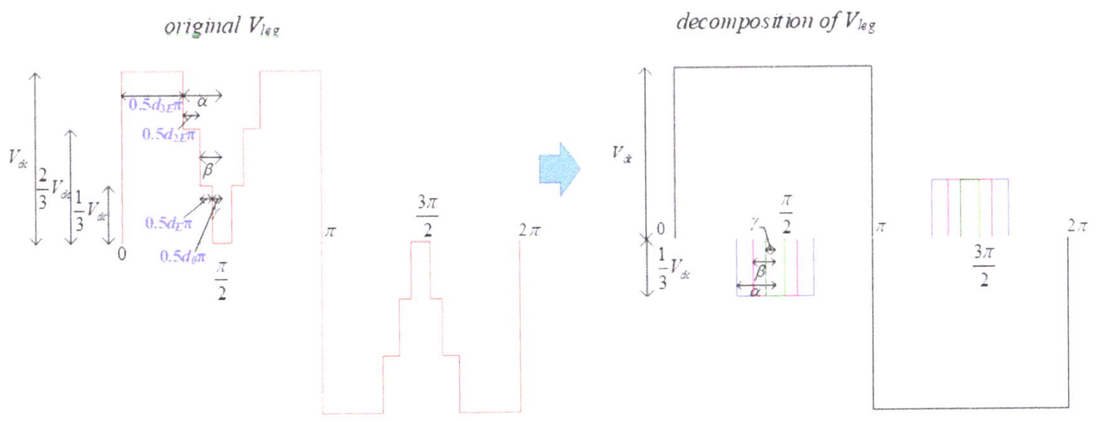

Figure 10. Decomposition of V_{leg} according to the voltage sag range.

5. Design Guideline

5.1. Design of the Resonant Tank

The proposed converter regulates the output voltage by modulating the magnitude of V_{leg} while fixing the operating frequency to the resonance point. Therefore, the voltage gain range can be determined from the length of the voltage sag section in the normal operating range as shown in Equation (10). If the load increases while the output voltage is fixed, the DC link capacitors' charging/discharging current increases and the duty compensation effort must be increased. Therefore, the period of the voltage sag section may be out of the normal range and may no longer be able to undertake voltage compensation. With fixed V_{dc} and V_O values, d_E decreases as n increases. At this time, if the load increases, d_E may have an abnormal value due to the increased duty compensation parameters; therefore, n must be appropriately designed considering the sag section under the maximum load condition. When n is determined, a Q factor must be selected. A high Q implies a large L_r and small C_r, leading to a large reactor size and increased withstand voltage in C_r. On the other hand, if Q is decreased, the shape of the resonance current deviates from a sinusoidal wave, increasing the root mean square (RMS) value of I_{Lr}. Thus, an appropriate value must be selected. C_r is determined as $C_r = 1/(2\pi{\cdot}f_r{\cdot}Q{\cdot}R_{ac})$, and L_r is determined by $L_r = 1/[(2\pi{\cdot}f_r)^2{\cdot}C_r]$. L_m is selected by determining the ratio k between L_m and L_r. Reducing L_m ensures ZVS operation while increasing circulating current and switch losses. Conversely, a large L_m reduces circulating current and switch losses, but ZVS operation may not be guaranteed. In addition, L_m affects the voltage gain in the region other than the resonance point. Therefore, an appropriate value of L_m must be selected.

5.2. Resonant Current and Voltage

In this section, the peak resonant current I_{Lr_pk} and voltage V_{Cr_pk} are calculated in order to select an appropriate device by determining the rated switch current and the withstand voltage of the resonance capacitor. When operating at the resonance point, I_{Lr} and I_{Lm} are in a steady state, as shown in Figure 11. Here, it is assumed that I_{Lr} and V_{Cr} are sinusoidal for convenience of calculation, as in Equation (11). The initial phase difference ϕ between the I_{Lr} and V_{leg} waveforms is determined as Equation (12). From the observation that the difference between I_{Lr} and I_{Lm} during a half cycle is equal to the output current divided by n, I_{Lr_pk} is determined to be Equation (13) [2]. I_{Lr_pk} is proportional to the load current and inversely proportional to L_m. This is consistent with the fact that as the L_m value decreases, the RMS value of I_{Lr} increases with the increased circulating current. From the energy balance principle, V_{Cr_pk} is calculated as Equation (14).

$$I_{Lr}(t) = I_{Lr\_pk}\sin(\omega t - \phi) \tag{11}$$

$$\phi = \sin^{-1}\left(\frac{nV_OT}{4I_{Lr\_pk}L_m}\right) \tag{12}$$

$$I_{Lr\_pk} = \frac{I_O\sqrt{\frac{n^4R_L{}^2}{f_r^2L_m{}^2} + 4\pi^2}}{4n} \tag{13}$$

$$V_{Cr\_pk} = I_{Lr\_pk} \cdot Z \tag{14}$$

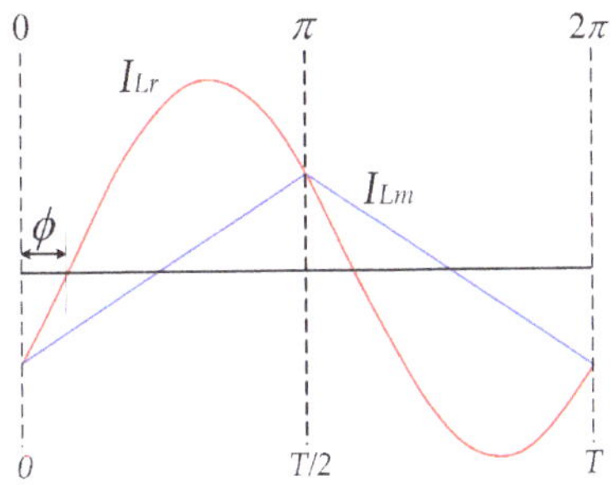

Figure 11. Resonant current waveform at $f_{sw} = f_r$.

5.3. Input and Output Capacitances

The input DC link capacitances can be calculated from the charging/discharging current of the capacitor and the duration time of the voltage sag period. MNRV DPWM determines the duration time of the reference vector to satisfy the magnitude of V_C^* on average during the unit switching period, as shown in Figure 12. Therefore, the relationship among α, β, and V_C^* can be derived as in Equation (15) (assuming γ = 0 when $|V_C^*| \geq 2/3 \times V_{dc}$).

In addition, given that the duty compensation parameters in the steady state are very small, we can assume that $d_E = d_{2E}$ and α = 2β. The input–output voltage relationship of Equation (10) at the resonance point can be simplified as Equation (16) in a high V_C^* case where α and β are small, and α can be expressed as Equation (17). Thus, d_E can be derived via Equation (18).

$$V_C^* = V_{dc} - \frac{2V_{dc}}{3\pi}(\alpha + \beta) = V_{dc}(1 - \frac{\alpha}{\pi}) \tag{15}$$

$$V_{dc} = \frac{nV_O}{1 - \frac{\sin\alpha + \sin\beta}{3}} \approx \frac{nV_O}{1 - \frac{\alpha}{2}} \tag{16}$$

$$\alpha = 2(1 - \frac{nV_O}{V_{dc}}) \tag{17}$$

$$d_E = 1 - \frac{V_C^*}{V_{dc}} = \frac{\alpha}{\pi} = \frac{2}{\pi}(1 - \frac{nV_O}{V_{dc}}) \tag{18}$$

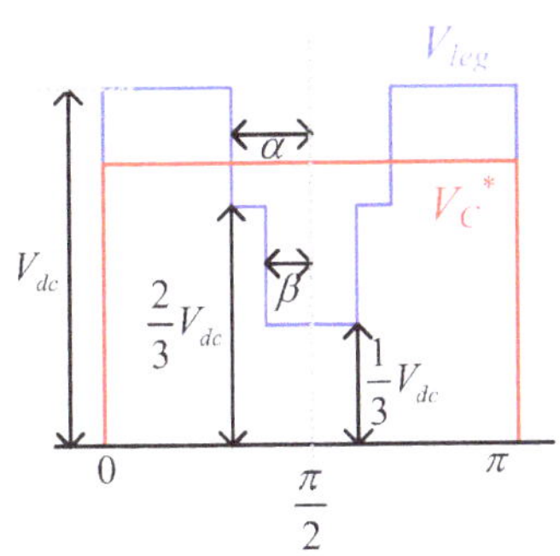

Figure 12. Relationship between V_C^* and V_{leg}.

Figure 13 describes the characteristics of the voltage change of the input DC link capacitors in the LCM. From Equation (19), the increment ΔV_{dc1} during the unit half cycle, which equals decrement ΔV_{dc3}, is related to C_{dc}, f_{sw}, I_{Lr_pk}, α, β, and ϕ. When the target voltage fluctuation amount $\Delta V_{dc,target}$ is determined, the C_{dc} value is given as in Equation (20) using the previously obtained values of I_{Lr_pk}, α, β, and ϕ. If CM alternates between the UCM and LCM at every resonant period in a steady state, the total voltage change per unit capacitor becomes $2 \times \Delta V_{dc}$.

$$\begin{aligned}\Delta V_{dc1} = \Delta V_{dc3} &= \frac{1}{C_{dc}}\left[\begin{array}{l}\int_{\frac{T}{4}-0.5(d_E+d_{2E})\frac{T}{2}}^{\frac{T}{4}-0.5d_E\frac{T}{2}} \frac{2}{3} I_{Lr\_pk}\sin(\omega t-\phi)dt + \int_{\frac{T}{4}-0.5d_E\frac{T}{2}}^{\frac{T}{4}+0.5d_E\frac{T}{2}} \frac{1}{3} I_{Lr\_pk}\sin(\omega t-\phi)dt \\ +\int_{\frac{T}{4}+0.5d_E\frac{T}{2}}^{\frac{T}{4}+0.5(d_E+d_{2E})\frac{T}{2}} \frac{2}{3} I_{Lr\_pk}\sin(\omega t-\phi)dt\end{array}\right] \\ &= \frac{I_{Lr\_pk}}{3\pi C_{dc} f_{SW}}(2\sin\alpha-\sin\beta)\cos\phi\end{aligned} \quad (19)$$

$$C_{dc} \approx \frac{I_O}{\Delta V_{dc,target}} \cdot \frac{\sqrt{\frac{n^4 R_L{}^2}{f_r^2 L_m{}^2}+4\pi^2}}{4n\pi f_r} \cdot \sqrt{1-\frac{n^2 V_O{}^2 T^2}{16 I_{Lr\_pk}{}^2 L_m{}^2}\cdot\left(1-\frac{nV_O}{V_{dc}}\right)} \quad (20)$$

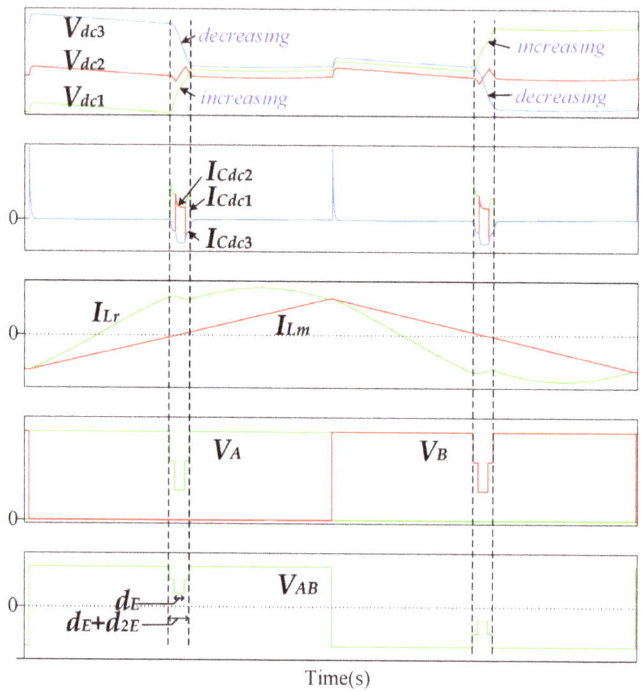

Figure 13. Voltage change characteristics of DC link capacitors in the LCM.

The output capacitor current I_{CO} in the steady state, assuming a continuous current mode, is shown in Figure 14. Because I_{CO} equals the output diode current (I_{DO1} or I_{DO2}) minus I_O, the net charge increment of C_O and Q_{CO} from t_1 and t_2, respectively, is calculated using Equation (21). Here, $t_1 = \sin^{-1}(2/\pi)/\omega$, $t_2 = [\pi - \sin^{-1}(2/\pi)]/\omega$. Thus, to satisfy the target output voltage fluctuation range $\Delta V_{Co,target}$, C_O is determined by Equation (22).

$$Q_{CO} = \int_{t_1}^{t_2} I_{CO}(t)dt = \int_{t_1}^{t_2} I_O\left(\frac{\pi}{2}\sin\omega t - 1\right)dt = 0.105\frac{I_O}{f_{sw}} \quad (21)$$

$$C_O = 0.105 \times \frac{I_O}{\Delta V_{Co,target} f_r} \quad (22)$$

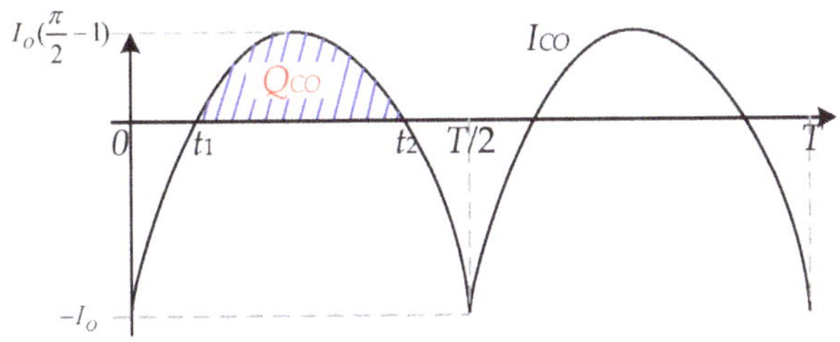

Figure 14. Current waveform of the output capacitor.

6. Four Representative Sag Types of V_{leg} According to the Carrier Deformation

For stable operation of the MNRV DPWM-based full-bridge diode-clamped multilevel resonant converter as described thus far, an appropriate carrier design is required. In DC/DC converters, it is important to maintain the symmetry of the voltage applied to the transformer. In order to maintain the symmetry of V_{leg}, which is the resonant tank input voltage, the carrier counting direction and initial value must be set appropriately according to CM. Figure 15 shows the representative V_{leg} that can be implemented according to the carrier deformation. V_{leg} basically takes the shape of a square wave for maximum power transfer. Additionally, at the middle, edge, or rear of the waveform, taking the form of a stepped sag, i.e., $V_{dc} \rightarrow 2/3 \times V_{dc} \rightarrow 1/3 \times V_{dc} \rightarrow 2/3 \times V_{dc} \rightarrow V_{dc}$, V_{leg} decreases and increases again. Compensation for the voltage deviations of the DC link capacitors is performed in that sag section. Depending on the location of the sag, we refer to these as the middle, edge, rear, and end sags, where sag sections occur respectively at the middle, edge, rear, and end of V_{leg}. In fact, the type described and interpreted thus far corresponds to the middle sag.

V_{leg} sag types are determined by the initial value and the counting direction of the carrier at the time the UCM and LCM start [29]. For example, when the UCM starts, if the carrier decreases from its maximum value and the carrier increases from its minimum value at the moment the LCM starts, it becomes a middle sag type with a dip in the middle. The remaining sag types can also be implemented by properly designing the initial value and the counting direction of the carrier. In addition to these four types, various V_{leg} waveforms can be created depending on the shape of the carrier. Each sag type has distinct characteristics. For the middle sag, the compensation ability for voltage deviation is excellent because the voltage-deviation-compensation operation occurs near the peak value of the resonance current, while the switching losses are greatest due to the large switching current. In addition, the sag occurs in the middle of the square wave, which has the greatest influence on the fundamental component of V_{leg}, meaning that the input C_O output voltage transfer gain is low. However, given that the period in which the output diode conducts is the longest, the RMS values of the output diode current and the resonance current are smallest, which is advantageous when conducting a large current. Regarding the edge sag, because the sag section appears at the edge, the ability to compensate for voltage deviations is somewhat lower, but the fundamental component of V_{leg} is larger than that in the middle sag case. In addition, unlike the middle sag, where all compensation operations for voltage deviation occur under hard switching conditions, the edge sag case can reduce the switching losses because part of the compensation operations for voltage deviation may occur in the ZVS region. The rear sag is characterized as intermediate between the middle sag and the edge sag. All three of these types utilize up and down carriers. On the other hand, the end sag shows the highest efficiency because the number of switching operations is reduced compared to other types because the sag occurs only at the end of the square wave and uses up or down carriers.

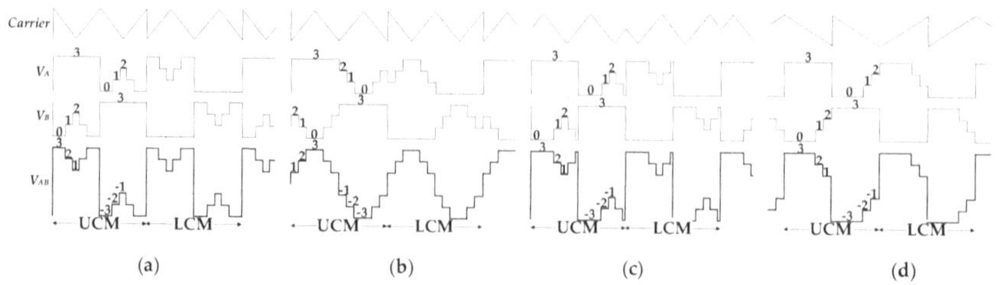

Figure 15. Four typical V_{leg} waveforms depending on the carrier wave shape: (**a**) middle, (**b**) edge, (**c**) rear, and (**d**) end sag types.

7. Simulation and Experiments

7.1. Simulation

To verify the operation of the proposed four-level *LLC* resonant converter, a simulation was conducted with the conditions shown in Table 1.

Table 1. Simulation condition.

P_O (kW)	V_{dc} (V)	V_O (V)	L_r (mH)	L_m (mH)	C_r (nF)	C_{dc} (μH)	C_O (μH)	n	f_{sw} (kHz)
1	700	350	1.5	4.28	168	100	11	1.68	10

Figure 16 presents the simulation results in a steady state for the middle, edge, rear, and end sags. From the top are V_C^*, CM, GateA, GateB, V_{dc}, I_{Cdc}, I_{Lr}, I_{Lm}, I_{DO}, V_A, V_B, I_{QA_upper}, I_{QA_lower}, and I_{DA}. Here, I_{Cdc} is the current flowing through each DC link capacitor, and I_{QA_upper}, I_{QA_lower}, and I_{DA} are the upper and lower switch currents and the clamping diode currents for leg A, respectively. Although the currents of the switches and clamping diodes for leg B are not shown here, their characteristics are symmetric with those of leg A.

All four results are consistent with those described above. In a steady state, CM alternates between the UCM and LCM for every cycle. For every half cycle, one leg is clamped with a positive or negative DC rail, and the switch gates of the other leg that is not clamped are turned on/off in a specific pattern determined by the PWM_CMDs and pre-defined carrier according to the sag types. The resonance current generated according to the magnitude of the fundamental component of the V_{leg} input to the resonance tank is transferred to the secondary side for constant control of the output voltage. In the sag section, voltage deviations among input DC link capacitors are reduced, and when the load or input voltage source changes, instead of sweeping f_{sw}, V_{ampl} changes to control the fundamental component of V_{leg}.

The voltage transfer gain decreased in the order of the edge, end, rear, and middle sag types. Therefore, the magnitude of V_C^* required to maintain a constant output voltage under identical conditions was found to be largest at the middle sag, as expected. On the other hand, the RMS values of the output diode and the resonance currents increased in the order of the middle, rear, end, and edge sag types. Owing to the reduced switching number in the end sag case, the overall efficiency was largest for the end sag. The compensation operation for the deviation of the capacitor voltage was performed in three steps during one cycle on average, and the switching step voltage of the four-level converter was lower by one third compared to that of a two-level converter. Therefore, the switching loss of the end sag type was expected to be nearly identical to that of the conventional two-level converter.

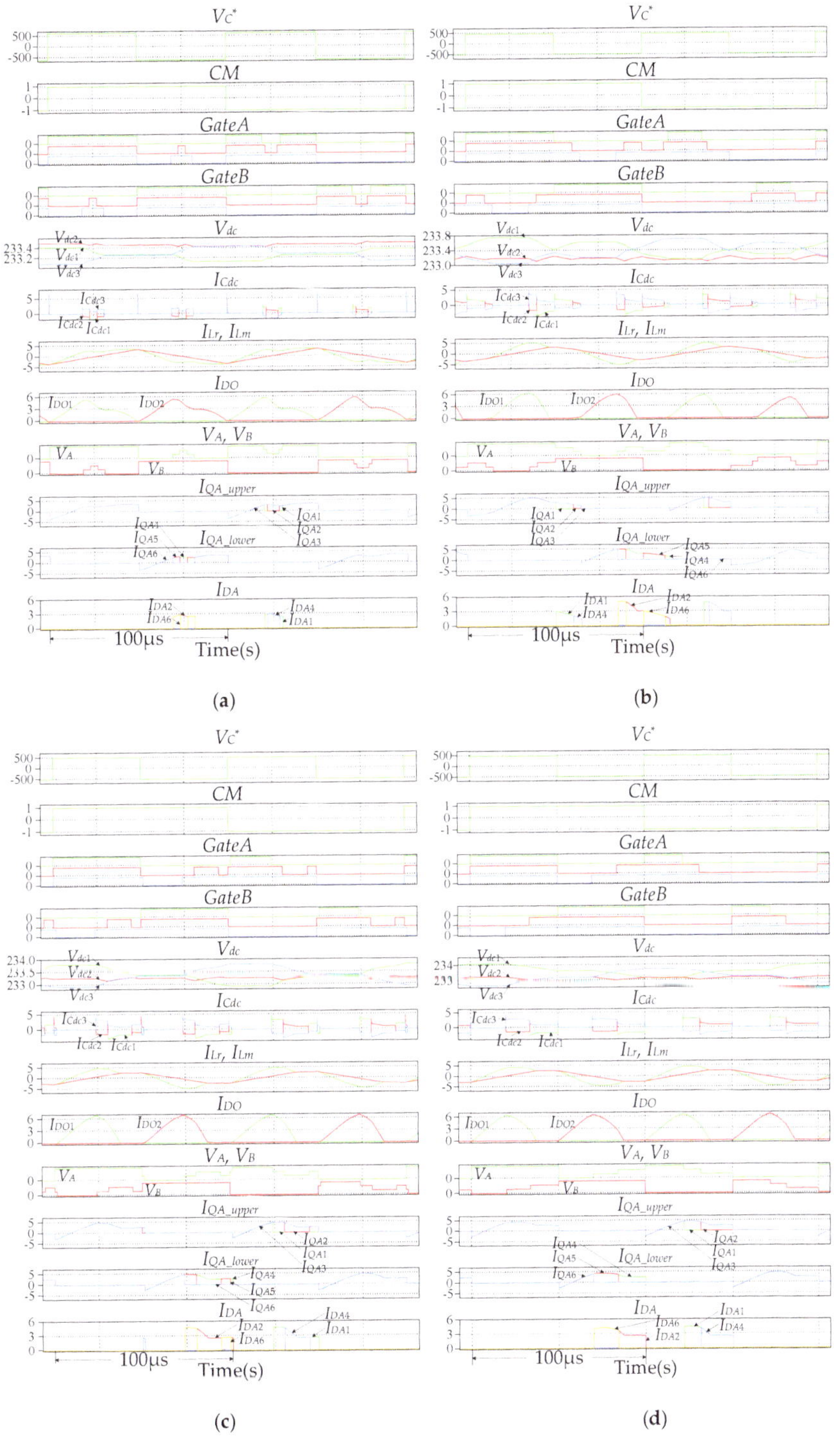

Figure 16. Simulation results for the (**a**) middle, (**b**) edge, (**c**) rear, and (**d**) end sags.

Figure 17 shows the loss analysis result of the switching devices according to the sag types when P_O = 1000 W. Losses were calculated by means of thermal loss modeling with PLECS software, utilizing the datasheets of the actual devices mentioned in Section 7.2 [32]. Here, $P_{cond,Q}$ and $P_{sw,Q}$ are conduction and switching losses of the main switches, respectively. $P_{cond,D}$ and $P_{sw,D}$ are the conduction and switching losses of the clamping diodes, respectively. Lastly, $P_{Cond,DO}$ denotes the conduction losses of the output diodes. As expected, the middle sag had largest switching losses compared to the other types while also showing the smallest conduction losses. Considering the winding losses of the magnetic material, it is predicted that the middle sag type would be more advantageous under a heavy load condition.

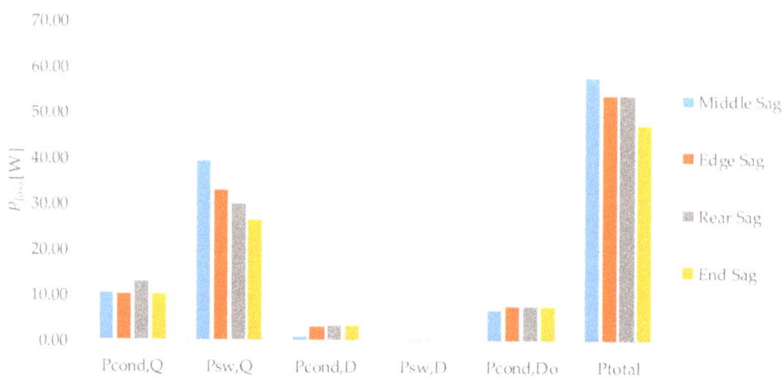

Figure 17. Switches loss analysis for the middle, edge, rear, and end sag types when P_O = 1000 W.

7.2. Experiments

A prototype circuit (Figure 18) was designed to confirm the operation of the proposed converter and to compare the performance outcomes according to the voltage sag types. The experimental conditions are identical to those in Table 1, except that here P_O varied between 500 W and 1500 W. The following devices were used in this experiment. The main switches were Fuji-Electric FGW35N60HD, the clamping diodes were ST-Microelectronic STTH15RQ06-Y, and the output diodes were Vishay VS-HFA06TB120S-M3. The center-tapped transformer had four stacked ferrite cores of the EE6565S with an air gap, and its turn ratio was 47:28:28. The resonant inductor consisted of four stacked ferrite cores of the EE4242S with an air gap. The controller was implemented using TI TMS320F28377D. Increased active switch control, voltage balancing of DC link capacitors, and CM selection increased the amount of computation in the MNRV DPWM compared to conventional two-level converters. Extra logic was also added to handle the clamped PWM CMDs every half cycle and the carrier whose phase is inverted whenever CM changes. It takes a lot of interrupt service routine (ISR) time to set and change the relevant registers of every switching pair that is different from each other. This increased ISR time limits the increase in the switching frequency. In this paper, the PWM switching part was implemented using the embedded ePWM module of the DSP due to the laboratory limitations. In order to overcome this limitation of increasing the switching frequency, it is necessary to implement a PWM switching part by using an FPGA with a parallel operation function and fast processing speed. This allows for a much higher switching frequency than in this experiment.

Figure 19 shows the V_A, V_B, V_{leg}, V_O, and I_{Lr} waveforms according to the voltage sag types when P_O = 1500 W. It can be seen that the output voltages held constant at 350 V in all four voltage sag types and that the V_{leg} values input to the resonance tank differed depending on the sag type, showing unique resonance current shapes. In addition, the selection of CM differed according to the magnitudes of V_{dc1} and V_{dc3}. On the other hand, the RMS value of I_{Lr} was lowest in the middle sag case because the conduction time of the output diode was longest for that sag type. However, because the voltage transfer gain was lowest in the middle sag case, the length of d_{3E} was longest for the middle sag at a higher V_{ampl}, whereas for the edge and end sags, it was conversely short.

Figures 20 and 21 show the waveforms when P_O = 1000 W and 500 W, respectively, showing results similar to those in the previous analysis. As the load decreased, it can be seen that conduction time of the output diode was shortened as the magnitude of the resonance current decreased. Figure 22 confirms the DC link capacitor voltage balancing capability for each voltage sag type. As described above, the voltage imbalance was eliminated by changing the CM based on the voltage deviation of the upper and lower capacitors. Meanwhile, the d_{comp} parameters in Equation (1) had the smallest value in the middle sag case, where the sag operation was performed at the peak point of the resonance current.

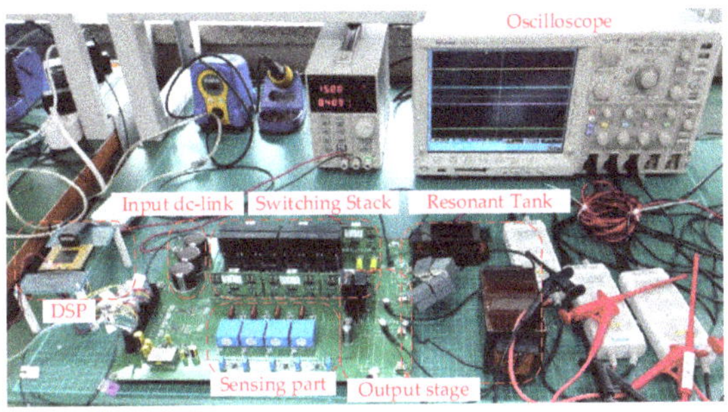

Figure 18. Prototype circuit of the proposed converter.

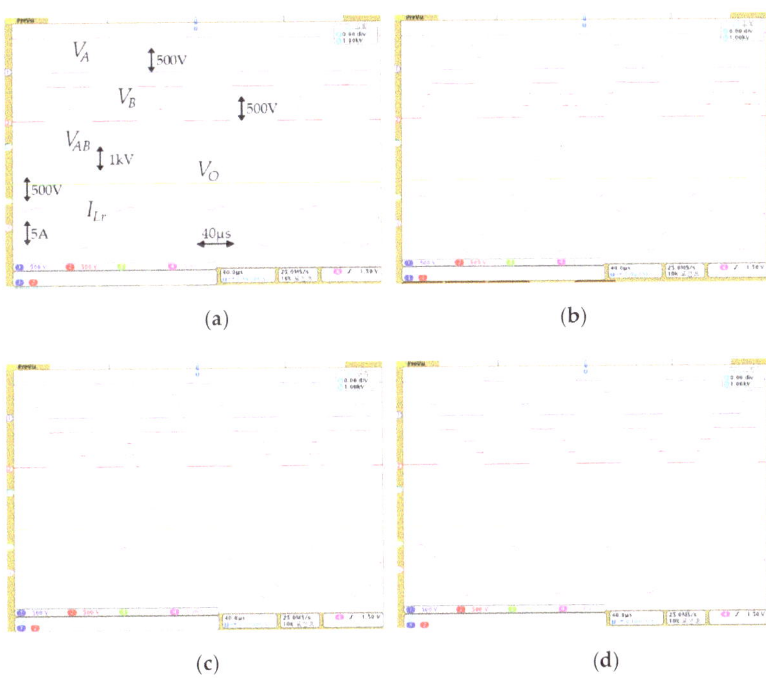

Figure 19. Experimental results for the (**a**) middle, (**b**) edge, (**c**) rear, and (**d**) end sags when P_O = 1500 W.

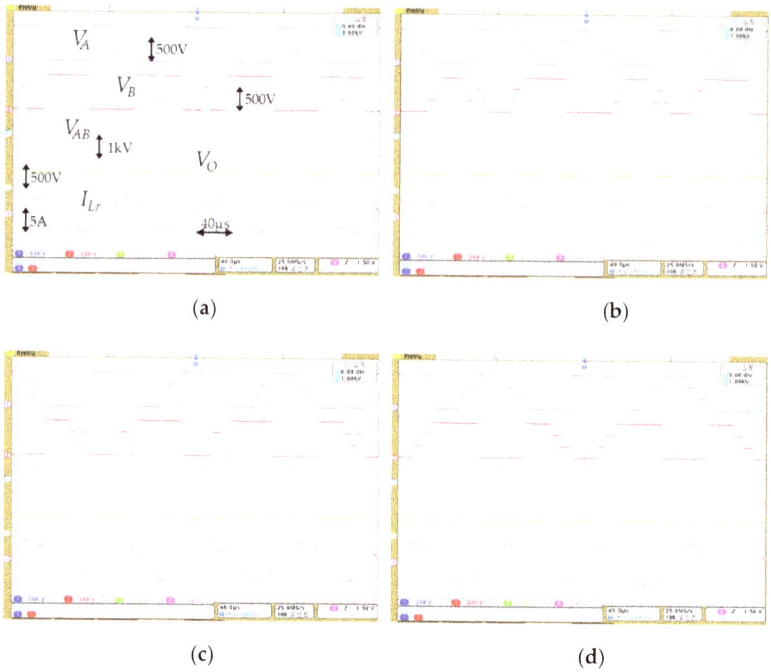

Figure 20. Experimental results for the (**a**) middle, (**b**) edge, (**c**) rear, and (**d**) end sags when P_O = 1000 W.

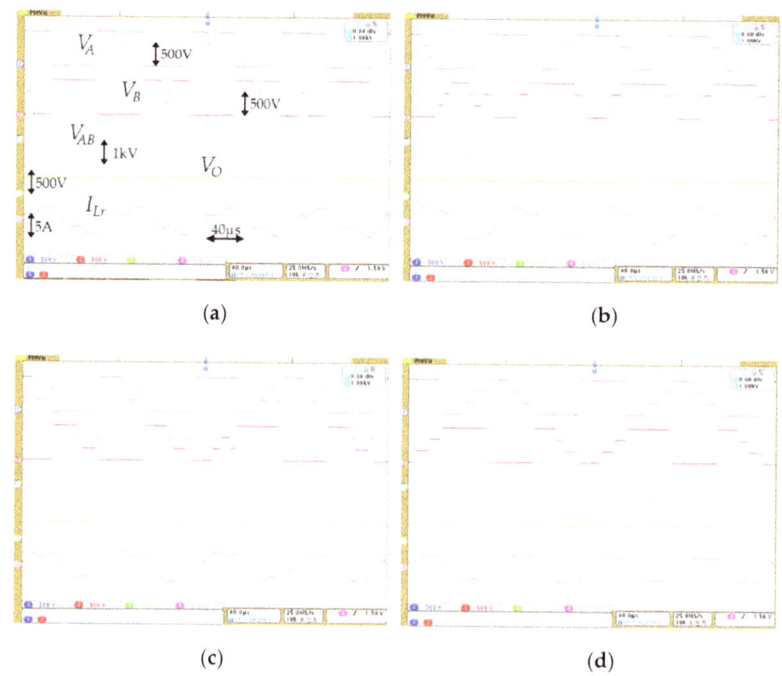

Figure 21. Experimental results for the (**a**) middle, (**b**) edge, (**c**) rear, and (**d**) end sags when P_O = 500 W.

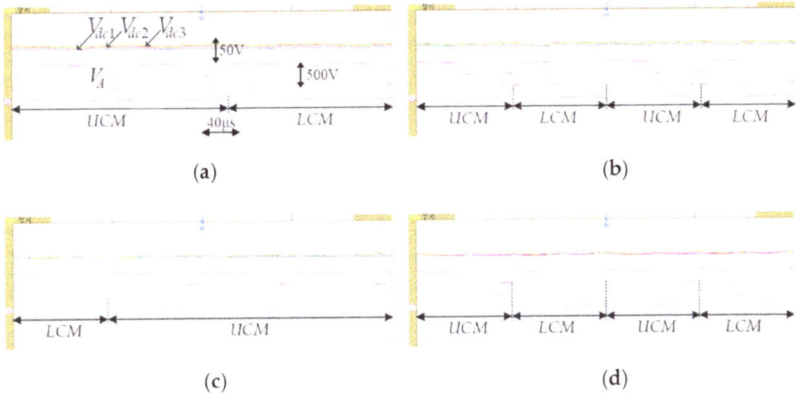

Figure 22. Voltage-balancing performances for the (**a**) middle, (**b**) edge, (**c**) rear, and (**d**) end sags when P_O = 1000 W.

Using a Yokogawa WT333E power meter, the power conversion efficiencies for each sag type were measured and compared as shown in Figure 23. When the load was small, the middle sag type with large switching loss showed the lowest efficiency (91.1% efficiency at P_O = 500 W), but as the load increased, the middle sag type with the smallest RMS values of the resonance current and output diode current showed the highest efficiency (95.2% efficiency at P_O = 1500 W). In fact, in this experiment, because the transformer turn ratio was designed to be small considering the middle sag with a small voltage transfer gain, the period of resonance section was reduced in the edge, rear, and end sag cases. Thus, the resonance currents and the output diode currents of edge, rear, and end sag types became large, resulting in high conduction losses of switches and high copper losses of the transformer and inductor. Therefore, if optimally designed for each sag type, the efficiency rate of the edge, rear, and end sag types can be improved compared to those in this experiment. With an optimal design of the transformer turn ratio for each sag type, the $V_C{}^*$ should be placed at a high level of LVR in its normal operating range.

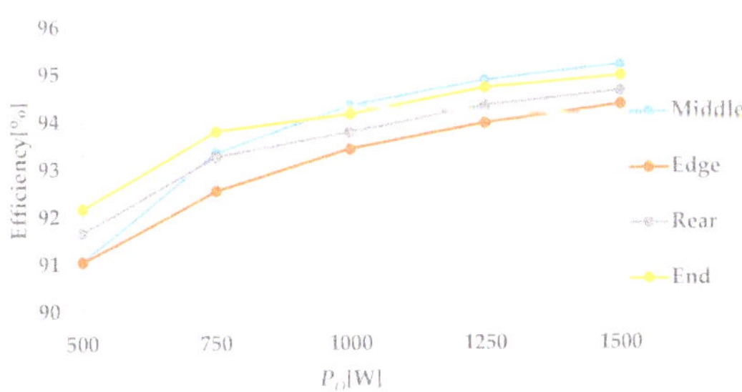

Figure 23. Efficiency comparison for the middle, edge, rear, and end sags.

8. Conclusions

In this paper, we proposed an amplitude-modulating full-bridge diode-clamped multilevel *LLC* resonant converter suitable for power conversion systems that use high input voltages. A new modulation scheme called MNRV DPWM was applied to eliminate voltage deviations in DC link capacitors connected in series. The proposed multilevel resonant converter operates by modulating the magnitude of the fundamental wave input to the resonant tank while fixing the operating frequency at the resonant point. The fixed operating point facilitates the design of passive devices and enables stable operation over a wide operating range with just one power conversion stage. In this paper, we proposed a control algorithm and analyzed the operating characteristics of the proposed diode-clamped four-level *LLC* resonant converter and presented design guidelines. The feasibility and effectiveness of the proposed converter were verified through a simulation and via experimentation with the prototype converter.

Author Contributions: Conceptualization, M.-S.S.; methodology, M.-S.S. and J.-B.L.; software, M.-S.S.; validation, M.-S.S. and J.-B.L.; formal analysis, M.-S.S.; experiments, M.-S.S. and J.-B.L.; writing—original draft preparation, M.-S.S.; writing—review and editing, M.-S.S. and J.-B.L. All authors have read and agreed to the published version of the manuscript.

Funding: This work is supported by the Korea Agency for Infrastructure Technology Advancement (KAIA) grant funded by the Ministry of Land, Infrastructure and Transportation (Grant RS-2022-00143566).

Institutional Review Board Statement: Not applicable.

Informed Consent Statement: Not applicable.

Data Availability Statement: Not applicable.

Conflicts of Interest: The authors declare no conflict of interest.

References

1. Yang, B.; Lee, F.C.; Huang, G. LLC resonant converter for front end dc/dc conversion. In Proceedings of the APEC, Seventeenth Annual IEEE Applied Power Electronics Conference and Exposition (Cat. No.02CH37335), Dallas, TX, USA, 10–14 March 2002.
2. Lu, B.; Liu, W.; Liang, Y.; Lee, F.C.; van Wyk, J.D. Optimal design methodology for LLC resonant converter. In Proceedings of the Twenty-First Annual IEEE Applied Power Electronics Conference and Exposition, Dallas, TX, USA, 19–23 March 2006.
3. Beiranvand, R.; Rashidian, B.; Zolghadri, M.R.; Alavi, S.M.H. Using LLC resonant converter for designing wide-range voltage source. *IEEE Trans. Ind. Electron.* **2010**, *58*, 1746–1756. [CrossRef]
4. Fu, M.; Fei, C.; Yang, Y.; Li, Q.; Lee, F.C. A GaN-based dc–dc module for railway applications: Design consideration and high-frequency digital control. *IEEE Trans. Ind. Electron.* **2019**, *67*, 1638–1647. [CrossRef]
5. Feng, J.; Chu, W.Q.; Zhang, Z.; Zhu, Z.Q. Power electronic transformer-based railway traction systems: Challenges and opportunities. *IEEE Trans. Emerg. Sel. Top. Power Electron.* **2017**, *5*, 1237–1253. [CrossRef]
6. Rodriguez, J.; Lai, J.-S.; Peng, F.Z. Multilevel inverter: A survey of topologies, controls, and applications. *IEEE Trans. Ind. Electron.* **2002**, *49*, 724–738. [CrossRef]
7. Kouro, S.; Malinowski, M.; Gopakumar, K.; Pou, J.; Franquelo, L.G.; Wu, B.; Rodriguez, J.; Pérez, M.A.; Leon, J.I. Recent Advances and Industrial Applications of Multilevel Converter. *IEEE Trans. Ind. Electron.* **2010**, *57*, 2553–2580. [CrossRef]
8. Gupta, K.K.; Ranjan, A.; Bhatnagar, P.; Sahu, L.K.; Jain, S. Multilevel Inverter Topologies with Reduced Device Count: A Review. *IEEE Trans. Power Electron.* **2015**, *31*, 135–151. [CrossRef]
9. Venkataramanaiah, J.; Suresh, Y.; Panda, A.K. A review on symmetric, asymmetric, hybrid and single DC sources based multilevel inverter topologies. *Renew. Sustian. Energy Rev.* **2017**, *76*, 788–812. [CrossRef]
10. Poorfakhraei, A.; Narimani, M.; Emadi, A. A Review of Modulation and Control Techniques for Multilevel Inverters in Traction Applications. *IEEE Access* **2021**, *9*, 24187–24204. [CrossRef]
11. Yazdani, A.; Iravani, R. A neutral-point clamped converter system for direct-drive variable-speed wind power unit. *IEEE Trans. Energy Convers.* **2006**, *21*, 596–607. [CrossRef]
12. Jounanne, A.V.; Dai, S.; Zhang, H. A multilevel inverter providing DC link balancing, ride-through enhancement, and common mode voltage elimination. *IEEE Trans. Ind. Electron.* **2002**, *49*, 739–745. [CrossRef]
13. Pan, Z.; Peng, F.Z.; Corzine, K.A.; Stefanovic, V.R.; Leuthen, J.M.; Gataric, S. Voltage balancing control of diode-clamped multilevel rectifier/inverter systems. *IEEE Trans. Ind. Appl.* **2005**, *41*, 1698–1706. [CrossRef]
14. Pou, J.; Pindado, R.; Boroyevich, D. Voltage-balance limits in four-level diode-clamped converters with passive front ends. *IEEE Trans. Ind. Electron.* **2005**, *52*, 190–196. [CrossRef]
15. Ajami, A.; Shokri, H.; Mokhberdoran, A. Parallel switch-based chopper circuit for DC capacitor voltage balancing in diode-clamped multilevel inverter. *IET Power Electron.* **2014**, *7*, 503–514. [CrossRef]
16. Sano, K.; Fujita, H. Voltage-Balancing Circuit Based on a Resonant Switched-Capacitor Converter for Multilevel Inverters. *IEEE Trans. Ind. Appl.* **2008**, *44*, 1768–1776. [CrossRef]
17. Cui, D.; Ge, Q. A Novel Hybrid Voltage Balance Method for Five-Level Diode-Clamped Converters. *IEEE Trans. Ind. Electron.* **2017**, *65*, 6020–6031. [CrossRef]
18. Monge, S.B.; Bordonau, J.; Boroyevich, D.; Somavilla, S. The nearest three virtual space vector PWM—A modulation for the comprehensive neutral-point balancing in the three-level NPC Inverter. *IEEE Trans. Power Electron.* **2004**, *2*, 11–15.
19. Monge, S.B.; Somavilla, S.; Bordonau, J.; Boroyevich, D. Capacitor voltage balance for the neutral-point-clamped converter using the virtual space vector concept with optimized spectral performance. *IEEE Trans. Power Electron.* **2007**, *22*, 1128–1135. [CrossRef]

20. Hu, C.; Yu, X.; Holmes, D.G.; Shen, W.; Wang, Q.; Luo, F.; Liu, N. An Improved Virtual Space Vector Modulation Scheme for Three-Level Active Neutral-Point-Clamped Inverter. *IEEE Trans. Power Electron.* **2016**, *32*, 7419–7434. [CrossRef]
21. Wang, K.; Zheng, Z.; Xu, L.; Li, Y. A generalized carrier-overlapped PWM method for neutral-point clamped multilevel converters. *IEEE Trans. Power Electron.* **2020**, *35*, 9095–9106. [CrossRef]
22. Wang, K.; Zheng, Z.; Xu, L.; Li, Y. Neutral-Point Voltage Balancing Method for Five-Level NPC Inverters Based on Carrier-Overlapped PWM. *IEEE Trans. Power Electron.* **2021**, *36*, 1428–1440. [CrossRef]
23. Jin, F.; Liu, F.; Ruan, X.; Meng, X. Multi-Phase Multi-Level LLC Resonant Converter with Low Voltage Stress on the Primary-Side Switches. In Proceedings of the 2014 IEEE Energy Conversion Congress and Exposition (ECCE), Pittsburgh, PA, USA, 14–18 September 2014.
24. Zumel, P.; Ortega, L.; Lazaro, A.; Fernandez, C.; Barrado, A. Control strategy for modular Dual Active Bridge input series output parallel. In Proceedings of the 2013 IEEE 14th Workshop on Control and Modeling for Power Electronics (COMPEL), Salt Lake City, UT, USA, 23–26 October 2013.
25. Haga, H.; Kurokawa, F. Modulation method of a full-bridge three level LLC resonant converter for battery charger of electrical vehicles. *IEEE Trans. Power Electron.* **2017**, *32*, 2498–2507. [CrossRef]
26. Wei, Y.; Luo, Q.; Mantooth, A. Overview of Modulation Strategies for LLC Resonant Converter. *IEEE Trans. Power Electron.* **2020**, *35*, 10423–10443. [CrossRef]
27. Song, M.-S. Multi-Neighboring Reference Vector Discontinuous PWM Based on Compensation for Voltage Deviation of the DC-link Capacitors Applicable for Single-Phase NPC Multi-level PWM Converter. *J. Korean Soc. Railw.* **2018**, *21*, 349–362. [CrossRef]
28. Song, M.-S. NPC Multi-level Propulsion System Using Multi-Neighboring Reference Vector Discontinuous PWM. *J. Korean Soc. Railw.* **2019**, *22*, 1–17.
29. Song, M.-S.; Cho, I.-H.; Lee, J.-B.; Ryu, J.-H. MN*R*V DPWM Based Diode-Clamped Multi-level LLC Resonant Converter for High Voltage Utilization. *J. Korean Soc. Railw.* **2021**, *24*, 288–302. [CrossRef]
30. Song, M.-S.; Cho, I.-H.; Lee, J.-B. $\pm 180^{\circ}$ Discontinuous PWM for Single-Phase PWM Converter for High-Speed Railway Propulsion System. *Energies* **2020**, *13*, 1550. [CrossRef]
31. Choi, H.S. Design consideration of half-bridge LLC resonant converter. *J. Power Electron.* **2007**, *7*, 13–20.
32. The Simulation Platform for Power Electronic Systems. Available online: https://www.plexim.com/plecs (accessed on 5 April 2022).

electronics

MDPI

Article

A Nonisolated Transformerless High-Gain DC–DC Converter for Renewable Energy Applications

Mohammad Zaid [1], Ifham H. Malick [1], Imtiaz Ashraf [1], Mohd Tariq [1,*], Basem Alamri [2] and Eduardo M. G. Rodrigues [3,4,*]

1 Department of Electrical Engineering, ZHCET, Aligarh Muslim University, Aligarh 202002, India; zaidahmadzhcet@gmail.com (M.Z.); ifhmal@gmail.com (I.H.M.); iashrafamu@gmail.com (I.A.)
2 Department of Electrical Engineering, College of Engineering, Taif University, P.O. Box 11099, Taif 21944, Saudi Arabia; b.alamri@tu.edu.sa
3 INESC-ID, Sustainable Power Systems Group, Av. Rovisco Pais 1, 1049-001 Lisboa, Portugal
4 Instituto Superior Tecnico, University of Lisbon, Av. Rovisco Pais 1, 1049-001 Lisboa, Portugal
* Correspondence: tariq.ee@zhcet.ac.in (M.T.); eduardo.g.rodrigues@tecnico.ulisboa.pt (E.M.G.R.)

Abstract: Dc–dc converters with a high gain, continuous input current, and common ground are usually employed in renewable energy applications to boost the generated output voltage of renewable energy sources. In this paper, a high-gain dc–dc converter comprising a voltage multiplier cell (VMC) and a common ground with continuous input current and low-voltage stress across semiconductor devices is proposed. The converter produces a voltage gain of about ten times compared to the conventional boost converter at a duty ratio of 50% by utilizing switched capacitors and switched inductors. The simultaneous operation of both the switches with the same gate pulse offers easy and simple control of the proposed converter with a wide range of operations. The boundary operation of the converter is analyzed and presented in both modes, i.e., continuous conduction mode (CCM) and discontinuous conduction mode (DCM). Ideal and nonideal analysis of the converter is carried out by integrating real models of passive elements and semiconductor devices by using PLECS software. The simulation is also used to calculate the losses and hence the working efficiency of the converter. The performance of the converter analyzed in the steady state is compared with various similar converters based on the voltage boosting capability and switching stresses. A hardware prototype is also developed to confirm and validate the theoretical analysis and simulation of the proposed converter.

Keywords: dc–dc converter; high gain; low-voltage stress; voltage multiplier cell; nonisolated converter

Citation: Zaid, M.; Malick, I.H.; Ashraf, I.; Tariq, M.; Alamri, B.; Rodrigues, E.M.G. A Nonisolated Transformerless High-Gain DC–DC Converter for Renewable Energy Applications. *Electronics* **2022**, *11*, 2014. https://doi.org/10.3390/electronics11132014

Academic Editors: Diego Bellan and Jelena Loncarski

Received: 4 April 2022
Accepted: 6 June 2022
Published: 27 June 2022

Publisher's Note: MDPI stays neutral with regard to jurisdictional claims in published maps and institutional affiliations.

1. Introduction

High-gain boost converters have gained prominence in recent times due to their suitability in various applications such as solar PV systems [1], switch-mode power supplies (SMPS), electric vehicles, and aerospace applications. A conventional boost converter has certain shortcomings such as discontinuous input and output currents and increased losses in the system at higher duty cycle operations, which leads to its limited solar PV and fuel cell applications. The traditional quadratic boost converter (TQBC) shown in Figure 1 uses a single switch of high voltage and a current rating which results in low efficiency at higher duty ratios. In the literature, many high-gain dc–dc converters are being published [2,3] to overcome the limitations of the conventional boost and quadratic boost converter. The use of coupled inductors, Z-source converters, switched inductors, switched capacitors, and voltage multiplier cells (VMCs) are popular methods to increase the gain of high-gain dc–dc converters.

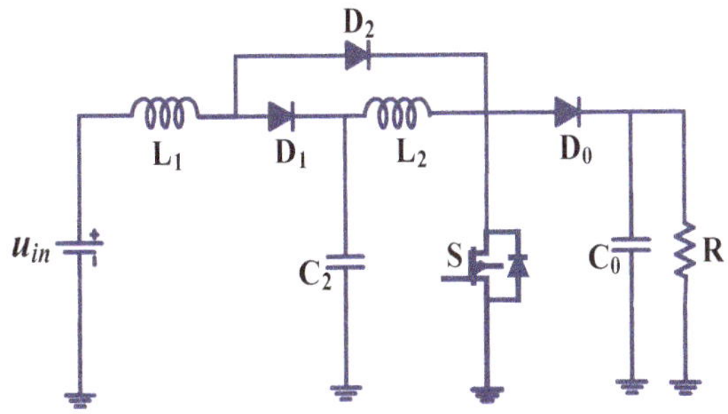

Figure 1. A traditional quadratic boost converter (TQBC).

Isolated step-up quadratic boost converters are implemented in [4,5] with coupled inductors that have a high-voltage gain and low switching stress. The converter in [4] has a voltage doubler cell and produces a higher gain as compared to [5]. Soft switched dc–dc converters eliminate the switching losses [6,7] however their hardware implementation and operation are difficult. Various Z-source converters [8–11] have been proposed in the literature that contains a unique impedance network. In [10], a quasi-Z source converter is implemented using a switched capacitor and switched inductor configuration for achieving a high-voltage gain along with low-voltage stress on capacitors and a continuous input current, but the topology lacks a common ground. In [11], the hybrid quasi-Z source converter is proposed by a combination of traditional Z source converters.

A set of nonisolated switched inductors and switched capacitor converters are discussed in [12–17]. The converter proposed in [12] has a switched capacitor and quadratic gain with a common ground and low input current ripples. The converter is lightweight due to the absence of a transformer and can be utilized for solar PV applications. In [13], the authors have proposed a switched inductor boost converter that has low current stress in inductors but suffers from a low-voltage gain and high-switch stresses. In [14], the twin duty cycle and three switches configuration add to the complexity of the circuit [14,15] as compared to [16] which has two switches following the same duty cycle. A non-isolated converter is proposed in [17] and has a simple design as compared to [15,16] along with a high gain and reduced voltage stresses on switches, diodes, and capacitors. It has a symmetrical structure that helps to ease the choice of the components and its design. In [18], a single switch nonisolated topology is implemented with low switch stress. In [19], the transformerless converter shown in Figure 2 is proposed. It consists of a VMC, with a gain of six times that of a conventional boost converter with a single switch and two inductors. The converter shown in Figure 2 has a single switch but the converter lacks a common ground. Moreover, the input current pulsates. The converter proposed in this paper has a common ground and continuous input current.

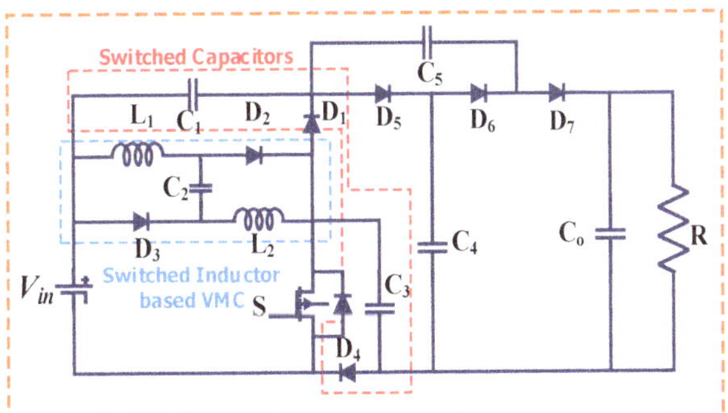

Figure 2. Converter proposed in [19].

In [20], a modified Ćuk converter is proposed with switched capacitors. The authors have proposed a buck-boost converter in [21] for renewable energy applications with continuous input. The converter exhibits high efficiency above 95% around the 56.5% duty cycle; however, the switching stress also increases drastically. In [22], an extendable boost converter employing active-passive inductor cells (APIC) is proposed. The efficiency and voltage gain at a lower duty ratio decrease drastically as the number of APICs is increased. The authors in [23] have introduced an extended boost converter by implanting a cell consisting of switched capacitors and inductors between a conventional boost converter. The converter has the same voltage gain for the complementary duty ratio, which is nearly constant, while the duty ratio is varied from 30% to 70%. A quadratic boost converter is proposed in [24]. The gain of the converter is lower as compared to similar topologies. A variety of quadratic boost converters are introduced in [25–30], each with its set of advantages and disadvantages. The authors in [25] have proposed a simple quadratic boost converter, but it lacks a common ground, whereas a modified quadratic boost converter in [26] is implemented, resulting in twice the gain as compared to the quadratic boost converter. The converter proposed in [27] has a common ground but a lower gain as compared to the topology proposed in [28], which lacks a common ground. The quadratic converter proposed by the authors in [29] has more components as compared to [30] but has a higher gain at the cost of a decreased efficiency due to increased heat losses.

In this paper, a new transformerless high-gain dc–dc converter is implemented that employs a VMC to boost the voltage. The attractive features of the proposed topology are

- The converter achieves a high gain of 10 times the conventional boost and 5 times that of the quadratic boost converter at a 50% duty ratio.
- The converter has low voltage stress of 5% of the output voltage across switch S_1.
- The converter uses the same gate signal for both switches, which leads to its easy operation.
- The topology has a continuous input current and a common ground, making it feasible for PV applications.

The paper discusses the structure of the proposed topology and its detailed ideal analysis in Section 2 followed by the nonideal analysis in Section 3. The comparison of the proposed topology with various similar converters is carried out in Section 4. The simulation and the experimental results are presented in Section 5 along with the efficiency of the proposed converter at different input voltages and power.

2. Structure of Proposed High-Gain DC–DC Converter Topology

The components of the proposed nonisolated dc–dc converter topology depicted in Figure 3 consist of a fixed DC input voltage source V_{in}; two switches (MOSFETs) S_1 and S_2; three inductors L_1, L_2, and L_3; seven diodes D_1, D_2, D_3, D_4, D_5, D_6, and D_O; five capacitors C_1, C_2, C_3, C_4, and C_O; and a resistive load R_O with an output voltage of V_O. The circuit also consists of a VMC with elements L_2, L_3, C_4, D_5, and D_6. The nonisolated converter does not include a high-frequency transformer and hence contributes to its low bulkiness, size, and cost reduction. This further aids in easier control of the topology as compared to isolated converters.

2.1. Operation of the Converter in Continuous Conduction Mode (CCM)

The proposed converter operates in two modes in each cycle while working in the CCM state. Each mode is discussed in detail by providing the voltage and current equations of each component in both modes.

Mode 1 ($0 \leq t \leq t_O$): In mode 1, both switches S_1 and S_2 are turned ON simultaneously, as shown in Figure 4, by applying the appropriate gate pulse. In this mode, diodes D_2, D_4, D_5, and D_6 are forward-biased, whereas the other diodes D_1, D_3, and D_O are reverse-biased. The inductor L_1 is charged through the input voltage source and the inductors L_2 and L_3 are charged through the capacitor C_3. The capacitor C_3 is discharged, while the capacitor C_5 is charged in this mode. The output capacitor C_O discharges and feeds the load to maintain a constant voltage V_O across the output load resistor.

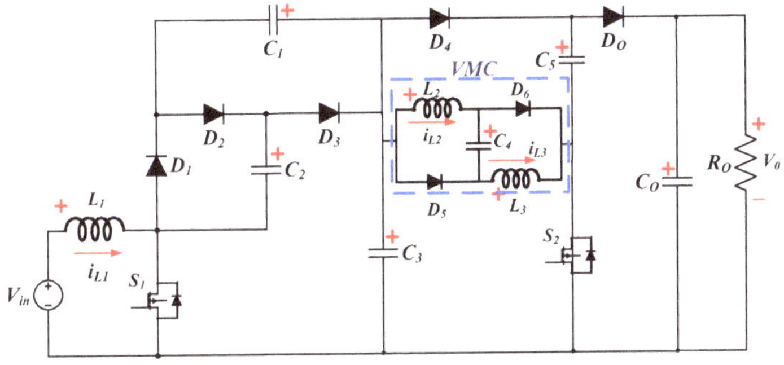

Figure 3. Proposed high-gain dc–dc converter topology.

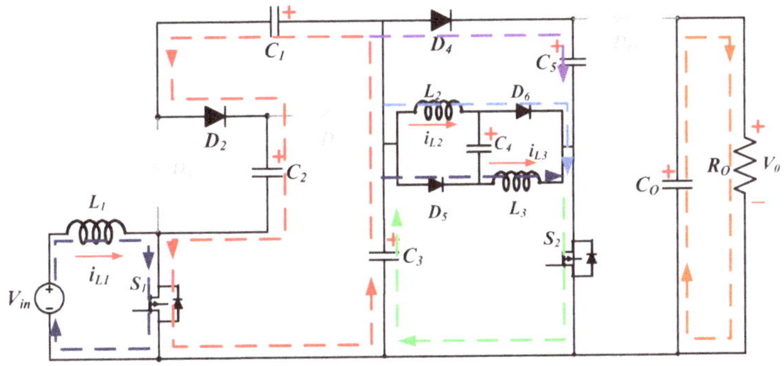

Figure 4. Conduction diagram of the converter in ON-state (Mode 1).

The voltage and current equations across the inductors and capacitors during mode 1 when both switches are ON can be expressed as (1) and (2), respectively.

$$\begin{cases} V_{L1} - V_{in} \\ V_{CO} = V_O \\ V_{C3} = V_{C1} + V_{C2} \\ V_{L2} = V_{L3} = V_{C3} = V_{C5} = -V_{C4} \end{cases} \tag{1}$$

$$\begin{cases} I_{L_1} = I_{in} \\ I_{C1} = I_{C2} = I_{D2} = I_{S1} - I_{in} \\ I_{S_2} = I_{in} + i_{C3} - I_{s1} = I_{C3} - I_{C1} \\ I_{D4} = I_{C5} \\ I_{CO} = I_O \\ I_{C4} = I_{L2} \end{cases} \tag{2}$$

Mode 2 ($t_O \leq t \leq T$): In mode 2, both switches S_1 and S_2 are turned OFF simultaneously. In Figure 5, it can be seen that diodes D_1, D_3 and D_O are forward-biased, whereas the remainder of diodes D_2, D_4, D_5 and D_6 are reverse-biased. In this mode capacitor, C_3 is charged through capacitors C_1 and C_2. Inductors L_2 and L_3 along with capacitor C_4 transfer power to load and charge, and capacitor C_O maintains a constant voltage V_O about the load. The voltage and current equations across the inductors and capacitors during Mode 2 when both switches are OFF can be expressed as (3) and (4), respectively.

$$\begin{cases} V_{C1} = V_{C2} \\ V_{L1} = V_{in} + V_{C1} - V_{C3} = V_{in} - V_{C1} \\ V_{C0} = V_0 \\ V_{C3} = V_{L2} + V_{L3} + V_{C4} - V_{C5} + V_{Co} = V_{L2} + V_{L3} + V_{C4} - V_{C5} + V_O \\ V_{L2} = V_{L3} = \frac{1}{2}(3\,V_{C3} - V_0) \end{cases} \tag{3}$$

$$\begin{cases} I_{C1} = I_{C2} = I_{D1} = I_{D2} = \frac{-I_{in}}{2} \\ I_{C4} = I_{C5} = I_{Do} = I_{in} + I_{C3} \\ I_{C0} = I_0 - I_{in} - I_{C3} \\ I_{L2} = I_{C4} \end{cases} \tag{4}$$

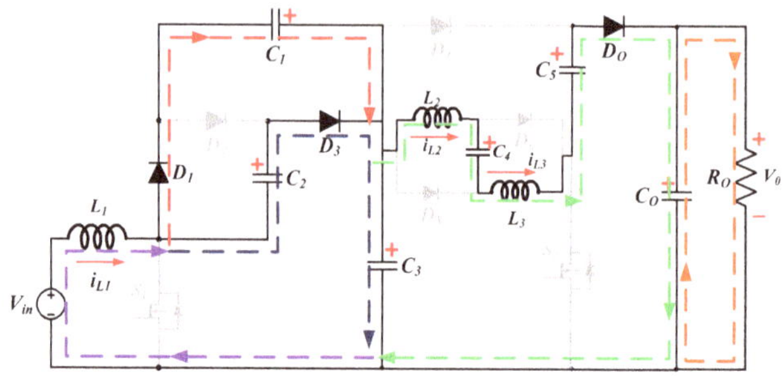

Figure 5. Conduction diagram of the proposed converter in OFF state (Mode 2).

The ideal voltage gain (M) can be calculated by using the volt-sec balance principle in an inductor in CCM operation as per the following relation given in (5),

$$\frac{1}{T}\left(\int_0^{DT} V_{LON} dt + \int_{DT}^{T} V_{LOFF} dt \right) = 0 \tag{5}$$

Applying the volt-sec principle in the inductor L_1

$$\begin{cases} \frac{1}{T}\left(\int_0^{DT} V_{in} dt + \int_{DT}^{T} (V_{in} - V_{C1}) dt \right) = 0 \\ V_{C1} = V_{C2} = \frac{V_{in}}{(1-D)} = \frac{V_{C3}}{2} \end{cases} \tag{6}$$

Applying the volt-sec balance principle to L_2

$$\begin{cases} \frac{1}{T}\left(\int_0^{DT} V_{C3} dt + \int_{DT}^{T} \frac{(3V_{C3} - V_O)}{2} dt \right) = 0 \\ V_{C3} = 2V_{C1} = \frac{(1-D)V_O}{(3-D)} \end{cases} \tag{7}$$

Equating Equations (6) and (7)

$$M_{CCM} = \frac{V_0}{V_{in}} = \frac{2(3-D)}{(1-D)^2} \tag{8}$$

where 'D' denotes the duty cycle for switches S_1 and S_2.

The voltages across the capacitors and inductors (shown in Figure 6) are represented in (9) and can be easily evaluated by applying KVL in respective loops as,

$$\begin{cases} V_{C1} = V_{C2} = \frac{(1-D)V_O}{2(3-D)} \\ V_{C3} = 2V_{C1} = \frac{(1-D)V_O}{(3-D)} \\ V_{C4} = -V_{C5} = -V_{C3} = \frac{-(1-D)V_O}{(3-D)} \\ V_{CO} = V_0 \\ V_{L1_{OFF}} = \frac{(-D)V_{in}}{(1-D)} = \frac{-D(1-D)V_O}{2(3-D)} \\ V_{L2_{OFF}} = V_{L3_{OFF}} = \frac{-(1+D)V_{in}}{2(3-D)} = \frac{-(1+D)V_O}{(1-D)^2} \end{cases} \quad (9)$$

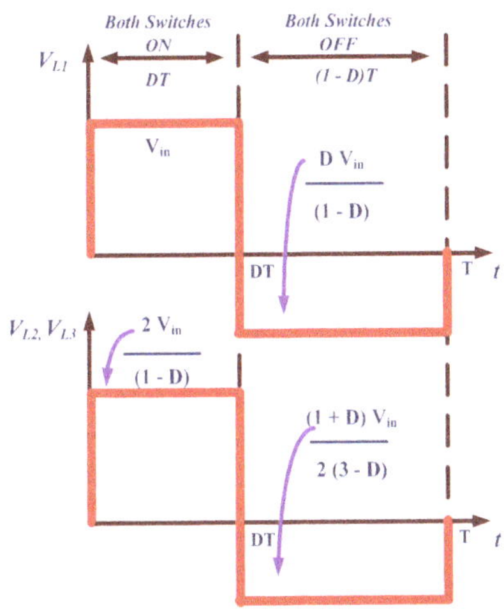

Figure 6. Inductor voltages.

2.1.1. Calculation of Voltage Stress

The voltage stress across the switches as shown in Figure 7 and the diodes shown in Figures 8 and 9 are evaluated at the instance when the switches are not conducting and are in the OFF state. The voltage stress particularly refers to the peak inverse voltage across the switch.

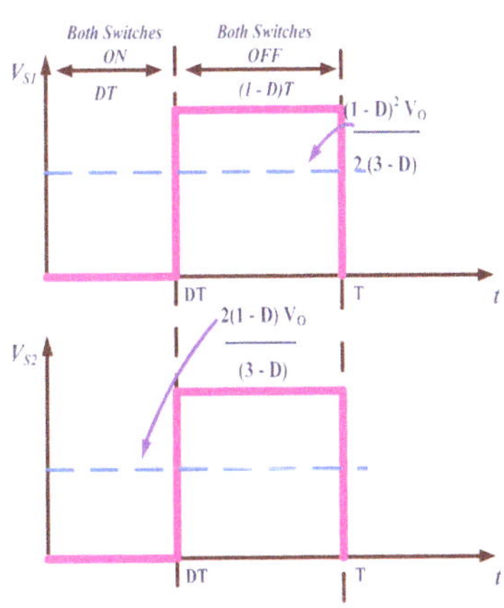

Figure 7. Switch voltages.

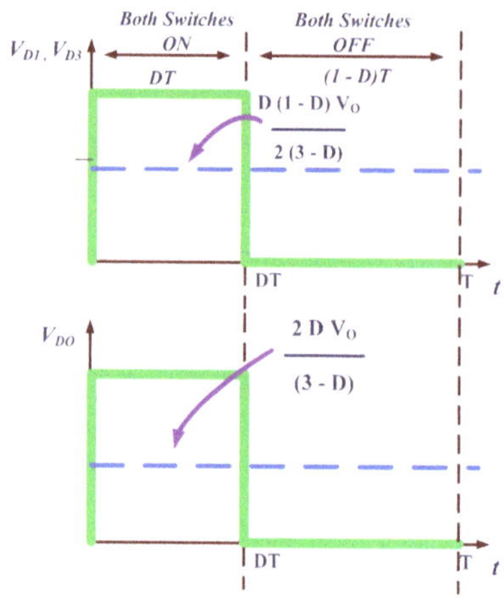

Figure 8. Diode voltages.

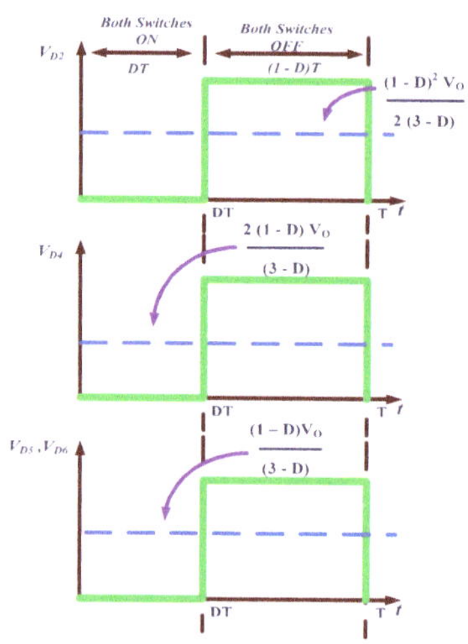

Figure 9. Diode voltages.

The voltage stresses across the switches in the OFF state are given as,

$$\begin{cases} V_{S1} = (1 - D)V_{C1} = \frac{(1-D)^2 V_o}{2(3-D)} \\ V_{S2} = (1-D)(V_{C5} - V_o) = \frac{-2(1-D)\ V_o}{(3-D)} \end{cases} \quad (10)$$

The voltage stress across different diodes when they are not conducting can be expressed as in (11):

$$\begin{cases} V_{D1} = V_{D3} = D\ V_{C1} = \frac{D(1-D)V_o}{2(3-D)} \\ V_{D2} = (1-D)V_{C1} = \frac{(1-D)^2 V_o}{2(3-D)} \\ V_{D4} = (1-D)(V_{C3} - V_O) = \frac{-2(1-D)\ V_o}{(3-D)} \\ V_{D5} = V_{D6} = \frac{-(1-D)V_o}{(3-D)} \\ V_{D0} = \frac{-2D\ V_o}{(3-D)} \end{cases} \quad (11)$$

2.1.2. Calculation of Average and RMS Currents

The average currents through the capacitors and switches can be easily determined by applying the current-sec balance principle in capacitors.

$$\frac{1}{T}\left(\int_{0}^{DT} I_{C_{ON}}dt + \int_{DT}^{T} I_{C_{OFF}}dt\right) = 0. \tag{12}$$

The average switch currents (in Figure 10) are given as,

$$\begin{cases} I_{S1_{avg}} = I_{L1} - I_{D1} = \frac{(3-D)(1+D)i_o}{D(1-D)^2} \\ I_{S2_{avg}} = 2I_{L2} = \frac{2i_o}{(1-D)} \end{cases} \tag{13}$$

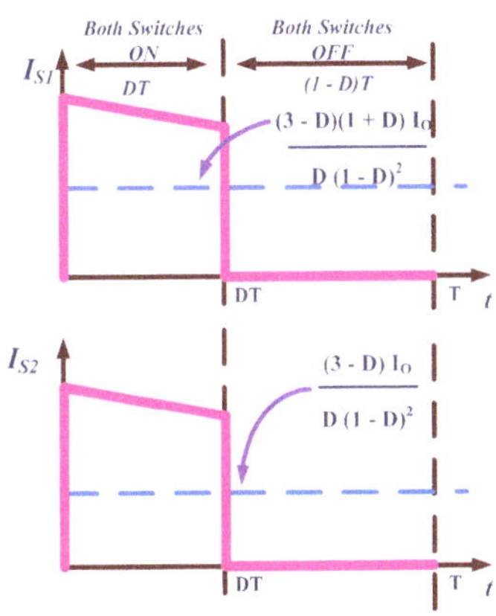

Figure 10. Switch currents.

The average inductor currents (in Figure 11) are given as,

$$\begin{cases} i_{L2_{avg}} = i_{L3_{avg}} = \frac{i_o}{(1-D)} \\ i_{DO_{avg}} = i_{D4_{avg}} = i_o \\ i_{L1_{avg}} = i_{in} = \frac{2(3-D)i_o}{(1-D)^2} \\ i_{D5_{avg}} = i_{D6_{avg}} = 2i_{L2_{avg}} = \frac{2i_o}{(1-D)} \\ i_{D1_{avg}} = i_{D2_{avg}} = i_{D3_{avg}} = \frac{(3-D)i_o}{D(1-D)} \end{cases} \tag{14}$$

The RMS values of the switch currents are given as,

$$\begin{cases} i_{S1_{rms}} = \frac{(3-D)(1+D)i_o}{D\sqrt{D}(1-D)^2} \\ i_{S2_{rms}} = \frac{2i_o}{\sqrt{D}(1-D)} \end{cases} \tag{15}$$

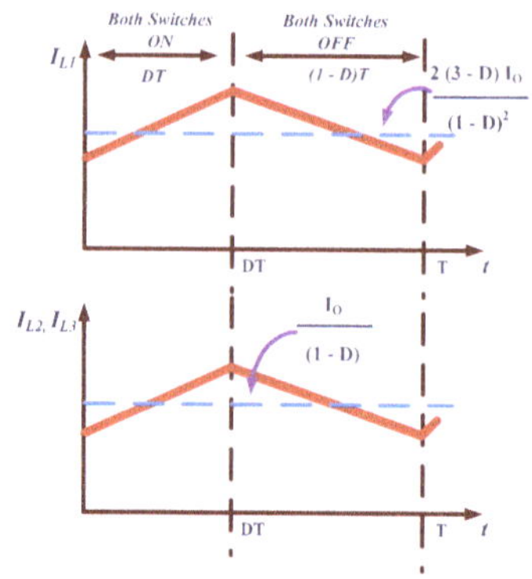

Figure 11. Inductor currents.

The RMS values of the various currents through the components are given as,

$$\begin{cases} i_{L2_{rms}} = i_{L3_{rms}} = \frac{i_o}{(1-D)} \\ i_{DO_{rms}} = i_{D4_{rms}} = \frac{i_o}{\sqrt{D}} \\ i_{L1_{rms}} = i_{in} = \frac{2(3-D)i_o}{(1-D)^2} \\ i_{D5_{rms}} = i_{D6_{rms}} = 2i_{L2_{rms}} = \frac{2i_{io}}{\sqrt{D}(1-D)} \\ i_{D1_{rms}} = i_{D2_{rms}} = i_{D3_{rms}} = \frac{(3-D)i_O}{D\sqrt{D}(1-D)} \\ i_{C1_{rms}} = i_{C2_{rms}} = \frac{3-D}{(1-D)^2}\sqrt{\frac{1-D}{D}}i_o \\ i_{CO_{rms}} = \sqrt{\frac{D}{1-D}}i_o \\ i_{C3_{rms}} = \frac{(5-D)}{(1-D)\sqrt{D\,(1-D)}}i_o \\ i_{C5_{rms}} = i_{C4_{rms}} = \frac{1}{\sqrt{D\,(1-D)}}i_o \end{cases} \quad (16)$$

2.1.3. Design of Inductors and Capacitors

The ripple current in the inductor L_1 can be found as in (17) and rearranged to obtain the critical value of the inductor for CCM operation.

$$\begin{cases} (\Delta i_{L1})_{ON} = \frac{V_{in}}{L_1}DT \\ L_{1_{Cri}} = \frac{V_{in}}{(\Delta i_{L1})_{ON} f_s}D = \frac{D(1-D)^2\,V_O}{2(3-D)\Delta i_{L1} f_s} \end{cases} \quad (17)$$

The ripple current and the critical value of the inductors L_2 and L_3 can be found as

$$\begin{cases} (\Delta i_{L2})_{ON} = \frac{V_{C3}}{L_2}(1-D)T \\ L_{2_{Cri}} = L_{3_{Cri}} = \frac{D(1-D)V_O}{(3-D)\Delta i_{L2} f_s} \end{cases} \quad (18)$$

The peak-to-peak ripple voltages across the capacitors can be given as

$$\begin{cases} \Delta V_{c1} = \Delta V_{c2} = \frac{(3-D)V_0}{(1-D)R_O C_1 f_s} \\ \Delta V_{c3} = \frac{(5-D)V_0}{(1-D)R_O C_3 f_s} \\ \Delta V_{c4} = \Delta V_{c5} = \frac{V_0}{R_O C_4 f_s} \\ \Delta V_{cO} = \frac{DV_0}{R_O C_O f_s} \end{cases} \quad (19)$$

2.2. Operation of Converter in Discontinuous Conduction Mode (DCM)

The proposed converter can also operate in DCM. This mode consists of three different sub-modes of operation, as depicted in Figure 12.

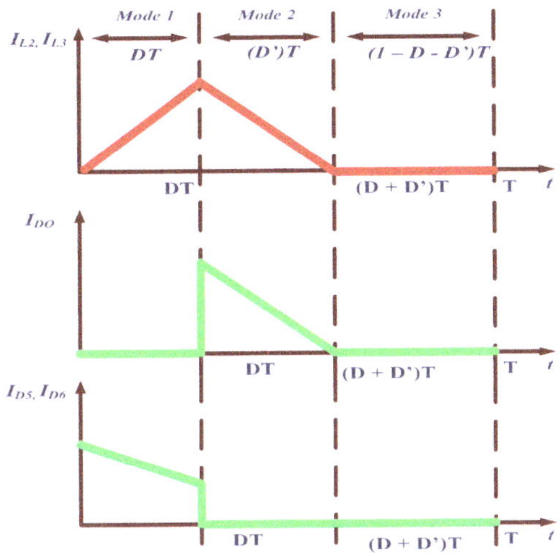

Figure 12. Typical waveforms during DCM.

- **(i)** *Mode 1*: In this mode, both switches are turned ON for duty cycle D as in the case of CCM.
- **(ii)** *Mode 2*: In this mode, both switches are turned OFF and the inductors start discharging. The inductors discharge through diodes D_1, D_3, and D_O for a duty cycle D′ and the mode ends at D + D′.
- **(iii)** *Mode 3*: In this mode, none of the switches or diodes conduct, and the load is fed entirely through the output capacitor C_O. The mode is operated for a duration of 1 − D − D′.

The inductors are charged in Mode 1 (from t = 0 to t = DT) and are discharged in Mode 2 (from t = DT to t = D′T). Hence, by applying volt-sec balance across the inductors, we obtain the voltage gain in DCM.

Applying volt-sec balance in inductor L_1, we obtain,

$$V_{c1} = \frac{(D+D')V_{in}}{D'^2} \tag{20}$$

Applying volt-sec in L_2, we obtain,

$$V_{c1} = \frac{D'V_O}{2(2D+3D')} \tag{21}$$

From Equating (20) and (21), we obtain,

$$M_{DCM} = \frac{V_O}{V_{in}} = \frac{2(D+D')(2D+3D')}{D'^2} \tag{22}$$

Rearranging Equation (22), we obtain,

$$D' = \frac{\sqrt{D^2+4M_{DCM}}-5D}{6-M_{DCM}} \tag{23}$$

Under DCM analysis for inductor L_2, we have,

$$\begin{cases} V_{L2} = L_2 \frac{\Delta i_{L2}}{DT} \\ \Delta i_{L2} = \frac{V_O D' DT}{L_2(2D+3D')} \end{cases} \tag{24}$$

For DCM operation,

$$I_{L2} = \frac{\Delta i_{L2}}{2} = I_O \tag{25}$$

From (24) and (25) we obtain,

$$D' = \frac{4\tau f_s D}{D - 6\tau f_s} \tag{26}$$

where τ is the time constant represented as $\frac{L_2}{R_O}$ and f_s is the switching frequency of the switches.

Equating (25) and (26), we obtain,

$$M_{DCM} = \frac{D(D - 2\tau f_s)}{4\,(\tau f_s)^2} \tag{27}$$

Let $K_e = \tau f_s$; then,

$$M_{DCM} = \frac{D(D - 2K_e)}{4\,(K_e)^2}. \tag{28}$$

2.3. Converter Operation at Boundary Conditions

The boundary condition represents the critical conduction state of the converter on the boundary of DCM and CCM. The relation for the boundary condition can be easily formulated by equating the voltage gains in the CCM and DCM operations, respectively. The boundary condition is determined by the boundary-normalized time constant of the inductor L_2, is as shown in Figure 13. The variation in the time constant of the inductor as a function of the duty cycle is represented as

$$K_{ec} = \frac{D(1-D)}{8(3-D)}\left(D + \sqrt{25 - 8D} - 1\right) \tag{29}$$

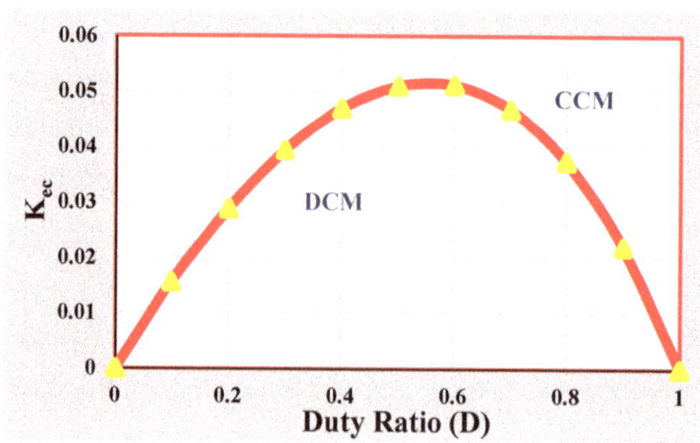

Figure 13. Boundary normalized inductor time constant versus the duty ratio.

The maximum value of K_{ec} is 0.00516 obtained at D = 0.55. The converter operates in DCM mode for a time constant less than K_{ec} and in CCM mode for a time constant more than K_{ec}.

3. Nonideal Analysis

The nonideal analysis of the converter accounts for the power loss analysis of the circuit. Thus far, only the ideal lossless analysis of the converter is being considered in the paper, ignoring the parasitic series resistances of the inductors, equivalent series resistance (ESR) of the capacitors, barrier potential voltage and leakage resistances of the diode, and the ON-state resistance of the switches (MOSFET), as shown in Figure 14. The loss analysis is proceeded by the inclusion of the above-mentioned parameters in the ideal circuit of the proposed topology. The nonidealities of the elements tend to decrease the voltage gain of the converter and also result in an increase in power losses. The total power loss across all the elements is given as,

$$P_{loss_{total}} = \sum_{i=1}^{2} P_{Si_{loss}} + \sum_{i=0}^{6} P_{Di_{loss}} + \sum_{i=0}^{5} P_{Ci_{loss}} + \sum_{i=1}^{3} P_{Li_{loss}} \tag{30}$$

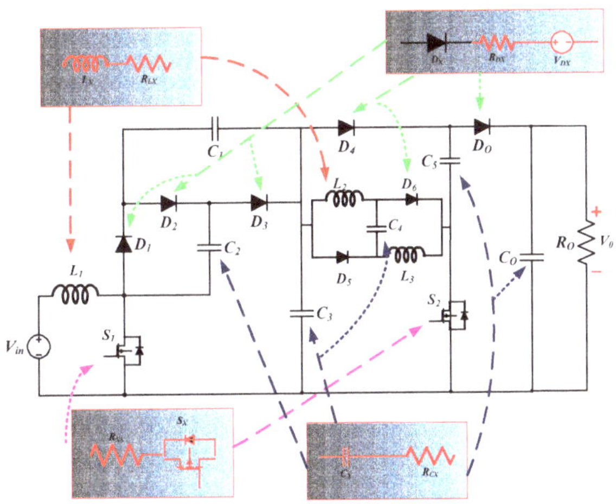

Figure 14. Nonideal realization of the proposed topology.

3.1. Calculation of Losses across Switches

The losses across switches can be either conduction losses that are encountered when the switch is ON and current flows through it, or they may be switching losses that are encountered when the switch is in the transition from the ON state to OFF state or vice-versa.

$$P^{S}_{loss_{total}} = P^{S_{1,2}}_{loss_{conduction}} + P^{S_{1,2}}_{loss_{switching}} \tag{31}$$

For the conduction loss calculations, the resistance of both switches is assumed to be the same, i.e., $r_{S_1} = r_{S_2} = r_S$.

$$\begin{cases} P^{S_{1,2}}_{loss_{conduction}} = i^2_{S1_{rms}} r_{S1} + i^2_{S2_{rms}} r_{S2} \\ P^{S_{1,2}}_{loss_{conduction}} = \left(\frac{(3-D)(3+D)i_o}{D\sqrt{D}(1-D)^2}\right)^2 r_{S1} + \left(\frac{2i_o}{\sqrt{D}(1-D)}\right)^2 r_{S2} \\ P^{S_{1,2}}_{loss_{conduction}} = \frac{1}{(D)^3(1-D)^4}((3-D)^2(1+D)^2 + (D)^2(1-D)^2)P_O\frac{r_S}{R_O} \end{cases} \tag{32}$$

For the calculation of switching losses for a switch operating at a switching frequency of f_s, the rise time (t_r) and fall time (t_f) of the gate pulses are considered. Then, the switching losses can be given as,

$$\begin{cases} P^{S_{1,2}}_{loss_{switching}} = \left(\frac{t_r + t_f}{2}\right) \times \left(i_{S1_{avg}} V_{S1_{avg}} + i_{S2_{avg}} V_{S2_{avg}}\right) f_s \\ P^{S_{1,2}}_{loss_{switching}} = \left(\frac{t_r + t_f}{2}\right) \times \left(\frac{(3-D)(1+D)i_o}{D(3-D)^2} \times \frac{(1-D)V_o}{2(3-D)} + \frac{2i_o}{(1-D)} \times \frac{2V_o}{(3-D)}\right) \times f_s \\ P^{S_{1,2}}_{loss_{switching}} = \left(\frac{t_r + t_f}{2}\right) \times \left(\frac{3 + 2D - D^2}{2D(1-D)(3-D)}\right) P_O \times f_s \end{cases} \tag{33}$$

3.2. Calculation of Losses across Diodes

For the calculation purpose, the cut-in voltage and resistance of all diodes are assumed to be the same, i.e., $V_{DO} = V_{D1} = V_{D2} = V_{D3} = V_{D4} = V_{D5} = V_{D6} = V_D$ and $r_{DO} = r_{D1} = r_{D2} = r_{D3} = r_{D4} = r_{D5} = r_{D6} = r_D$. The losses across different diodes are given as:

$$\begin{cases} P_{DO_{loss}} = i_{DO_{avg}} V_{DO} + i^2_{DO_{rms}} r_{DO} \\ P_{DO_{loss}} = i_o V_{DO} + \left(\frac{i_o}{\sqrt{D}}\right)^2 r_{DO} \\ P_{DO_{loss}} = \frac{V_D}{V_O} P_o + \left(\frac{1}{\sqrt{D}}\right)^2 \frac{r_D}{R_O} P_O \end{cases} \tag{34}$$

$$\begin{cases} P_{D1_{loss}} = i_{D1_{avg}} V_{D1} + i^2_{D1_{rms}} r_{D1} \\ P_{D1_{loss}} = \frac{(3-D)}{D(1-D)} \frac{V_D}{V_O} P_o + \left(\frac{(3-D)}{D\sqrt{D(1-D)}}\right)^2 \frac{r_D}{R_O} P_O \end{cases} \tag{35}$$

$$\begin{cases} P_{D2_{loss}} = i_{D2_{avg}} V_{D2} + i^2_{D2_{rms}} r_{D2} \\ P_{D2_{loss}} = \frac{(3-D)}{D(1-D)} \frac{V_D}{V_O} P_o + \left(\frac{(3-D)}{D\sqrt{D(1-D)}}\right)^2 \frac{r_D}{R_O} P_O \end{cases} \tag{36}$$

$$\begin{cases} P_{D3_{loss}} = i_{D3_{avg}} \times V_{D3} + i^2_{D3_{rms}} \times r_{D3} \\ P_{D3_{loss}} = \frac{(3-D)}{D(1-D)} \frac{V_D}{V_O} P_o + \left(\frac{(3-D)}{D\sqrt{D(1-D)}}\right)^2 \frac{r_D}{R_O} P_O \end{cases} \tag{37}$$

$$\begin{cases} P_{D4_{loss}} = i_{D4_{avg}} V_{D4} + i^2_{D4_{rms}} r_{D4} \\ P_{D4_{loss}} = \frac{V_D}{V_O} P_o + \left(\frac{1}{\sqrt{D}}\right)^2 \frac{r_D}{R_O} P_O \end{cases} \tag{38}$$

$$\begin{cases} P_{D5_{loss}} = i_{D5_{avg}} V_{D5} + i^2_{D5_{rms}} r_{D5} \\ P_{D5_{loss}} = \frac{2}{(1-D)} \frac{V_D}{V_O} P_o + \left(\frac{2}{\sqrt{D}\,(1-D)}\right)^2 \frac{r_D}{R_O} P_O \end{cases} \tag{39}$$

$$\begin{cases} P_{D6_{loss}} = i_{D6_{avg}} V_{D6} + i^2_{D6_{rms}} r_{D6} \\ P_{D6_{loss}} = \frac{2}{(1-D)} \frac{V_D}{V_O} P_o + \left(\frac{2}{\sqrt{D}\,(1-D)}\right)^2 \frac{r_D}{R_O} P_O \end{cases} \tag{40}$$

$$P_{D_{loss_{total}}} = P_{DO_{loss}} + P_{D1_{loss}} + P_{D2_{loss}} + P_{D3_{loss}} + P_{D4_{loss}} + P_{D5_{loss}} + P_{D6_{loss}} \tag{41}$$

3.3. Calculation of Losses across Capacitors

For the calculation of conduction loss in capacitors, the parasitic resistances of the capacitors r_{C1}, r_{C2}, r_{C3}, r_{C4}, and r_{C5} are assumed to be equal, while the resistance of the capacitor C_O is assumed to be r_{CO}

$$\begin{cases} P_{C_{loss_{total}}} = i^2_{CO_{rms}} r_{CO} + i^2_{C1_{rms}} r_{C1} + i^2_{C2_{rms}} r_{C2} + i^2_{C3_{rms}} r_{C3} + i^2_{C4_{rms}} r_{C4} + i^2_{C5_{rms}} r_{C5} \\ P_{C_{loss_{total}}} = \left(\sqrt{\frac{D}{1-D}}\right)^2 \frac{r_{CO}}{R_O} P_O + 2\left(\frac{3-D}{(1-D)^2}\sqrt{\frac{1-D}{D}}\right)^2 \frac{r_{C1}}{R_O} P_O + \\ \left(\frac{(5-D)}{(1-D)\sqrt{D\,(1-D)}}\right)^2 \frac{r_{C3}}{R_O} P_O + \left(\frac{1}{\sqrt{D\,(1-D)}}\right)^2 \frac{r_{C5}}{R_O} P_O \\ P_{C_{loss_{total}}} = \left(\frac{D}{(1-D)}\right) \frac{r_{CO}}{R_O} P_O + \left(\frac{44-24D+4D^2}{D(1-D)^3}\right) \frac{r_C}{R_O} P_O \end{cases} \tag{42}$$

3.4. Calculation of Losses across Inductors

The loss calculations for inductors are carried out by ignoring the ripple in the inductor current through the leakage resistances r_{L1}, r_{L2}, and r_{L3}. Inductors L_2 and L_3 have the same design values; hence, their parasitic resistances are assumed to be equal.

$$\begin{cases} P_{L_{loss_{total}}} = i^2_{L1_{rms}} r_{L1} + i^2_{L2_{rms}} r_{L2} + i^2_{L3_{rms}} r_{L3} \\ P_{L_{loss_{total}}} = \left(\frac{2(3-D)}{(1-D)^2}\right)^2 \frac{r_{L1}}{R_O} P_O + 2\left(\frac{1}{(1-D)}\right)^2 \frac{r_{L2}}{R_O} P_O \end{cases} \quad (43)$$

3.5. Calculation of Efficiency of the Converter in Nonideal Mode

The efficiency of the converter is given as the ratio of the output power transferred to the total input power fed to the circuit, which can be represented as the sum of the output power and the total losses in the converter. The efficiency (η) is given as

$$\begin{cases} \eta = \frac{P_0}{P_0 + P_{S_{loss}} + P_{D_{loss}} + P_{C_{loss}} + P_{L_{loss}}} \\ \eta = \frac{1}{1 + \frac{P_{S_{loss}} + P_{D_{loss}} + P_{C_{loss}} + P_{L_{loss}}}{P_0}} = \frac{1}{1+K} \end{cases} \quad (44)$$

where K is a constant given simplified as,

$$\begin{cases} K = \Big\{ \left[\frac{1}{D^3(1-D)^4}\left((3-D)^2(1+D)^2 + (D)^2(1-D)^2\right)\frac{r_S}{R_O}\right] \\ \quad + \left[\frac{(2D+3)(3-D)}{D(1-D)}\frac{V_D}{V_O}\right. \\ \quad + \left.\frac{27-18D+13D^2-4D^3+2D^4}{D^3(1-D)^2}\frac{r_D}{R_O}\right] \\ \quad + \left[\frac{D}{1-D}\frac{r_{CO}}{R_O} + \frac{35-18D+3D^2}{D(1-D)^3}\frac{r_C}{R_O}\right] \\ \quad + \left[\left(\frac{2(3-D)}{(1-D)^2}\right)^2 \frac{r_{L1}}{R_O} + 2\left(\frac{1}{1-D}\right)^2 \frac{r_{L2}}{R_O}\right]\Big\} \end{cases} \quad (45)$$

After substituting the value of K obtained in (45) in (44), we obtain,

$$\eta = \frac{D^3(1-D)^4}{D^3(1-D)^4 + \left[a\left(\frac{r_S}{R_O}\right) + b\left(\frac{r_D}{R_O}\right) + c\left(\frac{r_{CO}}{R_O}\right) + d\left(\frac{r_C}{R_O}\right) + e\left(\frac{r_{L1}}{R_O}\right) + f\left(\frac{r_{L2}}{R_O}\right) + g\left(\frac{V_D}{V_O}\right)\right]} \quad (46)$$

where,

$$\begin{cases} a = (9 + 12D - D^2 - 6D^3 + 2D^4) \\ b = (27 - 18D + 13D^2 - 4D^3 + 2D^4) \\ c = D^2(1-D)^4 \\ d = D(1-D)^2(35 - 18D + 3D^2) \\ e = 2D(3-D)^3 \\ f = 2D(1-D)^4 \\ g = D^3(1-D)^4(3-D)(2D+3) \end{cases} \quad (47)$$

3.6. Calculation of Nonideal Voltage Gain

For the derivation of nonideal voltage gain, all the lossy components of the elements such as the ON-state resistance of diodes and switches, ESRs of the inductor, and capacitor are considered for the analysis. Thus far, the input and output powers are equal due to lossless analysis, whereas, considering the above assumptions, the modified relation between the input and output power can be expressed as:

Input Power = Output Power + Losses across elements

3.7. Variation in Nonideal Voltage Gain

The nonideal gain (M_{actual}) of the proposed converter is derived as,

$$
\begin{cases}
M_{actual} = \frac{1}{1+K}(M_{CCM})_{ideal} = \eta(M_{CCM})_{ideal} \\
M_{\text{actual}} = \dfrac{2D^3(1-D)^2(3-D)}{\left(\begin{array}{c} D^3(1-D)^4 + \\ \left[a\left(\frac{r_S}{R_O}\right) + b\left(\frac{r_D}{R_O}\right) + c\left(\frac{r_{CO}}{R_O}\right) + d\left(\frac{r_C}{R_O}\right) + e\left(\frac{r_{L1}}{R_O}\right) + f\left(\frac{r_{L2}}{R_O}\right) + g\left(\frac{V_D}{V_O}\right)\right] \end{array}\right)}
\end{cases} \tag{48}
$$

From the voltage gain comparison of ideal and nonideal voltage gains of the proposed topology in Figure 15, it is observed that both gains have similar values up to the duty ratio of 0.4, and then the nonideal gain increases until D = 0.7 but deviate from the ideal characteristics. Thereafter, it gradually decreases to zero at the unity duty ratio.

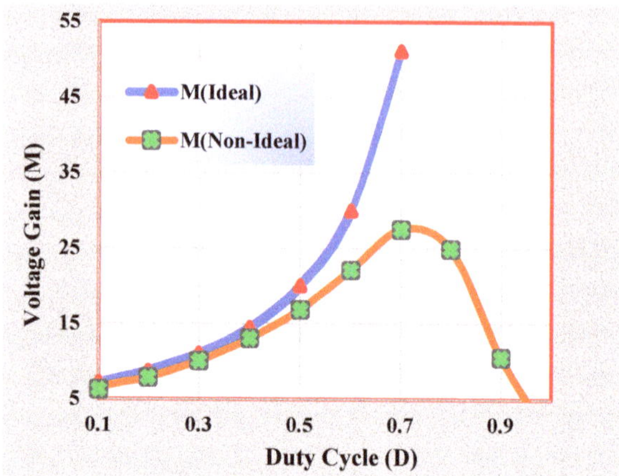

Figure 15. Comparison of calculated ideal and simulated nonideal voltage gain of the proposed converter.

The voltage gain of the converter can be affected by the parasitic resistances of different elements of the converter such as the parasitic resistance of inductors, capacitors, and switches. As the parasitic resistances of different elements increase, the voltage gain of the converter decreases gradually.

From Figure 16, it can be observed that, when the parasitic resistance of the switch is increased while other elements are kept ideal, the voltage gain of the converter decreases. For the parasitic resistance of 0.2% of the load resistance, the maximum gain is at the duty ratio of 0.7 with a gain of 24.18, while it decreases to 15.84 and 11.77 for 4% and 6% of the parasitic resistance, respectively, at D = 0.7.

From Figure 17, it can be observed that, when the parasitic resistance of the inductor is increased while other elements are kept ideal, the voltage gain of the converter decreases. For a parasitic resistance of 0.2% of the load resistance, the maximum gain is at the duty ratio of 0.75 with a gain close to 34, while it decreases to 24.88 and 19.80 for 4% and 6% of the parasitic resistance, respectively, at D = 0.7.

From Figure 18, it can be observed that, when the parasitic resistance of the capacitor is increased while other elements are kept ideal, the voltage gain of the converter decreases. For the parasitic resistance of 0.2% of the load resistance, the maximum gain is at the duty ratio of 0.8 with a gain of 63, while it decreases to around 41 and 21 for 4% and 6% of the parasitic resistance, respectively, at a near-duty ratio of 0.8.

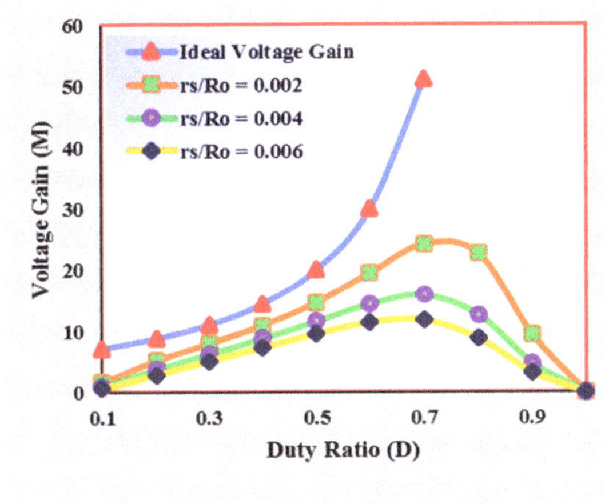

Figure 16. Voltage gain variation as per parasitic resistance of the switch.

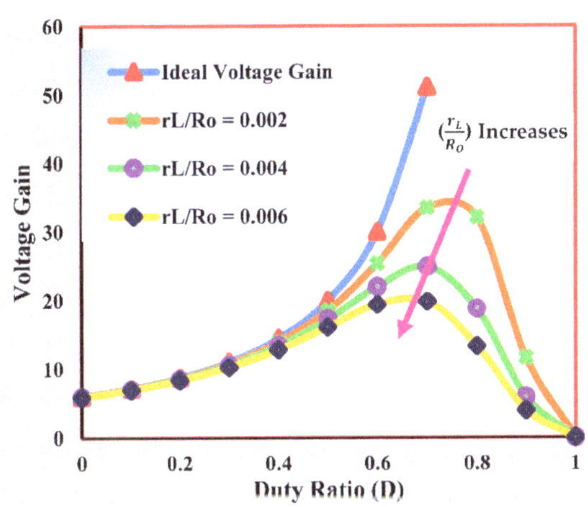

Figure 17. Voltage gain variation as per parasitic resistance of the inductor.

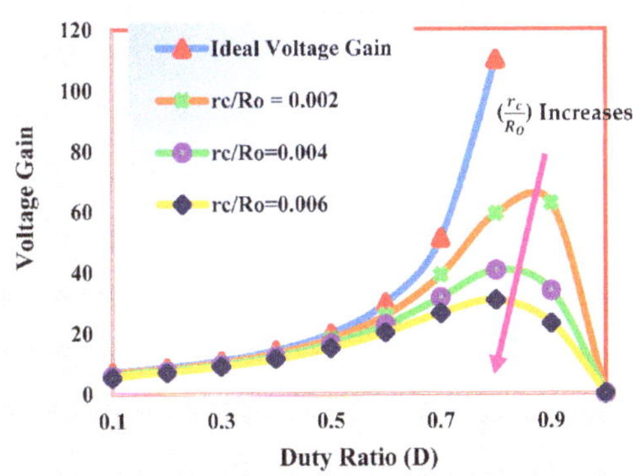

Figure 18. Voltage gain variation as per parasitic resistance of the capacitor.

4. Comparison of the Proposed Topology

In this section, a comparison between the proposed topology with two traditional topologies and recent topologies is discussed. The topologies are compared based on their voltage gain (M_{CCM}), number of inductors (N_L), number of capacitors (N_C) and number of diodes (N_D) in the converter. The voltage stress appearing across the switches (V_{si}/V_{in}) is also compared along with the availability of common ground.

For the comparison of the topologies, the graph between the ideal voltage gain and the duty cycle is plotted in Figure 19a. Figure 19b depicts the comparison of the nonideal gain of the converter with the ideal gain of some selected topologies whose gain intersects the nonideal gain plot of the proposed topology. The voltage stress across the switches in the compared converters is shown in Figure 20.

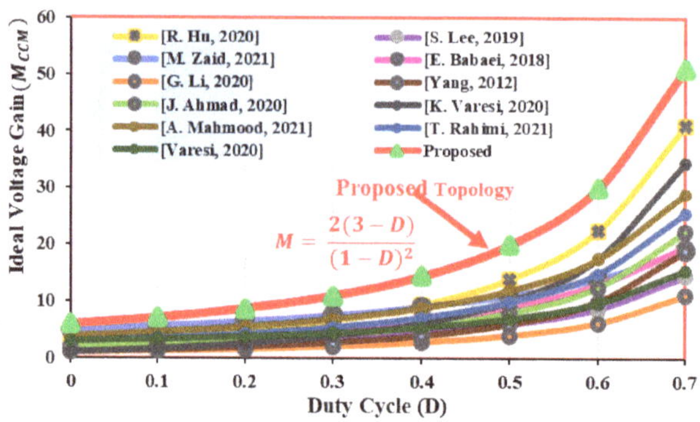

(a)

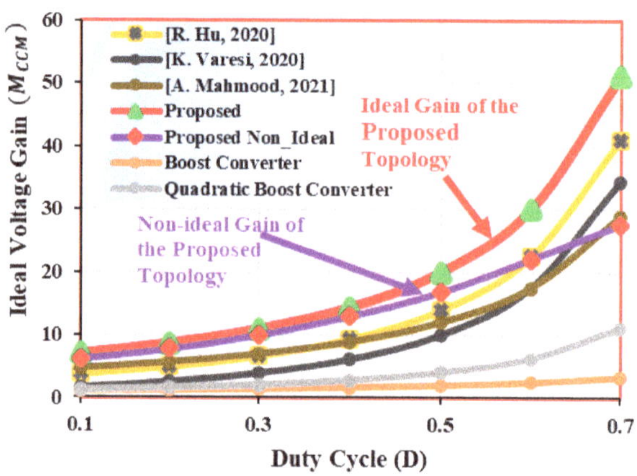

(b)

Figure 19. (**a**) Comparison of ideal voltage gain of various similar converters [4,5,16,22,24–30]. (**b**) Comparison of various topologies with the nonideal voltage gain of proposed topology [4,24,27,28].

It can be observed from Figure 19a that the proposed topology has the highest ideal gain among all the high-gain boost converters compared to all converters. An ideal voltage gain of above 14 is achieved by the proposed topology at a duty ratio of 40% and a gain of 20 is achieved at a duty ratio of 50%. The proposed converter also has a higher nonideal gain, as shown in Figure 19b, as compared to the ideal gain of the similar topologies, as presented in Table 1, until the duty ratio of 60%. The switching stress of the converter is very low as compared to the traditional topologies. In converters [4,5], even with the presence of an isolation transformer/coupled inductor operating at a transformation ratio of one, a lower gain than the proposed topology is produced, which works without a transformer or coupled inductor, hence reducing the overall cost of the topology.

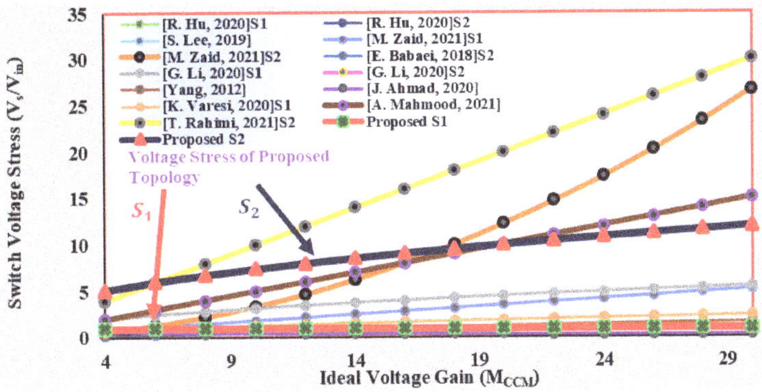

Figure 20. Comparison of calculated switch stress versus voltage gain [4,5,16,22,24–29].

For n = 1, the ideal gain of the converter in [4] when compared with the gain of the proposed converter is found to be half. Moreover, the voltage stress across the switch S_1 in [4] increases significantly as shown in Figure 20.

The topology proposed in [16] also has a VMC and two switches and produces a high gain at lower duty ratios with a lower number of components as compared with the proposed topology, but it decreases as the duty ratio is increased above 40%. The proposed converter despite having fewer inductors, switches and diodes than the converter [22] is capable of producing a much higher voltage gain. Although the switch stress across the converter [22] is very low as compared to the proposed one, the converter lacks a common ground that is available in the proposed topology. The converters in [24,25], despite having 2 and 3 inductors respectively and 2 switches which is the same as used in the proposed topology, produce a lower gain than the proposed topology and, concurrently, the voltage stress across the switch.

In the case of the topology shown in [25], the switch stress increases after a conversion gain ratio of 14. The converters in [26,27] also have three inductors but suffer from a low-voltage gain. The switch S_1 of [27] has high-voltage stress after a voltage gain of 8. The quadratic boost converter in [29] has the same inductors as the proposed converter, but the voltage gain is half as compared to the proposed converter, and it has a single switch only. The topology in [29] exhibits high-voltage stress across switch S_2 amongst all the compared topologies. The topology in [30] has the least voltage stress across its switches but suffers from a low ideal voltage gain, which is much lower than the nonideal voltage gain produced by the proposed converter.

From the comparison, it can be inferred that the topology has the highest voltage gain and the switch voltage gain across S_1 is very low. The switch voltage gain across the switch S_2 is found to be high for low-voltage gains, but it does not increase further as the voltage gain increases beyond 15. Hence, the proposed topology can be used at a voltage gain higher than 14, which is easily achieved at any duty ratio beyond 40%. Moreover, continuous input current and common ground are other advantages of the proposed converter.

Table 1. Comparison of proposed topology with certain similar topologies.

Topology	N_L	N_C	N_{SW}	N_D	M_{CCM}	M_{CCM} at D = 0.5	Common Ground	$S_{CCM} = \frac{V_{S_i}}{V_{in}}$
Boost Converter	1	1	1	1	$\frac{1}{(1-D)}$	2	Yes	$S = \frac{1}{1-D}$
Quadratic Boost Converter	2	2	1	3	$\frac{1}{(1-D)^2}$	4	Yes	$S = \frac{1}{(1-D)^2}$
[4]	1 + 1 coupled inductor	5	2	5	$\frac{2n+1+D}{(1-D)^2}$	14	Yes	$S_1 = \frac{1}{1-D}$ $S_2 = \frac{1+D}{(1-D)^2}$
[5]	1 + 1 coupled inductor	3	1	5	$\frac{1+n-D}{(1-D)^2}$	6	Yes	$S = \frac{2}{1-D}$
[16]	3	4	2	5	$\frac{5+D}{1-D}$	11	No	$S_1 = \frac{1}{1-D}$ $S_2 = \frac{1}{(1-D)^2}$
[22]	8	1	4	17	$\frac{1+7D}{(1-D)}$	9	No	$S_1 = 1$ $S = \frac{1+5D}{1+7D}$
[24]	2	2	2	2	$\frac{1}{(1-D)^2}$	4	Yes	$S_1 = \frac{1}{1-D}$ $S_2 = \frac{1}{(1-D)^2}$
[25]	3	4	1	4	$\frac{1+D}{(1-D)^2}$	6	Yes	$S = \frac{1}{(1-D)^2}$
[26]	3	3	1	5	$\frac{2}{(1-D)^2}$	8	Yes	$S = \frac{2}{(1-D)^2}$
[27]	3	5	2	4	$\frac{1+3D}{(1-D)^2}$	10	Yes	$S_1 = \frac{1}{1-D}$ $S_2 = \frac{1+D}{(1-D)^2}$
[28]	2	5	1	6	$\frac{2(2-D)}{(1-D)^2}$	12	No	$S = \frac{2-D}{(1-D)^2}$
[29]	3	6	1	6	$\frac{(3-D)}{(1-D)^2}$	10	Yes	$S_1 = \frac{3-D}{D^2(1-D)}$. $S_2 = \frac{3-D}{(1-D)^2}$
[30]	2	3	2	3	$\frac{D^2-3D+3}{(1-D)^2}$	7	Yes	$S_1 = \frac{1-D}{D^2-3D+3}$ $S_2 = \frac{1}{D^2-3D+3}$
Proposed Topology	**3**	**6**	**2**	**7**	$\frac{2(3-D)}{(1-D)^2}$	**20**	**Yes**	$S_1 = 1$ $S_2 = \frac{4}{(1-D)}$

5. Results

In this section, we discuss the simulation results performed on the **P**iecewise **L**inear **E**lectrical **C**ircuit **S**imulation (PLECS) and the experimental results obtained on hardware set-upfor the proposed topology.

5.1. Simulation Results

In this section, the simulation results of the proposed converter are presented. The simulation was carried out in PLECS software at an input voltage of 20 V and switching frequency of 50 kHz. The duty ratio was maintained at 0.4. The value of the inductor was 330 µΩ and its parasitic resistance was maintained at 0.12 Ω, while the value of capacitors was 47 µΩ and its parasitic resistance was kept at 0.1 Ω. The ON-state resistance of the switches was found to be 70 mΩ. With the above parameters, the simulation was performed and the output voltage was found to be $V_O = 248.08$ V at a 40% duty ratio with $V_{in} = 20$ V, which can be seen in Figure 21.

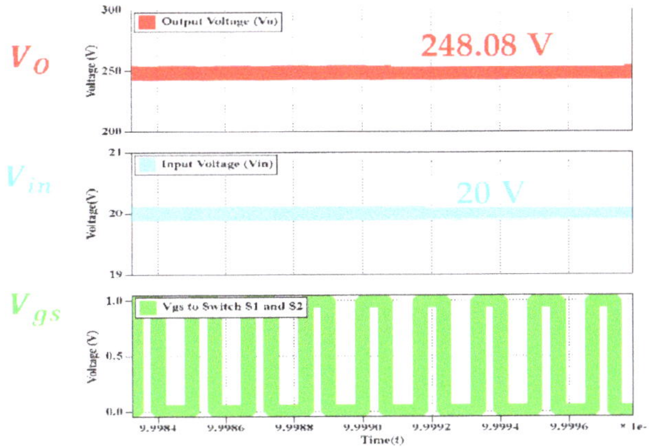

Figure 21. Simulated output voltage (V_O), input voltage (V_{in}), and duty cycle (D).

In Figure 22, the inductor current through inductor L_1 was found to be $I_{L1} = 5.06$ A and $I_{L2} = I_{L3} = 0.58$ A for inductor L_2. The ripple in the current through the inductor was lower and, hence, the average and RMS values were found to be the same.

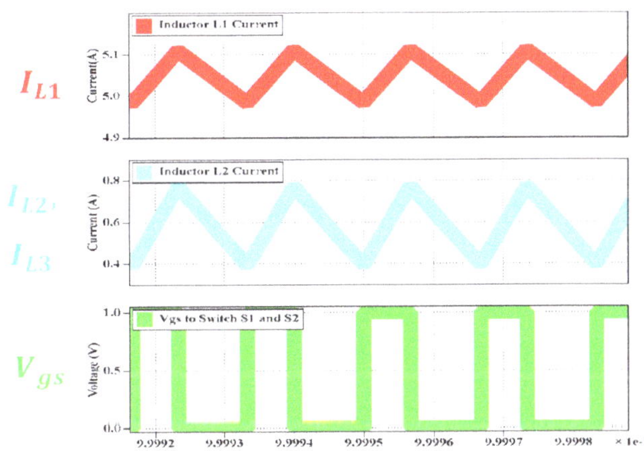

Figure 22. Simulated inductor currents (I_{L1}, I_{L2}, and I_{L3}) and duty cycle (D).

From Figure 23, the peak voltage across switches S_1 and S_2 was found to be $V_{S1(peak)} = 30.687$ V and $V_{S2(peak)} = 155.43$ V, respectively, while the RMS voltage was $V_{S1(rms)} = 23.77$ V. and $V_{S1(rms)} = 120.40$ V, respectively. From Figure 24, the average capacitor voltages in the simulation were found to be $V_{C1} = V_{C2} = 27.5209$ V and $V_{C3} = 58.65$ V, while from Figure 25, the capacitor voltages were $V_{C4} = -56.58$ V, $V_{C5} = 57.16$ V, and $V_{CO} = V_O = 248.08$ V. The ripple in capacitor voltages was negligible and, hence, the RMS and average voltages were found to be the same.

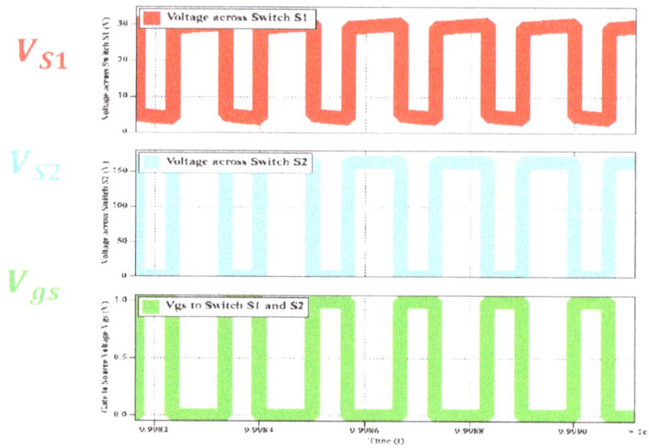

Figure 23. Simulated switch voltages and duty cycle (D).

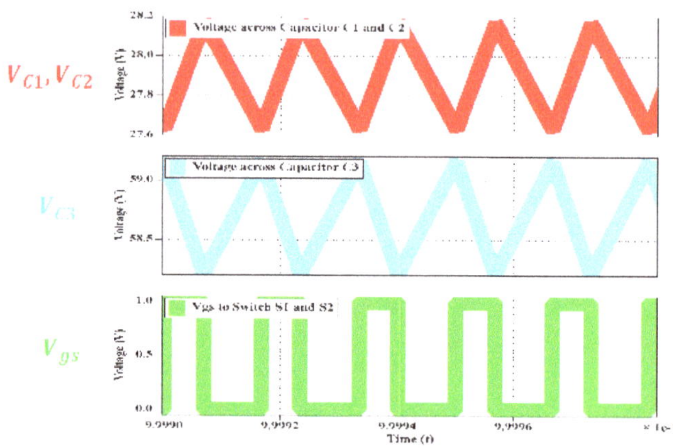

Figure 24. Simulated capacitor voltages and duty cycle (D).

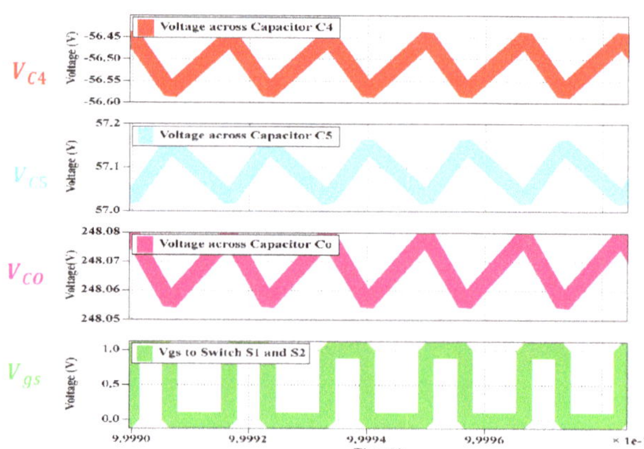

Figure 25. Simulated capacitor voltages and duty cycle (D).

5.2. Experimental Results

The experimental analysis of the proposed converter was carried out on the same parameters as the simulation procedure. To verify the overall voltage boosting, continuous current, and capacitor voltage handling capability of the proposed converter, a hardware prototype of the proposed converter was tested under standard laboratory conditions with the parameters mentioned in Table 2. To demonstrate the working of the topology, a converter with an output power of 200 W at a duty ratio of 40% was implemented with a load resistance of 300 Ω. For switches S_1 and S_2, Power MOSFET with part number SPW52N50C3 provided with a duty cycle (D) of 0.4 operating at a switching frequency (f_s) of 50 kHz was used, whereas diodes ($D_O - D_6$) with part number HER806 were used. From Figure 26, a 20 V DC supply was used at the input side and an output of 240 V was obtained for the same with V_{gs} as the gate drive signal.

Table 2. Hardware values of parameters.

Parameter	Symbol	Value
Input Voltage	V_{in}	20 V
Duty Cycle	D	0.4
Output Power	P_O	67 W
Load Resistance	R_O	850 Ω
Inductors	L_1	330 µH, ESR = 0.12 Ω
	L_2, L_3	0.4 mH ESR = 0.14 Ω
Capacitors	$C_1 - C_5$	47 µF, ESR = 0.1 Ω
	C_O	100 µF, ESR = 0.22 Ω
Diodes	D_O to D_6	HER806
Power MOSFETs	S_1, S_2	SPW52N50C3
Driver		TLP250H
Controller		STM32F334R8
Switching Frequency	f_s	50 kHz

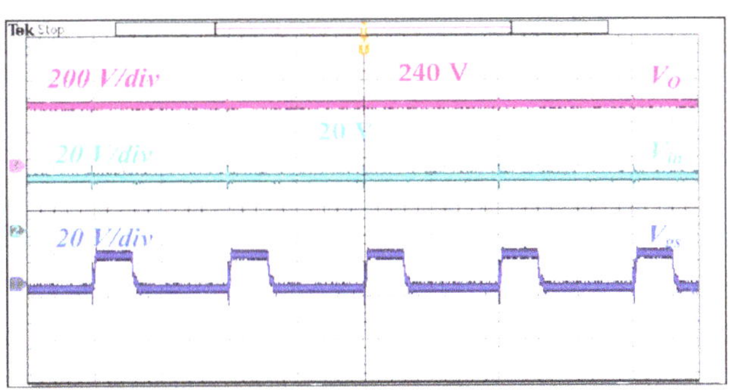

Figure 26. Top to bottom: experimental waveforms of output voltage (V_O), input voltage (V_{in}), and V_{gs} at D = 0.4.

From Figure 27, when the switches are ON, the inductors become charged and the current through them increases. For inductor L_1, it increases from an initial value of 4.8 A to a peak value of 5.2 A and it increases from an initial value of 0.7 A to a peak value of 1.0 A for inductors L_2 and L_3. When the switches are OFF, the inductors release their stored energy, and the currents through them decrease back to their initial values. During the OFF state, the switches are reversed-biased and block a peak voltage of 30 V and 160 V, respectively, in each cycle, which can be seen in Figure 28. The peak reverse blocking voltage is about 30 V for the switch S_1 and 160 V for switch S_2 for the output voltage of 240 V. Capacitor C_1 has a voltage of 30 V across it, while capacitors C_3 and C_4 have a voltage of 60 V with very-low-voltage ripples as in Figure 29. The experimental setup is shown in Figure 30.

Figure 31 shows the variation in efficiency with the output power as the input voltage is increased. It can be observed that as the input voltage increases, the efficiency of the converter increases. This occurs because with the increase in voltage, the current decreases to conserve the power; hence, the conduction losses across various elements of the converter including switches, diodes, and parasitic resistances of the inductor and capacitor are reduced.

The maximum efficiency of the converter is found as 97.85% at 30 V, 66 W followed by 96.5% for operation at 20 V for the same power. From Figure 32, the majority of the conduction losses i.e., around 39%, occur in the capacitors, out of which 58.5% of the losses in capacitors are due to capacitor C_3 itself. The switches and diodes contribute 40% of the total losses. The losses can be further reduced by using diodes and switches with low parasitic resistance.

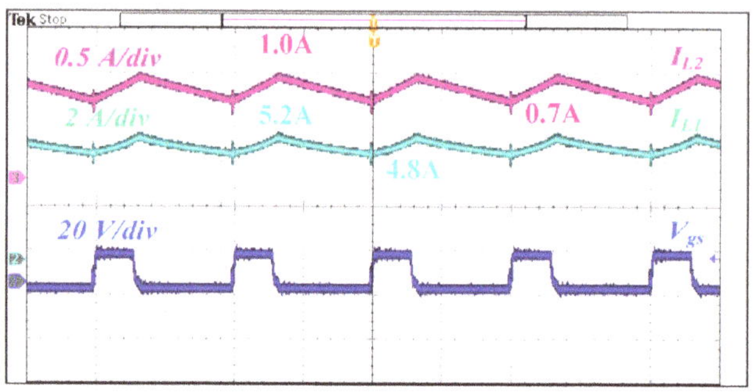

Figure 27. Top to bottom: experimental waveforms of current of inductor L_1 (i_{L1}), inductor L_2 (i_{L2}), and V_{gs} at D = 0.4.

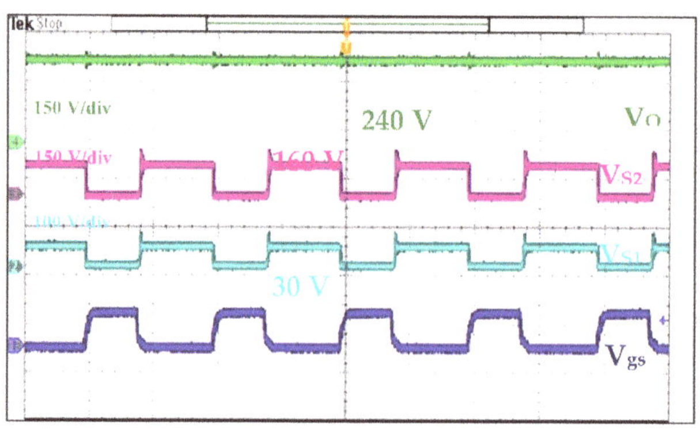

Figure 28. Top to bottom: experimental waveforms of output voltage (V_O), voltage across switch S_2 (V_{S2}), voltage across switch S_1(V_{S1}), and V_{gs} at D = 0.4.

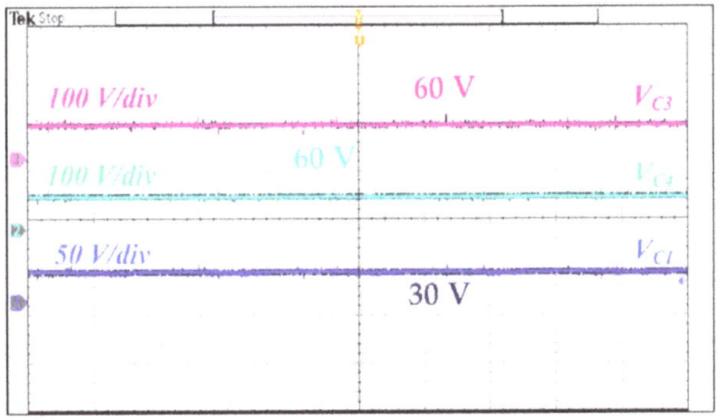

Figure 29. Top to bottom: experimental waveforms of voltage across capacitor C_3 (V_{C3}), the voltage across capacitor C_4 (V_{C4}), and the voltage across capacitor C_1 (V_{C1}) at D = 0.4.

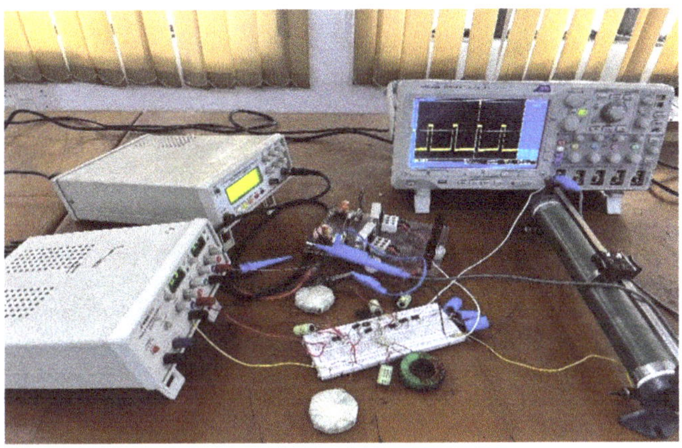

Figure 30. Experimental Set-up.

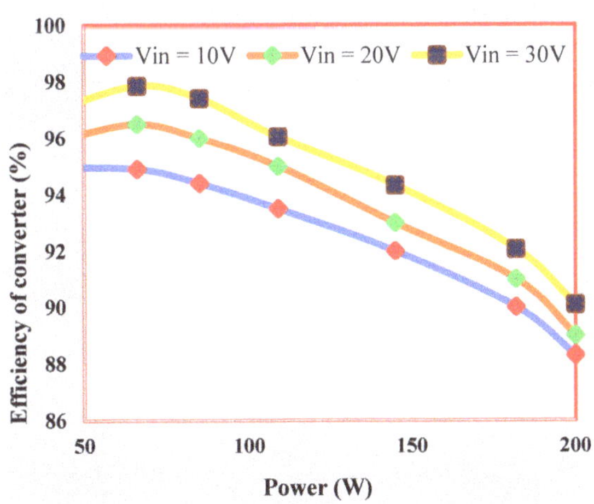

Figure 31. Simulated efficiency vs. output power comparison of the proposed converter at different input voltages.

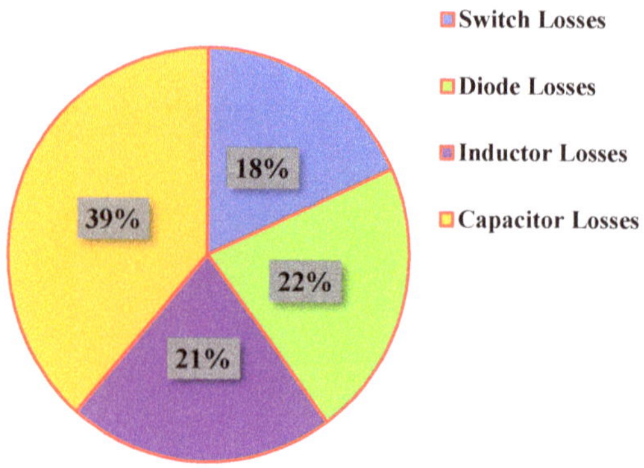

Figure 32. Distribution of losses across various elements of the converter.

6. Conclusions

The converter produces an ideal voltage gain of 11 times at a duty ratio of 30% with an ideal voltage gain of 14.44 at a duty ratio of 40%. The nonideal gain of the proposed converter at a D less than 60% is still found to be higher as compared to the ideal gains of the compared converters. In addition, the voltage stress across the switches is found to be lower than the output voltage and is also much lower than the compared topologies even at higher duty ratios. The voltage stress across S_1 is 6.92% of the V_o and across switch S_2, it is 46.15% of V_o at the duty ratio of 40%. The maximum efficiency of 97.85% is obtained at 66 W when the input voltage is 30 V, while it is 96.44% at a load of 38 W, keeping the input voltage at 20 V. The efficiency of the prototype model decreases with the power level due to the absence of galvanic isolation and parasitic resistances. The voltage and current stresses across the elements are low, which can be observed from the analysis and the hardware results. The converter has a common ground and operates at continuous input current, making it feasible for low- and medium-power solar and renewable energy PV applications.

Author Contributions: Conceptualization, writing—original draft preparation, M.Z., I.H.M. and M.T.; supervision, M.Z. and I.A.; writing—review and editing, B.A. and E.M.G.R.; funding acquisition, M.T., B.A. and E.M.G.R. All authors have read and agreed to the published version of the manuscript.

Funding: The authors would like to acknowledge the financial support from Taif University Researchers Supporting Project Number (TURSP-2020/278), Taif University, Taif, Saudi Arabia and from the collaborative research grant scheme (CRGS) project, Hardware-In-the-Loop (HIL) Lab, Department of Electrical Engineering, Aligarh Muslim University, India having project numbers CRGS/MOHD TARIQ/01 and CRGS/MOHD TARIQ/02.

Acknowledgments: The authors also acknowledge the support provided by the Hardware-In-the-Loop (HIL) Lab and Non-Conventional Energy (NCE) Lab, Department of Electrical Engineering, Aligarh Muslim University, India.

Conflicts of Interest: The authors declare no conflict of interest.

References

1. Forouzesh, M.; Siwakoti, Y.; Gorji, S.; Blaabjerg, F.; Lehman, B. Step-Up DC–DC Converters: A Comprehensive Review of Voltage-Boosting Techniques, Topologies, and Applications. *IEEE Trans. Power Electron.* **2017**, *32*, 9143–9178. [CrossRef]
2. Goyal, V.K.; Shukla, A. Isolated DC–DC Boost Converter for Wide Input Voltage Range and Wide Load Range Applications. *IEEE Trans. Ind. Electron.* **2021**, *68*, 9527–9539. [CrossRef]
3. Ayachit, A.; Reatti, A.; Kazimierczuk, M.K. Magnetising inductance of multiple-output flyback dc–dc convertor for discontinuous conduction mode. *IET Power Electron.* **2017**, *10*, 451–461. [CrossRef]
4. Hu, R.; Zeng, J.; Liu, J.; Guo, Z.; Yang, N. An Ultrahigh Step-Up Quadratic Boost Converter Based on Coupled-Inductor. *IEEE Trans. Power Electron.* **2020**, *35*, 13200–13209. [CrossRef]

5. Lee, S.; Do, H. Quadratic Boost DC–DC Converter with High Voltage Gain and Reduced Voltage Stresses. *IEEE Trans. Power Electron.* **2019**, *34*, 2397–2404. [CrossRef]
6. Das, P.; Mousavi, S.A.; Moschopoulos, G. Analysis and Design of a Nonisolated Bidirectional ZVS-PWM DC–DC Converter with Coupled Inductors. *IEEE Trans. Power Electron.* **2010**, *25*, 2630–2641. [CrossRef]
7. Da Cunha Duarte, C.; Barbi, I. A new family of ZVS-PWM active-clamping DC-to-DC boost converters: Analysis, design, and experimentation. *IEEE Trans. Power Electron.* **1997**, *12*, 824–831. [CrossRef]
8. Vinnikov, D.; Roasto, I.; Strzelecki, R.; Adamowicz, M. Step-up dc/dc converter switch cascaded quasi Z-source network. *IEEE Trans. Ind. Electron.* **2012**, *59*, 3727–3736. [CrossRef]
9. Liu, J.; Wu, J.; Qiu, J.; Jun, Z. Switched Z-Source/Quasi-Z-Source DC–DC Converters with Reduced Passive Components for Photovoltaic Systems. *IEEE Access* **2019**, *7*, 40893–40903. [CrossRef]
10. Haji-Esmaeili, M.; Babaei, E.; Sabahi, M. High Step-Up Quasi-Z Source DC–DC Converter. *IEEE Trans. Power Electron.* **2018**, *33*, 10563–10571. [CrossRef]
11. Shen, H.; Zhang, B.; Qiu, D. Hybrid Z-Source Boost DC–DC Converters. *IEEE Trans. Ind. Electron.* **2017**, *64*, 310–319. [CrossRef]
12. Karthikeyan, V.; Kumaravel, S.; Gurukumar, G. High step-up gain DC–DC converter with switched capacitor and regenerative boost configuration for solar PV applications. *IEEE Trans. Circuits Syst. II Express Briefs* **2019**, *66*, 2022–2026. [CrossRef]
13. Tang, Y.; Fu, D.; Wang, T.; Xu, Z. Hybrid Switched-Inductor Converters for High Step-Up Conversion. *IEEE Trans. Ind. Electron.* **2015**, *62*, 1480–1490. [CrossRef]
14. Samiullah, M.; Iqbal, A.; Ashraf, I.; Maroti, P. Voltage Lift Switched Inductor Double Leg Converter with Extended Duty Ratio for DC Microgrid Application. *IEEE Access* **2021**, *9*, 85310–85325. [CrossRef]
15. Bhaskar, M.S.; Meraj, M.; Iqbal, A.; Padmanaban, S.; Maroti, P.K.; Alammari, R. High gain transformer-less double-duty-triple mode DC/DC converter for DC microgrid. *IEEE Access* **2019**, *7*, 36353–36370. [CrossRef]
16. Zaid, M.; Khan, S.; Siddique, M.D.; Sarwar, A.; Ahmad, J.; Sarwer, Z.; Iqbal, A. A transformerless high gain dc–dc boost converter with reduced voltage stress. *Int. Trans. Electr. Energy Syst.* **2021**, *31*, e12877. [CrossRef]
17. Ahmad, J.; Siddique, M.; Sarwar, A.; Lin, C.; Iqbal, A. A high gain noninverting DC–DC converter with low voltage stress for industrial applications. *Int. J. Circuit Theory Appl.* **2021**, *49*, 4212–4230. [CrossRef]
18. Saravanan, S.; Babu, N. Design and Development of Single Switch High Step-Up DC–DC Converter. *IEEE J. Emerg. Sel. Top. Power Electron.* **2018**, *6*, 855–863. [CrossRef]
19. Khan, S.; Zaid, M.; Mahmood, A.; Nooruddin, A.S.; Ahmad, J.; Alghaythi, M.L.; Alamri, B.; Tariq, M.; Sarwar, A.; Lin, C.H. A New Transformerless Ultra High Gain DC–DC Converter for DC Microgrid Application. *IEEE Access* **2021**, *9*, 124560–124582. [CrossRef]
20. Andrade, A.; Faistel, T.; Toebe, A.; Guisso, R. Family of Transformerless Active Switched Inductor and Switched Capacitor Ćuk DC–DC Converter for High Voltage Gain Applications. *IEEE J. Emerg. Sel. Top. Ind. Electron.* **2021**, *2*, 390–398. [CrossRef]
21. Sarikhani, A.; Allahverdinejad, B.; Hamzeh, M. A Nonisolated Buck–Boost DC–DC Converter with Continuous Input Current for Photovoltaic Applications. *IEEE J. Emerg. Sel. Top. Power Electron.* **2021**, *9*, 804–811. [CrossRef]
22. Babaei, E.; MashinchiMaheri, H.; Sabahi, M.; Hosseini, S. Extendable Nonisolated High Gain DC–DC Converter Based on Active–Passive Inductor Cells. *IEEE Trans. Ind. Electron.* **2018**, *65*, 9478–9487. [CrossRef]
23. Shahir, F.M.; Babaei, E.; Farsadi, M. Extended topology for a boost DC–DC converter. *IEEE Trans. Power Electron.* **2019**, *34*, 2375–2384. [CrossRef]
24. Li, G. A Novel Quadratic Boost Converter with Low Inductor Currents. *CPSS Trans. Power Electron. Appl.* **2020**, *5*, 1–10. [CrossRef]
25. Yang, P.; Xu, J.; Zhou, G.; Zhang, S. A new quadratic boost converter with high voltage step-up ratio and reduced voltage stress. In Proceedings of the 7th International Power Electronics and Motion Control Conference, Harbin, China, 2–5 June 2012. [CrossRef]
26. Ahmad, J.; Zaid, M.; Sarwar, A.; Tariq, M.; Sarwer, Z. A New Transformerless Quadratic Boost Converter with High Voltage Gain. *Smart Sci.* **2020**, *8*, 163–183. [CrossRef]
27. Varesi, K.; Esfahlan, A.G. An Enhanced-Gain Quadratic-Boost DC–DC Configuration. In Proceedings of the 2020 28th Iranian Conference on Electrical Engineering (ICEE), Tabriz, Iran, 4–6 August 2020; pp. 1–5. [CrossRef]
28. Mahmood, A.; Zaid, M.; Ahmad, J.; Khan, M.A.; Khan, S.; Sifat, Z.; Lin, C.H.; Sarwar, A.; Tariq, M.; Alamri, B. A Non-Inverting High Gain DC–DC Converter with Continuous Input Current. *IEEE Access* **2021**, *9*, 54710–54721. [CrossRef]
29. Rahimi, T.; Ding, L.; Gholizadeh, H.; Shahrivar, R.S.; Faraji, R. An Ultra High Step-Up DC–DC Converter Based on the Boost, Luo, and Voltage Doubler Structure: Mathematical Expression, Simulation, and Experimental. *IEEE Access* **2021**, *9*, 132011–132024. [CrossRef]
30. Varesi, K.; Hassanpour, N.; Saeidabadi, S. Novel high step-up DC–DC converter with increased voltage gain per devices and continuous input current suitable for DC microgrid applications. *Int. J. Circ. Theor. Appl.* **2020**, *48*, 1820. [CrossRef]

energies

Article

DC-DC High-Step-Up Quasi-Resonant Converter to Drive Acoustic Transmitters

Emad Roshandel [1,*], Amin Mahmoudi [1,*], Solmaz Kahourzade [2] and Hamid Davazdah-Emami [3]

[1] College of Science and Engineering, Flinders University, Adelaide 5042, Australia
[2] STEM, University of South Australia, Adelaide 5001, Australia
[3] Research and Development Department, Eram Sanat Mooj Gostar Company, Shiraz 71976-87811, Iran
* Correspondence: emad.roshandel@flinders.edu.au (E.R.); amin.mahmoudi@flinders.edu.au (A.M.)

Abstract: This paper proposes a quasi-resonant step-up DC-DC converter to provide the DC-link voltage for piezoelectric transmitters (PZETs). The resonance not only provides a soft-switching condition for the converter switches, but also helps to decrease the converter element sizes for marine applications. Operation modes of the proposed converter are discussed. The current and voltage of the converter components are derived analytically, and hence the converter gain is obtained. The performance of the proposed high-step-up, high-power density converter is examined through a comprehensive simulation study. The simulation results demonstrate the soft-switching operation and short transient time required for the converter to reach the reference output voltage. The converter output voltage remains unchanged when an inverter with a passive filter is placed at its output while supplying the PZET. The proposed DC-DC converter is prototyped to validate the converter gain and soft-switching operation experimentally. The converter gain and soft-switching results in simulation are well matched with those of the experimental tests. The converter efficiency in three different switching frequencies is obtained experimentally. The power density of the proposed topology is determined via the designing of a printed circuit board. The experimental results demonstrate the appropriate gain and efficiency of the converter in the tested power range.

Keywords: acoustic transmitters; DC-DC power supply; high power density converter; high power step-up converter; PIEZO-electric drive; PIEZO-electric transmitters; resonant converter

Citation: Roshandel, E.; Mahmoudi, A.; Kahourzade, S.; Davazdah-Emami, H. DC-DC High-Step-Up Quasi-Resonant Converter to Drive Acoustic Transmitters. *Energies* **2022**, *15*, 5745. https://doi.org/10.3390/en15155745

Academic Editors: Mario Marchesoni and Vítor Monteiro

Received: 8 July 2022
Accepted: 6 August 2022
Published: 8 August 2022

1. Introduction

Acoustic transmitters have a significant role in the provision of a safe maritime trip. These transmitters use sound navigation and ranging (SONAR) technique to detect the seabed, near species, and possible obstacles during a voyage [1,2]. In addition, these waves warn the available fishes, whales, and all other living animals to move away from the moving vessels [3]. SONAR systems usually generate sound waves over a course of time, called pinging, and then stop generation to receive the reflected waves from the seabed and available obstacles [4]. Figure 1 shows the transmitting and receiving waves between the underwater transmitter and a sample object. In this regard, piezoelectric materials that can work in two operating areas (i.e., generating and motoring modes) are employed for pinging and receiving waves. The piezoelectric materials generate underwater sound during the motoring period when energized via an alternative current (AC) power supply. Piezoelectrics receive the reflected sounds when the energization time is concluded. In fact, they generate a low voltage power because of the received energy.

Piezoelectric transmitters (PZETs) require a high voltage level to work in motoring mode [5]. However, the available power source in ships and submarines, where the PZETs are located, is usually limited to a low-voltage battery package. It must be noted that the required voltage level for PZETs varies from 50 to 1000 volts, based on the operation of PZETs at different sea depths. If the applied voltage to the PZET is set to a higher level as required in the lower depths of the sea, the piezoelectric element will be consequently broken. Thus, a high-step-up power converter is needed to provide the precise voltage level for driving the PZET [6]. The availability of only a small space to place the transmitters and their power supply restricts the utilization of low-power-density converters for PZETs [7]. These converters are placed in harsh and noisy environments. Therefore, an analog robust controller is preferred instead of the digital controllers.

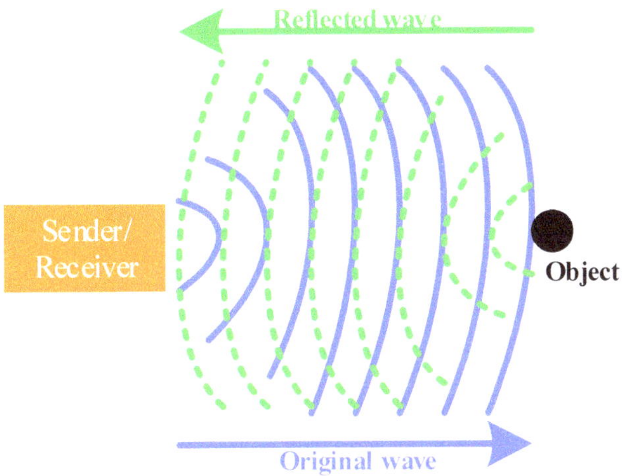

Figure 1. Graphical abstract of the operation of PZETs in underwater transmission systems.

Designers mostly used class A, B, and AB linear power amplifiers to supply PZETs [8]. The low efficiency of linear amplifiers not only demands large heat sinks because of the generation of the unwanted heat [9], but also increases the required size of the power source (e.g., battery packages). The gain of a class AB was enhanced by a class D switching in [10]. However, such a linear amplifier still suffers from low efficiency and large volume. Switching power converters are suitable substitutions for the linear power converters because of their higher power density and efficiency.

The DC-DC switching power converters are categorized into isolated and non-isolated families [11–13]. The isolated converters, which benefit from the availability of high-frequency transformers in their structures, provide an intrinsic input–output insulation [14]. The insulation is required in some high-power applications where engineers need to decrease the electric shocks of each side of the power supply [15]. For instance, it provides a layer of insulation in grid-tied converters [16]. However, they have a lower power density and efficiency compared to non-isolated power converters. Moreover, the isolation complicates the implementation of the control system. Thus, the non-isolated power supplies are more popular, especially in low-power applications [17].

Usually, non-isolated converters employ coupled inductors or resonance phenomena to reach a higher output voltage [18]. Switching power supplies equipped with coupled inductors have lower power densities due to the bulky components (e.g., inductors and capacitors) available in their structure [19,20]. Hence, a resonant power converter can be used in an application where the power density is a significant factor [21]. The resonance frequency in these types of converters is higher than the switching frequency, which leads to a reduction in the converter elements' size. Although the energy conversion through resonance creates higher voltage and current stress, it provides a higher gain for a converter. Therefore, the resonance-based converters have a higher gain and power density in comparison to other types of non-isolated converters [22]. Moreover, the lack of an isolating high-frequency transformer enables them to have higher efficiency compared to isolated converters.

Zero-voltage and current-switching strategies are two advanced switching methods of decreasing the switching losses of a converter and helping switches to work in their safe operating areas (SOAs). The conventional resistive-capacitive snubber was the earlier method of providing ZVS and ZCS for the switches of a converter [23]. The resistance of such topologies reduces the converter efficiency. Hence, the active-clamped circuits [24] and resonant snubbers [25] are more popular among engineers because of their lossless characteristic. The resonant-based power converters have a natural soft switching, which helps them to protect their components and gives them a high efficiency [26].

Although the literature does not consider the resonant converter as an interface circuit for PZETs, it seems logical to use the resonance phenomena in a PZET drive system to provide the DC voltage. It is obvious that the converter operating at a higher frequency requires smaller components, and is consequently a high-power density converter. Furthermore, the resonance not only assists in increasing the converter gain, but also provides the soft-switching condition for the converter switches. Therefore, the converter efficiency is increased because of the soft switching.

Usually, PZETs are energized over a small time span and receive the echo sounds over a greater time span [27]. Under this condition, the response time of the power supply is pivotal in the energization of the PZETs. Due to the resonance, the proposed converter operates rapidly enough to reach the required output voltage. As a non-isolated converter, the control system of the converter has much lower complexity than an isolated converter. It was also demonstrated that the proposed converter has a power density and gain comparable to previously introduced converters in literature.

Commercialized power supplies have a limited range of output voltage [28]. Hence, they are not suitable to be used as drivers of PZETs transmitters requiring a wide range of DC volage [5]. Furthermore, most commercialized converters have an isolated topology, complicating the implementation of the control system (i.e., provision of the feedback signal). The literature has introduced different topologies of high-step-up DC-DC converters for different applications [29–31]. However, they cannot operate in a wide range of duty cycles. It was shown that the proposed converter of this study not only offers the required gain over a wider range of duty cycles, but also has a higher power density as compared to those of commercially available converters. From the scientific point of view, the topology of the converter is new, and it offers comparable performance parameters as compared to the previously introduced converters discussed in literature [32–35]. In addition to the new topology, the mathematical model of the converter is provided, facilitating understanding of the converter's behavior.

The proposed converter operation modes and its mathematical modeling are described in Section 2. The design procedure is described at the end of this section. A 120 W converter is designed in the third section, and its performance is analyzed using the simulation analysis to validate the presented theory. The converter is prototyped to validate the circuit. The experimental results are presented in Section 4.

2. Proposed DC-DC Power Converter

A resonance circuit consisting of an inductor, a capacitor, and two power switches was added to the basic structure of the conventional boost converter to create the proposed converter (see Figure 2). AC voltages with high total harmonic distortion reduce PZET lifetime and lead to the breakage of PZETs. Thus, PZETs must be supplied by a semi-sinusoidal voltage to work safely. A package consisting of an H-bridge converter and LC filter was at the end of the proposed DC-DC converter to provide the required PZET voltage. In other words, the sharp edges of a rectangular pulse were obviated using the designed LC filter to protect the PZET. The converter switching strategy and the energy conversion through the resonance, which are explained in the next part, had crucial roles in the correct performance of the converter.

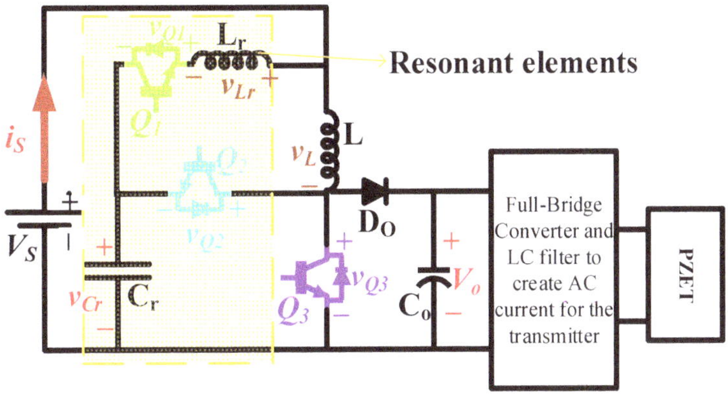

Figure 2. The proposed DC-DC power converter topology with highlighted resonant elements.

In this section, the ESR of the capacitor and resistance of the inductor are disregarded for the modeling and analysis. Additionally, it is assumed that the converter is supplied with a lossless battery package. The rising and falling time of the diode and switch conduction time are disregarded.

2.1. Converter Operation

All operating modes of the proposed DC-DC converter shown in Figure 3 are discussed in this section. Several selective waveforms, presented in Figure 4, are used to explain the operation of the proposed converter clearly. According to these figures, V_{Q1}, V_{Q2}, and V_{Q3} represent the voltage of the converter switches, while I_{Qk} (k = 1, 2, 3) is the related switch current. The current of the resonance inductor (L_r) is equal to I_{Q1}. According to Figure 4, the energy conversion period starts from t_0 when all switches are off, and the load is supplied by the output capacitor. Note that the transient operation of the converter ends at t_0, and it is the start of the steady-state operation.

In the steady-state operation of the converter, the resonance capacitor (C_r) has an initial charge (V_{int}) at t_1, which is stored from the resonance of the main inductor (L) with C_r. Thus, V_{int} is significantly greater than the source voltage (V_S). In the first time interval ($t_1 \leq t \leq t_2$), Q_1 and Q_3 are required to be turned on when the C_r initial voltage is greater than the source and the initial energy of both L and L_r is zero. According to Figure 3a, C_r depletes L_r through two different current paths, which charges L and the battery package. Because the source does not find time to energize the converter elements, both switches have time to start conduction in ZVS condition. Additionally, Q_2 can be turned on in the zero current switching (ZCS) condition when Q_3 is on, and C_r releases its energy. In this time interval, the power converter elements voltage and current vary based on the equations presented in (1):

$$\begin{cases} v_{Cr}(t) = (V_{int} - V_s)\cos(\omega_r t) \\ i_{Lr}(t) = \frac{V_{int}-V_s}{\omega_r L_r}\sin(\omega_r t) \\ i_L(t) = \frac{V_S}{L}t + i_{Lr}(t) \end{cases} \quad (1)$$

where

$$\omega_r = \frac{1}{\sqrt{L_r C_r}} \quad (2)$$

In the second time interval ($t_2 \leq t \leq t_3$), when v_{Cr} accedes to zero, the stored energy in L_r circulates through L while depleting into the battery package (Figure 3b). This time interval is concluded when all L_r stored energy is delivered to L and the battery package. Note that C_r is negatively charged when L_r delivers energy to the battery package. At the end of this time interval, as shown in Figure 3c, a small amount of the stored energy in L is delivered to C_r to satisfy the charge balance of the capacitor. t_2 can be easily derived from the resonance capacitor voltage equation in (1). The converter element energy conversion can be formulated as follows in this time interval:

$$\begin{cases} v_{Cr}(t_2) = 0 \rightarrow t_2 = \frac{\pi}{2\omega_r} \rightarrow i_L(t_2) = \frac{V_S}{L}\frac{\pi}{2\omega_r} \\ i_{Lr}(t_2) = \frac{V_{int} - V_s}{\omega_r L_r} \\ i_{Lr}(t) = i_{Lr}(t_2) - \frac{(V_S)}{L_r}(t - t_2) \\ i_L(t) = \frac{V_S}{L}(t - t_2) + i_L(t_2) \end{cases} \tag{3}$$

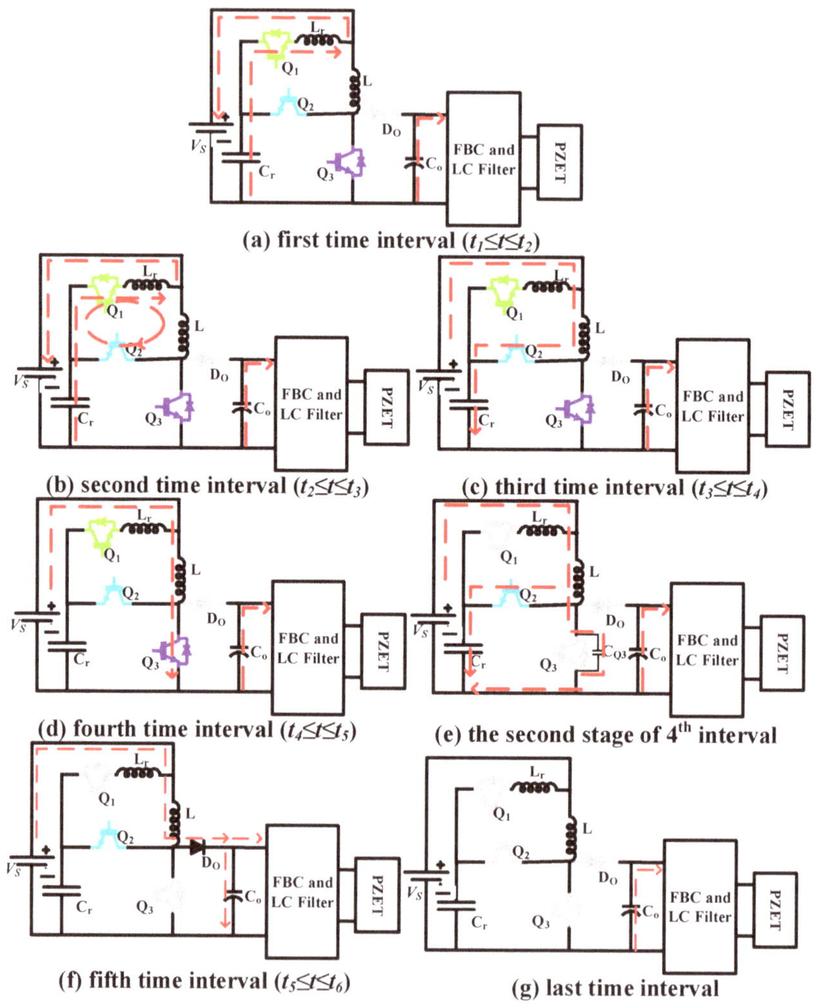

Figure 3. Circuit configurations of the proposed converter in different operation modes. The dashed line shows the current paths in each time interval. The green, blue, and purple are used to distinguish the Q_1, Q_2, and Q_3 from each other. Any gray switch shows that the switch is off in the mentioned time interval.

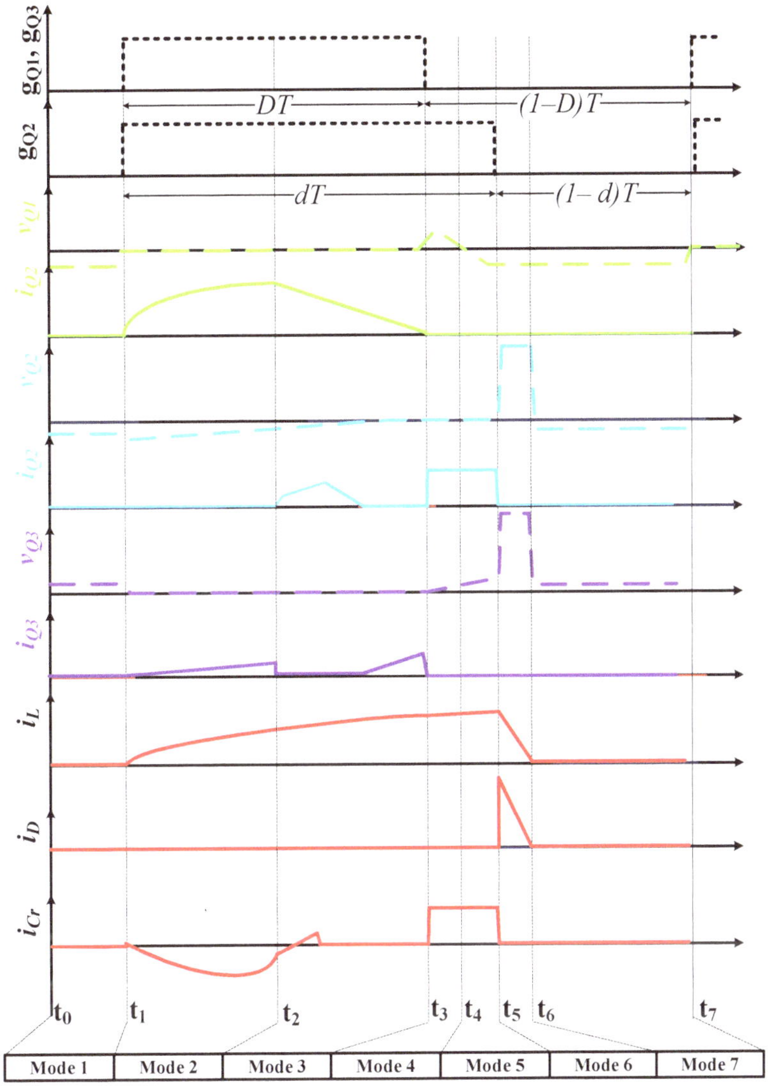

Figure 4. Selected waveforms of the converters in a switching period. The solid lines are used to represent the current waveforms of each component. The voltage waveforms are presented by the dashed lines. The gate pulses are shown by dotted lines. In this figure, g_{Q1}, g_{Q2}, and g_{Q3} show the gate pulses of Q_1, Q_2, and Q_3, respectively. The voltage and current of each switch are shown by v_{Qx}, and i_{Qx} where x is the switch number defined in Figure 2. The main inductor current, diode current (D_o), and resonant capacitor current are shown by i_L, i_D, and i_{Cr}, respectively.

The L_r current reaches zero at the end of this time interval. Then, t_3, as presented in (4), is calculated by means of the $I_{Lr}(t)$ equation from (3).

$$t_3 = \frac{i_{Lr}(t_2) L_r}{V_S} + t_2 \tag{4}$$

In the third time interval ($t_3 \leq t \leq t_4$), Q_2 is on and v_{Cr} is zero. Q_3 is already turned on in ZVS at t_1, and its voltage is zero. Then, Q_3 starts conduction in ZVS and L is charged by the input source (Figure 3d). This time interval is concluded when the turn-off gate commands are sent to Q_1 and Q_3. When the Q_1 current and v_{Cr} are zero because of the previous energy conversion during previous intervals, Q_1 and Q_3 are turned-off in the ZCS and ZVS conditions, respectively. The main inductor current equation is written in (5).

$$i_L(t) = \frac{V_S}{L}(t - t_3) + i_L(t_3) \tag{5}$$

The inductor current can be calculated by (5) where $t_4 = DT$.

$$i_L(t_4) = \frac{V_S}{L}(DT - t_3) + i_L(t_3) \tag{6}$$

In the fourth time interval ($t_4 \leq t \leq t_5$), Q_2 begins to conduct the input current in ZVS through a resonance between L and C_r. According to the Figure 3e, the parallel capacitor of Q_3 is charged during the L and C_r resonance and its voltage is equal to v_{Cr}. Therefore, Q_2 can be turned off in ZVS when the off command is sent to its gate. i_L and v_{Cr} can be calculated by (7) in this time interval:

$$\begin{cases} i_{Lr}(t) = 0 \\ i_L(t) = \frac{V_S}{\omega_m L}\sin(\omega_m(t - t_4) + i_L(t_4)) \\ v_{Cr}(t) = V_S(1 - \cos(\omega_m(t - t_4))) \end{cases} \tag{7}$$

where

$$\omega_m = \frac{1}{\sqrt{LC_r}} = 2\pi f_{sw}\,,\ f_{sw} = \frac{1}{T} \tag{8}$$

This interval is concluded when Q_2 is turning-off at dT. The calculated value for v_{Cr} at $t_5 = dT$ is equal to V_{int} defined in (1).

$$V_{\text{int}} = V_S(1 - \cos(\omega_m(dT - t_4))) \tag{9}$$

The main inductor stored energy is delivered to load through the output diode (D_O) at the fifth time interval presented in Figure 3f ($t_5 \leq t \leq t_6$). Note that the value of d, which determines Q_2 operating time, plays a prominent role in the determination of the converter gain. According to Figure 4, d must be greater than D. C_r voltage reaches the output voltage after a quarter of the resonance, which leads to D_O conduction. In this condition, D_O will conduct current to load inherently, which will be the highest gain of the converter. Therefore, Q_2 turn-off time must be smaller than a quarter of the resonance after Q_1 and Q_3 turn-off, where Q_2 is used to control the gain. Where a quarter of the resonance is calculated by (10), dT can be at most $0.25T$ greater than DT.

$$t_q = \frac{\pi}{2\omega_m} = \frac{\pi}{4\pi f_{sw}} = \frac{1}{4}T \tag{10}$$

Thus, $0.25T$ required time for maximum dT limits the maximum value of D to 0.75, which forces the proposed converter to work in discontinuous conduction mode (DCM) operation. Therefore, DCM operation is the appropriate mode for designing the value of converter elements.

In the last time interval, shown in Figure 3g, the output capacitor supplies the load with its stored energy.

The converter offers soft switching when the voltage second balance of the inductors and current balance of the capacitors are provided. According to (10), a quarter of the resonance should be completed to achieve this voltage and current balances in inductor and capacitors. Thus, the maximum duty cycle is 0.75. Additionally, a quarter of the resonance is required to charge the inductor L_r up to its peak current. Therefore, the minimum possible duty cycle for permeation of the soft switching is 0.25 if the switching frequency is set to be equal to the resonance frequency. The minimum value of the duty cycle can be reduced by increments in the switching frequency. For instance, the minimum duty cycle is 0.2 when switching frequency and resonance frequency are 100 kHz and 120 kHz.

2.2. Converter Gain

The converter gain is obtained using the voltage-balance equation of the main inductor presented in (11a). In (11a), inductance coefficient ($\frac{1}{L}$) is simplified from two sides of the equality. Then, the integral terms should be calculated to find the converter gain. The simplified term is shown in (11b):

$$\frac{1}{L}\left(\int_0^{DT} V_S dt + \int_{DT}^{t_5} (V_S - v_{Cr}) dt\right) = -\frac{1}{L}\int_{t_5}^{t_6} (V_S - V_O) dt \tag{11a}$$

$$\int_0^{DT} V_S dt + \int_{DT}^{t_5} (V_S - v_{Cr}) dt = -\int_{t_5}^{t_6} (V_S - V_O) dt \tag{11b}$$

where v_{Cr} is the resonance capacitor voltage equation presented in (7) and t_5 is equal to dT. The second term of the left side of (11b) is found by (12).

$$\begin{aligned} &\textstyle\int_{DT}^{t_5} (V_S - v_{Cr}) dt = \int_{DT}^{t_5} V_S - V_S(1 - \cos(\omega_m (t - t_4))) dt \\ &\overset{t_5 = dT}{\longrightarrow} \quad \textstyle\int_{DT}^{t_5} (V_S - v_{Cr}) dt = +\frac{V_S}{\omega_m} \sin(2\pi(d - D)) \end{aligned} \tag{12}$$

If the converter works in the boundary of DCM operation, D_O must conduct the main inductor current utmost until $t_6 = T$. Thus, the load will be energized by input source and L from dT to the end of the period (T). In this condition, the right side of (11) is calculated as follows:

$$\int_{t_5}^{t_6} (V_S - V_O) dt = \int_{dT}^{T} (V_S - V_O) dt = (V_S - V_O)(1 - d)T \tag{13}$$

Considering (11)–(13), the converter gain, which depends on the switching frequency, is calculated by (14).

$$\frac{V_O}{V_S} = \frac{1 + D - d + \sin(2\pi(d - D))}{1 - d} \tag{14}$$

2.3. Design Procedure

The design of the converter power components starts with the design of the main inductor. The value of the obtained inductor is used to find the resonance capacitor (C_r) based on the resonance, which occurs between these two elements in the fourth interval of energy conversion. The resonance inductor is obtained by consideration of the resonance happening between the C_r and L_r through the first interval of the energy conversion.

The voltage equation of the main inductor (L), when Q_3 is on, is used to calculate the value of the inductor as below:

$$V_S = V_L = L\frac{\Delta I_L}{DT} \rightarrow L = \frac{V_S DT}{\Delta I_L} \tag{15}$$

where ΔI_L is the maximum value of the inductor current. The ΔI_L value, which can be determined by the input and output power relationship presented in (16), is required to continue the calculation.

$$V_S I_{in} = V_O I_O \rightarrow I_{in} = \frac{V_O I_O}{V_S} \tag{16}$$

The average of the input current I_{in} must be equal to the average of the inductor current within a period to ensure the voltage-second balance of the inductor. The maximum value of ΔI_L is calculated at the DCM boundary. Figure 5 shows the main inductance current (I_L) and the battery current (I_{in}) at the DCM boundary. As demonstrated in (17), the maximum value of ΔI_L will be twice the average of input current.

$$\begin{cases} A_{tri} = \frac{\Delta I_L T}{2} \\ A_{rec} = I_{in} T \end{cases} \xrightarrow{A_{tri}=A_{rec}} I_{in} = \frac{\Delta I_L}{2} \tag{17}$$

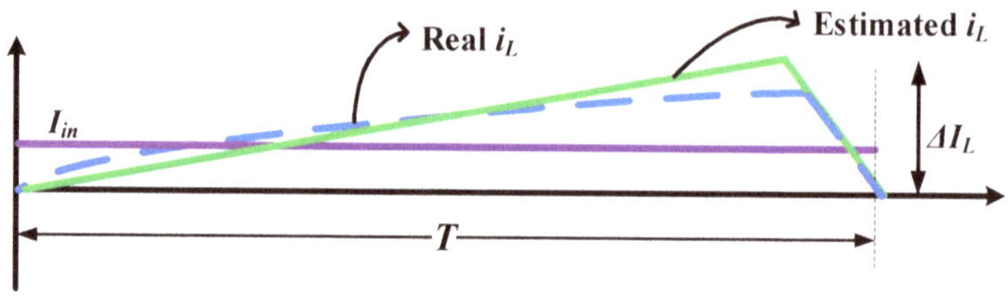

Figure 5. Estimation of I_L based on the average value of the input current.

Therefore, the maximum of ΔI_L is computed by (18) where I_{in} is replaced by its calculated value from (16):

$$\Delta I_L = \frac{2V_O I_O}{V_{S,min}} = \frac{2P_O}{V_{S,min}} \tag{18}$$

where P_O is the maximum output power. The voltage-second balance of L must be satisfied when the converter works on its maximum duty cycle (D_{max}). Therefore, D in (15) must be equal to the maximum required duty cycle, which can be derived from (14). For a certain converter, the ratio of V_O over V_s is given by the user. Thus, the maximum value of D using (14) is obtained when the maximum value of d (i.e., maximum dT = 0.25 T) is assumed at the DCM boundary. Consideration of D_{max} and substitution of the calculated ΔI_L in (15) results in L, as below:

$$L = \frac{2V_{s,min}^2 D_{max} T}{P_O} \tag{19}$$

Since the main inductance must be in resonance with C_r within a specific time interval, C_r can be calculated based on the switching frequency, as follows:

$$C_r = \frac{1}{\omega_m^2}\frac{1}{L}, \omega_m = 2\pi f_{sw}, f_{sw} = \frac{1}{T} \tag{20}$$

The resonance frequency of L_r with C_r is greater than the switching frequency, which is shown by f_r. When f_r is assumed by a designer, the value of L_r is calculated by (21).

$$L_r = \frac{1}{\omega_r^2}\frac{1}{C_r}, \omega_r = 2\pi f_r \tag{21}$$

The value of the output capacitor is determined in the last section of the design procedure. It is obvious that the output capacitor (C_O) must supply the load when energy conversion among the elements does not let the battery package supply the load. Therefore, the worst condition occurs in the maximum duty cycle when the maximum output current is obtained from C_O. According to the relationship of the capacitor voltage and current, C_O capacity can be calculated by (22):

$$\begin{gathered} i_c(t) = C\frac{dv_c(t)}{dt} \\ i_c(t) = C\frac{\Delta V_O}{D_{max}T} = I_{max} \rightarrow C_O = \frac{D_{max} T I_{O,max}}{\Delta V_O} \end{gathered} \tag{22}$$

where ΔV_O represents the maximum allowable output voltage level. In the next section, a simulation study is carried out to examine the performance of the proposed power converter.

3. Simulation Study

3.1. PZET Operation

PZETs must be supplied in their resonance frequency within a period to transmit data. After that, the power supply stops working, and the PZETs receive the reflected acoustic waves. For instance, a special PZET with the impedance of $Z = 330\angle-19^\circ\ \Omega$ was studied in [6], requiring an AC voltage with 40 kHz frequency within 15 milliseconds to send data. After this operating period, the supply was disconnected and the PZET needed 2 s to receive the reflected wave from the underwater objects. It is assumed that the proposed converter drives the aforementioned PZET in this simulation study.

3.2. Design of the Converter Elements

In this study, the converter was designed based on the maximum allowable duty cycle ($D_{max} = 0.75$) to examine the converter performance in all duty cycles. Additionally, it is supposed that the switching frequency (f_{sw}) and resonance frequency (f_r) are 100 kHz and 120 kHz, respectively. Therefore, the L value calculated by (19) is 0.86 µH, where the PZET consumes 100 W power in the maximum duty cycle and the converter is supplied by 2.4 volts rechargeable battery package. In this condition, C_r and L_r values extracted from (20) and (21) based on the obtained L are equal to C_r = 3.3 µF and L_r = 0.57 µH, respectively. Finally, the output capacitor of 1.2 µF is obtained by (22) when a 5-volt voltage ripple is considered as the applied restriction for ΔV_O.

Table 1, which compares the proposed converter elements' sizes with four other previously proposed high-step-up high-efficiency converters in an approximately equal power range, demonstrates the high-power density of the proposed converter. This table also shows the capability of the converter for operation over a wider/comparable range of duty cycles, allowing more flexibility to achieve different voltage levels.

Table 1. Comparison of the proposed DC-DC power converter element size vs. two other high-step-up converters in the literature.

	The Proposed Converter	Ref [32]	Ref [33]	Ref [34]	Ref [35]
Converter power	120 W	120 W	200 W	200 W	200 W
Switching frequency	25–100 kHz	100 kHz	60 kHz	50 kHz	30 kHz
Maximum Efficiency	95.8%	96.7%	94.5%	93.1%	92.5%
Operating duty cycle	0.2–0.75	0.3–0.7	0.5–0.9	0.1–0.75	0.05–0.3
Number of inductors	2	1	4	2	2
Inductors' value	L = 0.86 µH, L_r = 0.57 µH	L_m = 50 µH, L_k = 1.1 µH	$L_1 = L_2$ = 1.4 µH, $L_3 = L_4$ = 100 µH	L_m = 100 µH, L_k = 300 µH	$L_1 = L_2$ = 220 µH
Number of capacitors	2	5	4	3	4
Capacitors' value	C_r = 3.3 µF, C_o = 1.2 µF	C_1 to C_4 = 30 µF, C_o = 47 µF	$C_1 = C_2$ = 5 µF, $C_3 = C_4$ = 220 µF	C_1 = 470 µF, C_2 = 1000 µF, C_3 =100 µF	$C_1 = C_2 = C_3$ = 100 µF, C_4 = 220 µF
Number of diodes	1	5	4	3	4
Number of switches	3	1	2	2	2

Most of the commercialized converters include a high-frequency transformer, which leads to the isolation of input/output [28]. The transformer availability complicates the control system and increases the power density of the system. Furthermore, the output voltage range becomes limited to a smaller range. For instance, an H-bridge converter can only operate in smaller duty cycles in the range of 0.1–0.4.

3.3. Simulation Results

The simulation was run to validate the soft-switching operation of the proposed converter switches and its gain in the presence of a PIEZO-electric load model with impedance of $Z = 330\angle -19^{\circ}\ \Omega$.

The simulation was executed when the converter switches were commanded with a 50% duty cycle in 100 kHz operation frequency shown in Figure 6. Figure 7 shows the ZVS and ZCS operation of the proposed converter switches in a period when the considered gate commands were applied to them. The soft switching not only reduced the switching losses of the converter but also facilitated the operation of the SOA.

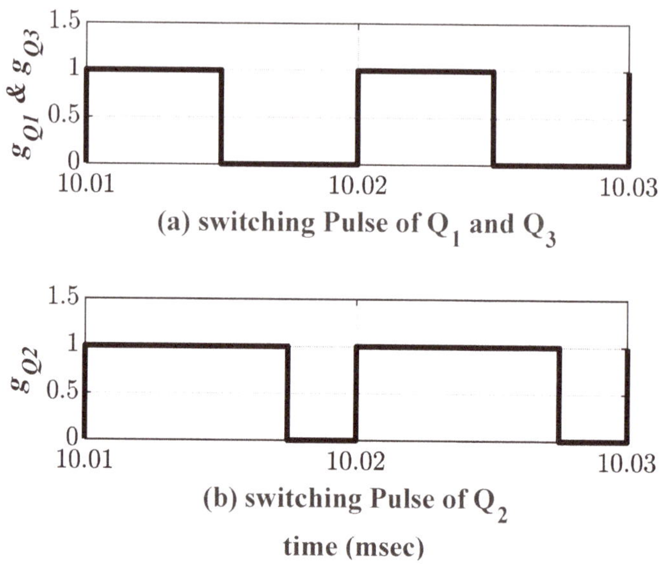

Figure 6. Gate pulses of the switches at the simulated test point.

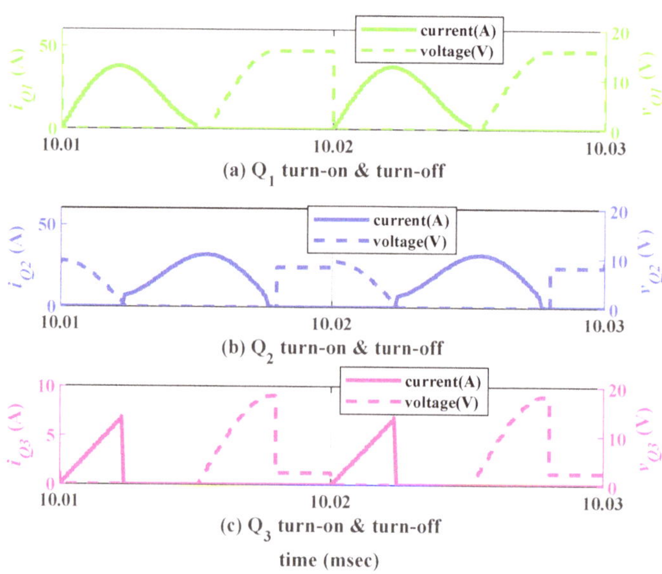

Figure 7. Soft-switching operation of the converter switches in simulation.

The input current, diode current, and current variation of the resonance elements in three switching periods are shown in Figure 8 to show the energy swing among the converter elements.

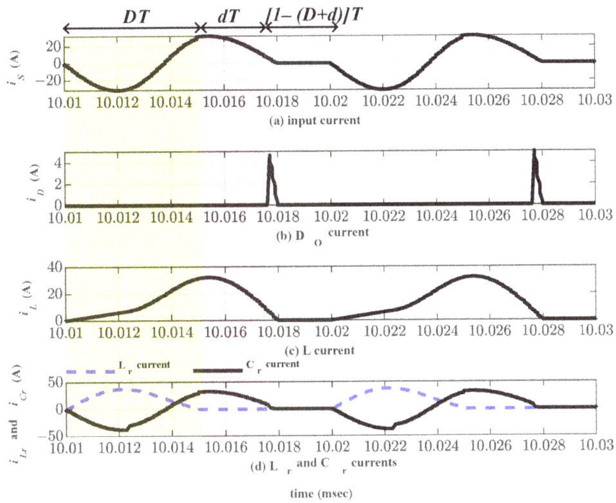

Figure 8. Simulated current waveforms of the resonance elements and output energizing current of the proposed converter.

The demonstration of soft switching enabled the checking of the converter gain. According to (14), the converter gain at $D = 0.5$ and $d = 0.75$ was about 8. Figure 9a demonstrates that the converter amplified 2.4 V input to 17.9 V in its steady-state operation which proved a gain of 7.45. Therefore, the calculated gain ratio for the converter is approximately correct, and the slight difference between the simulation and calculation analysis values came from the considered estimations during the analytical investigation. According to Figure 9a, the converter has a fast response time, which results from the availability of small size elements in the topology of the converter. The fast response enables the system controller to change output based on the user command as rapidly as possible. The proposed converter gain depends on both switching duty cycles (D and d). Figure 9b demonstrates the converter gain variations vs. sensitive parameters. According to Figure 9b, the increase in D leads to a higher gain, while d increments reduce the converter gain in a constant D.

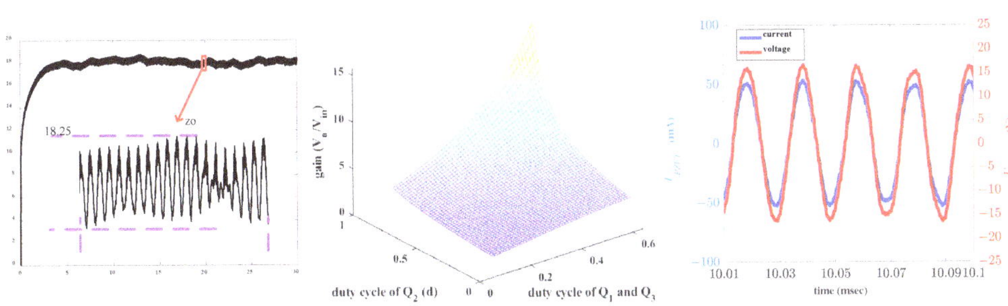

Figure 9. (**a**) Voltage response of the studied step-up DC-DC converter, which shows fast transients, high gain, and low ripple of the DC-link voltage in a simulation study. (**b**) Gain variations of the proposed DC-DC power converter against simultaneous changes of the switches' duty cycles in a simulation study. (**c**) PZET current and voltage waveforms in the simulation.

An H-bridge converter, which was controlled by a 40 kHz sinusoidal pulse width modulation (SPWM) command, was placed at the output of the proposed DC-DC converter to supply the understudy PZET through an LC filter. A regular sinusoidal voltage and currents of the PZET are shown in Figure 9c to prove the acceptable performance of the converter while it was connected to a DC-AC converter.

4. Experimental Results

Due to the lack of a high-voltage and high-power battery package in the available laboratory, the test setup was provided with a smaller rating power. According to earlier explanations, PZETs require about 50–1000 volts in different sea depths (greater sea depth requires higher voltage level). To realistically produce the power density of the industrial version of the studied converter, the element sizes were selected for such a voltage level. Lack of suitable experimental equipment limited us to testing the converter in industrial conditions. As such, a prototype of the proposed converter supplied by a 3.7 V 12.5 Ah Lithium-ion battery package was provided to experimentally validate the soft switching and the converter performance. The battery package included six parallel connected battery cells with 3.7 V and 2 Ah characteristics.

Arduino Mega2560 worked as a microprocessor to generate the required gate pulses presented in Figure 10a. The control pulses of Q_1 and Q_3 were generated using the first timer of the Mega2560. The required control pulse of Q_2 was generated by the second timer of this microprocessor. Although the metal–oxide–semiconductor field-effect transistors (MOSFETs) were the appropriate choice for 100 kHz switching frequency, an insulated-gate bipolar transistor (IGBT) was chosen to examine the converter behavior due to the ability of IGBTs to tolerate higher voltages. Because of the availability of three floating emitter IGBTs in the proposed topology, the gate pulses must command the gate switches through three different isolated ground MinMax power supply (MAU324). Thus, the MinMax power supply used at the output of three separated HCPL 3120 optocouplers provided 15 V voltage to supply the gates. The main structure of the converter included two SGH80N60UFD IGBTs, a FGA25N120ANTD IGBT, and a UF4007 ultrafast diode. Additionally, an electrolyte capacitor with a capacity of 2.2 µF was used to provide the DC-link. The inductors are constructed from the iron powder cores with a permeability of 7.5 µH/m and copper winding.

The images in Figure 10b,c show the 3D and printed circuit board (PCB) views of the converter. According to the presented 3D view, the power density of the converter based on the selected switches was about 13 W/cm^3 which enabled the converter to be classified as a high-power density converter. The PCB was designed in 2 layers, as shown in Figure 10c. A higher power density is achieved if the PCB is designed in more layers using surface-mount device (SMD) components. The schematic of this design is presented in Figure 11 to show the track paths of the considered PCB. In this figure, P1 and P2 are the input and output terminals of the circuit. The power required to supply the isolated power supplies was provided by P3. P4 was the terminal chosen to provide the gate commands from the controller circuit. R1, R2, and R3 were the input resistors of the opto-coupler. The 5 Ω gate resistance was selected at the output of the opto-couplers to drive the switches.

Measurement of signals in switching power converters is a difficult task. The availability of a two-channel oscilloscope limited us to using the differential measurement technique of signal measurement. In this paper, the gate drive signals were produced using a microcontroller in open-loop condition. It allowed measurement of the produced signal gates when the load was disconnected from the circuit. The currents of the switches were measured by the instalment of 0.3 Ω resistance in the collector emitter path. The grounds of both probes of the oscilloscope were connected to the collector. The voltage of the resistance and voltage of the collector-emitter were measured when the other sides of the probes were connected to the other sides of the series resistor (i.e., installed for current measurement) and emitter. Thus, the voltage and current of the switches were obtained together.

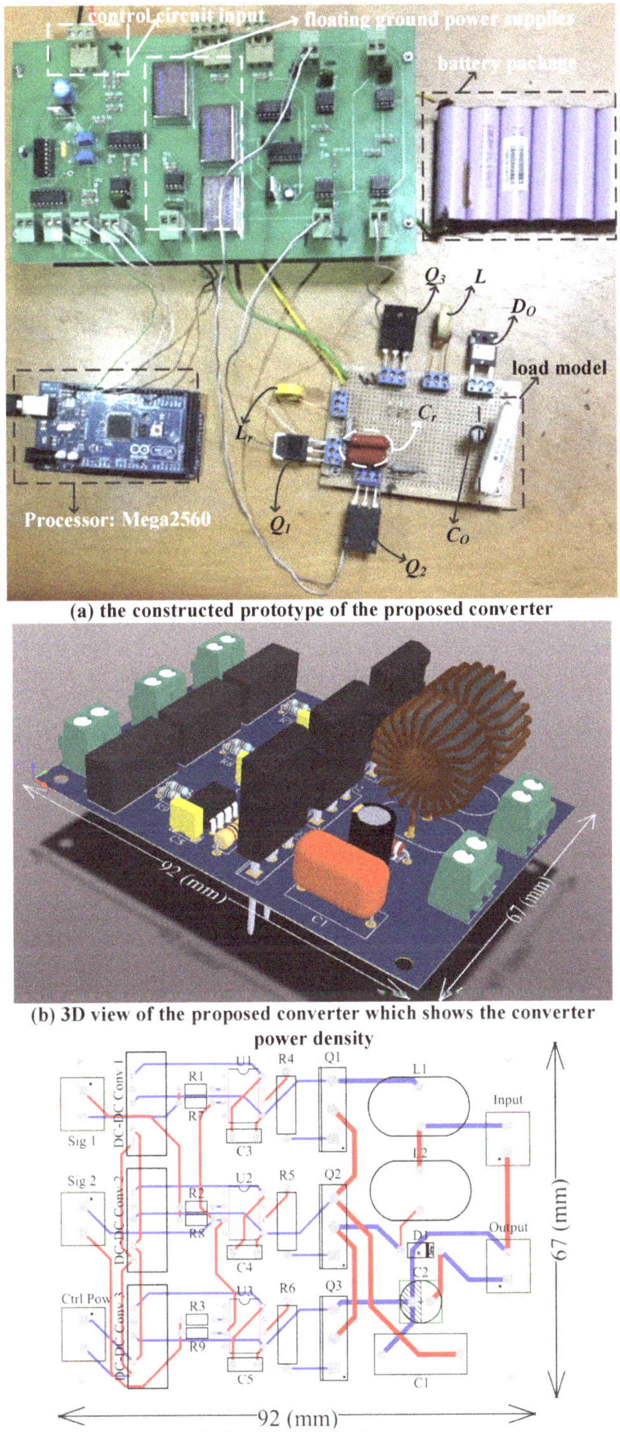

Figure 10. The constructed prototype and considered industrial version.

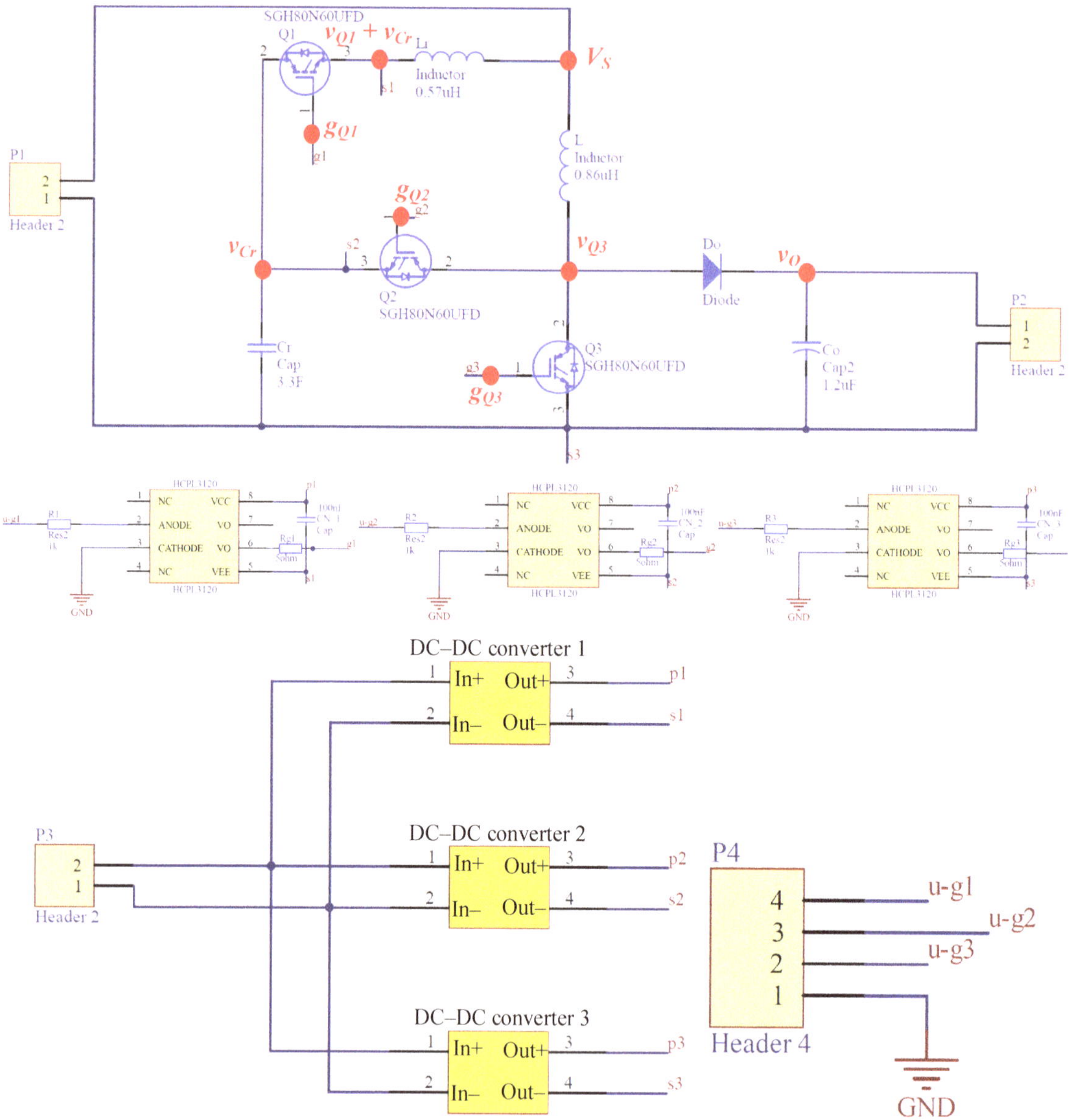

Figure 11. Schematic of the proposed converter.

The converter gain is the first factor of the proposed converter that must be examined practically. On the other hand, the proposed converter's operating duty cycles are limited to $0.2 \leq D \leq 0.75$. Hence, the experimental verification was performed at the converter operation boundary when the primary duty cycle (D) of the power converter at 100 kHz switching frequency was equal to 20%, as shown in Figure 12a. This figure shows the duty cycle of Q_2, which should be 0.25 larger at 45%. The ratio of the converter output to input voltages presented in Figure 12b was about 3.1 which verified the calculated converter gain ratio in (14). The input voltage of the converter shown in Figure 12b, which was provided by the battery package, demonstrated that a large amount of energy was obtained from the battery when the converter switches were turned-on. Therefore, it is evident that the battery voltage drops dramatically in a switching period. However, this energy is sent back to the battery in that period and its voltage is equal to its initial value before the start of the operation period.

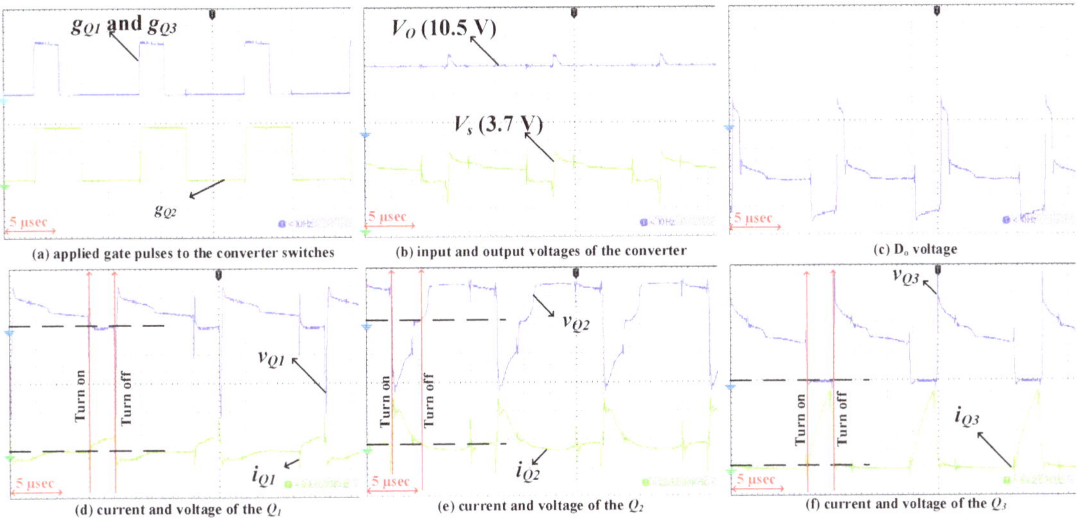

Figure 12. Experimental validation of the proposed converter gain and soft switching. The time division of all presented images in this figure is 2.5 μs. (**a**) Control signals (2 V/div); (**b**) output voltage (5 V/div), input voltage (2 V/div); (**c**) diode voltage (5 V/div); (**d**) Q_1 voltage (5 V/div), Q_1 current (1.66 A/div); (**e**) Q_2 voltage (2 V/div), Q_2 current (0.66 A/div); (**f**) Q_3 voltage (2 V/div), Q_3 current (0.66 A/div).

The resulting peak in the output voltage waveform, presented in Figure 12b, shows that the main inductor energy was delivered to the load and C_o at the end of each switching period. Since the proposed DC-DC converter works an interface circuit between the battery package and an inverter that supplies the PZETs, the charging spikes at the end of each period shown in the waveform of output voltage cannot damage the PZET. These spikes are eliminated by passing through an AC filter. Figure 12c shows D_o voltage variation in a switching period, in which the anode voltage is lower than the cathode in the rest of a switching period. Considering both output voltage and diode voltage together, the output is supplied by the converter when the main inductor (L) stored energy is delivered to the load. As shown in Figure 12d, Q_1 is turned on in the ZVS condition and the turned-off command meets its gate when its current is zero and enables it to be turned off when the transistor voltage is negative. Furthermore, the ZCS turn-on and ZVS turn-off of the Q_2 are presented in Figure 12e. The soft switching turn-on/off of the third switch (Q_3) is shown in Figure 12f. Figure 13 shows the simulation results in the same duty cycle and frequency as the experimental results. The small difference between the waveforms resulted from the gate drive circuit of the switches and the parasitic inductances of the wires in the prototyped circuit. The simulation results show the Q_1 and Q_3 conduct currents together in the same time interval. Q_3 starts conduction when Q_1 turns off. According to the results, all the switches operate in soft-switching condition. The soft-switching operation reduces the converter switching losses, which inherently increases the converter efficiency.

Figure 14a shows the energy conversion between the resonant capacitor and the main inductor. C_r voltage meets its negative peak when it is charged by the battery package through L. C_r voltage starts to increase from the negative peak when both Q_1 and Q_2 receive the turned-off command and L energy is delivered to C_r through Q_2. It should be highlighted that the voltage spike of the inductor occurs when the output diode (D_O)

starts conduction. The inductor and its resistances are small. The current of the inductor cannot change its direction instantly when the output voltage is connected to one side of the inductor. The small resistance of the inductor cannot provide a large time constant for damping the spike. For this reason, this spike is shown in the provided test setup. This issue is solved if the inductor is provided with a larger resistance.

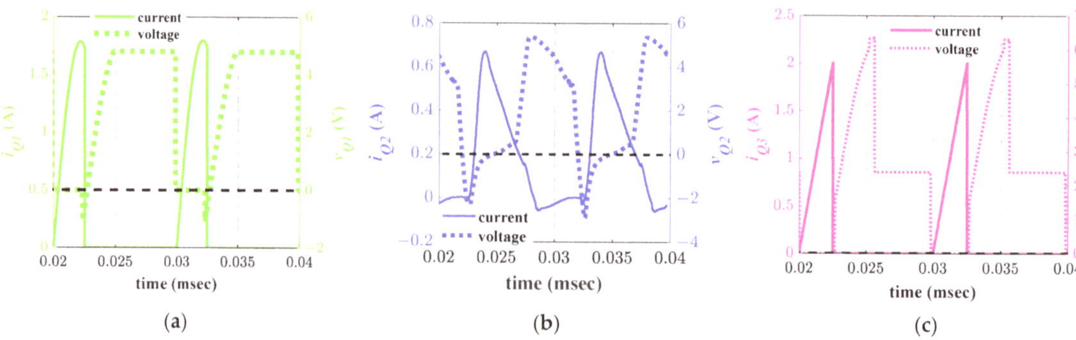

Figure 13. The simulation results showing the soft switching of the switches in the test conditions (D = 0.2 and d = 0.45). (**a**) Q_1 voltage and current; (**b**) Q_2 voltage and current; (**c**) Q_3 voltage and current.

Note that three 0.3 Ω resistors are placed in series with Q_1, Q_2, and Q_3 emitters to measure the switch currents during operation. For this reason, the voltage/division of the measured signal shown in Figure 12 must be mapped to Ampere/division by consideration of the value of the series resistors.

The proposed converter provides ZVS and ZCS switching for all switches in smaller switching frequencies up to 25 kHz. The resonances occur faster than the smaller switching frequency. Therefore, the energy conversion among the resonance elements occurs several times in a period of switching, which leads to a higher gain in comparison to the chosen based frequency (100 kHz) during the design procedure. In this condition, higher current cross through the switches and windings cause increments of conduction losses. In brief, although working in a lower switching frequency brings about a higher gain in a soft-switching operation, the converter efficiency is decreased because of the increase in conduction losses. To consider the conduction and switching losses in the efficiency analysis of the converter, the converter efficiency curve is extracted experimentally. The efficiency is defined by the ratio of the output active power over input active power. The input and output powers are calculated by multiplication of the input/output voltage and current measured by the digital amperemeters and voltmeters. Figure 14b shows the variation in the efficiency vs. output power in three different switching frequencies.

A similar procedure investigated the uncertainty of the measurement in reporting the results of an analysis on the repeatability, reproducibility, and stability of the measured results, as reported in [36,37]. The temperature variation was considered as an uncertain variable during the test.

According to the repeatability definition, the power converter was tested (turn/on and turn/off) under 100 W operation when the switch temperature was constant in each iteration. The input power was measured using the same means in each test. The collected data after 10 tests demonstrated the repeatability of the results with standard deviation of 0.0179 W. The reproducibility was studied by collecting another set of data on a 100 W experiment at the same switch temperature. The standard deviation of the second set of data was 0.0166 W. Thus, the measured data were reproducible because the difference of the calculated standard deviation was 0.0001 W. Finally, the converter was examined under 100 W loads on four consecutive days. The standard deviation of the collected data over four days was 0.0128 W, which showed the stability of the converter behavior.

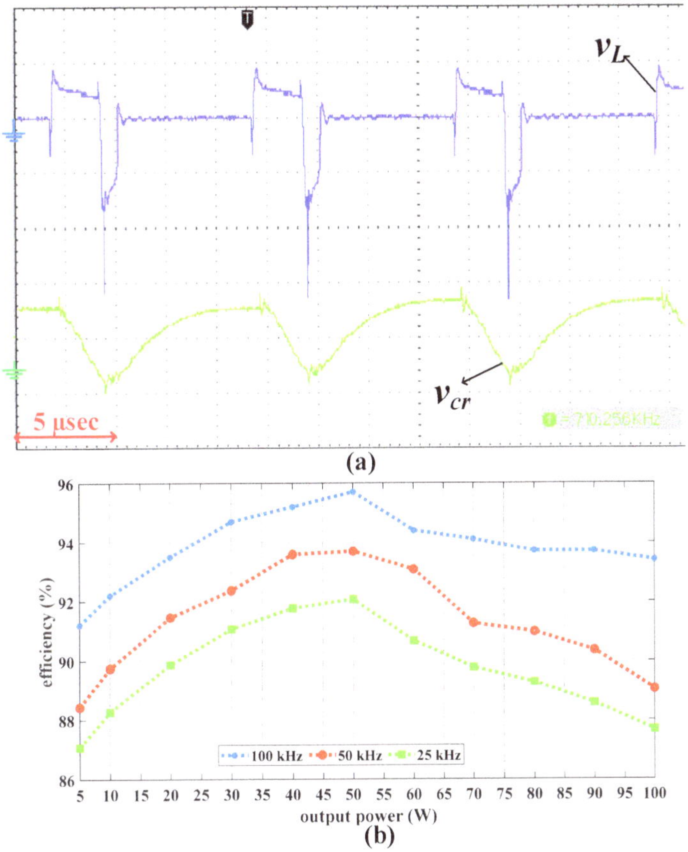

Figure 14. (**a**) Cr voltage (5 V/div) and L voltages (2 V/div) with time division of 2.5 µs. (**b**) Experimentally measured efficiency vs. output power of the prototyped DC-DC converter at different switching frequencies.

5. Conclusions

A high-step-up, high-efficiency, high-power density converter was proposed for driving PZETs. The availability of small-size power components in the structure of the proposed topology helped to decrease the required space of the converter. Because of the soft-switching operation of the converter switches, the converter has a higher efficiency in comparison to the most of the previously proposed high-step-up converters, which work in hard-switching conditions. Such a low-loss converter is suitable not only for driving PZETs but also for use in all other applications where a battery package is used as a power source.

Furthermore, the fast resonance increases the voltage build-up time at the output of the converter during transient conditions. The fast response of the converter is required when the energization time of a load is limited by the users. For instance, in PZETs, users need to energize the piezoelectric in a short time and stop energization to receive data.

Compared to the digital processors, controller designers prefer to use analog circuits in harsh noisy environments. The common input/output ground of the non-isolated converters eases the preparation of the feedback signals, which are required in a closed-loop control circuit. Therefore, the proposed converter is suitable to be utilized in the harsh conditions because of its non-isolated topology. The closed-loop controller design of the system and converter testing with higher power rating will be the subject of study in future works.

Author Contributions: Conceptualization, E.R.; methodology, E.R.; validation, E.R. and H.D.-E.; formal analysis, E.R. and H.D.-E.; investigation, E.R.; resources, S.K. and A.M.; data curation, E.R. and H.D.-E.; writing—original draft preparation, E.R.; writing—review and editing, A.M. and S.K.; visualization, E.R., H.D.-E. and A.M.; supervision, A.M.; project administration, E.R. and A.M. All authors have read and agreed to the published version of the manuscript.

Funding: This research received no external funding.

Institutional Review Board Statement: Not applicable.

Informed Consent Statement: Not applicable.

Data Availability Statement: Not applicable.

Conflicts of Interest: The authors declare no conflict of interest. The funders had no role in the design of the study; in the collection, analyses, or interpretation of data; in the writing of the manuscript, or in the decision to publish the results.

References

1. Yang, C.; Wang, T.; Li, Y.; Yang, H.; Li, J.; Qu, D.; Xu, B.; Yang, Y.; Daemen, J. Feasibility analysis of using abandoned salt caverns for large-scale underground energy storage in China. *Appl. Energy* **2015**, *137*, 467–481. [CrossRef]
2. Alanne, K.; Cao, S. An overview of the concept and technology of ubiquitous energy. *Appl. Energy* **2019**, *238*, 284–302. [CrossRef]
3. Deng, Z.D.; Lu, J.; Li, H.; Tian, C.; Myjak, M.J.; Bellgraph, B.J.; Cartmell, S.; Xiao, J. *Signal Transmitter and Methods for Transmitting Signals from Animals*; Pacific Northwest National Lab. (PNNL): Richland, WA, USA, 2019.
4. Zhang, W.; Han, G.; Wang, X.; Guizani, M.; Fan, K.; Shu, L. A node location algorithm based on node movement prediction in underwater acoustic sensor networks. *IEEE Trans. Veh. Technol.* **2020**, *69*, 3166–3178. [CrossRef]
5. Kiran, M.R.; Farrok, O.; Abdullah-Al-Mamun, M.; Islam, M.R.; Xu, W. Progress in piezoelectric material based oceanic wave energy conversion technology. *IEEE Access* **2020**, *8*, 146428–146449. [CrossRef]
6. Roshandel, E.; Namazi, M.M.; Saghaian-Nejad, S.M.; Ahn, J.-W. A Step-Up High-Power-Density DC/AC Converter to Drive Piezoelectric Transmitters. In Proceedings of the 2018 21st International Conference on Electrical Machines and Systems (ICEMS), Jeju, Korea, 7–10 October 2018; pp. 688–692. [CrossRef]
7. Song, H.; Zhang, Y.; Gao, J.; Zhang, Y. Small power ZVS circuits for the marine-controlled source electromagnetic transmitter. *J. Eng.* **2019**, *2019*, 4325–4330. [CrossRef]
8. Choi, H. Prelinearized class-b power amplifier for piezoelectric transducers and portable ultrasound systems. *Sensors* **2019**, *19*, 287. [CrossRef]
9. Smith, N.B.; Webb, A. *Introduction to Medical Imaging: Physics, Engineering and Clinical Applications*; Cambridge University Press: Cambridge, UK, 2010.
10. Benson, B.; Li, Y.; Kastner, R.; Faunce, B.; Domond, K.; Kimball, D.; Schurgers, C. Design of a low-cost, underwater acoustic modem for short-range sensor networks. In Proceedings of the OCEANS'10 IEEE Sydney, Sydney, Australia, 24–27 May 2010; pp. 1–9. [CrossRef]
11. Bhaskar, M.S.; Ramachandaramurthy, V.K.; Padmanaban, S.; Blaabjerg, F.; Ionel, D.M.; Mitolo, M.; Almakhles, D. Survey of DC-DC non-isolated topologies for unidirectional power flow in fuel cell vehicles. *IEEE Access* **2020**, *8*, 178130–178166. [CrossRef]
12. Lu, Y.; Wu, H.; Sun, K.; Xing, Y. A family of isolated buck-boost converters based on semiactive rectifiers for high-output voltage applications. *IEEE Trans. Power Electron.* **2015**, *31*, 6327–6340. [CrossRef]
13. Meraj, M.; Bhaskar, M.S.; Reddy, B.P.; Iqbal, A. Non-isolated DC–DC power converter with high gain and inverting capability. *IEEE Access* **2021**, *9*, 62084–62092. [CrossRef]
14. Zhao, X.; Chen, C.-W.; Lai, J.-S.; Yu, O. Circuit design considerations for reducing parasitic effects on GaN-based 1-MHz high-power-density high-step-up/down isolated resonant converters. *IEEE J. Emerg. Sel. Top. Power Electron.* **2019**, *7*, 695–705. [CrossRef]
15. Lee, S.-W.; Do, H.-L. Isolated High Step-Up Dual-Flyback DC–DC Converter with a Resonant Voltage Multiplier. *Electr. Power Compon. Syst.* **2020**, *48*, 871–880. [CrossRef]
16. Liu, C.; Li, X.; Zhi, Y.; Cai, G. New breed of solid-state transformer mainly combing hybrid cascaded multilevel converter with resonant DC-DC converters. *Appl. Energy* **2018**, *210*, 724–736. [CrossRef]
17. Moradzadeh, M.; Babaei, E.; Zamiri, E.; Hamkari, S. A new high step-up DC/DC converter structure by using coupled inductor with reduced switch-voltage stress. *Electr. Power Compon. Syst.* **2017**, *45*, 1705–1719. [CrossRef]
18. Veerachary, M.; Senjyu, T.; Uezato, K. Modeling of closed-loop voltage-mode controlled interleaved dual boost converter. *Comput. Electr. Eng.* **2003**, *29*, 67–84. [CrossRef]
19. Li, F.; Liu, H. A cascaded coupled inductor-reverse high step-up converter integrating three-winding coupled inductor and diode–capacitor technique. *IEEE Trans. Ind. Inform.* **2016**, *13*, 1121–1130. [CrossRef]
20. Babu, A.R.; Raghavendiran, T.A. High voltage gain multiphase interleaved DC-DC converter for DC micro grid application using intelligent control. *Comput. Electr. Eng.* **2019**, *74*, 451–465. [CrossRef]

21. Deng, C.; Yong, M.; Rodriguez, L.A.G.; Balda, J.C.; Li, R. Passive integration using FMLF technique for integrated boost resonant converters. *IEEE Trans. Ind. Electron.* **2019**, *67*, 3756–3766. [CrossRef]
22. Yuan, Y.; Zhang, Z.; Mei, X. Boost-integrated LCL resonant converter with high voltage gain. *IET Power Electron.* **2020**, *13*, 332–339. [CrossRef]
23. Zhu, B.; Ren, L.; Wu, X.; Song, K. ZVT high step-up DC/DC converter with a novel passive snubber cell. *IET Power Electron.* **2017**, *10*, 599–605. [CrossRef]
24. Lin, B.R.; Huang, C.L. Zero voltage switching active clamp buck-boost stage Cuk converter. *IET Electr. Power Appl.* **2007**, *1*, 173–182. [CrossRef]
25. Sayed, K.F.; Nakaoka, M.; Morimoto, K.; Kwon, S.K. New high-frequency linked half-bridge softswitching PWM DC-DC converter with input DC rail side active edge resonant snubbers. *IET Power Electron.* **2010**, *3*, 774–783. [CrossRef]
26. Li, Z.; Wu, T.; Zhang, G.; Yang, R. Hybrid modulation method combining variable frequency and double phase-shift for a 10 kW LLC resonant converter. *IET Power Electron.* **2018**, *11*, 2161–2169. [CrossRef]
27. Roshandel, E.; He, J. High Efficiency High Step-Up DC-DC Converter to Drive Piezo-Electric Transmitters. In Proceedings of the 2019 IEEE Industry Applications Society Annual Meeting, Baltimore, MD, USA, 29 September–3 October 2019; pp. 1–16. [CrossRef]
28. Mean-Well SD-100C-24, M. 100W Single Output DC-DC Converter. Available online: https://au.mouser.com/ProductDetail/MEAN-WELL/SD-100C-24?qs=sGAEpiMZZMtz8P%2FeuiupSWlCmLAwku1ozA57skhLwW4%3D (accessed on 7 July 2022).
29. Yu, D.; Yang, J.; Xu, R.; Xia, Z.; Iu, H.H.-C.; Fernando, T. A family of module-integrated high step-up converters with dual coupled inductors. *IEEE Access* **2018**, *6*, 16256–16266. [CrossRef]
30. Tang, Y.; Tong, H.; Afzal, R.; Guo, Y. High step-up ZVT converter based on active switched coupled inductors. *IEEE Access* **2020**, 1. [CrossRef]
31. Zhu, B.; Chen, S.; Zhang, Y.; Huang, Y. An interleaved zero-voltage zero-current switching high step-up DC-DC converter. *IEEE Access* **2020**, *9*, 5563–5572. [CrossRef]
32. Hassan, W.; Lu, D.D.-C.; Xiao, W. Single-switch high step-up DC–DC converter with low and steady switch voltage stress. *IEEE Trans. Ind. Electron.* **2019**, *66*, 9326–9338. [CrossRef]
33. Chen, Y.-T.; Lu, Z.-X.; Liang, R.-H. Analysis and design of a novel high-step-up DC/DC converter with coupled inductors. *IEEE Trans. Power Electron.* **2017**, *33*, 425–436. [CrossRef]
34. Rajabi, A.; Rajaei, A.; Tehrani, V.M.; Dehghanian, P.; Guerrero, J.M.; Khan, B. A Non-Isolated High Step-Up DC-DC Converter Using Voltage Lift Technique: Analysis, Design, and Implementation. *IEEE Access* **2022**, *10*, 6338–6347. [CrossRef]
35. Zhu, X.; Zhang, B.; Jin, K. Hybrid nonisolated active quasi-switched DC-DC converter for high step-up voltage conversion applications. *IEEE Access* **2020**, *8*, 222584–222598. [CrossRef]
36. Carobbi, C.F.M.; Lalléchère, S.; Arnaut, L.R. Review of uncertainty quantification of measurement and computational modeling in EMC part I: Measurement uncertainty. *IEEE Trans. Electromagn. Compat.* **2019**, *61*, 1690–1698. [CrossRef]
37. Cepova, L.; Kovacikova, A.; Cep, R.; Klaput, P.; Mizera, O. Measurement system analyses-gauge repeatability and reproducibility methods. *Meas. Sci. Rev.* **2018**, *18*, 20. [CrossRef]

energies

MDPI

Review

Review of Single-Phase Bidirectional Inverter Topologies for Renewable Energy Systems with DC Distribution

Meshari Alshammari [1,2] and Maeve Duffy [1,*]

1 Power Electronics Research Centre, School of Engineering and Ryan Institute, University of Galway, H91 HX31 Galway, Ireland
2 Electrical Engineering Department, College of Engineering, Jouf University, Sakaka 72388, Saudi Arabia
* Correspondence: maeve.duffy@universityofgalway.ie

Abstract: Recent developments in renewable energy installations in buildings have highlighted the potential improvement in energy efficiency provided by direct current (DC) distribution over traditional alternating current (AC) distribution. This is explained by the increase in DC load types and energy storage systems such as batteries, while renewable energy sources such as photovoltaics (PVs) produce electricity in DC form. In order to connect a DC distribution system to the alternating current grid (e.g., for backup, delivering energy storage to the grid) there is a need for a bidirectional inverter, which needs to operate over a wide range of source and load conditions and is therefore critical to the overall system performance. However, DC distribution in buildings is relatively new, with much of the research focused on the control of the DC bus connection between sources and loads, rather than on the grid connection. Therefore, this review aims to explore recent developments in bidirectional inverter technologies and the associated challenges imposed on grid-connected DC distribution systems. The focus is on small-scale building applications powered by photovoltaic (PV) installations, which may include energy storage in the form of batteries. An evaluation of existing inverter topologies is presented, focusing on semiconductor technologies, control techniques, and efficiency under variable source and load conditions. Challenges are identified, as are optimal solutions based on available technologies. The work provides a basis for future developments to address current shortcomings so that the full benefits of DC distribution can be achieved.

Keywords: bidirectional inverter; DC distribution system; grid integration; single-phase inverter; renewable energy integration

Citation: Alshammari, M.; Duffy, M. Review of Single-Phase Bidirectional Inverter Topologies for Renewable Energy Systems with DC Distribution. *Energies* **2022**, *15*, 6836. https://doi.org/10.3390/en15186836

Academic Editors: Jelena Loncarski and Diego Bellan

Received: 19 August 2022
Accepted: 11 September 2022
Published: 19 September 2022

1. Introduction

1.1. Background and Motivation

Renewable energy sources, including solar photovoltaics (PVs) and wind turbines, are considered the most dominant solutions to guarantee energy security, with solar PVs outweighing the advantages of other sources in terms of cost and environmental friendliness [1]. However, in order to maximize the supply of energy from renewable sources, the efficiency of the path from source to load needs to be optimized. For instance, the integration of a photovoltaic (PV) system with a conventional alternating current (AC) distribution system requires an inverter to convert the direct current (DC) electricity produced by PVs into a standard AC grid form. On the other hand, there is an ever-increasing range of domestic appliances and equipment that operate from a DC supply, e.g., computing and audiovisual equipment, cordless vacuum cleaners, etc., but they require an AC/DC rectifier stage to connect to the conventional AC distribution system (mains). In recognition of the improved efficiency provided by DC distribution between a DC source and DC loads (through the elimination of two complementary stages of power conversion), there has been significant growth in the range of appliances configured for supply from a DC distribution system, e.g., cooling, heating, lighting, refrigerator, washing machines, etc. [2]. Indeed, standards are being developed for DC-configured products under the EMerge Alliance [3], in which a high DC voltage level of 300–380 V is preferred in terms of distribution capability, lower equipment cost, and simplicity of integration with existing system infrastructure [4]. It is worth mentioning that such DC appliances correspond to over 60% of the total electricity consumption of householders in the U.S. [5].

In order to determine the benefit provided by DC distribution, recent studies have analyzed the relative efficiency of AC and DC systems [6,7]. Increased efficiency of up to 16% has been predicted for DC vs. AC when a PV installation and energy storage are utilized. A similar level of improvement was reported in [8] in which an additional energy source, such as a gas engine, was included with PVs, providing increased efficiency of 15%. Even in the absence of a PV source, in [9], the application of a DC distribution system in an office building was predicted to have lower power losses by up to 14.9% compared to the AC systems when only powered by the grid energy. In the case where renewable energy and energy storage are integrated with the utility grid through DC, optimal efficiency of up to 50% has been predicted for small-scale buildings [10,11]. This is encouraging for residential buildings which are likely to achieve self-sustainability compared to commercial buildings [10], thereby achieving the advantages of net-zero energy, such as resilience and reliability, and sustainability in buildings can be optimized.

However, when a DC distribution system is implemented and integrated with the AC grid, an inverter with bidirectional power flow is usually needed to feed the grid in the case of excess power from the PVs and to supply power from the grid to maintain the DC bus at a nominal voltage when the load demand is higher than PV generation. This compares with standard unidirectional inverters, which are normally used to feed PV energy into an AC distribution system. Bidirectional inverters have been widely used in higher power applications such as energy storage batteries and plug-in hybrid or fully electric vehicles. In electric vehicle (EV) applications, the bidirectional capability may be required to facilitate vehicle-to-grid (V2G) between the grid and the DC bus, although normally, only a unidirectional rectification stage is used to charge the EV battery. However, since EV battery charging has its own specific requirements in terms of battery voltage and charging algorithms, it is not considered further. Instead, this work concentrates on the bidirectional inverters in DC distribution systems integrated into residential buildings.

Despite their importance, the design and implementation of bidirectional inverters for DC distribution in buildings are relatively immature. As a result, there are several approaches described in the literature, but they have not been widely compared or assessed. This review takes the opportunity to address this gap so as to advance the understanding of the impact of bidirectional inverters in DC distribution systems, while also identifying technical challenges to be addressed in future research. The results will lead to the development of more efficient DC distribution systems towards enabling net-zero buildings, as predicted in the studies above.

1.2. Bidirectional Inverter Challenges

The interface between the DC bus and the AC grid is crucial because it can reduce the operation efficiency and stability of the overall system performance. Therefore, methods for increasing the efficiency of bidirectional inverters have received considerable attention because they relate to the return on investment of the DC system. High efficiency over a wide range of power levels has significant benefits for increasing DC system efficiency in buildings, especially small-scale domestic installations, where there is usually a mismatch between periods of PV production and energy consumption [12–14]. This is less relevant for commercial buildings.

Transformerless topologies are widely used to reduce losses (and costs) associated with transformers in isolated topologies. However, it is important to highlight that due to dynamic coupling between the shared AC and DC ground in a transformerless solution, the parasitic capacitance of PV panels causes a leakage current. This has a negative impact on the inverter efficiency and needs to be minimized to avoid significant loss [15]. The effect varies depending on the common mode voltage (CMV) across the parasitic capacitor, and therefore, modulation strategies can alleviate the issue [16]. Other solutions are discussed in more detail below.

High leakage current may lead to the additional distortion of the output grid current caused by high-frequency operation, where the utility grid standardization needs to be in compliance with the safety regulations of electrical equipment. On account of this, an EMI filter is required to attenuate the grid current to an optimal level of quality on the AC side in the case of inverter mode. However, while the ground of the AC and DC sides are shared through the distribution plant in the building, the common-mode (CM) and electromagnetic interference (EMI) noise-related issues will cause an impact on the DC side. The DC-bus voltage, as well as differential-mode (DM) and (CM) noise, must be kept within a small-voltage ripple to provide improved power quality, ensure reliable load operation, and reduce significant loss in the system. Different filtering schemes to address this are described later. It is shown that the additional number of passive components, as well as the complexity of the associated control approach, would reduce the efficiency of the design. As a consequence, a trade-off must be made in the development of a suitable EMI filter in order to reduce the aforementioned concerns.

Given the strong correlation between the leakage current and the EMI filter, it should be noted that the range of current that the filter inductor needs to support is another challenge due to the potential nonlinear characteristic of its core material. As a result, the tracking accuracy of the employed current controller to set the required grid current may be influenced by the inductor current ripple. Moreover, errors in zero-crossing detection for grid synchronization could be introduced due to insufficient magnetization of the inductor core material which has an impact on the grid current. This effect is important for both inversion and rectification modes in a grid-connected inverter. Indeed, in rectification mode, power factor correction (PFC) is required to reduce the consumers' load demand for reactive power. In addition, the harmonic regulation has to be fulfilled according to the required grid code. It is worth noting that the response to different weather conditions of renewable energy sources such as PVs can cause a wide voltage variation on the DC side of the bidirectional inverter. Under this circumstance, the authors of [17] offer buck/boost maximum power point tracking (MPPT) to mitigate the sudden change of voltage and associated stress applied to the DC bus interfaced with a grid-connected inverter. This is important to ensure that the bidirectional inverter supplies local load conditions as well as to the grid, efficiently and reliably. The power flow of the bidirectional inverter needs to be maintained based on the DC-bus voltage when subjected to variable source and load conditions [18]. The control capability also has to assure the stability of the entire system when subjected to a certain level of load demand due to shared power supply either from renewable energy or the grid. Moreover, the DC bus needs to be regulated at a nominal voltage to enable the operation of the battery charge controller and a reliable supply of DC system. On the other hand, the power delivered by the DC-bus in case of excess power from PVs has to meet the standard power regulations of the grid [19,20].

In addition to maintaining stable DC and AC voltage levels and grid synchronization [21,22], the DC-side interaction with the AC grid must be decoupled and isolated in the case of a fault or potential transient which can affect the dynamic operation of the system [23]. Furthermore, the phase shift during the transition period between inversion and rectification modes should not affect the DC system's reliability. Considering that certain bidirectional inverter types have two stages of power conversion, the control structure may become more complex if the transition between modes is performed effectively to achieve high performance levels. The control design of the DC-bus interface with renewable energy is discussed in more detail in Section 5.

1.3. Paper Structure

The paper consists of five sections. A brief overview of DC distribution system configurations involving a bidirectional inverter is presented in Section 2. Section 3 begins by laying out the historical overview and theoretical implementation of bidirectional inverter topologies and looks at how they are controlled with the utility grid. Section 4 presents a comparison of the performance between different topologies to highlight the optimal design. Section 5 presents a comparison and classification of the various techniques and parameters used in different control systems for bidirectional inverters. Finally, a conclusion summary is given in Section 6, which includes challenges and limitations associated with the bidirectional inverter and suggestions to be considered for future implementation.

2. DC Distribution System Configurations

With the development of efficient DC distribution systems in buildings in recent years, it has commonly been assumed that the combination of different energy sources could enhance performance. However, it has been shown that the DC system performance could vary in terms of efficiency depending on load profile, utilized renewable energy sources, and utility grid integration [10]. Figure 1 illustrates a general schematic configuration of DC distribution systems in buildings, including renewable energy sources (PV in this case), energy storage, and a mix of AC and DC loads [24–27]. The system contains a power optimizer known as a maximum power point tracking (MPPT) to maximize the PV output power under different weather conditions [28]. In addition, a step-up/down DC/DC power converter is implemented for charging/discharging the energy storage and to control the power flow to regulate the voltage level of the DC bus.

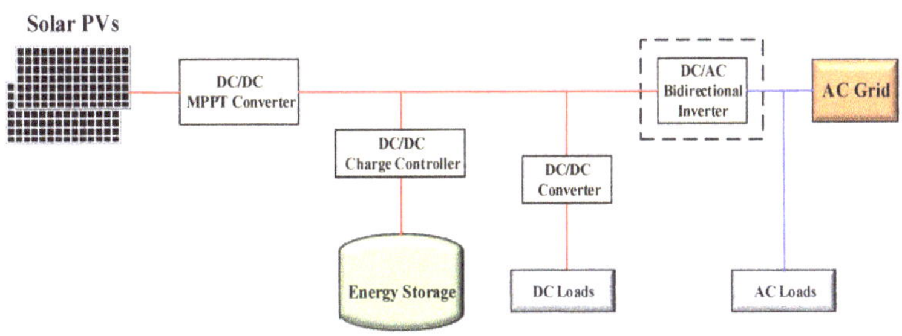

Figure 1. Schematic configuration of DC distribution systems integrated with a grid and a backup system.

The integration of grid power is required to assure the continuous operation of the system in supplying the DC loads in the case of insufficient power, either from renewable energy sources or energy storage. However, there are some configurations of DC systems that do not employ renewable energy sources, in which energy storage and/or a fossil fuel-based generator and grid power are utilized instead. Under these circumstances, energy storage could be beneficial to address the peak saving and ensure the security of supply, or for applying demand-side management when there are no alternative energy sources. On the other hand, regardless of the importance of energy storage, it is not always included in DC system configurations.

Despite the progress and optimal efficiency of DC distribution system implementation, there is still no consensus on standardization and regulation. The EMerge Alliance has developed a standard for 24 Vdc distribution for low power loads within inhabited areas, and several certified infrastructure solutions have successfully shown compliance with this standard [2]. However, 24 V is not compatible with typical battery and inverter voltages. In addition, the alliance has recommended the implementation of a 380 Vdc standard for use in higher power applications such as data centers and central telecom offices [2]. This standard might pave the way for DC distribution in residential and commercial applications at voltages greater than 24 V, which would be advantageous for both sectors. As a result, the academic literature on such systems integrated into buildings has revealed the emergence of several contrasting themes in terms of DC bus configuration and voltage level [29–31]. The purpose of a DC-distribution system is to eliminate some of the integrated conversion stages to increase the overall distribution system efficiency. However, depending on the range of loads to be supplied, there are two dominating DC bus structures, bipolar and unipolar, as presented in Figure 2. The principle of a bipolar type is that the DC bus has three lines, positive and negative lines, +Vdc, −Vdc, and the middle line is a neutral line [32–34], to allow the supply of loads with two available voltage ranges, as depicted in Figure 2a. In addition, the system has robust reliability to guarantee that loads are fed when an electrical fault occurs in either line [35]. Moreover, with a bipolar structure, the neutral point is grounded, which considerably reduces the line-to-ground safety risk by significantly reducing the highest allowable DC-line voltage relative to the ground.

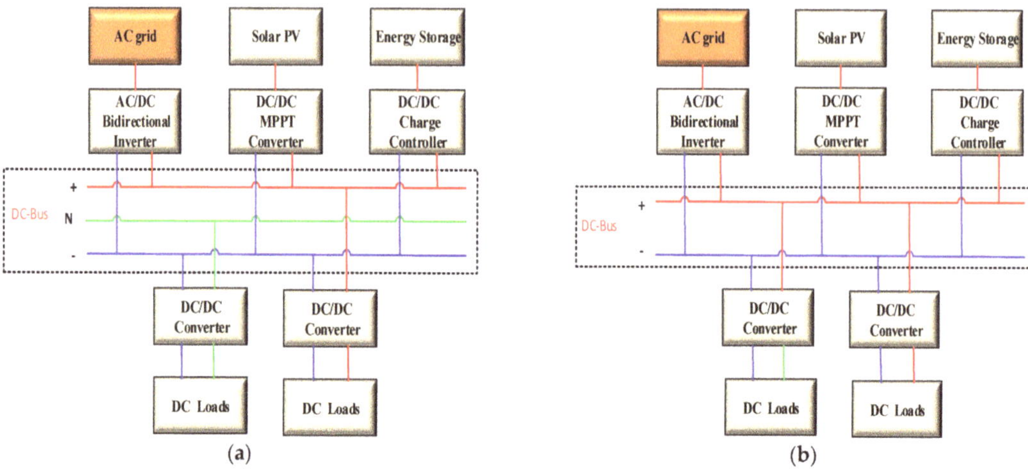

Figure 2. The structure of (**a**) bipolar DC bus, (**b**) unipolar DC bus.

Figure 2b shows the unipolar DC distribution system with one consistent voltage level, represented by the use of two lines, one labeled "+Vdc" and the other "−Vdc". The unipolar structure is more suitable for systems with low power loads due to the cost saving associated with reduced wiring compared to the bipolar type [36]. However, the lower voltage level of unipolar systems may limit the efficiency that can be achieved regardless of its cost benefits [37]. In terms of these DC bus structures integrating with the grid through a bidirectional inverter, there is one study implemented with a bipolar type (Figure 2a) [38]. This approach usually experiences additional costs in terms of control structure and computation, and ultimately the overall benefit is slightly higher when unipolar is used.

A comparison of the existing empirical literature on DC-bus voltage versus power integrated with a single-phase bidirectional inverter is illustrated in Figure 3. Based on the results, it may be observed that once the power level is high, the voltage typically corresponds to that of the peak mains, i.e., $\sqrt{2}$ times the root mean square (rms) voltage. It can be seen that the majority of the DC bus voltage levels fall somewhere in the region of 380 to 400 V. High-voltage operation (i.e., 600 V) is applied in cases where two voltage levels (600 V/300 V) may be required, allowing for the adaptability of DC load connection. Nevertheless, it is possible to conclude that the high power and voltage level (i.e., 10 kW and 600 V) is not appropriate for residential buildings from a safety perspective, despite the fact that it provides a high level of efficiency.

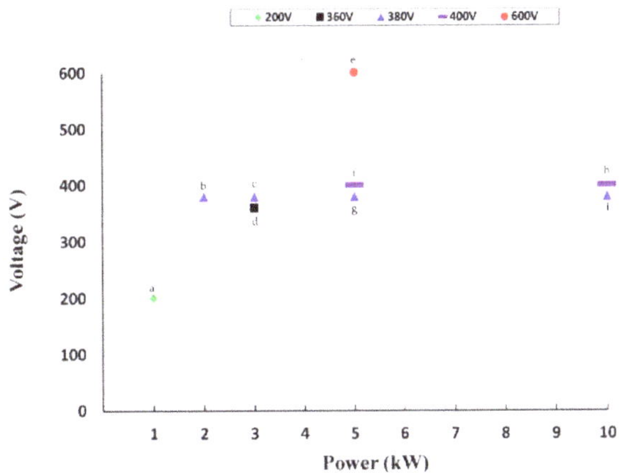

Figure 3. DC-bus voltage level integrated with a bidirectional inverter. a—[39,40]; b—[41]; c—[42]; d—[43]; e—[44]; f—[45–50]; g—[51–55]; h—[38,56,57]; i—[58].

3. Single-Phase Bidirectional Inverter Topologies

Single-phase inverters are generally classified into two types: voltage source (VS) and current source (CS) inverters. The VS inverter is widely used for PV grid-connected applications due to its advantages of high efficiency, economical cost, and the size of implementation [59,60]. It provides a good solution when the required voltage needs to be maintained regardless of the current variation. Meanwhile, the CS inverter is usually used in applications involved in controlling the torque, such as electrical vehicles (EVs), where the current needs to be controlled. This paper is mostly concerned with VS inverters.

Traditional inverters have unidirectional power flow, while in a DC distribution system, a bidirectional power flow interfaced between the DC bus and utility grid is usually required. A summary of the existing implemented bidirectional topologies used in DC distribution systems in buildings is illustrated in Figure 4. These may be categorized into transformer and transformerless inverters/converters. In turn, the transformerless topology can be grouped into four sub-groups based on their functionality of (a) voltage stress and decoupling between the DC and AC sides, i.e., common ground; (b) H-bridge; (c) H6; and (d) two-stage non-isolated topology [61,62]. In order to shed more light on each topology considering the main concerning factors such as leakage current (I_{cm}) and common mode voltage (CMV), the following sections provide an analysis of the topologies utilized, illustrating the efficiency and control challenges and optimal design aspects.

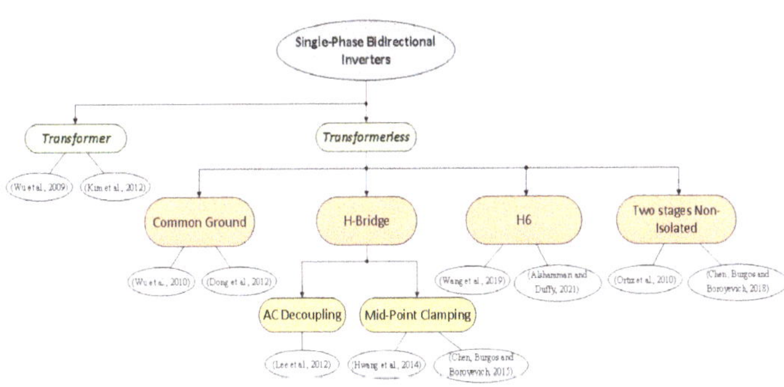

Figure 4. Classification of single-phase bidirectional inverter topologies integrated with a DC distribution system. (We et al., 2009)—[63]; (Kim et al., 2012)—[51]; (Wu et al., 2010)—[52]; (Dong et al., 2012)—[44]; (Lee et al., 2012)—[43]; (Hwarg et al., 2014)—[39]; (Chen, Burgos and

Boroyevich, 2015)—[56]; (Wang et al., 2019)—[53]; (Alshammari and Duffy, 2021)—[54]; (Ortiz et al., 2010)—[64]; (Chen, Burgos and Boroyevich, 2018)—[57].

3.1. Transformer Topologies

To overcome the potential effects of leakage current caused by parasitic capacitance between the PVs and the ground of the utility grid, a transformer may be used to provide galvanic isolation [65]. However, if operating at the mains/line frequency (LF), a large transformer size is needed to prevent saturation in the core, and that could decrease the efficiency and introduce further weight and cost to the system [66]. An alternative method is to resort to a high-frequency (HF) transformer to reduce the transformer size. However, this could lead to an increase in power loss and requires a relatively high level of system complexity, especially when integrating DC distribution systems, as the bidirectional power flow and regulation of the DC bus are required to be balanced when multiple power sources are applied, in addition to the bus being subjected to different load conditions.

The study of [63] has provided details on the adaption of the push-pull isolated topology for low-voltage DC bidirectional inverters integrated into the grid as shown in Figure 5a. This topology provides the ability to step up or down, which is important in applications where the DC bus voltage level is less than the peak mains. However, the topology needs to be operated at a relatively high frequency (20 kHz) to ensure that the transformer size is reasonable, and this limits its efficiency to some extent. The push-pull also provides the function of PFC in the case of the rectification mode, while using phase-shift control in the inversion mode. A notable feature of this topology is the benefit of a low number of switches and associated driver circuitry, which can avoid additional power losses compared to other transformer-based bidirectional inverters [67]. However, it seems that the current harmonic in PFC mode is relatively high when the system is operated for a wide range of power levels (1 kW to 5 kW in this case). Similarly, for the inversion mode, a high current harmonic distortion is introduced when the load demand is higher than 1 kW.

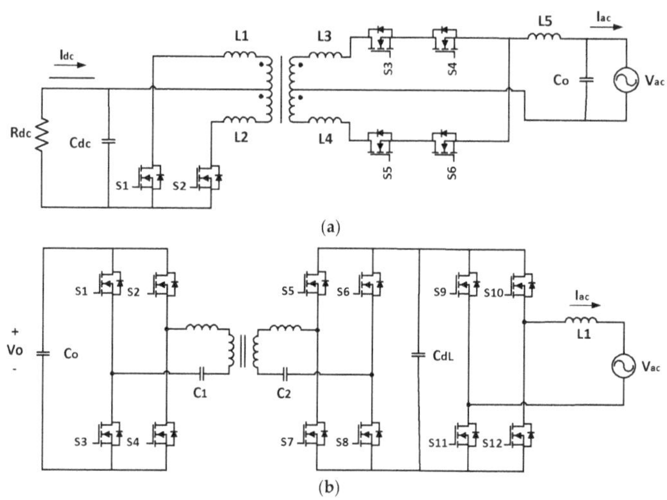

Figure 5. Illustration of bidirectional transformer topologies: (**a**) push-pull topology [63] and (**b**) isolated bidirectional boost rectifier AC–DC converter [51].

In recognition of their superior performance and efficiency, significant research on the application of DC/DC topologies to isolated bidirectional converters has recently been published [68]. Thus far, the development of a high-power isolated full-bridge boost rectifier with bidirectional capability is proposed [69]. However, as with all transformer-based solutions, leakage inductance causes high current spikes during switching and significant effort has been directed at addressing this through snubber circuits. Using an

RCD passive snubber to clamp the voltage is the most straightforward method; however, the resistor dissipation adds to power loss, resulting in low efficiency. Alternatively, active and passive clamping circuits may reduce voltage stress caused by the leakage inductance of current-fed inductors and isolation transformers [69]. However, due to the existing resonant current of the capacitor in some clamped circuits, the current stress on the switches is significantly increased, resulting in higher power loss and a corresponding decrease in efficiency. Therefore, a flyback snubber provided by the authors of [70] is used to recycle the energy that the clamping capacitor absorbs. Moreover, the clamping capacitor voltage can be controlled autonomously by the flyback snubber. The characteristic of zero-voltage transition (ZVT) was applied to a phase-shift full bridge at high frequency to achieve high efficiency for bidirectional capability [71]. The topology maintains a constant output voltage by decreasing the input voltage, which usually varies.

The bidirectional operation of an LLC resonant converter is introduced in [72] and has the ability to reduce the switches' voltage stress without any snubber circuity. However, it is possible that the wide operating range of the converter necessitates a compromise between the turn ratio of the transformer and the utilization of the resonant characteristics. Therefore, a high-efficiency isolated bidirectional inverter with two stages of power conversion was proposed by [51] to overcome the high switch conduction loss of the bidirectional boost rectifier, as shown in Figure 5b. However, the overall efficiency of this topology tends to be low at light loads.

3.2. Transformerless Topologies

The full-bridge (H4) transformerless topology, as shown in Figure 6a, is most commonly used for bidirectional conversion in renewable energy applications, e.g., [52]. Since there is no transformer, it generally provides a smaller solution, and the issue of transformer leakage current is eliminated. However, it has its own issues. The fluctuation of filter inductance due to non-linear core material properties over a wide current range causes current oscillation and significant current ripple, lowering current tracking accuracy [52]. As a consequence, the DC-bus and grid voltages will vary. In order to reduce this fluctuation in a single cycle, a closed-loop control system was designed that attenuated the required duty cycle to provide the desired inductor current while also reducing variations in the DC-bus and grid voltage. However, the efficiency of this topology is low.

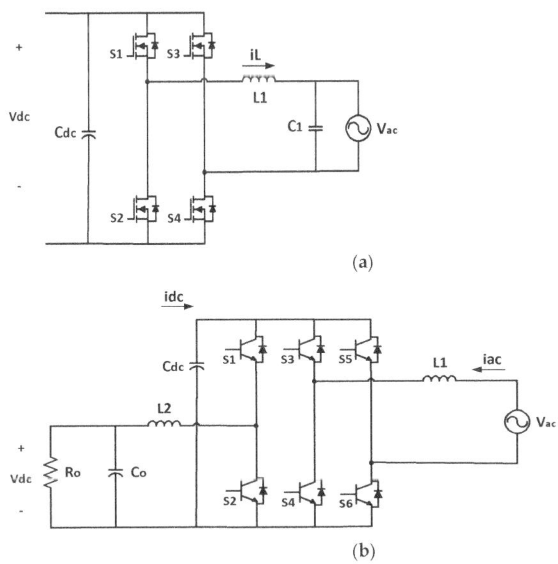

Figure 6. Bidirectional inverter (**a**) conventional H4 [52] and (**b**) two-stage topology [64].

To address the challenge of maintaining a stable DC bus voltage under transient load conditions, the work of [64] offered an improvement in which the stability of the DC-bus voltage is maintained and the DC-link capacitor is reduced by using a two-stage power conversion scheme where the H4 is employed in series with a DC/DC stage as shown in Figure 6b. As can be seen, the DC/DC acts as a buck during rectification and as a boost during inversion.

A modified H4 topology with an EMI filter connected to the grid is proposed in [44], as shown in Figure 7. The topology aims at reducing the leakage current on the DC side caused by high-frequency unipolar PWM on the AC side. It includes an output LCL filter with a common-mode choke in which the neutral point is connected directly to the mid-point of the DC link. This involves a split-phase DC line employing two extra DC capacitors. Consequently, this AC filter structure could add additional power loss, which limits efficiency. Therefore, there is a trade-off between CMV on the DC side and high efficiency, in which case the overall system performance, particularly at low consumption power, could be improved.

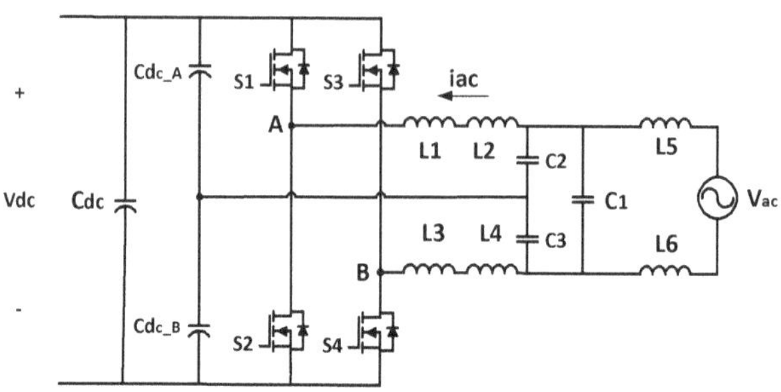

Figure 7. Modified H4 bidirectional full-bridge topology with EMI structure [44].

Another issue with the H4 topology is that it, too, suffers from power loss due to leakage current through the parasitic capacitance of the PVs [73]. In order to overcome this effect, the authors of [43] propose a solution that provides two additional current paths on the AC side, where the inverter output is driven into three levels of VAB (Vdc, 0, and –Vdc), as depicted in Figure 8. Since the rate of change in voltage across the parasitic capacitance (ground capacitor) is reduced, this design can limit high-frequency components and reduce leakage current. As a result, the inductor loss during freewheeling is reduced, while injected reactive power from the grid is reduced in the case of zero crossing. Due to the fact that the topology operates as a boost in rectification mode, the DC-link voltage limitation is required to be nearly twice as high as the power grid's peak voltage. However, there is no capacitor connected to the AC terminals, which impacts the EMI performance. High efficiency was achieved in this case but is partly owing to the use of oversized semiconductor devices relative to the design specification.

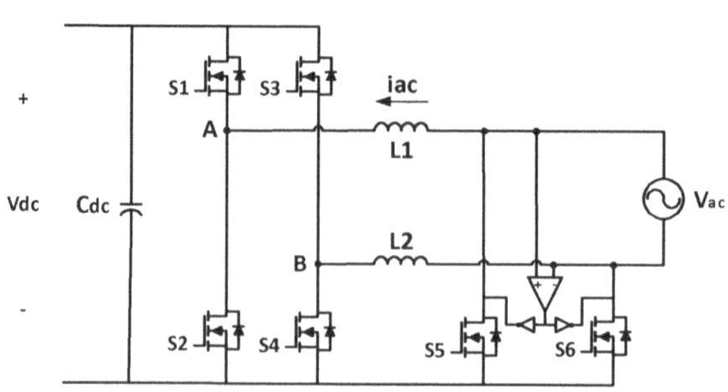

Figure 8. H4 bidirectional inverter full-bridge topology with additional paths [43].

Loss analysis of the H4 topology found that high losses are contributed by large inductor current ripple, in addition to the loss of the switches operating at high frequency [39]. Therefore, an interleaved three-leg full-bridge inverter was proposed as shown in Figure 9 in which high and low switching frequencies are applied for the middle and outer legs, respectively, where a predictive control strategy was implemented to synchronize the grid voltage and current. However, leakage current still presents a challenge in which the flow of current during different parts of the cycle is discontinuous. No details of the efficiency were provided in this case.

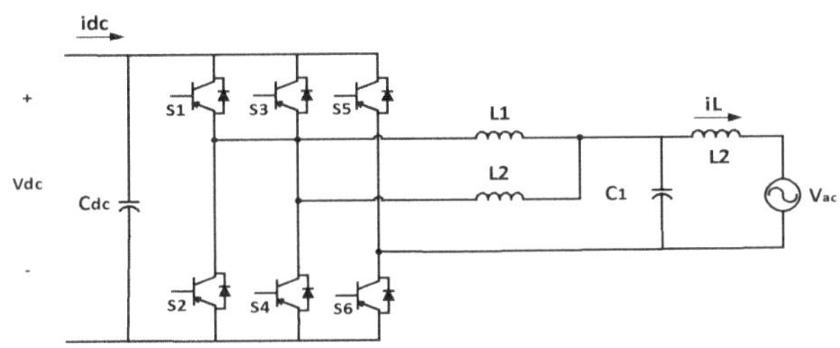

Figure 9. Interleaved single-phase three-leg of full bridge topology [39].

The study of [56] offers probably one of the most comprehensive empirical analyses of the transformerless H4 topology when employed in a two-stage interface between the grid and DC sources. The authors present an analytical loss breakdown for the H4 and propose a method for minimizing losses caused by the EMI and CM filters. Figure 10 shows the proposed 2-level (2L) full bridge with LCL filter in (a), while a parallel version is proposed in (b). The efficiency was highest when the 2L parallel structure was employed due to the reduction in conduction losses provided by paralleling and the use of SiC MOSFETs for decreasing the switching losses. Similarly, an emphasis on the performance of semiconductors device such as SiC MOSFET provides potential performance for enhancing the bidirectional inverter efficiency compared to ideal MOSFETs and IGBTs [41]. However, this topology does not address the issue of capacitive leakage current and associated CMV.

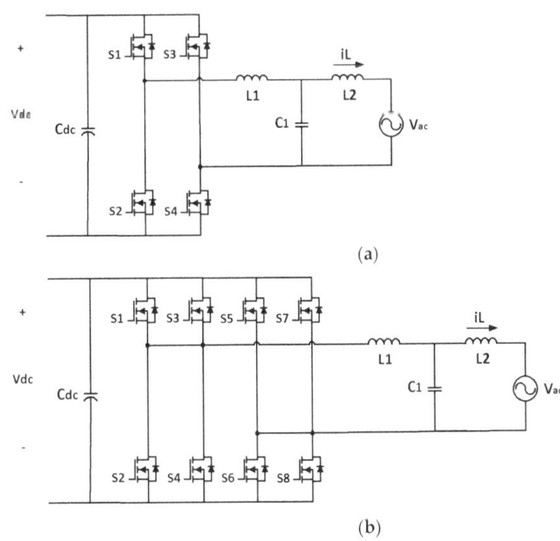

Figure 10. Bidirectional inverter topologies of (**a**) a 2L full bridge with LCL filter, (**b**) a 2L parallel full bridge [56].

Since solutions to the leakage current and associated CMV may require additional components and increase the cost and size of the bidirectional inverter, a new H6 transformerless inverter topology was proposed (firstly as a unidirectional inverter) in [74] that eliminates the leakage current, reduces the inductor harmonics current, and prevents the voltage spikes that occur in inductive loads. During the inductor free-wheeling interval in the positive half-cycle, the current flows through D1 and diode S5. This not only disconnects the DC side from the grid but also prevents the high-frequency voltage from building up in the parasitic capacitance, which lowers the leakage current. The same is applicable for the path through S6 and D2 during the negative half-cycle of the grid. Furthermore, its efficiency was improved in [75] with a modified PWM pattern when there is a phase shift between output current and voltage. Therefore, these features of the H6 outweigh other unidirectional transformerless topologies such as the H5, HERIC, etc.

The H6 was modified to operate with the capability of bidirectional power flow in [53], as shown in Figure 11a. However, it is indicated that under light load conditions (roughly 1.2 kW for a 5-kW rated system), a reduction in inverter efficiency was observed. This study also confirms that switches of the inverter could play an important role to maximize the overall efficiency. Therefore, a study of [54] proposed the synchronous H6 topology, as can be seen in Figure 11b, to reduce the greatest impact of the diode conduction loss and its related switching ON losses under full loads. Similarly, these dominant losses were reduced under rectification mode by using SiC MOSFETs instead of diodes D1 and D2.

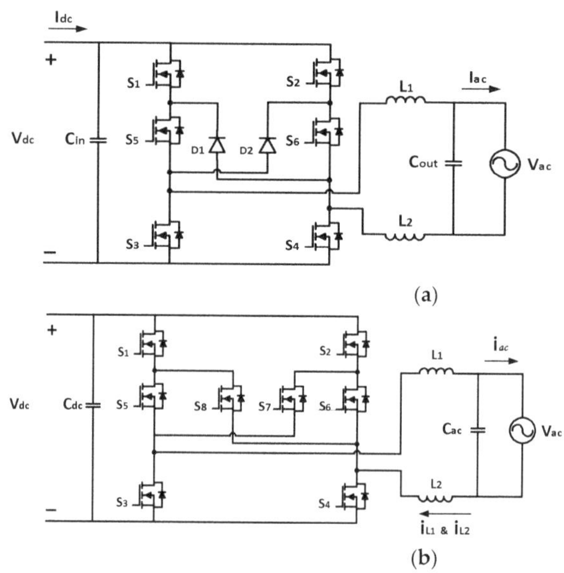

Figure 11. H6 bidirectional inverter topology: (**a**) standard H6 [53] and (**b**) synchronous H6 [54].

A small number of papers discuss bidirectional inverters for a bipolar DC configuration, in which the DC and low-frequency CM voltages need to be closely regulated to ensure symmetrical DC bus voltages and to reduce leakage current. The high-frequency CM noise can be filtered out by passive components, as with unipolar DC systems [23]. The authors of [76] proposed a control solution for grounded unidirectional inverter systems with power converters, which is based on an active common-mode duty cycle injection approach. As a result, a reduction in voltage ripple was achieved, which minimized the ground leakage current. However, this control technique does not apply to bidirectional operation. Therefore, the topology proposed in [57] and presented in Figure 12 was designed for enabling bidirectional capability and high frequency. Moreover, the leakage current was reduced due to the decoupling of the CM voltage, while the high-frequency noise was eliminated by the filter. Moreover, an interleaved schematic of this topology was introduced to mitigate the inductor choke ripple, which may lead to an additional loss, resulting in low efficiency in both stages. Furthermore, this topology requires extra switching devices and passive components to enable the converter to transition between the inverter and rectifier operation while maintaining DC-bus voltage. In addition, when the active damping method is implemented, the need for extra voltage and current sensors increases the overall cost of the system. Therefore, it is feasible that an alternative solution may be established to perhaps reduce the cost and size of the converter with respect to its level of efficiency.

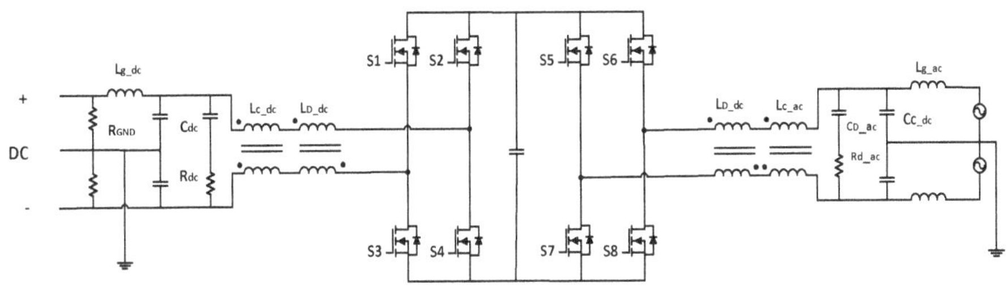

Figure 12. Two-stage bidirectional converter with CM decoupling [57].

4. Performance and Evaluation

The bidirectional inverter topologies considered in Section 3 are summarized and compared in this section. Table 1 compares the topologies in terms of controller implementation, switching frequency, and output filter.

Table 1. Control modulation and passive devices of transformer and transformerless bidirectional topologies for DC distribution systems.

Ref.	Topology	Power Level (W)	SPWM/ Implementation	Switching Frequency	Filter/EMI	DC-Bus Voltage	DC Link Capacitor	Line Voltage
[63]	Push-pull phase-shifted—transformer	600	DsPIC30F2020 at dc side Cycle converter at the ac side bipolar	20 kHz -	LC	-	Non	50 Vac/120 Hz
[52,77]	Full-bridge bidirectional inverter transformerless	5000	DsPIC30F2023	20 kHz	LC	340–380	3760 F	220 Vac/60 Hz
[45]	Full-bridge bidirectional inverter transformerless	5000	TMS320LF2406A	20 kHz IGBT	LCL	360–400	560 F × 10	220 Vrms/60 Hz
[78]	Full-bridge bidirectional inverter transformerless	-	DsPIC30F2023	- IGBT	LC	-	-	-
[46]	Full-bridge bidirectional inverter transformerless	5000	TMS320LF2406A Bipolar/Unipolar	20 kHz IGBT	LC	360–400	560 F ×10	220 Vrms/60 Hz
[47]	Full-bridge bidirectional inverter transformerless	5000	TMS320LF2406A	- IGBT	LC	360–400	-	-
[44]	Full-bridge bidirectional inverter transformerless	10,000	PWM Unipolar	20 kHz IGBT	parallel LCL	600	360 F	240 Vrms/60 Hz
[55]	AC–DC boost bidirectional rectifier	5000	TMS320F28335 Unipolar	13.8 kHz MOSFETs IGBT	LPF	380	2.7 mF	220 Vac/60 Hz
[51]	Full-bridge CLLC resonant bidirectional converter	5000	TMS320F28335 PFM	58–65 kHz MOSFETs	FIR	380	200 nF	220 Vac/60 Hz
[43]	Full-bridge bidirectional inverter transformerless	3000	PI	16 kHz MOSFETs	parallel L	360	2200 μF	220 Vrms/60 Hz

Table 1. *Cont.*

Ref.	Topology	Power Level (W)	SPWM/ Implementation	Switching Frequency	Filter/EMI	DC-Bus Voltage	DC Link Capacitor	Line Voltage
[48]	Full-bridge bidirectional inverter transformerless	5000	D-Σ digital PWM rectifier	20 kHz IGBT	LCL	360–400	-	220 Vrms/60 Hz
[79]	Bidirectional transformerless PV inverter based on HFLs	8000	Phase-shift	50 kHz IGBT MOSFETs	LC	380	6 × 1.0 mF	220 Vrms/50 Hz
[38,57]	Two-stage bidirectional AC–DC converter	10,000	Pluggable phase-leg module	40 kHz MOSFETs	DM	380 bipolar	100 nF	240 Vrms/60 Hz
[41]	Two-stage bidirectional AC–DC converter—NPC and FB	2000	Unipolar	20 kHz MOSFETs IGBT	LCL	380	-	127 Vrms/60 Hz
[53]	H6 bidirectional transformerless inverter with bridgeless PFC rectifier	5000	Unipolar	20 kHz MOSFETs	parallel L&C	380	-	220 Vac/50 Hz

What is notable is that digital rather than analog control is most commonly used. This is due to the better control precision and dynamic reactions of digital compared to analog implementations. According to Table 1, unipolar PWM is used for the majority of bidirectional inverters. Given the fact that unipolar leads to lower DC current ripple, it is possible to minimize the AC-side harmonics by a substantial amount in comparison to bipolar PWM. Using unipolar outperforms bipolar, which requires a large inductor to minimize current ripple due to the voltage peak value. However, PWM schemes create a high leakage current at a high frequency on the DC side that flows to the ground in transformerless topologies because of the different switching transitions and dead-times. Consequently, when high-efficiency operation is needed, it is critical to take this into consideration.

Almost all bidirectional inverter topologies were operated at 20 kHz due to the good trade-off between the inductor loss and switching loss of the employed semiconductor devices. Among these are SiC MOSFETs, which have a lower switching loss compared to Si MOSFETs. Conduction loss is a significant issue for IGBT switches, and they are not recommended for use in these bidirectional inverter topologies.

Another trend that is evident in Table 1 is the fact that LC and LCL filters are most common in transformerless topologies, which implies that the inductor design may contribute to minimizing the converter power loss. However, current distortion may occur at low power levels due to the impact of continuous inductor current flow, which may increase conduction loss and reduce overall efficiency.

The majority of bidirectional converter topologies have a power rating of 5 kW, indicating that they can be used in both commercial and residential buildings. While it is true that residential buildings are typically operated at a lower power level, this results in low efficiency for most topologies that are used.

A summary of semiconductor parameters implemented in the bidirectional topologies considered is given in Table 2. In terms of the technologies used, it was noticed that Si MOSFETs and IGBTs were the first to be developed, indicating that they were the primary switching devices available during this period of time. Indeed, such technology, particularly IGBT switches, is suitable for high voltage operation, which explains why they are the mainstream switches. However, as a result of advancements in semiconductor technology, SiC MOSFETs are capable of operating at higher voltages while also delivering higher efficiency when compared to Si MOSFETs and IGBTs. Moreover, because the cost of SiC MOSFETs is high, it is necessary to consider a trade-off between high efficiency and cost when designing a bidirectional inverter. Similarly, the diode used in some topologies, such as those in which the ideal Schottky diode can be replaced by an SiC Schottky diode due to its low power dissipation, can be justified in the same way.

Table 2. Semiconductor components of transformer and transformerless bidirectional topologies for DC distribution systems.

Ref.	Year of Publication	Device Type	Device Model	Manufacture	V Rated	I Rated	Diode Type	Diode Model
[45,46,49,52]	2010, 2011, 2012, 2016	IGBT	40N60A4	Fairchild Semiconductor	600 V	75 A	Ultrafast	RURG5060
[51]	2013	MOSFETs	IXKR47N60C5	Littelfuse	600 V	75 A	-	-
		MOSFETs	SPW47N60CFD	Infineon	600 V	75 A	Schottky	C3D20060D
[43]	2014	Cool MOSFETs	FCA 75N60N	IXYS Corporation	600 V	47 A		
		IGBT	IGW75N60T	Infineon	600 V	46 A		
[48]	2014	IGBT	G40N60A4	Gree	600 V	75 A	Schottky	C3D20060D
[38,57]	2017, 2019	SiC MOSFET	C2M0025120D	Wolfspeed	1200 V	90 A	-	-
[53]	2019	SiC MOSFET	C3M0120090D	Gree	900 V	23 A	SiC Schottky diode	CVFD20065A

The relative advantages and disadvantages of the different topologies are compared in Table 3. As compared to other transformerless topologies, the H6 topology claims to provide a high level of efficiency to support the premise that leakage current and high current distortion are kept to a minimum.

Table 3. Main comparison summary of bidirectional single-phase inverters for integration with DC distribution systems.

Ref.	Topology	Advantages	Disadvantages	Inverter Size	Maximum Efficiency	No. Conversion Stages	Leakage Current	Control Strategy
[63]	Push-pull transformer based	Low DC voltage ripple	High THD Hard switching increase loss	Large	89%	2	High	Simple
[51,55]	Isolated AC/DC converter	No CM effect	High level of complexity	Large	96%	2	High	Complex
[52,77]	Full-bridge	Small filter. Inductor variation minimized	High THD Voltage and current drop due to inductance change	Small	97.2%	1	Medium	Complex
[43]	Modified full-bridge	Low THD	Oversized switches High output voltage and current ripple	Medium	98.6%	1	Low	Medium
[39]	Interleaved three-leg full-bridge	Low switching loss	Medium THD due to inductor current ripple in respect invert size Expected CM fluctuation	Medium	-	1	High	Medium
[56]	2L full-bridge with paralleled SiC	Small filter	High conduction loss Increased switching loss as frequency increases	Small	93%	1	High	Medium

Table 3. *Cont.*

Ref.	Topology	Advantages	Disadvantages	Inverter Size	Maximum Efficiency	No. Conversion Stages	Leakage Current	Control Strategy
[53]	Standard H6	Low output current ripple and Icm Constant CM voltage	Diode conduction loss	Medium	97.5%	1	Low	Medium

Finally, the efficiency of the different topologies is compared in Figure 13 for both inversion and rectifier modes. These results have been extracted from the measurement data presented in the publications. For a fair comparison, ref [63] was excluded since its power is significantly lower (300 W). As can be seen, the majority of topologies provide relatively high efficiency at full load. It is worth noting that the highest efficiency of [43] comes from using 75 A oversized semiconductor devices at 3 kW. The next highest efficiency is achieved with a two-stage solution that includes CM decoupling, but it has high cost and size due to additional components. Next comes the H6 topology [48], but its efficiency drops rapidly at low power.

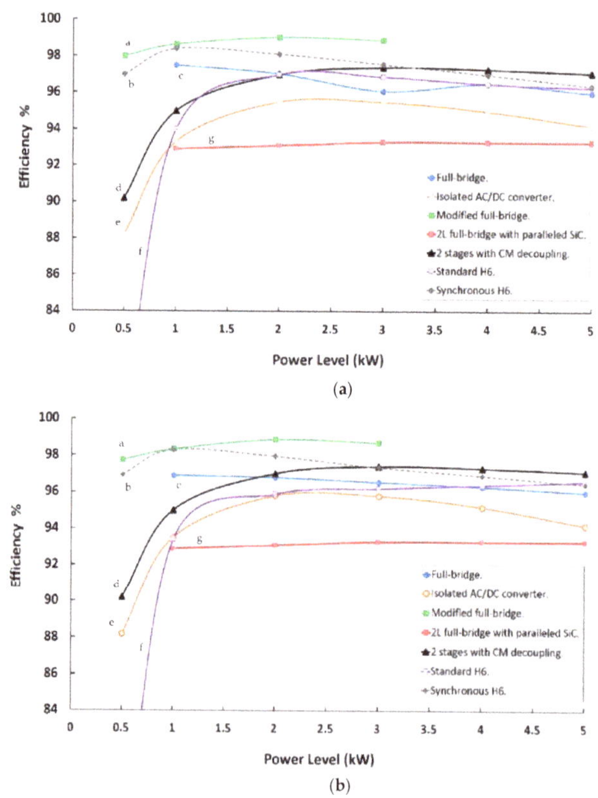

Figure 13. Bidirectional inverter efficiency comparison curves of (**a**) inversion mode, (**b**) rectification mode. a—[43]; b—[54]; c—[52]; d—[57]; e—[51]; f—[53]; g—[56].

However, a residential building will generally operate at a lower load than its maximum rated over the majority of the time. Therefore, bidirectional inverters with low efficiency at light loads would impact the overall system efficiency. The evaluation of the synchronous H6 using PSIM software with detailed analytic equations applied to predict component power loss for suitably sized semiconductors shows that the light load efficiency of the standard H6 can be addressed by replacing diodes with MOSFETs, as explained in [54].

5. Smart Grid Control

The control system is an important component of grid-connected power converters where it maintains the system efficiency and stability. Integrating a DC system into the grid may cause several issues, including grid instability and disruption, so several control solutions have been proposed to overcome grid distortions and fulfill the criteria of the grid's standard power [80,81].

The main control system requirement is as follows: phase estimation, which is known as a phase-locked loop (PLL), voltage control loop (VCL), and current control loop (CCL), as shown in Figure 14. The PLL is used to generate the grid reference current, which is synchronous with the grid voltage, where the most implemented type is the basic PLL in such applications. The DC-bus voltage is regulated by the use of a VCL that produces the required grid current amplitude to be compared with the grid current reference obtained from the PLL. The inductor current needs to be compared with the grid current to determine the duty cycle and track any sudden change through a CCL. The performance of the control system can best be treated according to the following tasks:

- Reactive power control for supplying the DC system.
- Active power control for feeding the grid.
- DC-bus voltage control.
- Grid synchronization.
- Dynamic operational modes control (inverter/rectifier).

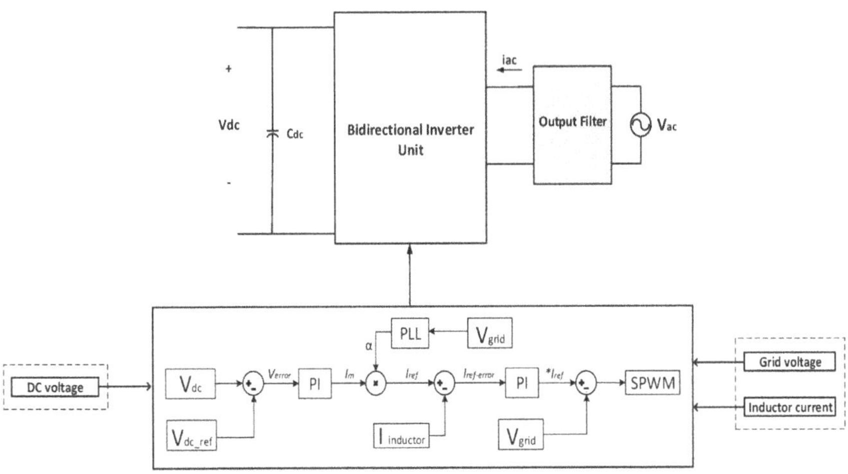

Figure 14. General configuration of the bidirectional inverter control system.

There is a possible classification of bidirectional inverter control systems based on a comparison of the following criteria: DC-bus voltage regulation, inductor current control, and bidirectional capability. This section provides a basis for comparison to enable the identification of the control system structure and to distinguish different levels of performance.

5.1. Controller Challenges

5.1.1. DC-Bus Voltage Regulation

The integration of a DC bus with different types of energy sources such as renewable energy and energy storage may result in system instability; therefore, the bus voltage needs to be retained at a nominal level to enable the reliable operation of the grid, battery, and DC loads. Regulating a DC bus is challenging due to the wide range of load and source conditions that it is expected to support, and the introduction of current harmonics through the grid connection. This issue was examined in [45], where fast regulation control is suggested to track any sudden change of bus voltage every quarter cycle. Likewise, the authors of [47,49] propose a quarter cycle approach (QLCRA) to regulate the variation in DC-bus voltage in situations where the integrated grid interface causes AC current distortion. The distorted line current was minimized when the inductor current was adjusted each quarter cycle instead of the conventional line cycle approach (OLCRA). The bus voltage also can vary when there are more input sources connected to the system.

The study of [42] applied predictive control to track an AC current reference by sensing the inductor current to obtain the desired output bus voltage. The stability of the bus voltage is achieved during both inversion and rectification modes. There is, however, a noticeable current distortion during the phase shift transition between these two modes, particularly at zero crossing, due to insufficient magnetizing inductor current. To achieve high power quality, two switching schemes were compared: unipolar and bipolar for delivering and consuming the utility grid power, respectively [46]. It was found that insufficient inductor current causes current distortion in unipolar operation, whereas bipolar operation decreases the distortion but induces higher current ripple and switching loss. The combined PWM schemes reduced the current distortion and at the same time relatively decreased the voltage ripple and switching loss. As a result, the unipolar system is deemed to be better in terms of lower voltage ripple and switching loss in both inversion and rectification modes.

However, a particular frequency response for designing the AC filter is required to eliminate the applied high frequency on switches, besides the switching loss when using a bipolar scheme. Alternatively, half of the switches in the bridge may be operated at high frequency, while low frequency is applied to the other switches with a hysteresis switching scheme [82]. It is worth mentioning that the switching loss for the hysteresis scheme has the same loss as the unipolar scheme [83].

Moreover, to ensure consistent frequency operation during zero-crossing, it is necessary to consider the DC-link voltage ripple and inductor current harmonics. In fact, high-frequency operation could result in a significant distortion of the delivered grid voltage caused by the dead time control aspect. As a result, the authors of [84] proposed a single switch to operate during the transient period when mode switching occurs in which a freewheeling path is provided to sustain the inductor current. This technique relies on the idea of miniaturizing the leakage current, which is still not optimal. Since the employed semiconductor devices were designed on the maximum operating current level, there is an additional potential cost-related consideration.

5.1.2. Inductor Nonlinearity

Recent developments in bidirectional inverters have heightened the need for controlling the nonlinearity of the inductor current that might occur in one operational cycle due to the inductance variation of different core materials over a wide range of currents. Consequently, a high current ripple has an impact on the accuracy of tracking the inductor current to generate the required duty ratio. Therefore, the authors of [52,77] proposed a predictive control technique to mitigate this variation by sensing the inductor current and using the grid and DC-bus voltage as iteration variables for determining the required duty ratio for the next cycle. However, it is expected that the proposed control is almost certainly limited when the system is subjected to a high load profile resulting in a voltage drop. In contrast to the earlier study, however, no evidence of reducing the current distortion was reported. According to [50,85] a fuzzy logic controller strategy is recommended to overcome the mentioned issue by increasing the inductor voltage to enable the fast-tracking of the current reference. It is believed that the fluctuation of the inductance current peak waveform was minimized; however, there is still some spiking in the zero-crossing point due to the line frequency operation. The strategies of applying low-frequency control to improve the inductor current oscillation might involve current sharing between the switches' parasitic capacitor which leads to a decrease in efficiency. Therefore, high-frequency operation can reduce the shared current in certain topology configurations that utilize middle switches (floating switches). In addition, the controller's bandwidth can be increased to enhance load transient response.

Despite this, little progress has been made with regard to deploying a high-resolution current for sensing the AC-side current during a rapid change of the inductor current. Analog-to-Digital converters (ADCs) can be adapted to achieve the required accuracy response under a wide range of the grid current. However, the implementation of ADCs suffers from higher complexity and higher cost. As a result, it has been shown conclusively that the continuous conducting mode (CCM) of the inductor current (rather than discontinuous conducting mode (DCM)) is desirable due to the simplicity of implementation in which the current measurement was replaced by a practical sensorless, average current regulation control algorithm model [86,87]. Nevertheless, it is important that other factors can affect the sensing procedure of the inner current loop control, such as inductance, inductor resistance, and related voltage conduction [88,89]. Therefore, the authors of [40] developed a sensorless current control with bidirectional capability to enhance the functionality of the PFC during the grid injection. It seems possible that a transient current is expected to contribute to the associated dynamic decoupling between DC and AC sides, in which a significant response time is required during a change of mode. Collectively, these studies outline a critical role for inductance variation on the system performance, which potentially needs to be taken into consideration.

5.1.3. Bidirectional Capability

The correlation of the inductor current and the DC-bus voltage has an influence on the transition region between inversion and rectification, as depicted in Figure 15. As can be seen, a linear relationship between the inductor current and DC-bus voltage determines the operation mode of the power flow. It is important to note that the DC-bus voltage should be operated in a range of ±10% Vdc in order to avoid frequent changes of mode during a sudden heavy load. As a result, it is possible that current distortion occurs during the transition period in the zero-crossing voltage. There is some evidence to suggest that PFC compensation could be utilized when the magnetization current is inadequate to mitigate the phase-shift region [46]. However, the time required to adjust the duty cycle for changing the switching mode is considered relatively slow, where a fast control response is needed in such conditions. This indicates a need to understand the various aspects of control ability that exist in two-stage power conversion topology, especially in the isolated DC/DC stage This may lead to an inappropriate escalation in uncontrolled power conversion mode when the load demand becomes negative, resulting in a significant increase in the delivered power to the output capacitor due to the combination of load and converter terminal.

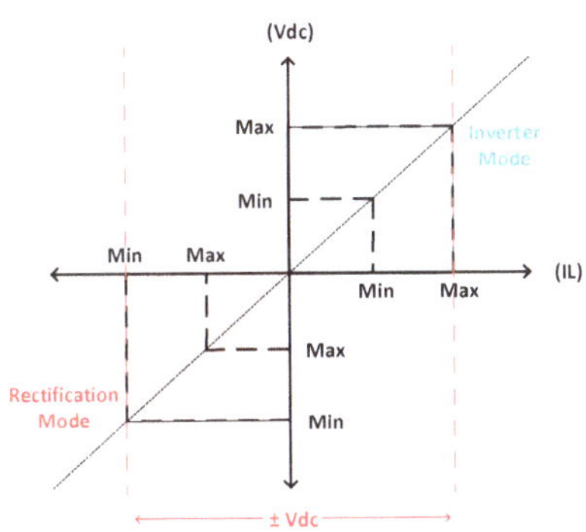

Figure 15. The regulation technique of DC-bus voltage (Vdc) with associated linearity of relationship with the inductor current (IL).

The authors of [51] set out a proposal of employing a dead-band control algorithm to smooth any abrupt state in performance where the DC-bus voltage is varying in two-stage systems. However, the utilized controller is not autonomous because an additional slave controller is required for the integrated power conversion stages. In contrast, the authors of [90] offer a fly-back bidirectional control capability using a single power conversion stage to overcome the complex implementation of controllers required in isolated two-stage converters. This is suitable only for integration between the energy storage and utility grid, which can be adapted for integration with DC distribution systems. Perhaps the most serious disadvantage of this technique is that a significant voltage and current stress is imposed on the converter semiconductors, which has been thought of as a key factor in system efficiency and performance. The distinctions between single-stage and two-stage bidirectional power conversion, with respect to which isolation's value is evaluated, are evidently a significant constraint on all the work discussed in this area.

A comparison summary of the implemented control is presented in Table 4.

Table 4. Summary of control implementation of transformer and transformerless bidirectional topologies for DC distribution systems.

Ref.	Advantages	Disadvantages	Controller Type
[51,55]	Fast dynamic response Low THD	Complex control structure	Hysteresis
[52,77]	Fast dynamic response Easy implementation	More calculation required	Predictive
[45]	Reactive power control High dynamic	Complex control structure High THD	Predictive
[46]	Simplicity	No full PFC Two controllers	-
[47,49]	Instantaneous current control Less circuitry	Complex transfer function	Predictive
[42]	High dynamic Low THD	Complex transfer function	Fuzzy logic
[50]	Lower inductor current ripple	High THD No full control of PFC	Fuzzy logic
[40]	One inductor filter	Slow tracking response No full control of PFC	Predictive
[58]	Fast tracking and dynamic response High gain	Limited operational range	Resonant

6. Conclusions

In this review, the aim is to assess the performance of existing bidirectional inverter topologies integrated with a DC distribution system in which renewable energy sources, energy storage, and DC loads are used. It was found that transformerless topologies outweigh transformer-based topologies due to higher efficiency and smaller size of the power converters. However, the primary issue with transformerless topologies is the connection between the utility grid and the DC bus, which results in high leakage current and high CMV, which have a negative effect on the overall efficiency. Freewheeling paths and two-stage power conversion are the prominent solutions for the drawbacks of transformerless topologies. It was found that the standard H6 topology provides the best performance in terms of high efficiency and cost-benefit related to the semiconductor devices and passive components required to eliminate leakage current and CMV. However, it is important to note that the efficiency of this topology decreases significantly when operating at less than 20% of its maximum rated power (5 kW) compared to other transformerless topologies.

The present study adds to the growing body of research regarding DC distribution systems that aim to enhance the system efficiency of power delivery between DC power sources, loads, and energy storage devices over AC distribution. It was shown that there is scope for further research on methods to improve the efficiency of bidirectional topologies under light-load conditions, and the dynamic response of control systems in two-stage bidirectional power converters. Furthermore, while the requirements of bidirectional inverters integrated with DC distribution in buildings may not be compatible for use in different applications such as (EV) and energy storage, the circuit topologies and control methods described may be adapted for other bidirectional applications. In conclusion, it is believed that this review will provide a reference for academics, engineers, manufacturers, and end-users interested in implementing DC distribution systems using bidirectional inverters with grid-connected and renewable energy systems.

Author Contributions: Conceptualization, M.A. and M.D.; methodology, M.A. and M.D.; software, M.A.; validation, M.A. and M.D.; formal analysis M.A.; investigation, M.A.; resources, M.A. and M.D.; data curation, M.A.; writing—original draft preparation, M.A.; writing—review and editing, M.D.; visualization, M.A. and M.D.; supervision, M.D.; project administration, M.A. and M.D.; funding acquisition, M.D. All authors have read and agreed to the published version of the manuscript.

Funding: This review paper was invited by the editor of Energies to the corresponding author.

Institutional Review Board Statement: Not applicable.

Informed Consent Statement: Not applicable.

Data Availability Statement: Not applicable.

Conflicts of Interest: The authors declare no conflict of interest.

References

1. Kannan, N.; Vakeesan, D. Solar energy for future world: A review. *Renew. Sustain. Energy Rev.* **2016**, *62*, 1092–1105. [CrossRef]
2. Garbesi, K.; Vossos, V.; Shen, H. *Catalog of DC Appliances and Power Systems*; Lawrence Berkeley National Laboratory: Berkeley, CA, USA, 2012; pp. 1–77. [CrossRef]
3. EMerge Alliance.pdf. Available online: https://www.emergealliance.org/standards/our-standards/ (accessed on 1 October 2021).
4. Sabry, A.H.; Shallal, A.H.; Hameed, H.S.; Ker, P.J. Compatibility of household appliances with DC microgrid for PV systems. *Heliyon* **2020**, *6*, e05699. [CrossRef] [PubMed]
5. U.S. Energy Information Administration. Available online: https://www.eia.gov/ (accessed on 1 October 2021).
6. Glasgo, B.; Azevedo, I.L.; Hendrickson, C. How much electricity can we save by using direct current circuits in homes? Understanding the potential for electricity savings and assessing feasibility of a transition towards DC powered buildings. *Appl Energy* **2016**, *180*, 66–75. [CrossRef]
7. Vossos, V.; Garbesi, K.; Shen, H. Energy savings from direct-DC in U.S. residential buildings. *Energy Build.* **2014**, *68*, 223–231. [CrossRef]
8. Kakigano, H.; Nomura, M.; Ise, T. Loss evaluation of DC distribution for residential houses compared with AC system. In Proceedings of the 2010 International Power Electronics Conference-ECCE Asia, Sapporo, Japan, 21–24 June 2010; pp. 480–486. [CrossRef]
9. Liu, Z.; Li, M. Research on Energy Efficiency of DC Distribution System. *AASRI Procedia* **2014**, *7*, 68–74. [CrossRef]
10. Alshammari, M.; Duffy, M. Feasibility Analysis of a DC Distribution System for a 6 kW Photovoltaic Installation in Ireland. *Energies* **2021**, *14*, 6265. [CrossRef]
11. Chauhan, R.K.; Rajpurohit, B.S. DC distribution system for energy efficient buildings. In Proceedings of the 2014 Eighteenth National Power Systems Conference (NPSC), Guwahati, India, 18–20 December 2014; pp. 1–6. [CrossRef]
12. Palaniappan, K.; Veerapeneni, S.; Cuzner, R.; Zhao, Y. Assessment of the feasibility of interconnected smart DC homes in a DC microgrid to reduce utility costs of low income households. In Proceedings of the 2017 IEEE Second International Conference on DC Microgrids (ICDCM), Nuremburg, Germany, 27–29 June 2017; pp. 467–473. [CrossRef]
13. Boeke, U.; Wendt, M. DC power grids for buildings. In Proceedings of the 2015 IEEE First International Conference on DC Microgrids (ICDCM), Atlanta, GA, USA, 7–10 June 2015; pp. 210–214. [CrossRef]
14. Alshammari, M.; Duffy, M. Comparative analysis between AC and DC distribution systems in a photovoltaic system: A case study of a school in Ireland. In Proceedings of the PCIM Europe 2019; International Exhibition and Conference for Power Electronics, Intelligent Motion, Renewable Energy and Energy Management, Nuremberg, Germany, 7–9 May 2019; pp. 1–8.

15. Estévez-Bén, A.A.; Alvarez-Diazcomas, A.; Macias-Bobadilla, G.; Rodríguez-Reséndiz, J. Leakage Current Reduction in Single-Phase Grid-Connected Inverters—A Review. *Appl. Sci.* **2020**, *10*, 2384. [CrossRef]
16. Guo, X.; Jia, X. Hardware-Based Cascaded Topology and Modulation Strategy with Leakage Current Reduction for Transformer-less PV Systems. *IEEE Trans. Ind. Electron.* **2016**, *63*, 7823–7832. [CrossRef]
17. Wu, T.-F.; Kuo, C.-L.; Sun, K.-H.; Chen, Y.-K.; Chang, Y.-R.; Lee, Y.-D. Integration and Operation of a Single-Phase Bidirectional Inverter with Two Buck/Boost MPPTs for DC-Distribution Applications. *IEEE Trans. Power Electron.* **2013**, *28*, 5098–5106. [CrossRef]
18. Ryu, M.-H.; Kim, H.-S.; Baek, J.-W.; Kim, H.-G.; Jung, J.-H. Effective Test Bed of 380-V DC Distribution System Using Isolated Power Converters. *IEEE Trans. Ind. Electron.* **2015**, *62*, 4525–4536. [CrossRef]
19. Yang, Y.; Wang, H.; Blaabjerg, F. Reactive power injection strategies for single-phase photovoltaic systems considering grid requirements. *IEEE Trans. Ind. Appl.* **2014**, *50*, 4065–4076. [CrossRef]
20. Robles, O.E.O.; Beristain, J.A.; Ramírez, J.P. Single-phase bidirectional high frequency link photovoltaic inverter with reactive power compensation function. In Proceedings of the 2015 IEEE Workshop on Power Electronics and Power Quality Applications (PEPQA), Bogota, Colombia, 2–4 June 2015; pp. 1–6. [CrossRef]
21. Vancu, M.F.; Soeiro, T.; Muhlethaler, J.; Kolar, J.W.; Aggeler, D. Comparative evaluation of bidirectional buck-type PFC converter systems for interfacing residential DC distribution systems to the smart grid. In Proceedings of the 2012 IECON Proceedings (Industrial Electronics Conference), Montreal, QC, Canada, 25–28 October 2012; pp. 5153–5160. [CrossRef]
22. Dong, D. Ac-dc Bus-interface Bi-Directional Converters in Renewable Energy Systems. Ph.D. Thesis, Virginia Polytechnic Institute and State University, Ann Arbor, MN, USA, 2012.
23. Dong, D.; Cvetkovic, I.; Boroyevich, D.; Zhang, W.; Wang, R.; Mattavelli, P. Grid-Interface Bidirectional Converter for Residential DC Distribution Systems—Part One: High-Density Two-Stage Topology. *IEEE Trans. Power Electron.* **2012**, *28*, 1655–1666. [CrossRef]
24. Rawat, G.S. Sathans Survey on DC microgrid architecture, power quality issues and control strategies. In Proceedings of the 2nd International Conference on Inventive Systems and Control (ICISC), Coimbatore, India, 19–20 January 2018; pp. 500–505. [CrossRef]
25. Rodriguez-Diaz, E.; Savaghebi, M.; Vasquez, J.C.; Guerrero, J.M. An overview of low voltage DC distribution systems for residential applications. In Proceedings of the 2015 IEEE 5th International Conference on Consumer Electronics—Berlin (ICCE-Berlin), Berlin, Germany, 6–9 September 2015. [CrossRef]
26. Ghareeb, A.T.; Mohamed, A.A.; Mohammed, O.A. DC microgrids and distribution systems: An overview. In Proceedings of the IEEE Power and Energy Society General Meeting, Vancouver, BC, Canada, 21–25 July 2013.
27. Liu, P.; Wu, Z.; Gu, W.; Lu, Y.; Yang, X.; Sun, K.; Sun, Q. Security-constrained AC–DC hybrid distribution system expansion planning with high penetration of renewable energy. *Int. J. Electr. Power Energy Syst.* **2022**, *142*, 108285. [CrossRef]
28. Subudhi, B.; Pradhan, R. A Comparative Study on Maximum Power Point Tracking Techniques for Photovoltaic Power Systems. *IEEE Trans. Sustain. Energy* **2012**, *4*, 89–98. [CrossRef]
29. Chauhan, R.K.; Rajpurohit, B.S.; Hebner, R.E.; Singh, S.N.; Gonzalez-Longatt, F.M. Voltage Standardization of DC Distribution System for Residential Buildings. *J. Clean Energy Technol.* **2015**, *4*, 167–172. [CrossRef]
30. Dragi˘, T.; Guerrero, J.M. DC Microgrids—Part II: A Review of Power Architectures, Applications, and Standardization Issues. *IEEE Trans. Power Electron.* **2016**, *31*, 3528–3549.
31. Dragičević, T.; Lu, X.; Vasquez, J.C.; Guerrero, J.M. DC Microgrids–Part I: A Review of Control Strategies and Stabilization Techniques. *IEEE Trans. Power Electron.* **2016**, *31*, 4876–4891. [CrossRef]
32. Jung, T.-H.; Gwon, G.-H.; Kim, C.-H.; Han, J.; Oh, Y.-S.; Noh, C.-H. Voltage Regulation Method for Voltage Drop Compensation and Unbalance Reduction in Bipolar Low-Voltage DC Distribution System. *IEEE Trans. Power Deliv.* **2017**, *33*, 141–149. [CrossRef]
33. Wu, T.-F.; Chang, C.-H.; Lin, L.-C.; Yu, G.-R.; Chang, Y.-R. DC-Bus Voltage Control with a Three-Phase Bidirectional Inverter for DC Distribution Systems. *IEEE Trans. Power Electron.* **2012**, *28*, 1890–1899. [CrossRef]
34. Riccobono, A.; Santi, E. Comprehensive Review of Stability Criteria for DC Power Distribution Systems. *IEEE Trans. Ind. Appl.* **2014**, *50*, 3525–3535. [CrossRef]
35. Kakigano, H.; Miura, Y.; Ise, T. Low-Voltage Bipolar-Type DC Microgrid for Super High Quality Distribution. *IEEE Trans. Power Electron.* **2010**, *25*, 3066–3075. [CrossRef]
36. Mishra, A.; Rajeev, K.; Garg, V. Assessment of 48 volts DC for homes. In Proceedings of the 2018 IEEMA Engineer Infinite Conference (eTechNxT 2018), New Delhi, India, 13–14 March 2018; pp. 1–6. [CrossRef]
37. Dastgeer, F.; Gelani, H.E. A Comparative analysis of system efficiency for AC and DC residential power distribution paradigms. *Energy Build.* **2017**, *138*, 648–654. [CrossRef]
38. Chen, F.; Burgos, R.; Boroyevich, D. A high-efficiency interleaved single-phase AC-DC converter with common-mode voltage regulation for 380 VDC microgrids. In Proceedings of the 2017 IEEE Energy Conversion Congress and Exposition (ECCE), Cincinnati, OH, USA, 1–5 October 2017; pp. 4128–4135.
39. Hwang, J.-C.; Chen, P.-C.; Liu, C.-S.; Chen, L.-R. A novel single-phase interleaved Bi-directional inverter for grid-connection control. *IEEE Int. Symp. Ind. Electron.* **2014**, *43*, 375–379. [CrossRef]
40. Chen, H.-C.; Liao, J.-Y. Bidirectional Current Sensorless Control for the Full-Bridge AC/DC Converter with Considering Both Inductor Resistance and Conduction Voltages. *IEEE Trans. Power Electron.* **2013**, *29*, 2071–2082. [CrossRef]

41. Sathler, H.H.; Marcelino, F.L.F.; de Oliveira, T.R.; Seleme, S.I.; Garcia, P.F.D. A comparative efficiency study on bidirectional grid interface converters applied to low power DC nanogrids. In Proceedings of the 14th Brazilian Power Electronics Conference COBEP 2017, Juiz de Fora, Brazil, 19–22 November 2017; pp. 1–6. [CrossRef]
42. Yu, G.-R.; Wei, J.-S. Modeling and control of a bi-directional inverter for DC microgrids. In Proceedings of the 2011 International Conference on System Science and Engineering (ICSSE), Macau, China, 8–10 June 2011; pp. 425–430. [CrossRef]
43. Lee, S.; Kim, K.; Kwon, J.; Kwon, B. Single-phase transformerless bi-directional inverter with high efficiency and low leakage current. *IET Power Electron.* **2014**, *7*, 451–458. [CrossRef]
44. Dong, D.; Luo, F.; Boroyevich, D.; Mattavelli, P. Leakage Current Reduction in a Single-Phase Bidirectional AC–DC Full-Bridge Inverter. *IEEE Trans. Power Electron.* **2012**, *27*, 4281–4291. [CrossRef]
45. Wu, T.-F.; Kuo, C.-L.; Sun, K.-H.; Chang, Y.-C. Dc-bus voltage regulation and power compensation with bi-directional inverter in dc-microgrid applications. In Proceedings of the 2011 IEEE Energy Conversion Congress and Exposition, Phoenix, AZ, USA, 17–22 September 2011; pp. 4161–4168. [CrossRef]
46. Wu, T.-F.; Kuo, C.-L.; Sun, K.-H.; Chang, Y.-C. Current distortion improvement for power compensation with bi-directional inverter in DC-distribution system. In Proceedings of the 2012 15th International Power Electronics and Motion Control Conference (EPE/PEMC), Novi Sad, Serbia, 4–6 September 2012; pp. 1–7. [CrossRef]
47. Wu, T.-F.; Kuo, C.-L.; Sun, K.-H.; Yu, G.-R. DC-bus voltage regulation with bi-directional inverter in DC distribution systems. In Proceedings of the IEEE Applied Power Electronics Conference and Exposition (APEC), Orlando, FL, USA, 5–9 February 2012; pp. 769–774. [CrossRef]
48. Chen, Y.-K.; Wu, T.-F.; Chang, Y.-C.; Lin, L.-C.; Yao, N. Extended Application of D-${\bf Σ}$ Digital Control to a Single-Phase Bidirectional Inverter With an LCL Filter. *IEEE Trans. Power Electron.* **2014**, *30*, 3903–3911.
49. Wu, T.-F.; Kuo, C.-L.; Lin, L.-C.; Chen, Y.-K. DC-Bus Voltage Regulation for a DC Distribution System with a Single-Phase Bidirectional Inverter. *IEEE J. Emerg. Sel. Top. Power Electron.* **2015**, *4*, 210–220. [CrossRef]
50. Yu, G.-R.; Lai, J.-J.; Liu, J.-Y. T-S fuzzy control of a single-phase bidirectional inverter. *Proc. IEEE Int. Conf. Ind. Technol.* **2016**, *2016*, 1462–1467. [CrossRef]
51. Kim, H.-S.; Ryu, M.-H.; Baek, J.-W.; Jung, J.-H. High-Efficiency Isolated Bidirectional AC–DC Converter for a DC Distribution System. *IEEE Trans. Power Electron.* **2012**, *28*, 1642–1654. [CrossRef]
52. Wu, T.-F.; Sun, K.-H.; Kuo, C.-L.; Chang, C.-H. Predictive Current Controlled 5-kW Single-Phase Bidirectional Inverter With Wide Inductance Variation for DC-Microgrid Applications. *IEEE Trans. Power Electron.* **2010**, *25*, 3076–3084. [CrossRef]
53. Wang, J.; Gao, S.; Sun, Y.; Ji, Z.; Cheng, L.; Li, L.; Gu, W.; Zhao, J. Single Phase Bidirectional H6 Rectifier/Inverter. *IEEE Trans. Power Electron.* **2019**, *34*, 10710–10719. [CrossRef]
54. Alshammari, M.; Duffy, M. A novel synchronous H6 for improving light load efficiency of bidirectional inverters in a DC distribution system. In Proceedings of the 2021 IEEE Applied Power Electronics Conference and Exposition (APEC), Phoenix, AZ, USA, 14–17 June 2021; pp. 1666–1673. [CrossRef]
55. Jung, J.-H.; Kim, H.-S.; Ryu, M.-H.; Kim, J.-H.; Baek, J.-W. Single-phase bidirectional AC-DC boost rectifier for DC distribution system. In Proceedings of the 2013 IEEE ECCE Asia Downunder-5th IEEE Annual International Energy Conversion Congress and Exhibition, IEEE ECCE Asia 2013, Melbourne, Australia, 3–6 June 2013; pp. 544–549. [CrossRef]
56. Chen, F.; Burgos, R.; Boroyevich, D. Efficiency comparison of a single-phase grid-interface bidirectional AC/DC converter for DC distribution systems. In Proceedings of the 2015 IEEE Energy Conversion Congress and Exposition (ECCE), Montreal, QC, Canada, 20–24 September 2015; pp. 6261–6268. [CrossRef]
57. Chen, F.; Burgos, R.; Boroyevich, D. A Bidirectional High-Efficiency Transformerless Converter with Common-Mode Decoupling for the Interconnection of AC and DC Grids. *IEEE Trans. Power Electron.* **2018**, *34*, 1317–1333. [CrossRef]
58. Chen, F.; Burgos, R.; Boroyevich, D.; Dong, D. Control loop design of a two-stage bidirectional AC/DC converter for renewable energy systems. In Proceedings of the 2014 IEEE Applied Power Electronics Conference and Exposition-APEC 2014, Fort Worth, TX, USA, 16–20 March 2014; pp. 2177–2183. [CrossRef]
59. Du, W.; Jiang, Q.; Erickson, M.J.; Lasseter, R.H. Voltage-Source Control of PV Inverter in a CERTS Microgrid. *IEEE Trans. Power Deliv.* **2014**, *29*, 1726–1734. [CrossRef]
60. Wiechmann, E.P.; Aqueveque, P.; Burgos, R.; Rodriguez, J. On the Efficiency of Voltage Source and Current Source Inverters for High-Power Drives. *IEEE Trans. Ind. Electron.* **2008**, *55*, 1771–1782. [CrossRef]
61. Singh, J.; Dahiya, R.; Saini, L.M. Recent research on transformer based single DC source multilevel inverter: A review. *Renew. Sustain. Energy Rev.* **2018**, *82*, 3207–3224. [CrossRef]
62. Khan, N.H.; Forouzesh, M.; Siwakoti, Y.P.; Li, L.; Kerekes, T.; Blaabjerg, F. Transformerless Inverter Topologies for Single-Phase Photovoltaic Systems: A Comparative Review. *IEEE J. Emerg. Sel. Top. Power Electron.* **2019**, *8*, 805–835. [CrossRef]
63. Wu, T.-F.; Wang, S.-A.; Kuo, C.-L.; Lee, K.-Y. Design and implementation of a push-pull phase-shifted bi-directional inverter with a dsPIC controller. In Proceedings of the International Conference on Power Electronics and Drive Systems (PEDS 2009), Taipei, Taiwan, 2–5 January 2009; pp. 728–733. [CrossRef]
64. Ortiz, G.; Biela, J.; Bortis, D.; Kolar, J.W. Converter for Renewable Energy Applications Converter for Renewable Energy Applications. In Proceedings of the 2010 International Power Electronics Conference-ECCE ASIA, Sapporo, Japan, 21–24 June 2010; pp. 3212–3219.

65. Sabry, A.H.; Mohammed, Z.M.; Nordin, F.H.; Ali, N.H.N.; Al-Ogaili, A.S. Single-Phase Grid-Tied Transformerless Inverter of Zero Leakage Current for PV System. *IEEE Access* **2019**, *8*, 4361–4371. [CrossRef]
66. Ge, W.; Wang, Y.; Zhao, Z.; Yang, X.; Li, Y. Residual Flux in the Closed Magnetic Core of a Power Transformer. *IEEE Trans. Appl. Supercond.* **2013**, *24*, 1–4. [CrossRef]
67. Otero-De-Leon, R.; Mohan, N. A push-pull DC-AC high frequency power electronics transformer for photovoltaic applications. In Proceedings of the IECON Proceedings (Industrial Electronics Conference) 2014, Dallas, TX, USA, 29 October–1 November 2014; pp. 1131–1136. [CrossRef]
68. Pan, X.; Li, H.; Liu, Y.; Zhao, T.; Ju, C.; Gae, A.K.R. An Overview and Comprehensive Comparative Evaluation of Current-Fed-Isolated-Bidirectional DC/DC Converter. *IEEE Trans. Power Electron.* **2019**, *35*, 2737–2763. [CrossRef]
69. Xue, F.; Yu, R.; Huang, A.Q. A 98.3% Efficient GaN Isolated Bidirectional DC–DC Converter for DC Microgrid Energy Storage System Applications. *IEEE Trans. Ind. Electron.* **2017**, *64*, 9094–9103. [CrossRef]
70. Wu, T.-F.; Chen, Y.-C.; Yang, J.-G.; Kuo, C.-L. Isolated Bidirectional Full-Bridge DC–DC Converter With a Flyback Snubber. *IEEE Trans. Power Electron.* **2010**, *25*, 1915–1922. [CrossRef]
71. Naayagi, R.T.; Forsyth, A.J.; Shuttleworth, R. High-Power Bidirectional DC–DC Converter for Aerospace Applications. *IEEE Trans. Power Electron.* **2012**, *27*, 4366–4379. [CrossRef]
72. Chen, W.; Rong, P.; Lu, Z. Snubberless Bidirectional DC–DC Converter With New CLLC Resonant Tank Featuring Minimized Switching Loss. *IEEE Trans. Ind. Electron.* **2009**, *57*, 3075–3086. [CrossRef]
73. López, O.; Freijedo, F.D.; Yepes, A.G.; Fernández-Comesaña, P.; Malvar, J.; Teodorescu, R. Eliminating ground current in a transformerless photovoltaic application. *IEEE Trans. Energy Convers.* **2010**, *25*, 140–147. [CrossRef]
74. Ji, B.; Wang, J.; Zhao, J. High-Efficiency Single-Phase Transformerless PV H6 Inverter With Hybrid Modulation Method. *IEEE Trans. Ind. Electron.* **2012**, *60*, 2104–2115. [CrossRef]
75. Wang, J.; Luo, F.; Ji, Z.; Sun, Y.; Ji, B.; Gu, W.; Zhao, J. An Improved Hybrid Modulation Method for the Single-Phase H6 Inverter with Reactive Power Compensation. *IEEE Trans. Power Electron.* **2017**, *33*, 7674–7683. [CrossRef]
76. Chen, F.; Burgos, R.; Boroyevich, D.; Zhang, X. Low-Frequency Common-Mode Voltage Control for Systems Interconnected with Power Converters. *IEEE Trans. Ind. Electron.* **2016**, *64*, 873–882. [CrossRef]
77. Wu, T.-F.; Sun, K.-H.; Kuo, C.-L.; Yang, M.-S.; Chang, R.-C. Design and implementation of a 5 kW 1φ bi-directional inverter with wide inductance variation. In Proceedings of the 2010 IEEE Energy Conversion Congress and Exposition, Atlanta, GA, USA, 12–16 September 2010; pp. 45–52. [CrossRef]
78. Yu, G.-R.; Wei, J.-S. Fuzzy control of a bi-directional inverter with nonlinear inductance for DC microgrids. In Proceedings of the 2011 IEEE International Conference on Fuzzy Systems, Taipei, Taiwan, 27–30 June 2011; pp. 1941–1945. [CrossRef]
79. Liu, C.; Wang, Y.; Cui, J.; Zhi, Y.; Liu, M.; Cai, G. Transformerless Photovoltaic Inverter Based on Interleaving High-Frequency Legs Having Bidirectional Capability. *IEEE Trans. Power Electron.* **2016**, *31*, 1131–1142. [CrossRef]
80. Nejabatkhah, F.; Li, Y.W.; Tian, H. Power Quality Control of Smart Hybrid AC/DC Microgrids: An Overview. *IEEE Access* **2019**, *7*, 52295–52318. [CrossRef]
81. Nejabatkhah, F.; Li, Y.W. Overview of Power Management Strategies of Hybrid AC/DC Microgrid. *IEEE Trans. Power Electron.* **2014**, *30*, 7072–7089. [CrossRef]
82. Li, R.T.H.; Chung, H.S.-H.; Lau, W.-H.; Zhou, B. Use of Hybrid PWM and Passive Resonant Snubber for a Grid-Connected CSI. *IEEE Trans. Power Electron.* **2010**, *25*, 298–309. [CrossRef]
83. Lai, R.; Ngo, K.D.T. A PWM method for reduction of switching loss in a full-bridge inverter. *IEEE Trans. Power Electron.* **1995**, *10*, 326–332.
84. Liao, Y.-H. A Novel Reduced Switching Loss Bidirectional AC/DC Converter PWM Strategy with Feedforward Control for Grid-Tied Microgrid Systems. *IEEE Trans. Power Electron.* **2013**, *29*, 1500–1513. [CrossRef]
85. Zheng, Z.; Zhang, T.; Xue, J. Application of Fuzzy Control in a Photovoltaic Grid-Connected Inverter. *J. Electr. Comput. Eng.* **2018**, *2018*, 3806372. [CrossRef]
86. Suzdalenko, A. Current Sensorless Control of Front-end Bidirectional AC/DC Converter Based on Half-bridge Topology. *Electr. Control Commun. Eng.* **2013**, *4*, 19–25. [CrossRef]
87. Suzdalenko, A. Development of single-switch model for current sensorless control of bidirectional half-bridge AC/DC converter. In Proceedings of the 2014 55th International Scientific Conference on Power and Electrical Engineering of Riga Technical University, RTUCON 2014, Riga, Latvia, 14 October 2014; pp. 38–42. [CrossRef]
88. Chen, H.-C. Single-Loop Current Sensorless Control for Single-Phase Boost-Type SMR. *IEEE Trans. Power Electron.* **2009**, *24*, 163–171. [CrossRef]
89. Azcondo, F.J.; De Castro, A.; López, V.M.; García, O. Power factor correction without current sensor based on digital current rebuilding. *IEEE Trans. Power Electron.* **2010**, *25*, 1527–1536. [CrossRef]
90. Jeong, Y.-S.; Lee, S.-H.; Jeong, S.-G.; Kwon, J.-M.; Kwon, B.-H. High-Efficiency Bidirectional Grid-Tied Converter Using Single Power Conversion with High-Quality Grid Current. *IEEE Trans. Ind. Electron.* **2017**, *64*, 8504–8513. [CrossRef]

World Electric Vehicle Journal

MDPI

Article

Analysis, Design, and Experimental Results for a High-Frequency ZVZCS Galvanically Isolated PSFB DC-DC Converter over a Wide Operating Range Using GaN-HEMT

Abdullah Eial Awwad

Department of Electrical Power Engineering and Mechatronics, Faculty of Engineering, Tafila Technical University, Tafila 66110, Jordan; abdullah.awad@ttu.edu.jo

Abstract: This paper investigates the potential of the emerging gallium nitride (GaN) high-electron mobility transistors (HEMT) power devices to meet certain power conversion challenges. The advantages of utilizing GaN HEMT transistors in a high-frequency, high-power isolated DC-DC topology are explored experimentally. Using the GaN HEMT's parasitic elements, e.g., output capacitance, and the leakage inductance of the transformer, a soft switching zero-voltage zero-current switching (ZVZCS) phase shift converter is proposed. Accordingly, the freewheeling current is terminated, and soft switching is realized for most of the primary and secondary active devices. Furthermore, without using any additional circuitry, the overshoot voltage across the bridges of active rectifier diodes is clamped at their voltage level. In addition, a high-frequency power transformer is optimized to minimize the overall transformer losses (e.g., winding and core losses). Combined the conductor types, e.g., litz wire and copper foil, shows good electrical and thermal performance by reducing the AC and DC resistance. Finally, a 5 kW, 100–250 kHz prototype is built and tested. The experimental results show a conversion efficiency of up to 98.18% for the whole converter.

Keywords: GaN HEMT; phase shift full bridge; isolated DC-DC converter; PSFB; high-frequency transformer; zero-voltage zero-current switching; soft switching; wide bandgap (WBG)

Citation: Eial Awwad, A. Analysis, Design, and Experimental Results for a High-Frequency ZVZCS Galvanically Isolated PSFB DC-DC Converter over a Wide Operating Range Using GaN-HEMT. *World Electr. Veh. J.* **2022**, *13*, 206. https://doi.org/10.3390/wevj13110206

Academic Editors: Diego Bellan and Jelena Loncarski

Received: 26 September 2022
Accepted: 29 October 2022
Published: 2 November 2022

1. Introduction

Currently, wideband gap (WBG) semiconductors are emerging candidates for high-power applications because of their lower on-state resistance and faster switching frequency, compared to conventional transistors, such as silicon (Si) MOSFETs and IGBTs [1–3]. Due to their superior characteristics, silicon carbide (SiC) and gallium nitride (GaN) high-electron mobility transistors (HEMT) have been the subject of extensive study and development. In terms of their application, the most advantageous features of GaN HEMT devices are their high breakdown field, high thermal conductivity, high-temperature operation, and minor on-state resistance [4–6].

Currently, the use of medium and high-power DC-DC converters is growing rapidly in different applications, including renewable power conversion [7], hybrid electric vehicles [8], power distribution in the micro-grid system [9], and storage systems [10]. The phase-shifted full bridge (PSFB) converter is commonly utilized as a DC-DC stage for high-power applications, and it is a feasible alternative for DC-DC converters.

In electric vehicle (EV) applications, the DC-DC converter is an essential part. Due to the extensive conversion ratio between the input and output, the galvanic isolation with specific turns ratio is preferred. A recent study has summarized the current advancements in GaN WBG semiconductor devices, particularly for EV applications [11]. The study investigates how GaN devices may be used to realize high-efficiency and high-power density converters in EV applications. Reference [12] proposed an integration method of auxiliary power modules and onboard battery chargers, which provide simultaneous operation without an extra circuit and ZVS. This improves the total efficiency, as well as volume and cost. A previous study investigated a modified multi-port DC-DC converter based on Z-Quasi Resonant for off-board EV battery charging [13]. A control strategy for charging and discharging EV batteries is proposed. Prior work introduced a multiple dual active bridge (DAB) DC-DC stage for EV fast charger. Generalized small-signal modeling and a generalized control approach were proposed to ensure equal power-sharing [14].

The advantages of applying the GaN HEMT devices in the PSFB converter have rarely been discussed in the literature [5,15]. However, the benefit of using SiC MOSFET in the PSFB converter has been introduced [16]. The experimental results for 10 kW, 250 kHz prototypes validate the feasibility of using SiC transistors. Indeed, clamping the overshoot voltage across the rectifier diodes and achieving soft switching in a wide load power range are the main benefits of using SiC MOSFETs in such topology. Using conventional transistors, different studies have focused on improving efficiency and achieving a wide soft switching operating range [17–19]. Other earlier studies have suggested adding snubber circuits and recovery clamp circuits as workable options to suppress the overvoltage across the output rectifier diodes [20–22]. These techniques have a large impact on the clamping of the voltage overshoot over the secondary side but increase the control complexity and add more losses. Furthermore, a previous study [23] used the capacitive output filter to minimize the diode rectifier ringing. This minimization of the diode rectifier ringing allows switching devices to operate at low switching losses under zero-voltage switching (ZVS).

Currently, the HF magnetic components are the bulkiest components of the high-frequency (HF) power converter. In the PSFB topology, the high-frequency transformer (HFT) is a design limitation to achieving high power density [24].

The transformer core and winding wire selection should be optimized for proper design to minimize the core and winding losses. In addition, the selection of the core material is critical in HFT design due to its impact on the cost, efficiency, and volume [3,25,26]. Moreover, the influence of wire types and winding arrangements are critical parameters for enhancing the dynamic performance of the HFT. To minimize winding losses, multi-strand HF litz wire or copper foil are frequently employed. According to previous work [24,27], to reduce the overall winding losses, the ideal number of litz wire strands should be chosen with an AC resistance (R_{AC}) that is almost twice as great as its DC resistance (R_{DC}). The skin and proximity effects are significantly reduced using litz wire. In contrast, greater thermal resistance is anticipated due to the several isolated strands, making it difficult to cool the litz wire.

Furthermore, the proximity effect is reduced by interleaving the primary and secondary windings, while optimizing the foil thickness effectively reduces the skin effect [24,27]. The copper foil plate is preferred over other materials for usage in high-current applications because of its larger width-to-thickness ratio. Due to this ratio, it offers higher heat conduction, making it possible to accomplish effective heat transmission to the surrounding area.

This work experimentally assesses the advantages of using GaN HEMT devices in the high-frequency, high-power DC-DC converter. First, the dynamic performance of the GaN HEMT is evaluated with different operating conditions. Then, the design optimization of the transformer is presented, and the impacts of different conductor types and winding arrangements are investigated. Finally, a soft switching zero-voltage zero-current switching (ZVZCS) PSFB converter is proposed. The parasitic output capacitance of the GaN HEMT and the power transformer's parasitic leakage inductance are used to achieve soft switching, which results in a low-cost, high-efficiency, lightweight, and compact converter design. A 5 kW, 100–250 kHz prototype experimentally demonstrates an efficiency of up to 98.18% for the full converter.

2. PSFB DC-DC Converter

The isolated PSFB DC-DC converter (Figure 1) consists of a controlled GaN HEMT full-bridge inverter at the input (GaN Systems GS66508T), an HF transformer providing galvanic isolation and voltage level conversion, and a Schottky barrier diode (SBD) bridge rectifier (ROHM SCS230AE2) as the output stage.

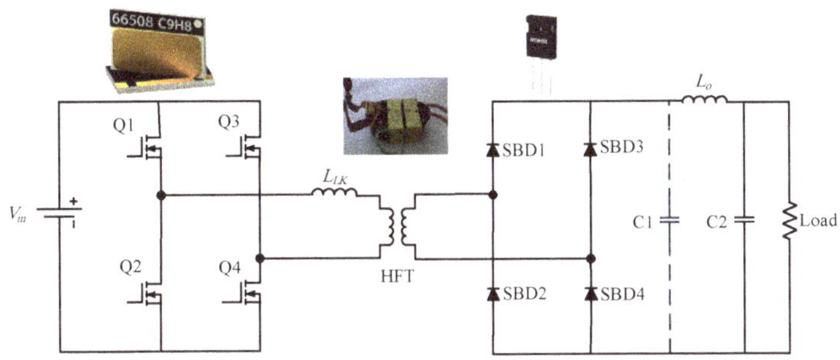

Figure 1. Topology of the GaN HEMT PSFB converter.

In this paper, the circuit parasitics, such as the output parasitic capacitance of the GaN devices (C_{oss}) and the transformer leakage inductance (L_{LK}), are used to achieve ZVZCS. To ensure resonant discharge of the C_{oss}, the two legs of the controlled GaN HEMT full bridge operate in phase-shift mode. The ZVS of the leading leg (Q1 and Q2) is achieved with help of energy storage in both the transformer leakage inductor and the output filter inductor (L_o). The zero-current switching (ZCS) can be achieved for the lagging-leg (Q3 and Q4) during the freewheeling diode transition by forcing the primary current (I_p) to zero.

The simulated key waveforms of the ZVZCS PSFB converter are shown in detail in Figure 2. As presented, the lagging leg is turned off with ZCS by forcing the primary current to reduce to zero at t_3. At this time, Q1 is turned off, and the primary current charges the output capacitance of Q1 (C_{oss1}) and discharges the output capacitance of Q2 (C_{oss2}). Furthermore, the leading leg always achieves ZVS; indeed, even a minimum primary current amplitude is reached with the help of the energy storage (E_L) in the transformer leakage inductance ($E_L \propto I_p^2$).

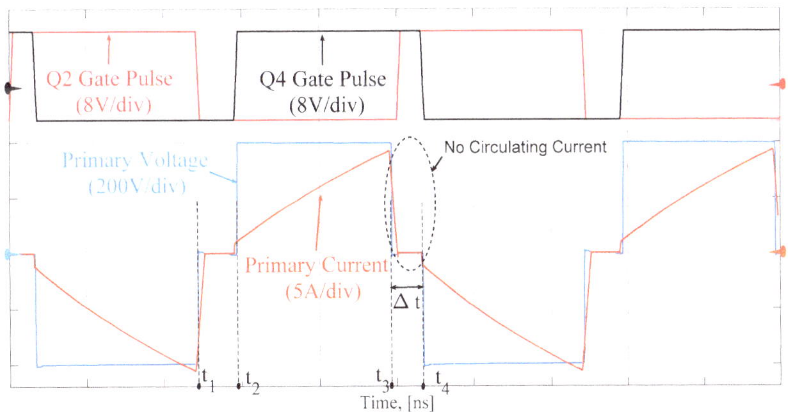

Figure 2. Main waveforms of the PSFB converter.

However, one of the main advantages of this topology is that as the primary current reaches zero, no more power is supplied to the load. This phenomenon mainly occurs because the SBD1 and SBD2 become reverse biased, and, hence, the switching losses of the output rectifier diodes and the reverse recovery losses are eliminated. To ensure a continuous current in the load side during the interval (Δt; Figure 2), the output filter capacitance (C2) is used to supply the load current. In this context, contrary to the Si-based converter, the circulating current in the primary side is eliminated during the freewheeling period, Δt.

Conversely, the output capacitor (C1) is used to suppress the voltage spikes generated during the power transmission stage that appear across the rectifier bridge.

Moreover, at t_4, as the primary current is zero, and due to the limitation of the current slope caused by the transformer leakage inductance, the switch Q3 can be turned on with ZVS. Conversely, at t_2, either ZVS or a hard-switching (HS) state with a relatively small current will cause Q4 to turn on. The switching states of the proposed topology are summarized in Table 1.

Table 1. Switching states of the proposed topology.

		Turn on	Turn off
GaN HEMTBridge	Q1	ZVS	HS with a relatively small turn-off current
	Q2	ZVS	HS with a relatively small turn-off current
	Q3	ZVS or HS with a relatively small turn-on current	ZCS
	Q4	ZVS or HS with a relatively small turn-on current	ZCS

3. Design Considerations

3.1. Switching Device Selection

Fast switching is present in GaN HEMT devices, and the process of turning on and off is accompanied by an unwanted current and voltage ringing. Therefore, in order to achieve the optimal utilization of these devices, it is necessary to completely comprehend the switching behavior and, hence, determine the limiting factors in dynamic processes, e.g., turn-on and turn-off transitions.

In this paper, the commercial GaN HEMT transistor (GaN Systems GS66508T) is used as an active device. To characterize the switching performance of the device under test (DUT), the standard double-pulse test (DPT) circuit with inductive load was set up (see Figure 3a).

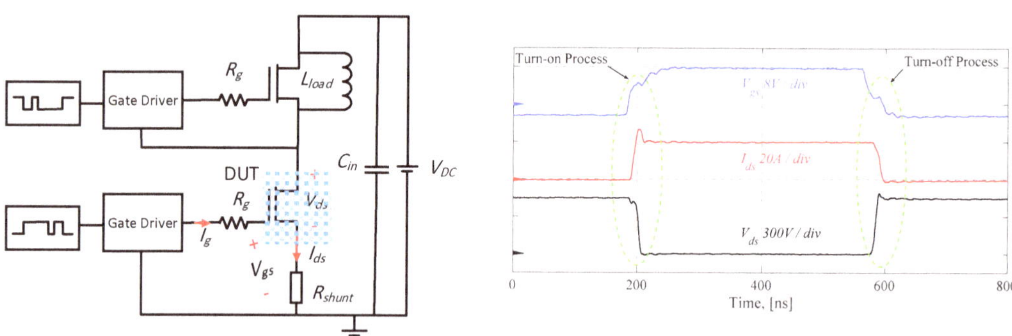

(a) **(b)**

Figure 3. Double-pulse test circuit: (**a**) schematic of the DPT and (**b**) typical waveforms.

Figure 3b depicts the typical inductive double-pulse switching waveforms, the drain-source voltage (V_{ds}), the drain-source current (I_{ds}), and the gate-source voltage (V_{gs}). As shown in the figure, at the end of the first pulse and the beginning of the second pulse, the device's switching dynamic waveforms are recorded.

The switching energy loss (E_{loss}) is calculated directly from the measurement, according to the following equation:

$$E_{Loss} = \int_t v_{ds}(t) i_{ds}(t).dt \quad (1)$$

The switching experiments were conducted up to 400 V_{DC}, with different case temperatures (T_c) and varying load currents from 2 A to 20 A. Figure 4 illustrates the measured turn-on (E_{on}) and turn-off (E_{off}) switching energy losses at two different case temperatures.

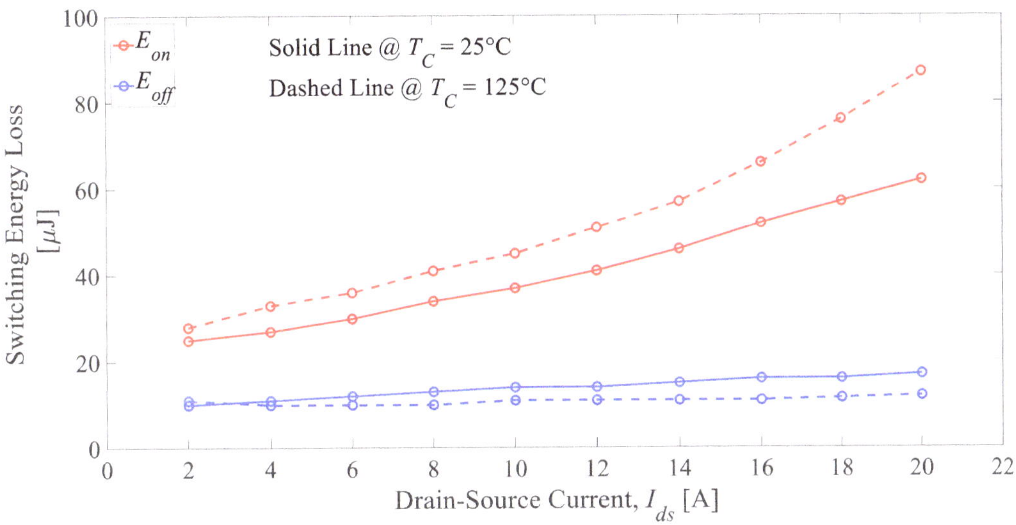

Figure 4. Switching energy loss as a function of the drain-source current at 400 V and two different case temperatures.

As expected, with increased current levels, the overall switching losses increase. In detail, as shown in Figure 4, the turn-off loss drops slightly as the temperature increases. This is mainly due to the fact that the plateau voltage increases as the temperature increases. A higher plateau voltage means there is a faster voltage decrease from the plateau to the threshold voltage. A high rate of change of v_{ds} (dv_{ds}/dt) reduces the overlap interval between the channel current and the applied voltage, thus, minimizing the loss interval [28].

Conversely, the turn-on loss is more sensitive to temperature. As observed, at I_{ds} = 20 A, the E_{on} increases by 40.3% when varying the temperature from 25 °C to 125 °C.

To explain this phenomenon, the transfer characteristic of the DUT was examined with different temperatures. Figure 5 depicts the measured transfer characteristics with different T_c at a constant V_{ds} of 9 V. As shown, in the saturation region, the current is reduced accordingly with the increase in temperature, resulting in a reduction in the average transconductance (g_m; Figure 6). As observed, the transconductance decreases by 59% when increasing the temperature from 25 °C to 125 °C. The decrease in g_m is mainly due to the decrease in both the electron mobility and the electron velocity, and thus limits the displacement current, which results in an increase in the turn-on switching time, and hence the loss [28].

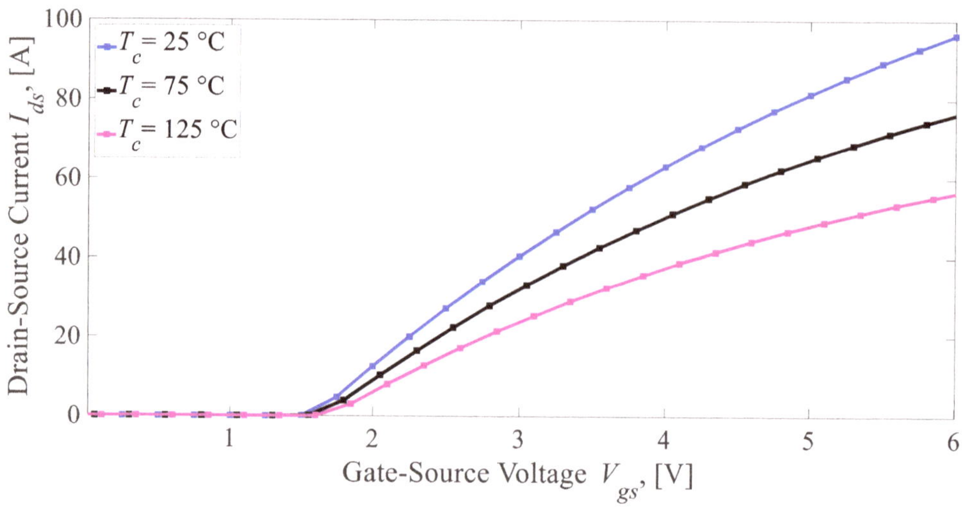

Figure 5. Transfer characteristics measured at different case temperatures.

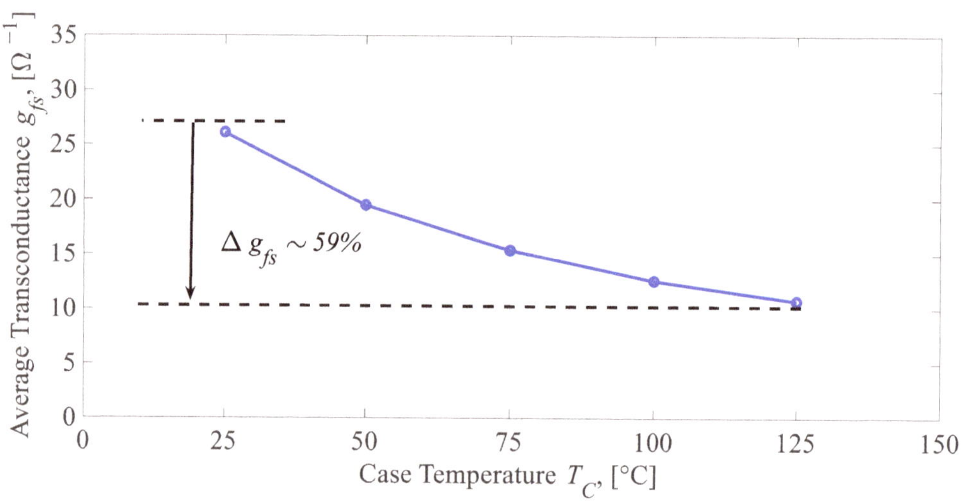

Figure 6. Transconductance as a function of case temperature.

3.2. Switching Frequency and Thermal Limitation

As previously mentioned, GaN HEMT devices operate under high temperatures and HF conditions, thus, reducing the overall size and cost of the passive components and resulting in an increase in total power density. Conversely, the thermal design becomes more difficult, necessitating the use of a sophisticated cooling system. Indeed, the cooling system represents a key factor limiting the utilization of the potential advantages of GaN HEMT transistors in high-power converter applications.

In this context, the switching frequency and thermal limitation factors are elucidated to design a high switching frequency power converter that operates continuously with 5 kW output power.

The maximum power dissipated (P_{DMAX}) in the high-power semiconductors is calculated as follows:

$$P_{DMAX} = \frac{T_j - T_c}{R_{th}} \tag{2}$$

$$R_{th} = R_{th-jc} + R_{th-ca} \tag{3}$$

where T_j is the junction temperature, R_{th} is the total thermal resistance, R_{th-jc} (=0.5 °C/W) is the thermal resistance from the junction to the case, and R_{th-ca} (=0.2 °C/W) is the thermal resistance from the case to the ambient temperature.

The junction temperature is a function of the total power losses and total thermal resistance:

$$T_j = T_c + R_{th} P_{Loss}(T_j) \tag{4}$$

$$P_{Loss}(T_j) = P_{cond}(T_j) + P_{sw}(T_j) \tag{5}$$

where the $P_{cond}(T_j)$ and $P_{sw}(T_j)$, respectively, are the conduction and switching losses as a function of the junction temperature, which is given as follows:

$$P_{cond}(T_j) = D[1 + \alpha_1(T_j - T_o)] R_{on1} I^2_{ds-RMS} \tag{6}$$

$$P_{sw}(T_j) = \left[E_{on}(V_{ds}, I_{ds}, T_j) + E_{off}(V_{ds}, I_{ds}, T_j)\right] f_{sw} \tag{7}$$

where R_{on1} is the on-state resistance at room temperature (T_o) (e.g., for the GS66508T device, the measured R_{on1} = 55 mΩ), α_1 is the on-state resistance coefficient (α_1 = 0.64 mΩ/°C), $I_{ds\text{-}RMS}$ is the drain-source RMS current calculated over the on-time duty cycle, D, and f_{sw} is the switching frequency.

The maximum power dissipated in the DUT is calculated using Equation (2). The converter is forced-air cooled, giving an estimated case temperature of 70 °C for a power dissipation of 114 W. To ensure safe operation, a safety margin (β) of 20% is assumed.

Therefore, the relationship between the maximum allowable switching frequency and the total losses should satisfy the following constraint:

$$P_{Loss}(T_j) \leq (100\% - \beta) P_{DMAX} \tag{8}$$

$P_{Loss}(T_j)$ can be rewritten as follows:

$$P_{Loss}(T_j) = P_{cond} + f_{sw}(E_{on} + E_{off}) \tag{9}$$

Therefore, the maximum permitted switching frequency (f_{sw-MAX}) can be calculated as follows:

$$f_{sw-MAX} = \frac{(100\% - \beta)P_{DMAX} - P_{cond}}{P_{sw}} \tag{10}$$

As a result, the relationship between the maximum permitted switching frequency and the possible output power can be simply calculated. In this work, to prevent overheating of the core and windings and to ensure the safe operation of the GaN HEMT and diode bridges under the use of the forced-air cooling system, the switching frequency was limited to 250 kHz.

3.3. Design and Optimization of the HF Transformer

The transformer provides galvanic isolation and the required voltage matching in isolated DC-DC converters. In this work, in order to minimize the core and winding losses, the selection of the transformer core and winding wire was optimized. Table 2 shows the operating conditions of the HF transformer.

Table 2. DC-DC converter design parameters.

Output Power, P_{out}	5 kW
Input Voltage, V_{in}	400 V
Output Voltage, V_{out}	500–600 V

Table 2. *Cont.*

Switching Frequency, f_{sw}		100–250 kHz
Turns Ration, n		1:1.5
Deadtime (t_{dead})		70 ns
GaN HEMTs	GS66508T	– Drain-Source Blocking Voltage (BV_{DS}) = 650V – On-state resistance ($Ron1$) = 55/118 mΩ (Measured at 25/125 °C).
Rectifier Diodes	SCS230AE2	– Reverse Voltage (VR) = 650 V – Continuous Forward Current (IF) = 30 A
	High-Frequency Transformer	
Ferrite Material		N87
Leakage Inductance, L_{LK}		1.9 μH
Primary Turns, np		10
Primary Windings		4 Litz wires (parallel 100 × Ø 0.05 mm)
Secondary Turns, ns		15
Secondary Windings		Copper Foil (36 mm × 0.15 mm)
DC Resistance, R_{DC}		17 mΩ
AC Resistance, R_{AC}, 100/250 kHz		57/71 mΩ

The design methodology proposed in this paper is devoted to minimizing the total power loss, including the core (P_{core}) and winding (P_w) losses, and hence improving the overall converter performance.

The core loss can be expressed as follows [24]:

$$P_{core} = K f_{sw}^{\alpha} \Delta B^{\beta} A_c l_m \tag{11}$$

where K is the core loss coefficient, ΔB is the peak AC flux density, A_c is the core cross-section area, l_m is the core mean magnetic path length, and α and β are the core loss exponents.

In terms of the winding losses, the following expression can be used [24]:

$$P_w = K_r \frac{\lambda^2 \rho(MLT)}{4 A_{win} k_u A_c^2} \left(\frac{1}{\Delta B}\right)^2 I_{RMS}^2 \tag{12}$$

where K_r is the normalized value of the AC resistance at HF, λ is the applied primary volt-sec, ρ is the resistivity of copper, MLT is the mean length per turn, k_u is the fill factor, A_{win} and A_c are the window and cross-section areas of the core, respectively, and I_{RMS} is the RMS winding current.

Furthermore, the flux density and the number of turns are considered key parameters in the transformer design. The optimal flux density ($B_{optimal}$) occurs when the first derivative of the total loss (P_{tot}) is zero [24]:

$$\frac{dP_{tot}}{d(\Delta B)} = \frac{dP_{core}}{d(\Delta B)} + \frac{dP_w}{d(\Delta B)} = 0 \tag{13}$$

Therefore, the minimum loss can be achieved (Figure 7) if the core operating is designed at the optimal flux density $B_{optimal}$ [24]:

$$\Delta B_{optimal} = \left(\frac{K_r I_{rms}^2 \rho \lambda^2 MLT}{4K_u A_{win} A_c^3 l_m f_{sw}^{\alpha}} \right)^{\frac{1}{\beta+2}} \quad (14)$$

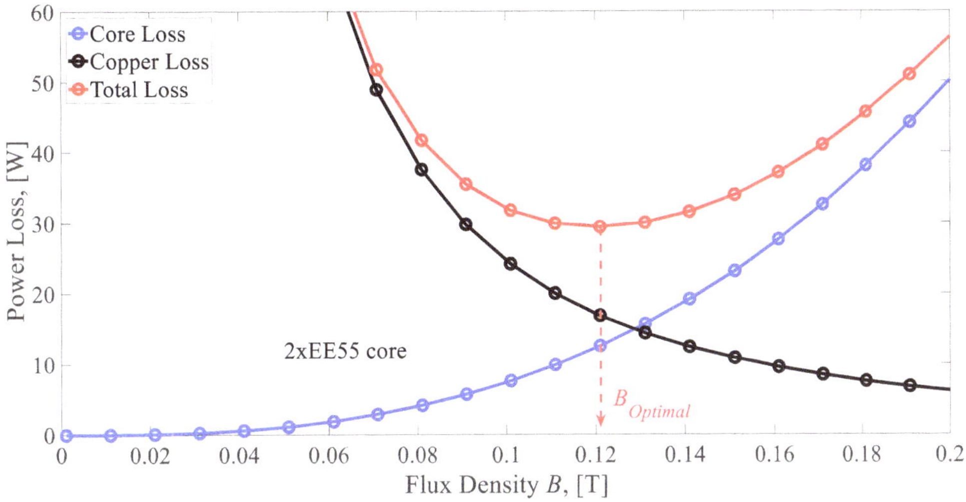

Figure 7. The optimal flux density versus the total power loss of the high-frequency transformer.

Moreover, by solving $\left(\frac{dP_{tot}}{dn} = 0\right)$, the optimal number of turns ($n_{optimal}$) that minimize the total transformer power loss can be determined:

$$n_{optimal} = \left(\frac{\beta K f_{sw}^{(\alpha-\beta)} \left(\frac{V_{in}}{2A_c}\right)^{\beta} A_c l_m k_u A_{win}}{2\rho K_r MLT I_{rms}^2} \right)^{\frac{1}{\beta+2}} \quad (15)$$

In this work, to avoid the expected high surface temperature rise, a double EE55/28/21 core using N87 material was used in the design. As shown in Figure 7, using Equation (14), the optimal flux density can be calculated as 121 mT and the optimal number of turns as 10. This results in 17 W of winding losses and 15 W of core losses at an f_{sw} of 100 kHz.

The copper foil plate is chosen over other materials for usage in high-current applications because of its larger width-to-thickness ratio, which means better heat transfer to the ambient temperature can be obtained. In addition, selecting the optimal foil thickness directly reduces the skin effect, and applying the interleaving method to the primary and secondary windings minimizes the proximity effect.

Conversely, using litz wires effectively reduces the skin and proximity effects, but a large core window area is required. Further to this, higher thermal resistance is expected because of the multi-isolated strands, and thus the cooling system becomes inactive.

According to previous work [3,27], combining different conductor types (litz wire and copper foil; Figure 8) provides good electrical and thermal characteristics. With this approach (Table 1), the copper foil is applied to the low-voltage, high-current side, with a cross-section area of 5.4 mm^2 (width: 36 mm × thickness: 0.15 mm). In contrast, the litz wire is applied to the high-voltage side with four parallel 100 strands with a diameter of 0.05 mm.

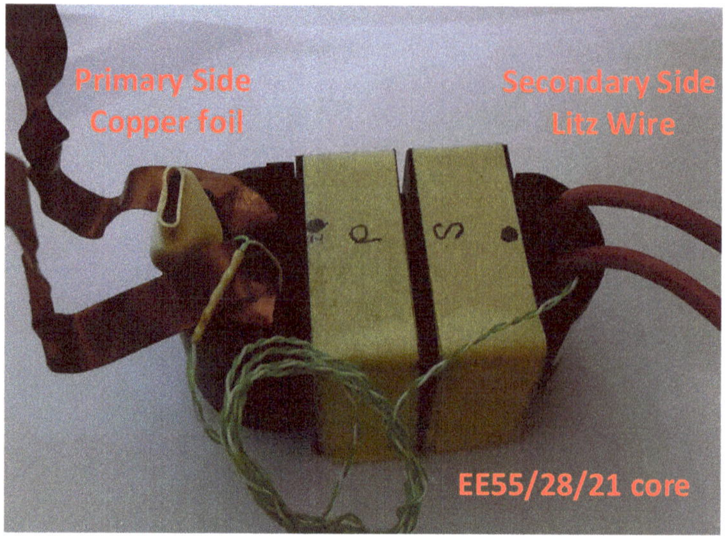

Figure 8. The 5 kW ferrite core (N87) high-frequency transformer prototype.

4. Experimental Results and Discussion

In this study, a prototype of a 5 kW ZVZCS PSFB DC-DC converter has been built and tested. A switching frequency in the range of 100–250 kHz is applied. Operating at different switching frequencies provides a clear picture of the optimum efficiency and power density design. Additionally, to drive and control the high-speed power devices, a microcontroller from STMicroelectronics based on the ARM Cortex architecture (STM32F429, clock rate up to 180 MHz) is used.

The converter specifications used in this work are shown in Table 1. The key waveforms of the proposed ZVZCS converter at different loads with a switching frequency of 100 kHz are illustrated in Figure 9.

It is clear that the proposed converter is able to operate with ZVZCS even at light load, without any additional auxiliary circuit. As shown, the freewheeling current mode is removed, increasing the overall efficiency. Furthermore, the capacitive output filter clamped the overshoot in the primary and secondary voltages to the appropriate voltage level, eliminating the need for an additional circuit.

Additionally, the proposed converter was investigated at a higher switching frequency. Figure 10 illustrates the key waveforms of the proposed ZVZCS at different loads and a fixed switching frequency of 250 kHz.

It can be concluded that, even at a higher switching frequency, the ZVZCS can be achieved using only the small parasitic leakage inductance of the HFT. This achievement of ZVZCS leads to improvements in overall efficiency for a wide power range.

The measured efficiencies of the overall proposed converter at 100 kHz, 150 kHz, and 250 kHz are compared in Figure 11. The design and optimization methodology of HFT, the application of ZVZCS, the optimization of the conductor types (e.g., mixing the litz wire and copper foil), and the superior features of the GaN HEMT devices lead to an efficiency of up to 98.18% for the whole converter.

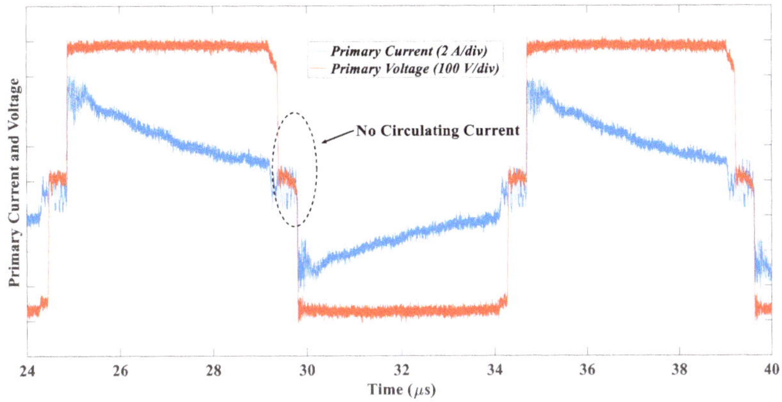

(a)

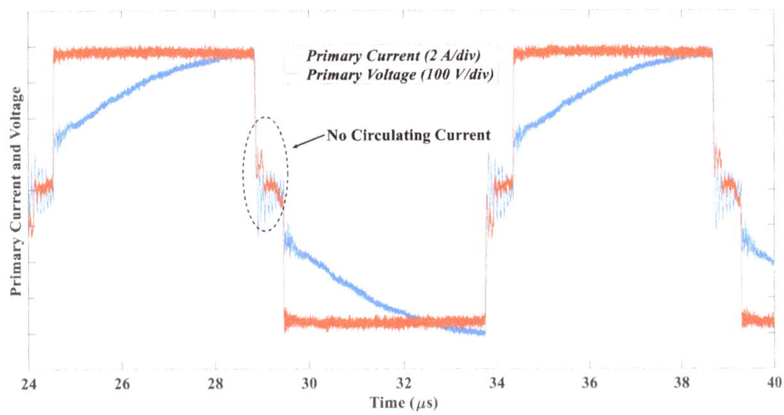

(b)

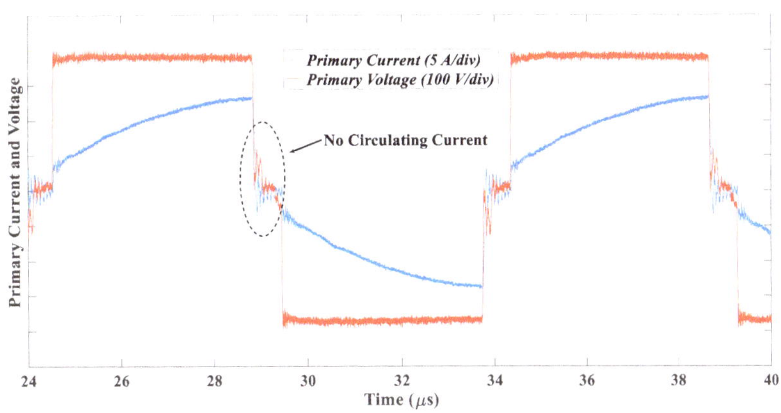

(c)

Figure 9. Measured operation waveforms at 100 kHz and output powers of (**a**) 800 W, (**b**) 3 kW, and (**c**) 5 kW.

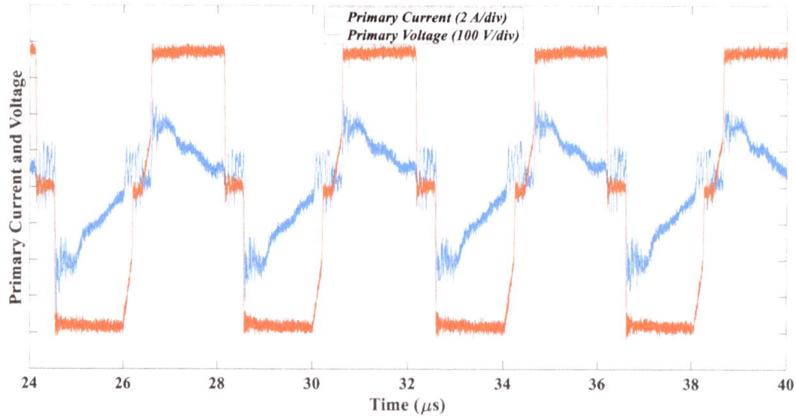

(**a**)

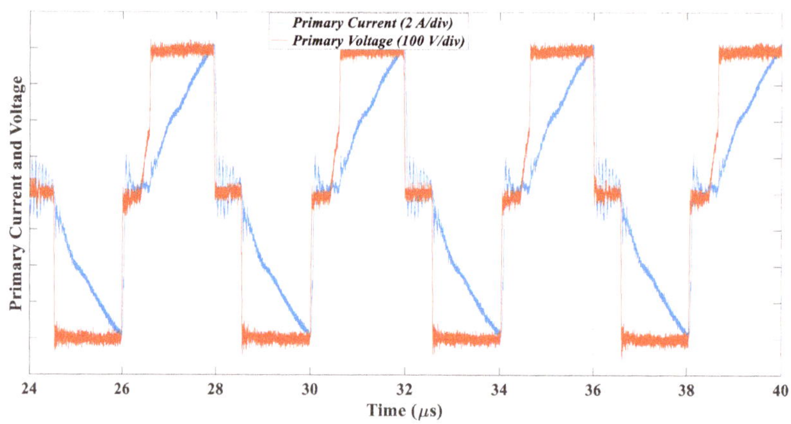

(**b**)

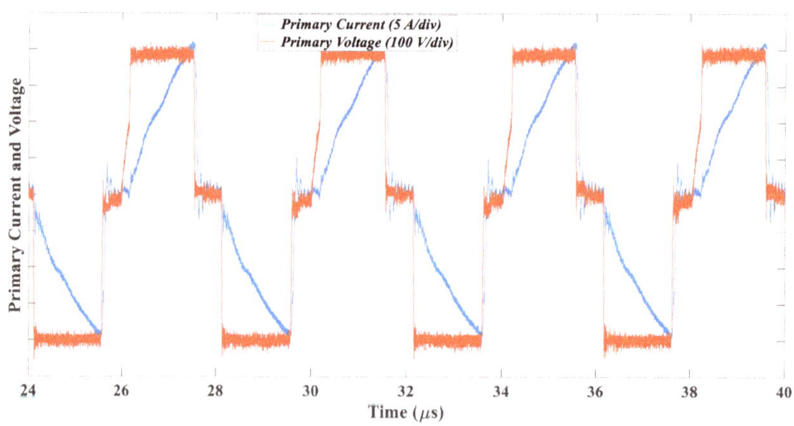

(**c**)

Figure 10. Measured operation waveforms at 250 kHz and output powers of (**a**) 800 W, (**b**) 3 kW, and (**c**) 5 kW.

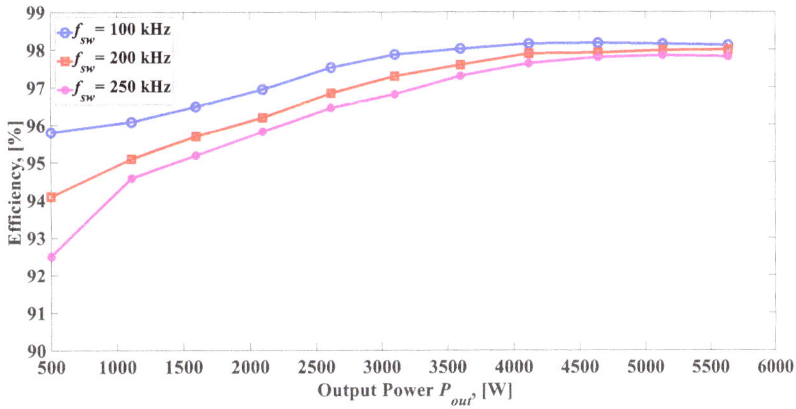

Figure 11. Efficiency comparison of the whole proposed converter at different switching frequencies.

The measured power losses are separated into GaN HEMT power loss (e.g., conduction and switching) and other losses (e.g., core and winding losses of the HFT, rectifier diodes, filter, DC-link, etc.; Figure 12). To extract the power losses, the input and output power are measured using a power analyzer (Zimmer LMG671). The high-speed current and voltage switching of the GaN HEMT and HFT are captured with high-bandwidth meters of 200–2000 MHz. Furthermore, the total equivalent series resistance (ESR) of the HFT is measured as a function of frequency. Specifically, the switching losses of the GaN HEMT devices at given turn-on and turn-off currents are extracted based on the measured DPT results shown in Figure 4. The conduction loss is estimated using the measured on-state resistance. Finally, the core and winding losses are calculated based on Equations (11) and (12), respectively.

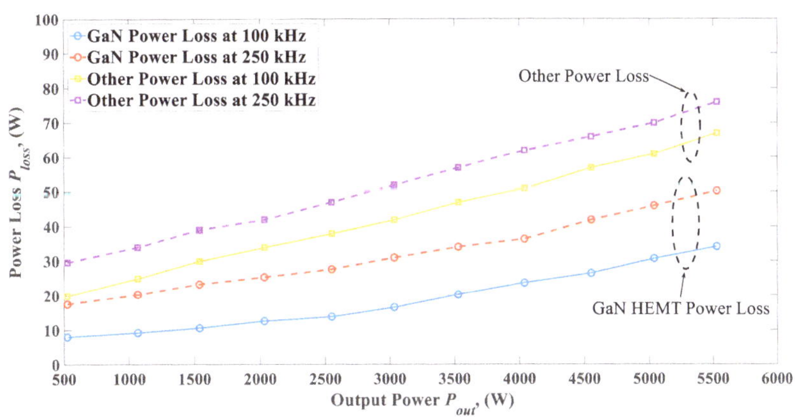

Figure 12. Loss distribution of the proposed converter at different output powers and two different switching frequencies.

Additionally, for the other losses, e.g., conduction and switching losses of leading and lagging legs, the losses of the output rectifier diodes, the loss breakdown methodology described in [16] is adaptive and applied in this work.

As an example, Figure 13 shows the loss distribution at 5 kW and different switching frequencies. As demonstrated, at f_{sw} = 100 kHz, the full bridge GaN HEMT loss is about 32% of the overall power losses (23% conduction loss and 9% switching loss of the overall power losses). It is worth mentioning that even working at high switching frequency

(f_{sw} = 250 kHz), the GaN HEMT devices still operate with 40% of the maximum power dissipated (Figure 13).

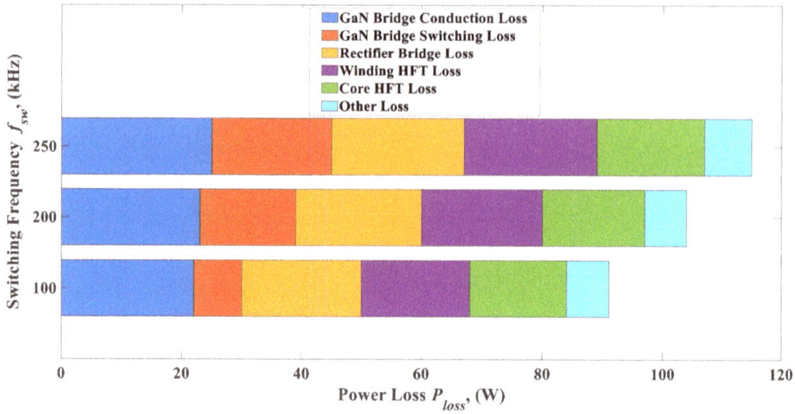

Figure 13. Detailed loss distribution of the proposed converter at 5 kW and different switching frequencies.

5. Conclusions

In this paper, a design methodology for achieving a soft switching phase shift full bridge converter using GaN HEMT devices is presented. Using the parasitics capacitance of the GaN HEMT, a soft switching ZVZCS PSFB converter is proposed using the power transformer's parasitic leakage inductance.

The main benefits of using the GaN HEMT device in such topology are the elimination of the freewheeling circulating current, the clamping of the overshoot voltage across the primary and secondary bridges, and the obtaining of soft switching in a wide load power range without using additional auxiliary circuits or an additional high-frequency inductor. Furthermore, in this study, the design and optimization of a high-frequency, high-power transformer are presented. In this context, using different conductor types (litz wire and copper foil) shows good thermal and electrical properties.

A 100–250 kHz, 400/600 V, 5 kW GaN-based PSFB prototype was built and tested, which showed a good dynamic performance in the wide load power range and an efficiency of up to 98.18% for the whole converter.

Funding: This research received no external funding.

Data Availability Statement: Not applicable.

Conflicts of Interest: The authors declare no conflict of interest.

References

1. Kimoto, T.; Cooper, J.A. *Fundamentals of Silicon Carbide Technology: Growth, Characterization, Devices and Applications*; Wiley: Singapore, 2014; ISBN 1118313550.
2. Suganuma, K. *Wide Bandgap Power Semiconductor Packaging: Materials, Components, and Reliability*; Woodhead Publishing: Oxford, UK, 2018; ISBN 9780081020944.
3. Eial Awwad, A. *On the Perspectives of SiC MOSFETs in High-Frequency and High-Power Isolated DC/DC Converters*; Universitätsverlag der TU Berlin: Berlin, Germany, 2020; ISBN 3798330964.
4. Morya, A.K.; Gardner, M.C.; Anvari, B.; Liu, L.; Yepes, A.G.; Doval-Gandoy, J.; Toliyat, H.A. Wide Bandgap Devices in AC Electric Drives: Opportunities and Challenges. *IEEE Trans. Transp. Electrific.* **2019**, *5*, 3–20. [CrossRef]
5. Ma, C.-T.; Gu, Z.-H. Review of GaN HEMT Applications in Power Converters over 500 W. *Electronics* **2019**, *8*, 1401. [CrossRef]
6. Bu, H.; Cho, Y. GaN-based Matrix Converter Design with Output Filters for Motor Friendly Drive System. *Energies* **2020**, *13*, 971. [CrossRef]
7. Eial Awwad, A. Dynamic Performance Enhancement of a Direct-Driven PMSG-Based Wind Turbine Using a 12-Sectors DTC. *WEVJ* **2022**, *13*, 123. [CrossRef]

8. Janabi, A.; Wang, B. Switched-Capacitor Voltage Boost Converter for Electric and Hybrid Electric Vehicle Drives. *IEEE Trans. Power Electron.* **2020**, *35*, 5615–5624. [CrossRef]
9. Shamsi, P.; Fahimi, B. Stability Assessment of a DC Distribution Network in a Hybrid Micro-Grid Application. *IEEE Trans. Smart Grid* **2014**, *5*, 2527–2534. [CrossRef]
10. Avila, A.; Lucu, M.; Garcia-Bediaga, A.; Ibarguren, U.; Gandiaga, I.; Rujas, A. Hybrid Energy Storage System Based on Li-Ion and Li-S Battery Modules and GaN-Based DC-DC Converter. *IEEE Access.* **2021**, *9*, 132342–132353. [CrossRef]
11. Yadlapalli, R.T.; Kotapati, A.; Kandipati, R.; Balusu, S.R.; Koritala, C.S. Advancements in energy efficient GaN power devices and power modules for electric vehicle applications: A review. *Int. J. Energy Res.* **2021**, *45*, 12638–12664. [CrossRef]
12. Yu, G.; Choi, S. An Effective Integration of APM and OBC With Simultaneous Operation and Entire ZVS Range for Electric Vehicle. *IEEE Trans. Power Electron.* **2021**, *36*, 10343–10354. [CrossRef]
13. Harini, S.; Chellammal, N.; Chokkalingam, B.; Mihet-Popa, L. A Novel High Gain Dual Input Single Output Z-Quasi Resonant (ZQR) DC/DC Converter for Off-Board EV Charging. *IEEE Access.* **2022**, *10*, 83350–83367. [CrossRef]
14. ElMenshawy, M.; Massoud, A. Modular Isolated DC-DC Converters for Ultra-Fast EV Chargers: A Generalized Modeling and Control Approach. *Energies* **2020**, *13*, 2540. [CrossRef]
15. Joo, D.-M.; Lee, B.-K.; Kim, D.-S.; Kim, J.-S.; Kim, H.-J. Analysis of GaN HEMT-based phase shifted full bridge dc-dc converter. In Proceedings of the 2015 IEEE International Telecommunications Energy Conference (INTELEC), Osaka, Japan, 18–22 October 2015; pp. 1–5. [CrossRef]
16. Awwad, A.E.; Badawi, N.; Dieckerhoff, S. Efficiency analysis of a high frequency PS-ZVS isolated unidirectional full-bridge DC-DC converter based on SiC MOSFETs. In Proceedings of the 2016 18th European Conference on Power Electronics and Applications (EPE'16 ECCE Europe), Karlsruhe, Germany, 5–9 September 2016; pp. 1–10. [CrossRef]
17. Kim, Y.-D.; Cho, K.-M.; Kim, D.-Y.; Moon, G.-W. Wide-Range ZVS Phase-Shift Full-Bridge Converter with Reduced Conduction Loss Caused by Circulating Current. *IEEE Trans. Power Electron.* **2013**, *28*, 3308–3316. [CrossRef]
18. Ruan, X.; Yan, Y. A novel zero-voltage and zero-current-switching PWM full-bridge converter using two diodes in series with the lagging leg. *IEEE Trans. Ind. Electron.* **2001**, *48*, 777–785. [CrossRef]
19. Shi, K.; Zhang, D.; Zhou, Z.; Zhang, M.; Gu, Y. A Novel Phase-Shift Dual Full-Bridge Converter with Full Soft-Switching Range and Wide Conversion Range. *IEEE Trans. Power Electron.* **2016**, *31*, 7747–7760. [CrossRef]
20. Ilic, M.; Maksimovic, D. Phase-Shifted Full Bridge DC-DC Converter with Energy Recovery Clamp and Reduced Circulating Current. In Proceedings of the APEC 07-Twenty-Second Annual IEEE Applied Power Electronics Conference and Exposition, Anaheim, CA, USA, 25 February–1 March 2007; pp. 969–975. [CrossRef]
21. Mweene, L.H.; Wright, C.A.; Schlecht, M.F. A 1 kW 500 kHz front-end converter for a distributed power supply system. *IEEE Trans. Power Electron.* **1991**, *6*, 398–407. [CrossRef]
22. Cho, J.-G.; Baek, J.-W.; Jeong, C.-Y.; Yoo, D.-W.; Joe, K.-Y. Novel zero-voltage and zero-current-switching full bridge PWM converter using transformer auxiliary winding. *IEEE Trans. Power Electron.* **2000**, *15*, 250–257. [CrossRef]
23. Gautam, D.S.; Musavi, F.; Eberle, W.; Dunford, W.G. A Zero-Voltage Switching Full-Bridge DC—DC Converter with Capacitive Output Filter for Plug-In Hybrid Electric Vehicle Battery Charging. *IEEE Trans. Power Electron.* **2013**, *28*, 5728–5735. [CrossRef]
24. Erickson, R.W.; Maksimović, D. *Fundamentals of Power Electronics*, 2nd ed.; Kluwer Academic: New York, NY, USA, 2001; ISBN 9780306480485.
25. Cremasco, A.; Rothmund, D.; Curti, M.; Lomonova, E.A. Voltage Distribution in the Windings of Medium-Frequency Transformers Operated with Wide Bandgap Devices. *IEEE J. Emerg. Sel. Top. Power Electron.* **2022**, *10*, 3587–3602. [CrossRef]
26. Liu, R.; Mi, C.C.; Gao, D.W. Modeling of Eddy-Current Loss of Electrical Machines and Transformers Operated by Pulsewidth-Modulated Inverters. *IEEE Trans. Magn.* **2008**, *44*, 2021–2028. [CrossRef]
27. Kazimierczuk, M.K. *High-Frequency Magnetic Components*, 2nd ed.; Wiley: Chichester, UK, 2014; ISBN 9781118717790.
28. Lidow, A.; de Rooij, M.; Strydom, J.; Reusch, D.; Glaser, J. *GaN Transistors for Efficient Power Conversion*, 3rd ed.; Wiley: Hoboken, NJ, USA, 2020; ISBN 9781119594376.

energies

Article

Performance Analysis and Evaluation of Multiphase Split-Source Inverters

Sherif M. Dabour [1,2,*], Ahmed A. Aboushady [2], I. A. Gowaid [2,3], Mohamed. A. Elgenedy [2] and Mohamed E. Farrag [2]

1 Department of Electrical Power and Machines Engineering, Faculty of Engineering, Tanta University, Tanta 31733, Egypt
2 Electrical Power Engineering Department, School of Computing, Engineering and Built Environment, Glasgow Caledonian University, Glasgow G4 0BA, UK
3 Electrical Engineering Department, Faculty of Engineering, Alexandria University, Alexandria 21544, Egypt
* Correspondence: sherif.dabour@gcu.ac.uk

Abstract: Due to their many advantages over their counterparts, such as Z-source inverters (ZSIs), split-source inverters (SSIs) have recently received much attention as single-stage boost inverters. This paper discusses a multiphase version of the SSI topology for the first time. Among multiphase systems undergoing a revolution in the research area, five-phase motor drives are a relatively practical selection in industrial applications. Therefore, this paper focuses on a five-phase SSI as an example. The topology, operating principles, modulation techniques, and performance analysis of the analyzed topology are introduced. A modified space-vector modulation (MSVM) scheme is developed to eliminate low-frequency ripples in the input current. There is also a detailed analysis and graphical evaluation of the output currents ripples using the space-vector approach. It is evident that multiphase SSI is suitable for motor drives, especially when a high-output voltage gain is required. In addition to having a nearly identical ripple in output current to a conventional VSI, it has the benefit of performing the boosting action in a single stage with fewer passive components and a low ripple in input current. Finally, the simulation and experimental results have been conducted to demonstrate the viability of the multiphase SSI studied in the theoretical study and analysis.

Keywords: split-source inverter; multiphase systems; five-phase inverters; modified space-vector modulation; DC and AC ripples analysis

Citation: Dabour, S.M.; Aboushady, A.A.; Gowaid, I.A.; Elgenedy, M.A.; Farrag, M.E. Performance Analysis and Evaluation of Multiphase Split-Source Inverters. *Energies* **2022**, *15*, 8411. https://doi.org/10.3390/en15228411

Academic Editors: Diego Bellan and Jelena Loncarski

Received: 16 October 2022
Accepted: 7 November 2022
Published: 10 November 2022

1. Introduction

Recently, multiphase (more than three-phase) machines have begun to replace their conventional AC counterparts because of their high reliability, inherent torque ripple reduction, and resulting reduced power electronic devices required for operation [1]. As a result of these advantages, they are great candidates for a variety of critical applications, including variable speed drives (VSDs), wind energy, and ship propulsion [1,2]. Amongst multiphase machines, the five-phase motor may produce the lowest copper loss and total loss [3]. In most cases, voltage source (VSIs) or current source inverters (CSIs), along with matrix converters (MCs), are employed as power supplies for five-phase motor drives, as shown in Figure 1 [4–6].

As VSIs and matrix converters only operate in bucking mode, additional circuits are required to increase voltage gain [4,6]. As a result, these circuits increase the cost of the system and reduce its overall efficiency [5]. CSIs operate in boosting mode, but this is complex to implement and requires more passive components [5]. As a result of their superior performance in terms of system complexity, size, weight, and the ability to perform both bucking and boosting actions, single-stage topologies have replaced conventional architectures. Five-phase Z- and quasi-Z-source inverters (ZSI/qZSI), shown in Figure 2, have been gaining increasing attention in recent years [7,8]. This overcomes the conceptual and theoretical shortcomings of the traditional VSIs and CSIs. Compared to VSD systems that are based on VSIs and CSIs, the ZSI offers additional advantages [9,10]:

- Regardless of the input voltage, the voltage gain, *G*, can be improved by producing any desired AC output voltage, even that greater than the line voltage.
- No additional circuits are needed to provide ride-through during voltage sags.
- This reduces the line's common-mode voltage and harmonic current while improving the power factor.

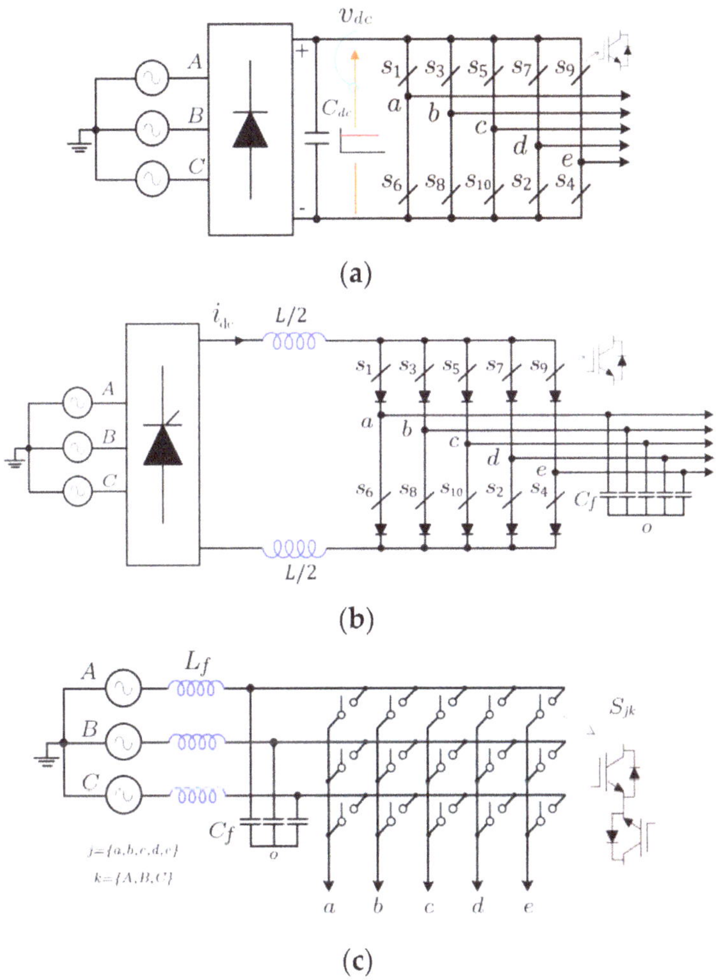

Figure 1. Basic power converters for five-phase drives. (**a**) Five-phase voltage-source inverter (VSI) [4]. (**b**) Five-phase current-source inverter (CSI) [5]. (**c**) Three-phase to five-phase matrix converter (MC) [6].

Many publications discuss the modulation techniques of ZSIs for the different number of phases and different applications [11–14]. It can be mainly classified into three methods: (1) simple-boost (SB), (2) maximum-constant-boost (MCB), and (3) maximum-boost (MB) control methods. The SB control scheme produces lower voltage gains than the MB control scheme. Meanwhile, the MB method requires more passive elements in the impedance network because the shoot-through (ST) duty ratio periodically varies. Furthermore, this variation increases the peak inductor current and capacitor voltage. An improved maximum-constant boost (MCB) modulation method obtains constant ST variation with a higher boosting factor than SB modulation. Despite the merits of multiphase ZSIs, their output voltage magnitude varies accurately in a narrow range by controlling the shoot-through (ST) duty ratio.

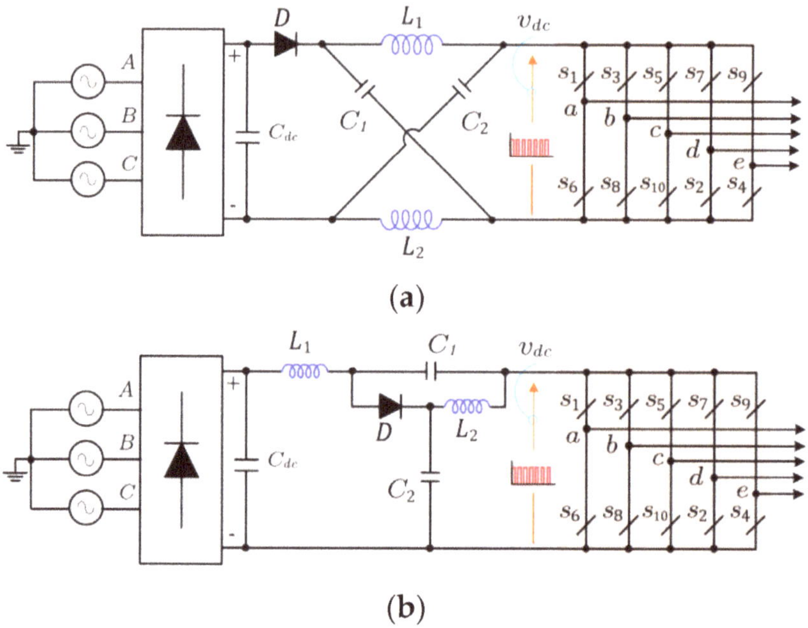

Figure 2. Z-source topologies for five-phase drives. (**a**) Basic five-phase Z-source inverter [7]. (**b**) Five-phase quasi-Z-source inverter [8].

Recently, the split-source inverter (SSI) represents an alternative to the basic ZSI and qZSI topologies. It has been first invented by [15] for a single-phase system and developed in [16] for the three-phase topology, as shown in Figure 3. The potential advantages of SSI over ZSI can be summarized as follows [9]:

- It has a continuous input current and DC-link voltage.
- It has lower voltage stresses across the switches for high-voltage gains.
- It requires a lower passive component count.
- It uses the same switching states as the VSI, which uses a different state for ST.
- Compared with the two-stage architecture, no additional active switch is required in the SSI.

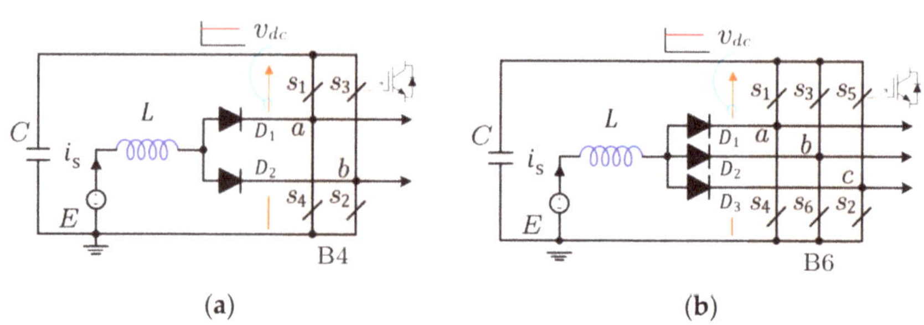

Figure 3. Basic SSI topology, (**a**) single-phase SSI [15], and (**b**) three-phase SSI [16].

Several topologies have been developed with the idea of the basic SSI to achieve higher boosting capabilities [17–19], multilevel technologies [20–22], and multiport output inverters [23,24]. Although the reduction in the number of switch counts and passive elements in these architectures leads to a reduction in the volume and weight of the inverters, it suffers from some demerits, such as rapid commutations for the input diodes and higher current stresses for the lower switches [16,19]. The authors of this paper, however, have not conducted any extensive research to explore and extend the concept of basic SSI to multiphase systems based on their best knowledge. Therefore, this paper aims to fill this gap in the multiphase drive architecture by extending the basic SSI to n-phase designs. A five-phase SSI topology is fully explored in this paper by analytical, simulation, and experimental studies. Moreover, a modified space-vector modulation (MSVM) scheme is proposed to ensure a boosting action with sinusoidal output voltages and to improve the switching characteristic of both the input and output sides of the SSI. The proposed MSVM scheme produces constant charging and discharging intervals for certain modulation indexes to minimize the input current ripples. Accordingly, the required passive elements (inductor and capacitor) for the boosting action are reduced. This paper also assesses the quality of output current waveforms due to the MSVM scheme and compares it with the classical VSI. It is important to note that the output current ripple affects the motor copper losses at the switching frequency, the motor torque ripple, the load current total harmonic distortion (THD), and the insulation breakdown. It should be minimized to improve the drive-system efficiency and lifetime [12].

2. Review of Three-Phase Split-Source Inverters

2.1. Topology

The three-phase SSI topology is shown in Figure 3b [16]. This topology is acquired by combining a DC–DC boost converter into the VSI bridge (B6). This can be done by connecting a boost inductor, L, to the midpoint of each leg of the inverter (a, b, and c) via the forward diodes (D_1–D_3).

2.2. Operation

Since the three-phase SSI utilizes the B6 bridge of the standard VSI, it uses the same switching vectors ($V_o - V_7$), shown in Figure 4. The inductor, L, and the diodes of Figure 3 are used to boost the supply voltage to the desired DC-link voltage at the capacitor, C_{dc}. It is important to note that the three-phase SSI uses the switching vectors $V_o - V_6$ to charge the input inductor, while the remaining zero vector V_7 is used to discharge the inductive energy to the DC link capacitor. The equivalent circuits of both inductive charging and discharging modes are shown in Figure 5.

(1) *Inductive Charging Mode*: The charging mode can be obtained by switching ON at least one of the lower switches in the B6 bridge. In this case, the voltage at the output of the DC-side is zero (i.e., $v_{inv} = 0$) due to the short circuit, as shown in Figure 5a. As a result, the inductor L is charged during an interval of T_{ch}.

(2) *Inductive Discharging Mode*: The discharging mode occurs when all of the upper switches in the B6 bridge are turned ON (i.e., $V_7\{111\}$). In this case, the inductor L will release its energy in the capacitor C via the antiparallel diodes of the upper switches during an interval of d_{dis}, where $T_{dis} = T_7 = T_s - T_{ch}$, T_s is the sampling time. Hence, the voltage at the DC side equals the capacitor voltage (i.e., $v_{inv} = v_{dc}$), as shown in Figure 5b.

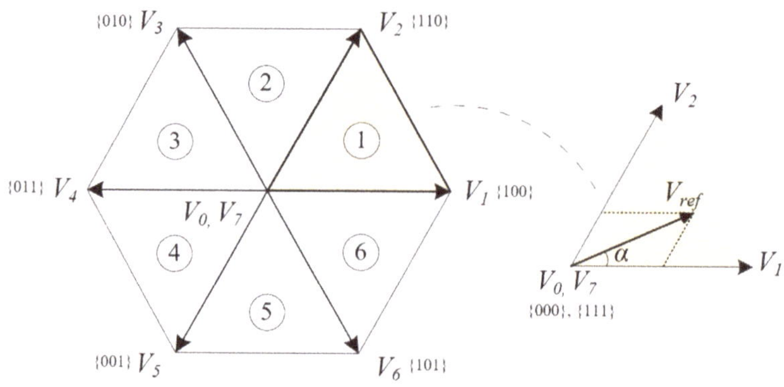

Figure 4. Switching state representation of three-phase SSI based on space-vector approach.

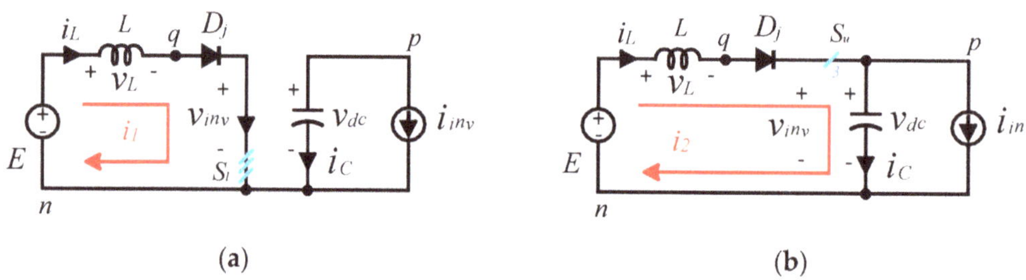

Figure 5. Mode of operation of three-phase SSI [14]. (**a**) Basic SSI topology (for states V_0–V_6). (**b**) Inductive discharging (for state V_7).

2.3. DC Boosting Factor

The DC-boosting factor, β, which represents the ratio between the DC-link voltage, v_{dc} (that is, the capacitor voltage), to the supply voltage, β, can be expressed as follows [16]:

$$\beta = \frac{1}{(1-D_{av})} \tag{1}$$

where D_{av} is the average value of the charging duty cycle ($d_{ch} = T_{ch}/T_s$), it can be determined from $D_{av} = 1/T_s \int_0^{T_s} T_{ch} dt$.

2.4. Modulation

Several modulation schemes have been presented with the idea of level shifting to obtain the boosting and inversion processes for three-phase SSIs [16–19]. The level shifting leads to wider boosting capabilities [19]. The commonly used schemes for SSI are reviewed and compared in this section. The schemes are level-shifted sinusoidal PWM (SM), third-harmonic injection PWM (THM), symmetrical PWM (SYM), and modified space-vector PWM (MSVM), which are fully discussed in [16,25].

(1) *Mathematical Relations*: Considering the linear modulation operating mode, a general representation of the modulating signals, u_k^*, for all of the schemes using the carrier-based PWM modulators is governed by

$$u_k^* = u_k + u_{zss} \tag{2}$$

where u_{zss} is the zero-sequence signal that gives us the degree of freedom to obtain the different modulation schemes; $k \in \{a, b, c\}$ and u_k are the basic sinusoidal three-phase modulating signals, and they can be derived from

$$u_k = \begin{bmatrix} u_a \\ u_b \\ u_c \end{bmatrix} = M \begin{bmatrix} \sin(\vartheta) \\ \sin(\vartheta - 2\pi/3) \\ \sin(\vartheta + 2\pi/3) \end{bmatrix} \tag{3}$$

where $\vartheta = \omega t$ is the fundamental angular frequency, and M is the modulation index, which can be determined from

$$M = \mu \hat{V}_o / V_{dc} \quad \forall \quad 0 < M < 1 \tag{4}$$

where $\hat{V}_o$ is the maximum value of the fundamental component of the output-phase voltage, V_{dc} is the average DC-link voltage (capacitor voltage), and μ is a modulation coefficient, which equals 2 for SM and $\sqrt{3}$ for THM, SYM, and MSVM schemes.

The duty cycles of the reference modulating signal, d_k^*, which are compared with the up–down carrier wave to obtain the gating pulses of the SSI, can be determined from

$$d_k^* = 1/2(1 + u_k^*) \quad \& \quad 0 \le d_k^* \le 1. \tag{5}$$

Figure 6 shows the duty cycles d_k^* of the selected modulation schemes. It can be observed that the charging duty cycle, d_{ch}, is time-variant in the SM and SYM schemes, while it is fixed during the switching period in the MSVM and RSVM schemes. As a result, ripples with low-order harmonics will appear in the DC-link voltage and inductor current when SM and SYM are employed. So, from the point of view of the input current and DC-link voltage ripples, the MSVM and RSVM schemes are the best choices for SSIs.

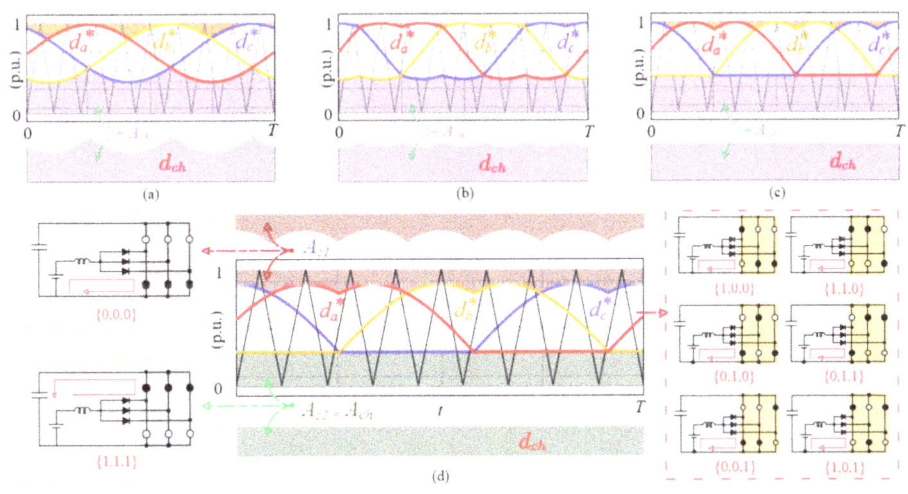

Figure 6. Modulation of three-phase split-source inverter based on level-shifted approach. (**a**) SM, (**b**) SYM, (**c**) MSVM, (**d**) and RSVM scheme. The superscript * denotes the reference waveform.

2.5. Design Procedure

The inductor current and capacitor voltage ripples (ΔI_L, ΔV_{dc}) are crucial factors in the design procedure. When using MSVM and RSVM schemes, there is only a high-switching ripple based on the switching frequency, and the value of L and C are determined from [19]

$$L = \frac{d_{ch} E}{f_{sw} \Delta I_L} \quad \& \quad C = \frac{(1 - d_{ch}) I_L}{f_{sw} \Delta V_{dc}} \tag{6}$$

where f_{sw} is the switching frequency.

It is worth noting that the value of the required passive elements L and C are increased in the case of SM and SYM schemes to mitigate the effect of low-order harmonics and obtain the same current ripples.

After reviewing the three-phase SSI, in the next sections, this paper extends the concept of the basic three-phase SSI to the five-phase case as an example of the n-phase SSI topologies.

3. Proposed Five-Phase Split-Source Inverters

The five-phase SSI topology is shown in Figure 1. In this case, five forward diodes are used to combine the DC-boosting inductance into the multiphase inverter bridge, as shown in Figure 7. The multiphase SSI topology is used here to supply a star-connected balanced inductive load, as shown in Figure 7. The circuit is supplied by a DC-supply voltage e, which is supposed to be almost constant ($e \approx E$).

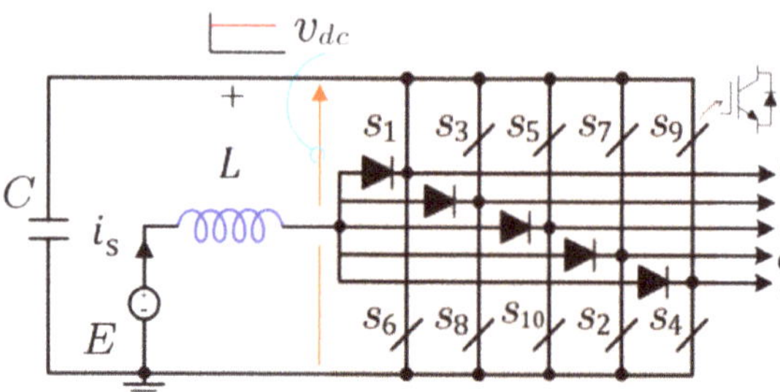

Figure 7. Five-phase SSI topology.

As known, there are 32 (2^n, $n = 5$) switching action combinations ($V_o - V_{31}$) for the 5-phase VSI, from which two are the zero vectors (V_o and V_{31}) and the remaining are active, as shown in Figure 8. In the presented SSI topology, the switching vectors $V_o - V_{30}$ are used for inductive charging, while the remaining zero vector V_{31} is used to release inductive energy. The equivalent circuits of inductive charging and discharging modes are shown in Figure 9a,b.

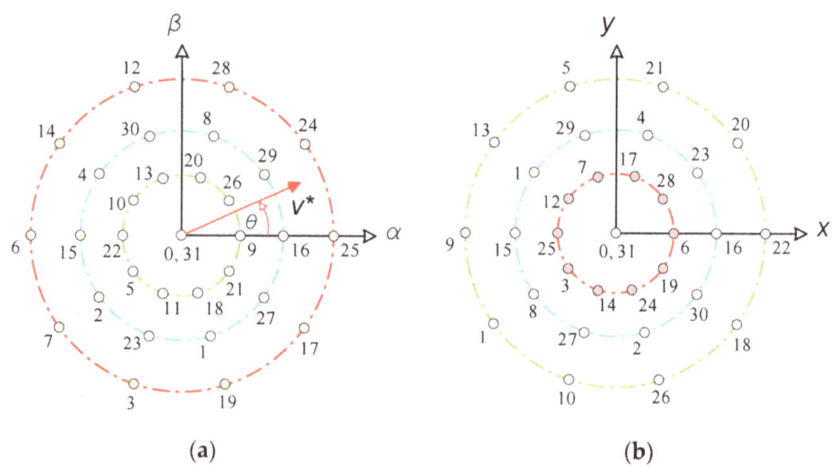

Figure 8. Space-vector model five-phase SSI. (**a**) $\alpha - \beta$ subspace. (**b**) $x - y$ subspace.

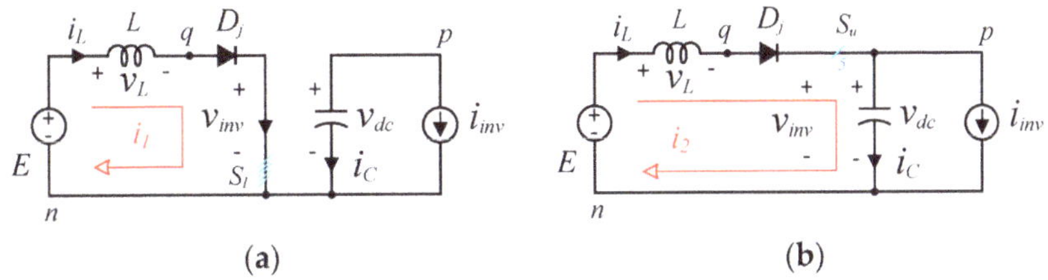

Figure 9. Operating modes of five-phase SSI. (**a**) Inductive charging (for states 0–30). (**b**) Inductive discharging (for state 31).

Among the best-known and reviewed PWM schemes of the 3-phase SSI is the MSVM scheme, which is developed here to control the operation of the 5-phase SSI. The modulating signals for the case of five-phase SSI can be mathematically represented as

$$v_j = v_j^* + v_{zss} \tag{7}$$

where v_j is the duty cycles of the modulating signals, and $j = a, b, \ldots, e$ and v_j^* are the sinusoidal reference signals, which are shifted by $2\pi n/5$ as given in (4)

$$v_j^* = k_2 M \cos\left(\omega t - \frac{2\pi n}{5}\right),\ n \in \{0-4\} \tag{8}$$

where k_2 is a constant equal to $1/(2\sin(2\pi/5))$ and ω is the line frequency in rad/s, while v_{zss} in (7) denotes the zero-sequence signal (ZSS) that is selected to obtain different SVM schemes, such as sinusoidal modulation (SM), fifth harmonic injection modulation (FHM), symmetrical SVM (SYM), and MSVM. The ZSS, in each case, is governed by

$$v_{zss} = \begin{cases} D_{av} & \forall \text{ SM} \\ D_{av} - k_2 M \cos\omega t & \forall \text{ FHM} \\ D_{av} - \frac{1}{2}k_2(v_{\max} + v_{\min}) & \forall \text{ SYM} \\ D_{av} - k_2 v_{\min} & \forall \text{ MSVM} \end{cases} \tag{9}$$

where v_{max} and v_{min} are the maximum and minimum envelop of all the sinusoidal reference signals of (9), respectively, while M is the modulation index. It is worth noting that the modulation index of any modulation technique should be limited by one.

This paper focuses on the MSVM scheme, which leads to a constant DC-side current ripple. For the sake of illustration, Figure 10 shows the duty cycles of the modulating signals of the MSVM scheme of five-phase SSI, which is governed by (7)–(9) for M = 0.75. As observed in the first sector of Figure 10, $v_a > v_b > v_e > v_c > v_d$. Therefore, the minimum voltage is v_d. Thus, the five-phase duty cycles are given by

$$\begin{cases} v_a = k_2\left(v_a^* - v_d^*\right) + 1 - M \\ v_b = k_2\left(v_b^* - v_d^*\right) + 1 - M \\ v_c = k_2\left(v_c^* - v_d^*\right) + 1 - M \\ v_d = 1 - M \\ v_e = k_2\left(v_e^* - v_d^*\right) + 1 - M \end{cases} \tag{10}$$

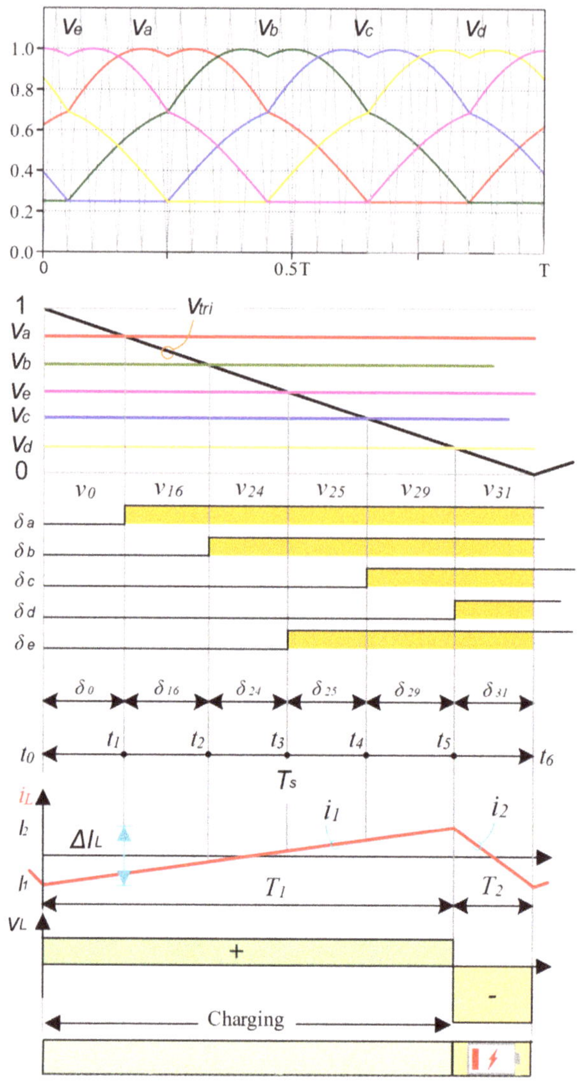

Figure 10. Duty cycles of the modulating signals of 5-phase SSI for $M = 0.75$ for one complete period and inductor current and voltage waveforms for sector 1.

Figure 10 also shows the generation mechanism of the switching pattern in sector 1 of the five-phase SSI. Due to the symmetry, only the first half of the switching period is explained. The duty cycles of the resulting active (δ_{16}, δ_{24}, δ_{25}, and δ_{29}) and zero vectors, δ_z (δ_0 and δ_{31}) in this switching pattern can be determined based on the volt-second balance concept by (11). Figure 11 shows the variations of these duty cycles for sector 1.

$$\begin{cases} \delta_0 = M(1 - \sin(\omega t + 2\pi/5)) \\ \delta_{16} = k_2 LM \sin(\pi/5 - \omega t) \\ \delta_{24} = M \sin(\omega t) \\ \delta_{25} = M \sin(\pi/5 - \omega t) \\ \delta_{29} = k_2 LM \sin(\omega t) \\ \delta_{31} = 1 - M \\ \delta_z = \delta_0 + \delta_{31} \end{cases} \tag{11}$$

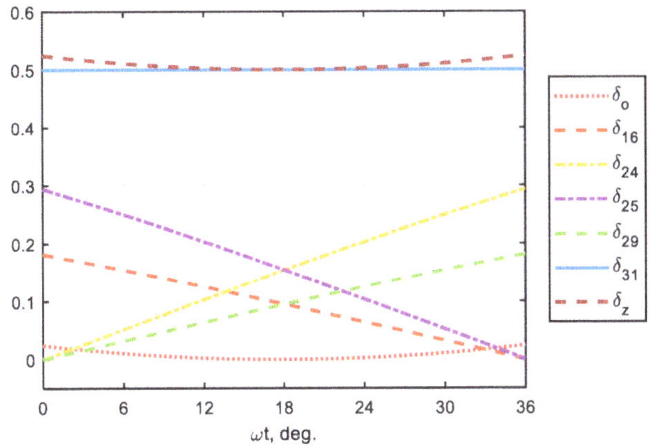

Figure 11. The variation of duty cycles during sector −1 for M = 0.5.

It is worth seeing from (11) and Figure 10 that the duty cycle of the discharging mode, δ_{31}, is constant during the sector. This will also lead to obtain a constant duty cycle of the charging, D, which is equal to the sum of all remaining duty cycles in a switching period and can be determined by

$$D = \delta_0 + \delta_{16} + \delta_{24} + \delta_{25} + \delta_{29} = 1 - \delta_{31}. \tag{12}$$

Hence,

$$D_{av} = D = M. \tag{13}$$

From (13), the DC-boosting and AC inverter gain are

$$\beta = \frac{1}{1-M} \quad \& \quad G = \frac{M}{2\sin(2\pi/5)(1-M)} \tag{14}$$

Figure 12 shows a graphical correlation between the modulation index and the DC-boosting factor for the same voltage gains. It can be observed that the five-phase SSI will operate at a wide range of modulation indices and has a directly proportional relationship between the gain and modulation index, i.e., a higher modulation index leads to greater gains. Having a lower THD of the output is one of the significant advantages of this inverter.

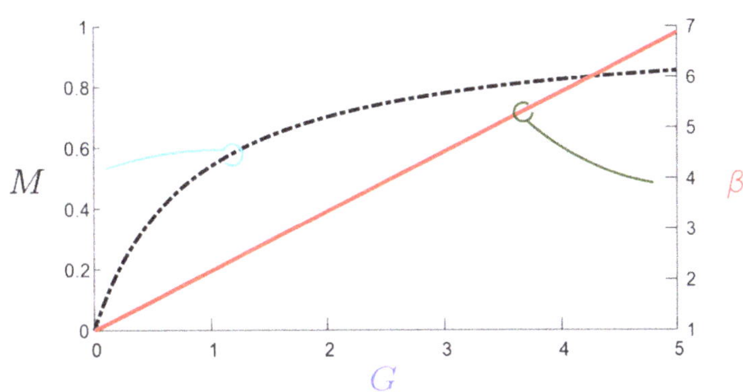

Figure 12. Modulation index and boosting factor versus the AC-gain of the MSVM scheme for the five-phase SSI.

4. Current Ripple Analysis of Five-Phase SSI

This section introduces an analytical analysis of the DC-input current ripple, error-voltage vector, and output harmonic distortion flux (HDF), which is related to the output current ripple of five-phase SSI based on the MSVM scheme.

4.1. DC Ripple and Design of Passive Components

The inductor currents ripple of the five-phase SSI topology can be evaluated as in the basic DC–DC boost-converter. This is owed to fixing the charging and discharging intervals (T_1 and T_2) of a given modulation index, which is obtained by applying the MSVM scheme. During the inductive charging mode of Figure 9a, the inductor current is given by

$$i_1 = \frac{E}{L} \cdot t + I_1 \tag{15}$$

where I_1 is the initial current in charging mode. During this mode, the inductor current rises, and the inductor voltage will be positive, as shown in Figure 10.

The inductor current during the discharging mode of Figure 9b is given by

$$i_2 = \frac{(E - v_{dc})}{L} \cdot t + I_2, \tag{16}$$

where I_2 is the initial current in the discharging mode. For a stable operation, the inductor current must fall, and therefore the inductor voltage goes to a negative value, as shown in Figure 10.

Assuming the high-frequency switching operation, the peak-to-peak inductor current ripple, and the DC-link capacitor voltage ripples as follows:

$$\begin{cases} \Delta I_L = I_2 - I_1 = \frac{E}{L}\frac{D_{av}}{f_s} = \frac{E}{L}\frac{M}{f_s}, \\ \Delta V_C = I_{dc}\frac{1-D_{av}}{Cf_s} = I_{dc}\frac{1-M}{Cf_s}. \end{cases} \tag{17}$$

It can be observed from (17) that the inductor current and the capacitor voltage ripples are inversely proportional to their respective (L/C) values and the switching frequency.

4.2. AC Current Ripple

The AC ripples can be evaluated using both the time domain and the space-vector domain [26,27]. In the time-domain analysis, the output current ripple, Δi_o, is determined from

$$\Delta i_o(t) = i_o(t) - i_{o1}(t) \tag{18}$$

where i_o is the actual output current and i_{o1} is the fundamental component of the sinusoidal output current. Figure 13 shows the graphical waveforms of these currents.

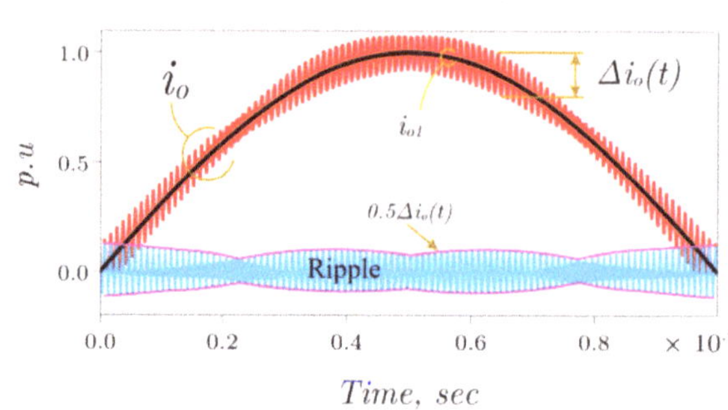

Figure 13. Evaluation of AC output current ripple using the time-domain analysis.

On the other hand, the analysis of the peak current ripples in this paper is based on the space-vector approach, which gives a quantitative evaluation of the current ripple [4,26,27]. Due to symmetry, only the first sector in the $\alpha\beta$ plane is examined. Figure 14a,b show how the error-voltage vectors between the reference $\overline{v}^*$ and the actual space vectors, $\overline{v}_k$ (k is the vector number), are generated in both $\alpha\beta$ and xy subspaces. It should be noticed that a zero-reference space vector is assumed in the xy plane to eliminate the low-order harmonics from the output voltage waveform. Additionally, it can be observed that the error vectors depend on the index, M, and the angle θ and are defined as [4]

$$\Delta\overline{v}_k(M,\,\theta) = \overline{v}_k - \overline{v}^*(M,\,\theta) \tag{19}$$

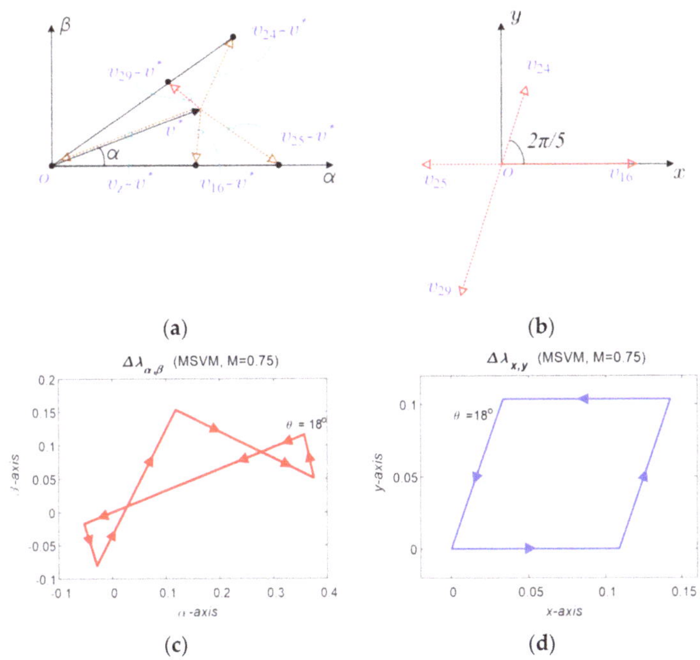

Figure 14. Generation of error voltage vector and harmonic-flux trajectory during the half of the switching cycle of five-phase SSI using MSVM scheme [26]. (**a**) Generated error-voltage vectors in $\alpha\beta$ subspace. (**b**) Generated error-voltage vectors in xy subspace. (**c**) Harmonic-flux trajectory for M = 0.75, $\theta = 18^\circ$ in $\alpha\beta$ plane. (**d**) Harmonic-flux trajectory for M = 0.75, $\theta = 18^\circ$ xy plane.

In this approach, the output current ripple, Δi_o, is evaluated from the harmonic distortion flux (HDF), $\Delta\overline{\lambda}$, as follows [4]

$$\Delta i_o = (1/L_\sigma)\Delta\overline{\lambda} \tag{20}$$

where L_σ is the equivalent motor inductance, which can be determined in the case of induction motor drives from the stator leakage inductance, L_1; the rotor leakage inductance, L_2; and the magnetization inductance, L_m, as follows [26]:

$$L_\sigma = L_1 + \frac{L_2 L_m}{L_2 + L_m} \tag{21}$$

The HDF, $\Delta\overline{\lambda}$, is defined as the error vector between the reference and actual space vectors during one switching period. It can be also called the harmonic-flux trajectory and is determined from

$$\Delta\overline{\lambda_j} = \Delta\overline{v}_k(M,\,\theta)\cdot\Delta t \tag{22}$$

where Δt is the switching period and $\overline{v}^*$, $\overline{v}$ are the reference and actual switching space vectors, respectively. It is worth noting that $\Delta\overline{\lambda}$ is time-dependent, and it is a function of M and θ. In the case of five-phase inverters, the HDF should be calculated for both $\alpha\beta$ and xy subspaces.

Assuming zero initial conditions for the harmonic flux at the beginning of the switching period, Figure 14c,d show the harmonic-flux trajectory in both subspaces for M = 0.75 and $\theta = 18^\circ$. To give a fair comparison, Figure 15 shows the trajectory of the harmonic flux for both five-phase VSI and SSI. It is worth noting that the VSI is modulated by the conventional SVM, while SSI is modulated by the MSVM scheme. Based on (22), the mean-square value of the HDF in the $\alpha\beta$ and xy planes over a sampling interval ($0 \leq t \leq T_s$) is

$$\Delta\overline{\lambda}^2_{abcde-MS}(M,\ \theta) = \int_0^{T_s} \Delta\overline{\lambda}_j^2\ dt \tag{23}$$

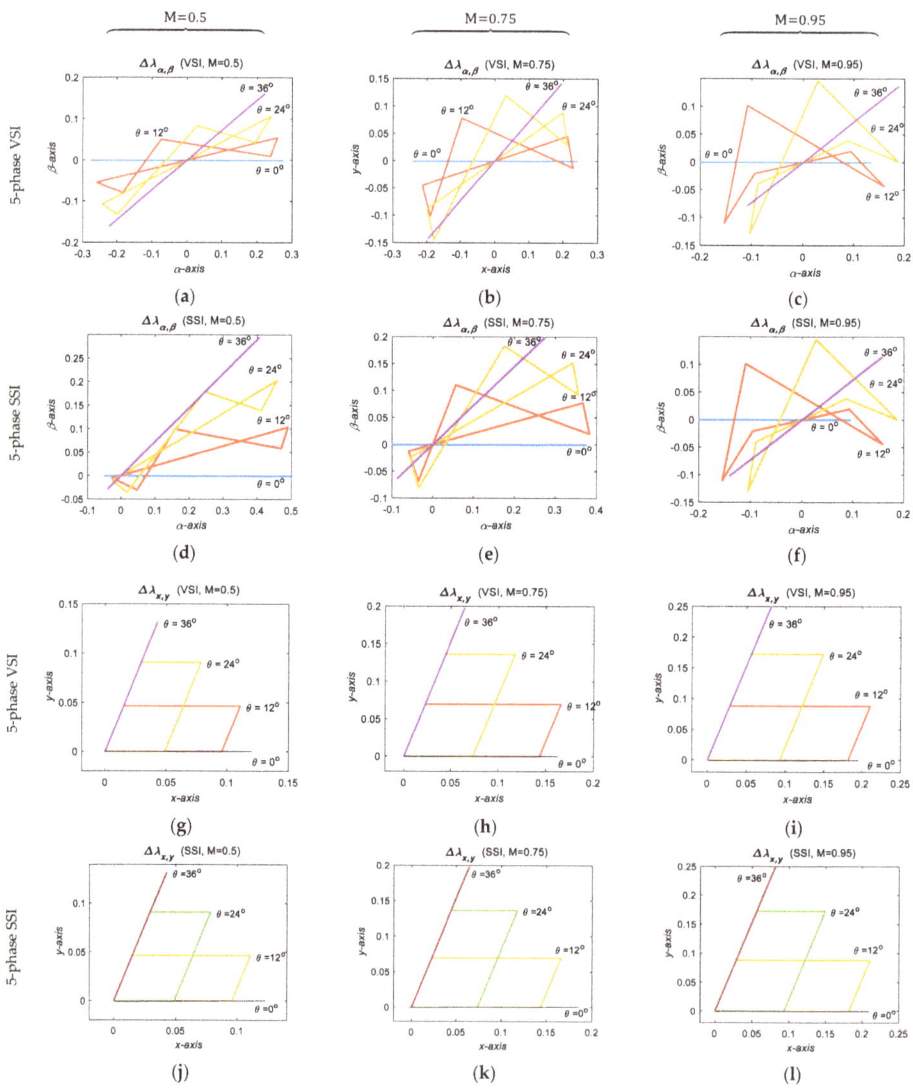

Figure 15. Error-voltage vector and harmonic-flux trajectory during the half of the switching cycle for VSI and SSI [26].

Hence, the per-fundamental cycle HDF, $\Delta\lambda^2_{abcde-MSf}$, is

$$\Delta\lambda^2_{abcde-MSf}(M) = \frac{5}{\pi}\int_0^{\pi/5} \Delta\lambda^2_{abcde-MS}\, d\theta. \tag{24}$$

The HDF for both the five-phase VSI and SSI topologies is shown in Figure 16. The HDF is highly affected by the modulation index and modulation scheme. Based on Figures 15 and 16, the following conclusions can be drawn:

- Due to the symmetrical distribution of the zero vectors in the classical SVM of VSI, symmetrical harmonic flux trajectories are observed from the $\alpha\beta$ plane of Figure 15a–c, while there are some asymmetries in the MSVM for SSI because of the level shift approach, as shown in Figure 15d–f. However, the same trajectories are obtained in the xy plane, as shown in Figure 15g–l. This is owed to the zero references on this plane.
- As shown in Figure 16a,b, the HDF is significantly greater for SSI than for VSI in the $\alpha\beta$-plane, particularly for low-modulation indices. However, in the xy-plane, the HDF is the same.
- Due to the elimination of the level shift effect on SSI modulating waveforms at M = 1, both VSI and SSI have approximately equal HDFs, as shown in Figure 16c.
- The SSI produces higher current ripples than the traditional VSI, but since the SSI also boosts the voltage, it does not require an additional DC booster.

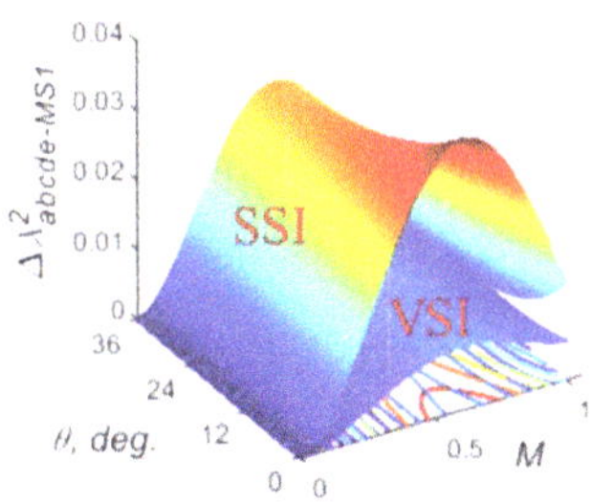

(a)

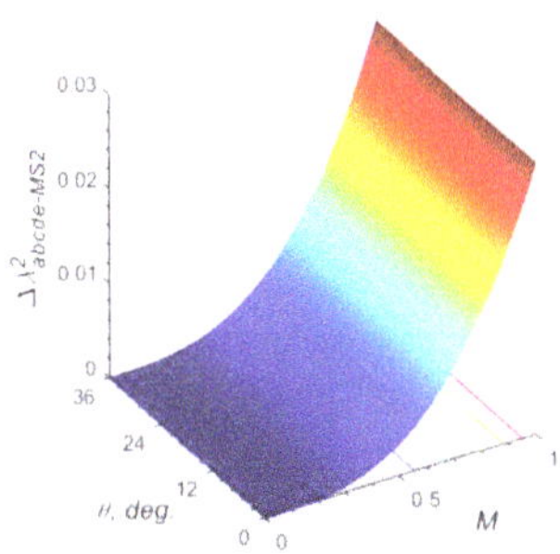

(b)

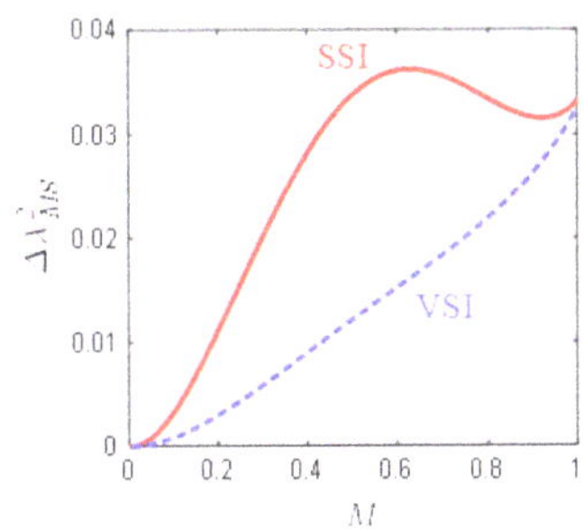

(c)

Figure 16. Harmonic distortion flux for both five-phase VSI and SSI [26]. (**a**) HDF in $\alpha\beta$ plane for both VSI and SSI. (**b**) HDF in xy plane for both VSI and SSI. (**c**) Total HDF for both VSI and SSI.

5. Simulation and Experimental Results

Based on the analysis, a PLECS simulation model has been developed for a five-phase SSI topology. The topology is powered by a fixed DC source of 45 V and loaded by a five-phase inductive load. Firstly, the simulation study aims to prove the concept of five-phase SSI by showing the input and output waveforms of the voltages and currents. The parameters of the simulation are listed in Table 1.

Table 1. Simulation parameters for five-phase SSI.

Parameters	Values	Parameters	Values
Input voltage	45 V	Input inductance	1.28 mH
DC-link capacitor	480 µF	RL-load	4.7 Ω, 5 mH
Modulation index	0.5	Switching frequency	15 kHz

Figure 17 shows the output-phase voltage, output currents, DC-link voltage, and inductor current waveforms for the case study. Additionally, the zoom window of Figure 17 illustrates these results from a microscopic perspective to illustrate the concept and the inductor current and capacitor voltage ripples.

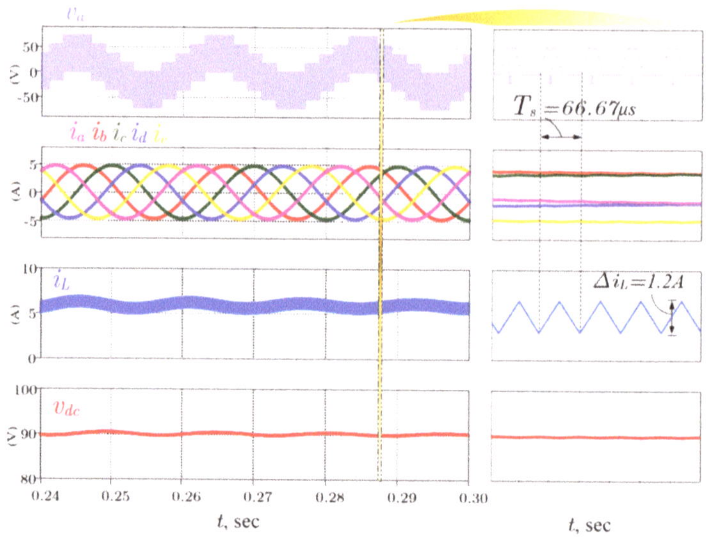

Figure 17. Simulated voltages and current waveforms of the five-phase SSI under the case study using PLECS.

Based on the simulation results, it can be observed that

- The simulated waveforms and measured values from PLECS for the five-phase SSI show good agreement with that estimated from the analysis. For example, the average DC-link voltage is 90V, and the input current ripple 1.2 A agrees with that calculated from (13) and (16), respectively, at the given condition of the case study.
- The output-phase currents are sinusoidal waves, and the input current is completely free from the low-order harmonics. Moreover, the DC-link capacitor voltage exhibits near-constant value.

Furthermore, the proposed five-phase SSI is compared with the standard two-stage five-phase inverter shown in Figure 18, which is composed of a boost converter followed by a five-phase VSI. To obtain a fair comparison, it is important to provide the same DC-link voltage to the inverter bridge for both topologies. Therefore, the voltage supplied to VSI is determined by the expected average voltage of the DC-link capacitor in the 5-phase SSI ($V_{dc} = \beta E$). In this case, the modulation index is set to 0.5 for both topologies, and the duty cycle for the boost converter stage in the conventional architecture is set to 0.5. An evaluation in terms of phase voltage and current ripples is introduced in Figure 19. As expected from the current ripple analysis of Section 4, the conventional VSI has a lower current ripple at $M < 1$ than the proposed SSI. However, the conventional VSI needs an extra converter stage.

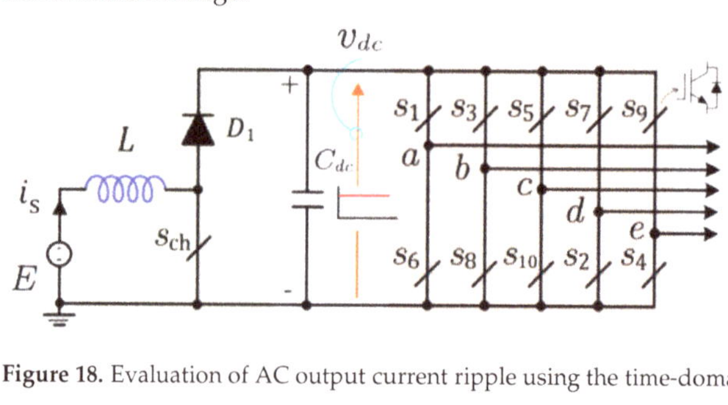

Figure 18. Evaluation of AC output current ripple using the time-domain analysis.

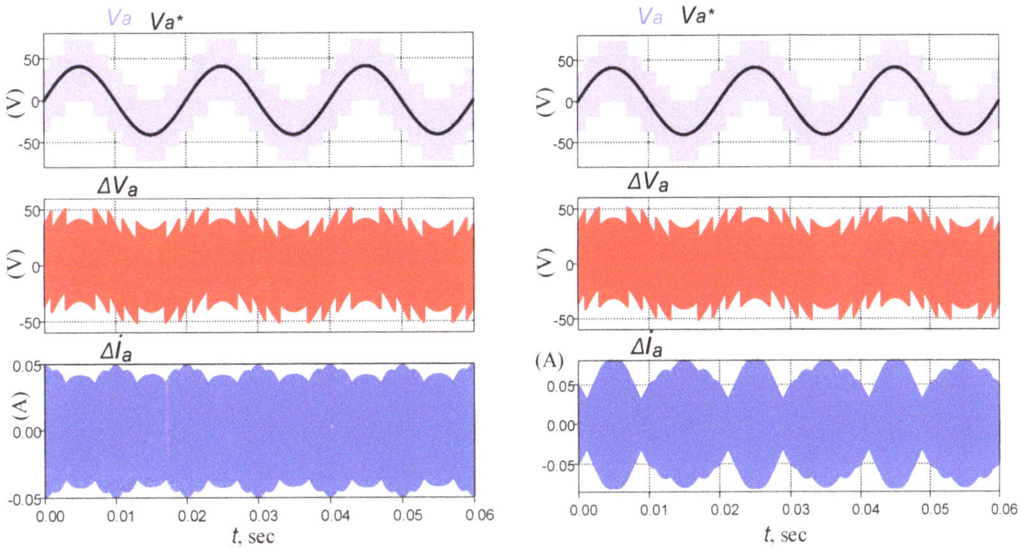

(**a**) (**b**)

Figure 19. The simulation waveform of output-phase voltage, error voltage, and the corresponding current ripples for both VSI and SSI topologies. (**a**) Boost-converter followed by five-phase VSI. (**b**) Proposed five-phase SSI.

6. Experimental Results

This section describes the experimental results conducted on the laboratory prototype, shown in Figure 20a, which is designed according to the presented procedure. A five-phase star-connected inductive load of 4.7 ohms and 5 mH is connected to the inverter output terminals. The whole experimental setup is shown in Figure 20b. During this study, a low-cost LaunchPad Delfino development kit (F28379D) was utilized to implement the modulation technique. To generate symmetrical triangular signals, the kit was programmed using the PLECS coder with a UP/DOWN counter at 15 kHz, while modulation signals have a modulation index of 0.5 and a frequency of 50 Hz.

(**a**) (**b**)

Figure 20. Photographs of the five-phase SSI prototype and experimental setup. (**a**) SSI. (**b**) Setup.

The measured voltage and current waveforms are recorded via a multi-channel digital oscilloscope from ROHDE&SCHWARZ (R&S RTM3004) [28] and shown in Figures 21–23. The upper trace of Figure 21 shows the output voltage, output current, and supply current waveforms for the case study, while the lower trace shows a zoom of the measured waveforms. Moreover, Figure 22 shows the DC-link or capacitor-voltage waveform. In addition, the inductor voltage and the current stress on one of the forward diodes are shown in the upper trace of Figure 23; moreover, a zoom of the inductor voltage is shown in the down trace.

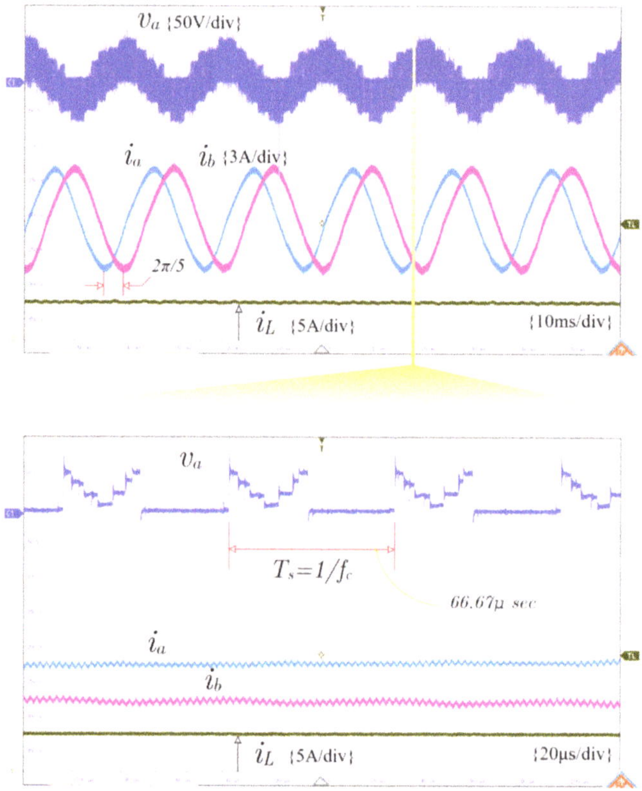

Figure 21. Experimental waveforms of output-phase voltage, two adjacent phase currents, and the supply or inductor current for the proposed five-phase SSI with the modified space-vector modulation scheme.

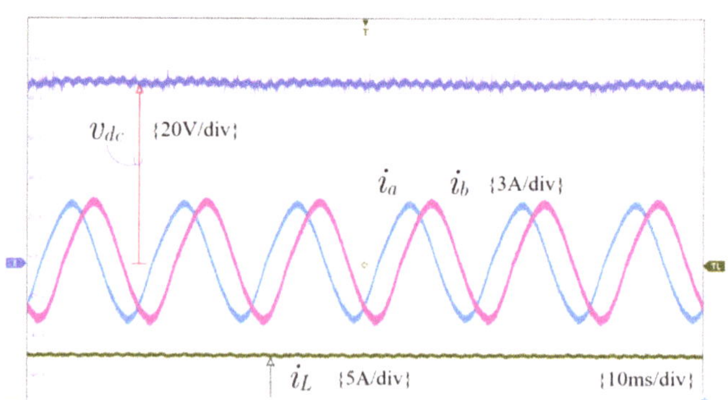

Figure 22. DC-link voltage and input/output current waveforms.

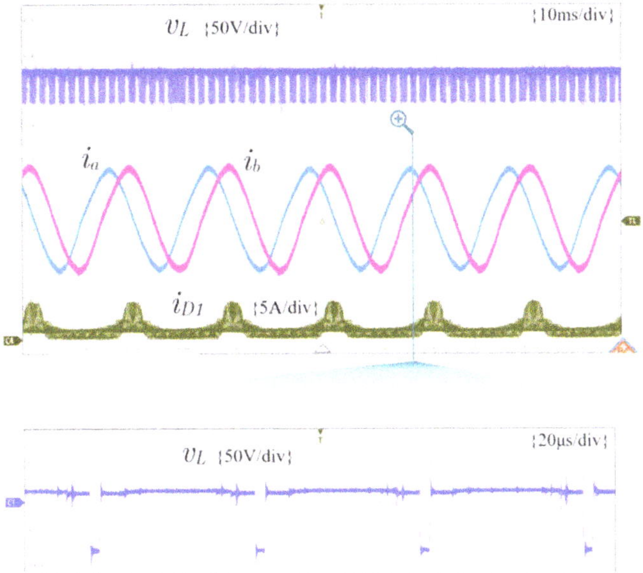

Figure 23. Five-phase SSI prototype.

From these results, the following observation can be made:

- Based on Figure 21, the waveforms of the output currents are near sinusoidal waveforms with low distortion and displaced by $2\pi/5$. These results are close to the simulation results. As can be observed, the pulse width modulated output voltage is chopped according to the proposed scheme with a frequency, f_c, of 15 kHz.
- From the inductor (supply) current waveform shown in Figure 21, it is worth noting that the inductor current is free from the low-order harmonics with the proposed PWM scheme. This is owed to utilizing fixed charging and discharging duty cycles in the proposed MSVM scheme. As you can see, the inductor current is continuous, with an average value of 5.3 A.
- The measured average voltage at the DC link from Figure 22 is 89 V. It agrees with the simulation and the mathematical analysis.
- Moreover, it can be observed from Figure 23 that the average voltage across the inductor is zero due, which proves the volt-second balance concept.

From the analytical, simulation, and experimental studies presented in this paper, the proposed five-phase split-source inverter possesses all of the advantages of the basic three-phase topology, and it is promising for applications involving five-phase motor drives, particularly if a high-output voltage gain is required. It presents a good alternative to the counterparts of the boosting inverters.

7. Conclusions

A five-phase split-source inverter topology is presented in this paper as an example of the multiphase architecture of the basic split-source inverter. This topology shows a good alternative for the low-voltage-supplied five-phase drive system, which is based on the basic two-stage DC–DC–AC converter and Z-source inverters. The paper discusses in detail the operating principles and modulation technique of the proposed topology. Furthermore, a detailed analysis of the ripples in the input and output currents is presented along with graphical representations. In addition, the results of the theoretical study are verified and validated through a simulation study using PLECS. An experimental prototype is built to validate the theoretical analysis. Simulation and experimental results are clearly in agreement, and experimental differences can be attributed to deadtime parasitic effects.

To conclude the findings behind this work, the five-phase SSI topology has some merits over its counterparts, which are used particularly if a high-output voltage gain is required. In summary, these merits are as follows:

(a) It has a continuous input current free of low-order harmonics;
(b) The AC-output ripple is like that of the conventional space vector modulated five-phase VSIs; therefore, these ripples are smaller than those of Z-source inverters.
(c) The boosting action is performed in a single stage, with fewer passive elements than Z-source inverters and without active switches as in two-stage inverters.
(d) In addition, it uses the same standard modulation schemes as the five-phase inverter and there is no need for additional shoot-through pulses.

Moreover, it is also possible to apply the same concept of this study to the other multiphase SSI topologies such as six-phase, seven-phase, and nine-phase architectures, and the same conclusions can be drawn.

Author Contributions: Conceptualization, software, methodology, validation, and formal analysis, S.M.D.; investigation, A.A.A.; resources, M.E.F.; experimental work, M.A.E.; writing—original draft preparation, S.M.D., I.A.G.; writing—review and editing, A.A.A. and I.A.G.; visualization, supervision, M.E.F.; project administration, S.M.D.; All authors have read and agreed to the published version of the manuscript.

Funding: This research received no external funding.

Institutional Review Board Statement: Not applicable.

Informed Consent Statement: Not applicable.

Data Availability Statement: Not applicable.

Conflicts of Interest: The authors declare no conflict of interest.

References

1. Kindl, V.; Cermak, R.; Ferkova, Z.; Skala, B. Review of Time and Space Harmonics in Multi-Phase Induction Machine. *Energies* **2020**, *13*, 496. [CrossRef]
2. Laksar, J.; Cermak, R.; Hruska, K. Challenges in the Electromagnetic Design of Multiphase Machines: Winding and Equivalent Circuit Parameters. *Energies* **2021**, *14*, 7335. [CrossRef]
3. Dabour, S.M.; Rashad, E.M.; Abdel-khalik, A.; Ahmed, S.; Massoud, A. A new fifteen-switch inverter topology for two five-phase motors drive. In Proceedings of the 2016 Eighteenth International Middle East Power Systems Conference (MEPCON), Cairo, Egypt, 27–29 December 2016; pp. 729–734. [CrossRef]
4. Dabour, S.M.; Abdel-Khalik, A.S.; Massoud, A.M.; Ahmed, S. Analysis of Scalar PWM Approach With Optimal Common-Mode Voltage Reduction Technique for Five-Phase Inverters. *IEEE J. Emerg. Sel. Top. Power Electron.* **2019**, *7*, 1854–1871. [CrossRef]
5. Elgenedy, M.A.; Elserougi, A.A.; Abdel-Khalik, A.S.; Massoud, A.M.; Ahmed, S. A Space Vector PWM Scheme for Five-Phase Current-Source Converters. *IEEE Trans. Ind. Electron.* **2016**, *63*, 562–573. [CrossRef]
6. Dabour, S.M.; Abdel-Khalik, A.; Ahmed, S.; Massoud, A. Performance of a three-to-five matrix converter fed five-phase induction motor under open-circuit switch faults. In Proceedings of the ISCAIE, Penang, Malaysia, 30–31 May 2016; pp. 159–164.
7. Ellabban, O.; Abu-Rub, H. Field-oriented control of a five-phase induction motor fey by a Z-source inverter. In Proceedings of the 2013 IEEE International Conference on Industrial Technology (ICIT), Cape Town, South Africa, 25–28 February 2013; pp. 1624–1629. [CrossRef]
8. Abduallah, A.A.; Iqbal, A.; Meraj, M.; Ben-Brahim, L.; Alammari, R.; Abu-Rub, H. Discontinuous space vector pulse width modulation techniques for a five-phase quasi Z-source inverter. In Proceedings of the IECON 2015-41st Annual Conference of the IEEE Industrial Electronics Society, Yokohama, Japan, 9–12 November 2015; pp. 004205–004210. [CrossRef]
9. Ellabban, O.; Abu-Rub, H. Z-Source Inverter: Topology Improvements Review. *IEEE Ind. Electron. Mag.* **2016**, *10*, 6–24. [CrossRef]
10. Kavya Santhoshi, B.; Mohana Sundaram, K.; Padmanaban, S.; Holm-Nielsen, J.B. Critical Review of PV Grid-Tied Inverters. *Energies* **2019**, *12*, 1921. [CrossRef]
11. Abdelhakim, A.; Blaabjerg, F.; Mattavelli, P. Modulation Schemes of the Three-Phase Impedance Source Inverters—Part I: Classification and Review. *IEEE Trans. Ind. Electron.* **2018**, *65*, 6309–6320. [CrossRef]
12. Yuan, J.; Yang, Y.; Blaabjerg, F. A Switched Quasi-Z-Source Inverter with Continuous Input Currents. *Energies* **2020**, *13*, 1390. [CrossRef]
13. Yi, F.; Cai, W. A Quasi-Z-Source Integrated Multiport Power Converter as Switched Reluctance Motor Drives for Capacitance Reduction and Wide-Speed-Range Operation. *IEEE Trans. Power Electron.* **2016**, *31*, 7661–7676. [CrossRef]

14. Dabour, S.M.; Abdel-Khalik, A.S.; Ahmed, S.; Massoud, A.M. A Family of Discontinuous PWM Strategies for Quasi Z-Source Nine-Switch Inverters. *IEEE Access* **2021**, *9*, 169161–169176. [CrossRef]
15. Ribeiro, H.; Pinto, A.; Borges, B. Single-stage DC-AC converter for photovoltaic systems. In Proceedings of the IEEE Energy Conversion Congress and Exposition, Atlanta, GA, USA, 12–16 September 2010; pp. 604–610.
16. Abdelhakim, A.; Mattavelli, P.; Spiazzi, G. Three-Phase Split-Source Inverter (SSI): Analysis and Modulation. *IEEE Trans. Power Electron.* **2016**, *31*, 7451–7461. [CrossRef]
17. Akbar, F.; Cha, H.; Ahmed, H.F.; Khan, A.A. A Family of Single-Stage High-Gain Dual-Buck Split-Source Inverters. *IEEE J. Emerg. Sel. Top. Power Electron.* **2020**, *8*, 1701–1713. [CrossRef]
18. Chen, M.; Yin, C.; Loh, P.C. Magnetically Coupled High-Voltage-Boost Split Y-Source Inverter Without Leakage-Induced Voltage Spikes. *IEEE Trans. Ind. Electron.* **2020**, *67*, 5444–5455. [CrossRef]
19. Dabour, S.M.; Alotaibi, M.A.; Abd-Elaziz, A.A.; Alshahat, M.A.; Abdallah, M.; Eltamaly, A.M.; Abdel-Khalik, A.S.; Massoud, A.M.; Ahmed, S. Modeling and Control of Single-Stage Quadratic-Boost Split Source Inverters. *IEEE Access* **2022**, *10*, 24162–24180. [CrossRef]
20. AbdulSalam, M.; Dabour, S.M.; Rashad, E.M. Cascaded Multilevel Split-Source Inverters: Analysis and Modulation. In Proceedings of the 2019 21st International Middle East Power Systems Conference (MEPCON), Cairo, Egypt, 17–19 December 2019; pp. 1204–1209. [CrossRef]
21. Montazeri, S.H.; Milimonfared, J.; Zolghadri, M. A New Modeling and Control Scheme for Cascaded Split-Source Converter Cells. *IEEE Trans. Ind. Electron.* **2022**, *69*, 7618–7628. [CrossRef]
22. Montazeri, S.H.; Milimonfared, J.; Zolghadri, M.R. Multidimensional Pulsewidth Modulation for Cascaded Split-Source Inverter. *IEEE Trans. Ind. Electron.* **2023**, *70*, 137–146. [CrossRef]
23. Dabour, S.M.; Abdel-Khalik, A.S.; Ahmed, S.; Massoud, A. An Optimal PWM Technique for Dual-Output Nine-Switch Boost Inverters With Minimum Passive Component Count. *IEEE Trans. Power Electron.* **2021**, *36*, 1065–1079. [CrossRef]
24. Pinjala, M.K.; Bhimasingu, R. Improving the DC-Link Utilization of Nine-Switch Boost Inverter Suitable for Six-Phase Induction Motor. *IEEE Trans. Transp. Electrif.* **2020**, *6*, 1177–1187. [CrossRef]
25. Abdelhakim, A.; Mattavelli, P.; Boscaino, V.; Lullo, G. Decoupled Control Scheme of Grid-Connected Split-Source Inverters. *IEEE Trans. Ind. Electron.* **2017**, *64*, 6202–6211. [CrossRef]
26. Dabour, S.M.; Aboushady, A.A.; Elgenedy, M.A.; Gowaid, I.A.; Farrag, M.E. Current Ripple Evaluation of Space Vector Modulated Five-Phase Split-Source Inverters. In Proceedings of the 2022 57th International Universities Power Engineering Conference (UPEC), Istanbul, Turky, 30 August–2 September; pp. 1–6. [CrossRef]
27. Dujic, D.; Jones, M.; Levi, E. Analysis of Output Current Ripple RMS in Multiphase Drives Using Space Vector Approach. *IEEE Trans. Power Electron.* **2009**, *24*, 1926–1938. [CrossRef]
28. Rohde & Schwarz R&S®RTM3000 Oscilloscope. Available online: https://www.rohde-schwarz.com/uk/products/test-andmeasurement/oscilloscopes/rs-rtm3000-oscilloscope_63493-427459.html (accessed on 10 October 2021).

electronics MDPI

Article

Design Optimization of an Efficient Bicolor LED Driving System

Fouzia Ferdous and A.B.M. Harun-ur Rashid *

Department of Electrical and Electronic Engineering, Bangladesh University of Engineering and Technology, Dhaka 1205, Bangladesh
* Correspondence: abmhrashid@eee.buet.ac.bd

Abstract: There are some challenges involved in the design of a multicolor LED driver, such as the precise control of color consistency, i.e., maintaining the correlated color temperature (CCT) and luminous intensity. CCT deviation causes a color shift of composite light. This paper approaches the method of nonlinear optimization of the LED currents of two LED sources to achieve the desired CCT. A bicolor blended-shade white LED system is formed by using a warm color LED source of 1000 K CCT and a cool color LED source of 6500 K CCT. By using a nonlinear optimization methodology, the reduced deviation of the blended CCT and optimum LED currents are obtained. The optimized currents in the two LED strings are maintained by the control circuit of the single-ended primary inductor converter (SEPIC). The obtained reduced deviation of the CCT is 43 K, and the precision is 99.15%. Again, harmonics in the input current hamper power quality, i.e., reduce the power factor and increase power loss. This paper proposes the harmonic reduction technique to achieve the lowest value of total harmonic distortion (THD) through the nonlinear parametric optimization of the SEPIC. Measured THD = 4.37%; PF = 0.96; and efficiency = 92.8%. The system stability was determined and found to be satisfactory.

Keywords: LED; color consistency; CCT; THD; bicolor

Citation: Ferdous, F.; Harun-ur Rashid, A.B.M. Design Optimization of an Efficient Bicolor LED Driving System. *Electronics* **2022**, *11*, 3984. https://doi.org/10.3390/electronics11233984

Academic Editors: Diego Bellan and Jelena Loncarski

Received: 1 November 2022
Accepted: 27 November 2022
Published: 1 December 2022

Publisher's Note: MDPI stays neutral with regard to jurisdictional claims in published maps and institutional affiliations.

1. Introduction

Multicolor LED lights can provide a large range of colors and quality lighting with a high color rendering index (CRI), and they can be applied in many areas [1]. As red, green, and blue LEDs have different thermal and luminous characteristics, variation in one color can have a significant effect on the targeted color point (CCT). Color control in LED systems is a multidisciplinary subject involving electric power, circuit topology, thermal management, and optical performance [2,3]. There is a lack of measurement results that describe the luminous and power parameters of multicolor LED systems. This paper presents an LED driving system which describes essential luminous and power measurements.

The color recognition of the LED light is a crucial research area in multicolor LED systems [4,5]. For the smooth intensity control of an LED light, optical regulation can be conducted either by changing the amplitude level or the duty cycle, or by changing both of the currents flowing through the LEDs [6,7]. White and other different colors can be generated using LED lights. Currently, LED-based white light is generated from a blue or a near-ultraviolet emitting chip coated with yellow phosphor, or by mixing red, green, and blue light [8]. Light from monochromatic red-green-blue (RGB) LEDs [9–12] or red-green-blue-amber (RGBA) LEDs were blended to obtain different shades of white. However, these types of LED systems require three or more pulse-width modulated (PWM) output channels. The RGB [10,13] or RGBA [14] LED driver uses independent converters for its power supply. These are redundant. The proposed bicolor LED driver uses a single power converter to drive both LED strings.

There are some challenges in obtaining a high-quality multicolor LED light source, such as junction temperature variations caused by the self-heating of the LEDs and disturbances in ambient temperatures [10,15]. These factors affect the LED current and hamper color consistency, i.e., CCT. The researchers proposed different feedback control structures based on LED junction temperature [13,15] to stabilize the luminous intensity and color consistency. However, the use of sensors not only increases the complexity but also increase the expense of the system. The proposed LED driver does not require sensors.

Therefore, in the design of LED drivers for diversified applications, constituents such as color consistency (CCT) and color intensity are needed along with other key requirements such as low THD, high PF, and high efficiency. Several attempts have been made to achieve low harmonics in the input current [16–19]. The resulting aim would be to achieve an LED driver that can maintain a constant CCT with constant intensity along with low THD and high luminous efficiency. High efficiency enables a low consumption of electrical energy. However, in most cases, the efficiency of LED drivers is low [20–24], and they cause harmonics in the input current and a low power factor [25–27]. The proposed LED driver has overcome these THD and power factor issues. This paper has achieved a low THD value by using a non-linear optimization methodology on SEPIC parameters. There was a bit of a discrepancy between the measured results and the calculated results for the THD. A high power factor was achieved by using a proper control circuit.

The precise color control of the multicolor LED system is still a challenging task for the chromaticity (CCT) shift due to the junction temperature variation and aging [28–33]. Efforts are ongoing to maintain CCT control. Papers [34–38] used the linear method for the CCT control of bicolor LEDs, and they considered the CCT constant, which is inaccurate. The CCT assumption fails to account for the temperature-dependent color temperature of LEDs and the nonlinearity of LEDs. It employs duty cycle control to mix color. The proposed LED driver is based on the use of nonlinear optimization to find the least CCT deviation. CCT and dimming control are performed in [39] using a microcontroller, where CCT deviation varies from 145 K to 510 K for a CCT range 2700 K to 6500 K. This large CCT deviation will cause color shifts. Using a microcontroller enhances complexity and cost. An open-loop system is used in [40] for CCT control, and the deviation of CCT is relatively higher. In [39,41,42], the authors focus on color control based on PWM switched systems. Refs. [39,41] uses one AC–DC converter and two LED drivers. The closed-loop control system is highly dependent on color sensors. MCU is used to generate PWM. All these increase complexity and cost. Paper [42] worked on CRI optimization with CCT, but there was no option to mitigate junction temperature and aging. They worked only on the color control of LEDs. The papers did not mention anything about the power source, power quality, and power related issues, etc. On the other hand, our proposed LED driver gives the full driver circuit design while considering color and power issues. The proposed driver has mitigated the current rising with the junction temperature and the aging effect by using a proper color feedback circuit. The driver shows a CCT deviation range of 43 K to 234 K, which cannot be seen by the naked eye. A microcontroller is not required in the proposed driver, and this reduces cost and complexity.

In previous studies [40,41] of bicolor LEDs, researchers have worked with only bicolor LEDs, whereas we have not only worked with bicolor LEDs but we have also designed a bicolor composite LED driver. In [43], the authors propose a general-purpose LED driver, and in [44], the authors describe a bicolor LED driver. In both cases, the power parameters are good. In [45], the authors worked with a bicolor LED driver based on luminous flux as a function of the LED voltages, but luminous flux is a function of LED current, which is more accurate. They did not mention the power quality parameters. In the proposed paper, we designed a bicolor LED driver for which the CCT was maintained at two color temperatures, namely, 1000 K and 6500 K, with a blended-shade white composite LED light through controlling the LED string current. At the same time, we achieved the least THD and a high power factor by optimizing the SEPIC parameters and using a proper control circuit, respectively.

In the proposed LED driver, the bicolor LED string's current is regulated by a single power converter. The combined CCT of the composite LED light is determined. CCT deviation is minimized through the constrained nonlinear optimization of the LED currents of the two LED strings. A nonlinear constrained parametric optimization methodology is also applied to refine the parameters of the SEPIC to obtain the lowest THD value. The proposed driver also boosts the efficiency and power factor. The SEPIC is operated in DCM and hence a control circuit improves the power factor. Thus, the LED driver maintains the desired CCT with minimal THD. Nowadays, people prefer a diversity of light color, e.g., shopping centers, garden lights, and spotlights where colors change with different control techniques. The nonlinear optimization model can be simplified to achieve practical solutions which are then transformed into real life applications. The result is more accurate than those obtained with linear approaches. The idea can be applied to LED systems with multiple color temperatures and is not limited to white LEDs.

The remainder of the paper is organized as follows: the basic configuration and driver topology are described in Section 2, the nonlinear optimization approach is laid out in Section 3, the nonlinear optimization of the reduction in CCT deviation is shown in Section 4, the analysis of the color-control scheme is described in Section 5, the proposed parametric optimization methodology for harmonic reduction is detailed in Section 6, the stability studies are listed in Section 7, the results are discussed in Section 8, and a conclusion is provided in Section 9.

2. Basic Configuration and Driver Topology

The bicolor LED driver circuit consists of two parts. The first is the power stage, and the second is the color management stage. The power stage comprises a bridge rectifier and a SEPIC to regulate the LED outputs. The SEPIC is chosen for PFC due to its superior characteristics: positive DC output, high–low output level, low switching stress, and low output ripple current. The total LED current regulation is carried out using the control circuit of the SEPIC. The block diagram in Figure 1 shows the power stage. The color controller controls the LED current in each of the LED strings.

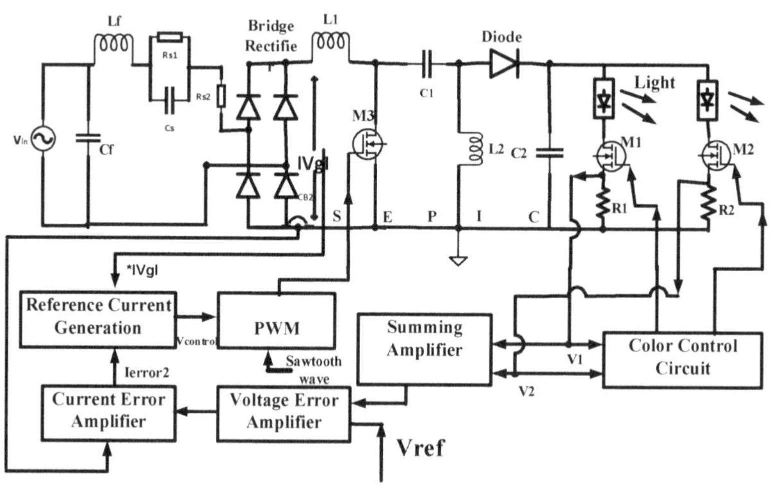

Figure 1. Power stage (basic SEPIC) and color control system.

The CCT and THD are controlled in the bicolor LED driver circuit simultaneously. The composite CCT is obtained by blending the colors of the LEDs of two parallel branches. By maintaining optimized LED currents in both branches, the minimization of CCT deviation can be obtained. In the color control scheme, the required current in each LED string flows individually through each color control feedback loop to achieve the target CCT with color stability. Reduced THD can be obtained by using optimized SEPIC parameters with the LED driver. The circuit parameters are achieved by performing constrained nonlinear optimization within the workable boundary.

The SEPIC provides a constant current, which is equal to the sum of the desired currents in the two LED strings, i.e., the SEPIC output current is divided into two LED strings according to the equivalent resistor of each branch. For the proposed converter topology, bicolor LED strings with string voltages (V1 and V2) act as the load. The sensed voltages of the two LED strings are summed and then passed through a voltage error amplifier. This generates an error signal, which is the reference current of the current error amplifier. The inductor current (I_L) is sensed and compared with the reference. It performs a proportional and integral function. The output of the current error amplifier V_{error2} is multiplied by the input sinusoidal voltage template, and the resulting signal forms the reference to generate the PWM signal. This helps to improve the power factor and reduce the harmonics in the input current. The PWM signal is generated by comparing a sawtooth carrier signal of 20 kHz with the output of a multiplier. The LED currents are as follows:

From ref. [43], the average SEPIC output current is:

$$I_0 = \frac{DD_1V_g}{2fLeq} \tag{1}$$

I_0 has been divided into R_{01} and R_{02} (R_{01} and R_{02} are the equivalent resistances of string1 and string2).

The average LED currents are now:

$$I_{av1} = \frac{R_{02}}{R_{01}+R_{02}}\frac{DD_1V_g}{2fL_{eq}},\quad I_{av2} = \frac{R_{01}}{R_{01}+R_{02}}\frac{DD_1V_g}{2fL_{eq}} \tag{2}$$

$$I_{peak1} = \frac{I_{av1}}{D_{M1}},\quad I_{peak2} = \frac{I_{av2}}{D_{M2}} \tag{3}$$

D_{M1} *and* D_{M2} are the duty cycles of MOSFETs M1 and M2.

Finally, the PWM signal is fed into the gate of SEPIC MOSFET M_3, and thus the LED current is maintained. An RC circuit is incorporated before the bridge rectifier to suppress the inrush current and voltage spike. The RC circuit is also shown in Figure 1. The details of the color control scheme are provided later.

3. Nonlinear Optimization Approach

Problem-based nonlinear constrained optimization methods are effective for fixing the parameters of analogue circuits. This optimization method is applied to find the optimum value of the LED current to achieve the desired CCT, and to find the lowest THD value with the optimum parameters for SEPIC L_1, L_2, and C_1 and the optimum filter parameters for L_f and C_f. Minimized CCT deviation and minimal THD are both desirable for maintaining color consistency and power quality in a multicolor LED driver, respectively. Figure 2 shows a functional block diagram of the CCT optimization process.

Non-linear optimization can be conceptually divided into the following distinct parts: (i) define the objective function or cost function, (ii) create constraints, and (iii) solve the problem iteratively within these constraints. The objective function is a mathematical model quantitatively describing the desired behavior of the circuit. We chose to define the circuit performance metric in order to minimize the cost function. The solution of the cost function achieves the lowest value among all the combinations of possible variables.

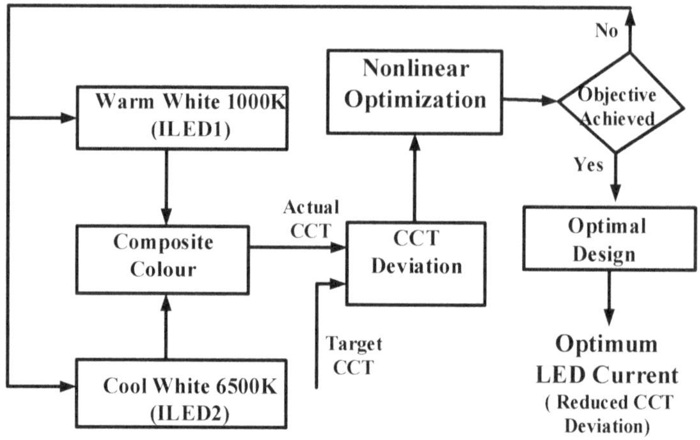

Figure 2. Functional block diagram of the CCT optimization.

Constraints of optimization constitute a hard boundary placed on the value of a variable, which prevents transition to unrealistic values. In nonlinear constrained optimization, the error function is calculated and then the parameters are adjusted to minimize the error function, causing the actual circuit behavior to approach the target function as closely as possible within the boundary limit.

4. Nonlinear Optimization for Reduction of CCT Deviation

Color temperature is a characteristic of visible light. Changing the color consistency causes adverse effects on human perception and comfort. Therefore, it is important to choose the right color temperature based on the desired mood. The International Commission for Illumination (CIE) has regulated and evolved multiple color spaces for the computation, assessment, and demonstration of visualizable colors [8]. In this paper, the system is configured based on two different color temperatures. The combined CCT of the composite LED light is determined. The CCT deviation is minimized through the nonlinear optimization of the LED currents of the two LED strings.

Creation of Optimization Problem

For the two LED modules with color temperatures of 1000 K and 6500 K, the chromaticity coordinates of the emitted light according to CIE 1931 are, respectively, (x_1, y_1) and (x_2, y_2) with luminous fluxes φ_1 and φ_2. Referring to Figure 1, the currents that flow through the mentioned LED strings are I_{LED1} and I_{LED2}, and their corresponding voltages are V_{LED1} and V_{LED2}, respectively. In order to find the desired CCT within the infinite number of possible solutions, an objective function is introduced to the system of equations. Equations (4) and (5) illustrate the mathematical model of the CCT. The CCT deviation is:

$$\Delta CCT = CCT_{target} - CCT_{est} \tag{4}$$

According to McCamy's approx. algorithm [1], the estimated CCT of synthesized light is:

$$CCT_{est} = 449n_{est}^3 + 3525n_{est}^2 + 6823.3n_{est} + 5520.33 \tag{5}$$

where $n_{est} = \frac{x-0.3320}{0.1858-y}$.

The chromaticity coordinate of the light synthesized by the two LED channels is based on [1]. Each color LED features a different spectral distribution and luminous output. If P is the consumed electrical power of each LED string and ε_i is the luminous efficacy of the

LED light source, then $\varphi_i = \varepsilon_i P_{eleci} = \varepsilon_i V_{LEDi} I_{LEDi}$, where I = 1,2 for the bicolor LED string. By calculating the values of x, y, and φ_i, n is estimated as follows:

$$n_{est} = \frac{\frac{\frac{x_1}{y_1}\varepsilon_1 V_{LED1} I_{LED1} + \frac{x_2}{y_2}\varepsilon_2 V_{LED2} I_{LED2}}{\frac{1}{y_1}\varepsilon_1 V_{LED1} I_{LED1} + \frac{1}{y_2}\varepsilon_2 V_{LED2} I_{LED2}} - 0.3320}{0.1858 - \frac{\varepsilon_1 V_{LED1} I_{LED1} + \varepsilon_2 V_{LED2} I_{LED2}}{\frac{1}{y_1}\varepsilon_1 V_{LED1} I_{LED1} + \frac{1}{y_2}\varepsilon_2 V_{LED2} I_{LED2}}} \tag{6}$$

The target CCT is obtained from the (x, y) coordinate of the blackbody curve. This CCT_{est} model is used in order to obtain the least CCT deviation.

Formation of Constraint with boundary—LEDs are current-driven devices with low equivalent dynamic resistance. The LED model is shown in Equation (7). Figure 3 represents the graph of LED load model. The constraint equation represents the I-V relationships of the LEDs.

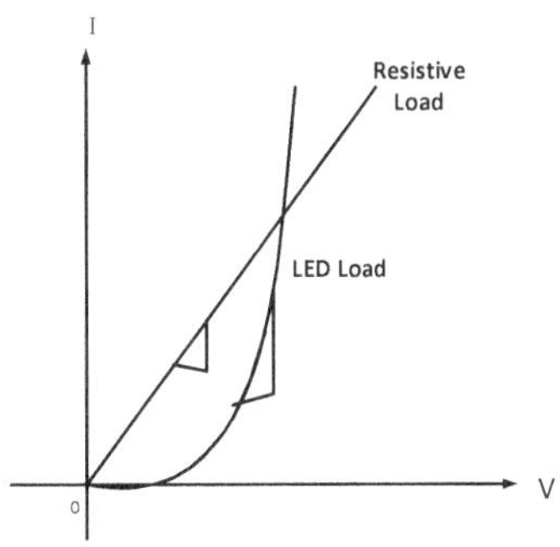

Figure 3. LED load model.

The LED current is:

$$I_{LED} = I_0\left(e^{\frac{V_{LED}}{nV_T}} - 1\right) \tag{7}$$

Rearranging

$$e^{\frac{V_{LED}}{nV_T}} = \frac{I_{LED}}{I_O} + 1$$

The reverse saturation current Io can be practically considered as 10 µA. The ideality factor n = 2 denotes the direct semiconductor (GaAs and InP). The room temperature thermal voltage V_T is considered to be 0.025 V.

The reduction in CCT deviation reduces color shift. Theoretically, various combinations are possible for CCT minimization. However, CCT deviations, i.e., color shifts, will only be minimized if the LED currents are within the specified ranges and the other parameters are meaningful. We measured the I-V characteristics of LEDs of 1000 K (LED_1) and 6500 K (LED_2), which were confined, according to their observed I-V characteristics. We considered the current limit of I_{LED1} to be from 220 mA to 320 mA when the LED_1 voltage varies between 23 volts and 25 volts, and the current I_{LED2} rises from 510 mA to 610 mA when the LED_2 voltage varied from 23 volts to 25 volts. For the change in the AC input voltage, the SEPIC output changed within the limit for feedback control.

Luminous efficacy differs with color variation. The efficacy of the LED was measured at 10 lm/watt and 33.7 lm/watt for LED_1 and LED_2, respectively. The minimized CCT deviation was obtained by solving the nonlinear optimization problem in MATLAB. We obtained ΔCCT = 87.7 K for two optimized LED currents, I_{LED1} – 290 mA and I_{LED2} = 610 mA. In the bicolor LED driver, the color control circuit should be designed in such a way that the 290 mA and 610 mA currents will flow through the LED_1 and LED_2 strings, respectively, to ensure minimal CCT deviation.

Some other constraints are the luminous intensity, efficacy, temperature, and electrical power of LEDs. All these depend upon LED currents. Since LED sources have a narrow spectral range in comparison with the visible spectrum, composite LED systems are highly sensitive to variations in temperature and peak wavelengths. As the junction temperature is not controllable, we can compensate for its effect with a color feedback circuit. Again, variations in circumambient luminosity and LED lifetime will lead to even more significant deviations in the real flux and CCT.

5. Proposed Color Control Scheme

Luminous output changes with temperature, and it changes over time in a way that cannot be accurately prognosticated. These factors influence the CCT, light intensity, and color point. This problem can be solved by applying a proper current control scheme.

The CCT and luminance set points are provided to the system in a convenient form using two series resistors, R_1 and $R_{2,}$, connected with both LED strings, respectively. The bicolor LED driver with a color controller is shown in Figure 4. The LED string current is sensed from series resistors R_1 and R_2. The current error amplifier maintains the LED current according to the reference value. This reference value is tuned to achieve the value of the current obtained from the MATLAB optimization program. It maintains the CCT of the corresponding color point. The control voltage is the output of the current error amplifier. The PWM signal is generated by comparing the control voltage with the sawtooth wave of the same clock frequency (20 kHz). The PWM signal generated from each color control loop is fed into the gate of the MOSFET of both LED strings and maintains the desired LED current individually. Thus, the corresponding color point and CCT of the respective LED currents remain stable. The switching state transition times of M_1 and M_2 are very short (a few ns), so it is undetectable to humans. The dimming capability can be achieved by changing the reference voltage. For the color controller, if one LED current is controlled instead of two, under the premise of controlling the sum of the two LED currents, one LED current will be discontinuous, and the other will be continuous. For currents of a different nature, smooth color control becomes difficult. This is why two-color control loops are used.

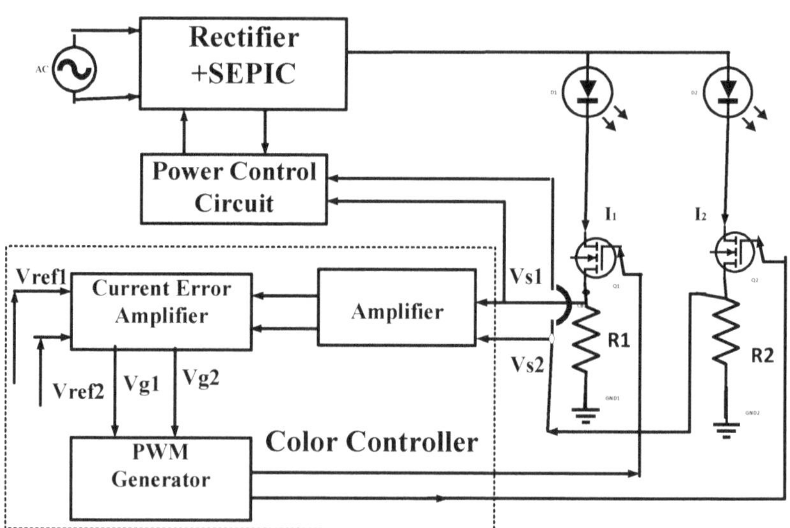

Figure 4. Bicolor LED driver with a feedback loop to regulate each LED string current to maintain CCT.

6. Proposed Parametric Optimization Method for Harmonic Reduction

The parameters that influence harmonics in the input current are the value of the input inductor L_1, input capacitor C_1, and output inductor L_2, and the filter components L_f and C_f. The nonlinear optimization method is applied in order to minimize the THD, where the parameters of the SEPIC and input filter parameters are optimized.

6.1. Expression for the Optimization Problem

We have applied the problem-based nonlinear constrained optimization methodology by using MATLAB. For this purpose, the SEPIC parameter optimization problem is formulated first. THD is defined in the optimization problem. Next, the necessary constraints are set and then the nonlinear optimization algorithm is executed iteratively to minimize the cost function by varying the parameters within the range set by the constraints.

To minimize the THD, the objective functions is obtained from Equations (9) and (10) of reference [43].

$$THD = \sqrt{\frac{1}{32I_{1rms}{}^2}\left(\frac{DD_1L_fC_f}{fL_1L_2C_1}\right)^4 R^2V_m{}^2 - 1} \quad (8)$$

6.2. Formation of Constraint with Boundary

To achieve minimal THD, the SEPIC parameters L_1, L_2, and C_1 and the filter parameters L_f and C_f are optimized using nonlinear optimization in MATLAB, and the constraint boundary limit is defined in the practical domain. Here, we first consider the limit of L_1. The value of L_1 is defined in Equation (9):

$$L_1 = \frac{V_{in}.D}{\Delta I_{L1}.f_s} \quad (9)$$

Here, L_1 depends upon ΔI_{L1}. L_1 must be high enough for ΔI_{L1} to be significantly minimized. In the practical circuit, L_1 has some limitations, such as size, cost, and the availability of component values. It should also be cost effective. Considering the above, L_1 is varied from 1 mH to 10 mH. L_2 is defined as the equivalent resistance of L_{eq} and L_1. L_{eq} is dependent on the dynamic resistance, voltage conversion ratio, and switching frequency. Considering the transfer of energy in the SEPIC, it is feasible to restrict L_2 to a moderate range, i.e., to the micro-henry range.

It is confined in the range of 1 μH to 100 μH. Now, the constraint parameter C_1 depends upon the resonance frequency f_r,,which should be within the line frequency and switching frequency. The resonance frequency is considered to be 3 kHz. C_1 is the input energy transfer capacitor. Therefore, it should be a small value for the smooth transfer of energy from the input to the output, and for maintaining proper circuit operation according to the constraint of Equation (10).

$$f_r = \frac{1}{2\pi\sqrt{C_1(L_1 + L_2)}} \quad (10)$$

For convenience, the range for C_1 is from 0.001 μF to 0.01 μF. For the input filter, high impedance is used to restrict the unwanted frequency in the input terminal. It should also be noted that to restrict the high starting current and voltage spike, an RC soft-start is incorporated in the circuit after the LC filter in the input. The effect of these RC components is also considered while calculating L_f and C_f. Thus, undesired signals, especially high frequency signals, will be blocked from the power grid.

$$L_{feq} = L_f + \frac{R_{s1}}{\omega^2 R_{s1} C_s + \omega} + \frac{R_{s2}}{\omega} \quad (11)$$

$$f_r = \frac{1}{2\pi\sqrt{C_f L_{feq}}} \quad (12)$$

Here, L_{feq} is the equivalent inductance of the filter components. By evaluating a realistic situation, C_f is restricted in the range of 0.1 μF to 1.0 μF and L_f is restricted in the range of 1 mH to 10 mH, according to the constraints (12). The minimized THD is obtained from the solution of the nonlinear optimization by using MATLAB. THD = 2.1% is obtained for the SEPIC parameters L_1 = 10.0 mH, L_2 = 40.0 μH, C_1 = 0.01 μF, L_f = 10.0 mH, and C_f = 1 μF. The measured values are provided in Section 8.

7. Stability Studies

A state space-averaged model for the SEPIC-based LED driver during different modes of operation has been developed [43]. Following this methodology, the factors of the bicolor LED system are considered, and the overall open-loop transfer function of the System is obtained as follows:

$$Trg = \frac{6.761s^4 + 250.4s^3 - 1.287s^2 - 3.206e^{-5}s + 4.322e^{-14}}{s^5 + 0.05135s^4 + 3.739e^{-6}s^3 + 4.77e^{-14}s^2 + 1.634e^{-18}s} \quad (13)$$

From the step response shown in Figure 5, it is found that for the compensated model, the system reaches the steady state within only 1 ms.

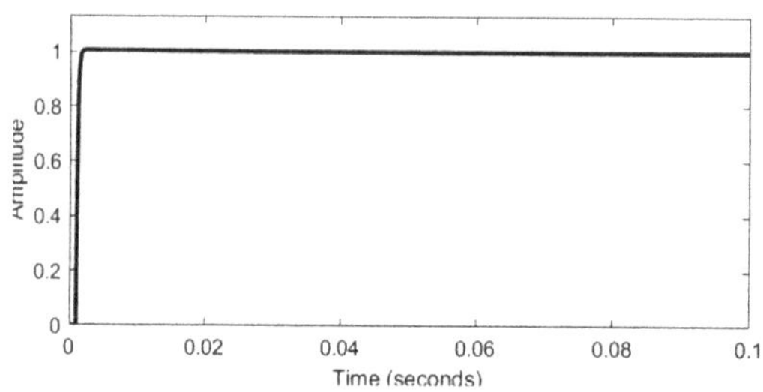

Figure 5. Steady state is achieved at 1 ms (step response) of the compensated SEPIC.

8. Results

We have designed a nonlinear optimization scheme in MATLAB for bicolor LED combinations. For both cases, CCT and THD, to achieve the lowest value, nonlinear optimization was performed by refining the boundary limits to obtain optimum performance. In the bicolor system, both the LED strings were of 24 volts, and LED_1 (2 × 12 V) and LED_2 (3 × 8 V) were mounted on the heatsink. The experimental setup consisted of the SEPIC circuit and a control circuit. The control circuit consisted of a summing amplifier, voltage error amplifier, current error amplifier, and multiplier, and PWM generators were constructed using the OPAMP, diode, capacitor, and resistors. We determined the different luminous and electrical characteristics of 1000 K and 6500 K LEDs.

By applying the proposed closed-loop control scheme, we regulated the specified currents, 290 mA and 610 mA, found in the nonlinear optimization algorithm to flow through the two LED strings at an input voltage of 220 V. The CCT, CRI, and luminous flux of the composite LED light were measured with an LED current from the Lab-sphere Photometric Measurement System PM-105-16 at different input voltages. Figure 6 shows the CCT and flux vs. the LED current curve for both LEDs. The CCT was almost constant, but the flux increased with the increase in LED current. The measured composite CCT was 5043 K, which was very close to the desired CCT of 5000 K. The CCT deviation was 43 K at an operating voltage of 220 V, which is within the non-perceivable (+/−283 K) color limit [26]. The measured and target CCT values with the flux of the bicolor LED light are shown in Figure 7. The experimental results show that, despite variations in light intensity, the bicolor LED lamp with a closed-loop control remained stable with the strict control of the CCT.

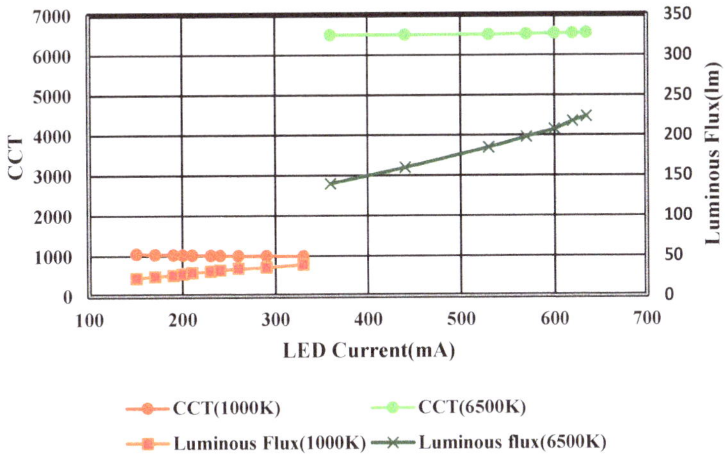

Figure 6. Measured CCT and luminous flux vs. LED current for CCT 1000 K and 6500 K LEDs.

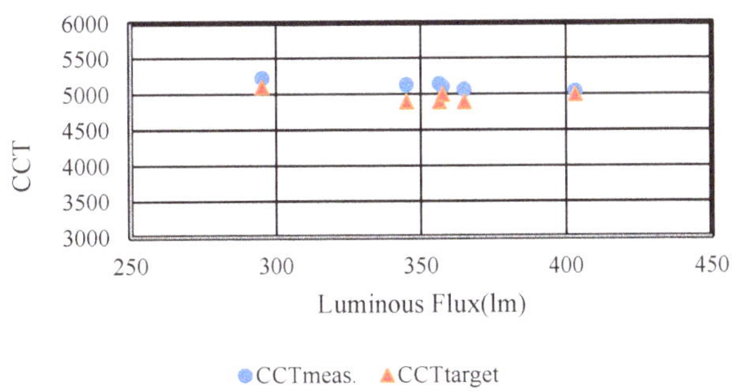

Figure 7. Variation in CCT with the change in luminosity of the composite LED light.

Table 1 compares the target CCT with the corresponding measured CCT for the blended LED system. The minimum and maximum percentage errors are 0.86% and 5.08%, respectively. This shows that the measured CCT values from the color sphere conform closely with the corresponding target values. Figure 8a shows the spectral distribution of the composite color LED light with the wavelength. This allows us to determine the nature of the composite light at different wavelengths. The spectral flux vs. the wavelength of the composite LED light (LED_2, wavelength of 500–565 nm with a CCT of 6500 K; LED_1, wavelength of 625–740 nm with a CCT of 1000 K) is shown in the curve. In the wavelengths of the spectrum, only certain wavelengths (the spikes) are strongly present. From the SPD curve, it can be seen that the peaks of the green (LED_2) region are greater than those of the red (LED_1) region. These spikes indicate the dominance of the rendering of color for objects illuminated by the light source. The measured CCT of 5043 K leads to a CRI value of 57.32% in this region. It produces a light similar to daylight (5000 K). Its ability to render color across the spectrum is not bad, but it is certainly much worse than daylight. We found from the measured data that the CCT varies from around 4967 K to 5228 K with the variation in the LED current.

Table 1. CCT measurements.

LED Current (mA)		CCT Target	CCT Meas.	CCT Dev.	% Error	CRI
I_{LED1}	I_{LED2}					
220	512	5100	5228	128	2.50	61.34
250	520	4900	5134	234	4.78	62.17
270	543	4900	5149	149	5.08	60.46
278	563	4900	4967	67	1.36	61.14
282	610	5000	5107	107	2.14	59.73
290	610	5000	5043	43	0.86	57.32

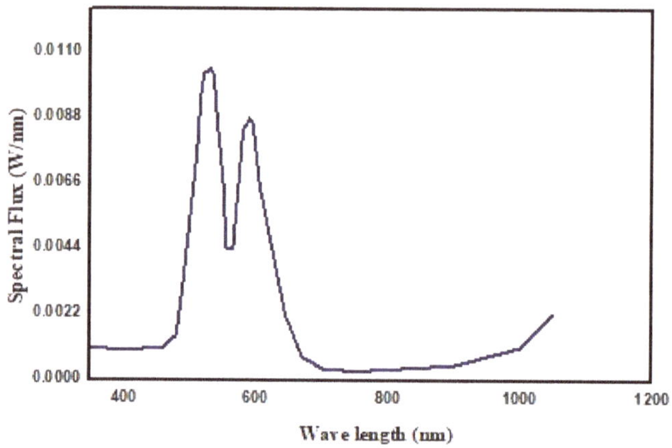

(a)

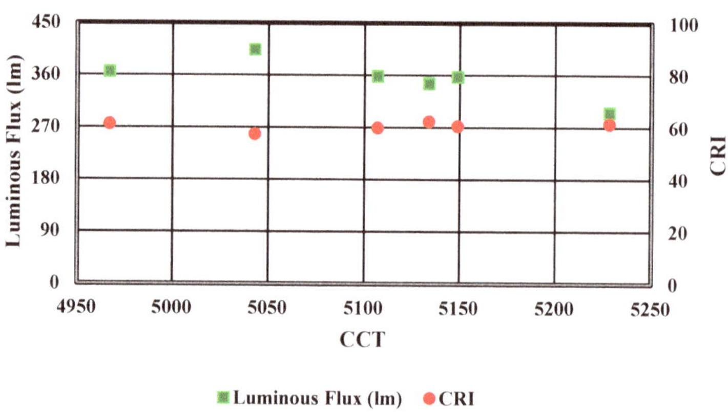

(b)

Figure 8. (**a**) Spectral distribution of composite R-G color measured using the Lab-sphere Photo metric Measurement System. (**b**) Luminous flux and CRI vs. CCT variation.

Figure 8b shows that the luminous flux changes (increases and decreases) very little with the increase in CCT. The luminous flux is proportional to the LED current. However, when the LED voltage and efficacy change, the flux no longer remains proportional. The CRI is independent of the CCT.

The duty cycle of the SEPIC is defined as:

$$D_{max} = \frac{V_{output}}{V_{output} + V_{input(min)}}$$

where $V_{input(min)} = 90$ V.

It is found that an increase or decrease in the duty cycle will result in an increase or decrease in the flux value. Thus, the overall CCT value increases or decreases. For instance, a larger D_1 value will cause a decrease in the CCT value, i.e., it will become more reddish white. Similarly, a larger D_2 value will increase the CCT value, i.e., it will become more bluish white. Because the duty ratio changes, the LED current changes from the optimized values, and the CCT also changes. The composite CCT vs. duty ratios (D_1, D_2) are graphically illustrated in Figure 9. We find that the CCT deviation shown in Table 1 is within the allowable CCT tolerance. Therefore, it is experimentally verified that the CCT values continue to track closely with the target values, regardless of the variations in the luminous flux. This prevents a color shift in the composite color LED driver. Table 2 represents the optimized parameters of prototype of LED driver.

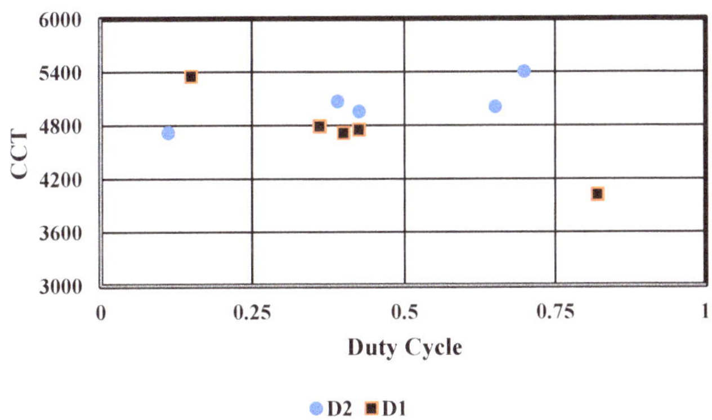

Figure 9. Composite CCT variation with change in duty cycle.

Table 2. Optimized parameters used in prototype.

Parameters	Values
Input inductor, L1	10 mH
Output inductor, L2	100 µH
Input capacitor, C1	0.01 µF/600 V
Input Filter Inductor, Lf	10 mH
Input Filter Capacitor, Cf	0.94 µF
Duty Cycle, Dmax	0.21
Power Diode	1000 V/10 A
Power MOS, M3	12 N60 (600 V/12 A)
Operating Voltage	220 V
MOS Switch, M1, M2	K2098

The measured power loss shown is 7.2%, the efficiency is 92.8%, and the THD is 4.37% at 220 V. The measured power losses are distributed component-wise, as shown in Figure 10. Here, R-sense stands for the sense resistor, V-divider denotes the resistance for the template of the input voltage, and R-diode and S-diode denote the rectifier and SEPIC diode, respectively.

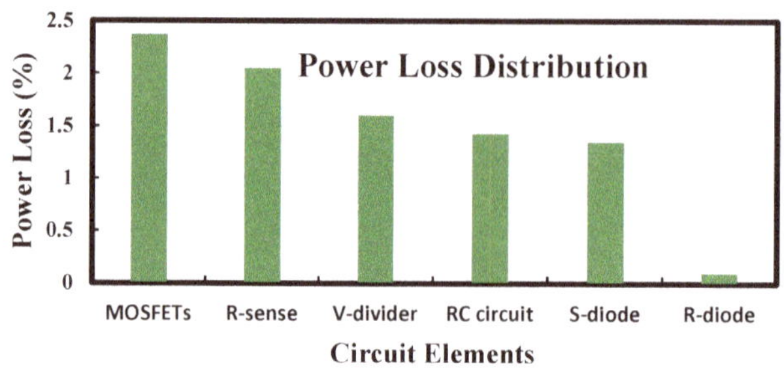

Figure 10. Power loss components in LED driver.

The power consumed in the DC bias circuit is P_{DCbias} = 0.30 watt. It can be seen that the largest power loss occurs in the MOSFETs, accounting for approximately 2.37% of the total power consumption. The proposed LED driver shows low THD (4.37%) and high PF (0.96) at 220 V due to the optimization of the SEPIC parameters and the well-designed feedback control circuit. A crest factor of 1.4 is obtained.

The average measured LED voltage and LED current of the two branches are 24 V, 290 mA and 24.4 V, 610 mA, respectively. Figure 11 shows the waveforms of the input voltage (v_s) and the input current (i_s) in the steady state condition. From the results, we find that the input current follows the input voltage, and therefore the P.F is high.

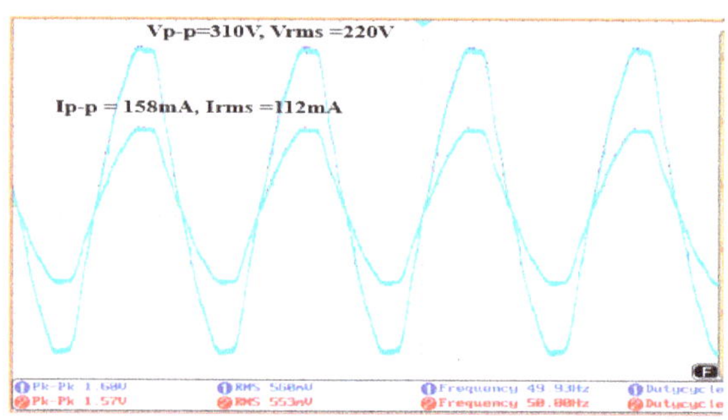

Figure 11. The measured input voltage: 220 V rms (50 V/division); the input current: 112 mA rms (7 mA/division). THD = 4.37% and P.F = 0.96.

The currents of both LED strings were measured by varying the respective duty cycles (D_1 and D_2) of MOSFETs M_1 and M_2. Figure 12a,b shows the PWM and corresponding LED voltage and LED current at Vs = 220 V in a steady-state condition at a duty cycle D_1 of 42.55% and 63% for LED_1. The same curves are shown in Figure 13a,b for LED_2 by varying the duty cycle D_2 to 11.61% and 36.65%. For Figure 12a,b and Figure 13a,b, we find that with dimming, the LED voltages and LED currents vary. The pulse width of the LED current changes with the change in duty cycle. However, the obtained LED voltages are smooth and constant due to the use of a capacitor of 1000 µF (ref. Figure 1) parallel to both LED strings. The effect of dimming on the luminosity data are shown in Table 3.

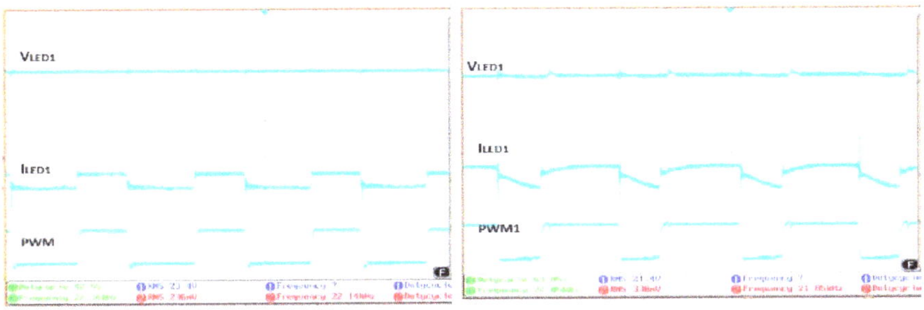

(a) (b)

Figure 12. LED voltage and current at duty cycles of (**a**) 42.55% and (**b**) 63% with corresponding gate pulse for string 1.

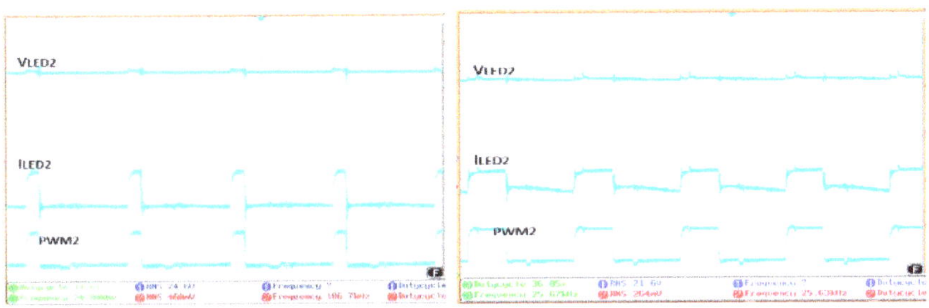

(a) (b)

Figure 13. LED voltage and current with duty cycles (D_2) of (**a**) 11.61% and (**b**) 36.65% in LED string 2.

Table 3. Current, voltage, and luminosity data variation with dimming.

No. of String	Duty Cycle (%)	LED Current (mA)	LED Voltage (V)	CCT	% CCT Deviation	Luminous Flux	CRI
1	42.55	236	23.4	4955	0.90	381.69	58.21
	63	330	24.4	5104	2.08	401.69	49.81
2	11.61	264	24.6	4719	5.62	379.63	62.64
	36.65	460	24.6	4790	4.20	300.07	63.21

Table 3 shows the LED current, voltage, and luminosity data variations with dimming. Here, we see that, in spite of the variation in the duty cycle, the CCT deviates only from 0.90% to 5.62%, i.e., from 45 K to 281 K, a change which cannot be recognized by the human eye.

Table 4 shows the measured power parameters of the designed bicolor LED driver. By the nonlinear optimization of the parameters of the SEPIC, a minimized THD was obtained. For efficiency enhancement, we have chosen circuit components with loss-related parameters of minimum values. Figure 14a shows the efficiency, THD, and P.F. variation of the LED driver with the variation in the duty cycle. From the curve, it can be seen that the efficiency and the THD decrease and that the power factor increases with the increase in the duty cycle. Figure 14b shows the efficiency, THD, and P.F. variation of the LED driver with the variation in input voltage.

Table 4. The measured power parameters of the proposed bicolor LED drivers.

Vs (Volt)	Is (mA)	P.F.	THD	V_{LED1} (Volt)	I_{LED1} (mA)	V_{LED2} (Volts)	I_{LED2} (mA)	Efficiency %
90	206	0.988	1.27	23.1	220	23.8	512	91.6
110	190	0.981	2.40	23.8	250	24.0	520	90.7
150	150	0.980	5.40	24.0	270	24.5	543	88.5
180	125	0.977	4.36	24.0	278	24.5	563	91.8
200	115	0.960	4.15	24.6	282	24.8	580	95.3
220	110	0.960	4.37	24.0	290	24.4	610	92.8

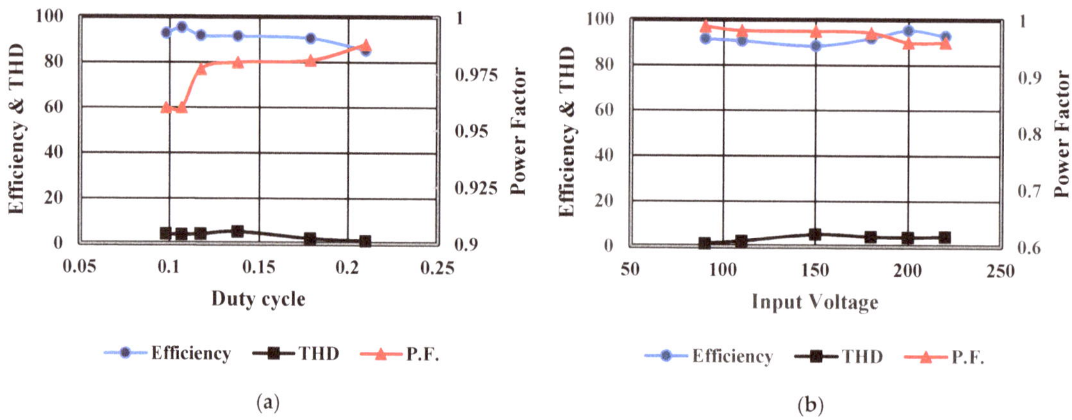

Figure 14. (**a**) THD, P.F., and efficiency vs. duty cycle graph; (**b**) THD, P.F., and efficiency vs. input voltage graph.

The curve shows that efficiency and THD increase and that the power factor decreases with the increase in input voltage. The performance in Figure 14a,b justifies the electrical characteristics of an LED driver. Table 5 shows comparisons between the luminous flux and CCT of the proposed system and those of the bicolor LED systems presented in [39] and [41]. The proposed LED driver shows luminous bicolor LED system features that are comparable with those obtained using the more accurate nonlinear approach, but exhibits better features than those obtained using the linear approach [39]. Ref. [41] presents an LED driver which shows a CCT less than 50 K, but nothing is said about the luminous flux. Neither of these papers mention anything about the power quality parameters.

Table 6 shows the comparison of the power parameters with those of [44,45]. Our designed LED driver shows better performance in THD and efficiency than [41], but on the other hand it shows comparable performance in P.F. The author of [45] has worked with bicolor LED drivers that show CCT variations of 0.73% to 2.58%, but they do not mention power quality parameters. In the proposed bicolor LED driver, the CCT varies from 0.86% to 5.08%. The result is comparable. Ref. [45] uses a single-inductor dual-output converter with a 12 V DC input, whereas the proposed LED driver is SEPIC based and has a 220 V AC input. Figure 15 shows the prototype of the bicolor LED driver with the experimental setup.

Table 5. Comparison of luminosity parameters with those of prior art [39,41].

Bicolor White LED Light	Luminous Flux (lm)	CCT Measured	CCT Desired	CCT Deviation
CCTC = 5000 K	100	2754	3300	546
CCTw = 2700 K	289	3814	4000	186
(Linear approach) [39]	455	2798	3300	502
CCTC = 6500 K	274.57	4713	4900	187
CCTw = 1000 K	354.86	5070	5000	70
(Proposed)	401.69	5404	5300	104
CCTC = 5000 K				
CCTw = 2700 K	95	3250	3300	50
(Non-linear approach)	296	5515	5400	115
[39]	562	3392	3300	92
CCTC = 7500 K	-	3050	3000	50
CCTw = 2700 K	-	3410	3400	10
[41]	-	3505	3500	05

Table 6. Comparison of power quality parameters with those of [44,45].

Item	LED Driver in [44]	Designed LED Driver	LED Driver in [45]
Input Voltage (V)	220	220	12 V(DC)
MOS Switches	3	3	3
Converter	SIMO	SEPIC	SIDO
Total Current (mA) ($I_{LED1} + I_{LED2}$)	411.9	810	800
Transformer	1	No need	No need
Frequency (kHz)	50	20	200
Power Factor	0.982	0.96	-
THD	7.15%	4.37%	-
Efficiency	85.56%	92.8%	Not mentioned

Figure 15. Prototype of the bicolor LED driver with Lab-sphere Photometric Measurement System PM-105-16.

9. Conclusions

In this paper, a bicolor composite LED driver is designed based on nonlinear optimization for maintaining color consistency and regulating THD. Though the empirical model of the CCT depends upon junction temperature, ambient temperature, ambient luminance and aging for a proper color feedback circuit, the CCT deviation remains within the limit. The experimental results are: CCT deviation = 43 K, error = 0.86%, and CRI = 57.32%. Thus, the proposed approach is accurate and tightly controls CCT. By determining the proper THD model, choosing the constraints accounting for the appropriate functionality and availability of components, and applying nonlinear optimization, the simulated THD was 2.1% and the experimentally obtained THD was 4.37%. The Power factor was increased by using a proper feedback circuit (reference current generation block). The nonlinear optimization method was also appropriate for achieving lower THD. The other obtained power quality parameters were: CF = 1.41%, P.F. = 0.96, and efficiency = 92.8%. These are satisfactory. The cost of our proposed AC–DC bicolor LED driver is BDT 750 (USD 7.14), whereas a traditional 22-watt AC–DC driver costs BDT 440 (USD 4.2). The traditional driver's efficiency is 86%, whereas we have designed 22-watt driver with 92% efficiency. The simplicity of the generation of precise and consistent light output from the proposed bicolor LED system makes it an attractive solution.

Author Contributions: Conceptualization, A.B.M.H.-u.R.; data curation, F.F.; formal analysis, F.F.; investigation, F.F.; methodology, F.F.; resources, A.B.M.H.-u.R.; supervision, A.B.M.H.-u.R. All authors have read and agreed to the published version of the manuscript.

Funding: This research received no funding.

Conflicts of Interest: The authors declare no conflict of interest.

Abbreviations

CCT	Corelate color temperature
CRI	Color rendering index
THD	Total harmonic distortion
P.F.	Power factor
I_O	Output current
D	Duty cycle
V_g	Rectified input voltage
ε_i	Luminous efficacy
φ_1, φ_2	Luminous flux of LED 1 and 2
R_{01}, R_{02}	Equivalent resistance of strings 1 and 2
SEPIC	Single-ended primary inductor converter
D_{M1}, D_{M2}	Duty cycle of MOS of 1st and 2nd LED strings
$\boldsymbol{CCT_{target}}$	CCT of target point
$\boldsymbol{CCT_{est}}$	Estimated CCT
$\boldsymbol{\Delta CCT}$	Deviation between target and estimated CCT
R_1, R_2	Sense resistor of LED strings 1 and 2
PWM	Pulse width modulation
ΔI_{L1}	Inductor current increment
$\boldsymbol{R_{s1}}$, $\boldsymbol{R_{s2}}$	Series resistors 1 and 2 of RC circuit

References

1. Zhan, X.; Wang, W.; Chung, H. A novel Color Control Method for Multi-colour LED Systems to Achieve High Colour Rendering Indexes. *IEEE Trans. Power Electron.* **2018**, *33*, 8246–8258. [CrossRef]
2. Chen, H.T.; Tan, S.C.; Hui, S.Y.R. Colour variation reduction of GaN-based white light-emitting diodes via peak-wavelength stabilization. *IEEE Trans. Power Electron.* **2014**, *29*, 3709–3719. [CrossRef]
3. Hui, S.Y.R.; Qin, Y.X. A general photo-electro-thermal theory for light-emitting-diode (LED) systems. *IEEE Trans. Power Electron.* **2009**, *24*, 1967–1976. [CrossRef]
4. Paschotta, R. *Encyclopedia of Laser Physics and Technology*; Wiley: New York, NY, USA, 2008.

5. Mills, P.R.; Tomkins, S.C.; Schlangen, L.J.M. The effect of high correlated color temperature office lighting on employee wellbeing and work performance. *J. Circadian Rhythm.* **2007**, *5*, 1–9. [CrossRef]
6. Yau, T.Y.; Hwu, K.I.; Liu, K.J. AC–DC Flyback Dimmable LED Driver with Low-Frequency Current Ripple Reduced and Power Dissipation in BJT Linearly Proportional to LED Current. *J. Energies* **2020**, *13*, 4270. [CrossRef]
7. Nassary, M.; Orabi, M.; Ahmed, E.M.; Hasaneen, E.-S.; Gaafar, M. Modified Harmonic Injection Technique for Electrolytic Capacitor-Less LED Driver. In Proceedings of the Nineteenth International Middle East Power Systems Conference (MEPCON), Cairo, Egypt, 19–21 December 2017; pp. 1459–1464.
8. Ferdous, S.M.; Oninda, M.A.M.; Sarowar, G.; Islam, K.K.; Hoque, M.A. Non-Isolated Single Stage PFC based LED Driver with THD Minimization using Cúk Converter. In Proceedings of the 9th International Conference on Electrical and Computer Engineering, Dhaka, Bangladesh, 16 February 2017; pp. 471–474.
9. Yoomak, S.; Ngaopitakkul, A. The Study of Harmonic Reduction in Light Emitting Diode (LED) Roadway Lighting System. In Proceedings of the 17th International Scientific Conference on Electrical Power Engineering (EPE), Prague, Czech Republic, 16–18 May 2016; pp. 1–5.
10. Miao, C.H. Colour Temperature Adjustable Lamp. U.S. Patent US8159125B2, 17 April 2012.
11. Jonsson, K. Light with Changeable Color Temperature. U.S. Patent US20120146505A1, 20 November 2012.
12. Schubert, E.F. *Light-Emitting Diodes*, 2nd ed.; Cambridge University Press: Cambridge, UK, 2016; pp. 332–346.
13. Muthu, S.; Schuurmans, F.J.; Pashley, M.D. Red, green, and blue LEDs for white light illumination. *IEEE Trans. J. Sel. Top. Quantum Electron.* **2002**, *8*, 333–338. [CrossRef]
14. Wang, F.C.; Tang, C.-W.; Huang, B.J. Multivariable robust control for a red-green-blue LED lighting system. *IEEE Trans. Power Electron.* **2012**, *25*, 417–428. [CrossRef]
15. Tang, C.W.; Huang, B.J.; Ying, S.-P. Illumination and color control in red -green-blue light-emitting diode. *IEEE Trans. Power Electron.* **2014**, *29*, 4921–4937. [CrossRef]
16. Zhao, H.; Lee, S.W.R. Determination of Driving Current of RGB LEDs for White Light Illumination. In Proceedings of the 13th International Conference on Electronic Packaging Technology & High Density Packaging, Guilin, China, 13–16 August 2012; pp. 1540–1545.
17. Qu, X.H.; Wong, S.C.; Tse, C.K. Temperature measurement technique for stabilizing the light output of RGB LED lamps. *IEEE Trans. IEEE Trans. Instrum. Meas.* **2010**, *59*, 661–670.
18. Gilman, J.M.; Miller, M.E.; Grimaila, M.R. A simplified control system for a daylight-matched LED lamp. *Light. Res. Technol.* **2013**, *45*, 614–629. [CrossRef]
19. Available online: https://docplayer.net/203504867-White-paper-color-management-of-a-red-green-and-blue-led-combinational-light-source-abstract-using-rgb-leds.html (accessed on 30 October 2022).
20. Górecki, K.; Ptak, P. Modeling power supplies of LED lamps. *Int. J. Circuit Theory Appl.* **2018**, *46*, 629–663. [CrossRef]
21. Rata, G.; Rata, M. The study of harmonics from dimmable LED lamps, using CompactRIO. In Proceedings of the 2016 International Conference on Development and Application Systems (DAS), Suceava, Romania, 19–21 May 2016; pp. 180–183.
22. Pollock, A.; Pollock, H.; Pollock, C. High Efficiency LED Power Supply. *IEEE J. Emerg. Sel. Top. Power Electron.* **2015**, *3*, 617–623. [CrossRef]
23. Uddin, S.; Shareef, H.; Mohamed, A.; Hannan, M.A. Harmonics and thermal characteristics of low wattage LED lamps. *Przegląd Elektrotechniczny* **2012**, *88*, 266–271.
24. Lho, Y.H.; Lee, S.Y. A Study on methodology to improve the power factor of high-power LED module. In Proceedings of the 2014, 14th International Conference on Control, Automation and Systems (ICCAS 2014), Seoul, Republic of Korea, 22 October 2014; pp. 1404–1406.
25. Górecki, K. The influence of power supply voltage on exploitive parameters of the selected lamps. *Inf. MIDEM-J. Microelectron. Electron. Compon. Mater.* **2013**, *43*, 193–198.
26. Uddin, S.; Shareef, H.; Mohamed, A.; Hannan, M. An analysis of harmonics from dimmable LED lamps. In Proceedings of the 2012 IEEE International Power Engineering and Optimization Conference, Melaka, Malaysia, 6–7 June 2012; pp. 182–186.
27. Ionescu, C.; Dima, M.; Bonfert, D. Flicker distortion power factor analysis in lighting LED's. In Proceedings of the 2017 IEEE 23rd International Symposium for Design and Technology in Electronic Packaging (SIITME), Constanta, Romania, 26–29 October 2017; pp. 280–285.
28. Li, Y.C. A novel control scheme of quasi-resonant valley-switching for high-power-factor AC-to-DC LED drivers. *IEEE Trans. Ind. Electron.* **2015**, *62*, 4787–4794. [CrossRef]
29. Liu, P.J.; Hsu, Y.-C. Boost converter with adaptive reference tracking control for dimmable white LED drivers. *Microelectron* **2015**, *46*, 513–518. [CrossRef]
30. Liu, P.J.; Peng, K.L. Adaptive Driving Bus Voltage and Energy Recycling Control Schemes for Low- Power AC–DC RGB-LED Drivers. *IEEE Trans. Power Electron.* **2017**, *64*, 7741–7748. [CrossRef]
31. Zhang, Y.; Chen, H.; Dongsheng, M. A V_O-Hopping Reconfigurable RGB LED Driver with Automatic Detection and Predictive Peak current Control. *IEEE J. Solid-State Circuits* **2015**, *50*, 1287–1298. [CrossRef]
32. Zhan, X.; Chung, H.; Zhang, R. Investigation into the Use of Single Inductor for Driving Multiple Series-Connected LED Channels. *IEEE Trans. Power Electron.* **2017**, *32*, 3034–3050. [CrossRef]

33. Ng, S.K.; Loo, K.H.; Lai, Y.M.; Tse, C.K. Color Control System for RGB LED with Application to Light Sources Suffering from Prolonged Aging. *IEEE Trans. Ind. Electron.* **2014**, *61*, 1784–1798. [CrossRef]
34. Lee, A.; Sin, J.; Chan, P. Scalability of quasi-hysteretic FSM based digitally controlled single-inductor dual-string buck LED driver to multiple strings. *IEEE Trans. Power. Electron.* **2014**, *29*, 501–513. [CrossRef]
35. Kim, H.C.; Yoon, C.S.; Ju, H.; Jeong, D.-K.; Kim, J. An AC-powered, flicker-free, multi-channel LED driver with current-balancing SIMO buck topology for large area lighting applications. In Proceedings of the 2014 IEEE Applied Power Electronics Conference and Exposition-APEC 2014, Fort Worth, TX, USA, 16–20 March 2014; pp. 3337–3341.
36. Zhang, W.; Sutardja, P. Correlated Color Temperature Control Methods and Devices. U.S. Patent US20130020956 A1, 24 January 2013.
37. Wang, H.-C.; Chen, Y.T.; Lin, J.T.; Chiang, C.P.; Cheng, F.H. Enhanced visualization of oral cavity for early inflamed tissue detection. *Optics Express* **2010**, *18*, 11800–11809. [CrossRef] [PubMed]
38. Chen, Y.S.; Lin, C.Y.; Yeh, C.M.; Kuo, C.T.; Hsu, C.W.; Wang, H.C. Anti-glare LED lamps with adjustable illumination light field. *Optics Express* **2014**, *22*, 5. [CrossRef] [PubMed]
39. Chen, H.T.; Tan, S.C. Nonlinear Dimming and Correlated Color Temperature Control of Bi-Color White LED Systems. *IEEE Trans. Power Electron.* **2014**, *30*, 6934–6947. [CrossRef]
40. Malik, R.; Ray, K.; Mazumdar, S. Wide range, Open-loop, CCT and Illuminance Control of an LED Lamp using Two-component Colour Blending. *IEEE Trans. Power Electron.* **2017**, *33*, 9803–9818. [CrossRef]
41. Lee, A.T.L.; Chen, H.; Tan, S.C.; Hui, S.Y. Precise Dimming and Colour Control of LED Systems Based on Color Mixing. *IEEE Trans. Power Electron.* **2016**, *31*, 65–80. [CrossRef]
42. Lohaus, L.; Leicht, E.; Dietrich, S.; Wunderlich, R.; Heinen, S. Advanced Color Control for Multicolor LED Illumination Systems with Parametric Optimization. In Proceedings of the IECON 2013-39th Annual Conference of the IEEE Industrial Electronics Society, Vienna, Austria, 10–13 November 2013; pp. 3305–3310.
43. Ferdous, F.; Rashid, A.B.M.H. Design of a High-Performance AC-DC LED Driver Based on SEPIC Topology. *Int. J. Power Electron. Drive Syst. (IJPEDS)* **2021**, *12*, 870–885. [CrossRef]
44. Pal, S.; Singh, B.; Chandra, A. A Novel Ambiance Changing Energy Saving LED Lighting System with PFM Operated Zeta Converter. In Proceedings of the 2015 IEEE Industry Applications Society Annual Meeting, Addison, TX, USA, 18–22 October 2015; pp. 1–8.
45. Germaine, C.P.; Lee, A.T.; Tan, S.C.; Hui, S.R. Precise Luminous Flux and Color Temperature Control of Dimmable Bi- Colour White Light-Emitting Diode Systems. In Proceedings of the 2022 IEEE Applied Power Electronics Conference and Exposition (APEC), Houston, TX, USA, 20–24 March 2022.

World Electric Vehicle Journal

MDPI

Article

The Suppression of Modular Multi-Level Converter Circulation Based on the PIR Virtual Impedance Strategy

Chun Wang, Wenxu Yan *, Wenyuan Wang, Hongyu Ni and Jie Chu

School of Internet of Things, Jiangnan University, Wuxi 214000, China
* Correspondence: ywx01@jiangnan.edu.cn

Abstract: In recent years, with the rise of the electric vehicle industry, there has been significant research on charging and power supply vehicle technologies for electric vehicles. In terms of the corresponding converter usage, modular multi-level converters (MMCs) are also increasingly used in the field of electric vehicle power supply research because of their unique advantages. However, the circulating current problem of MMCs has not been effectively addressed in existing domestic and international studies. In this paper, we propose a proportional-integral resonant (PIR) control method combined with virtual impedance for the optimal suppression of the MMC internal circulating current problem based on the comparison and generalization of the existing methods. Based on the analysis of the working principle of MMCs, this paper proposes and adopts the control strategy of combining virtual impedance and proportional-integral resonance to suppress the circulating current and builds a simulation model in MATLAB to verify that the control strategy proposed in this paper is feasible.

Keywords: MMC; harmonic suppression; circuit circulation; PIR control; virtual impedance; MATLAB/Simulink simulation

1. Introduction

In recent years, with the promotion of new energy use, more and more fields are using new energy as energy supplies. Because of its unique characteristics of new technology, the electric vehicle has many links to new energy storage-related fields for its charging and discharging. The charging and discharging process of an EV battery is inseparable from the converter, and there have been many power electronics pieces of research on the interaction between EVs, power supplies, and grids. Modular multi-level converters (MMCs) have received significant attention since they were proposed in 2002 and have been widely used in power electronic transformers, energy routers, and the energy internet due to their easy scalability, modularity, and applicability to large-capacity applications. Due to their special cascade topology, they have a smoother output waveform and lower switching losses with good application prospects in the field of power converters for electric vehicles [1–3].

In the use of electric vehicle power systems, modular multi-level converters have numerous advantages over conventional converter topologies. MMC circuits are modularly cascaded, have strong scalability, and meet the requirements of various voltage-level applications by cascading different numbers of submodules in the MMC. The MMC avoids the series connection between quantitative power-switching devices, which greatly reduces the converter manufacturing difficulty of the converter [4]. However, the MMC topology also has certain drawbacks compared to the conventional converter. Because of the unbalanced energy distribution between phases, MCCs could cause the formation of circulating currents inside the structure, resulting in more considerable losses inside the converter due to the bridge arm current distortion. It was found that, on the one hand, the presence of circulating harmonics increases the bridge arm current distortion rate, boosts the rated current capacity of the switching components, and increases the system cost; on the other hand, the circulating current increases system losses, leads to higher battery cell temperatures, and accelerates battery aging [5–7].

Citation: Wang, C.; Yan, W.; Wang, W.; Ni, H.; Chu, J. The Suppression of Modular Multi-Level Converter Circulation Based on the PIR Virtual Impedance Strategy. *World Electr. Veh. J.* **2023**, *14*, 17. https://doi.org/10.3390/wevj14010017

Academic Editor: Joeri Van Mierlo

Received: 7 December 2022
Revised: 26 December 2022
Accepted: 27 December 2022
Published: 5 January 2023

In the current use and research of MMC devices in various countries, many scholars have studied the internal circulating current of MMCs. From the perspective of passive hardware improvement, some scholars have proposed the method of improving the hardware circuit by increasing the bridge arm reactor to suppress the circulating current, but this increases the overall volume of the MMC, increasing the cost; the system would generate uncontrollable energy internally [8,9]. From the perspective of active control, some scholars have used the traditional proportional-integral (PI) control to achieve the active suppression of the circulating current, controlling the sum of the upper and lower bridge arm voltages as the direct flow without even harmonics, thus eliminating the even harmonics in the circulating current. However, the classical PI controller has limited control gain for AC signals and thus is only suitable for tracking DC signals and cannot completely eliminate a series of even harmonic components in the circulating current [10]. Some scholars have compensated the oscillations of the arm capacitor voltage by compensating the modulation with calculations, but this method requires additional energy to control the loop to achieve system balance [11,12].

This paper proposes a suppression strategy based on MMC harmonic circulation by combining PIR and virtual impedance for the typical MMC system. First, this paper analyzes the topological structure and working principle of MMCs from the objective analysis of suppressing harmonic circulation. Then, we propose a strategy combining virtual impedance and proportional-integral resonance control circulation suppression strategy based on the mathematical analysis of calculating MMC harmonic circulation. Finally, we built a typical MMC circuit simulation model to verify the effectiveness of the strategy of even-harmonic reduction.

2. MMC Analysis

In order to suppress the circulating harmonic current inside the MMC circuit, this section first models and analyzes the MMC circuit. This chapter explains the operation principle of MMCs in the three-phase balanced state and calculates and analyzes the main internal variables of the internal circulating current.

2.1. MMC Model

2.1.1. MMC Basic Circuit

As shown in Figure 1, a typical three-phase MMC circuit topology consists of a three-phase, six-bridge arm with an n submodule SM, a bridge arm reactor L_m, and bridge arm resistance R_m. The three-phase bridge arm on the DC side of the system is jointly connected to the HVDC DC bus, where U_{dc} and I_{dc}, are the current DC bus voltage and current, respectively. The AC side of the system includes equivalent AC reactance L_s and equivalent resistance R_s. The AC side load could be a resistance load, power grid, or a high-voltage motor [13–15].

When the MMC circuit operates in the inverter state, the multi-level voltage waveform is synthesized on the MMC AC side by controlling the input and cutting out of each submodule in the bridge arm, thus emitting active and reactive power to the AC side. The public DC bus of the MMC would also charge and discharge the DC capacitor of the submodule to maintain the stability of the module capacitance-voltage. In Figure 1, u_{uj} and u_{lj} are the module port voltage of the upper and lower bridge arms of the phase. j, i_{uj}, and i_{lj} are the current flowing through the upper and lower bridge arms of the phase. j and U_{sj} are the AC side voltage and the number of phases is $j = a, b, c$.

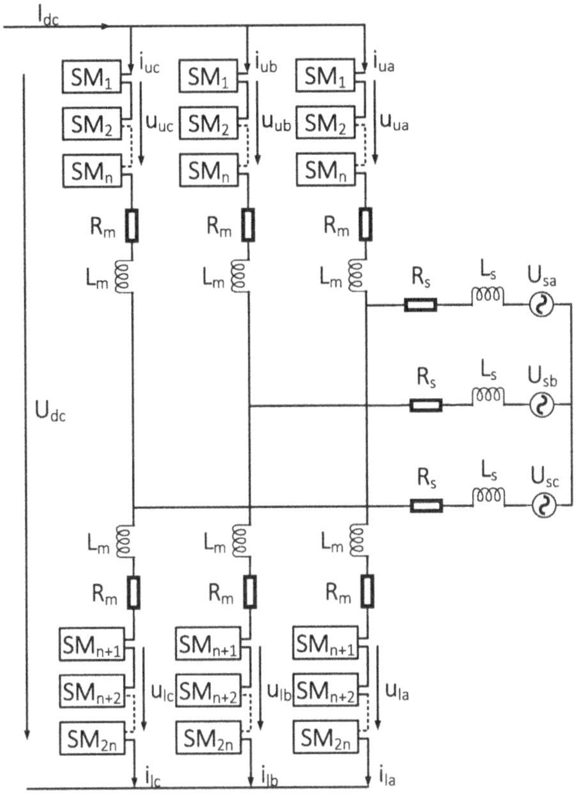

Figure 1. Typical MMC circuit structure.

Under the normal operation of an MMC, the total number of the connecting modules of the upper and lower bridge arm is to ensure the stable voltage of the DC side. The amplitude and the waveform of the output voltage could be changed by changing the difference between the number of input submodules in the upper and lower bridge arms.

Each *SM* in the circuit is an independent submodule, and the MMC submodule has two kinds of half-bridge structures and full-bridge structures [16]. Among them, the half-bridge structure submodule is more widely used due to the small number of devices and better economy. The commonly used MMC half-bridge submodule topology is shown in Figure 2, consisting of two power-switching tubes with anti-parallel diodes and a capacitor.

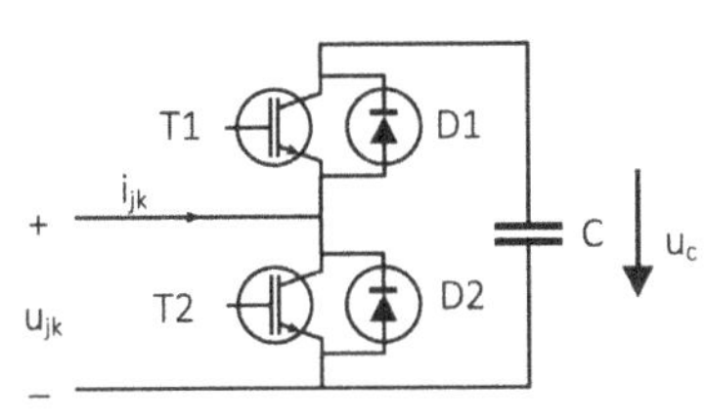

Figure 2. MMC half-bridge submodule structure.

During normal operation, the driving signals of the two power-switch tubes are opposite to avoid a short circuit of the capacitance. The bridge arm flow direction is the internal positive direction of the current i_{jk}, and the working state of the submodule *T*1 is *S*. According to the power-switch tube drive signal and the corresponding current flow direction, there are four modes of flow channels inside the submodule, as shown in Figure 3.

(1) When the drive signal of the upper switch T_1 is placed on the submodule, T_1 is in the input state $S = 1$. When the bridge arm current i_{jk} is positive, the current charges the capacitor through the anti-parallel diode $D1$ on the switch $T1$, the capacitance-voltage u_C rises, and the output voltage u_{jk} of the submodule is equal to the capacitance-voltage u_C, as shown in Figure 3a. When the bridge arm current i_{jk} is negative, the current discharges the capacitance through the upper switch tube $T1$, then the capacitance-voltage u_C drops and the output voltage u_{jk} of the submodule is equal to the capacitance-voltage u_C, as shown in Figure 3b.

(2) When the drive signal of the switch $T1$ is set to 0, the submodule $T1$ is in the bypass state $S = 0$. When the bridge arm current i_{jk} is positive, the current bypasses the capacitance through $T2$, the capacitance-voltage u_C remains unchanged, and the output voltage u_{jk} of the submodule is 0, as shown in Figure 3c. When the bridge arm current i_{jk} is negative, the current bypasses the capacitance through $D2$, and then the capacitance-voltage u_C remains unchanged, and the output voltage u_{jk} of the submodule is 0, as shown in Figure 3d.

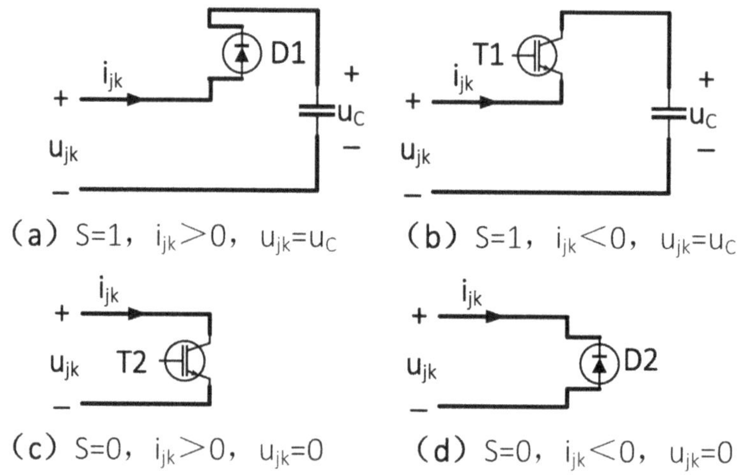

Figure 3. Working principle of an MMC half-bridge submodule. (**a**) Working condition in S = 1, $i_{jk} > 0$. (**b**) Working condition in S = 1, $i_{jk} < 0$. (**c**) Working condition in S = 0, $i_{jk} > 0$. (**d**) Working condition in S = 0, $i_{jk} < 0$.

Therefore, the relationship between the submodule output voltage and the internal capacitance-voltage of the submodule could be expressed as:

$$u_{jk} = S \times u_C \tag{1}$$

2.1.2. MMC Equivalent Circuit

For the above typical MMC circuit, the following basic assumptions are stated to simplify the analysis:

(1) The system is in a symmetrical three-phase normal operation state; (2) the AC current on the network side is divided symmetrically and evenly between the upper and lower bridge arms; (3) ignore the fluctuation of the DC bus voltage U_{dc}.

Based on the above assumptions, the upper and lower arm voltage of each phase could be equivalent to the AC voltage source u_{uj}, u_{lj}, and the upper and lower arm current could be represented by i_{uj},i_{lj}, as shown in Figure 4. Then, we built a three-phase equivalent circuit model of a typical MMC topology and performed an internal voltage and current analysis [17–19].

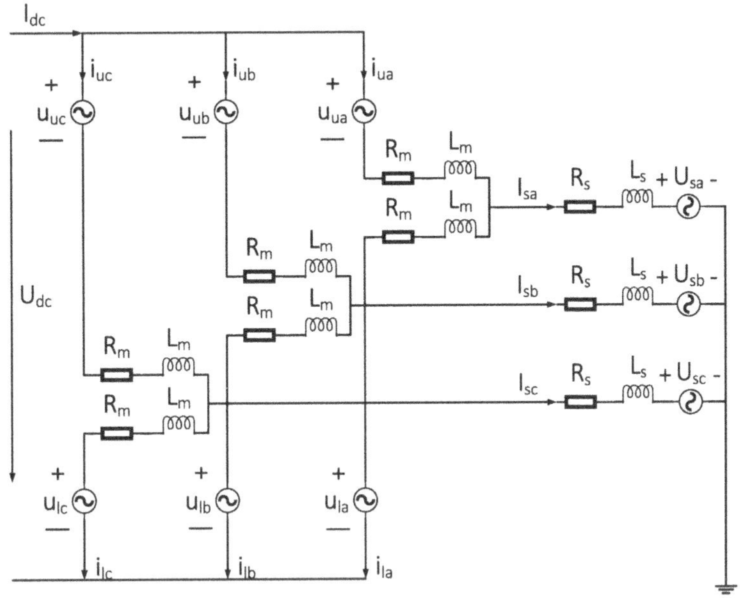

Figure 4. MMC three-phase equivalent model.

Without considering the bridge arm current harmonic component, the three-phase MMC equivalent circuit model is obtained with Kirchhoff's theorem, and the current of the upper and lower bridge arm in each phase is, respectively:

$$\begin{cases} i_{uj} = \frac{I_{dc}}{3} + \frac{I_{sj}}{2} \\ i_{lj} = \frac{I_{dc}}{3} - \frac{I_{sj}}{2} \end{cases} \tag{2}$$

In the equation, I_{dc} is the current in the DC bus, and I_{sj} is the phase current on the network side. Suppose that the number of submodules of the upper and lower bridge arm input conduction is i, j, and since the voltage needs to be kept stable on the DC side, generally:

$$i + j = n. \tag{3}$$

Assume that each submodule has uniform voltage distribution, then the sum of the output voltage of the upper and lower bridge arm submodules u_{uj},u_{lj} is, respectively:

$$\begin{cases} u_{uj} = \sum\limits_{i=1}^{n} u_{jk} \\ u_{lj} = \sum\limits_{j=1}^{n} u_{jk} \end{cases} \tag{4}$$

In the equation, u_{jk} is the output AC voltage of the upper and lower bridge arm submodule port.

From the MMC equivalent circuit of Figure 4, the upper and lower bridge arms could be regarded in parallel from the AC side, so the characteristic equation of the external characteristics is:

$$U_{sj} = \frac{(u_{uj} - u_{lj})}{2} + (R_s + \frac{R_m}{2})I_{sj} + (L_s + \frac{L_m}{2})\frac{dI_{sj}}{dt} \tag{5}$$

2.2. Circulation Analysis

According to the above analysis, the capacitor of each submodule is evenly divided and balanced at all times, and each capacitance voltage is $U_{dc}/(i+j)$ in the ideal working state. However, in the actual working conditions, the capacitance parameter in each submodule cannot be completely consistent, and the small error of the switch pipe conduction time would lead to the charge and discharge time of each input submodule not being fully synchronized; the capacitance-voltage in each submodule would produce a small deviation, and cannot always be fully balanced. The capacitance-voltage imbalance in the phase bridge arm submodule and the switching is not a completely ideal device and would cause the MMC to produce an alternating harmonic circulation inside [20–22].

Define U_{acj} as the j phase of the AC side's voltage, and the MMC output voltage modulation ratio is m. In practical engineering uses, $m \in (0,1)$ is generally selected:

$$m = \frac{2U_{acj}}{U_{dc}} \tag{6}$$

The power P_{uj},P_{lj} of each phase's upper and lower bridge arm could be calculated as:

$$\begin{cases} P_{uj} = u_{uj}i_{uj} = \frac{1}{2}U_{dc}[1 - m\sin(\omega t)]\left[\frac{1}{3}I_{dc} + \frac{1}{2}I_{sj}\sin(\omega t + \varphi)\right] \\ P_{lj} = u_{lj}i_{lj} = \frac{1}{2}U_{dc}[1 + m\sin(\omega t)]\left[\frac{1}{3}I_{dc} - \frac{1}{2}I_{sj}\sin(\omega t + \varphi)\right] \end{cases} \tag{7}$$

after calculating, the formula is expanded to obtain the following results:

$$\begin{cases} P_{uj} = \frac{1}{6}U_{dc}I_{dc} - \frac{1}{8}mU_{dc}I_{sj}\cos\varphi - \frac{1}{6}U_{dc}I_{dc}m\sin(\omega t) \\ \quad +\frac{1}{4}U_{dc}I_{sj}\sin(\omega t + \varphi) + \frac{1}{8}mU_{dc}I_{sj}\cos(2\omega t + \varphi) \\ P_{lj} = \frac{1}{6}I_{dc}U_{dc} - \frac{1}{8}mU_{dc}I_{sj}\cos\varphi + \frac{1}{6}U_{dc}I_{dc}m\sin(\omega t) \\ \quad -\frac{1}{4}U_{dc}I_{sj}\sin(\omega t + \varphi) + \frac{1}{8}mU_{dc}I_{sj}\cos(2\omega t + \varphi) \end{cases} \tag{8}$$

It could be seen that the power flowing through the upper and lower bridge arms includes the DC component, base frequency, and double frequency components. The external output power of the phase j unit P_{out} and the circulation power inside the ring P_{in} are, respectively:

$$\begin{cases} P_{out} = P_{uj} - P_{lj} = -\frac{1}{3}U_{dc}I_{dc}m\sin(\omega t) + \frac{1}{2}U_{dc}I_{sj}\sin(\omega t) \\ P_{in} = P_{uj} + P_{lj} = \frac{1}{3}U_{dc}I_{dc} - \frac{1}{4}mU_{dc}I_{sj}\cos\varphi + \frac{1}{4}mU_{dc}I_{sj}\cos(2\omega t + \varphi) \end{cases} \tag{9}$$

It could be analyzed that the output power of the MMC to the AC network's side only contains the base wave components; without the DC and harmonic components, the internal circulation power contains a double frequency component. By dividing the circulation power by the voltage, we see that the internal circulation has a second-order harmonic component.

Through the above derivation process, it could be concluded that the MMC bridge arm circulation and the capacitor voltage in the submodule affect each other, the second-order harmonic component of the voltage in each phase cause second-order harmonic circulation, and the second-order circulation further leads to the third-order capacitor's voltage fluctuations in the submodule, and further causes forth-order circulation, which produces messy harmonic components in the MMC. Since the odd harmonics of the upper and lower bridge arms could always cancel each other, only the even harmonic components flow internally. They have opposite symbols, and the phase difference is 120° and in the negative order.

Define the harmonic component of the circulating current present in the bridge arm of phase j as i_{cirj}, so Equation (2) could be modified and rewritten as:

$$\begin{cases} i_{uj} = \frac{I_{dc}}{3} + \frac{I_{sj}}{2} + i_{cirj} \\ i_{lj} = \frac{I_{dc}}{3} - \frac{I_{sj}}{2} + i_{cirj} \end{cases} \tag{10}$$

By defining i_{zj} as the circulation current on the bridge arm and adding i_{uj},i_{lj} in Equation (10) to offset the AC side I_{sj}, we obtain i_{zj} as:

$$i_{zj} = \frac{i_{uj} + i_{lj}}{2}. \tag{11}$$

According to Equation (10), i_{zj} is composed of direct flow and harmonic superposition so that it could be written as:

$$i_{zj} = \frac{I_{dc}}{3} + i_{cirj} \tag{12}$$

2.3. Modulation Method

In this paper, the nearest level modulation (NLM) was used to perform the submodule modulation of the MMC circuit [23,24]. The nearest level modulation approximates the modulating waveform steps and renders the output waveform produced by the converter similar to the modulating waveform by controlling the input and cut-out of each submodule in the circuit by superimposing the voltage of each submodule. The number of input submodules for the upper and lower bridge arms is as follows:

$$\begin{cases} i = \frac{n}{2} - \text{round}\left[\frac{n}{2} m \sin(\omega t)\right] \\ j = \frac{n}{2} + \text{round}\left[\frac{n}{2} m \sin(\omega t)\right] \end{cases}. \tag{13}$$

where $round[f(x)]$ is the integer function of approximation and the integer with the smallest absolute value of the difference with $f(x)$ is obtained.

3. MMC Circulation Control Strategy

This chapter introduces the MMC circuit harmonic loop suppressor designed in this paper, the principles of the main components of the controller, and the structure of the overall controller.

3.1. Controller Analysis

3.1.1. PIR Control

According to the previous introduction and analysis of the internal working principle of an MMC, it could be learned that the second and fourth-order harmonics are the most important components of the circulating harmonic current, and the high-order current is triggered by the low-order current. Therefore, by reducing the size of the secondary and fourth harmonics, the circulation of each harmonic in the circuit could be effectively suppressed.

The proportional-integral (PI) controller part could realize the unstated difference tracking of the DC components in the circulation, but the AC signal cannot be tracked without the unstated difference. The proportional resonance (PR) controller part could achieve infinite gain at the selected frequency so that it could control a certain frequency, but it needs tedious coordinate transformation and decoupling work before the control process. The proportional-integrated resonance (PIR) controller in the composite current controller could improve the harmonic characteristics and eliminate the derived specific harmonic components so that the controller could achieve a better control effect [25,26].

The traditional PR controller is developed from the internal mode principle, and the general transfer function of the resonant frequency ω_0 is:

$$G_{PR}(s) = K_p + \frac{K_i s}{s^2 + \omega_0^2}. \tag{14}$$

The ideal PR controller could have an infinite gain at the resonant frequency, but there is no gain on either side of this frequency. If the power grid frequency offset occurs, it would directly affect the stability of the system and seriously affect the effect of the controller.

An optimized quasi-proportional-integral resonant controller is proposed in order to maintain the advantages of the PR controller in terms of its high gain while effectively reducing the effects of grid frequency shifts. For the suppression of MMC circulation in this paper, the controller takes second and forth-order frequencies as the main control objects, so the PIR controller transmission function is:

$$G_{PIR}(s) = K_p + \frac{K_i}{s} + \frac{K_{r1}s}{s^2 + \omega_c s + (2\omega_0)^2} + \frac{K_{r2}s}{s^2 + \omega_c s + (4\omega_0)^2} \tag{15}$$

where K_p is the proportional coefficient, K_i is the integral coefficient, K_{r2} and K_{r4} are the resonance control gain of the binary and quadruple frequency, respectively, ω_0 is the base wave frequency, and ω_c is the resonant bandwidth parameter, which controls the range of the resonant frequency (bandwidth), and has the general value 5–15 rad/s.

3.1.2. Virtual Impedance

On the basis of the PIR controller, the additional virtual bridge arm impedance link could effectively undertake the circulation high-order harmonic component suppression function so as to achieve better results. The method of virtual impedance could inhibit many high harmonics with less content in the circulation so that not only the virtual impedance has a small partial pressure on the system, but it also could effectively inhibit the circulation [27,28].

As shown in Figure 5, L_v and R_v constitute the virtual impedance part together, with input i_{cir} as the bridge arm harmonic circulation and output u_v as the correction component of the modulated wave.

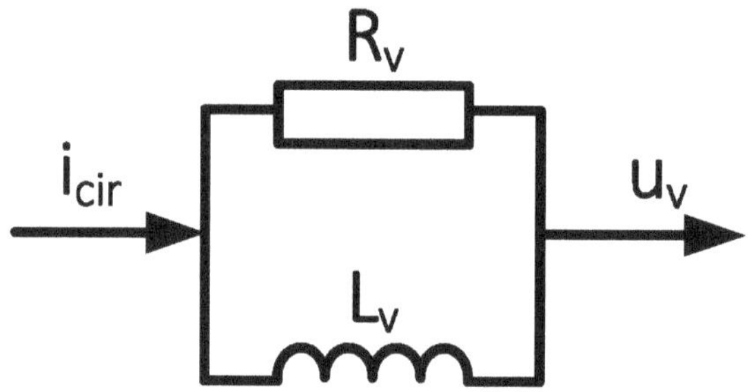

Figure 5. Virtual impedance equivalence diagram.

In the controller, the virtual impedance link is the first-order inertial link, whose transfer function is:

$$G_{vir}(s) = \frac{R_v L_v s}{L_v s + R_v} \tag{16}$$

3.2. Controller Instruction

Based on the synthesis of the two controllers, this paper proposed that, based on the PIR virtual impedance control strategy, the control logic is as follows: the bridge arm circulation i_{zj} is through the PIR controller adjustment control modulation wave component u_{pir}, then the harmonic component in the circulation goes through the virtual impedance first-order inertia correction u_{vir}, combining the two together to obtain the modulation wave voltage of the harmonic circulation suppression controller u_h. The overall control block diagram is shown in Figure 6:

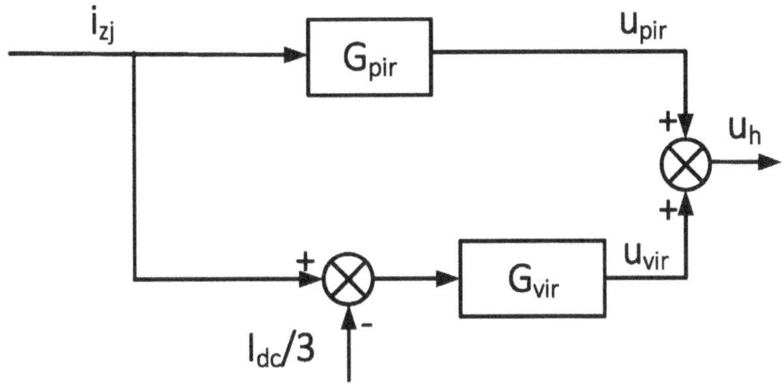

Figure 6. Structural diagram of the PIR virtual impedance controller.

4. Simulation Analysis

In order to verify the effect of the PIR virtual impedance composite control method proposed in this paper, the corresponding model was built in MATLAB for simulation verification (Supplementary Materials: Simulation files S1).

4.1. Simulation Parameters

In this paper, the simulation platform selected the MATLAB2020a Simulink environment and built a three-phase 22 submodule MMC circuit. The circuit adopted the double closed-loop control of the outer loop power control and the inner loop current control; the modulation method adopted the nearest level modulation (NLM) technology. The selected simulation parameters are shown in Table 1, and the selected controller parameters are shown in Table 2 [29,30].

Table 1. Simulation parameters.

Simulation Variable	Parameter Setting
DC bus voltage/kV	5.5
Number of single bridges Arm submodules *n*	22
Submodule capacitance/mF	7
Bridge arm reaction/mH	8
Switching frequency/kHz	10
Modulation ratio	0.9
Load resistance/Ω	1

Table 2. Controller parameters.

Controller Parameters	Values
K_p	40
K_i	600
K_{r1}	1000
K_{r2}	1000
ω_c	5
R_v/Ω	1
L_v/mH	10

4.2. Simulation Result

4.2.1. Output for the Steady-State Operation

To verify the feasibility of the control strategy used in this paper, the main output quantities in the circuit could be observed so that the steady-state operation control effect of the overall MMC circuit could be observed [31]. The simulation was set to start from t = 0 s, and the active power disturbance was added on the AC side at t = 0.15 s during the simulation to verify the robustness of the circuit against the power fluctuation in the grid-connected operation. The three-phase voltages and currents are shown in their respective pictures using red, blue, and orange.

The voltage and current waveforms of the three-phase AC side output are, respectively, shown in Figure 7a,b. It was observed that before t = 0.15 s, the voltage and current were running in balance; after t = 0.15 s, the voltage and current were also running in balance; at the moment of t = 0.15 s, due to the increase in power, the current automatically increased in order to ensure constant output voltage. The fast Fourier transform (FFT) analysis result shows that the total harmonic distortion (THD) of the three-phase output voltage and current were 1.9% and 0.94%, respectively, as shown in Figure 8a,b. It was observed that the output waveforms worked well with low distortion rates and exhibited good robustness when experiencing power disturbances.

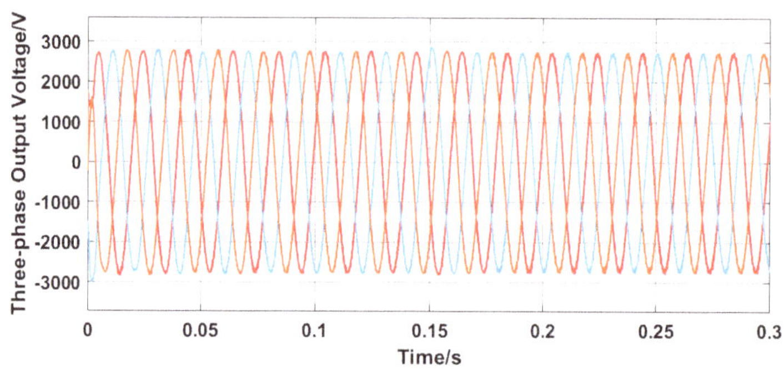

(**a**)

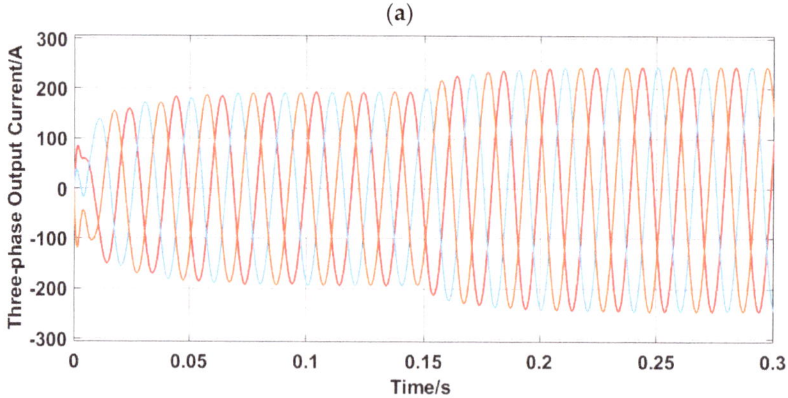

(**b**)

Figure 7. (**a**) Three−phase output voltage. (**b**) Three−phase output current.

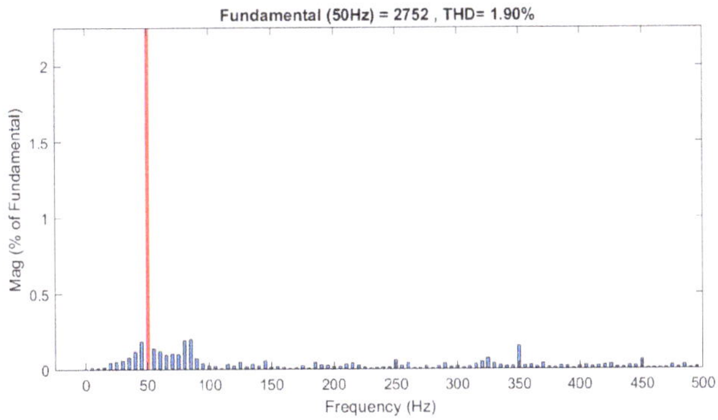

(**a**)

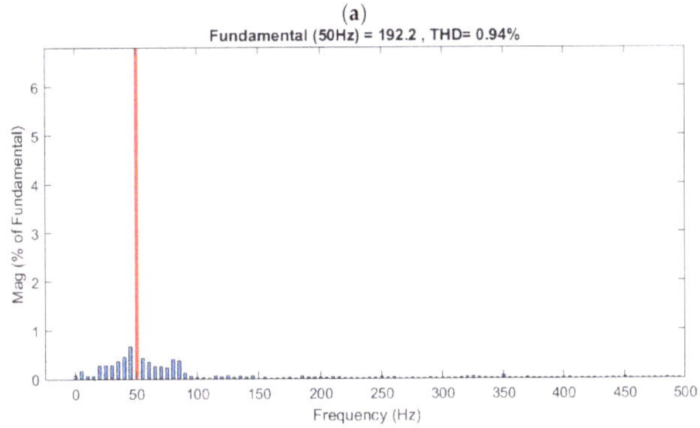

(**b**)

Figure 8. (**a**) FFT analysis of the three−phase output voltage. (**b**) FFT analysis of the three−phase output current.

The DC side's current is shown in Figure 9. When the DC side's current reached the steady state, it was controlled to fluctuate within a small range, which was allowed with the normal operating conditions.

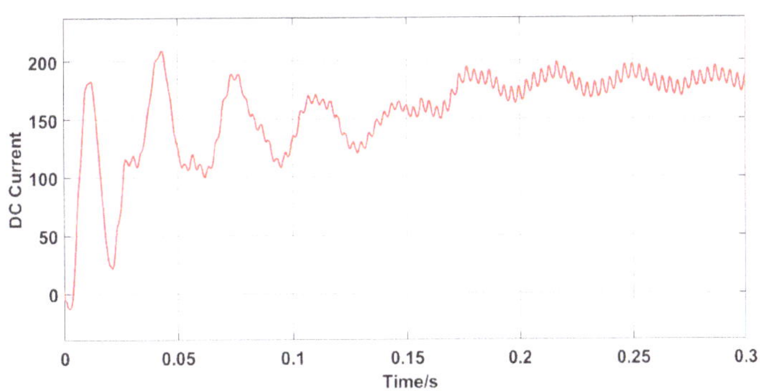

Figure 9. DC side's current.

The modulation process is shown in Figure 10, which shows the modulation waves of the upper and lower bridge arms of phase a. It was observed as a step wave similar to the sine waveform with a high degree of symmetry.

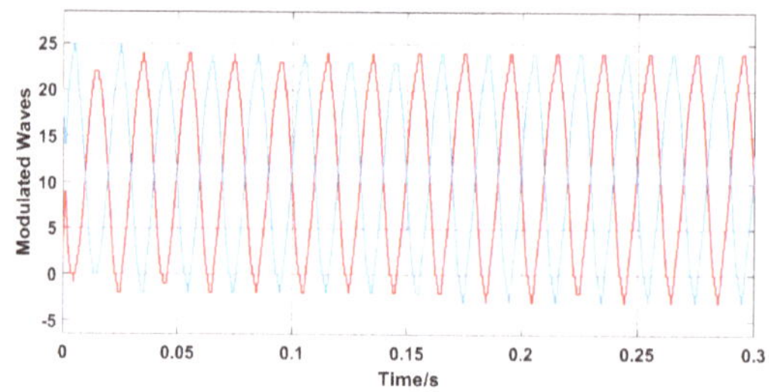

Figure 10. Modulated waves.

According to the analysis above, it is seen that the MMC circuit works stably with good output waveform, and the system has certain robustness under the use of the controller in this paper.

4.2.2. Comparison of the Control Effects

In order to verify the effectiveness of the harmonic circulating current suppression of the MMC circuit in this paper, the effect of the circulating current suppression was compared with that of the open-loop strategy, conventional PI control strategy, and QPR control strategy. The waveform and amplitude of the harmonic loop current component of the MMC circuit during steady-state operation could be observed. The current of the upper bridge arm of phase a was selected as the reference object, and the total harmonic distortion under different control methods was compared with FFT analysis.

The controller, without adding the circulating current suppressor, is shown in Figure 11. The current waveform of the harmonic current component is shown in Figure 11a, and the THD of the arm current was 30.04%, as shown in Figure 11b.

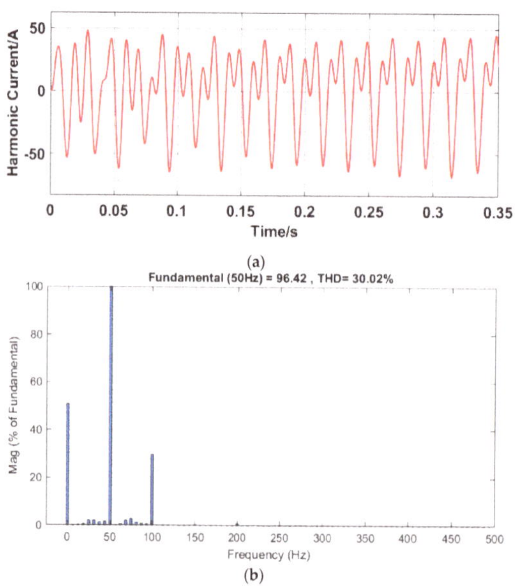

Figure 11. (**a**) Harmonic current without the suppressor. (**b**) FFT analysis of the bridge arm current without the suppressor.

The case of using the PI control circulating current suppressor is shown in Figure 12. The harmonic current component waveform is shown in Figure 12a, and the THD of the arm current was 5.23%, as shown in Figure 12b.

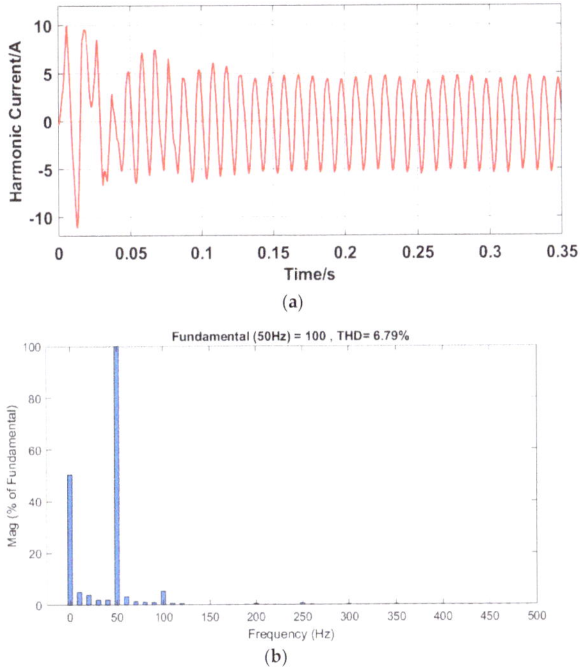

Figure 12. (**a**) Harmonic current with the PI control suppressor. (**b**) FFT analysis of the bridge arm current with the PI control suppressor.

The case of using the QPR control circulating current suppressor is shown in Figure 13. The harmonic current component waveform is shown in Figure 13a, and the THD of the arm current was 4.48%, as shown in Figure 13b.

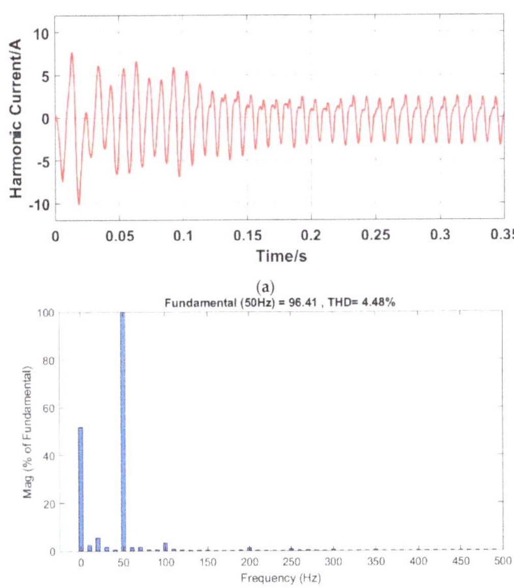

Figure 13. (**a**) Harmonic current with the QPR control suppressor. (**b**) FFT analysis of the bridge arm current with the QPR control suppressor.

The case of using the PIR virtual impedance composite suppressor designed in this paper is shown in Figure 14. The upper bridge arm current waveform is shown in Figure 14a, and the THD of the arm current was 1.68%, as shown in Figure 14b.

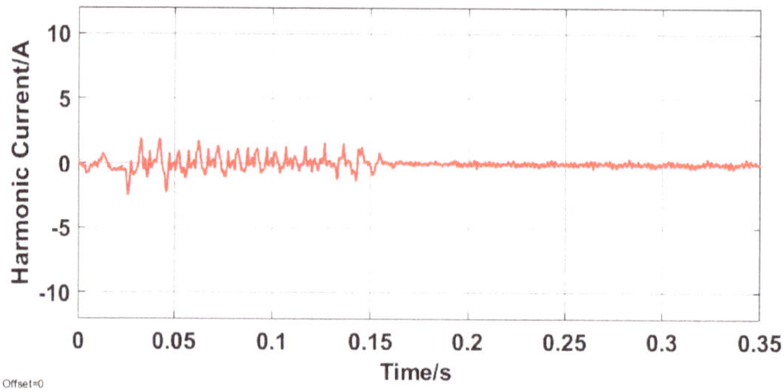

(a)

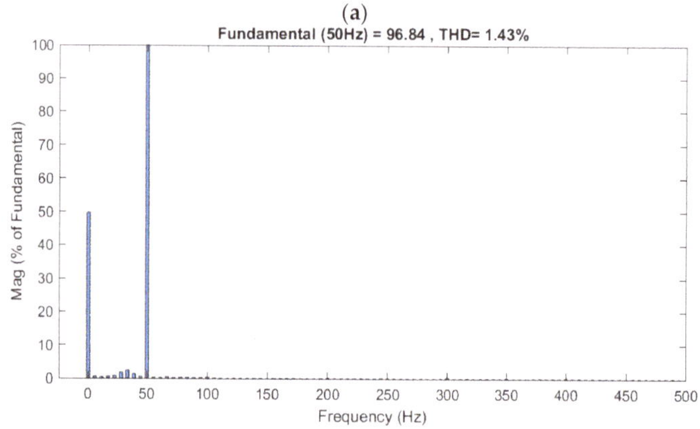

(b)

Figure 14. (**a**) Harmonic current with the PIR virtual impedance composite suppressor. (**b**) FFT analysis of the bridge arm current with the PIR virtual impedance composite suppressor.

Table 3 displays the cross-sectional comparisons of the THD of the bridge arm current for the different control methods.

Table 3. Comparisons of the THD using different methods.

Controller Use	THD of the Bridge Arm Current (%)
Without adding the current suppressor	30.02
PI control suppressor	6.79
QPR control suppressor	4.48
PIR virtual impedance composite suppressor	1.43

It is seen from the comparison that the current distortion rate of the bridge arm is very high without adding the circulation control, and there is still a large THD with the addition of the PI circulation control or QPR circulation control. When controlled by the PIR virtual impedance composite harmonic circulating current strategy proposed in this paper, the obtained harmonic current waveform flattens, and the distortion rate is lower, which proves a good control effect.

5. Discussion

With the control strategy used in this paper, the harmonic loops in the circuit could be effectively suppressed, which could reduce component losses in the circuit and extend the component's life. In the field of electric vehicles, this move could lead to optimized output voltage and current, more efficient application of charging lines, and better power quality.

In addition to the work already conducted in this paper, there is still some incompleteness, and these could be considered as subsequent research directions. Since there is no well-built physical experimental platform in this lab, experimental verification has not yet been completed, which would be a long-term work direction to follow. The control strategy proposed in this paper is mainly based on a three-phase balanced circuit, but the operation mode and fault ride-through capability of the circuit under unbalanced conditions were not thoroughly studied.

6. Conclusions

In this manuscript, the internal variables of the MMC circuit were analyzed and explained in detail based on the analysis of its topology in the second chapter. Through circuit derivation, its equivalent circuit was provided, the composition of the internal circulating current was further analyzed, and its harmonic components were explained. In the third chapter of this paper, based on the controller principle, this paper proposed a control method based on the combination of a proportional-integral resonant controller and virtual impedance, which was used to effectively suppress the internal harmonic circulations during the operation of the circuit. In the fourth chapter, the proposed control method was fully simulated through the MATLAB simulation platform. On the one hand, the stability of the operation was analyzed, and it was observed that the three-phase output voltage and current were better, with distortion rates of 1.9% and 0.94%, respectively, and still maintained good robustness when the rated active power fluctuated. On the other hand, the effect of the controller was verified by comparing it with some common circulating current control strategies. The a-phase upper bridge arm current distortion rate was 30.02%, 6.79%, and 4.48% under the traditional open-loop controller, PI controller and QPR controller, respectively, and the distortion rate was 1.43% when the selected controller was used. In summary, it could be concluded that the controller proposed in this paper could effectively suppress the internal harmonic loop current with good results while ensuring the overall stable operation of the circuit.

Supplementary Materials: The following supporting information can be downloaded at: https://www.mdpi.com/article/10.3390/wevj14010017/s1, Simulation files S1: MMC_PIRV.zip.

Author Contributions: Conceptualization, C.W. and W.Y.; Data curation, C.W.; Formal analysis, C.W. and W.W.; Funding acquisition, H.N. and J.C.; Investigation, C.W.; Methodology, C.W.; Resources, C.W.; Software, C.W.; Validation, C.W.; Writing–original draft, C.W.; Writing–review and editing. All authors have read and agreed to the published version of the manuscript.

Funding: This research was funded by the State Grid Zhejiang Province Technology Project, grant number 5211SX220003.

Institutional Review Board Statement: Not applicable.

Informed Consent Statement: Not applicable.

Data Availability Statement: Not applicable.

Conflicts of Interest: The authors declare no conflict of interest.

References

1. Gan, C.; Sun, Q.; Wu, J.; Kong, W.; Shi, C.; Hu, Y. MMC-Based SRM Drives With Decentralized Battery Energy Storage System for Hybrid Electric Vehicles. *IEEE Trans. Power Electron.* **2018**, *34*, 2608–2621. [CrossRef]
2. Guidi, G.; D'Arco, S.; Nishikawa, K.; Suul, J.A. Load Balancing of a Modular Multilevel Grid-Interface Converter for Transformer-Less Large-Scale Wireless Electric Vehicle Charging Infrastructure. *IEEE J. Emerg. Sel. Top. Power Electron.* **2020**, *9*, 4587–4605. [CrossRef]
3. Mao, M.; Ding, Y.; Chang, L.; Hatziargyriou, N.D.; Chen, Q.; Tao, T.; Li, Y. Multi-Objective Power Management for EV Fleet With MMC-Based Integration Into Smart Grid. *IEEE Trans. Smart Grid* **2017**, *10*, 1428–1439. [CrossRef]
4. Wang, D.; Liu, J.; Piegari, L.; Song, S.; Chen, X.; De Simone, D. A Battery Lifetime Improved Control Strategy of Modular Multilevel Converter for Electric Vehicle Application. In Proceedings of the 2019 IEEE 10th International Symposium on Power Electronics for Distributed Generation Systems (PEDG), Xi'an, China, 3–6 June 2019; pp. 594–598. [CrossRef]
5. Gumrukcu, E.; Asadollahi, E.; Joglekar, C.; Ponci, F.; Monti, A.; Guidi, G.; D'Arco, S.; Suul, J.A. Optimal Management for Megawatt Level Electric Vehicle Charging Stations With a Grid Interface Based on Modular Multilevel Converter. *IEEE Access* **2021**, *10*, 258–270. [CrossRef]
6. Wang, W.V.; Thirmawithana, D.J.; Riar, B.; Zane, R. A Novel Integrated Boost Modular Multilevel Converter for High Power Wireless EV Charging. In Proceedings of the 2018 IEEE Energy Conversion Congress and Exposition (ECCE), Portland, OR, USA, 23–27 September 2018; pp. 81–88.
7. Steinsland, V.; Kristensen, L.M.; Arghandeh, R.; Zhang, S. Design of Modular Multilevel Converters for the Shipnet in Medium Voltage DC All-Electric Ships. In Proceedings of the 2020 IEEE 21st Workshop on Control and Modeling for Power Electronics (COMPEL), Aalborg, Denmark, 9–12 November 2020; pp. 1–8.
8. Far, A.J.; Jovcic, D. Small-Signal Dynamic DQ Model of Modular Multilevel Converter for System Studies. *IEEE Trans. Power Deliv.* **2015**, *31*, 191–199. [CrossRef]
9. Bergna, G.; Suul, J.A.; D'Arco, S. State-space modelling of modular multilevel converters for constant variables in steady-state. In Proceedings of the IEEE Workshop Control Modeling for Power Electronics, Trondheim, Norway, 27–30 June 2016; pp. 1–9.
10. Belayneh, N.B.; Park, C.H.; Kim, J.M. Compensation of Arm Current Sensor Errors in Modular Multilevel Converter. *IEEE Trans. Ind. Appl.* **2019**, *55*, 5005–5012. [CrossRef]
11. Sharifabadi, K.; Harnefors, L.; Nee, H.; Norrga, S.; Teodorescu, R. *Design, Control, and Application of Modular Multilevel Converters for HVDC Transmission Systems (series Wiley-IEEE)*; Wiley: Hoboken, NJ, USA, 2016.
12. Prieto-Araujo, E.; Gross, D.; Dörfler, F.; Gomis-Bellmunt, O. Optimal Multivariable MMC Energy-Based Control for DC Voltage Regulation in HVDC Applications. In Proceedings of the 2020 IEEE Power & Energy Society General Meeting (PESGM), Montreal, QC, Canada, 2–6 August 2020; p. 1.
13. Li, H.; Fan, X.; Liu, S. Small-Signal Stability Modeling for MMC-Based DC Grids With Voltage Slope Control and Influence Analysis of Parameters. *IEEE Access* **2021**, *10*, 4686–4698. [CrossRef]
14. Wang, W.; Hang, L.; Qiu, J.; Lu, H.; Zeng, P. Application of MMC based on state observer in DC integration of renewable energy. In Proceedings of the 8th Renewable Power Generation Conference (RPG 2019), Shanghai, China, 24–25 October 2019; pp. 1–6.
15. Bahrani, B.; Debnath, S.; Saeedifard, M. Circulating Current Suppression of the Modular Multilevel Converter in a Double-Frequency Rotating Reference Frame. *IEEE Trans. Power Electron.* **2016**, *31*, 783–792. [CrossRef]
16. Yang, S.; Wang, H.; Chen, H.; Song, W.; Wang, T. Probability-Based Modeling and Analysis for PS-PWM in an MMC Distributed Control System With Sub-Module Asynchrony. *IEEE Trans. Power Electron.* **2019**, *34*, 10392–10397. [CrossRef]
17. Cheng, T.; Zhong, J.; Yang, B.; Wei, Y.; Xu, J.; Zhang, Z. Improved Circulation Suppression Strategy for MMC Considering Low Frequency Oscillation. In Proceedings of the 2021 11th International Conference on Power and Energy Systems (ICPES), China, Shanghai, 18–20 December 2021; pp. 57–61.
18. Bhesaniya, M.M.; Shukla, A. Norton Equivalent Modeling of Current Source MMC and Its Use for Dynamic Studies of Back-to-Back Converter System. *IEEE Trans. Power Deliv.* **2016**, *32*, 1935–1945. [CrossRef]
19. Gao, S.; Ye, H.; Liu, Y. Accurate and Efficient Estimation of Short-Circuit Current for MTDC Grids Considering MMC Control. *IEEE Trans. Power Deliv.* **2019**, *35*, 1541–1552. [CrossRef]
20. Liu, Z.; Zhao, J. Disturbance Interaction Analysis and Suppression Strategy of MMC-HVDC Systems Considering Sub-Module Capacitor Voltage Ripples. *IEEE Trans. Power Syst.* **2020**, *36*, 235–247. [CrossRef]
21. Ji, H.; Chen, A.; Liu, Q.; Zhang, C. A new circulating current suppressing control strategy for modular multilevel converters. In Proceedings of the 2017 36th Chinese Control Conference (CCC), Dalian, China, 26–28 July 2017; pp. 9151–9156.
22. Jia, G.; Chen, M.; Tang, S.; Zhang, C.; Zhao, B. A Modular Multilevel Converter With Active Power Filter for Submodule Capacitor Voltage Ripples and Power Losses Reduction. *IEEE Trans. Power Electron.* **2020**, *35*, 11401–11417. [CrossRef]
23. Qu, L.P.; Lu, Z.; Liu, C.J.; He, C.L. A Sort Method of Balancing Capacitor Voltage of MMC. In Proceedings of the 2019 18th International Symposium on Distributed Computing and Applications for Business Engineering and Science (DCABES), Wuhan, China, 8–10 November 2019; pp. 241–244.
24. Jin, Y.; Wang, S.; Xiao, Q.; Liu, Y.; Mu, Y.; Ji, Y.; Chaudhary, S.K.; Teodorescu, R. A Novel Harmonic Control Method for MMC Combining Improved Nearest Level Control and Selective Harmonic Elimination method. In Proceedings of the 2019 IEEE Energy Conversion Congress and Exposition (ECCE), Baltimore, MD, USA, 29 September–3 October 2019; pp. 6368–6375.

25. Zuo, Y.; Zhu, J.; Yuan, X.; Lee, C.H.T. Two-degree-of-freedom Quasi-PIR Controller for Smooth Speed Control of Permanent Magnet Vernier Machine. In Proceedings of the 2021 IEEE Energy Conversion Congress and Exposition (ECCE), Vancouver, BC, Canada, 10–14 October 2021; pp. 5022–5028.
26. Jun, S.; Zhen, X. An Improved Control Strategy for Double Feed Induction Generator in Low Frequency Resonant Power Grid. In Proceedings of the 2018 37th Chinese Control Conference (CCC), Wuhan, China, 25–27 July 2018; pp. 7460–7464.
27. Wai, R.J.; Zhang, Q.Q.; Wang, Y. A Novel Voltage Stabilization and Power Sharing Control Method Based on Virtual Complex Impedance for an Off-Grid Microgrid. *IEEE Trans. Power Electron.* **2019**, *34*, 1863–1880. [CrossRef]
28. Micallef, A.; Apap, M.; Spiteri-Staines, C.; Guerrero, J.M. Mitigation of Harmonics in Grid-Connected and Islanded Microgrids Via Virtual Admittances and Impedances. *IEEE Trans. Smart Grid* **2017**, *8*, 651–661. [CrossRef]
29. Isik, S.; Alharbi, M.; Bhattacharya, S. An Optimized Circulating Current Control Method Based on PR and PI Controller for MMC Applications. *IEEE Trans. Ind. Appl.* **2021**, *57*, 5074–5085. [CrossRef]
30. Ming, G.; Xiaohong, W.; Linping, W.; Daliang, W.; Weicheng, M.; Dong, Y. Research on Improved Circulation Suppression Strategy of MMC Based on Quasi-PR Controller. In Proceedings of the 2021 IEEE 5th Information Technology, Networking, Electronic and Automation Control Conference (ITNEC), Xi'an, China, 15–17 October 2021; pp. 106–109.
31. Ren, J.; Li, W.; Zhang, S. Study on AC Harmonic Characteristics of Netherland ±525 kV/2 GW MMC-HVDC System for Offshore Wind Farms Integration. In Proceedings of the 2021 International Conference on Power System Technology (POWERCON), Haikou, China, 8–9 December 2021; pp. 1534–1541.

energies

MDPI

Article

A High-Efficiency QR Flyback DC–DC Converter with Reduced Switch Voltage Stress Realized by Applying a Self-Driven Active Snubber (SDAS)

Jeong-Sang Yoo [1,*], Jong-Ok Baek [2] and Tae-Young Ahn [3]

1 Department of Electronics Engineering, Cheongju University, Cheongju 28503, Republic of Korea
2 R&D Division, Zinnos Inc., Busan 49108, Republic of Korea
3 Department of Electrical & Control Engineering, Cheongju University, Cheongju 28503, Republic of Korea
* Correspondence: jeongsangyoo@gmail.com

Abstract: In this paper, a QR flyback converter using a self-driven active snubber (SDAS) was proposed to solve the problem of voltage surge in the switch of QR flyback converters. In the proposed converter, the SDAS consisting of a clamping capacitor and an active switch can be configured in parallel with the main switch or transformer to reduce the voltage surge in the switch. To confirm the steady-state characteristics of the QR flyback converter to which the proposed SDAS is applied, equivalent circuits for each state were constructed, and the equations and characteristics for each state were determined. A 60 W class small AC–DC adapter was constructed to confirm the effectiveness of the proposed converter and the control circuit method, and the experimental results were analyzed. The size of the experimental AC–DC adapter was 74 × 29 × 23 mm, and it had a high power density of 20 W/in^3 or more. The experimental circuit was limited to the high power conversion efficiency of up to 91.56%, and the maximum voltage surge in the switch was approximately 450 V. One of the reasons for such high efficiency is the SDAS circuit, which sufficiently reduces the voltage surge of the QR flyback switch, compared with the RCD clamp circuit, and does not consume power in principle.

Keywords: quasi-resonant flyback converter; synchronous rectifier; active clamping switch; RCD snubber; valley switching; active snubber; self-driven active snubber

Citation: Yoo, J.-S.; Baek, J.-O.; Ahn, T.-Y. A High-Efficiency QR Flyback DC–DC Converter with Reduced Switch Voltage Stress Realized by Applying a Self-Driven Active Snubber (SDAS). *Energies* **2023**, *16*, 1068. https://doi.org/10.3390/en16031068

Academic Editors: Diego Bellan and Jelena Loncarski

Received: 21 December 2022
Revised: 12 January 2023
Accepted: 13 January 2023
Published: 18 January 2023

1. Introduction

Recently, due to the rapid development of the IT field, low-power energy consumption and high-efficiency power conversion devices are required. Following the trend of miniaturization and lightweight electronic devices, research is being actively conducted to increase the efficiency of power converters and reduce their weight. To reduce the size and weight of power supply devices, high-frequency switching has become an important design criterion. A switching power supply is a device that converts an input voltage into a stable output voltage, and appropriate circuit development is required depending on the application field and output capacity. For output powers of less than 150 W, flyback converters are most commonly used to achieve miniaturization and low manufacturing cost. In addition, since a transformer is used, multiple outputs can be comprised, and input and output isolation is possible [1,2].

Recently, with the introduction of the quasi-resonant (QR) flyback control method, the switching loss is reduced, and efficiency is increased compared with conventional flyback converters, resulting in a broader range of applications. The power conversion efficiency can be increased by reducing the turn-on loss to reduce the switching loss. However, in the case of an isolated switching power supply such as a QR flyback converter, a turn-off surge in voltage simultaneously occurs across the switch due to internal parasitic components. This high-voltage surge increases the rated voltage of the switch and becomes a major cause of deterioration in the reliability of the product. To reduce high-voltage switch surges, an

RCD-structured snubber using a resistor–capacitor–diode can be used across the primary and secondary sides of the transformer or the switch. The RCD snubber has the advantage of having a simple structure and easy design, but it also generates heat in the resistance element and lowers the power conversion efficiency of the converter because the entire energy resulting from the high surge in voltage is consumed in the resistor.

Another way to reduce the voltage surge in the switch is to apply an active clamp circuit by configuring the auxiliary switch in series with the clamp capacitor separately from the main switch. In this case, one of the methods to clamp the main switch voltage to a constant voltage is by introducing a signal in the drive signal of the main switch that is opposite to that of the auxiliary switch. However, this active clamp method cannot be applied to all flyback converter methods and only applies to converters operating in the boundary current mode (BCM), in which the transformer magnetizing inductor current is at the boundary at zero but not applicable to the quasi-resonant (QR) flyback method, which operates in discontinuous conduction mode (DCM) [3,4].

Figure 1 shows the basic circuit of a QR flyback converter with an RCD snubber. Figure 1a shows a converter in which the secondary side of the transformer is composed of a diode rectifier, while Figure 1b shows a converter composed of a synchronous rectifier circuit using MOSFET instead of a diode. In Figure 1a, the main switch of the converter is S, and the resistance R_s, capacitor C_s, and diode D_s of the RCD snubber are in parallel with the transformer T_1, and the surge energy due to the resonance of the transformer leakage inductor L_K and the switch parasitic capacitor C_R is consumed in the RCD snubber circuit. In Figure 1b, the main switch of the converter is S_1, and the synchronous rectifier switch S_R replaces the diode rectifier on the secondary side of the transformer T_1. Figure 2 shows the operating waveform in a steady state. Figure 2a is the voltage and current waveform of the diode rectifier, and Figure 2a is the driving signal and the voltage and current waveform of the MOSFET synchronous rectifier. In the QR flyback converter, the magnetizing current of the transformer increases linearly from zero during the turn-on period of the switch because it operates at the DCM and the continuous conduction mode (CCM). During the turn-off period, a voltage surge is generated during damping resonance due to the transformer leakage inductor and the parasitic capacitor of the switch. Although the RCD snubber circuit absorbs some of the surge current, the remaining current still generates a voltage surge in the switch. At this time, the voltage surge of the switch reaches maximum point, and the voltage surge can cause breakdown of the switch [5]. The secondary-side rectifier current linearly decreases, and after it becomes zero, the resonance resulting from magnetizing inductance L_M and the switch parasitic capacitor C_R starts, and the main switch is turned on at the valley, where the switch resonance voltage is lowest owing to the QR flyback control signal. In the case of the synchronous rectifier, the synchronous rectifier switch S_R is turned on while the rectifier current flows. For the synchronous rectifier switch S_R, a MOSFET with low on-resistance is selected to reduce the conduction loss. As shown in Figure 2, since the main switch voltage of the QR flyback converter is turned on when it reaches the lowest point due to resonance, the turn-on loss is reduced, thus reducing switching loss and enabling high-efficiency power conversion. However, since a voltage surge is generated when the switch S is turned off, this problem cannot be fundamentally solved by adding an RCD snubber circuit [6–8].

In this paper, a QR flyback converter using a self-driven active snubber (SDAS) is proposed to mitigate the problem of the high switch-turn-off voltage in the QR flyback converter. The proposed converter clamps the voltage surge of the switch to a constant voltage during turn-off operation owing to the SDAS composed of a clamping capacitor and an active switch, and since the active switch drive signal is controlled by the secondary-side current of the transformer, a separate control circuit is not required. In addition, the proposed converter has the advantage of using a switch with a low voltage rating due to the reduced turn-off voltage surge and using the conventional QR flyback control IC. For the design of its major components, the steady-state operation characteristics were analyzed, and the simulation and experimental results were compared. To confirm the effectiveness of the proposed converter and the control circuit method, a 60 W class miniaturized AC–DC adapter was constructed, and the experimental results were analyzed. The size of the experimental AC–DC adapter was 74 × 29 × 23 mm, the power density was about 20 W/in^3, and the target maximum efficiency was over 90% [9–11].

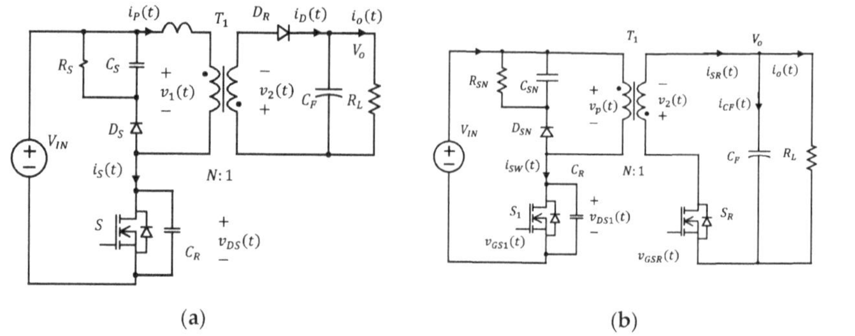

Figure 1. Basic circuit of QR flyback converter with RCD snubber: (a) QR flyback converter with diode rectifier; (b) QR flyback converter with synchronous rectifier.

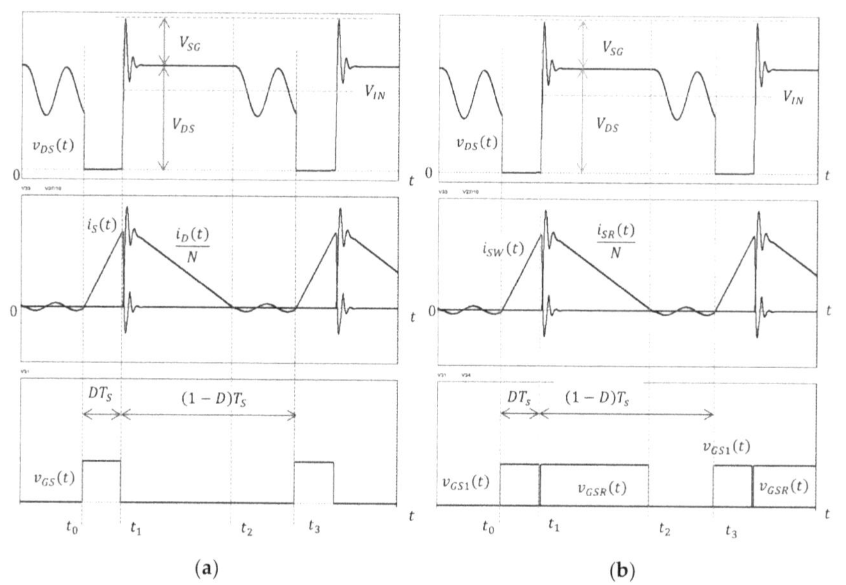

Figure 2. Steady-state operating waveform of QR flyback converter: (a) steady-state operating waveform of QR flyback converter with diode rectifier; (b) steady-state operating waveform of QR flyback converter with synchronous rectifier.

2. QR Flyback DC–DC Converter with SDAS

2.1. Basic Circuit Structure of QR Flyback Converter with SDAS

Figure 3 shows the basic circuit of the QR flyback converter with the SDAS proposed in this paper. In Figure 3a, the SDAS is configured in parallel with the main switch, while in Figure 3b, the SDAS is configured in parallel with the transformer. In this figure, the main switch is S_1, and the synchronous rectifier switch S_R replaces the existing diode rectifier on the secondary side of the transformer T_1. The SDAS circuit consists of a clamp switch S_2 and a clamp capacitor C_L in series. Transformer T_2 is a current transformer (CT) that detects secondary-side current and generates driving signals for the synchronous rectifier switch S_R and the clamp switch S_2. Therefore, since the two switches S_R and S_2 are turned on and off by the secondary-side current signal of the transformer, they are driven independently of the external PWM control signal. The clamp capacitor C_L absorbs the surge energy caused by the resonance of the transformer leakage inductor L_K and the switch parasitic capacitor C_R. In the conventional active clamp-type flyback converter, the location of the main switch S_1, the clamp switch S_2, and the clamp capacitor C_L are the same as the circuit proposed in this paper, but the clamp switch S_2 serves as a complementary switch to the main switch, so the QR flyback characteristics are not satisfied. Therefore, the conventional active-clamp flyback converter has different steady-state operation characteristics from the QR flyback converter configured with the SDAS in parallel. Since QR flyback converters are relatively widely applied in the design and manufacturing of small-capacity converters, several manufacturers have released control IC models. In this paper, while maintaining the basic characteristics and design environment of the existing QR flyback converter, we proposed a circuit method capable of high-efficiency power conversion by actively removing the voltage surge of the switch to lower the rated voltage of the switch and reducing the conduction loss of the switch [12–16].

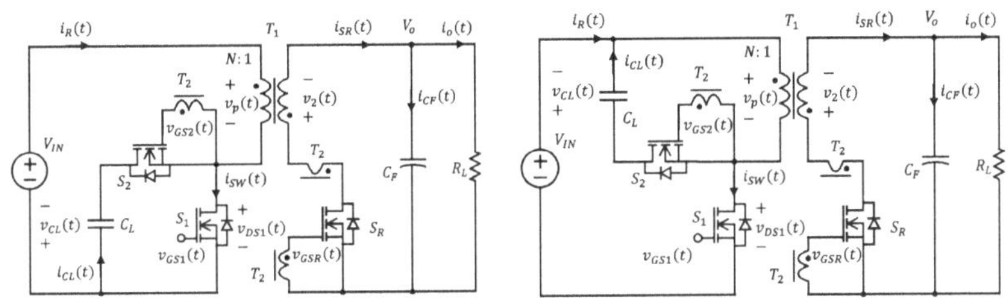

(a) (b)

Figure 3. QR flyback DC–DC converter with SDAS: (**a**) QR flyback converter with SDAS in parallel with main switch; (**b**) QR flyback converter with SDAS in parallel with transformer.

Figure 4 is an equivalent circuit to that in Figure 3a with the transformer T_1 removed. Figure 4a is the equivalent circuit with the clamp capacitor C_L, and Figure 4b is the equivalent circuit converted to the clamping constant voltage V_{CL} instead of the clamp capacitor. Transformer T_1 is represented by the leakage inductor L_K and magnetizing inductor L_M, and all switches are assumed to be ideal switches without loss. The resonant capacitor C_R placed in parallel with the main switch S_1 is the parasitic capacitor of the switch. The output load resistance and the output capacitor are assumed to be constant voltage in a steady state. In addition, in this study, to simplify the analysis and design, the equivalent circuit in Figure 4a was used as a basis for the steady-state analysis and design of major components [16,17].

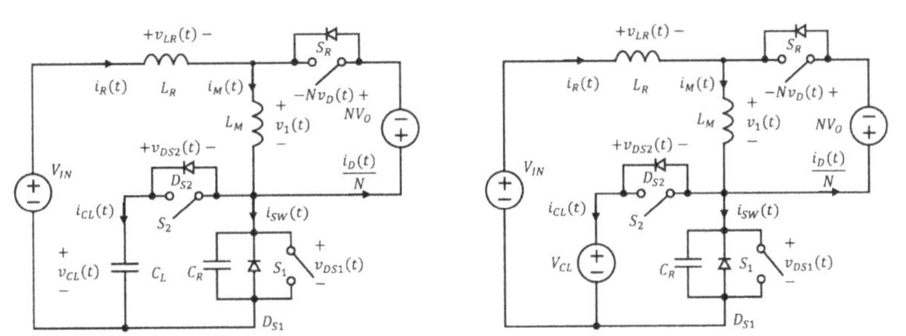

(a) (b)

Figure 4. Equivalent circuit of QR flyback converter with self-driving active snubber: (**a**) equivalent circuit of QR flyback converter with diode rectifier; (**b**) equivalent circuit of QR flyback converter with synchronous rectifier.

2.2. Steady-State Analysis

Figure 5 shows the operating waveform for the equivalent circuit in Figure 4. Figure 5a shows the switch voltage of the converter turned on in the second valley, whereas Figure 5b shows the switch voltage turned on in the first valley. As can be seen, when the converter operates in a steady state, its state is divided into six operating states for one cycle. In this case, the turn-off timing of the switch is different, but the waveform for each state is the same. The operation mechanism in each state of the converter operating under a steady state is described below [18,19].

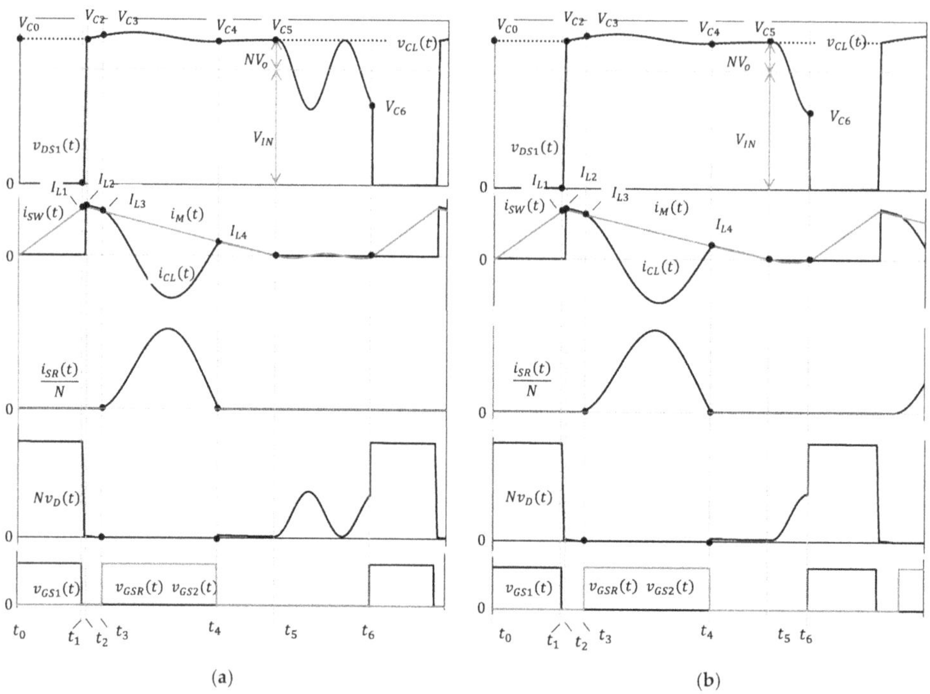

Figure 5. Steady-state operating waveform of QR flyback converter with SDAS: (**a**) steady-state operating waveform of a QR flyback converter with two valley voltages; (**b**) steady-state operating waveform of a QR flyback converter with one valley voltage.

First, in state 1, the main switch S_1 is turned on at time t_0, the secondary-side current of the transformer becomes zero, and the auxiliary switch S_2 and the synchronous rectifier switch S_R are turned off. The equivalent circuit of state 1 is shown in Figure 6. As can be seen in Figures 5 and 6, the input voltage is applied to the magnetizing inductor L_M and the leakage inductor L_K, and the current constantly increases from zero, as shown in Equation (1). Assuming that the main switch voltage is equal to that in Equation (2), and the clamp capacitance is large, the clamp voltage is equal to the turn-off steady-state voltage of the main switch shown in Equation (3). The switch voltage of the synchronous rectifier is determined using Equation (4), and the current is determined with Equation (5). When the main switch S_1 is turned off at time t_1, the magnetizing current i_M becomes I_{L1}, and state 1 ends [20,21].

$$i_{sw}(t) = i_M(t) = i_R(t) = \left(\frac{V_{IN}}{L_R + L_M}\right)t \tag{1}$$

$$v_{DS1}(t) = 0 \tag{2}$$

$$v_{CL}(t) = V_{C0} = V_{IN} + NV_o \tag{3}$$

$$Nv_D(t) = -(V_{IN} + NV_o) \tag{4}$$

$$\frac{i_{SR}(t)}{N} = 0 \tag{5}$$

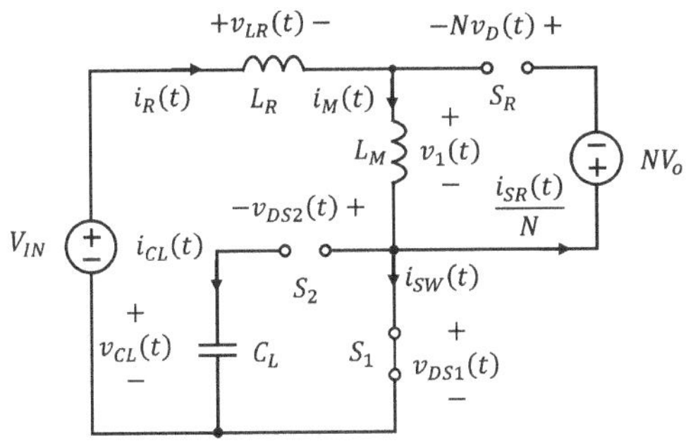

Figure 6. Equivalent circuit in state 1.

When the main switch S_1 is turned off at time t_1, state 2 starts, and the equivalent circuit is shown in Figure 7. At time t_1, the initial value of the magnetizing current i_M is I_{L1}, and the magnetizing inductor L_M and the switch equivalent capacitor C_R resonate, as shown in Equation (6). The switch resonant voltage is determined with Equation (7). At this time, the characteristic impedance Z_r and the resonant angular velocity ω_r are calculated using Equation (8). When the resonance voltage of the main switch S_1 becomes equal to the clamp capacitor voltage V_{CL}, the time becomes t_2, the maximum value of the magnetizing current i_M becomes I_{L2}, and state 2 ends.

$$i_M(t) = i_R(t) = \left(\frac{V_{IN}}{Z_r}\right)\sin\omega_r t + I_{L1}\cos\omega_r t \tag{6}$$

$$v_{DS1}(t) = I_{L1}Z_r\sin\omega_r t + V_{IN}(1 - \cos\omega_r t) \tag{7}$$

$$Z_r = \sqrt{\frac{L_M}{C_R}} \quad \omega_r = \frac{1}{\sqrt{L_M C_R}} \tag{8}$$

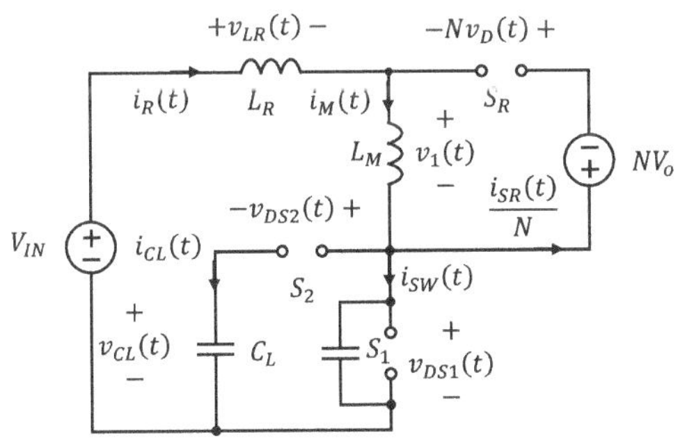

Figure 7. Equivalent circuit in state 2.

At time t_2, when the magnetizing current L_M flows through the internal diode of the clamp switch, state 3 starts, and the equivalent circuit in this state is shown in Figure 8. At time t_2, the initial value of magnetizing current L_M is I_{L2}, the magnetizing inductor L_M and the clamp capacitor C_L resonate, and the magnetizing current i_M is determined with Equation (9). The clamp capacitor resonant voltage is provided by Equation (10). At this time, the characteristic impedance Z_R and the resonant angular velocity ω_R are determined using Equation (11). When the clamp current i_{CL} becomes larger than the magnetizing current i_M, the magnetizing current becomes I_{L3}, the time becomes t_3, and state 3 ends.

$$i_{CL}(t) = i_M(t) = i_R(t) = \left(\frac{V_{IN} - V_{C2}}{Z_R}\right)\sin\omega_R t + I_{L2}\cos\omega_R t \quad (9)$$

$$v_{CL}(t) = I_{L2}Z_R\sin\omega_R t + V_{IN}(1 - \cos\omega_R t) + V_{C2} \quad (10)$$

$$Z_R = \sqrt{\frac{L_M}{C_L}} \quad \omega_R = \frac{1}{\sqrt{L_M C_L}} \quad (11)$$

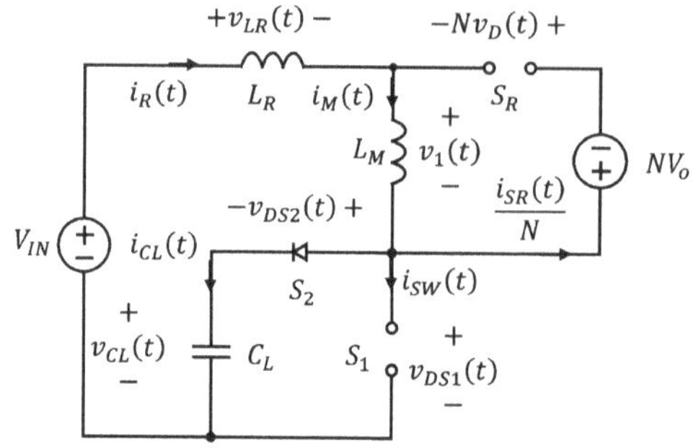

Figure 8. Equivalent circuit in state 3.

At time t_3, as the resonance current of the clamp current i_{CL} becomes larger than the magnetizing current i_M, the remaining current flows through the rectifier on the secondary side of the transformer. When the synchronous rectifier switch S_R and the clamp auxiliary switch S_2 are turned on with the rectifier current, state 4 starts, and the equivalent circuit in this state is shown in Figure 9. At time t_3, the initial value of the resonance current i_{CL} is I_{L3}, the leakage inductor L_K and the clamp capacitor C_L resonate, and the clamp capacitor current is as shown in Equation (12). The clamp capacitor resonant voltage is calculated using Equation (13). The secondary-side current starts at zero due to resonance and performs zero-current switching (ZCS); thus, its waveform is a sinusoidal wave reflecting the resonance. At this time, the characteristic impedance Z_O and the resonant angular velocity ω_O are expressed in Equation (15). When the magnetization current i_M and the clamp capacitor current i_{CL} become equal, the secondary rectifier current becomes zero, and the time becomes t_4, and state 4 ends [22,23].

$$i_{CL}(t) = i_R(t) = \left(\frac{V_{IN} + NV_O - V_{C3}}{Z_o}\right)\sin\omega_o t + I_{L3}\cos\omega_o t \quad (12)$$

$$v_{CL}(t) = I_{L3}Z_o\sin\omega_o t - (V_{IN} + NV_O - V_{C3})\cos\omega_o t + (V_{IN} + NV_O) \quad (13)$$

$$\frac{i_{SR}(t)}{N} = I_{L3}(1 - \cos\omega_o t) - \frac{NV_o}{L_M}t - \left(\frac{V_{IN} + NV_O - V_{C3}}{Z_o}\right)\sin\omega_o t \quad (14)$$

$$Z_o = \sqrt{\frac{L_R}{C_L}} \quad \omega_o = \frac{1}{\sqrt{L_R C_L}} \quad (15)$$

At time t_4, the magnetizing current i_M and the clamp current i_{CL} become equal, the secondary rectifier current becomes zero, and state 5 begins. The equivalent circuit in this state is shown in Figure 10. When the synchronous rectifier switch S_R and the clamp auxiliary switch S_2 are turned off with the rectifier current, the initial value of the resonance current i_{CL} is I_{L4} at time t_4. The magnetizing inductor L_M and the clamp capacitor C_L resonate, as shown in the figure, and the clamp capacitor current is calculated with Equation (16). The clamp capacitor resonant voltage is calculated using Equation (17). At this time, the characteristic impedance Z_R and the resonant angular velocity ω_R are expressed in Equation (18). When the magnetizing current i_M and the clamp current i_{CL} become zero, the time becomes t_5, and state 5 ends.

$$i_{CL}(t) = i_M(t) = i_R(t) = \left(\frac{V_{IN} - V_{C4}}{Z_R}\right)\sin\omega_R t + I_{L4}\cos\omega_R t \tag{16}$$

$$v_{CL}(t) = I_{L4} Z_R \sin\omega_R t + V_{IN}(1 - \cos\omega_R t) + V_{C4} \tag{17}$$

$$Z_R = \sqrt{\frac{L_M}{C_L}} \quad \omega_R = \frac{1}{\sqrt{L_M C_L}} \tag{18}$$

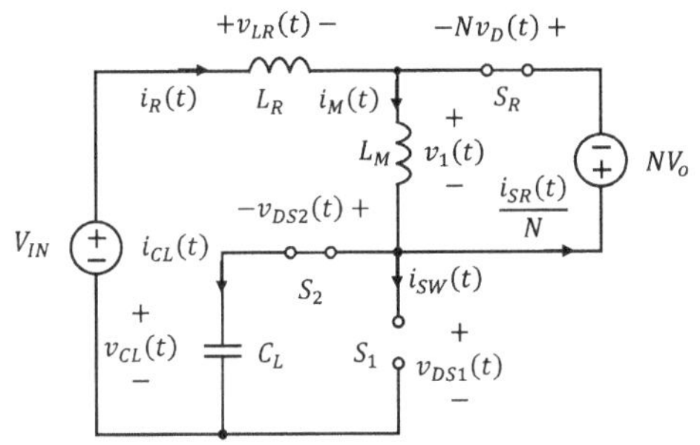

Figure 9. Equivalent circuit in state 4.

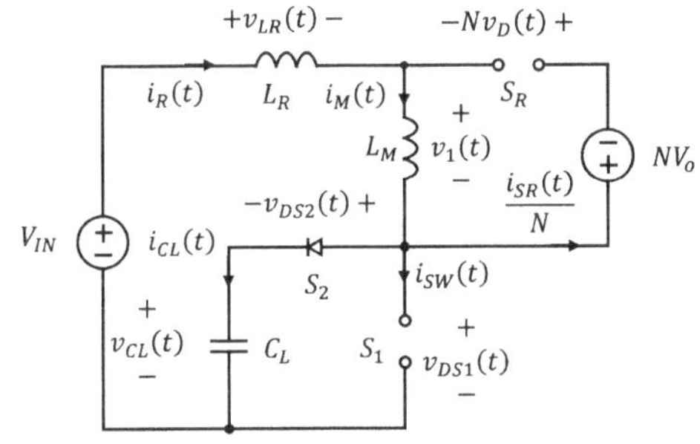

Figure 10. Equivalent circuit in state 5.

At time t_5, when the magnetizing current i_M, the clamp current i_{CL}, and the secondary rectifier current become zero, state 6 starts, and its equivalent circuit is shown in Figure 11. At time t_5, the initial value of the turn-off voltage of the main switch is V_{C5}, and the magnetizing inductor L_M and the switch equivalent capacitor C_R resonate, as shown in the figure. The magnetizing current is determined using Equation (19), and the switch resonant voltage is expressed in Equation (20). At this time, the characteristic impedance Z_r and the resonant angular velocity ω_r are found using Equation (21). When the switch voltage reaches the lowest point V_{C6} due to resonance, it becomes time t_6, and at this time, if the switch is turned on by the control circuit, the turn-on loss can be minimized.

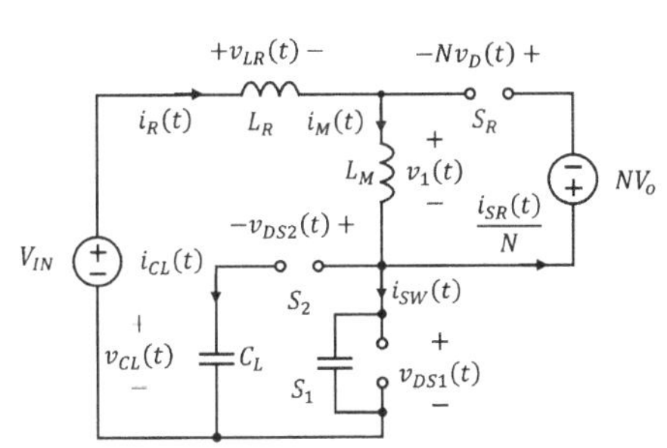

Figure 11. Equivalent circuit in state 6.

Figure 12 shows a simplified resonant equivalent circuit with the components constituting the resonant circuit for each state equivalent circuit. The initial current and voltage values for each resonant equivalent circuit are shown [24,25].

$$i_M(t) = i_R(t) = i_{SW}(t) = \left(\frac{V_{IN} - V_{C5}}{Z_r}\right)\sin\omega_r t \tag{19}$$

$$v_{DS1}(t) = (V_{C5} - V_{IN})\cos\omega_r t + V_{IN} \tag{20}$$

$$Z_r = \sqrt{\frac{L_M}{C_R}} \quad \omega_r = \frac{1}{\sqrt{L_M C_R}} \tag{21}$$

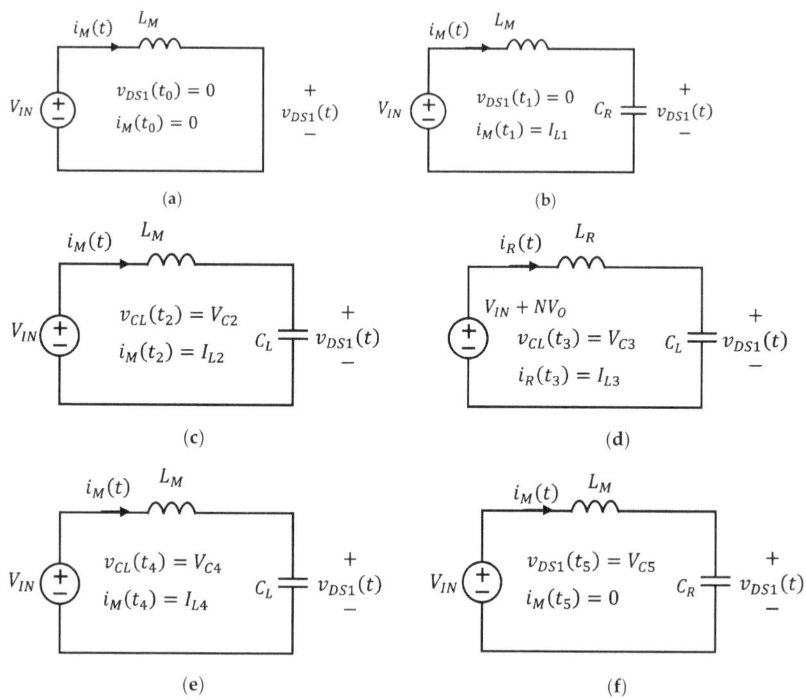

Figure 12. Equivalent circuit of resonance circuit for each state: (**a**) equivalent circuit of resonant circuit in state 1; (**b**) equivalent circuit of resonant circuit in state 2; (**c**) equivalent circuit of resonant circuit in state 3; (**d**) equivalent circuit of resonant circuit in state 4; (**e**) equivalent circuit of resonant circuit in state 5; (**f**) equivalent circuit of resonant circuit in state 6.

To find the appropriate value of the clamp capacitor, the voltage of the magnetizing inductor L_M is obtained using Equation (22). During states 4 and 5, the voltage value is NV_o, and the maximum current value is I_M^+. Assuming that the time of state 2–5 is twice the resonance time of the leakage inductor L_K and the clamp capacitor C_L, Equation (23) is used. The clamp capacitor value can be obtained from Equation (24). At this time, the maximum voltage value of the clamp capacitor is determined using Equation (25).

$$v_M(t) = L_M \frac{di_M(t)}{dt} \tag{22}$$

$$NV_o = L_M\left(\frac{I_M^+}{2 \times 2\pi\sqrt{L_K C_L}}\right) \tag{23}$$

$$C_L = \frac{1}{L_K}\left(\frac{L_M I_M^+}{4\pi N V_o}\right)^2 \quad (24)$$

$$V_{CL} = V_{IN} + NV_o \quad (25)$$

3. Design of QR Flyback DC–DC Converter with SDAS

3.1. Optimal Design Process for Major Devices

The electrical specifications shown in Table 1 were determined for the design of the SDAS-applied QR flyback converter proposed in this paper. The input voltage was AC 90–264 V, the output voltage was set at 12 V, and the maximum output was set at 60 W. Since the AC–DC power supply of less than 75 W is not subject to the IEC 61000-3-2 limitation of current harmonics, the PFC circuit was not used. The size of the power supply device was 29 mm in width, 23 mm in height, and 74 mm in length, and the volume at this time was less than 50 cc.

Table 1. Design conditions for QR flyback converter with SDAS.

Parameter	Symbol	Value	Unit
Input voltage range	V_{IN}	90–264	Vac
Output voltage	V_o	12	Vdc
Maximum output power	$P_{o\ max}$	60	W
Maximum output current	$I_{o\ max}$	5.0	A
Case size	$W \times H \times L$	29 × 23 × 74	mm
Efficiency	η	90	%

The winding ratio of the transformer is calculated with Equation (26), and V_c is the voltage that is converted from the output voltage to the primary-side voltage. K_c is the design margin coefficient. In this design, the secondary winding of the transformer was determined as the minimum natural number, so 5 was selected. Since the QR flyback converter always operates in a discontinuous current mode, the entire energy stored in the magnetizing inductor L_M is delivered to the output P_o. Therefore, the magnetizing inductor L_M is expressed using Equation (26). The maximum duty rate d_{max} appears at the lowest input voltage and is shown in Equation (28). The RMS values of the current in the primary and secondary windings of the transformer are determined with Equations (29) and (30). Equation (24) can be used for the clamp capacitor, and the result is shown in Equation (31). For the clamp capacitor value, 68 nF was selected, which is the closest standard value [26–30].

$$N = \frac{N_1}{N_2} = \frac{K_c V_c}{V_o} = \frac{1.5 \times 50}{12} = 6.3 \quad (26)$$

$$L_M = \frac{2\,P_o}{I_R^2\,f_{sw}} = \frac{2 \times 70}{2.5^2 \times 83k} = 260\ \mu\text{H} \quad (27)$$

$$d_{max} = \frac{I_{pk} L_M f_{sw}}{V_{IN\ min}} = \frac{2.5 \times 269\mu \times 83k}{90\sqrt{2}} = 0.42 \quad (28)$$

$$I_{1\ rms} = I_{pk}\sqrt{\frac{d_{max}}{3}} = 2.5\sqrt{\frac{0.42}{3}} = 0.94\ \text{A} \quad (29)$$

$$I_{2\ rms} = NI_{pk}\sqrt{\frac{1-d_{max}}{3}} = 2.5 \times 6.3\sqrt{\frac{1-0.42}{3}} = 6.93\ \text{A} \quad (30)$$

$$C_L = \frac{1}{5.3\mu}\left(\frac{260\mu \times 2.5}{4\pi \times 6.4 \times 13}\right)^2 = 73\ \text{nF} \quad (31)$$

3.2. Circuit Simulation Results

To verify the design results of the QR flyback converter using the SDAS designed in the previous section, a circuit simulation was performed using PSIM 11.0, and the simulation circuit diagram is shown in Figure 13. In the figure, the previously designed values were used for the main components of the converter, and the main switches and transformers were modeled as ideal components. To stabilize the output voltage, a valley-switching PWM control circuit suitable for the QR flyback converter was designed. Figure 14 is the waveform results of the PSIM simulation when the converter operated in a steady state. In Figure 14a, the input voltage was DC 320 V, and the load resistance was 5 Ω; in Figure 14b, the input voltage was DC 320 V, and the load resistance was 2.4 Ω. These values correspond to approximately 30 W and 60 W of output power [31].

Figure 13. PSIM simulation schematic of QR flyback converter with SDAS.

In the figure, shown from the top, the operating waveforms refer to the main switch voltage v_{DS1}, the magnetizing inductor current i_M, the clamp capacitor current i_{CL}, the switch drive voltage $v_{GS1,2}$, the transformer secondary rectifier current i_{SR}, and the switch current i_S. As can be seen from this figure, when the DC input voltage was 320 V, the switch was turned on at the lowest point of the valley. As the power load increased, the number of valleys decreased, and the switch was turned on in the first valley at full load. The driving voltage of the clamp and the synchronous rectifier switch was controlled by the secondary-side rectifier current of the transformer [32]. As a result, it can be seen that the SDAS circuit consisting of the clamp switch and the clamp capacitor operated properly, and the voltage surge was eliminated during the turn-off period of the main switch. Specifically, the rectifier current on the secondary side of the transformer is in the form of a half-wave rectified sinusoidal wave occurred by the resonance, and it showed ZCS characteristics and reduced the switching loss and voltage surge, lowering the rectifier voltage rating [33,34].

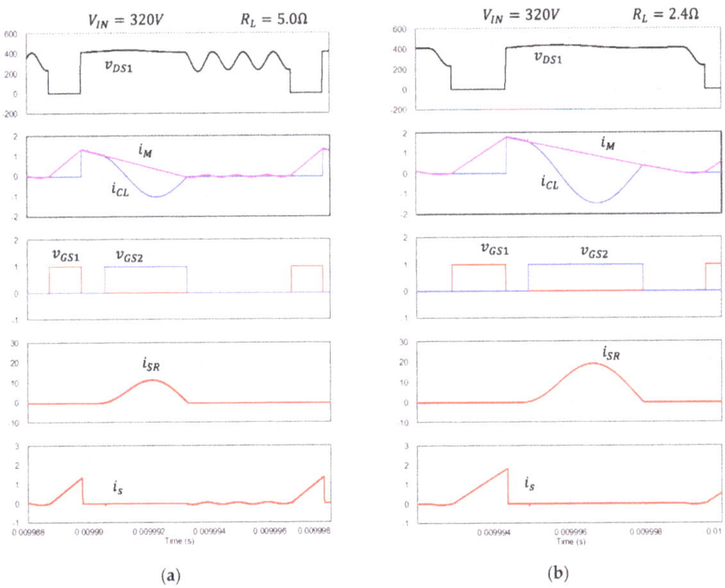

Figure 14. Steady-state waveform result of QR flyback converter with SDAS with PSIM simulation: (**a**) under low power load; (**b**) under large power load.

In Figure 15, the PSIM simulation results are compared with the experimental results when a 60 W class QR flyback converter with the SDAS was used under a steady state. Shown from the top, the operating waveforms refer to the switch drive voltage v_{GS1}, the main switch voltage v_{DS1}, the switch current i_s, the clamp capacitor current i_{CL}, the transformer secondary rectifier current i_{SR}, the clamp switch drive voltage v_{GS2}, the clamp capacitor voltage v_{CL}, and the transformer primary voltage v_1. It can be seen that the PSIM simulation results and the experimental waveforms are in relatively good agreement, and they are the same as the operation of each state described above. Therefore, it was confirmed that both the analysis of the steady-state operation and the design process were valid [35].

3.3. Compact AC–DC Adapter Design

Figure 16 shows the basic structure including the control circuit to construct the AC–DC adapter designed above. The basic circuit in the figure is for a QR flyback converter with the SDAS composed of existing commercialized QR flyback control components and an isolation circuit using a photocoupler. As shown in Figure 16, a synchronous rectifier driver component was used to control the synchronous rectifier switch S_R, and a photo driver was used to deliver a driving signal to the clamp switch S_2 on the primary side of the transformer. As a result, the operation of the synchronous rectifier and the clamp switch was always the same [36,37].

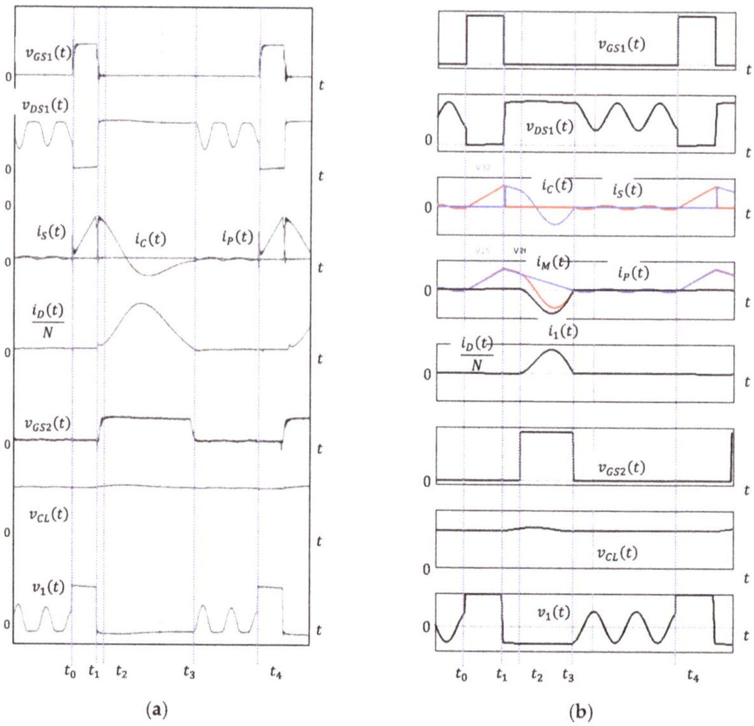

Figure 15. Comparison of operating waveforms of PSIM simulation and experimental circuit: (**a**) operating waveform of the experimental circuit; (**b**) operating waveform of PSIM simulation.

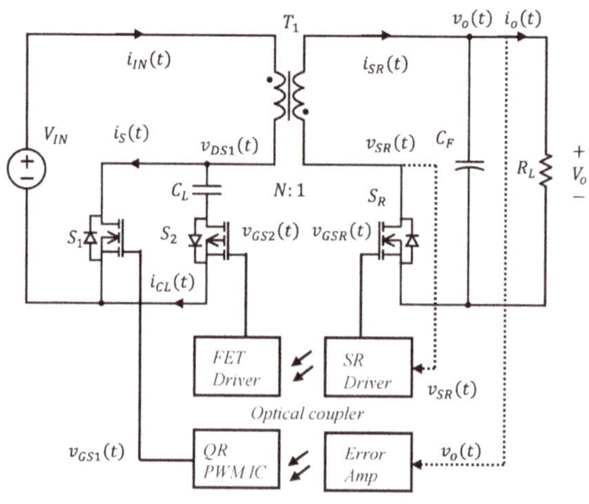

Figure 16. Basic circuit and control structure of QR flyback converter with SDAS.

Table 2 shows the main components and their electrical characteristics constituting the experimental AC–DC adapter circuit. For the main switch, a MOSFET with a relatively low voltage rating of 600 V was used, as it could reduce the on-resistance compared with other MOSFETs with a similar current rating. A 500 V class P-channel MOSFET was used for the clamp switch and a surface-mounted device (SMD) MOSFET with 6.3 mΩ on-resistance was used for the synchronous rectifier switch to minimize the conduction loss. A QR flyback control IC, NCP1380, was used to control the main switch, TEA1791 was used as the synchronous rectifier driver, and TLP118 was used as the photo driver. Table 3 shows the characteristics of the transformer used in the experimental AC–DC adapter circuit. The magnetic core size of the transformer was TDK's RM7. For the primary winding, Litz wires comprising a bundle of 30 copper wires with 0.1 mm thickness were used to reduce high-frequency loss, and Litz wires comprising 60 copper wires were used for the secondary winding to deal with a large RMS current value. The measured magnetizing inductance was 260 μH, and the leakage inductance was 5.3 μH. The clamp capacitor was a 68 nF film capacitor rated at 650 V. Figure 17 shows the experimental circuit diagram. In the circuit diagram, the AC–DC rectifier was omitted for convenience, but a general full-wave rectifier was applied in the experiment [38–40].

Table 2. Main devices and electrical ratings used in the experimental AC–DC adapters.

Parameter	Name	Model	Specifications	Shape
Main switch	S_1	TK31V60X	600 V, 30 A, 78 mΩ	DFN8 N-ch
Clamp switch	S_2	FQP3P50	−500 V, −2.7 A, 4.9 Ω	TO-220 P-ch
SR switch	S_R	BUK9Y8R5-80E	80 V, 100 A, 6.3 mΩ	LFPAK56 N-ch
QR controller	IC_1	NCP1380	28 V, 500 mA	SO-8
SR controller	IC_2	TAE1791	120 V, 3 A	SO-8
Photo driver	IC_3	TLP118	3750 V, 60 ns	SO-6 GaAlAs IRED

Table 3. Transformer characteristics of experimental AC–DC adapters.

Parameter	Name	Unit	Value	Spec.
Core size	T_1		RM7	
Primary winding	N_1	Turn	32	0.1 mm × 30 p
Secondary winding	N_2	Turn	5	0.3 mm × 60 p
Auxiliary winding	N_A	turn	5	0.1 mm × 30 p
Magnetizing inductance	L_M	μH	260	
Leakage inductance	L_K	μH	5.3	

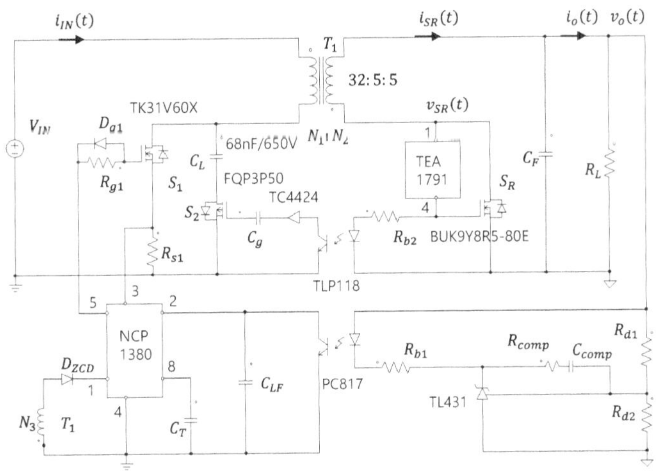

Figure 17. Schematic of QR flyback converter with SDAS used in the experiments.

Figure 18 shows the external appearance of the experimental AC–DC adapter. Figure 18a is the assembly drawing of the outer case, and Figure 18b is the inside and outside size and shape of the constructed experimental circuit. The AC–DC adapter was a 74 mm long rectangular cylinder case with a hollow inside and a rectangular lid attached to both ends. The AC terminal was integrated with the lid on one side. The volume of the adapter was 50 cc, and the power density was approximately 20 W/in^3. Figure 19 is the instrument composition and test environment to test the experimental circuit. The power supply used in the experiment was a PCR400LE, the electronic load was a PLZ1004WH from Kikusui (Yokohama, Kanagawa, Japan), the power analyzer was a WT1600 from YOKOGAWA (Tokyo, Japan), and the oscilloscope was an HDO6104 model from LeCroy (Chestnut Ridge, NY, USA). To measure the waveform of the experimental circuit in a steady state, the test was carried out without the case.

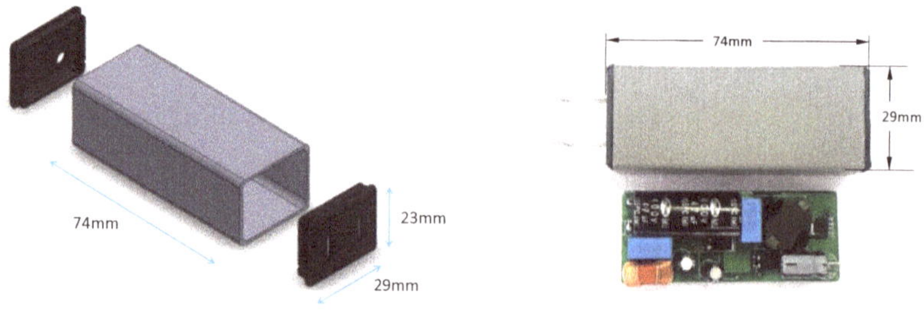

(a) (b)

Figure 18. External shape and size of AC–DC adapter using QR flyback converter with SDAS: (**a**) external shape and size of AC–DC adapter; (**b**) internal PCB and case of AC–DC adapter for experimentation.

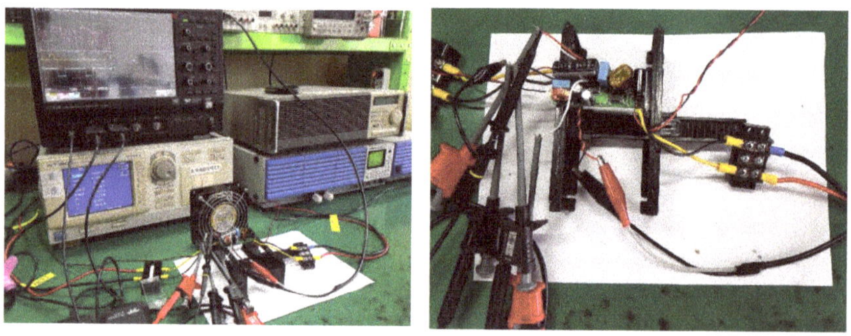

(a) (b)

Figure 19. Experimental circuit and its configuration: (**a**) load characteristic test of AC–DC adapter used in the experiment; (**b**) AC–DC adapter and its measurement setup.

4. Experimental Results

To verify the effectiveness of the QR flyback converter with the SDAS proposed in this paper, a 60 W AC–DC adapter was tested. Figure 20 is the steady-state operating waveform of the experimental circuit when the input voltage was 310 V. In Figure 20a, the output current was 0.3 A, the voltage v_{DS} of the main switch is shown on the top, and the switch current i_S is shown on the bottom. The power load was approximately 3.6 W, and the switch voltage showed a small duty cycle under a low load current. At this time, the turn-off voltage of the switch was clamped by the SDAS circuit and limited to 420 V. As the load current increased, the operating waveform changed from Figure 20b to Figure 20f. Figure 20f shows when the load current was 5 A, and the output power was 60 W. The maximum switch current was 2.4 A, and the maximum switch voltage was limited to about 450 V. It can be seen from this figure that depending on the load, the peak value of the switch voltage did not significantly increase and was clamped to a constant level [41].

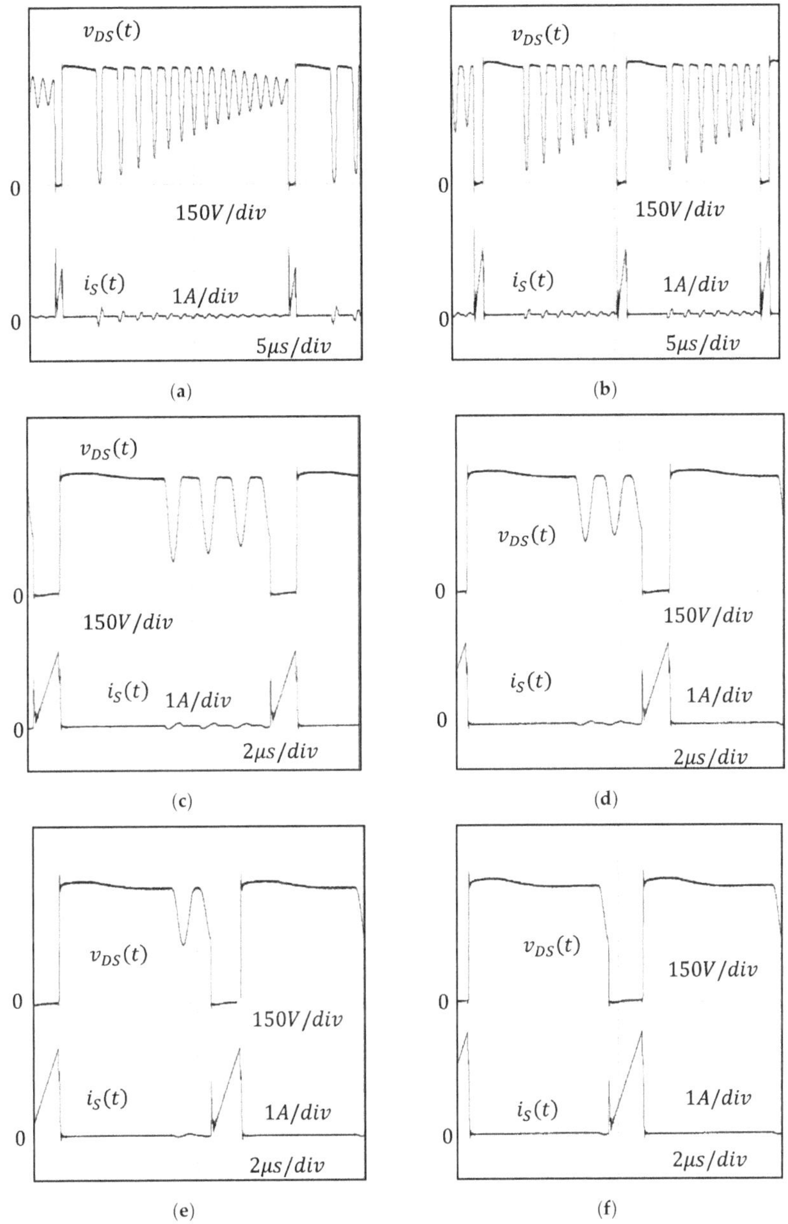

Figure 20. Steady-state waveforms of main switch voltage and current of the experimental circuit when the input voltage was 310 V: (**a**) 0.3 A load current; (**b**) 1.0 A load current; (**c**) 2.0 A load current; (**d**) 3.0 A load current; (**e**) 4.0 A load current; (**f**) 5.0 A load current.

Figure 21 shows a comparison of the switch voltage and current waveforms of a conventional QR flyback converter and the proposed QR flyback converter with the SDAS. The input voltage was 310 V, the load current was 3.5 A, and the output power was 42 W. Figure 21a shows the switch voltage and current waveforms of the experimental circuit using the existing QR flyback converter and RCD snubber, and Figure 21b is the switch voltage and current waveform of the QR flyback converter with the SDAS. As shown in

Figure 21a, the voltage surge of the switch occurred at the turn-off moment and had a peak value of up to 550 V. After that, the voltage surge was limited to about 500 V, there was a damping resonance period, and then the voltage surge stabilized at about 410 V, which was the turn-off steady-state voltage. The switch turned on in the second valley, and the switch current had a peak value of about 2 A. At this time, the switch-turn-off voltage also influenced the current during the damping resonance period, and therefore the current resonated simultaneously. As shown in Figure 21b, the turn-off voltage of the switch was clamped by the SDAS circuit and had a peak value of 430 V, and the peak value of the switch current was about 1.9 A. As can be seen in the figure, compared with the RCD clamp circuit, the SDAS circuit reduced the voltage surge of the QR flyback switch by up to 120 V or more. In addition, compared with RCD snubbers, which consume the voltage surge power in resistors, SDAS circuits do not consume power and, therefore, have higher efficiency characteristics in principle [42].

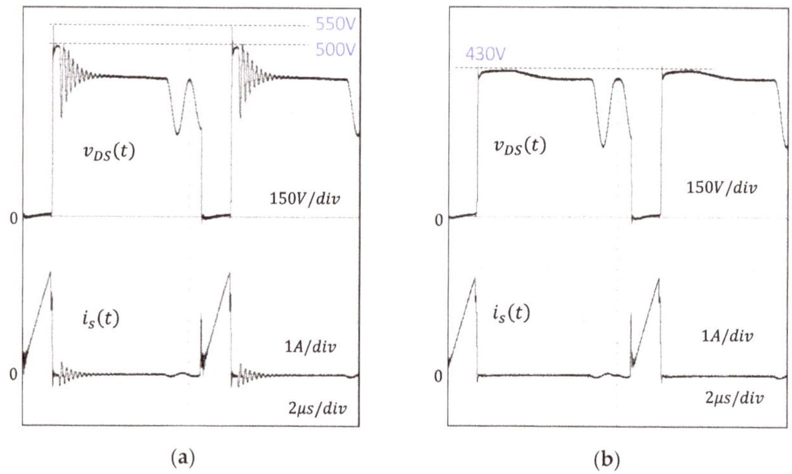

Figure 21. Comparison of steady-state switch voltage and current of QR flyback converter: (**a**) steady-state switch voltage and current waveforms of a QR flyback converter with RCD snubber; (**b**) steady-state switch voltage and current waveforms of a QR flyback converter with active snubber.

Figure 22 shows the power conversion efficiency of the experimental circuit at an input voltage of 310 V. The efficiency was over 85% when the output power was more than 5 W, and the efficiency was as high as 90% at the maximum power load of 60 W. Figure 23 shows the internal power dissipation of the adapter. At a power load of 10 W, there was a power loss of approximately 1.5 W, and at a maximum power load of 60 W, an internal loss of 6.7 W was observed. Figure 24 shows the output voltage relative to the power load. As evident in the figure, the output voltage remained stable, and a maximum voltage change of less than 5 mV was observed with the change in power load. Figure 25 shows the power conversion efficiency measured at the output power of 20 W, 40 W, and 60 W relative to the input voltage. The efficiency increased at a higher output and with a decrease in the input voltage. Specifically, the maximum power conversion efficiency was achieved when the input voltage was 140 V, and the output was 60 W, at 91.56%. When the output was 60 W, the power conversion efficiency was over 90% in all rates of input voltage.

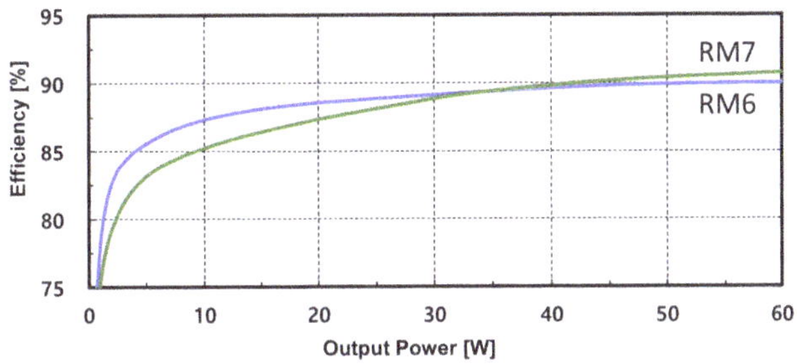

Figure 22. Power conversion efficiency characteristics versus load current of a QR flyback converter with SDAS at an input voltage of 310 V.

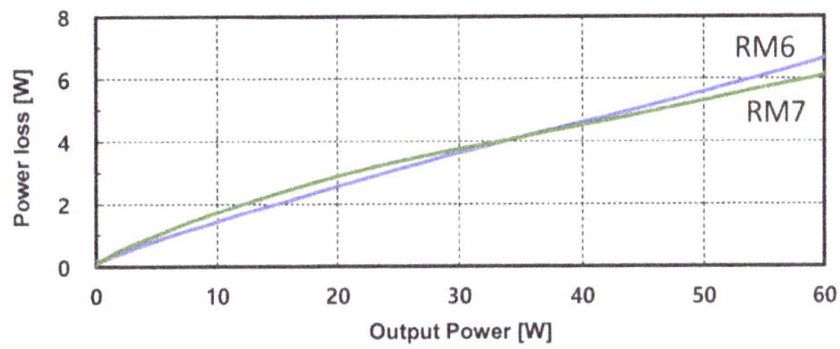

Figure 23. Characteristics of internal power dissipation versus load current of a QR flyback converter with SDAS at an input voltage of 310 V.

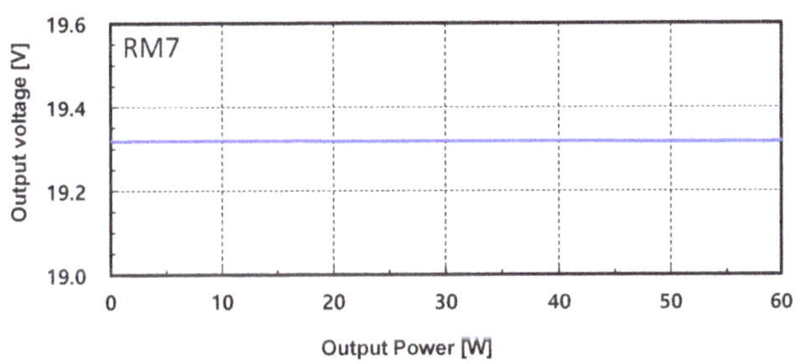

Figure 24. Output voltage characteristics versus load current of a QR flyback converter with SDAS at an input voltage of 310 V.

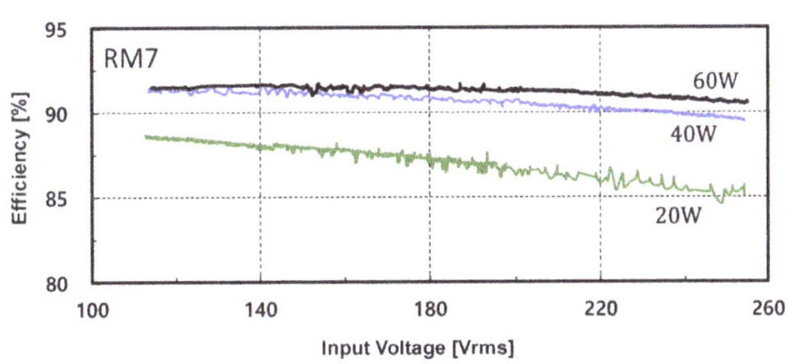

Figure 25. Power conversion efficiency characteristics versus input voltage of a QR flyback converter with SDAS at 20 W, 40 W, and 60 W power load.

These experimental results highlight the following findings: The AC–DC adapter of the proposed QR flyback converter with the SDAS showed a high power density of more than 20 W/in^3 and a high power conversion efficiency of up to 91.56%. Therefore, the proposed SDAS method can be used as the basic circuit method for high-efficiency power supply. One of the reasons for such high efficiency is due to the SDAS circuit, which sufficiently reduces the voltage surge of the QR flyback switch compared with the RCD clamp circuit and does not consume power in principle. Specifically, a separate control circuit is not required to drive the auxiliary switch for the clamp, and the circuit structure is simplified because it is self-driven inside the converter [43].

5. Conclusions

In this paper, a QR flyback converter using a self-driven active snubber (SDAS) was proposed to solve the problem of the switch voltage surge in the QR flyback converter. In the proposed converter, the SDAS consisting of a clamping capacitor and an active switch was configured in parallel with the main switch or transformer to reduce the voltage surge in the switch. The clamp switch was synchronized with the drive signal of an internal synchronous rectifier and a photo driver or pulse transformer, so no separate control circuit was required.

To confirm the steady-state characteristics of the proposed QR flyback converter with SDAS, equivalent circuits for each state were constructed, and the equations and characteristics for each state were determined. Using the main components for each state, the resonant circuit was separately arranged, and the components involved in resonance and the initial values of each state were identified. The values of the main elements were designed using the steady-state analysis results, and the optimal values of the clamp capacitor were derived. To compare the steady-state analysis results, a simulation using the PSIM program was performed. The simulation results and the theoretical waveforms were relatively well matched, and they showed a very similar shape to the experimental waveforms.

To confirm the effectiveness of the proposed converter and control circuit method, a 60 W class small AC–DC adapter was constructed, and the experimental results were presented. The size of the experimental AC–DC adapter was 74 × 29 × 23 mm and had a high power density of 20 W/in^3 or more. The experimental circuit was limited to the high power conversion efficiency of up to 91.56%, and the maximum voltage surge of the switch was approximately 450 V. Taken together, these results highlight the applicability of the proposed converter as the basic circuit method for achieving high-efficiency power supply. One of the reasons for such high efficiency is due to the SDAS circuit, which sufficiently reduces the voltage surge of the QR flyback switch compared with the RCD clamp circuit and does not consume power in principle.

Author Contributions: Conceptualization, T.-Y.A. and J.-O.B.; Methodology, J.-S.Y., J.-O.B. and T.-Y.A.; Validation, J.-S.Y., J.-O.B. and T.-Y.A.; Formal analysis, T.-Y.A.; Investigation, J.-S.Y. and J.-O.B.; Resources, T.-Y.A.; Data curation, J.-O.B. and T.-Y.A.; Writing—original draft preparation, J.-S.Y., J.-O.B. and T.-Y.A.; Writing—review and editing, J.-S.Y., J.-O.B. and T.-Y.A.; Visualization, J.-S.Y. and T.-Y.A.; Supervision, J.-O.B. and T.-Y.A.; Project administration, J.-O.B. and T.-Y.A.; Funding acquisition, T.-Y.A. All authors have read and agreed to the published version of the manuscript.

Funding: This work was supported by the Technological Innovation R&D Program (S3106696, Distributed power smart grid conversion module with GaN power semiconductor) funded by the Ministry of SMEs and Startups (MSS, Republic of Korea).

Data Availability Statement: Not applicable.

Conflicts of Interest: The authors declare no conflict of interest.

References

1. Park, H.-P.; Jung, J.-H. Design Methodology of Quasi-Resonant Flyback Converter with a Divided Resonant Capacitor. *IEEE Trans. Ind. Electron.* **2020**, *68*, 10796–10805. [CrossRef]
2. Wang, C.; Xu, S.; Shen, W.; Lu, S.; Sun, W. A Single-Switched High-Switching-Frequency Quasi-Resonant Flyback Converter. *IEEE Trans. Power Electron.* **2018**, *34*, 8775–8786. [CrossRef]
3. Baek, J.-O. A Study on New Active Clamp Circuit and Control Method for High Efficiency Quasi Resonant Flyback Converter. Ph.D. Thesis, Cheongju University, Graduate School of Cheongju University, Cheongju, Republic of Korea, 2018.

4. Wang, J.-M.; Lin, C.-W.; Huang, K.-Y.; Wong, J.-S. The Novel Quasi-Resonant Flyback Converter with Autoregulated Structure for Parallel/Serial Input. *IEEE Trans. Ind. Electron.* **2019**, *67*, 992–1004. [CrossRef]
5. Mauromicale, G.; Raciti, A.; Rizzo, S.A.; Susinni, G.; Fusillo, F.; Palermo, A.; Scrimizzi, F.; Scollo, R. Si and GaN Devices in Quasi Resonant Flyback converters for Wall Charger Applications. In Proceedings of the 2019 IEEE Energy Conversion Congress and Exposition (ECCE), Baltimore, MD, USA, 29 September–3 October 2019. [CrossRef]
6. Halil, O.; Sueel, V.; Hava, A. A comprehensive loss analysis of quasi resonant flyback converter for design purpose. In Proceedings of the PCIM Asia 2018 International Exhibition and Conference for Power Electronics, Intelligent Motion, Renewable Energy and Energy Management, Shanghai, China, 26–28 June 2018.
7. Xu, S.; Wang, C.; Qian, Q.; Zhu, J.; Sun, W.; Li, H. A single-switched high-switching-frequency quasi-resonant flyback converter with zero-current-switching and valley-switching. In Proceedings of the 2019 IEEE Applied Power Electronics Conference and Exposition (APEC), Anaheim, CA, USA, 17–21 March 2019. [CrossRef]
8. Min, G.-H.; Ha, J.-I. Inner Supply Data Transmission in Quasi-Resonant Flyback Converters for Li-Ion Battery Applications Using Multiplexing Mode. *IEEE Trans. Power Electron.* **2018**, *34*, 64–73. [CrossRef]
9. Cheng, M.; Liu, L.; Liao, Z.; Yu, J.; Peng, S.; Zhang, Z.; Ye, T.; Wang, N. On Silicon-Integrated Micro-Transformers and Their Applications in Quasi-Resonant Flyback Converters. *IEEE Trans. Magn.* **2022**, *58*, 8401209. [CrossRef]
10. Kang, B.; Low, K.-S.; Ng, K.J. Capacitor Charging by Quasi-Resonant Approach for a Pulsed Plasma Thruster in Nano-Satellite. *IEEE Trans. Plasma Sci.* **2020**, *48*, 1271–1278. [CrossRef]
11. Vračar, D.; Pejovic, P.V. Active-Clamp Flyback Converter as Auxiliary Power-Supply of an 800 V Inductive-Charging System for Electric Vehicles. *IEEE Access* **2022**, *10*, 38254–38271. [CrossRef]
12. Yao, J.; Li, Y.; Wang, S.; Huang, X.; Lyu, X. Analysis and Reduction of Radiated EMI in High-Frequency GaN IC-based Active Clamp Flyback Converters. In Proceedings of the 2020 IEEE Applied Power Electronics Conference and Exposition (APEC), New Orleans, LA, USA, 15–19 March 2020. [CrossRef]
13. Wang, C.; Sun, D.; Zhang, X.; Hu, J.; Gu, W.; Gui, S. A Constant Current Digital Control Method for Primary-Side Regulation Active-Clamp Flyback Converter. *IEEE Trans. Power Electron.* **2020**, *36*, 7307–7318. [CrossRef]
14. Kuo, C.-C.; Lee, J.-J.; He, Y.-H.; Wu, J.-Y.; Chen, K.-H.; Lin, Y.-H.; Lin, S.-R.; Tsai, T.-Y. A dynamic resonant period control technique for fast and zero voltage switching in GaN-based active clamp flyback converters. *IEEE Trans. Power Electron.* **2020**, *36*, 3323–3334. [CrossRef]
15. Xue, L.; Zhang, J. Design considerations of highly-efficient active clamp flyback converter using GaN power ICs. In Proceedings of the 2018 IEEE Applied Power Electronics Conference and Exposition (APEC), San Antonio, TX, USA, 4–8 March 2018. [CrossRef]
16. Hu, J.; Wen, B.; Burgos, R.; Kang, Y. Design of a Wide-Input-Voltage PCB-Embedded Transformer Based Active-Clamp Flyback Converter Considering Permeability Degradation. *IEEE Trans. Power Electron.* **2021**, *36*, 10355–10365. [CrossRef]
17. Ma, Z.; Yao, J.; Wang, S.; Sheng, H.; Lakshmikanthan, S.; Osterhout, D. Radiated EMI Reduction with Double Shielding Techniques in Active-clamp Flyback Converters. In Proceedings of the 2021 IEEE International Joint EMC/SI/PI and EMC Europe Symposium, Raleigh, NC, USA, 26 July–13 August 2021. [CrossRef]
18. Wu, S.-T.; Cheng, Y.-T. Design and Implementation of a Single-Stage PFC Active-Clamp Flyback Converter with Dual Transformers. *Electronics* **2021**, *10*, 2588. [CrossRef]
19. Chen, M.; Xu, S.; Huang, L.; Sun, W.; Shi, L. A Novel Digital Control Method of Primary-Side Regulated Flyback with Active Clamping Technique. *IEEE Trans. Circuits Syst. I Regul. Pap.* **2020**, *68*, 950–962. [CrossRef]
20. Xu, S.; Qian, Q.; Tao, T.; Lu, S.; Sun, W. Small Signal Modeling and Control Loop Design of Critical Conduction Mode Active Clamp Flyback Converter. *IEEE Trans. Power Electron.* **2020**, *36*, 7250–7263. [CrossRef]
21. Nasiri, A.; Banaei, M.R.; Abadi, A.S.S. Phase-shifted active clamp flyback converter for driving a magnetron. In Proceedings of the 2019 27th Iranian Conference on Electrical Engineering (ICEE), Yazd, Iran, 30 April–2 May 2019.
22. Nasiri, A.; Abadi, A.S.S. A New Driving Method for a Magnetron Using a Soft Switching Active Clamp Fly-back Converter. In Proceedings of the 2019 10th International Power Electronics, Drive Systems and Technologies Conference (PEDSTC), Shiraz, Iran, 12–14 February 2019. [CrossRef]
23. Li, H.; Li, S.; Xiao, W.; Hui, S.Y.R. A Modulation Method for Capacitance Reduction in Active-Clamp Flyback-Based AC–DC Adapters. *IEEE Trans. Power Electron.* **2022**, *37*, 9455–9467. [CrossRef]
24. Gao, S.; Song, H.; Wang, Y.; Xu, R.; Xu, D. A Secondary-Resonance MHz Active-Clamp Flyback Converter with Partial Power Processing. *IEEE Trans. Ind. Appl.* **2022**, *58*, 7988–7997. [CrossRef]
25. Gu, D.; Xi, J.; He, L. Digitally controlled GaN-based MHz active clamp flyback converter with dynamic dead time optimisation for AC–DC adapter. *IET Power Electron.* **2020**, *13*, 3777–3786. [CrossRef]
26. Xu, S.; Qian, Q.; Ren, B.; Liu, Q. An accurate small signal modeling and control loop design of active clamp flyback converter. In Proceedings of the 2019 10th International Conference on Power Electronics and ECCE Asia (ICPE 2019-ECCE Asia), Busan, Republic of Korea, 27–30 May 2019.
27. Tang, C.-Y.; Lin, W.-Z.; Tan, Y.-C. An Active Clamp Flyback Converter with High Precision Primary-Side Regulation Strategy. *IEEE Trans. Power Electron.* **2022**, *37*, 10281–10289. [CrossRef]
28. Xu, S.; Qian, Q.; Shi, R.; Shah, S.S.; Liu, Q.; Lu, S.; Sun, W. Sampled-Data Modeling for PCM and ZVS Controlled Critical Conduction Mode (CrCM) Active Clamp Flyback (ACF) Converter at Variable Switching Frequency. *IEEE Trans. Circuits Syst. I: Regul. Pap.* **2020**, *67*, 3588–3600. [CrossRef]

29. Liu, Y.-C.; Huang, B.-S.; Lin, C.-H.; Kim, K.A.; Chiu, H.-J. Design and implementation of a high power density active-clamped flyback converter. In Proceedings of the 2018 International Power Electronics Conference (IPEC-Niigata 2018-ECCE Asia), Niigata, Japan, 20–24 May 2018.
30. Zaman, A.; Radic, A. How to Design and Implement an Adapter Power Supply with Active Clamp Flyback: An All Silicon Design Methodology. *IEEE Power Electron. Mag.* **2020**, *7*, 36–43. [CrossRef]
31. Dey, S.; Ray, M.B.; Soni, H.; Ghosh, R.; Shah, M. Comparison between Quasi-Resonant and Active Clamp Flyback topologies for GaN-based 65 W Wall Charger Application. In Proceedings of the 2021 IEEE Applied Power Electronics Conference and Exposition (APEC), Phoenix, AZ, USA, 14–17 June 2021.
32. Hosseinabadi, F.; Soleymani, B.; Adib, E.; Islam, R.; Farzanehfard, H.; Kouzani, A.Z.; Mahmud, M.A.P. A soft-switching inverting high step-down converter with a pair of coupled inductors and self-driven synchronous rectifier. *IET Power Electron.* **2021**, *14*, 2441–2451. [CrossRef]
33. Mahdi, G.J.; Allahyari, H.; Shoaei, A.; Adib, E.; Bahrami, H.; Zadeh, M.A.L.; Fahimifar, M.H. A self-driven synchronous rectification wide-range ZVS single-switch forward converter controlled using the variable inductance. *IET Power Electron.* 2022; *preprint*. [CrossRef]
34. Liu, C.; Zhang, Z.; Liua, Y.; Si, Y.; Wang, M.; Lei, Q. A Universal Block of Series-Connected SiC MOSFETs Using Current-Source Gate Driver. *IEEE J. Emerg. Sel. Top. Power Electron.* **2021**, *10*, 3066–3086. [CrossRef]
35. Li, X.; Ma, H.; Yi, J.; Lu, S.; Xu, J. A Comparative Study of GaN HEMT and Si MOSFET-Based Active Clamp Forward Converters. *Energies* **2020**, *13*, 4160. [CrossRef]
36. Soleymani, B.; Adib, E. A High Step-Down Buck Converter with Self-Driven Synchronous Rectifier. *IEEE Trans. Ind. Electron.* **2019**, *67*, 10266–10273. [CrossRef]
37. Li, M.; Ouyang, Z.; Andersen, M.A.E.; Zhao, B. Self-Driven Gate Driver for *LLC* Synchronous Rectification. *IEEE Trans. Power Electron.* **2020**, *36*, 56–60. [CrossRef]
38. Di, Z.; Donglai, Z.; Zicai, W.; Hua, Z.; Ying, Z. Self-driven Active-Clamped Forward Power Converter with Output Ripple-Cancellation for Aerospace Applications. In Proceedings of the 2019 4th International Conference on Power and Renewable Energy (ICPRE), Chengdu, China, 21–23 September 2019. [CrossRef]
39. Guan, Y.; Hu, X.; Wang, Y.; Xu, D.; Wang, W. A Single-Switch Self-Driving High Frequency Converter Based on Optimal Feedback Network. In Proceedings of the 2018 IEEE Energy Conversion Congress and Exposition (ECCE), Portland, OR, USA, 23–27 September 2018. [CrossRef]
40. Konishi, A.; Umetani, K.; Hiraki, E. High-frequency self-driven synchronous rectifier controller for WPT systems. In Proceedings of the 2018 International Power Electronics Conference (IPEC-Niigata 2018-ECCE Asia), Niigata, Japan, 20–24 May 2018.
41. Bahrami, H.; Allahyari, H.; Adib, E. A Self-Driven Synchronous Rectification ZCS PWM Two-Switch Forward Converter with Minimum Number of Components. *IEEE Trans. Ind. Electron.* **2021**, *69*, 12842–12850. [CrossRef]
42. Nourieh, N.; Sun, Y.; Simpson, O. A novel resonant ZVS power converter with self-driven synchronous rectifier for low-voltage high-current applications. *IET Power Electron.* **2021**, *14*, 1397–1408. [CrossRef]
43. Scortegagna, R.G.; Gules, R. Self-Driven Current-Doubler Synchronous Rectifier and Design Tuning for Maximizing Efficiency in IPT Systems. *IEEE J. Emerg. Sel. Top. Power Electron.* **2021**, *10*, 1007–1016. [CrossRef]

electronics

MDPI

Article

Cascaded AC-DC Power Conversion Interface for Charging Battery

Jinn-Chang Wu [1], Hurng-Liahng Jou [2], Fu-Zen Chen [2,*] and Jung-Peng Li [1]

[1] Department of Microelectronics Engineering, National Kaohsiung University of Science and Technology, 142 Haijhuan Road, Kaohsiung 81143, Taiwan
[2] Department of Electrical Engineering, National Kaohsiung University of Science and Technology, 415 Chien-Kung Road, Kaohsiung 80778, Taiwan
* Correspondence: chenfuzen@nkust.edu.tw

Abstract: This paper develops a cascaded AC-DC power conversion interface (CADPCI) to convert AC power to charge the battery set. The proposed CADPCI is composed of a cascaded converter (CC) and a dual-input buck converter (DIBC). The CC is formed by connecting a full-bridge converter (FBC) and a bridgeless rectifier (BLR) in series. The CADPCI generates an 11-level input voltage and performs unity power factor correction. The switching loss is reduced because only the FBC with a lower DC port voltage is switched at a high frequency. The DIBC uses a buck converter and a selection switch set to generate a two-level DC voltage on the DC port of the BLR. By controlling the DC input voltage of the buck converter, the injected power of the BLR can match the input power of the utility. Therefore, the FBC does not require to handle the real power, saving an isolated converter for regulating the DC port voltage of the FBC, thus simplifying the power circuit of the CC. The buck converter also acts as a DC active filter to filter out low-frequency ripples of the charging current. A prototype is constructed to verify the performance of the proposed CADPCI.

Keywords: cascaded power converter; bridgeless rectifier; DC active filter

Citation: Wu, J.-C.; Jou, H.-L.; Chen, F.-Z.; Li, J.-P. Cascaded AC-DC Power Conversion Interface for Charging Battery. *Electronics* **2023**, *12*, 1192. https://doi.org/10.3390/electronics12051192

Academic Editors: Diego Bellan and Jelena Loncarski

Received: 10 February 2023
Revised: 26 February 2023
Accepted: 27 February 2023
Published: 1 March 2023

1. Introduction

Electronic equipment is widely used in industry, commerce, and households. It is generally powered by the utility through an AC-DC power conversion interface. Over the last two decades, batteries have started from powering our portable electronics to powering our vehicles and household equipment, such as robot vacuums, electric bikes, mowing machines, etc. With the development of robots and electric vehicles, the number of battery-powered devices has increased dramatically. Those batteries have to be charged from the utility through an AC-DC power conversion interface.

In order to keep good power quality for electronic equipment and ensure the performance of battery sets, an AC-DC power conversion interface must provide a stable and controllable voltage/current at the DC port and produce a sinusoidal input current at the AC port with nearly a unit power factor. Diode rectifiers do not meet these requirements and cannot work as an AC-DC power conversion interface alone [1]. Due to the price decrease in the power semiconductor components and advanced switching power supply technology, active power factor correction (PFC) rectifiers have been well-developed. Conventionally, a switching power converter, such as a boost, buck, or buck–boost converter, is connected to the DC port of a diode rectifier for a single-phase PFC circuit [1–6]. The diode rectifier converts the utility voltage into a rectified voltage, which is then further converted into a controllable DC voltage/current through a switching power converter. The switching power converter also shapes the input current for a good power factor and low current harmonic distortion.

Different from discontinuous input current driven by buck converter or buck–boost converter, boost converter drives its input current continuously, which eliminates the large-capacity input filter and reduces electromagnetic interference (EMI). In order to improve the power efficiency of PFC circuit, the bridgeless PFC circuit integrates the diode bridge and the boost converter to eliminate one rectifier diode voltage drop at the forward path [7,8]. However, the DC output voltage for a PFC circuit that uses a boost converter is always greater than the amplitude of the utility voltage. Hence, a buck converter with a high step-down ratio is required for low-voltage applications, and the overall power efficiency is low. To reduce leakage current and increase the step-down ratio, an isolated converter, such as a flyback or forward converter, is connected to the diode rectifier in the PFC circuit [9–11]. However, this produces a discontinuous input current and results in high voltage stress for power electronic switches. Moreover, the use of a transformer also decreases the power efficiency and induces spike voltages.

Power semiconductor components are worked as switches for power conversion applications. Their non-ideal switching characteristics induce switching losses during the switching turned-on and turned-off transitions. The switching loss is highly dependent on the transition time, transition voltage/current level, and semiconductor switching characteristics. For the switching power converter used in the conventional PFC circuit, the power semiconductor components are operated in hard switching with a high transition voltage level, resulting in a large switching loss. Soft switching technologies, which take advantage of the LC resonant to turn the switch components at nearly zero transition voltage/current level, can reduce switching losses dramatically [12–14]. However, most soft switching technologies significantly increase the complexity of the controller and power circuit design. In addition, the resonant performance is heavily affected by the drift of passive components.

Multi-level converters (MLCs) reduce the transition voltage level of the switching of power electronic switches, so both the switching harmonics and the switching loss are reduced. Therefore, the capacities of both passive filter components and heat dissipation components can be effectively reduced. The diode-clamped MLCs employ a number of clamped diodes as the conduction paths to generate more output voltage levels and reduce the transition voltage level of the power electronic switches [15,16]. However, these clamped diodes have higher voltage ratings and result in larger power losses. The flying-capacitor MLCs generate more output voltage levels and reduce the switching voltages of power electronic switches by inserting capacitors into the conduction path [17,18]. However, these capacitors enlarge the circuit volume. In addition, the issues of voltage balance in both diode-clamped MLCs and flying-capacitor MLCs have to be concerned [15–18]. The cascaded bridge MLCs connect several full-bridge converters (FBCs) to generate more output voltage levels and reduce the switching voltages of power electronic switches [19–23]. An independent DC source is necessary for each FBC, which is the primary consideration for cascaded bridge MLCs. Although most MLCs are applied in DC-AC power conversions, many AC-DC power conversion applications have been developed in recent years [24–27].

In order to keep the advantage of the continuous input current of the boost-type PFC circuits but eliminate the drawback of the high step-down ratio of the second stage and high transition voltage level, this paper proposes a cascaded AC-DC power conversion interface (CADPCI) to convert AC power from the utility into stable DC power to charge the battery set. The proposed CADPCI is composed of a cascaded converter (CC) and a dual-input buck converter (DIBC). The major contributions of the proposed CADPCI are listed as follows.

1. The CC uses an FBC and a bridgeless rectifier (BLR), connected in series, to generate an 11-level input voltage and perform a unity power factor. Only six power electronic switches are used in the CC.
2. The switching loss of the CC is reduced significantly because only the FBC with a lower DC port voltage is switched at a high frequency.
3. The DIBC controls the DC port voltage of the BLR to achieve a power balance between the BLR and the utility. The FBC does not handle real power to save an isolated converter for regulating the DC port voltage of the FBC, thus simplifying the power circuit of the CC.

4. DIBC further realizes the function of a DC active filter (DAF) with no additional circuit.

This paper is organized as follows: Section 2 reviews the cascaded converter in rectifier applications, followed by the principle of the proposed cascaded power conversion interface in Section 3. Then, Section 4 describes the operation of the proposed dual-input buck converter, and Section 5 explains the operation of the cascaded converter. The last two sections show the experimental results and conclude this paper.

2. Cascaded Converter

The CC is configured by connecting several FBCs in series, as shown in Figure 1. According to the DC port voltage of different FBCs, the CC is divided into symmetrical CC and asymmetrical CC. The DC port voltages of the FBCs in a symmetrical CC are always the same. The symmetrical CC has $2n + 1$ voltage levels at the AC port, where n is the number of FBCs [19,20]. The benefits of symmetrical CC include easily modulated and even distribution of the power losses. On the other hand, the DC port voltages of the FBCs in an asymmetrical CC are usually in multiple relationships. Accordingly, the asymmetric CC generates more levels of AC port voltage compared to the symmetrical CC [21–23]. The CCs with two FBCs are given as examples. A two-FBCs symmetrical CC generates five voltage levels at the AC port. A two-FBC asymmetrical CC, which has a 1:2 voltage ratio at their FBC DC port voltages, generates seven voltage levels at the AC port [22]. A two-FBC asymmetrical CC, which has a 1:3 voltage ratio at their FBC DC port voltages, generates nine voltage levels at the AC port [23].

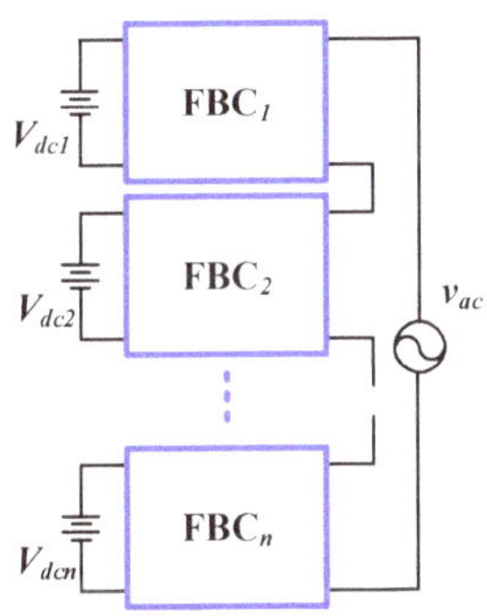

Figure 1. Topology of the CC.

Regardless of symmetrical or asymmetrical CC, each FBC requires an independent DC power supply, which increases the complexity of the DC power processing circuit.

3. Principle of Proposed Cascaded Power Conversion Interface

The power circuit of the proposed CADPCI is shown in Figure 2. The proposed CADPCI is composed of a CC and a DIBC. The CC combines an FBC and a BLR in a series connection, as shown in Figure 3. The BLR replaces the diode rectifier of the PFC circuit to decrease conduction loss [7,8], but an extra switch component is necessary. The FBC uses unipolar pulse width modulation (PWM) to control power switches S_{f1}–S_{f4} to generate a three-level pulse voltage at the AC port of the FBC (v_f). The DC port of the BLR is connected to the DIBC. The DIBC integrates a buck converter and a selector switch (S_{d1}) to provide a two-level DC voltage to the DC port of the BLR. The power switches S_{b1} and S_{b2} of BLR are switched synchronously with the utility voltage to generate a five-level step-wave voltage at the AC port of the BLR (v_b). The two-level DC voltage for the DC port of the BLR is two or four times the DC port voltage of the FBC (V_{Cf}). Therefore, the proposed CC can synthesize an 11-level input voltage by cascading the ac port voltages of FBC and BLR. Compared to the asymmetrical CC with two FBCs, the proposed CC can generate more voltage levels. The CC also produces a sinusoidal input current to perform a unity power factor. The FBC with a low DC port voltage is the only part switching in high frequency.

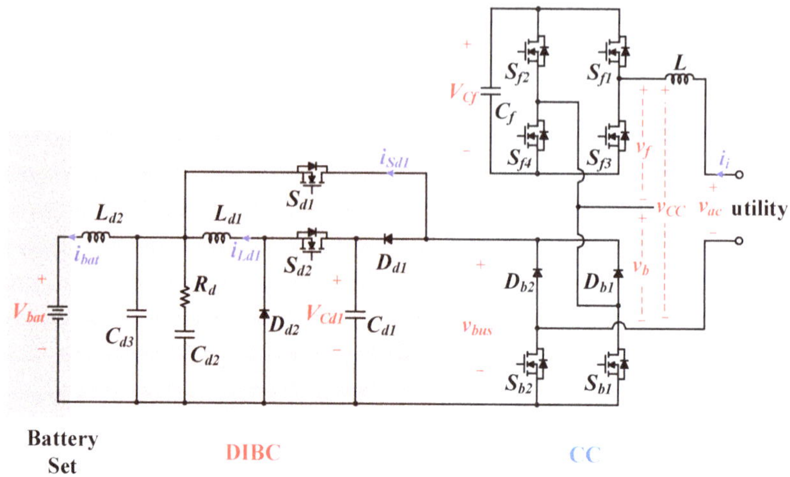

Figure 2. Power circuit of the proposed CADPCI.

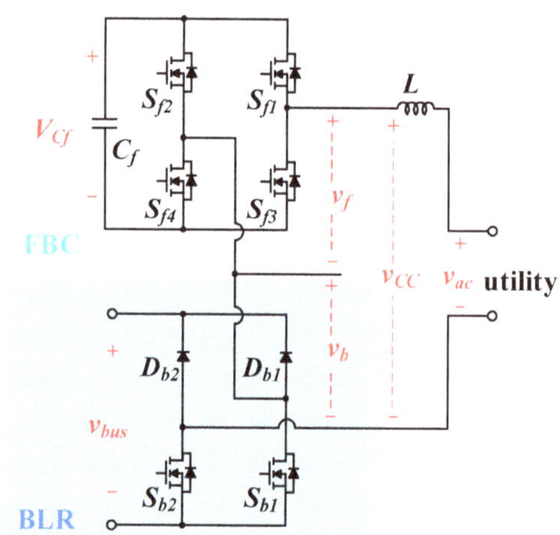

Figure 3. CC circuit of the proposed CADPCI.

The DIBC also performs the function of DAF with no additional circuit to filter out low-frequency ripple of the charging current. A low-pass filter that is configured using C_{d2}, C_{d3}, R_{d}, and L_{d2} is used to filter out the switching ripple of the charging current.

4. Operation of Dual-Input Buck Converter

Although the CADPCI performs unity power factor correction, the instantaneous input power from the utility includes not only a DC power but also a twice-utility-frequency AC power. The twice-utility-frequency AC power results in a ripple current on the DC side of the CADPCI that charges the battery set. This ripple current may reduce the life of the battery set [28,29]. The novelty of the DIBC is that the buck converter in the DIBC not only regulates the DC port voltage of the BLR but also acts as a DAF to filter out low-frequency ripple of the charging current for the battery set.

The input voltage of the DIBC has two levels: the voltage of the battery set (V_{bat}) and the input voltage of the buck converter (V_{Cd1}), depending on the operation of the selector switch S_{d1}. The DIBC operates in two modes, according to the selector switch, as shown in Figure 4.

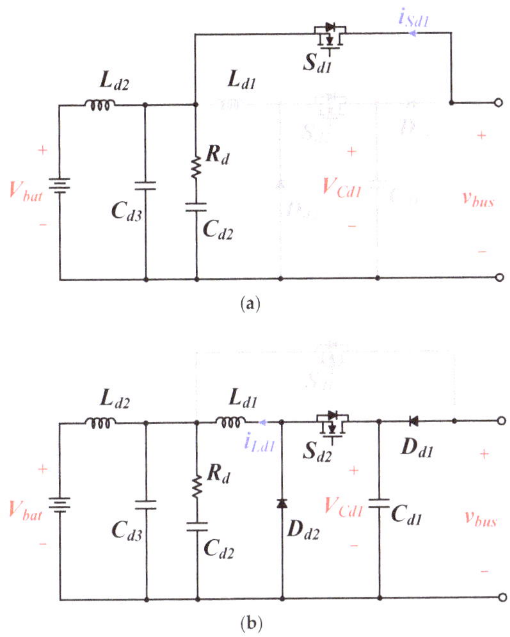

Figure 4. Operation of the proposed DIBC, (**a**) mode DI, (**b**) mode DII.

- Mode DI:
  Figure 4a shows the operation of this mode, S_{d1} is turned on, and D_{d1} is turned off. The battery set is directly charged from the utility through only the CC. The input voltage of the DIBC is equal to the voltage of the battery set.
- Mode DII:
  Figure 4b shows the operation of this mode, S_{d1} is turned off, and D_{d1} is turned on. The battery set is charged from the utility through both the CC and buck converter. The input voltage of the DIBPC is equal to the input voltage of the buck converter, which is higher than the voltage of the battery set due to the operation of the buck converter.

The buck converter adopts a current mode control to control the current of inductor L_{d1}, and its operation can be divided into two modes. When S_{d2} is turned on, the voltage v_{Dd2} across D_{d2} is V_{Cd1}. V_{Cd1} is higher than the voltage of the battery set, so the current of the inductor L_{d1} is increased. When S_{D2} is turned off, D_{d2} is conducted. Consequently, the voltage v_{Dd2} across D_{d2} is 0, and the current of the inductor L_{d1} is decreased. By controlling S_{D2} in PWM switching, the voltage v_{Dd2} across D_{d2} is a pulse voltage that varies between V_{Cd1} and 0, which can control the current of the inductor L_{d1} increasing or decreasing to follow its reference current. The reference current includes a DC component and an AC component. The DC component is used to regulate the input voltage of the buck converter. The AC component is calculated by extracting the ripple of the charging current to perform the function of DAF. Therefore, the DIBC can perform the function of DAF with no additional circuit.

5. Operation of Cascaded Converter

The CC in Figure 2 comprises an FBC and a BLR that are connected in series. The circuit for the BLR operates in three modes, as shown in Figure 5.

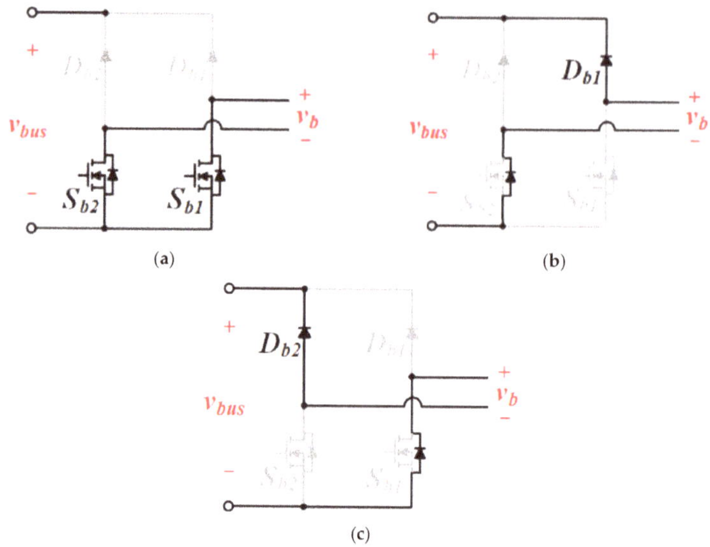

Figure 5. Circuit operation of the BLR: (**a**) mode RI, (**b**) mode RII, (**c**) mode RIII.

- Mode RI:
  As can be seen in Figure 5a, S_{b1}, and S_{b2} are turned on, and the current path is bidirectional. The AC port voltage of the BLR is:

$$v_b = 0 \quad (1)$$

- Mode RII:
  This mode is operated during the positive half cycle of the utility voltage, as shown in Figure 5b. The input current is positive. S_{b1} and S_{b2} are turned off, and D_{b1} and the body diode of S_{b2} conduct. The AC input port voltage of the BLR is:

$$v_b = v_{bus} \quad (2)$$

  where v_{bus} is the DC port voltage of BLR.
- Mode RIII:
  This mode is operated during the negative half cycle of the utility voltage, as shown in Figure 5c. The input current is negative. S_{b1} and S_{b2} are turned off, and D_{b2} and the body diode of S_{b1} conduct. The AC port voltage of the BLR is:

$$v_b = -v_{bus} \quad (3)$$

The DC port voltage of the BLR is generated by the DIBC and has two levels: the voltage of the battery set and the input voltage of the buck converter, depending on the selector switch. The low-pass filter, configured by C_{d2}, C_{d3}, R_d, and L_{d2}, is used to filter out the switching ripple, and its gain is close to unity for DC voltage. As a result, the output voltage of the buck converter is almost equal to the voltage of the battery set. S_{b1} and S_{b2} in the BLR are switched synchronously with the utility voltage, so the BLR generates a five-level stepped voltage at the AC port. The five levels are V_{Cd1}, V_{bat}, 0, $-V_{bat}$, and $-V_{Cd1}$.

The unipolar PWM control is adopted in the FBC. The FBC generates a three-level pulse voltage at the AC port. The three levels are V_{Cf}, 0, and $-V_{Cf}$.

Using asymmetric voltage technology for a CC increases the number of voltage levels. Therefore, the DC port voltage of the FBC, the voltage of the battery set, and the input voltage of the buck converter are designed in the ratio of 1:2:4 to allow the CC to generate an 11-level input voltage. The operation voltage for the CC is shown in Figure 6. Figure 6a shows that the BLR generates a five-level step-wave voltage, which contains voltage levels of $4V_{Cf}$, $2V_{Cf}$, 0, $-2V_{Cf}$, and $-4V_{Cf}$. Figure 6b shows that the FBC generates a three-level pulse voltage, where the three voltage levels are V_{Cf}, 0, and $-V_{Cf}$. The AC port voltage of the CC integrates the AC port voltages of the FBC and the BLR to generate an 11-level pulse voltage. The voltage levels are 5 V_{ice}, 4 V_{Cf}, 3 V_{Cf}, 2 V_{Cf}, V_{Cf}, 0, $-V_{Cf}$, $-2\ V_{Cf}$, $-3\ V_{Cf}$, $-4\ V_{Cf}$, and $-5\ V_{Cf}$, as illustrated in Figure 6c.

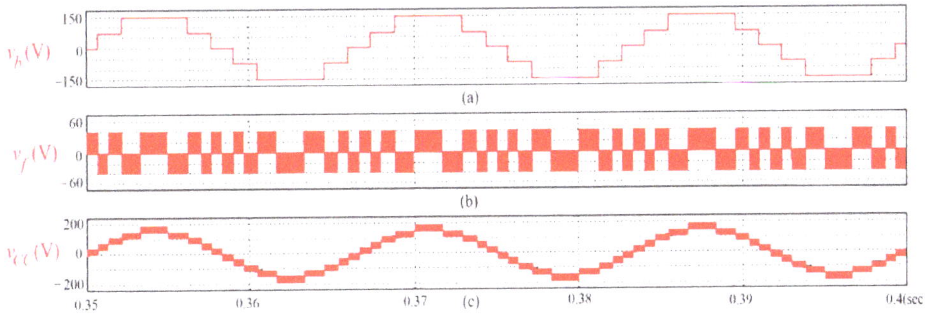

Figure 6. Simulation results on AC port voltages: (**a**) BLR; (**b**) FBC; (**c**) CC.

The difference in each voltage level is only V_{Cf}, and the ripple of the input current is written as:

$$\Delta i_i = \frac{V_{Cf}}{L \cdot f_p} D \cdot (1 - D), \tag{4}$$

where D is the duty cycle of the FBC and f_p is the frequency of the pulse voltage for the FBC. The FBC uses unipolar PWM control, so f_p is twice the switching frequency of the power electronic switches. Since the difference in each voltage level is greatly reduced, and f_p is multiplied, the filter inductor is significantly reduced in the proposed CC. Moreover, only the FBC is switched in high frequency. The DC port voltage of the FBC is about one-fifth that of a conventional PFC circuit; hence, the switching loss for the proposed CC is reduced significantly. The conduction resistance of MOSFET is proportional to its voltage rating. Hence, the conduction loss of FBC is low. Moreover, the input voltage of the buck converter is four-fifths as compared with that of the conventional PFC circuit, so the switching loss of the buck converter in the DIBC is also reduced. Although the proposed CADPCI uses a large number of components, its power efficiency is better than that of the conventional PFC circuit. In addition, the capacity of the passive filter and the EMI of the proposed CADPCI is significantly reduced as compared with the conventional PFC circuit. The DC port voltage of FBC in the proposed CC is also less than that in the asymmetrical CC with a DC port voltage ratio of 1:3, which is a quarter of the DC port voltage for the conventional PFC circuit. Therefore, the switching losses, the capacity of the passive filter, and the EMI of the proposed CC can be further reduced as compared with the asymmetrical CC with a DC port voltage ratio of 1:3. Figure 7 shows the percentage loss for the power semiconductor components in the CADPCI by using a thermal module of PSIM. For CC, the switching loss is significantly reduced. The largest loss in the CADPCI is the switching loss and conduction loss of S_{d2} of the DIBC.

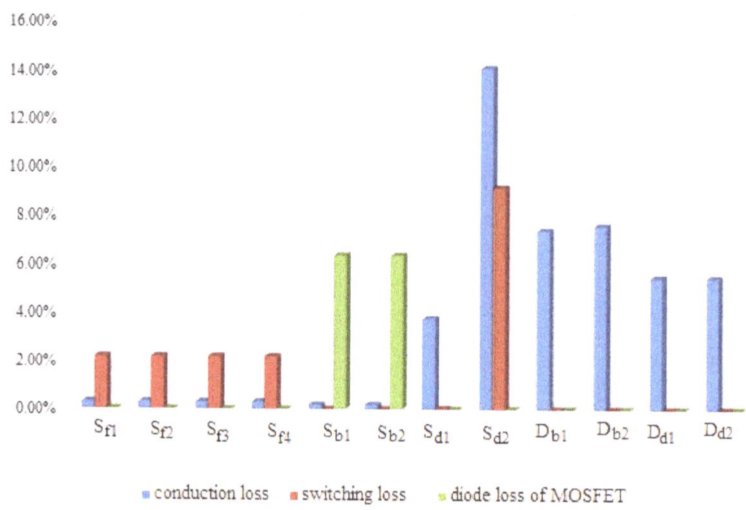

Figure 7. Percentage loss for the power semiconductor components in the CADPCI.

The major disadvantage of CC is that the DC ports for FBCs do not have a common ground, so several independent power supplies or isolated DC-DC power converters must be used to process the DC power for the FBCs. The power balance theory is used in the proposed CC, and only a capacitor C_f is used to be an energy buffer in the FBC, and an independent power source or an isolated DC-DC power converter is removed. The AC port voltage of the BLR (v_b) in Figure 6b can be written as:

$$v_b(\mathrm{t}) = \begin{cases} V_{Cd1}, & \theta_2 \le \omega t \le \pi - \theta_2 \\ V_{bat}, & \theta_1 \le \omega t \le \theta_2,\ (\pi - \theta_2) \le \omega t \le (\pi - \theta_1), \\ 0, & 0 \le \omega t \le \theta_1,\ \pi - \theta_1 \le \omega t \le \pi + \theta_1, 2\pi - \theta_1 \le \omega t \le 2\pi \\ -V_{bat}, & \pi + \theta_1 \le \omega t \le \pi + \theta_2,\ 2\pi - \theta_2 \le \omega t \le 2\pi - \theta_1 \\ -V_{Cd1}, & \pi + \theta_2 \le \omega t \le 2\pi - \theta_2 \end{cases} \tag{5}$$

where

$$\theta_1 = \sin^{-1}\left(\frac{V_{bat}}{V_{AC}}\right) \tag{6}$$

$$\theta_2 = \sin^{-1}\left(\frac{V_{Cd1}}{V_{AC}}\right) \tag{7}$$

and V_{AC} is the amplitude of the utility voltage. The Fourier series for the AC port voltage of the BLR can be expressed as:

$$V_b(t) = V_{b0} + \sum_{n=1}^{\infty} V_{bn} \sin(n\omega t + \varphi_n) \tag{8}$$

where V_{b0} is the average value, and it is 0. V_{bn} is the amplitude of the n-th harmonic, which is

$$V_{bn} = \frac{4}{n\pi}(V_{bat}(\cos(n\theta_1) - \cos(n\theta_2)) + V_{Cd1}\cos(n\theta_2)),\ n = 1,3,5,\cdots \tag{9}$$

The utility voltage is written as:

$$v_{ac}(t) = V_{AC}\ \sin(\omega t) \tag{10}$$

If the input current of the CC is controlled to be sinusoidal and the power factor is unity, it is written as:

$$i_i(t) = I_i \sin(\omega t) \tag{11}$$

The input real power for the CC is written as:

$$P_i = \frac{1}{2} V_{AC} I_i \tag{12}$$

The input real power for the BLR is derived as:

$$P_b = \frac{4I_i}{\pi}[V_{bat}(\cos(\theta_1) - \cos(\theta_2)) + V_{Cd1}\cos(\theta_2)] \tag{13}$$

The input real power for the CC is the sum of the input real powers for the BLR and the FBC, which is:

$$P_i = P_f + P_b \tag{14}$$

As can be seen in (13), the voltage of the battery set cannot be controlled; hence, the input real power of the BLR is controlled by the input voltage of the buck converter, V_{Cd1}. When the input real power for the BLR is rendered equal to the input real power of the CC by adjusting the input voltage of the buck converter, no real power is injected into the FBC. Therefore, an isolated DC-DC power converter is not required to convert the real power from the FBC to charge the battery set. If the DC port voltage of the FBC is less than its set value, the input voltage of the buck converter must be reduced so that the input real power for the BLR is less than that for the CC. At this time, the input real power for the FBC is positive and is used to charge the capacitor C_f. If the DC port voltage of the FBC is greater than its set value, the input voltage of the buck converter must be increased so that the input real power for the BLR is greater than that for the CC. Therefore, the input real power for the FBC has a negative value, and the capacitor C_f is discharged.

6. Control Block

Figure 8 shows the control block of the DIBC. The selector switch is controlled by comparing the absolute value of the utility voltage with the voltage of the battery set and the dc port voltage of the FBC. When the absolute value of the utility voltage is between the dc port voltage of the FBC and the input voltage of the buck converter, a control signal is generated to turn S_{d1} on.

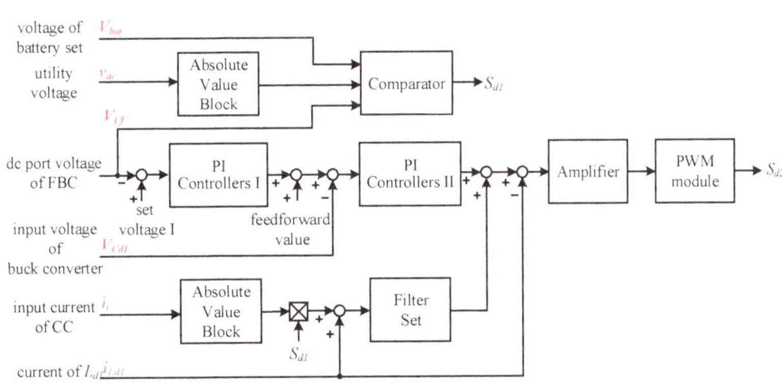

Figure 8. Control block of the proposed DIBC.

The buck converter performs two functions. The first function is a power balance control to control the input voltage such that the input real power of the BLR is equal to the input real power of the CC. The output from the buck converter is connected to the battery set, so the output voltage of the buck converter cannot be controlled. The duty of

S_{d2} is used to control the input voltage. The second function is DAF, which filters out the low-frequency ripple of the charging current for the battery set by controlling the output current of the buck converter.

In order to control the power balance, the DC port voltage of FBC must be regulated. The DC port voltage of the FBC is detected and compared with the set voltage I, and then the compared result is sent to the PI controller I. A feedforward value is added to the output of PI controller I. This sum is the set value for the input voltage of the buck converter. The feedforward value is four times that of the set voltage I. The input voltage of the buck converter is detected and compared with its set value, and then the compared result is sent to PI controller II. The output of PI controller II is the power balance control signal. Because two PI controllers form a dual-loop to control the DC port voltage of the FBC and the input voltage of the buck converter, the bandwidth of two control loops must be designed to differ at least four times to avoid oscillation. Since the set value for the input voltage of the buck converter is mainly determined by the feedforward value, the PI controller I only makes fine adjustments. Therefore, the response speed of the PI controller I is designed to be slower. For controlling the input current of CC to be sinusoidal, the level voltages for the AC port of the CC should overlap slightly. Considering the fluctuation of the battery set voltage, the set voltage I is slightly higher than one-fifth of the DC port voltage of the conventional PFC circuit to ensure level voltage overlap for the AC port of the CC.

In order to realize the function of DAF, the charging current of the battery set is calculated. As seen in Figure 2, the charging current of the battery set is the sum of the inductor current i_{Ld1} of the buck converter and the current i_{Sd1} of selection switch S_{d1}. The inductor current i_{Ld1} is measured directly using a current detector. The current i_{Sd1} is calculated by multiplying the absolute value of the input current for the CC by the control signal S_{d1}. The inductor current i_{Ld1} is added to the current i_{Sd1} to calculate the charging current of the battery set. The calculated charging current of the battery set is sent to a filter set to extract the low-frequency components. The filter set includes band-pass filters for 120 Hz, 240 Hz, and 360 Hz and a high-pass filter. The gains of the band-pass filters and the high-pass filter are assigned, respectively, to determine the attenuation rate for each ripple component of the charging current. Since the magnitude of the low-frequency components of the charging current is inversely proportional to their frequency, the gains for the 120 Hz, 240 Hz, and 360 Hz band-pass filter and the high-pass filter also decrease sequentially. The output of the filter set is the DAF control signal. The current reference signal is obtained by adding the power balance control signal and the DAF control signal. The current reference signal is compared with the detected inductor current iLd1, and the compared result is sent to an amplifier. The output of the amplifier is sent to a PWM module to generate the control signal of S_{d2}.

Figure 9 shows the control block of the CC. The control target of the CC is the input current. The detected utility voltage is sent to a sine-wave generator to generate a sine-wave signal with a unit amplitude that is in phase with the utility voltage. The sine-wave signal is multiplied by an amplitude signal to give the current reference signal. The amplitude signal is controlled by a constant current/constant voltage (CC/CV) charging strategy for the battery set. The battery set is charged in the CC mode, and then it is charged in the CV mode while the battery voltage reaches the floating charging voltage. The input current of the CC is detected and compared with its reference signal, and the compared result is sent to the current controller. The output of the current controller is added to a feedforward signal to give a modulation signal. The feedforward signal (v_{ff}) is written as:

$$v_{ff} = \left(\frac{v_{ac} - v_b}{V_{Cf}}\right) V_{tri} \tag{15}$$

where V_{tri} is the amplitude of the carrier signal for the PWM module. The modulation signal is sent to the PWM module. The PWM module uses unipolar PWM technology to generate the control signals of S_{f1}–S_{f4} for the FBC. The absolute value of the utility voltage is compared with the DC port voltage of the FBC, and the compared result is used to generate the control signals for S_{b1} and S_{b2} for the BLR.

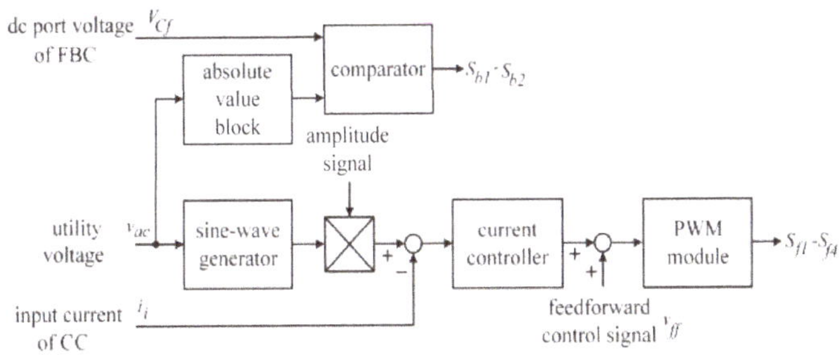

Figure 9. Control block of the CC.

7. Experimental Results

To verify the performance of the proposed CADPCI, an 800 W prototype was developed. Figure 10 shows the photo of the prototype. The circuit parameters of the prototype are shown in Table 1. The CADPCI is connected to a single-phase utility of 110 V and 60 Hz, and six batteries are connected in series to form the battery set. Considering the level voltage overlap for the AC port of the CC, the set voltage I for the DC port voltage of the FBC is 43 V.

Figure 10. Photo of the prototype: (**a**) power supply; (**b**) digital signal processor board; (**c**) phase-lock loop board; (**d**) current-detection board; (**e**) voltage-detection board; (**f**) driver board; (**g**) power circuit board.

Table 1. Circuit parameters of prototype.

		CC	
Inductor L	0.5 mH	Switching frequency	20 kHz
Capacitor Cf	2200 μF		
		DIBP	
Inductor L_{d1}	0.3 mH	Switching frequency	20 kHz
Capacitor C_{d1}	2200 μF	Capacitor C_{d2}, C_{d3}	14.1 μF
Inductor L_{d2}	0.3 mH	Resistor R_d	8 Ω

Figure 11 shows the experimental results for the AC side of the CC in the steady state. Figure 11a,b show that the AC port voltage of the CC is an 11-level voltage and is synchronized with the utility voltage. Figure 11b,c show that the input current of the CC is a sine-wave current that is in phase with the utility voltage, so the power factor is close to unity. Figure 12 shows the voltage waveform and frequency spectrum for the AC port voltage of the CC. The dominant harmonics for the AC port voltage of the CC appear at around 40 kHz, which is twice the switching frequency of the FBC. The AC port voltage of the CC is an 11-level voltage, so the amplitude of the dominant harmonics is very small. Therefore, the filter inductor in the prototype is very small. Figure 13 shows the total harmonic distortion (THD) of the input current of the CC. The THD of the input current of the CC is only 3.7%. Figure 14 shows the power factor of the CC. The power factor of the CC is close to unity.

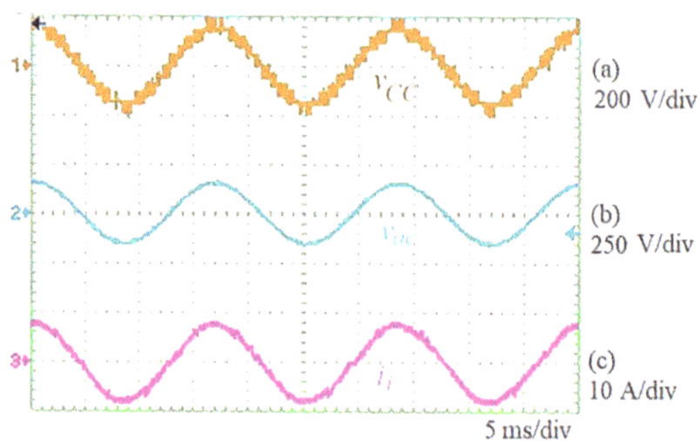

Figure 11. Experimental results for the AC side of the CC: (**a**) AC port voltage of CC; (**b**) utility voltage; (**c**) input current.

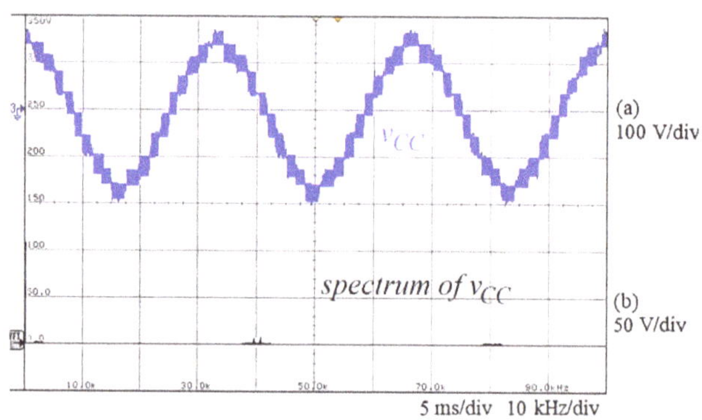

Figure 12. Voltage waveform and frequency spectrum for the AC port voltage of the CC: (**a**) voltage waveform; (**b**) frequency spectrum.

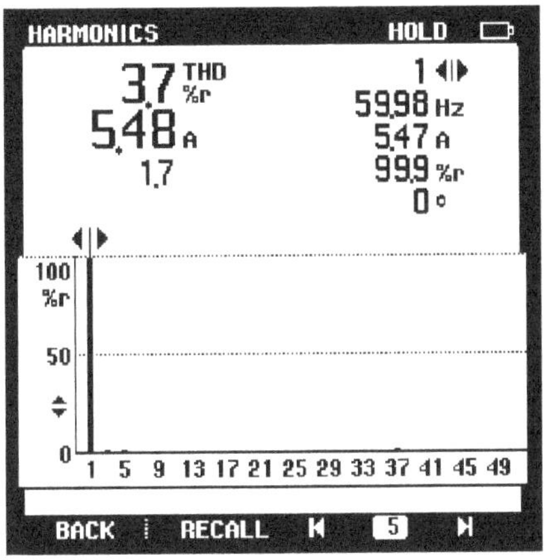

Figure 13. THD of the input current of the CC.

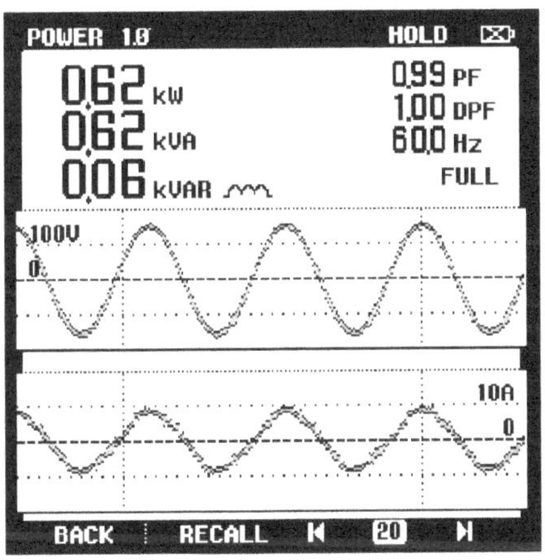

Figure 14. Power factor of the CC.

Figure 15 shows the experimental results for the voltages at the AC side of the CC. S_{b1} and S_{b2} are switched synchronously with the utility voltage, so the AC port voltage of the BDR, which is shown in Figure 15c, is a five-level step-wave voltage. The control for the FBC uses unipolar PWM. Figure 15b shows that the FBC generates a three-level high-frequency pulse voltage. The AC port voltage of the CC is the summation of the AC port voltages of the BLR and the FBC, so an 11-level AC voltage is generated, as shown in Figure 15a.

Figure 16 shows the experimental results for the DC side of the CC. The DC port voltage of the FBC is stabilized at about 43 V, and the input voltage of the buck converter is regulated at about 150 V. Therefore, it verifies that the DC port of the FBC only needs a capacitor to act as an energy buffer to stabilize the voltage, which can eliminate the need for an isolated DC-DC power converter.

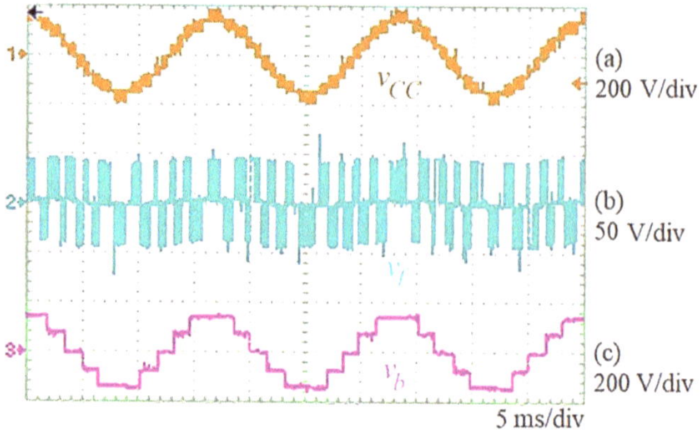

Figure 15. Experimental results for the voltage at the AC side of the CC: (**a**) AC port voltage of CC; (**b**) AC port voltage of FBC; (**c**) AC port voltage of BLR.

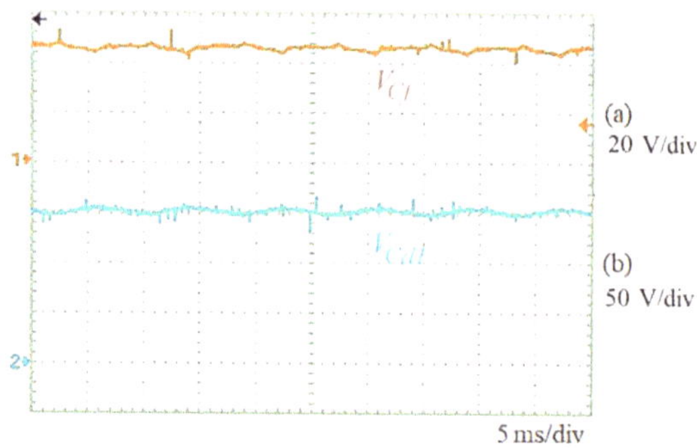

Figure 16. Experimental results for the DC side of the CC: (**a**) DC port voltage of FBC; (**b**) input voltage of the buck converter.

Figures 17 and 18 show the experimental results for the DIBC with and without the function of DAF. As can be seen in Figures 17c and 18c, it verifies that the DIBC, with the function of DAF, can effectively suppress the low-frequency ripple of the charging current for the battery set. Figure 19 shows the power efficiency of the CADPCI under the different output voltages. The higher the input voltage is, the higher the power efficiency will be. The maximum power efficiency of CADPCI is 96.68%.

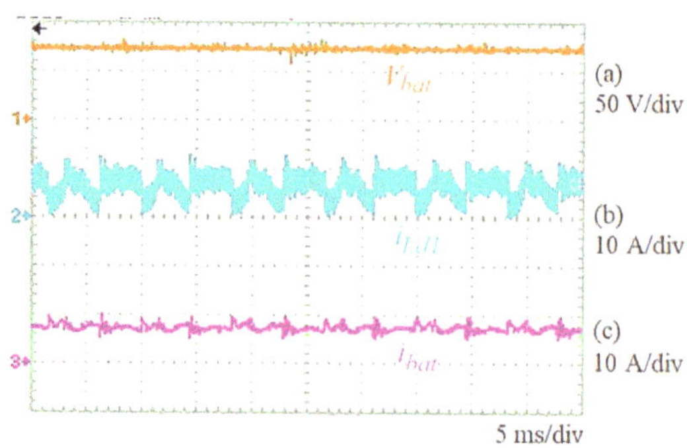

Figure 17. Experimental results for the proposed DIBC with the function of DAF: (**a**) voltage of the battery set; (**b**) inductor L_{d1} current; (**c**) charging current of the battery set.

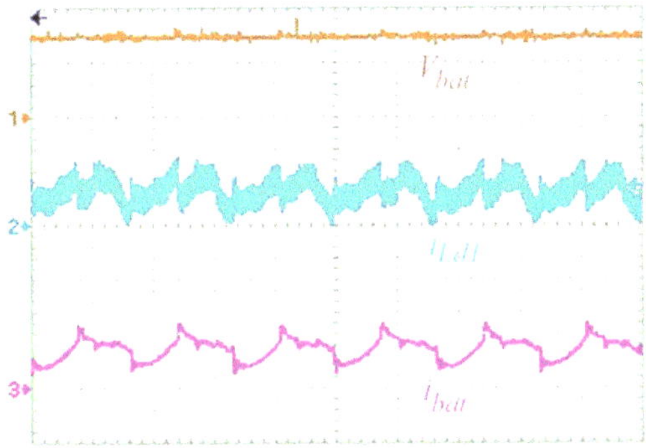

Figure 18. Experimental results for the proposed DIBC without the function of DAF: (**a**) voltage of the battery set; (**b**) inductor L_{d1} current; (**c**) charging current of the battery set.

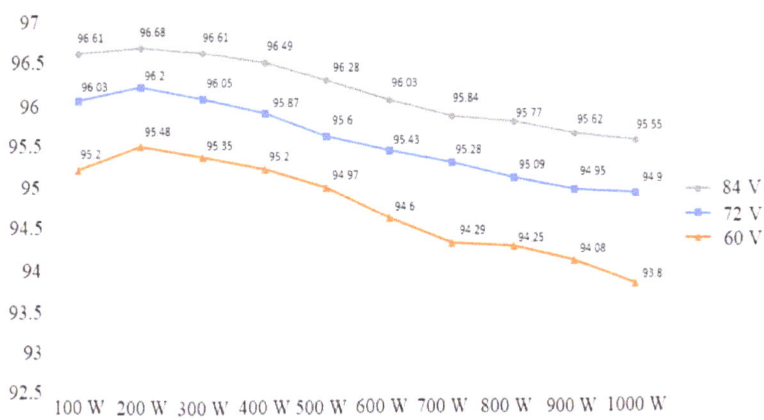

Figure 19. Power efficiency of the CADPCI under different battery voltages.

According to the experimental results, the merits of the prototype are summarized in Table 2.

Table 2. Merits of the proposed CADPCI of prototype.

AC Port Voltage	11-Level
dominant harmonics for the AC port voltage	double the switching frequency
power factor	close to unity
THD	less than 5%
low-frequency ripple of charging current	small
Maximum power efficiency	96.68%

8. Conclusions

An AC-DC power conversion interface with a stable and controllable DC voltage/current and unity power factor correction is expected to improve the power quality of electronic equipment and the performance of a battery set. A CADPCI is proposed to convert AC power from the utility into stable DC power to charge a battery set.

The experimental results show that the CC generates an 11-level voltage at the AC port and performs unity power factor correction. The dominant harmonics of the AC port voltage for the CC occur at around 40 kHz, and the amplitude is very small due to eleven voltage levels. The DC port of the FBC only needs a capacitor to stabilize the voltage, so there is no need for an isolated DC-DC power converter. The DIBC can effectively suppress the low-frequency ripple of the charging current for the battery set.

The proposed CADPCI has the advantages of higher power efficiency, the lower capacity of the passive filter, and the EMI. Hence, the proposed CADPCI is suitable for charging the battery of electric vehicles, robots, and home-based battery energy storage systems.

Author Contributions: Conceptualization, J.-C.W. and H.-L.J.; Validation, J.-P.L.; Writing—original draft, J.-C.W.; Writing—review and editing, H.-L.J. and F.-Z.C. All authors have read and agreed to the published version of the manuscript.

Funding: This research was funded by the Ministry of Science and Technology of Taiwan and ABLEREX Electronics Co., Ltd. (Taipei, Taiwan), grant number MOST 110-2622-E-992-018.

Data Availability Statement: The data used to support the findings of the study are available within the article.

Acknowledgments: The authors are grateful to the Ministry of Science and Technology of Taiwan and ABLEREX Electronics Co., Ltd. (Taipei, Taiwan) for financial support for this paper.

Conflicts of Interest: The authors declare no conflict of interest.

References

1. Al-Mawsawi, S.A. Efficiency Improvement of Computer Power Supplies Utilizing Active Power Factor Corrector. *Electr. Power Compon. Syst.* **2008**, *36*, 1310–1317. [CrossRef]
2. Tung, C.P.; Fan, J.W.T.; Chow, J.P.W.; Relekar, A.; Chan, W.T.; Ho, K.W.; Wang, K.W.; Chung, H.S.H. 1-kW Boost-type PFC Using a Low-Voltage Series Pass Module for Input Current Shaping. *IEEE Trans. Power Electron.* **2020**, *35*, 7596–7611. [CrossRef]
3. Kayisli, K.; Tuncer, S.; Poyraz, M. A Novel Power Factor Correction System Based on Sliding Mode Fuzzy Control. *Electr. Power Compon. Syst.* **2017**, *45*, 430–441. [CrossRef]
4. Das, P.; Pahlevaninezhad, M.; Drobnik, J.; Moschopoulos, G.; Jain, P.K. A Nonlinear Controller Based on a Discrete Energy Function for an AC/DC Boost PFC Converter. *IEEE Trans. Power Electron.* **2013**, *28*, 5458–5476. [CrossRef]
5. Singh, B.; Shrivastava, A. Buck Converter-Based Power Supply Design for Low Power Light Emitting Diode Lamp Lighting. *IET Power Electron.* **2014**, *7*, 946–956. [CrossRef]
6. Lee, S.W.; Lee, D.J.; Lee, S.H.; Do, H.L. Buck-Boost AC-DC LED Driver for Lamp with Visible Light Communication Module. *Electr. Power Compon. Syst.* **2019**, *47*, 372–381. [CrossRef]
7. Chen, Y.; Zhou, J.; Dai, W.P.; Hu, E. Application of Improved Bridgeless Power Factor Correction Based on One-cycle Control in Electric Vehicle Charging System. *Electr. Power Compon. Syst.* **2014**, *42*, 112–123. [CrossRef]
8. Valipour, H.; Mahdavi, M.; Ordonez, M. Resonant Bridgeless AC/DC Rectifier with High Switching Frequency and Inherent PFC Capability. *IEEE Trans. Power Electron.* **2020**, *35*, 232–246. [CrossRef]
9. Singh, B.; Singh, S.; Chandra, A.; Al-Haddad, K. Comprehensive Study of Single-Phase AC-DC Power Factor Corrected Converters with High-Frequency Isolation. *IEEE Trans. Ind. Inform.* **2011**, *7*, 540–556. [CrossRef]
10. Moon, S.; Koo, G.B.; Moon, G.W. A New Control Method of Interleaved Single-Stage Flyback AC–DC Converter for Outdoor LED Lighting Systems. *IEEE Trans. Power Electron.* **2013**, *28*, 4051–4062. [CrossRef]
11. Duarte, J.; Lima, L.; Oliveira, L.; Michels, L.; Rech, C.; Mezaroba, M. Single-Stage High Power Factor Step-Up/Step-Down Isolated AC/DC Converter. *IET Power Electron.* **2012**, *5*, 1351–1358. [CrossRef]
12. Cetin, S. Power-Factor-Corrected and Fully Soft-Switched PWM Boost Converter. *IEEE Trans. Ind. Appl.* **2018**, *54*, 3508–3517. [CrossRef]
13. Hanson, A.J.; Perreault, D.J. A High-Frequency Power Factor Correction Stage with Low Output Voltage. *IEEE J. Emerg. Sel. Top. Power Electron.* **2020**, *8*, 2143–2155. [CrossRef]
14. Hosseinabadi, F.; Adib, E. A Soft-Switching Step-Down PFC Converter with Output Voltage Doubler and High Power Factor. *IEEE Trans. Power Electron.* **2019**, *34*, 416–424. [CrossRef]
15. Umbría, F.; Gordillo, F.; Salas, F. Modeling and Full Decoupling Control of a Grid-Connected Five-Level Diode-Clamped Converter. *Electr. Power Compon. Syst.* **2017**, *45*, 1585–1598. [CrossRef]
16. Cui, D.; Ge, Q. A Novel Hybrid Voltage Balance Method for Five-Level Diode-Clamped Converters. *IEEE Trans. Ind. Electron.* **2018**, *65*, 6020–6031. [CrossRef]
17. Manai, L.; Armi, F.; Besbes, M. Flying Capacitor Multilevel Inverter Control Considering Lower Order Harmonics Elimination Based on Newton-Raphson Algorithm. *Electr. Power Compon. Syst.* **2017**, *45*, 1918–1928. [CrossRef]
18. Ramos, R.; Serrano, D.; Alou, P.; Oliver, J.A.; Cobos, J.A. Control Design of a Single-Phase Inverter Operating with Multiple Modulation Strategies and Variable Switching Frequency. *IEEE Trans. Power Electron.* **2021**, *36*, 2407–2419. [CrossRef]

19. Wanjekeche, T. Modeling, Control and Experimental Investigation of a Cascaded Hybrid Modular Inverter for Grid Interface Application. *IEEE Access* **2018**, *6*, 21296–21313. [CrossRef]
20. Odeh, C.I.; Agu, V.N. Cascaded Symmetrical Series Half-Bridge Cells Multi-Level Inverter. *Electr. Power Compon. Syst.* **2016**, *44*, 1418–1427. [CrossRef]
21. Chittathuru, D.; Padmanaban, S.; Prasad, R. Design and Implementation of Asymmetric Cascaded Multilevel Inverter with Optimal Components. *Electr. Power Compon. Syst.* **2021**, *49*, 361–374. [CrossRef]
22. Wu, J.C.; Jou, H.L.; Huang, P.H. Seven-Level Power Conversion System for Solar Power Generation System. *IET Renew. Power Gener.* **2020**, *14*, 1387–1394. [CrossRef]
23. Wanjari, K.D.; Salodkar, P.A.; Waghmare, M.A. Analysis of Multicarrier Switching Techniques for Nine Level Inverter. In Proceedings of the 2018 Second International Conference on Computing Methodologies and Communication (ICCMC), Erode, India, 15–16 February 2018; pp. 577–581. [CrossRef]
24. Wu, J.C.; Jou, H.L.; Jhang Jian, J.Y. Single-Phase Multi-Level AC–DC Power Conversion Interface. *IET Power Electron.* **2016**, *9*, 449–456. [CrossRef]
25. Teixeira, C.A.; Holmes, D.G.; McGrath, B.P. Single-Phase Semi-Bridge Five-Level Flying-Capacitor Rectifier. *IEEE Trans. Ind. Appl.* **2013**, *49*, 2158–2166. [CrossRef]
26. Norões Maia, A.C.; Jacobina, C.B. Single-Phase Ac–Dc–Ac Multilevel Five-Leg Converter. *IET Power Electron.* **2014**, *7*, 2733–2742. [CrossRef]
27. Wang, C.; Zhuang, Y.; Jiao, J.; Zhang, H.; Wang, C.; Cheng, H. Topologies and Control Strategies of Cascaded Bridgeless Multilevel Rectifiers. *IEEE J. Emerg. Sel. Top. Power Electron.* **2017**, *5*, 432–444. [CrossRef]
28. Elrajoubi, A.M.; Ang, S.S.; George, K. Design and Analysis of a New GaN-Based AC/DC Converter for Battery Charging Application. *IEEE Trans. Ind. Appl.* **2019**, *55*, 4044–4052. [CrossRef]
29. Cho, S.Y.; Lee, I.O.; Baek, J.I.; Moon, G.W. Battery Impedance Analysis Considering DC Component in Sinusoidal Ripple-Current Charging. *IEEE Trans. Ind. Electron.* **2016**, *63*, 1561–1573. [CrossRef]

electronics

Article

Large- and Small-Signal Modeling and Control Strategy for Multi-Output Welding Power Supply

Jingzhang Zhang, Shujun Chen, Yue Yu, Hongyan Zhao * and Mingyu Liu

Faculty of Materials and Manufacturing, Beijing University of Technology, Beijing 100124, China
* Correspondence: zhaohy@bjut.edu.cn

Abstract: Welding technology is a key aspect of the manufacturing industry. With the application of aluminum alloy expanding to more applications, the demand for advanced welding technology for aluminum alloy is constantly increasing. The aluminum alloy welding process requires a multi-output welding power supply to improve aluminum alloy welding quality. In this paper, we design a novel type of multi-output converter that outputs dual DC or single AC according to different welding process requirements and analyze its working principles in detail. Considering the influence of load current variation on the circuit's performance, we established small-signal models of the DC and AC working modes and large-signal models of the system. Based on these models, we designed a control algorithm for the proposed multi-electrode arc welding power supply. Finally, we constructed an experimental prototype and demonstrated the feasibility of the control strategy. Based on the welding power supply designed in this paper, the welding process control of aluminum alloy can be more accurate, resulting in better welding quality.

Keywords: large-signal; small-signal; welding power supply; control strategy

Citation: Zhang, J.; Chen, S.; Yu, Y.; Zhao, H.; Liu, M. Large- and Small-Signal Modeling and Control Strategy for Multi-Output Welding Power Supply. *Electronics* **2023**, *12*, 1800. https://doi.org/10.3390/electronics12081800

Academic Editors: Diego Bellan and Jelena Loncarski

Received: 18 February 2023
Revised: 4 April 2023
Accepted: 6 April 2023
Published: 11 April 2023

1. Introduction

Welding constitutes a significant manufacturing process in modern industry. Aluminum or magnesium alloys are commonly used in the industrial production of the aviation and aerospace fields due to their high specific modulus and strength, strong corrosion resistance, and other advantages. However, oxidation of the aluminum alloy leads to the formation of a dense oxide film, which has a higher melting point and faster heat dissipation than the aluminum alloy. This results in faster heat input and dissipation of the substrate during welding and leads to defects such as pores or deformation in the weld beam [1]. To reduce these weld defects, researchers proposed to weld the aluminum alloy by multi-electrode arc welding, where more than two electrodes are utilized to generate welding current and the output current of one electrode is used to clear the oxide film. After clearing the oxide film, the substrate is preheated to ensure that it can be quickly melted during welding. Simultaneously, the output current of the other electrode is used to melt the welding wire, whose droplets deposit onto the weld, forming a complete weld or accumulation layer [2].

Multi-electrode arc welding technology requires a multi-electrode arc welding power supply that has more than two current outputs. The power supply can adjust the output current in real-time according to the requirements of the welding process. The output current can be AC or DC, constant current or pulse current. Therefore, the power supply must have a rapid response speed for the welding power system to meet different welding process requirements [3–6].

There are many combinations of current multi-electrode welding processes, such as tandem welding, dual tungsten argon arc welding, bypass electrode GTAW welding, dual wire bypass coupled arc welding, arcing wire GTAW welding, and dual wire indirect arc alternating bypass welding. These welding processes are composed of three or more electrodes and DC or AC current. According to different process requirements, different welding power sources are combined with different welding processes.

Currently, the multi-output arc welding power supply is composed of several individual power supply sections, where each power supply has the complete rectifier, inverter, transformer, and output rectifier unit. It is crucial that the multi-electrode arc welding power supply has several electrodes, and the electrodes should coordinate with each other. In some welding processes, the output of multiple electrodes must be synchronized or have a certain phase difference, but the current multi-electrode arc welding power supply cannot meet those requirements. Therefore, it is necessary to develop the circuit of the multi-electrode arc welding power supply and design the control strategy on this basis, such as the transient response characteristics under the multi-electrode arc welding process.

The power supply for multi-electrode arc welding is a multi-output current power system. Studies have been conducted on multi-output current power supplies, whereas there has been extensive research performed on multi-output voltage power supplies. Therefore, the power structure of the multi-output voltage power supply can be used as guidance to design the multi-output current power supply. The architecture of the multi-output voltage power source is shown in Figure 1, which is referred to as the intermediate bus type. The power supply in this architecture has a high overall efficiency, and its volume and cost are low. Thus, this architecture is a suitable multi-output power supply architecture.

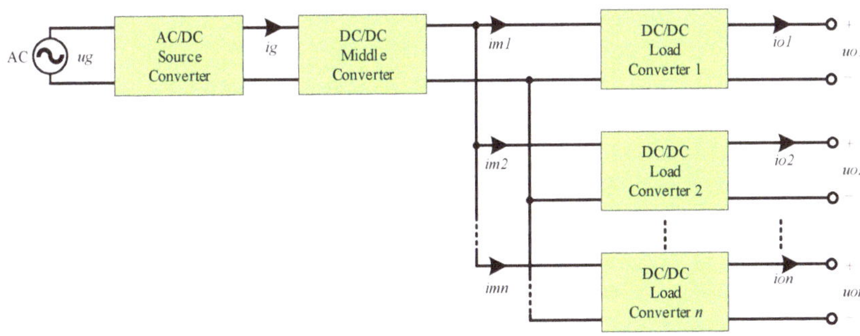

Figure 1. Intermediate bus power architecture.

The control strategy of the multi-electrode arc welding power supply is the most important aspect of realizing multi electrode welding. During the welding process, according to the welding process requirements, the current will be in a stable state or change rapidly at a certain time. The current change can be sorted into large-signal process and small-signal process according to different welding processes. In the large-signal process, the current changes over a large range, whereas in the small-signal process, it is dynamically adjusted within a small range. Therefore, the large- and small-signal modeling should be established before the control strategy is designed.

There are numerous small-signal modeling methods and many control methods based on the small-signal model. For example, the PID control method is based on the small-signal model. Ang et al. [7] studied the small-signal model of the PID controller based on the simulator for the DC-DC converter. This small-signal model describes the PID control process, but the model is not combined with the converter. The large-scale application of digital devices realizes the real-time tuning PID control method. On this basis, numerous researchers studied the adaptive PID control method [8,9], which can adjust the required duty cycle value in advance according to the established small-signal model. Based on the adaptive adjustment, Cao et al. [10] proposed the delay feedback automatic adjustment PID controller, which reduces the complexity of the control and shortens the PID controller's intervention and adjustment times. However, it is not suitable for large-scale adjustment. To make the PID small-signal model suitable for large-scale regulation, Du et al. [11]

proposed a PID controller based on the nonlinear small-signal model, which can complete the adjustment of some large signals but increases the complexity of the control system.

Small-signal control can only maintain stability at a steady operating point; however, the stability of the system under large-signal adjustment cannot be guaranteed, in which case the traditional linearized state space method is no longer applicable [12–14]. Therefore, it is necessary to establish a large signal model to determine the feasibility of the system.

Large-signal modeling methods are still under research. The traditional modeling method uses nonlinear switches to establish the large-signal model of the system [15]. However, this method is not suitable for the large signal model under different working conditions. The behavior description method is likewise used in large signal modeling. In addition, these two models are mostly used in simulation research [16]. Liu et al. [17] used the ideal transformer model to replace the switch to establish the large signal model; however, such a model can only be established in CCM (continuous conduction mode) mode. The traditional reduced-order simplified model substantially simplifies the system-level simulation and retains some module characteristics, which has evident advantages in simulation. However, this method cannot provide theoretical guidance to analyze the system [18]. Therefore, it is necessary to establish a new type of large signal model to describe the large signal change process in the welding process so as to apply the multi-electrode arc welding power supply.

Current multi-output welding power supplies involve combining multiple sources for different processes, using three or more electrodes with DC or AC current. This requires numerous power sources, increasing cost and impacting production efficiency. There is a need for a flexible, cost-effective multi-electrode power supply suitable for various processes. A power supply meeting the combination requirement can be achieved through multiple electrode combinations. Characterized by ease of use, affordability, and efficient control, this power supply holds significant potential. Research on multi-electrode arc welding power supplies is in its early stages, with no suitable method for the main circuit and control strategy.

In this study, we designed a topological structure for a multi-electrode arc welding power supply, dividing the current change process into large and small-signal change processes. By establishing models and analyzing the converter's working principle, we designed a control method and constructed an experimental prototype. The resulting power supply offers improved response time and cost-effectiveness compared to traditional multi-output welding power supplies, increasing its applicability, response time, and cost-effectiveness as compared to conventional multi-output welding power supplies, therefore providing a viable option for broader applicability.

2. Topology of Multi-Output Arc Welding Power Supply

The multi-output arc welding power supply proposed in this study is shown in Figure 2. i_{g1} and i_{g2} are the target current values controlled by the controller; i_{o1} and i_{o2} are the actual values.

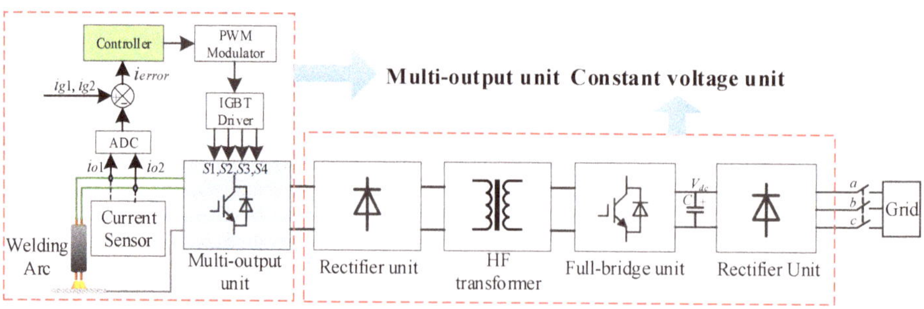

Figure 2. Topology of multi-electrode arc welding power supply.

The converter includes a three-phase rectifier unit, a full-bridge inverter unit, a high-frequency transformer, a rectifier unit, and a multi-channel output unit. The three-phase rectifier unit converts the 380 VAC to 540 VDC, which is converted to a high-frequency pulse with a frequency of 20 kHz after passing through the full-bridge inverter unit. The turn ratio of the high-frequency transformer is 14:3. After passing through the transformer, the voltage of the high-frequency pulse decreases and the current rises. After the output rectifier unit, the voltage becomes 50 VDC. This voltage stays constant during welding and is not affected by current output; hence, it can be omitted from the analysis in this study.

The multi-output unit is the most important component of the converter designed in this study, as shown in Figure 3. S_1–S_4 are the IGBTs in the circuit; L_1 and L_2 are the inductors for stable current in the circuit; R_1 and R_2 represent the welding arcs. In the stable welding process, the welding current and voltage basically remain constant and only change within a small range. I arc can be regarded as the resistance to analysis. I_{R1}, I_{R2}, V_{R1}, and V_{R2} are the welding currents and welding voltages, respectively.

The multi-output unit is composed of the buck chopper circuit, and according to the different output, this circuit can be converted into a full-bridge inverter circuit. We use different colors to represent different output patterns: In Figure 3a, the red and blue lines represent two DC currents; in Figure 3b, the red and blue lines represent different current directions in AC mode. Figure 4 illustrates the working principle of the IGBT in each mode, and the color representation is the same as in Figure 3. The multi-output unit can output dual DC or single AC, depending on the requirement.

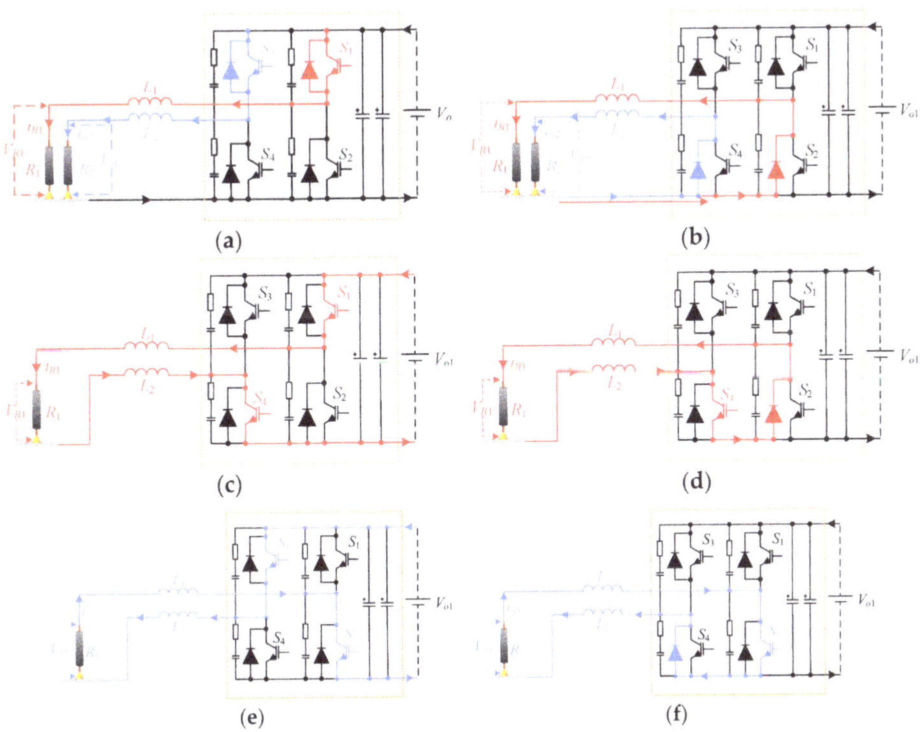

Figure 3. Different work modes. (**a**) IGBTs switch in DC mode (**b**) Continuous Current in DC mode; (**c**) IGBTs switch in AC forward mode (**d**) Continuous Current in AC forward mode; (**e**) IGBTs switch in AC reverse mode (**f**) Continuous Current in AC reverse mode.

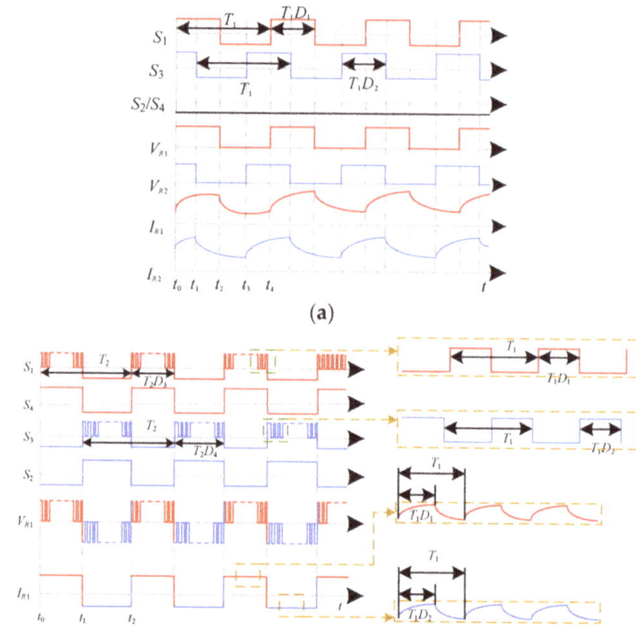

(a)

(b)

Figure 4. Principle of converter. (**a**) Principle of DC mode, (**b**) Principle of AC mode.

In the welding process, an inductor is used in the converter to maintain a stable current, and its operation mode is always CCM mode. Therefore, during the welding period, the output current and voltage remain constant. In the analysis, the welding arc is regarded as a resistance, whose magnitude changes slightly within a certain range.

As shown in Figure 4, S_1, S_2, S_3, and S_4 present the drive waveform of each IGBT. V_{R1} and V_{R2} are the voltages of the welding arc. I_{R1} and I_{R2} are the currents of the welding arc. S_2 and S_4 are keep in turn off state. When S_1 and S_3 are switched on, the currents flow through the IGBTs as shown in Figure 3a, and when S_1 and S_3 are switched off, the currents flow through the internal diode in S_2 and S_4 as shown in Figure 3b. So, in DC mode, when S_1 and S_3 are switched on and off, the DC voltage is chopped, inductors L_1 and L_2 are in CCM mode, and the current through the resistance is DC current. This working principle is shown in Figure 3a,b, and the specific waveform is shown in Figure 4a,b. In Figure 4a,b, T_1 = 50 µs, D_1 and D_2 are the duty cycles of the driving waveform of the IGBT. The duty cycles are changed according to the setting current.

At time T_0, S_1 and S_3 are switched off; the current on resistor R_1 flows from V_{o1+} through S_1, L_1, and R_1 to V_{o1-}, and the current on resistor R_2 flows from voltage V_{o1+} through S_2, L_2, and R_2 to V_{o1-}. At this time, inductors L_1 and L_2 store energy, and the current continues to increase.

At time T_1, S_1 is switched on and S_3 is switched off; the inductor L_1 continues to store energy, and the current on R_1 continues to increase. Since S_3 is off, L_2 discharges to stop current changes, and the current flowing through R_2 decreases. At this time, the current loop flowing through R_2 passes through the internal reverse diodes of S_4, R_2, and L_2.

At time T_2, S_1 and S_3 are switched off. The energy stored in the inductor L_1 starts discharging, and the energy in the inductor L_2 continues to discharge to maintain the current flowing across the resistors R_1 and R_2. At this time, the current loop flowing through R_1 passes through the internal reverse diodes of S_2, R_1, and L_1.

At time T_3, S_3 is switched on, S_1 is switched off, inductor L_2 is charged, and the current flowing through resistor R_2 starts to increase. At this point, the current across resistor R_1 continues to decrease.

At time T_4, S_1 and S_3 are switched on, returning to time T_0, as described above.

As shown in Figure 4b, S_1, S_2, S_3, and S_4 present the drive waveform of each IGBT. V_{R1} is the voltage of the welding arc. I_{R1} is the current of the welding arc. When the converter is in AC mode, the welding power supply can output forward and reverse currents. In the welding process, the frequency of AC current is generally 40~50 Hz. Therefore, in AC mode, an AC cycle can be regarded as the combination of two DC modes, which are the forward and reverse DC modes, respectively. Another difference with DC mode is that when the S_1 or S_3 is switched on or off, the S_2 or S_4 are kept in the on or off state to provide the circuit with continuous current. The specific working waveform is shown in Figure 4b, where T_2 is the cycle of AC welding and D_3 and D_4 are the forward and reverse current duty cycles during AC welding, respectively. In contrast with the DC mode, in the inductor discharge stage, assume that the loop controlled by S_1 and S_4 is the forward current as shown in Figure 3c, and the loop controlled by S_2 and S_3 is the reverse current as shown in Figure 3e. When the converter works in forward AC mode, S_1 is switched on or off. In this mode, S_4 is kept in turn-on mode, while S_2 and S_3 are kept in turn off states, and the continuous current can flow through S_4 and the internal diode in S_2 as shown in Figure 3d. When the converter works in reverse AC mode, S_3 is switched on or off. In this mode, S_2 is kept in turn-on mode, while S_1 and S_4 are kept in turn-off states, and the continuous current can flow through S_2 and the internal diode in S_4 as shown in Figure 3f. The discharge loop in the forward current is through L_1, R_1, L_2, S_4, and internal reverse diodes in S_2. The discharge circuit for the reverse current is through internal reverse diodes of S_4, R_1, L_1, S_2, and L_2.

At time T_0, S_1 and S_4 are switched on simultaneously, and the current flows through V_{o1+}, S_1, R_1, and S_4 to V_{o1-}. During the AC forward period, S_1 is switched off and on at a frequency of 20 kHz, while S_4 is always on. Therefore, this mode can be considered to be the same as the DC mode, where one IGBT is used for chopping while the other switch in the loop is always on.

At time T_1, in AC mode, the current changes from forward to reverse, and S_1 and S_4 are switched off. Then, after a certain dead time, S_2 and S_3 are switched on. The current flows through V_{o1+}, S_3, R_1, and S_2 to V_{o1-}. In the AC reverse mode, S_3 is switched off and on at a frequency of 20 kHz, and S_2 is consistently on. The mode in this case can be regarded as the reverse DC mode.

At time T_2, S_2 and S_3 are switched on, and S_1 and S_4 are switched off, as at time T_0. At this time, a complete AC cycle has been completed.

In AC mode, the dead time should be considered. For the converter in this paper, there is no dead time in DC mode. In AC mode, there is dead time when the forward and reverse currents are switched. In this paper, the difference with the full-bridge inverter is that there is no secondary side of the transformer to provide a continuous flow circuit to release the energy of the primary side when switching. Therefore, when the voltage on the inductor and welding arc drops to the constant voltage, the internal diode of the opposite IGBT would be turned off by a hard switch. Since the constant voltage is 50 V, the voltage on the IGBT is relatively low, and the effect on the IGBT is small due to the protection of the RC absorption circuit and the high voltage tolerance of the IGBT module. Due to inertia, the change in the arc lags behind the change in the current, so when the current drops to zero, the arc does not disappear.

3. Modeling

3.1. Small-Signal Modeling

There are numerous small-signal modeling methods, such as the state-space averaging method [12], the time averaging method [13,14], and the description function method [15]. In this study, a small-signal model of a multi-electrode arc welding power source is established based on the time averaging method. According to the time-average equivalent circuit principle [14], the small-signal equivalent circuit of the multi-electrode arc welding power supply can be obtained, as shown in Figure 5.

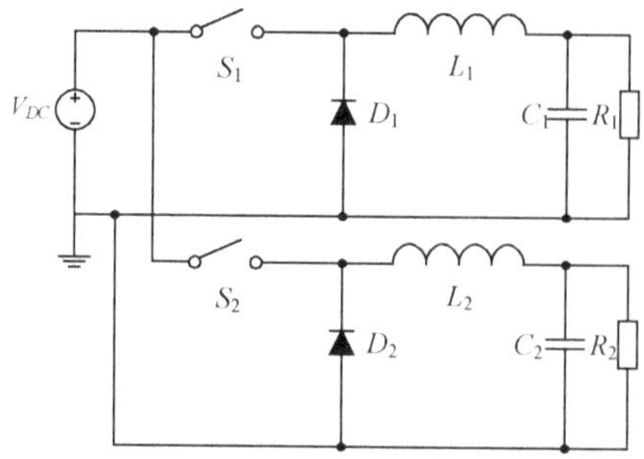

Figure 5. Simplified model of converter.

Taking the output loop controlled by S_1 as an example, the following switching states can be obtained:

Switching state 1: S_1 is switched on, diode D_1 is switched off; the inductor L_1 and capacitor C_1 are charged by the input power supply, and power is distributed to the load R_1. The inductor current i_{L1} rises linearly with slope e_1. Let the state variable $x = [i_{L1}\ v_a]^T$, then the equation of state is:

$$\begin{cases} \dot{i}_{L1} = -v_a/L_1 + v_i/L_1 \\ \dot{v}_a = i_{L1}/C_1 - v_a/(R_1C_1) \end{cases} \tag{1}$$

Switch state 2: switch S_1 is switched off, diode D_1 is switched on, inductor L_1 discharges to load R_1 and capacitor C_1, and inductor current i_{L1} linearly decreases with slope $-e_2$, then the state equation is:

$$\begin{cases} \dot{i}_{L1} = -\frac{1}{L_1}v_a \\ \dot{v}_a = -\frac{1}{R_1C_1}v_a \end{cases} \tag{2}$$

The rising slope e_1 and falling slope e_2 of inductor current are respectively:

$$\begin{cases} e_1 = (v_i - v_a)/L_1 \\ e_2 = v_a/L_1 \end{cases} \tag{3}$$

Additionally, since the two groups of parallel chopper circuits have the same structure and device parameters, the state equations of the control loop of S_2 can be obtained as follows:

$$\begin{cases} \dot{i}_{L2} = -v_b/L_2 + v_i/L_2 \\ \dot{v}_b = i_{L2}/C_2 - v_b/(R_2C_2) \end{cases} \tag{4}$$

$$\begin{cases} \dot{i}_{L2} = -\frac{1}{L_2}v_b \\ \dot{v}_b = -\frac{1}{R_2C_2}v_b \end{cases} \tag{5}$$

The rising slope e_3 and falling slope e_4 of inductor current are, respectively:

$$\begin{cases} e_3 = (v_i - v_b)/L_2 \\ e_4 = v_b/L_2 \end{cases} \tag{6}$$

According to the converter principle introduced in Figure 3, when the converter outputs two DC currents, the current is controlled by regulating the duty cycle of the S_1 and S_2 driving voltages. According to the different current values of I_A and I_B, the converter can be divided into four states, as shown below:

Thus, combining Figure 6 with Equations (1)–(6), letting the state variable $x = [i_{L1}\ v_a\ i_{L2}\ v_b]^T$, the model of the converter in DC mode can be obtained by using the state space average method:

$$\dot{\overline{x}} = A\overline{x} + Bv_i \tag{7}$$

Among them,

$$A = \begin{bmatrix} 0 & -\frac{1}{L_1}(1-d_1) & 0 & 0 \\ \frac{1}{C_1}d_1 & -\frac{1}{R_1C_1} & 0 & 0 \\ 0 & 0 & 0 & -\frac{1}{L_2}(1-d_2) \\ 0 & 0 & \frac{1}{C_2}d_2 & -\frac{1}{R_2C_2} \end{bmatrix} \tag{8}$$

$$B = \begin{bmatrix} \frac{1}{L_1}d_1 & 0 & \frac{1}{L_2}d_2 & 0 \end{bmatrix}^T \tag{9}$$

When the converter outputs AC, the simplified model is shown as Figure 7,

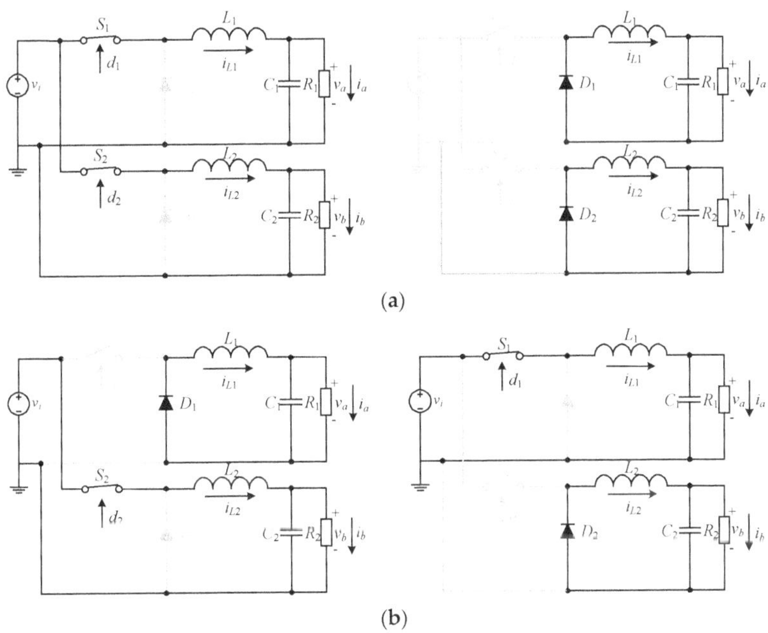

Figure 6. Switch status of S_1 and S_2 in DC mode. (a) $i_a = i_b$, S_1 and S_2 are switched on at the same time, (b) $i_a \neq i_b$, S_1 and S_2 are switched off at the same time.

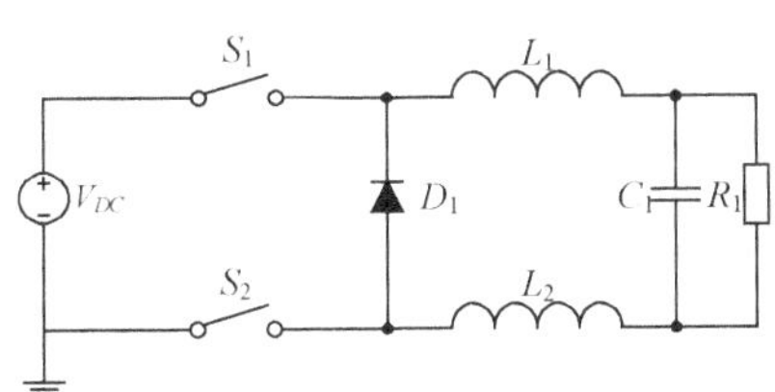

Figure 7. Simplified model.

By analyzing the working mode in the AC mode, the following switch states can be obtained:

Switching state 1: S_1 and S_2 are switched on, diode D_1 is switched off, the inductor L_1 and capacitor C_1 are charged by the input power supply, and output power is sent to the load R_1. The inductor current i_{L1} rises linearly with slope e_1. Let the state variable $x = [i_{L1}\ v_a]^T$, then the equation of the state is:

$$\begin{cases} \dot{i}_{L1} = -v_a/(L_1+L_2) + v_i/(L_1+L_2) \\ \dot{v}_a = i_{L1}/C_1 - v_a/(R_1C_1) \end{cases} \tag{10}$$

Switch state 2: S_1 is switched off, diode D_1 is switched on; inductor L_1 discharges to load R_1 and capacitor C_1, and inductor current i_{L1} decreases linearly with slope $-e_2$, then the state equation is as follows:

$$\begin{cases} \dot{i}_{L1} = -\frac{1}{L_1+L_2}v_a \\ \dot{v}_a = -\frac{1}{R_1C_1}v_a \end{cases} \tag{11}$$

The rising slope e_1 and falling slope e_2 of the inductor current are:

$$\begin{cases} e_1 = (v_i - v_a)/(L_1+L_2) \\ e_2 = v_a/(L_1+L_2) \end{cases} \tag{12}$$

In the AC state, the converter can be divided into four states, as shown below:

Therefore, combining Figure 8 and Equations (10)–(12), the state variable $x = [i_{L1+}\ v_{a+}\ i_{L1-}\ v_{a-}]^T$, the state space averaging model of the converter in AC mode can be obtained as follows:

$$\dot{\overline{x}} = A\overline{x} + Bv_i$$

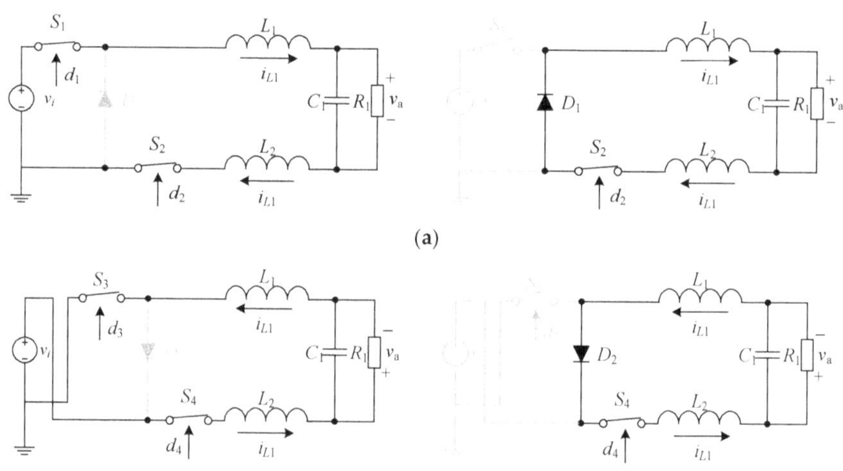

Figure 8. Switch status in AC mode. (**a**) AC mode (+) switch status, (**b**) AC mode (−) switch status.

Among them:

$$A = \begin{bmatrix} 0 & -\frac{1}{L_1+L_2}(1-d_1) & 0 & 0 \\ \frac{1}{C_1}d_1 & -\frac{1}{R_1C_1} & 0 & 0 \\ 0 & 0 & 0 & -\frac{1}{L_1+L_2}(1-d_3) \\ 0 & 0 & \frac{1}{C_1}d_3 & -\frac{1}{R_1C_1} \end{bmatrix} \tag{13}$$

$$B = \begin{bmatrix} \frac{1}{L_1+L_2}d_1 & 0 & \frac{1}{L_1+L_2}d_3 & 0 \end{bmatrix}^T \tag{14}$$

Small-signal disturbances are applied to the state space average model as shown in Equations (8), (9), (13) and (14), namely:

$$\begin{cases} i_{L1} = I_{L1} + \hat{i_{L1}}, i_{L2} = I_{L2} + \hat{i_{L2}} \\ v_a = V_a + \hat{v}_a, v_b = V_b + \hat{v}_b \\ d_1 = D_1 + \hat{d}_1, d_2 = D_2 + \hat{d}_2, d_3 = D_3 + \hat{d}_3, d_4 = D_4 + \hat{d}_4 \end{cases} \tag{15}$$

In this form, I_{L1}, I_{L2}, V_a, V_b, D_1, D_2, D_3, and D_4 are the steady values, and the "^" symbol variable is the small-signal disturbance.

Therefore, the expression of the small-signal model in the DC mode is as follows:

$$\begin{cases} \hat{i}_{L1} = \frac{1}{sL_1}[-(1-D_1)\hat{v}_a + (V_a + V_i)\hat{d}_1 + D_1\hat{v}_i] \\ \hat{v}_a = \frac{1}{sC_1}(D_1\hat{i}_{L1} + I_{L1}\hat{d}_1 - \hat{i}_{L1}) \\ \hat{i}_{L1} = \frac{1}{sL_2}[-(1-D_2)\hat{v}_b + (V_b + V_i)\hat{d}_2 + D_2\hat{v}_i] \\ \hat{v}_b = \frac{1}{sC_2}(D_2\hat{i}_{L2} + I_{L2}\hat{d}_2 - \hat{i}_{L2}) \end{cases} \tag{16}$$

The expression of the small-signal model of the AC mode is as follows:

$$\begin{cases} \hat{i}_{L+} = \frac{1}{s(L_1+L_2)}[-(1-D_1)\hat{v}_{a+} + (V_{a+} + V_i)\hat{d}_1 + D_1\hat{v}_i] \\ \hat{v}_{a+} = \frac{1}{sC_1}(D_1\hat{i}_{L+} + I_{L+}\hat{d}_1 - \hat{i}_{L+}) \\ \hat{i}_{L-} = \frac{1}{s(L_1+L_2)}[-(1-D_3)\hat{v}_{a-} + (V_{a-} + V_i)\hat{d}_3 + D_3\hat{v}_i] \\ \hat{v}_{a-} = \frac{1}{sC_1}(D_3\hat{i}_{L-} + I_{L-}\hat{d}_3 - \hat{i}_{L-}) \end{cases} \tag{17}$$

According to the small-signal model of the converter shown in Equations (16) and (17), the small-signal model of the converter shown in Figure 9 can be obtained when the input voltage disturbance is ignored.

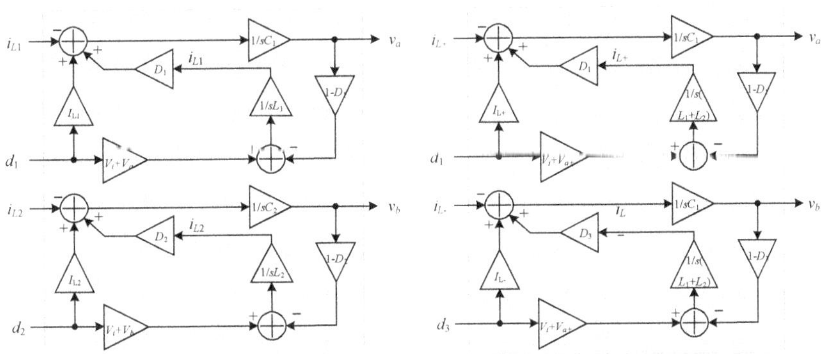

Figure 9. Small-signal model. (**a**) Small-signal model in DC mode (**b**) Small-signal model in AC mode.

After obtaining the transfer function of the equivalent circuit, the equivalent small-signal block diagram of the converter can be determined, as shown in Figure 10. Herein, $G(s)$ is the transfer function between duty cycle $D(s)$ and output $I_o(s)$, and $E(s)$ is the error transfer function between $I_o(s)$ and I_g. $G_m(s)$ is the transfer function of the PWM generator, $H(s)$ is the transfer function of the sampling, and $G_{vd}(S)$ is the transfer function of the controller.

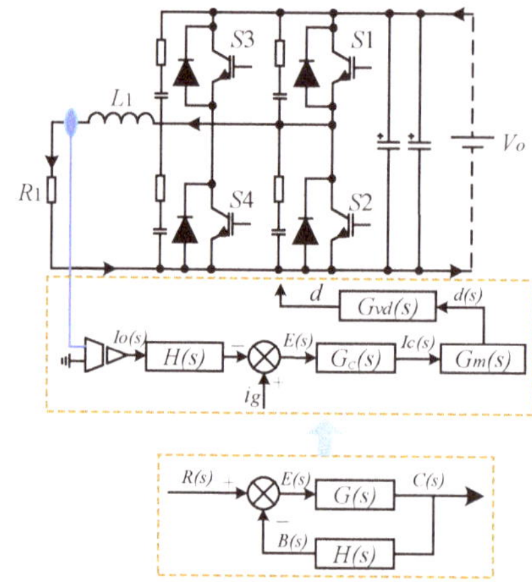

Figure 10. Small-signal block diagram of the converter.

The final small-signal model is as follows:

$$G_{DC}(s) = G_c(s)\frac{U_o}{d_1}\frac{1}{1 + s\frac{L_1}{R_1} + S^2 L_1 C_1} \tag{18}$$

$$G_{AC}(s) = G_c(s)\frac{U_o}{d_1}\frac{1}{1 + s\frac{L_1+L_2}{R_1} + S^2 (L_1 + L_2) C_1} \tag{19}$$

3.2. Large Signal Modeling

Since the small-signal working state is a response to the small-signal disturbance, when the power supply output changes rapidly, it will be in the large-signal working state, and a large-signal model is necessary to describe the transformation process of the converter [18–20]. During the large-signal process, the welding current changes rapidly, and the current control strategy must be fast and accurate to ensure that the current can be adjusted to the given value. In this section, we establish a large signal model to describe the working state of the converter in the large signal process. According to the small-signal establishment process described in Section 3.1, the averaged global state space model can be generally described as follows:

$$\begin{cases} \dot{x} = A(x,u)x + B(x,u)u \\ y = C(x,u)x + D(x,u)u \end{cases} \tag{20}$$

where x is the system state variable; U is the input variable; Y is the output variable; A, B, C, and D are the system matrixes of the response. According to the working principle of the converter mentioned above, the converter for welding processes is a non-linear system; if using the same large-signal control strategy for all large current changes, the current response time would be different. Therefore, the large-signal working phase should be divided into n sub-phases.

$$\begin{cases} \dot{x} = \sum\limits_{i=1}^{n} w_i(X_i, U_i)[A_i(X_i, U_i)x + B(X_i, U_i)u] \\ y = \sum\limits_{i=1}^{n} w_i(X_i, U_i)[C_i(X_i, U_i)x + D(X_i, U_i)u] \end{cases} \tag{21}$$

Among them, X_i and U_i are the state and input variables of the model, respectively; W_i is the weight function corresponding to the local model, which can be a Gaussian function [21], a trigonometric function [22], an exponential function [23–25], etc., due to the requirements.

$$\begin{cases} 0 \le w_i(X_i, U_i) \le 1 \\ \sum\limits_{i=1}^{n} w_i(X_i, U_i) = 1 \end{cases} \tag{22}$$

Because the output current range of the converter is large, an individual weight function cannot cover the entire working range, so the global working area of the converter is divided into three local working areas according to the actual working conditions of the converter.

Thus, the large-signal model of each sub-phase is obtained, after which the weight function is designed, finally yielding the global model. After the welding arc is generated, the voltammetry characteristics of the current and voltage are specific. For example, in the TIG welding process, the voltammetry characteristics are $U = 10 + 0.04i$. When the current changes, the voltage changes along with it. As shown in Figure 11, the current is limited to 300 A, and the work phase is divided into three segments, namely, 0–100, 101–200, and 201–300 A. Therefore, the benchmark operating points of the large signal are (14, 100), (18, 200), and (22, 300), respectively. In different working periods, the coefficients of the weight function are different, and the function of the large signal changes accordingly.

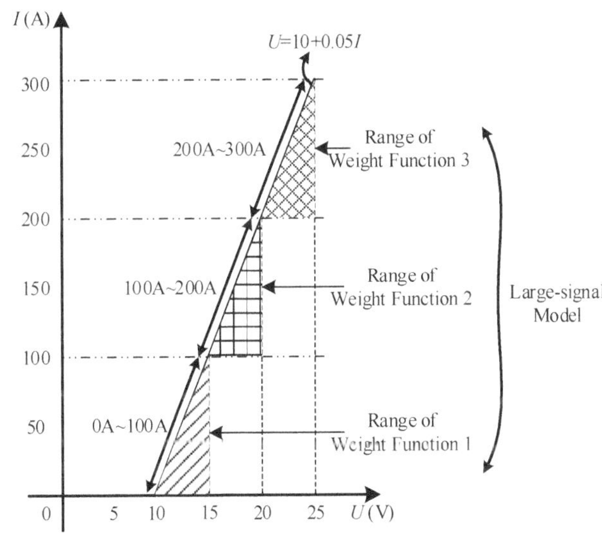

Figure 11. Region divided by large signal model.

Considering the complexity and accuracy, the Gaussian function is selected as the weight function:

$$w_i(x) = \frac{\mu_i(x)}{\sum\limits_{j=1}^{r} \mu_j(x)} \tag{23}$$

where

$$\mu_i(x_1, x_2) = \prod_{i=1}^{3} \left(\frac{1 - 1/\exp[m_i(x_i - x_{i0} - b)]}{1 + \exp[m_i(x_i - x_{i0} - b)]} \right) \tag{24}$$

where m_i, x_{i0} and b are constant terms.

When the input voltage and load current changes are considered separately, the weight function can be designed as follows:

$$
\begin{aligned}
\mu_i(i_g) &= \frac{1-1/\exp[-m_{i_g}(i_g-i_{g0i}-a)]}{1+\exp[m_{i_g}(i_g-i_{g0i}-a)]} \\
\mu_j(i_{load}) &= \frac{1-1/\exp[-m_{i_{load}}(i_{load}-i_{load0j}-b)]}{1+\exp[m_{i_{load}}(i_{load}-i_{load0j}-b)]}
\end{aligned}
\tag{25}
$$

wherein

$$
m_{i_{g0}} = \frac{13}{\Delta i_{g0}}; a = \frac{\Delta i_{g0}}{2}; m_{i_{load0}} = \frac{13}{\Delta i_{load0}}; b = \frac{\Delta i_{load0}}{2} \tag{26}
$$

Therefore, the large signal model of the converter can be approximated as:

$$
G_{vd}(s) = w_1(x)\frac{G_{vd01}(1+s/w_z)}{1+s/w_{p1}} + \sum_{2}^{3}[w_i(x)\frac{1}{1+sL/R+s^2LC}] \tag{27}
$$

4. Control Strategy Design

The above analysis shows the small-signal and large-signal models of the multi-output power supply; based on these models, the adaptive control strategy of the converter system can be constructed, as shown in Figure 12. After the load current is measured, we can determine whether the power supply is in the small-signal or the large-signal work process. If it is in the small-signal work process, then we adopt the small-signal control strategy. If it is in the large-signal work process, different weight functions must be selected according to the current value. The transfer function of the output to the control signal is as follows:

$$
\frac{v_{ctrl}}{i_o} = \sum_{i=1}^{n} w_i(I_g, i_{load})G_{ci}(I_g, i_{load}) \tag{28}
$$

According to Equation (28), the controller can be described as follows: when the working conditions of the converter (I_g and i_{load}) in a large range of current change, the weight function would monitor the change of working state in real-time and be adaptive to adjust the action of each local controller weight coefficient to ensure the system has good global control performance of the work area and a good transient response.

Therefore, during the welding process, the method proposed in this paper can be divided into large signal processing and small signal processing. The limitation of traditional PID control methods is that the PID adjusts using the same K_p and K_i parameters for all current modulations, which results in inconsistent results for small and large currents. Therefore, in this research, the current is divided into three ranges and combined with different weight functions and PIDs in each range, thus the adjustment effect is consistent in different current ranges. Based on the above theory, when the current is switched from a small to a large value, the weight function will be switched first so that the current will quickly rise to the required value.

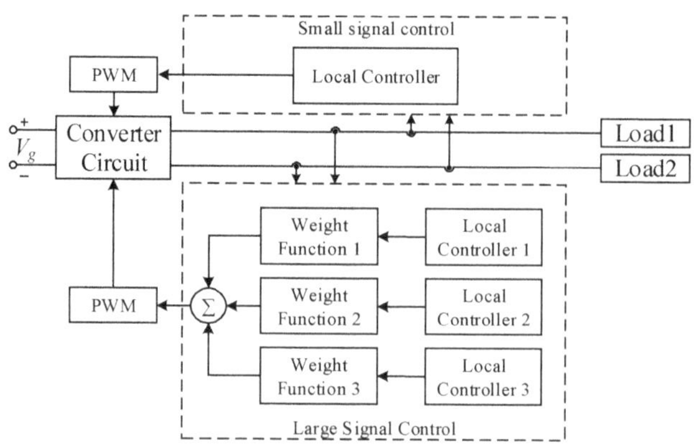

Figure 12. Control strategy.

5. Experimental Verification

To testify to the above theoretical analysis, an experimental prototype is built; the experimental prototype is shown in Figure 13. The HMI for multi-electrode output is designed, and the current and voltage can be set by the touch screen. STM32F407 is used as the controller to output the PWM to control the multi-channel output unit. The load box is used to simulate the arc, and the resistance of the load box ranges from 0.05 to 4 Ω. The bus capacitor is placed behind the constant-voltage output unit to stabilize the voltage during welding. A total of six 2200 uF electrolytic capacitors are used.

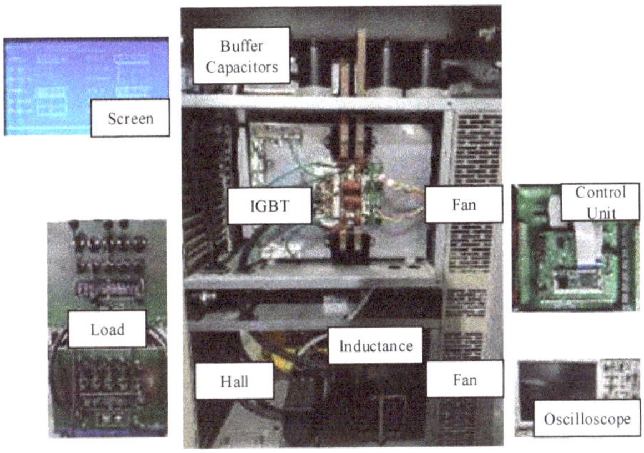

Figure 13. Experimental prototype.

In order to prevent the IGBT from being misdirected by dv/dt when the IGBT is turned off, a reverse turn-off voltage is applied to the IGBT in the drive circuit. The IGBT drive circuit designed in this paper has a +15 V/−8 V drive waveform, as shown in Figure 14.

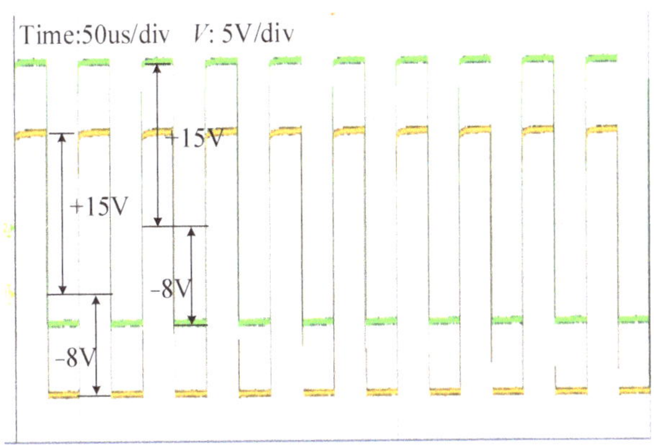

Figure 14. The drive waveform of IGBTs.

The rapid change of current will produce high voltage at both ends of the IGBT, so it is necessary to set up a protection circuit to protect the IGBT and to choose IGBTs with a high withstand voltage as well. In the circuit, the IGBT modules with a high withstand voltage value of 1200 V are selected. Moreover, an RC circuit is connected in parallel with the drain and source of the IGBT to absorb the peak voltage caused by the fast shutdown. The RC absorption circuit in this paper has a resistance of 6.2 ohms and a capacitor of 10 nF.

Experiments are designed to verify the performance of the converter. First, the experiment is conducted to check the current output of the converter in DC mode and AC mode.

In DC mode, the converter could generate two current outputs of different values. In AC mode, the converter could output the same forward and reverse currents. In addition, in AC mode, the times for forward current and reverse current are 21 ms and 4 ms, respectively.

The small-signal model and the large-signal model are also testified to. To verify the performance of the control strategy, the adjustment time could be observed and compared with the traditional PID closed-loop control strategy.

The experimental waveform in steady state is shown in Figure 15a,b indicate that the output current in DC or AC is relatively stable and the inductor is working in CCM mode. In steady-state operation, the output current is stable, indicating satisfactory system stability.

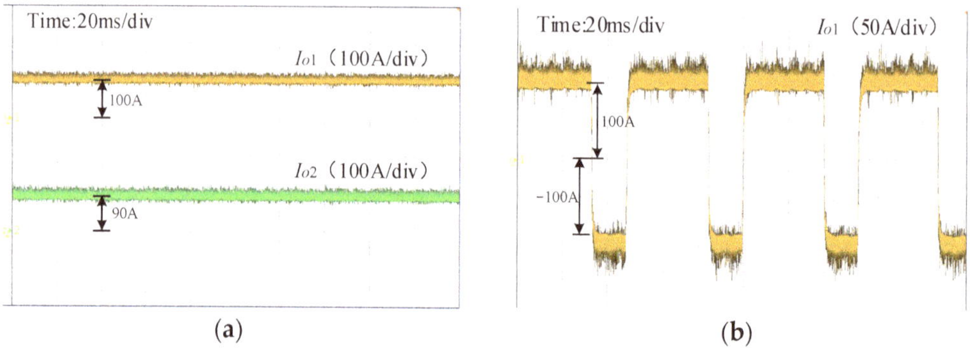

Figure 15. Steady state output waveform. (**a**) Dual DC currents (**b**) AC current.

In the working state of the small-signal process, the duty cycle change will not be drastic. When the input voltage changes drastically, it can be quickly adjusted to the given current. As shown in Figure 16, compared with the traditional control system with PI regulation only, the current system changes steadily and adjusts quickly. During welding, the stable current can keep the welding arc stable.

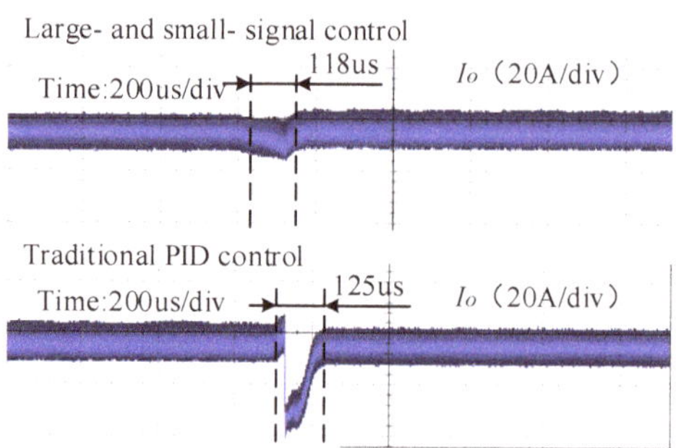

Figure 16. Small-signal control output waveform.

By observing the conversion between large current and small current, it can be seen that the large-signal control strategy designed in this paper can realize fast adjustment. As shown in Figure 17, only 187 us are taken when the current rises from 50 A to 250 A, and only 245 us are taken when the current falls from 250 A to 50 A. The response time is short, and the system has a small overshoot.

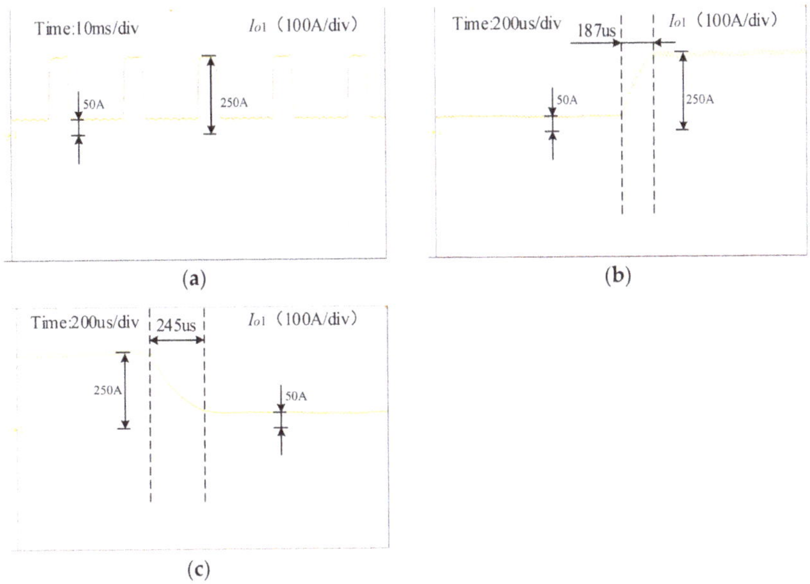

Figure 17. Large current and small current response waveform. (**a**) Large current and small current (**b**) Small current rises to large current (**c**) Large current drops to small current.

In the welding process, the large signal generally also appears when the current changes significantly. Therefore, the current direction change in the AC state is the most typical application scenario. As shown in Figure 18, compared with the traditional PID, the current changes faster in the AC state, and a higher welding frequency can be achieved in AC welding.

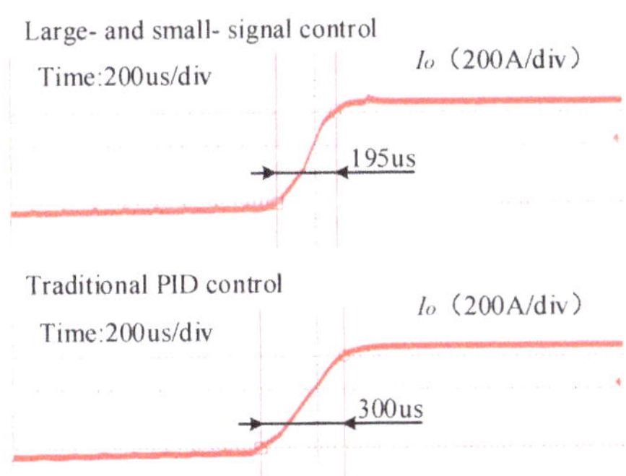

Figure 18. Large signal control output waveform in AC mode.

6. Conclusions

In this paper, based on the intermediate bus power architecture, we designed a converter suitable for the multi-output arc welding power supply and introduced its topology. Based on the designed topology, different IGBTs can be controlled to realize dual DC current output or single AC current output. When two IGBTs are controlled to switch on or off, the converter can generate dual DC current output. When an IGBT is controlled to switch on or off or to stay in the on state, the converter can output AC current. The working principle of the converter is analyzed in this paper and can be easily applied in a converter.

The small-signal model of the converter in DC and AC modes and the large-signal model of the system are established. The small-signal model is based on the time-averaging method, while the large-signal model is based on geometric segmentation. Based on the established small- and large-signal models, different current changes would be controlled by different control strategies to realize the rapid current adjustment.

An experimental prototype was designed to verify the model we proposed. The results showed that the converter designed in this study has good stability, and the control strategy can rapidly respond and generate the output current compared with the traditional PID control in both small-signal adjust and large-signal adjust, such that the welding current and adjustment ability are significantly improved. Furthermore, the proposed converter shows the availability of higher-frequency welding.

For future work, the research on multi-output power supplies for multi-electrode arc welding will focus on improving the current response speed and current output synchronization at high speeds to achieve higher AC current frequencies, as well as continuing to advance the control method of multi-output power supplies.

Author Contributions: Conceptualization, S.C.; Methodology, Y.Y.; Validation, M.L.; Writing—original draft, J.Z.; Funding acquisition, H.Z. All authors have read and agreed to the published version of the manuscript.

Funding: This research was funded by the National Natural Science Foundation of China under grant number 52205325.

Data Availability Statement: The data presented in this study are available in article.

Conflicts of Interest: The authors declare no conflict of interest.

References

1. Chen, S.; Feng, L.; Li, L.; Yin, S.; Zhang, S. Principle and parameter design of arc welding inverter using soft switching technique. *China Weld.* **1998**, *1*, 63–70.
2. Chen, J.S.; Lu, Y.; Li, X.R.; Zhang, Y.M. Gas tungsten arc welding using an arcing wire. *Weld. J.* **2012**, *91*, 261–269.
3. Chen, S.J.; Zhang, L.; Wang, X.P.; Wang, J. Stability of the cross-arc process—A preliminary study. *Weld. J.* **2015**, *94*, 158–168.
4. Lee, D.Y.; Leifsson, L.; Kim, J.; Lee, S.H. Optimization of hybrid tandem metal active gas welding using Gaussian process regression. *Sci. Technol. Weld. Join.* **2020**, *25*, 208–217. [CrossRef]
5. Zhang, Y.; Zhang, S.; Jiang, M.; Wu, L. Keyhole double-sided arc welding process. *J. Mater. Sci. Technol.* **2001**, *17*, 159–160. [CrossRef]
6. Li, K.H.; Chen, J.S.; Zhang, Y. Double-electrode GMAW process and control. *Weld. J.* **2007**, *86*, 231.
7. Ang, K.H.; Chong, G.; Li, Y. PID control system analysis, design, and technology. *IEEE Trans. Control Syst. Technol.* **2005**, *13*, 559–576.
8. Arias, M.; Diaz, M.F.; Lamar, D.G.; Linera, F.M.; Sebastian, J. Small-signal and large-signal analysis of the two-transformer asymmetrical half-bridge converter operating in continuous conduction mode. *IEEE Trans. Power Electron.* **2014**, *29*, 3547–3562. [CrossRef]
9. Bahrami, H.; Sepehrian, H.; Park, C.S.; Rusch, L.A.; Shi, W. Time-domain large-signal modeling of traveling-wave modulators on SOI. *J. Light. Technol.* **2016**, *34*, 2812–2823. [CrossRef]
10. Cao, S.G.; Rees, N.W.; Feng, G. Quadratic stability analysis and design of continuous-time fuzzy control systems. *Int. J. Syst. Sci.* **1996**, *27*, 193–203. [CrossRef]
11. Du, X.; Helaoui, M.; Jarndal, A.; Liu, T.; Hu, B.; Hu, X.; Ghannouchi, F.M. ANN-based large-signal model of AlGaN/GaN HEMTs with accurate buffer-related trapping effects characterization. *IEEE Trans. Microw. Theory Tech.* **2020**, *68*, 3090–3099. [CrossRef]
12. Zurbriggen, I.G.; Ordonez, M.; Anun, M. PWM-geometric modeling and centric control of basic DC-DC topologies for sleek and reliable large-signal response. *IEEE Trans. Ind. Electron.* **2015**, *62*, 2297–2308. [CrossRef]
13. Gregorcic, G.; Lightbody, G. Local model network identification with Gaussian processes. *IEEE Trans. Neural Netw.* **2007**, *18*, 1404–1423. [CrossRef]
14. Iqbal, M.T.; Maswood, A.I. An explicit discrete-time large- and small-signal modeling of the dual active bridge DC–DC converter based on the time scale methodology. *IEEE Trans. Emerg. Sel. Ind. Electron.* **2021**, *2*, 545–555. [CrossRef]
15. Kapat, S.; Krein, P.T. A tutorial and review discussion of modulation, control and tuning of high-performance DC-DC converters based on small-signal and large-signal approaches. *IEEE Open. J. Power Electron.* **2020**, *1*, 339–371. [CrossRef]
16. Li, C.; Du, Z. A novel method for computing small-signal stability boundaries of large-scale power systems. *IEEE Trans. Power Syst.* **2013**, *28*, 877–883. [CrossRef]
17. Liu, S.; Zhou, L.; Lu, W. Simple analytical approach to predict large-signal stability region of a closed-loop boost DC–DC converter. *IET Power Electron.* **2013**, *6*, 488–494. [CrossRef]

18. Molinas, M.; Sun, J. Guest editorial: Special issue on power electronics and systems: Modeling, analysis, control, and stability. *IEEE J. Emerg. Sel. Top. Power Electron.* **2017**, *5*, 1425–1426. [CrossRef]
19. Youssef, N.B.H.; Al-Haddad, K.; Kanaan, H.Y. Large-signal modeling and steady-state analysis of a 1.5-kW three-phase/switch/level (Vienna) rectifier with experimental validation. *IEEE Trans. Ind. Electron.* **2008**, *55*, 1213–1224. [CrossRef]
20. Prasad, A.; Fager, C.; Thorsell, M.; Andersson, C.M.; Yhland, K. Symmetrical large-signal modeling of microwave switch FETs. *IEEE Trans. Microw. Theory Tech.* **2014**, *62*, 1590–1598. [CrossRef]
21. Takagi, T.; Sugeno, M. Fuzzy identification of systems and its applications to modeling and control. *IEEE Trans. Syst. Man Cybern.* **1985**, *15*, 116–132. [CrossRef]
22. Wang, P.; Chen, X.; Tong, C.; Jia, P.; Wen, C. Large- and small-signal average-value modeling of dual-active-bridge DC–DC converter with triple-phase-shift control. *IEEE Trans. Power Electron.* **2021**, *36*, 9237–9250. [CrossRef]
23. Zhang, K.; Shan, Z.; Jatskevich, J. Large- and small-signal average-value modeling of dual-active-bridge DC–DC converter considering power losses. *IEEE Trans. Power Electron.* **2017**, *32*, 1964–1974. [CrossRef]
24. Khan, S.; Rahman, K.; Tariq, M.; Hameed, S.; Alamri, B.; Babu, T.S. Solid-State Transformers: Fundamentals, Topologies, Applications, and Future Challenges. *Sustainability* **2022**, *14*, 319. [CrossRef]
25. Iqbal, M.T.; Maswood, A.I.; Tariq, M.; Iqbal, A.; Verma, V.; Urooj, S. A Detailed Full-Order Discrete-Time Modeling and Stability Prediction of the Single-Phase Dual Active Bridge DC-DC Converter. *IEEE Access* **2022**, *10*, 31868–31884. [CrossRef]

World Electric Vehicle Journal

MDPI

Article

Experimental Design of an Adaptive LQG Controller for Battery Charger/Dischargers Featuring Low Computational Requirements

Jhoan Alejandro Montenegro-Oviedo [1], Carlos Andres Ramos-Paja [2], Martha Lucia Orozco-Gutierrez [1,*], Edinson Franco-Mejía [1] and Sergio Ignacio Serna-Garcés [3]

[1] Escuela de Ingeniería Eléctrica y Electrónica, Universidad del Valle, Cali 760042, Colombia; alejandro.montenegro@correounivalle.edu.co (J.A.M.-O.); edinson.franco@correounivalle.edu.co (E.F.-M.)
[2] Facultad de Minas, Universidad Nacional de Colombia, Medellin 050041, Colombia; caramosp@unal.edu.co
[3] Departamento de Electrónica y Telecomunicaciones, Instituto Tecnológico Metropolitano, Medellin 050013, Colombia; sergioserna@itm.edu.co
* Correspondence: martha.orozco@correounivalle.edu.co

Abstract: The growing use of DC/DC power converters has resulted in the requirement that their complex controllers be cheaper and smaller, thus using cost-effective implementations. For this purpose, it is necessary to decrease the computational burden in controller implementation to minimize the hardware requirements. This manuscript presents two methods for tuning an adaptive linear–quadratic–Gaussian voltage controller for a battery charger/discharger, implemented with a Sepic/Zeta converter, to work at any operating point. The first method is based on a lookup table to select, using the nearest method, both the state feedback vector and the observer gain vector, solving the Riccati's differential equation offline for each practical operating point. The second method defines a polynomial function for each controller element that is based on the previous data corresponding to the system operating points. The adaptability of the two controllers to fixed voltage regulation and reference tracking was validated using simulations and experimental tests. The overshoot and settling time results were lower than 11% and 3.7 ms, which are in the same orders of magnitude of a control approach in which the equations are solved online. Likewise, three indices were evaluated: central processing unit capacity, cost, and performance. This evaluation confirms that the controller based on polynomial interpolation is the best option of the two examined methods due to the satisfactory balance between dynamic performance and cost. Despite the advantages of the controllers in being based on a lookup table and polynomial interpolation, the adaptive linear–quadratic–Gaussian has the benefit of not requiring an offline training campaign; however, the cost saving obtained with the lookup table controllers and polynomial interpolation controllers, due to the possible implementation on small-size microcontrollers with development tool simple and easy maintenance, will surely be desirable for a large number of deployed units, ensuring that those solutions are highly cost-effective.

Keywords: adaptive controller; battery charger/discharger; Sepic/Zeta converter; DC bus regulation; cost-effective; comparison of performance

Citation: Montenegro-Oviedo, J.A.; Ramos-Paja, C.A.; Orozco-Gutierrez, M.L.; Franco-Mejía, E.; Serna-Garcés, S.I. Experimental Design of an Adaptive LQG Controller for Battery Charger/Dischargers Featuring Low Computational Requirements. *World Electr. Veh. J.* **2023**, *14*, 142. https://doi.org/10.3390/wevj14060142

Academic Editor: Joeri Van Mierlo

Received: 29 April 2023
Revised: 25 May 2023
Accepted: 26 May 2023
Published: 28 May 2023

1. Introduction

Power processing is the most critical activity in electric and electronic systems, and one device most used for this purpose is the DC/DC power converter. Power converters are used to drive DC sources, storage devices, and loads. Moreover, power converters require a controller for safely performing power processing tasks. Due to the wide range of controllers reported in the literature, evaluating functional and non-functional characteristics to choose the most suitable one for an industrial system is of significant relevance [1].

It is crucial to have performance criteria for controllers because they provide an objective way to assess the quality of their work. The evaluation criteria for DC/DC power converter controllers include factors such as precision of voltage regulation, system stability, efficiency, load-carrying capacity, and dynamic response [2–5]. They can also have advanced features such as overcurrent and overvoltage protection to ensure safe and reliable system operation. Moreover, some non-functional characteristics such as size, cost, and computational burden should be considered.

The viability of implementing a solution in control theory is a problem that can be solved if an algorithm with moderate computational requirements exists [6]. The moderate computational requirements mean that the computational resources (fundamentally processor and memory) that need to solve the problem are within the acceptable costs. Being inside these limits facilitates implementation and helps in reaching the system performance specifications [7]. The acronyms used here are summarized in Table 1.

Table 1. Acronyms.

Acronyms	Significance
LQG	linear–quadratic–Gaussian
MPC	model predictive control
DSP	digital signal processor
LQR	linear quadratic regulator
LQI	linear quadratic integral
PID	proportional–integral–derivative
RMSE	root mean square error
ADCs	analog-to-digital converters
MIPS	millions of instructions per second

The high complexity of the current systems requires the development of advanced control strategies to achieve the performance demanded by the application. These control strategies, in turn, require devices with high power processing, thus raising energy consumption and increasing the heat dissipation and costs. In order to develop cheaper solutions, with small and efficient implementations, it is necessary to reduce the computational burden of the control strategies without significantly impacting the response accuracy; for this, it is useful to simplify the models to reduce the number of differential equations, replace some dynamic equations with lookup tables, linearize some nonlinear relations, or eliminate redundant calculations. In the particular case of power converter control, strategies frequently used, apart from classical proportional–integral–derivative (PID) topologies, are model predictive control (MPC), robust control, and adaptive control, whose implementations are generally complex and produce a significant computational burden. Some works found in reviews of the state of the art implement those controllers, but there are few reported attempts to reduce the computational burden.

For example, in [8], the computational burden of a conventional finite control set MPC applied to a T-type 3P-3L (three-phase three-level) converter is reduced by implementing three improvements: a better optimization method to determine the vector for current tracking without vector enumeration; an algorithm for capacitor charge balance with the aim of obtaining the vector for neutral point potential balance, thus eliminating weight factor; and, finally, introducing a hybrid vector output mode to give full consideration to current tracking and neutral point potential balance. With these changes in the algorithm, it an improvement factor of 1.6 was achieved in the execution time of the implementation with C-language on the DSP TMS320F28335. However, although the system performance is improved in the solution, the complexity and processor requirements are high. In [9], the authors implemented a finite control set MPC variant, named fast MPC, applied to a multilevel inverter in a photovoltaic system. In this work, the change in the algorithm avoids testing all possible states of the multilevel converter and evaluate a cost function, reducing the six-factor computation time without effecting the dynamic response (the controller was implemented on a TMS320F28335 DSP). However, the finite control set MPC proposed in that work still executes complex operations, such as Lagrange extrapolation, that still overload the processor.

A method that has recently had a huge boom is deterministic artificial intelligence, which has shown tremendous improvement of performance in systems control compared to nonlinear adaptive methods. For example, some works present the control of the DC motor or the actuator of an unmanned underwater vehicle, where it is evident that the performance improvements are not due to the coefficient estimation but to the modeling methods [10] and the discretization [11–14]. This is why the deterministic artificial intelligence method is highly recommended for applications that require high levels of accuracy, consistency, and precision, such as in power electronics, although in the reviewed works, the controller does not directly control the DC motor driver, namely the power converter. However, these works do not report performance metrics related to computational burden.

A sliding mode controller that requires a processor with high performance is presented in [15]. The strategy consists of an adaptive-gain/super twisting sliding mode controller applied to two electro-pneumatic actuators, which is a slow system. This work does not evaluate the computational complexity; however, it requires a DS1104 board from DSpace to implement the control law, which is a costly device. Other works show some practical applications utilizing bidirectional buck–boost converters: a high-gain sliding controller in a supercapacitor in [16], a sliding-mode controller of the electro-mechanical actuator current of an aircraft powered by a supercapacitor in [17], and in [18–23], a current sliding-mode controller accompanied by supervisory control that carries the charge state of the battery, its charge/discharge current, the regulation of the high-voltage side, or maximizes generator current. These works highlight the advantages of these nonlinear controllers, but they do not feature their requirements for hardware, lower [19] when implemented using an M4 processor, nor their computational burden.

A methodology for combined design of controllers, both adaptive and linear quadratic regulator (LQR), is presented in [24]. The adaptive controller portion deals with the unknown dynamics of the plant. At the same time, the LQR control, based on a Kalman filter, is used to minimize the effects of the external disturbances. Unfortunately, this work only shows simulations without experimental validation, and there are neither simplifications nor strategies that could decrease the computational burden. Similarly, proposed in [25] is a combination of adaptive and LQR control for a grid-connected inverter. The current control is frequency-adaptive, while the optimal LQR selects desirable gains for the full-state feedback controller and full-state observer. However, the computational burden of the implementation is high, requiring a DSP TMS320F28335 with a sampling period of 100 μs, which is after decreasing the computational requirements by designing a discrete-time observer in the stationary reference frame.

A useful strategy to significantly reduce the computational burden is to perform any optimization process offline; this is, for example, defining the controller adjustments in several operation points before the dynamic behavior takes it there. Therefore, when the system evolves to one of those operations points, the controller parameters must be selected from a table and not dynamically calculated. This solution is reported in some practical cases. For example, the use of a lookup table for power converter control is reported in [26]; this work proposes two new lookup table methods for a three-level AC/DC converter supplied from the grid. The first method improves the nonlinear direct power control by rotating the orientation of the coordinates; the second method is an improvement of the first in which a new optimal division of the current error plane is introduced, which allows for better voltage vector assignment. Another example is reported in [27], which proposes a direct power control strategy using a switching table that can simultaneously control both active and reactive power in an electrical grid. It uses an extended p-q theory better suited for unbalanced grid voltages than the original p-q theory. The extended p-q theory is combined with voltage-oriented control to mitigate the influence of grid voltage unbalances. Nevertheless, the disadvantages of direct power control are evident: small scalability, requirement for high synchronization, does not work well with interference, and limited range [28,29].

Adaptive lookup table-based variable on-time control is proposed in [30] for critical mode boost power factor corrector converters, which combines the advantages of both the conventional lookup table method and the real-time calculation method. The paper realizes the adaptability of proposed control within the universal AC input, and the entire load range through a simple linear calculation. Furthermore, the proposed controller reduces the required memory space for variable on-time tables and significantly shortens the computation time, leading to low system cost.

In general, the parameters of PID controllers for nonlinear industrial systems are dynamically tuned using lookup tables. A method based on a cost function to tune the lookup table parameters used in gain-scheduled PID control is proposed in [31], where the L2 norm is adopted to prevent overlearning. This strategy allows the controller to work at any operation point.

Presented in [32] is a lookup table of maximun power point tracking (MPPT) algorithm for boost converters in photovoltaic systems, which has lower accuracy than a perturbed and observer (P&O) algorithm but a better response to sudden changes in temperature or irradiance. Instead, the works reported in [33,34] describe a digital PID for regulating the voltage on class E buck–boost converter and buck converter, respectively. The compensator is based on a lookup table and implemented with multi-phase digital pulsewidth modulation.

Lookup tables are also used in MPC implementations. For example, ref. [35] proposes an offline optimization and online lookup table for a two-layer model predictive control to reduce the computational burden. The work examines the properties of computation complexity, steady-state operation, and robustness of the proposed strategy through detailed simulations.

Although the works reported in [26–35] use controllers based on a lookup table, none is applied on an adaptive LQR controller nor for a Sepic/Zeta converter-like charger/discharger battery.

A longitudinal movement/optimal controller for heavy-duty vehicles is solved offline in [36] using a mixed integer quadratic program. Then, if the position is known, those trajectories can be used as lookup tables to find the reference for both velocity and freewheeling. However, this solution does not directly act on the actuators in the vehicle, limiting its accuracy. Another work that does not involve the actuator was reported [37], in which an adaptive lookup table algorithm is introduced that can accurately and automatically update the static table of an air compressor system. The algorithm has better dynamic performance, such as shorter regulating time and lower overshoot in comparison with the traditional control method (PI + static feedforward strategy).

Presented in this paper are similar strategies to reduce the computational burden of complex controllers applied to battery charger/dischargers. In particular, this works focuses on the battery charger/discharger discussed in [38], which is based on a Sepic/Zeta converter interfacing a battery with the DC bus of a microgrid. The implementation consists of a linear–quadratic–Gaussian (LQG) controller designed to adapt to any operation, with the aim of regulating the DC bus voltage to ensure safe microgrid operation. This controller uses an adaptive law to adjust the controller parameters depending on the operating point; it also requires a full-state observer to reduce the number of sensors, thus reducing the cost. The implementation reported in [38] has a high computational burden due to the online calculation of both the adaptive part and the observer in which a DSP TMS320F28379D is required for experimental deployment.

Therefore, this work proposes using an offline process to calculate the adaptive controller described in [38], thus preserving the required operation range. Two methods were designed to calculate the state matrix, calculate the input matrix, and to solve the Ricatti differential equations: the lookup table method and the polynomial interpolation method. The first method requires calculating the matrices, solving the Ricatti equation once for each operating point, and storing the controller parameters in a lookup table. The second method defines a polynomial function representing the changes in the state feedback vector and the observer gain vector. The polynomial degree is defined to achieve a significantly small root mean square error. Both solutions are evaluated using numerical metrics, circuital simulations, and experimental implementations. Moreover, the performance/cost balance is also analyzed and contrasted with the original implementation reported in [38], which provides a selection guide depending on the charger/discharger requirements in terms of dynamic performance and implementation cost.

This manuscript is divided into five parts: Section 2 presents the power stage and the mathematical model; Section 3 presents the design of the adaptive LQG voltage controller, describing the controller processing using both the lookup table and the polynomial interpolation methods. Section 4 presents a design example and simulation results, and Section 5 reports the experimental validation, comparison, and discussion of the two solutions. Finally, the conclusions of the work are reported in Section 6.

2. Modeling of the Power Stage

The battery charger/discharger based on a bidirectional Sepic/Zeta converter is shown in Figure 1. This charger/discharger circuit is discussed in [38], where deep circuit and mathematical analyses are provided. The variables used for the circuit analysis are summarized in Table 2.

Table 2. Variables.

Variable	Significance
i_o	Output current, DC bus current of a microgrid
v_{dc}	Output voltage, DC bus voltage of a microgrid
$v_{dc_{ref}}$	Desired DC bus voltage
v_b	Battery side voltage
v_{ci}	Coupling capacitor voltage
v_{ci_e}	Coupling capacitor voltage on stationary state
i_{L1}	Inductor current of battery side
i_{L2}	Inductor current of DC bus side
i_{L1_e}	Inductor current on stationary state of battery side
i_{L2_e}	Inductor current on stationary state of DC bus side
u	Binary control signal
d	Continuous duty cycle
d_e	Continuous duty cycle on stationary state
T_{sw}	Switching period

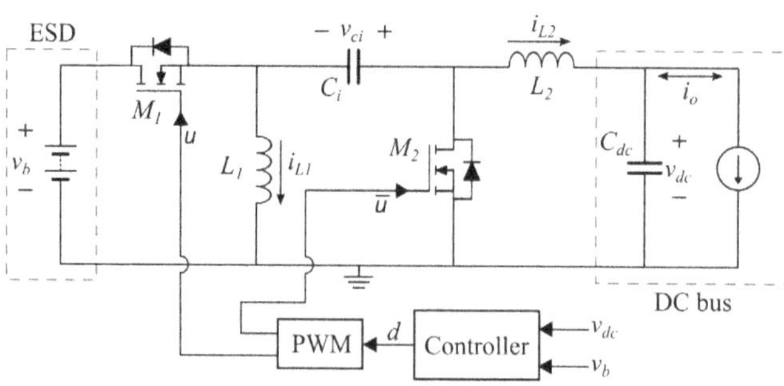

Figure 1. Sepic/Zeta-based circuital interface for bidirectional power flow.

In this circuit, the main objective is to regulate the output voltage v_{dc}, which corresponds to the voltage at the DC bus of a microgrid. In this way, the charger/discharger ensures the safe operation of all the devices connected to the DC bus.

For realistic purposes, the parasitic resistances in both the MOSFETs and inductors are considered for the analysis. A control-oriented model for this charger/discharger circuit is provided in [38]. In summary, the modeling procedure consists of obtaining the differential equations for the inductor currents and capacitor voltages, which are extracted from circuital analyses (charge and volt-second balances). Such a process is performed when the MOSFET control signal has ON ($u = 1$) and OFF ($u = 0$) states; then, those equations are combined using the control signal u to obtain the following switched model:

$$\frac{di_{L1}}{dt} = \frac{(v_b - (i_{L1} + i_{L2}) \cdot R_{on1}) \cdot u + (-v_{ci} - (i_{L1} + i_{L2}) \cdot R_{on2}) \cdot \bar{u} - i_{L1} \cdot R_{L1}}{L_1} \tag{1}$$

$$\frac{di_{L2}}{dt} = \frac{(v_{ci} - (i_{L1} + i_{L2}) \cdot R_{on1} + v_b) \cdot u - (i_{L1} + i_{L2}) \cdot R_{on2}\bar{u} - v_{dc} - i_{L2} \cdot R_{L2}}{L_2} \tag{2}$$

$$\frac{dv_{ci}}{dt} = \frac{i_{L1} \cdot \bar{u} - i_{L2} \cdot u}{C_i} \tag{3}$$

$$\frac{dv_{dc}}{dt} = \frac{i_{L2} - i_o}{C_{dc}} \tag{4}$$

It is difficult to use the previous switched model in the design of continuous controllers. Therefore, such a discontinuous model is averaged within the switching period T_{sw}, which is performed by replacing the binary control signal u by the continuous duty cycle d. Moreover, for the sake of simplicity, the ON resistances of the MOSFETs are considered equal ($R_{on1} = R_{on2} = R_{on}$). The resulting averaged model is nonlinear; thus, it is described in terms of small-signal variables in state space representations, which will be used to design the controller. Applying the previous procedure leads to the small-signal model given in (5) and (6), where A_m (state matrix), B_m (input matrix), and C_m (output matrix) must be evaluated in the desired operating point. The quantities $\tilde{x}$, $\tilde{y}$, and $\tilde{d}$ are the small-signal changes in the state vector, output, and duty cycle, respectively. Finally, the state vector variables are $\tilde{x}_1 = \tilde{i}_{L1}$, $\tilde{x}_2 = \tilde{i}_{L2}$, $\tilde{x}_3 = \tilde{v}_{ci}$ and $\tilde{x}_4 = \tilde{v}_{dc}$.

$$\dot{\tilde{x}} = \underbrace{\begin{bmatrix} -\frac{R_{on}+R_{L1}}{L_1} & -\frac{R_{on}}{L_1} & \frac{-(1-d_e)}{L_1} & 0 \\ -\frac{R_{on}}{L_2} & -\frac{R_{on}+R_{L2}}{L_2} & \frac{d_e}{L_2} & \frac{-1}{L_2} \\ \frac{(1-d_e)}{C_i} & \frac{-d_e}{C_i} & 0 & 0 \\ 0 & \frac{1}{C_{dc}} & 0 & 0 \end{bmatrix}}_{A_m} \cdot \tilde{x} + \underbrace{\begin{bmatrix} \frac{v_b+v_{cie}}{L_1} \\ \frac{v_b+v_{cie}}{L_2} \\ \frac{-i_{L1e}-i_{L2e}}{C_i} \\ 0 \end{bmatrix}}_{B_m} \cdot \tilde{d} \tag{5}$$

$$\tilde{y} = \underbrace{\begin{bmatrix} 0 & 0 & 0 & 1 \end{bmatrix}}_{C_m} \cdot \tilde{x} \tag{6}$$

The previous small-signal model is used in the following section to design the controller, aimed at regulating the DC bus voltage in a microgrid.

3. Adaptive LQG Voltage Controller

The controller aims to regulate the DC bus voltage under any operation condition, which are battery charge and discharge and standby modes. Moreover, the control system must remain stable over the entire operating range, and this means for different relations among battery and DC bus voltages. To achieve this performance, a linear–quadratic–Gaussian (LQG) controller with adaptive parameters is proposed. The LQG consists of three fundamental blocks: a linear quadratic integral (LQI) to regulate the DC bus, an optimal state observer to estimate the system states, and a dynamic process to determine the LQG parameters according to the operating point. Two methods are proposed to calculate the parameters values of controller in any operating condition: lookup table and polynomial interpolation.

3.1. LQI Controller Design

The regulation of the DC bus is defined as an optimal control problem, and its analytical design procedure was developed in [38]. In short, the quadratic cost function given in (7) is defined, which is solved using the linear quadratic regulator (LQR) approach [39], and an integrator is added to eliminate the steady-state error.

$$J = \int_0^{t_f} \frac{1}{2} \cdot (\tilde{z}^T Q z + r\tilde{d}^2) \cdot dt \tag{7}$$

In Equation (7) r is a positive scalar value, Q is 5×5 positive matrix, and $\tilde{z}$ is the extended state vector resulting from the inclusion of the error integral, named $\tilde{x}_i$. The state feedback law is defined in (8), where K_x is a gain vector that multiplies the states $\tilde{x}$, and K_i is a scalar that multiplies $\tilde{x}_i$. Then, the extended system is defined as reported in (9).

$$\tilde{d} = -K \cdot \tilde{z} = -\begin{bmatrix} K_x & K_i \end{bmatrix} \cdot \begin{bmatrix} \tilde{x} \\ \tilde{x}_i \end{bmatrix} \tag{8}$$

$$\dot{\tilde{z}} = \begin{bmatrix} \dot{\tilde{x}} \\ \dot{\tilde{x}}_i \end{bmatrix} = \underbrace{\begin{bmatrix} A_m & 0 \\ -C_m & 0 \end{bmatrix}}_{A_w} \cdot \underbrace{\begin{bmatrix} \tilde{x} \\ \tilde{x}_i \end{bmatrix}}_{\tilde{z}} + \underbrace{\begin{bmatrix} B_m \\ 0 \end{bmatrix}}_{B_w} \cdot \tilde{d} \tag{9}$$

The optimization problem is solved using the Hamiltonian matrix, the equation of states and co-states, and the stable state values, all of which are described in [38]. In such a procedure, the optimal value of $\tilde{d}^*$ is calculated as given in (10).

$$\tilde{d}^* = -r^{-1} \cdot B_w^T \cdot S_m \cdot \tilde{z} \tag{10}$$

Then, K is defined as (11); to perform the calculation of K, the S_m matrix must be solved from the Riccati differential equation reported in (12).

$$K = -r^{-1} \cdot B_w^T \cdot S_m \tag{11}$$

$$Q + A_w^T \cdot S_m + S_m \cdot A_w - S_m \cdot B_w \cdot r^{-1} \cdot B_w^T \cdot S_m + \dot{S}_m = 0 \tag{12}$$

Finally, this problem is formulated as an optimal regulator of infinite time, hence S_m corresponds to the steady-state solution of (12).

3.2. State Observer

In order to reduce the number of sensors required in the state feedback control system, the observer described by (13) and (14) is used. The design of the observer considers the linear model of the system (5) and (6) and the state feedback control law (8) [40]. The outputs of the observer are the estimated states $\hat{x}$, and the inputs of the observer are y and d, which correspond to the measured DC bus voltage v_{dc} and the duty cycle, respectively.

$$\dot{\hat{x}} = (A_m + \ell \cdot C_m) \cdot \hat{x} + B_m \cdot d - \ell \cdot y \tag{13}$$

$$\hat{y} = C_m \cdot \hat{x} \tag{14}$$

The observer gain ℓ is calculated using an optimization process, which guarantees the convergence of the estimated states to the real values. The calculation of ℓ, given in (15), requires solving the Riccati differential equation reported in (16). In such an expression, S_o

corresponds to the optimal error covariance matrix of the state observer, and γ is a scalar used to adjust the convergence speed of the observer states as explained in [41].

$$\ell = S_o \cdot C_m^T \cdot \gamma \quad (15)$$

$$A_m \cdot S_o + S_o \cdot A_m^T - S_o \cdot C_m^T \cdot \gamma^{-1} \cdot C_m \cdot S_o + B_m \cdot B_m^T + \dot{S}_o = 0 \quad (16)$$

3.3. *Adaptive Parameters of the LQG Controller*

In this section, two adaptive methods to adjust the control parameters under any operating condition are proposed. Unlike the approach presented in [38], which consists of online iterative calculation of the parameters, in this work the parameters are selected from a lookup table or calculated using polynomial interpolation. Since the parameters are calculated offline, both proposed methods require lower computational resources than the method presented in [38], as no online solutions of matrix differential equations are needed.

3.3.1. Adaptive Lookup Table Method

This method consists of calculating both the state feedback vector and the observer gain vector, solving the Riccati's differential equation offline for each practical operating point (v_{dc_ref} and v_b combinations). Subsequently, the controller parameters are stored in a lookup table to be inserted into the implementation device. The selection of the parameters from the lookup table consists of searching the closest value (nearest method) according to the measurements of both the bus reference v_{dc_ref} and the battery voltage v_b.

3.3.2. Adaptive Polynomial Interpolation Method

This method consists of defining a polynomial function that represents the variation of each element of both the state feedback and the observer gain vectors, being in agreement with the operating point. This polynomial interpolation is performed using the data calculated in the lookup table method. The order of the polynomial functions is defined in such a way that the root mean square error (RMSE) is significantly small. As a result, after calculating the value of the polynomial coefficients, functions depending on v_b and v_{dc_ref} are obtained for each adaptive parameter, which are subsequently implemented in a control device.

3.4. *Structure of the Adaptive LQG Voltage Controller*

The control scheme is shown in Figure 2, where the adaptive tuning block provides the matrices of the LGQ controller (A_m, B_m), the observer gain (ℓ), and the control parameters (K_x), all according to the operating point (v_{dc_ref}, v_b). Then, the LQG controller uses those parameters to calculate the optimal duty cycle d in real time.

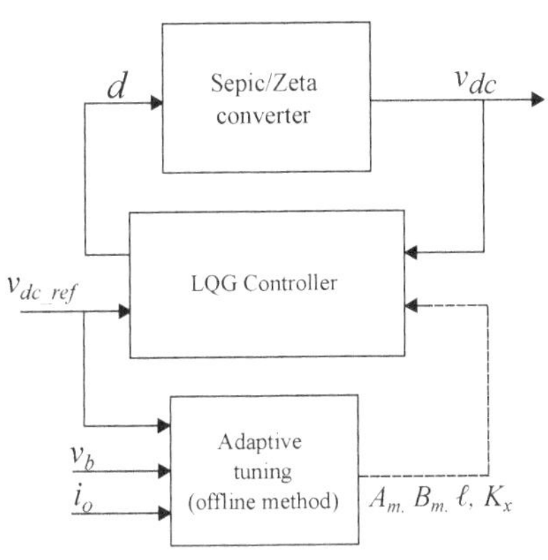

Figure 2. Structure of the adaptive LQG voltage controller.

4. Design Example and Simulation Results

This section carries out a design example, applying the proposed strategies to reduce the computational burden of adaptive controller implementation and verifying those solutions using detailed circuital simulations. The application example considers a power converter of 26 W, a switching frequency of 40 kHz, and battery and DC bus voltages between 10 and 26 volts. The application also requires a minimum efficiency of 90%, a maximum ripple of 20% on the inductor currents, and a maximum bus voltage ripple of 1%. The converter parameters to achieve those requirements are inductors (L_1, L_2) of 680 μH and capacitors (C_i, C_{dc}) of 330 μF. Those values are the same ones calculated in the design procedure reported in [38], with the objective of performing a comparison under the same operating conditions.

4.1. Calculation of Controller Parameters

The values of the matrices A_m and B_m of the observer are computed offline using Equation (5) of the state space system. The controller parameters K and ℓ are also calculated offline for both the lookup table and polynomial interpolation methods.

4.1.1. Adaptive Lookup Table

To generate the data for the lookup table, the controller parameters were obtained by varying the battery voltage and the DC bus voltage inside the established operating range. This calculation was performed using the LQR function of the Control System toolbox package from Matlab, and the controller parameters were calculated by varying v_b and v_{dc_ref} every 2 V. The adopted search algorithm is nearest, which returns the value of the table corresponding to the closest data according to the combination of the inputs (v_b, v_{dc_ref}). The data obtained for this method are reported in Appendix A.

4.1.2. Adaptive Polynomial Interpolation

To establish the polynomial functions that represent the variation of the state feedback vector, polynomial interpolation was performed using the data generated in the lookup table method. Therefore, for each element of the vector, a function of the voltages v_b and v_{dc_ref} is generated. This process was conducted with the Curve Fitter tool from Matlab, which calculates the coefficients according to the polynomial degree selected for the inputs. In this application, the linear least squares method was selected to calculate those coefficients. Moreover, the results show that a fourth order polynomial for the state feedback vector and a third order polynomial for the observer gain vector were sufficient in such a way that a maximum RMSE equal to 0.026 was obtained. This means that the resulting polynomial functions represent the state feedback vector in the operation range with satisfactory accuracy.

Figure 3 shows an example of the polynomial fit for the third element of the state feedback vector (K_3). It is evident that the resulting surface adequately represents the set of points generated for all the combinations of v_b and v_{dc_ref} in the defined range.

The previous procedure is also executed to calculate the coefficients of the polynomial functions related to the observer gain vector. Figure 4 shows the polynomial fit for the third element of the observer gain vector (ℓ_3) as an example, obtaining a surface that adequately represents the variation of this observer parameter. Finally, the polynomials and coefficients for this method are reported in Appendix A.

4.2. Simulations Results

The charger/discharger and control system performance in the regulation of the DC bus voltage is tested using circuital simulations. The tests are conducted at multiple operating points within the defined operating range. The simulations were executed in Simulink® from Matlab® using the Simscape™ Electrical™ toolbox [42]. The adaptive parameter methods were implemented using a three-dimensional lookup table (first method), and using Matlab® functions for the polynomial interpolation (second method). The simulations evaluate the solution performance in two scenarios: voltage regulation for a fixed reference value, which is the most common application; and reference tracking, which introduces a strong perturbation to test the system stability.

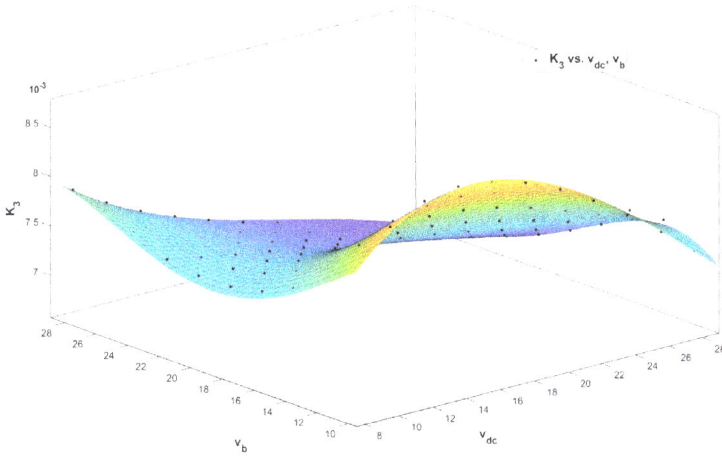

Figure 3. Fit plot of K_3.

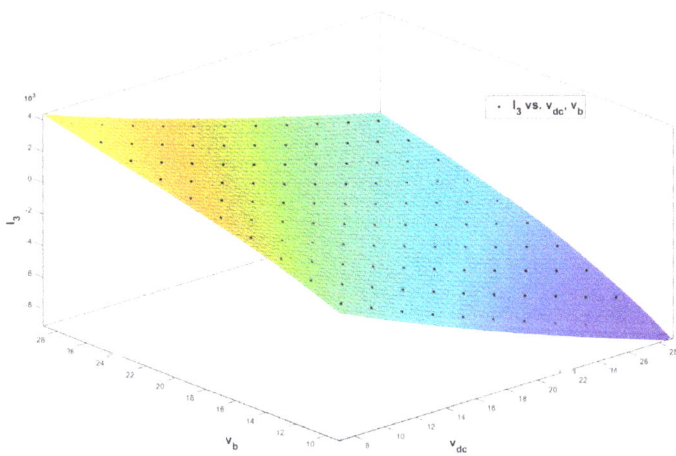

Figure 4. Fit plot of ℓ_3.

4.2.1. Bus Voltage Regulation for a Fixed Reference Value

This is the most common operation of any microgrid (fixed DC bus voltage), where the disturbances in the bus are generated by changes in the current (and power) profiles of both the loads and sources connected to the bus. Therefore, the simulations consider a current profile imposing the different modes of operation to the batteries, including charging, discharging, and standby. In addition, the tests are conducted with one and two batteries connected in series to test the scalability of the solution.

The first tests are carried out using the adaptive adjustment method based on the lookup table. Figures 5–7 show the performance of the lookup table solution considering a 12 V battery. Those simulation consider step changes in bus current every 50 ms: first, it starts in standby mode ($i_o = 0$ A), then switches to discharge mode ($i_o > 0$ A), followed by switching to charge mode ($i_o < 0$ A), and finally returning to standby mode. The simulations confirm that the adaptive LQG with lookup table provides the desired performance, in terms of voltage regulation, under any battery operating condition. Similarly, such results also illustrate the bus voltage stability in any operation condition. Moreover, the tests also evaluate the solution under different ratios between the batteries and the DC bus voltages (buck, boost, unitary gain), where satisfactory (and stable) performance is also evident. Finally, the lookup table solution does not saturate the duty cycle, thus ensuring the correct operation of system.

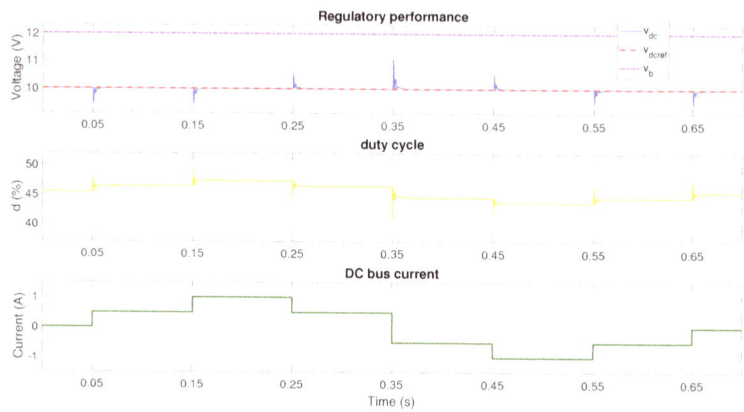

Figure 5. Performance with $v_b = 12$ V and reference voltage $v_{dc_{ref}} = 10$ V using adaptive LQG with lookup table (buck mode).

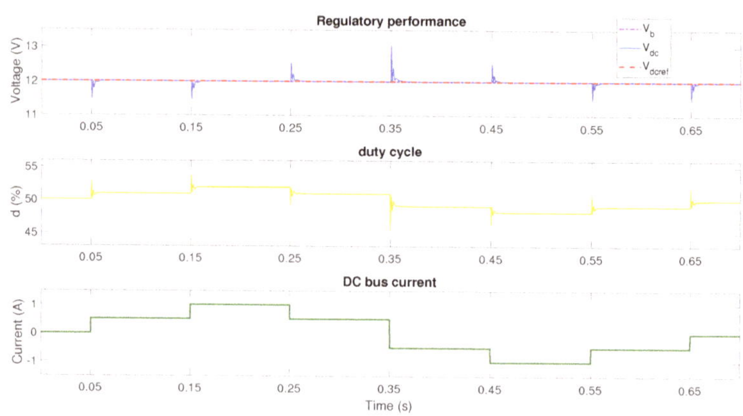

Figure 6. Performance with $v_b = 12$ V and reference voltage $v_{dc_{ref}} = 12$ V using adaptive LQG with lookup table (unitary gain mode).

In order to test the adaptability of the controller to large changes in battery voltage, the tests are repeated considering two batteries connected in series, thus $v_b = 24$ V. In this case, to validate the buck, boost, and unity gain modes, the bus voltage references are set to 20 V, 24 V, and 26 V. The results of the simulations are shown in Figures 8–10, where satisfactory DC bus voltage regulation is achieved. In this way, the adaptability of the controller is verified under any condition, and it can thus be used in applications with different numbers of batteries as long as the voltage of the batteries is considered for the lookup table calculation. Therefore, it is useful to perform such a lookup table calculation for the rated limits of the power converter.

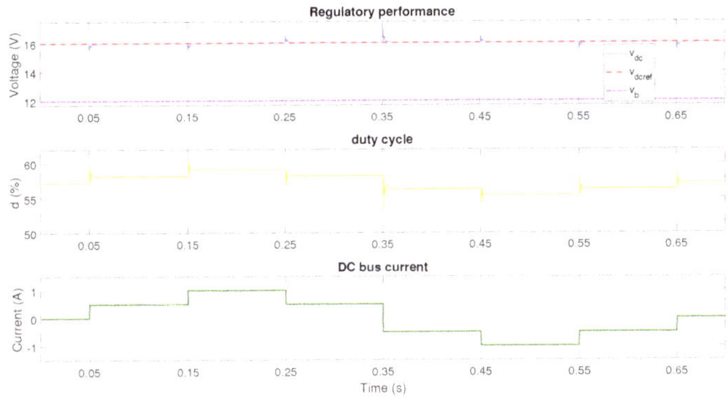

Figure 7. Performance with $v_b = 12$ V and reference voltage $v_{dc_{ref}} = 16$ V using adaptive LQG with lookup table (boost mode).

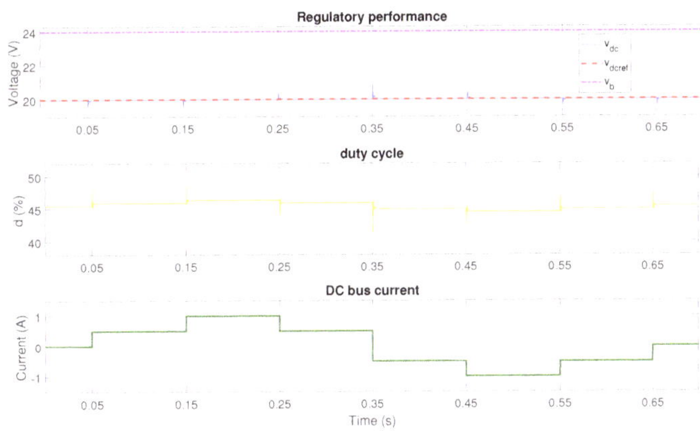

Figure 8. Performance with $v_b = 24$ V and reference voltage $v_{dc_{ref}} = 20$ V using adaptive LQG with lookup table (buck mode).

Figure 9. Performance with $v_b = 24$ V and reference voltage $v_{dc_{ref}} = 24$ V using adaptive LQG with lookup table (unitary gain mode).

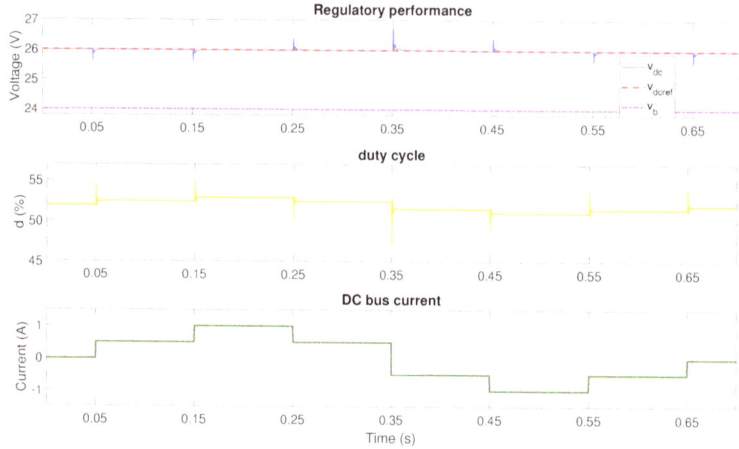

Figure 10. Performance with $v_b = 24$ V and reference voltage $v_{dc_{ref}} = 26$ V using adaptive LQG with lookup table (boost mode).

The same tests are performed for the adaptive parameter method based on polynomial interpolation. The tests are carried out with one and two batteries connected in series and testing the different modes of the converter. The simulation results are reported in Figure 11 (buck mode), Figure 12 (unitary gain mode) and Figure 13 (boost mode) for a single battery ($v_b = 12$ V) and in Figure 14 (buck mode), Figure 15 (unitary gain mode), and Figure 16 (boost mode) for two series-connected batteries ($v_b = 24$ V).

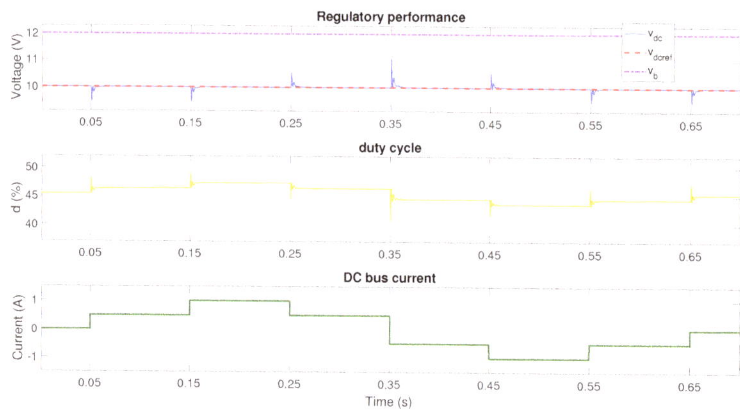

Figure 11. Performance with $v_b = 12$ V and reference voltage $v_{dc_{ref}} = 10$ V using adaptive lqg with polynomial interpolation (buck mode).

The tests of the adaptive controller with polynomial interpolation also show satisfactory voltage regulation in response to perturbations in the DC bus current in all the different modes of both the converter and battery. Hence, there are no significant differences in the transient behavior in contrast with the Lookup table method. Nevertheless, the choice of the method may be due to the limitations of the hardware to be implemented, since the lookup table method consumes more memory, but the polynomial interpolation method consumes more mathematical operations.

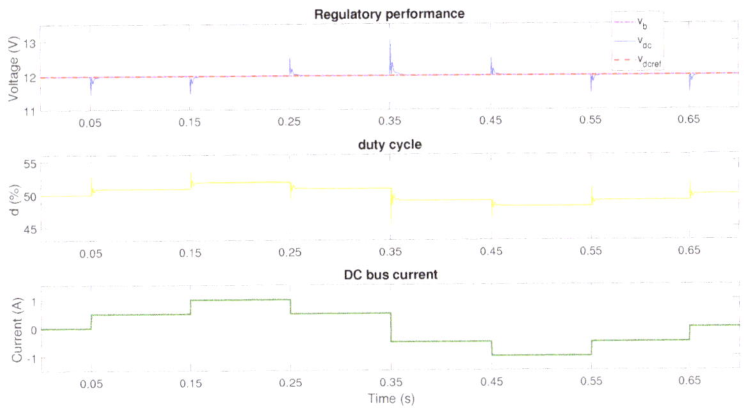

Figure 12. Performance with $v_b = 12$ V and reference voltage $v_{dc_{ref}} = 12$ V using adaptive LQG with polynomial interpolation (unitary gain mode).

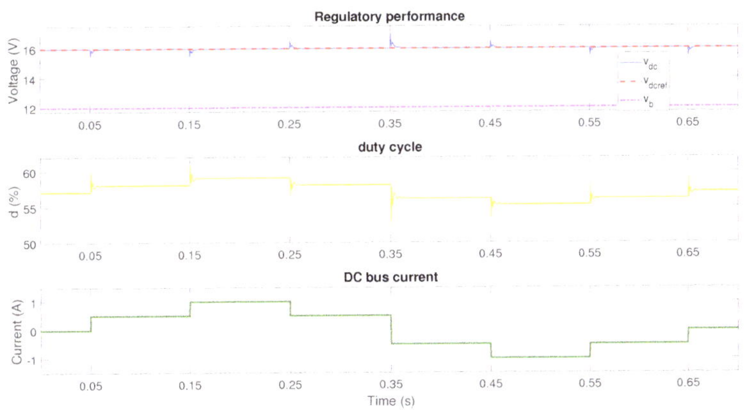

Figure 13. Performance with $v_b = 12$ V and reference voltage $v_{dc_{ref}} = 16$ V using adaptive LQG with polynomial interpolation (boost mode).

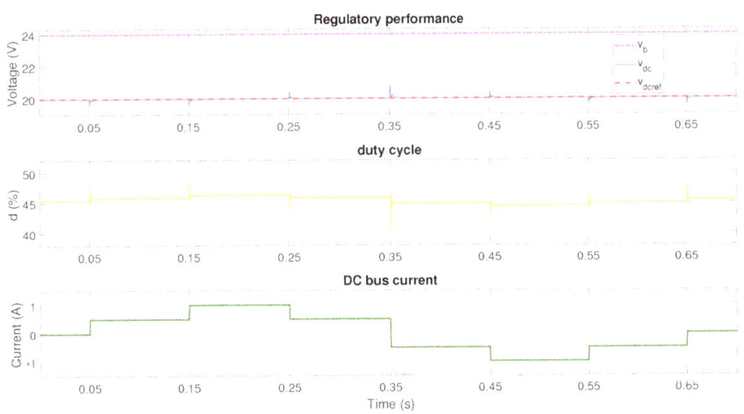

Figure 14. Performance with $v_b = 24$ V and reference voltage $v_{dc_{ref}} = 20$ V using adaptive LQG with polynomial interpolation (buck mode).

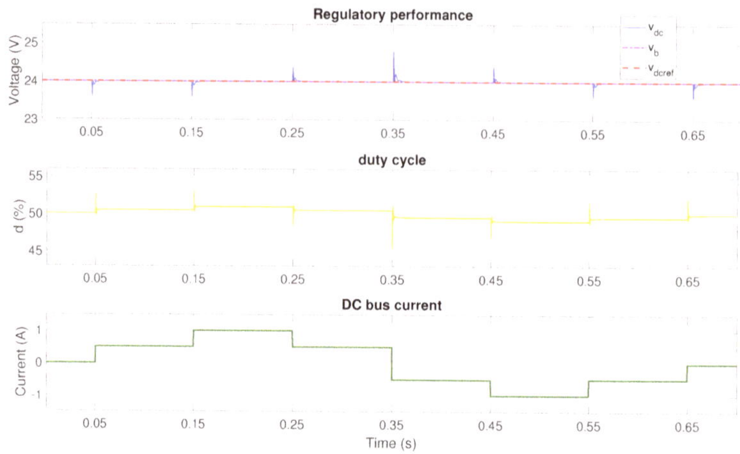

Figure 15. Performance with $v_b = 24$ V and reference voltage $v_{dc_{ref}} = 24$ V using adaptive LQG with polynomial interpolation (unitary gain mode).

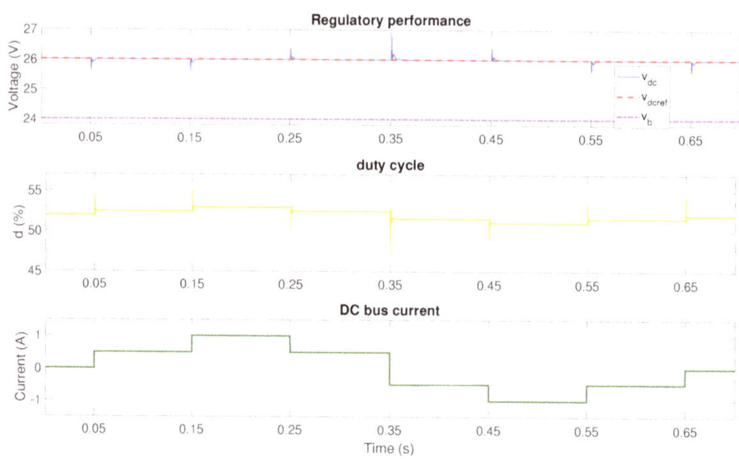

Figure 16. Performance with $v_b = 24$ V and reference voltage $v_{dc_{ref}} = 26$ V using adaptive LQG with polynomial interpolation (boost mode).

Table 3 summarizes the dynamic performance results of the tests in Figures 5–16. In addition, the table also reports the results of the online adaptive method reported in [38], which enables a fair comparison since all the simulations are conducted in the same operation conditions: the results show that the online solution exhibits a maximum overshoot of 9.7% and settling times under 10 ms in all cases, while the maximum overshoot is 10.9% and the settling times are under 10 ms for the offline adaptive LQG methods; hence, all the solutions fulfill the design requirements. It is noted that the online adaptive LQG method provides lower overshoot in the tests at $v_b = 12$ V in comparison with the offline methods, and there is also a longer settling time. This is caused by small tolerances in the controller parameters obtained with the offline adaptive methods; thus, an error in the settling time is expected. However, the offline methods provide a close performance in terms of both overshoot and settling time and, hence, the lookup table and polynomial interpolation methods are suitable for practical (experimental) applications.

Table 3. Summary of the dynamic performance for bus voltage regulation.

Test Conditions		Overshoot (%)			Settling Time (ms)		
v_b (V)	v_{dc} (V)	Lookup Table	Polynomial Interp.	LQG Online	Lookup Table	Polynomial Interp.	LQG Online
12	10	10.9	10.85	9.7	3.3	3.65	3.83
	12	8.8	8.76	8.75	3.2	3.45	3.46
	16	6.25	6.21	6.19	2.9	3	3.02
24	20	4.13	4.1	4.1	0.75	0.75	0.75
	24	3.3	3.29	3.29	0.671	0.672	0.67
	26	3	2.98	3	0.63	0.624	0.63

On the other hand, mean error and relative standard deviation percentages of the DC voltage of the tests in Figures 5–16 are presented in Table 4. It is observed that the lowest error is obtained by the polynomial interpolation method at reference DC voltage of 16 V. Moreover, the mean percentage error of polynominal interpolation is lower than those of lookup table method for all tested cases. In terms of relative standard deviation, the two proposed methods present similar values, which means that a low number of points of the dynamic voltage response deviate from the desired value.

Table 4. Mean error and relative standard deviation of the tests.

v_b (V)	v_{dc} (V)	Mean Error (%)		Relative Standard Deviation (%)	
		Lookup Table	Polynomial Interp.	Lookup Table	Polynomial Interp.
12	10	1.33×10^{-5}	1.04×10^{-5}	0.5674	0.5720
	12	6.02×10^{-5}	4.69×10^{-5}	0.4632	0.4667
	16	4.91×10^{-6}	9.09×10^{-7}	0.3440	0.3460
24	20	1.05×10^{-5}	8.43×10^{-6}	0.1797	0.1812
	24	6.21×10^{-5}	5.93×10^{-5}	0.1443	0.1455
	26	6.75×10^{-6}	5.16×10^{-6}	0.1367	0.1379

4.2.2. Reference Tracking

To evaluate the tracking of the bus voltage references and thereby evaluate the robustness of the solutions to large changes in the bus voltage value, changes were made at a rate of 60 V/s regarding the battery voltage. First, the system starts in discharge mode with a bus voltage higher than the battery voltage, then the reference is changed to a voltage lower than the battery voltage, then returning back to the initial voltage. At 35 ms, the DC bus current is switched from 1 A to −1 A, which forces switching between the discharge and charge modes of the batteries. Finally, the test conditions are repeated to test the reference tracking in both charge and discharge modes, and in both boost and bock modes.

Figure 17 shows the simulation results for the lookup table method considering a singe battery and a reference profile changing between $v_{dc} = 16$ V and $v_{dc} = 10$ V. The tests show adequate tracking of the reference in all the conditions; however, unexpected small transients occur when the reference changes. This is caused by the sudden changes in the controller parameters when the operating point changes significantly; this is a consequence of the discretization introduced by the nearest method.

Next, the behavior of the lookup table solution is evaluated with the connection of two batteries in series, and in this case the voltage references changes between $v_{dc} = 20$ V and $v_{dc} = 26$ V. The results, reported in Figure 18, confirm that the controller ensures satisfactory tracking of the reference at any condition, but there are unexpected transients in the ramp zones, thus exhibiting the same discretization observed in the previous test, i.e., of a single battery.

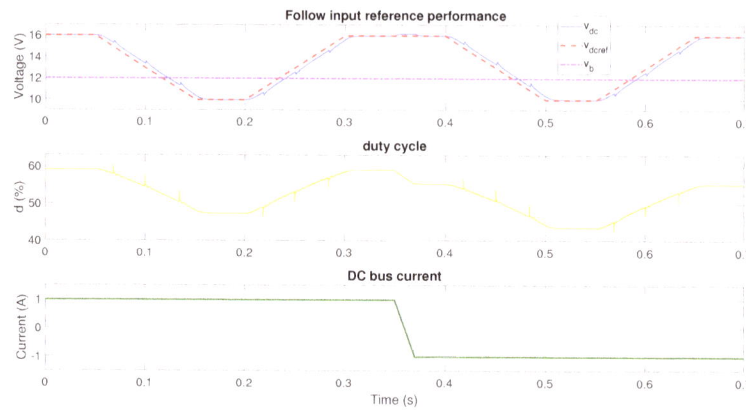

Figure 17. Reference tracking performance at $v_b = 12$ V using adaptive LQG with lookup table.

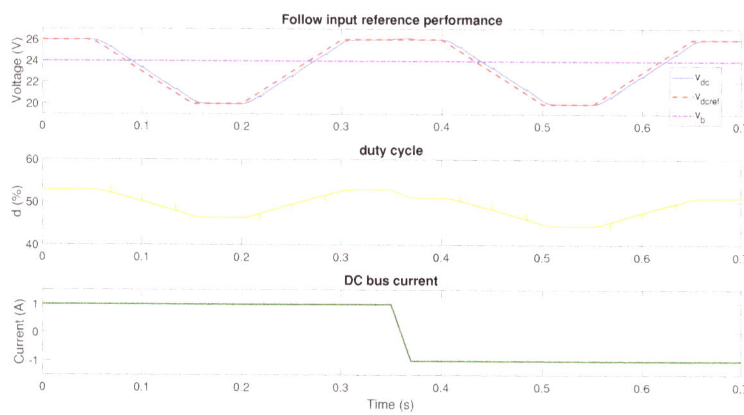

Figure 18. Reference tracking performance at $v_b = 24$ V using adaptive LQG with lookup table.

The next two simulations evaluate the performance of the adaptive method based on the polynomial interpolation. Figures 19 and 20 show the results for one and two batteries, respectively, where the polynomial interpolation method provides correct tracking of the reference in the different operation conditions. It is observed that, for this method, there are no unexpected transients in the ramp zones. This is achieved since the adaptive adjustment of the parameters correspond to non-discontinuous functions, unlike the lookup table method.

The tests performed on both offline adaptive methods, and reported in Figures 17–20, confirm the correct reference tracking under changes in the operating conditions. Finally, the following section validates the previous simulation results using an experimental prototype.

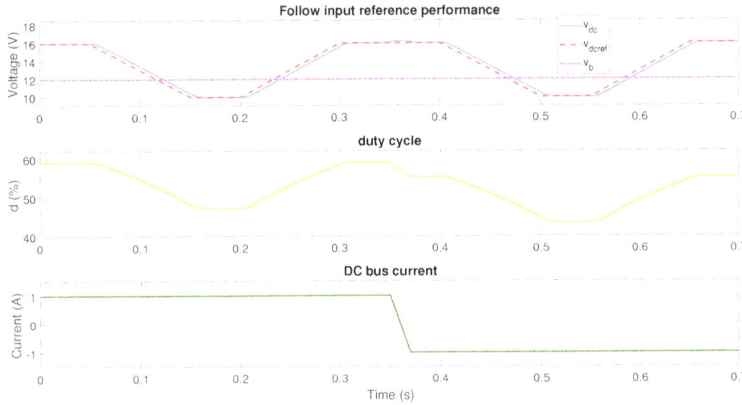

Figure 19. Reference tracking performance at $v_b = 12$ V using adaptive LQG with polynomial interpolation.

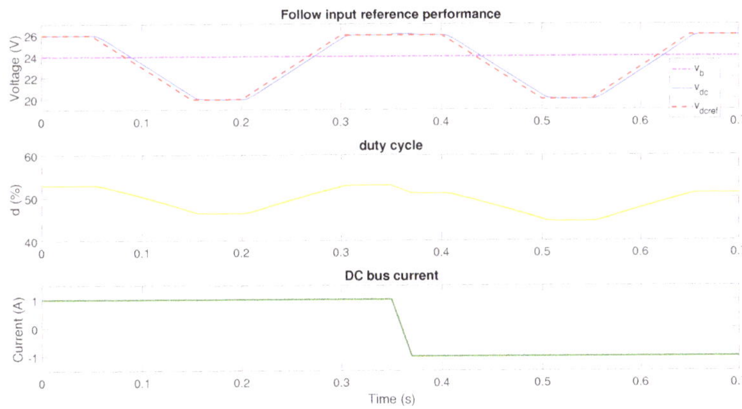

Figure 20. Reference tracking performance at $v_b = 24$ V using adaptive LQG with polynomial interpolation.

5. Experimental Validation

This section reports validations of the previous simulations using an experimental prototype of the Sepic/Zeta charger/discharger and adaptive control systems.

5.1. Experimental Implementation

The experimental scheme used to validate the proposed offline solutions is reported in Figure 21. The system includes one or two commercial batteries (12 V or 24 V), the Sepic/Zeta bidirectional converter, and a DC bus emulator constructed with a DC electronic load and electronic power supply, emulating the loads and sources connected to the microgrid bus, respectively. The battery and DC bus voltages are measured with LV25P sensors and signal conditioning circuits. The control card Delfino TMS320F28379D is used to implement the control system and adaptive method. It is worth noting that the control card uses an internal analog-to-digital converter (ADC) to digitize the measurements, and it uses an internal PWM module to produce the control signal for the MOSFETS drivers.

The experimental implementation of both the adaptive controller and charger/discharger is shown in Figure 22, where the Sepic/Zeta converter, sensors, and control card are highlighted. Figure 23 shows the complete experimental test bench, which also exhibits the commercial batteries and the practical implementation of the DC bus emulator.

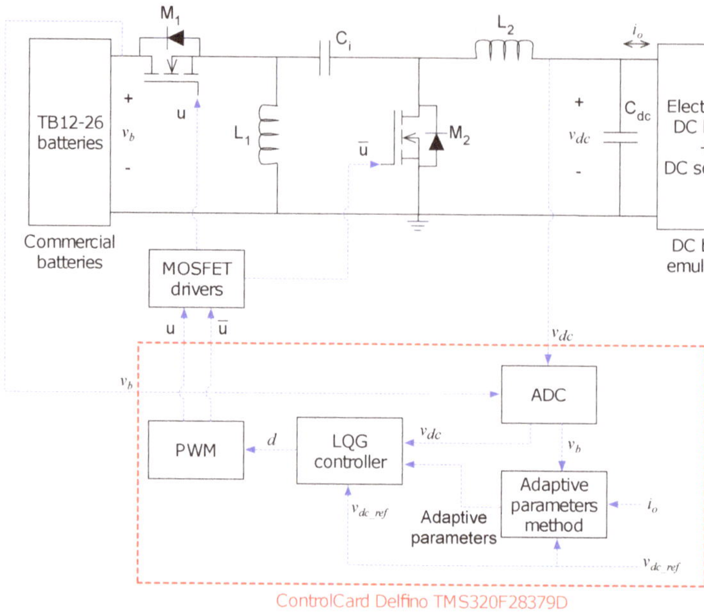

Figure 21. Scheme of the experimental test.

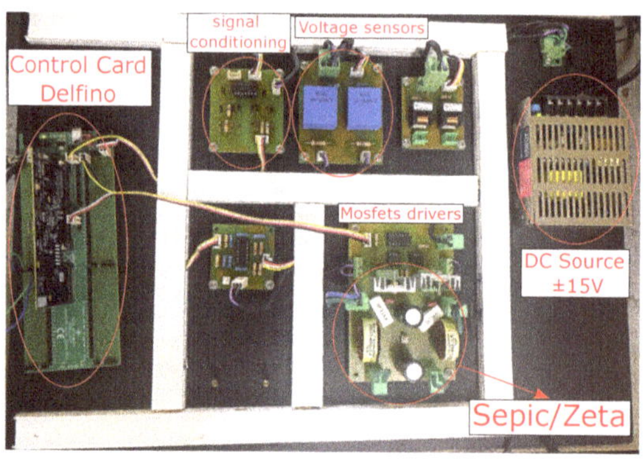

Figure 22. Experimental implementation.

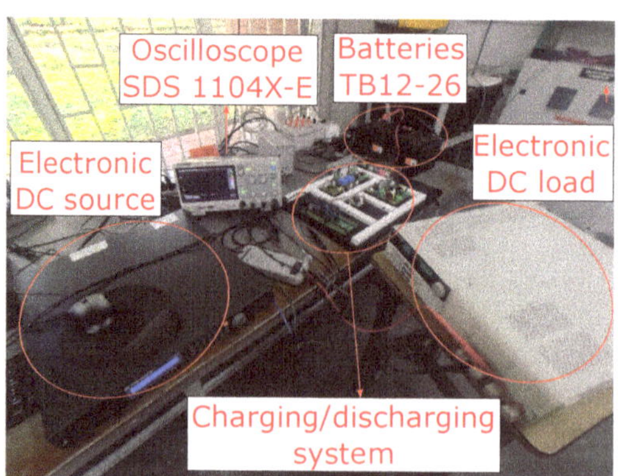

Figure 23. Experimental test bench.

5.2. Experimental Validation of the Bus Voltage Regulation

The same test conditions adopted for the simulations are implemented in this experimental validation. Thus, the DC bus current profile includes increments and decrements and shifts between charge, discharge, and standby modes. Figure 24a,b show the performance of the adaptive LQG with both the lookup table and polynomial interpolation methods. In this experiment, the bus voltage is regulated at 10 V (buck mode) for the charging, discharging, and standby modes of the battery. These experiments demonstrate the correct operation of the two offline parameter adjustment methods in terms of regulation, and no appreciable differences are observed between them in terms of their performance.

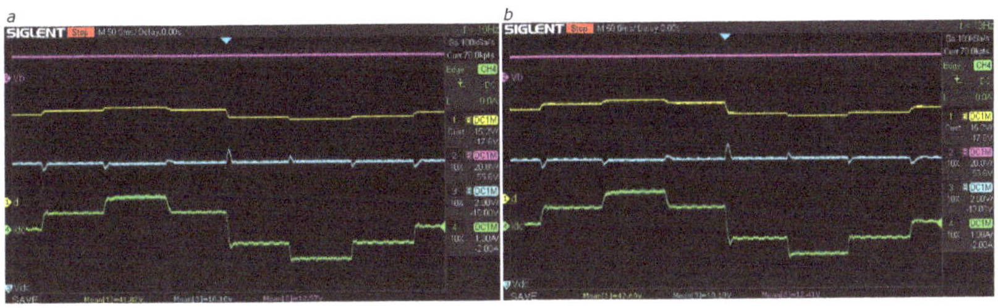

Figure 24. Bus regulation with v_b = 12 V and reference $v_{dc_{ref}}$ = 10 V (buck mode) using a LQG by (**a**) lookup table and (**b**) polynomial interpolation. Magenta: battery voltage v_b; yellow: duty cycle d; cyan: bus voltage v_{dc}; green: bus current i_o.

The previous tests were executed with bus voltages equal to 12 V (unitary gain mode) and 16 V (boost mode). The experimental data, reported in Figures 25 and 26 for both offline methods, confirm the adequate performance under different operation conditions of the DC bus current (green trace) and for both unitary gain and boost modes.

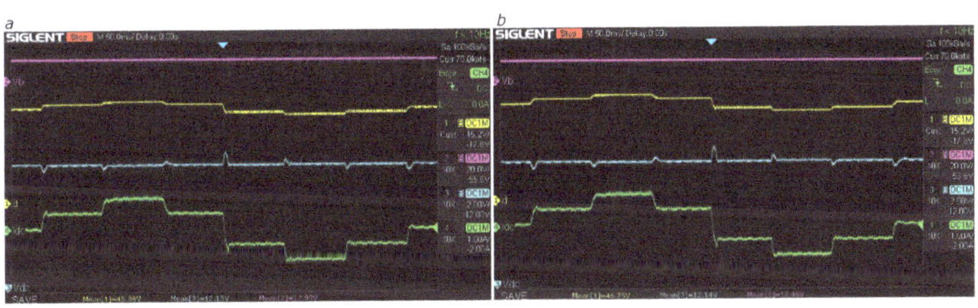

Figure 25. Bus regulation with v_b = 12 V and reference $v_{dc_{ref}}$ = 12 V (unitary gain mode) using adaptive LQG by (**a**) lookup table and (**b**) polynomial interpolation. Magenta: battery voltage v_b; yellow: duty cycle d; cyan: bus voltage v_{dc}; green: bus current i_o.

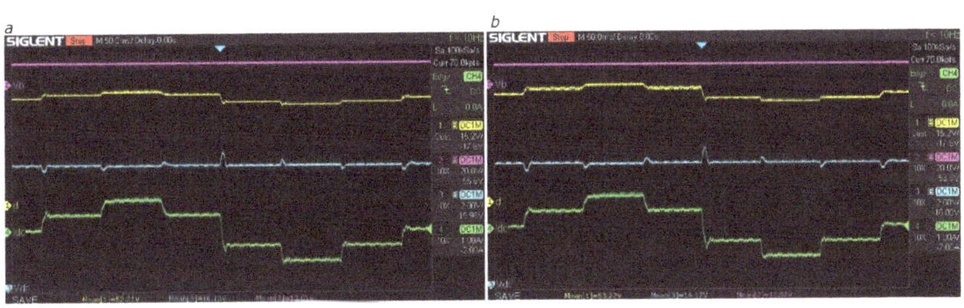

Figure 26. Bus regulation with v_b = 12 V and reference $v_{dc_{ref}}$ = 16 V (boost mode) using adaptive LQG by (**a**) lookup table and (**b**) polynomial interpolation. Magenta: battery voltage v_b; yellow: duty cycle d; cyan: bus voltage v_{dc}; green: bus current i_o.

In the same way, Figures 27–29 report a good performance in terms of DC bus regulation at 20 V (buck mode), 24 V (unitary gain mode) and 26 V (boost mode), respectively, considering two series-connected commercial batteries ($v_b = 24$ V). Therefore, the robustness of both offline methods is confirmed for large changes in the battery voltage.

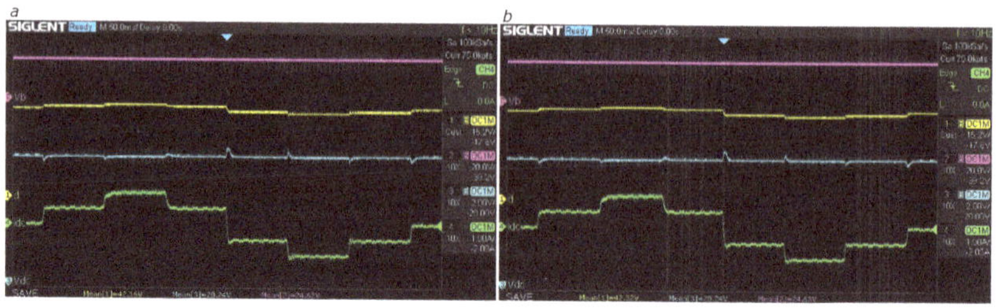

Figure 27. Bus regulation with $v_b = 24$ V and reference $v_{dc_{ref}} = 20$ V (buck mode) using adaptive LQG by (**a**) lookup table and (**b**) polynomial interpolation. Magenta: battery voltage v_b; yellow: duty cycle d; cyan: bus voltage v_{dc}; green: bus current i_o.

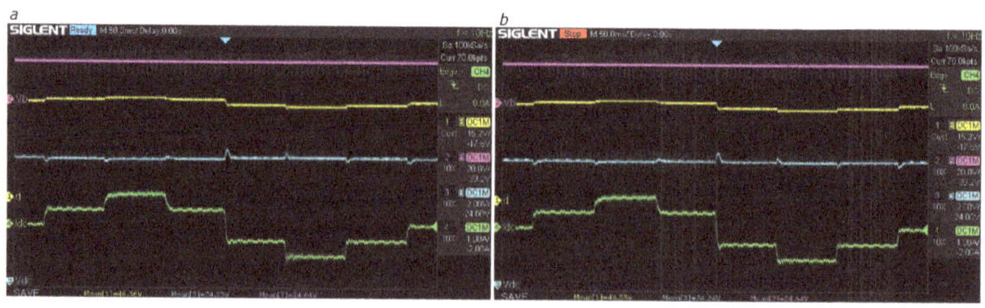

Figure 28. Bus regulation with $v_b = 24$ V and reference $v_{dc_{ref}} = 24$ V (unitary gain mode) using adaptive LQG by (**a**) lookup table and (**b**) polynomial interpolation. Magenta: battery voltage v_b; yellow: duty cycle d; cyan: bus voltage v_{dc}; green: bus current i_o.

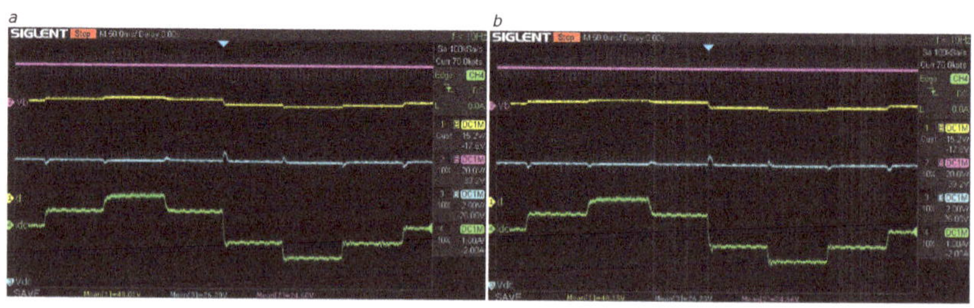

Figure 29. Bus regulation with $v_b = 24$ V and reference $v_{dc_{ref}} = 26$ V (boost mode) using adaptive LQG by (**a**) lookup table and (**b**) polynomial interpolation. Magenta: battery voltage v_b; yellow: duty cycle d; cyan: bus voltage v_{dc}; green: bus current i_o.

In general, in the experimental data concerning DC bus voltage regulation, both offline adaptation methods exhibit correct performance under charge, discharge, and standby conditions under different voltage conversion ratios (buck, boost, and unitary gain modes). In addition, the voltage variation of the battery pack enable the stability of both solutions at different operating points to be confirmed.

Table 5 summarizes the performance criteria (overshoot and settling time) obtained with the proposed adaptive offline methods. For comparative purposes, the adaptive online LQG approach presented in [38] was also implemented and experimentally tested, and its performance criteria are included in Table 5. The results with a single battery ($v_b = 12$ V) show that the adaptive online LQG exhibits lower overshoot and shorter stabilization time than the offline methods. However, in the test considering two batteries connected in series

($v_b = 24$ V), the proposed offline methods provide better dynamic performance, except for the lookup table method in standby mode. Those experiments confirm that the proposed offline methods are a suitable option to adapt the LQG parameters in practical application, since the difference with the online option is small.

Table 5. Dynamic performance in bus voltage regulation tests.

Test Conditions		Overshoot (%)			Settling Time (ms)		
v_b (V)	v_{dc} (V)	Lookup Table	Polynomial Interp.	LQG Online	Lookup Table	Polynomial Interp.	LQG Online
12	10	9.28	9.45	8.75	9.25	9.16	9.00
	12	8.07	7.91	7.00	9.03	8.93	8.80
	16	6.48	6.09	5.85	9.19	8.85	8.70
24	20	2.85	2.80	3.16	6.04	6.47	7.00
	24	2.68	2.33	2.63	6.37	5.80	6.43
	26	2.30	2.23	2.63	5.31	4.76	6.00

5.3. Experimental Validation of Reference Tracking

Several experiments were conducted to validate the adaptability of the offline methods to changes in the operating point by considering dynamic changes in the reference value. The changes in the bus reference are given in such a way that the converter changes between buck and boost modes. In the same way, the tests were repeated for the charge, discharge, and standby modes of the batteries. The experimental data of the reference tracking, for both lookup table and polynomial interpolation methods, are shown in the Figure 30, where a single battery ($v_b = 12$ V) is considered in those experiments. The proposed offline adaptive methods guarantee correct tracking of the references under different operating points for different modes of the converter (buck and boost modes).

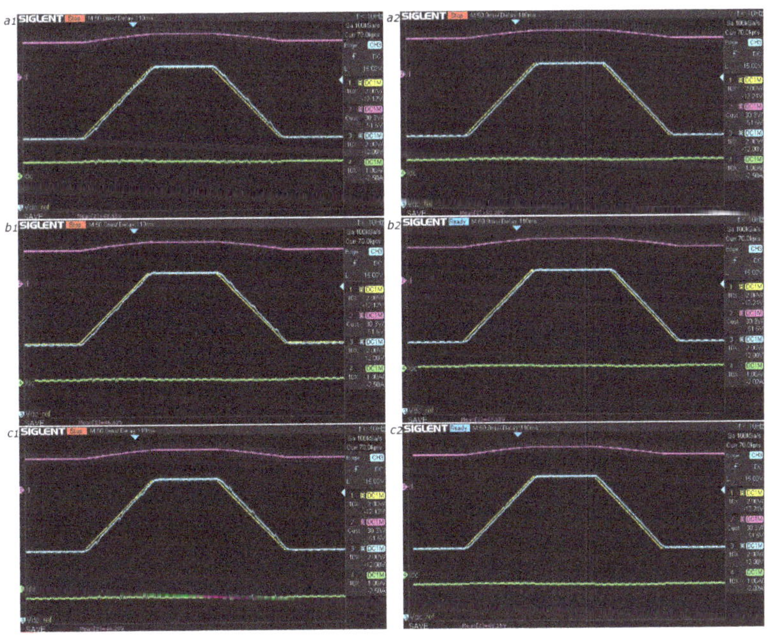

Figure 30. Reference tracking performance at $v_b = 12$ V using adaptive LQG with lookup table: (**a1**) discharge mode, (**b1**) standby mode, and (**c1**) charge mode; and with polynomial interpolation: (**a2**) discharge mode, (**b2**) standby mode, and (**c2**) charge mode. Magenta: duty cycle d; yellow: bus voltage reference v_{dcref}; cyan: bus voltage v_{dc}; green: bus current i_o.

In addition, the tracking performance was evaluated with two batteries connected in series ($v_b = 24$ V). For this experiment, the bus voltage reference starts at 20 V (buck mode), changes to 26 V (boost mode), and finally returns to 20 V. Figure 31 shows the experimental behavior of both proposed offline methods; similarly to the previous experiments, the results indicate the correct performance of the adaptive methods for different operating points.

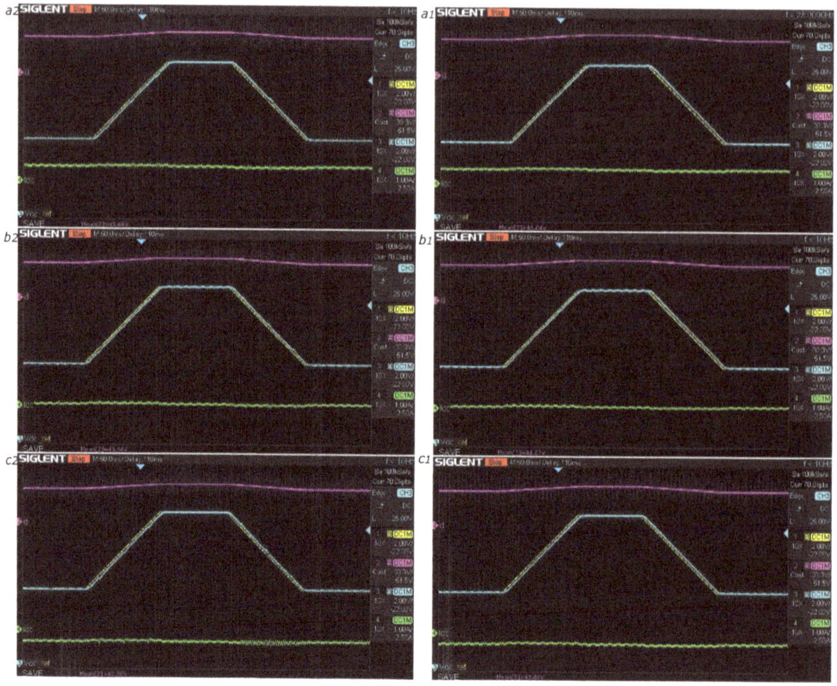

Figure 31. Reference tracking performance at $v_b = 24$ V using adaptive LQG with lookup table: (**a1**) discharge mode, (**b1**) standby mode, and (**c1**) charge mode; and with polynomial interpolation: (**a2**) discharge mode, (**b2**) standby mode, and (**c2**) charge mode. Magenta: duty cycle d; yellow: bus voltage reference v_{dcref}; cyan: bus voltage v_{dc}; green: bus current i_o.

In agreement with the simulation results, the lookup table method exhibits unexpected voltage transients when the reference voltage changes continuously, which is caused by the discretization of the method. However, these transients are more notable for a battery of 12 V, since the bus voltage is lower. This phenomenon can be reduced by increasing the resolution of the data stored in the lookup table; however, this increases the memory required in the control device and, also, the parametrization time for the method.

5.4. *Performance and Cost Comparison*

Since the main objective of the proposed offline methods is to reduce the computational resources required for the implementation of the adaptive LQG controller, this section evaluates the computational burden of the new solutions.

The control card TMS320F28379D has two cores, each providing 400 millions of instructions per second (MIPS), which means that the total capacity of the card is 800 MIPS. In the implementation, each adaptive offline proposed method is required to use a core at a time. Using the Matlab® report, the processor load for each of the adaptive solutions was extracted, which includes the online approach reported in [38]. This information is summarized in Table 6, where the two proposed (offline) solutions require almost the same

CPU utilization (MIPS required). Instead, the online method [38] requires between 7 and 8 times the MIPS for the adaptive process.

Table 6. Overall CPU utilization.

Adaptive Method	Max CPU Utilization (MIPS)
Lookup table	94
Polynomial interpolation	93
Online LQG	714

Taking into consideration the maximum MIPS required by proposed method, a suitable control card available in the market is selected to evaluate the implementation cost. Control cards from TI C2000 MCU series are suitable due to the available PWM, ADC modules, and high processing capacity. For example, the TMS320F2800157 control card provides 120 MIPS, which is enough to implement the adaptive offline methods. Moreover, this control card also provides the additional hardware required to execute the control strategy for the battery charger/discharger, such as programming ports and communication ports, among others. The cost of this card is USD 59; instead, the control card required to implement the online method (with 714 MIPS required) is the TMS320F28379D, which costs USD 159 and is thus significantly more expensive.

The previous practical considerations were analyzed by proposing three indices: designed CPU capacity index (I_p), cost index (I_{cost}) and performance index (I_{cpu}). The I_p index is calculated from the performance data reported in Table 5 using Equation (17), where a higher I_p implies greater performance. In such an expression, $E_{x(i)}$ is the overshoot or stabilization time, and $E_{ref(i)}$ corresponds to the lowest value of the data under comparison. In this way, the solution with the lower overshoot or shortest stabilization time will always have $I_p(i) = 1$, and the other solutions will have lower I_p value. Finally, the $I_p(i)$ values for each method are averaged to obtain a general performance index I_p for each adaptive method.

$$I_p(i) = \frac{E_{x(i)}^{-1}}{E_{ref(i)}^{-1}} \tag{17}$$

The I_{cpu} index uses the data reported in Table 6 and Equation (18), where the higher I_{cpu}, the lower the required MIPS. In (18), cpu_x and cpu_{ref} correspond to the MIPS required by the solution and the minimum CPU consumption of methods under comparison, respectively. In this way, the method with the lower MIPS requirement will have $I_{cpu} = 1$, while the other methods will exhibit a lower I_{cpu} value.

$$I_{cpu} = \frac{cpu_x^{-1}}{cpu_{ref}^{-1}} \tag{18}$$

Finally, the I_{cost} index is given in Equation (19), where $cost_x$ is the price of the corresponding control card, and $cost_{ref}$ is the price of the cheaper control card for all the implementations. As in the previous cases, the higher the index, the lower the required cost, where the solution with the lower cost will have an index equal to 1.

$$I_{cost} = \frac{cost_x^{-1}}{cost_{ref}^{-1}} \tag{19}$$

The results of applying three proposed indices to the solutions under comparison are reported in Table 7. A higher index value of I_p implies a better dynamic performance, a higher value of I_{cpu} indicates lower computational requirement, and a higher value of I_{cost} means lower cost of the control card required to implement the adaptive method. It is observed that the polynomial interpolation method exhibits a better overall performance, also requiring the lowest CPU consumption and cost; therefore, that method is the best alternative for commercial implementation of the adaptive LQG controller for battery charger/dischargers.

Table 7. Comparative summary of the indices.

Adaptive Method	Suitable Control Card	Performance Index	CPU Utilization Index	Cost Index
Lookup table	TMDSCNCD2800157	0.934	0.989	1
Polynomial interpolation	TMDSCNCD2800157	0.970	1	1
Online LQG	TMDSCNCD28379D	0.946	0.13	0.371

The previous data can be further processed to contrast both the offline (proposed) and online solutions. For this purpose, combined indicators are proposed: the performance–cost, which adds both I_p and I_{cost}; and the performance–CPU, which adds both I_p and I_{cpu}. Table 8 reports both combined indicators, where the polynomial interpolation method provides the best relation between performance of the charger/discharger and implementation cost. The second best method in this category is the lookup table, and the last is the online method reported in [38]. Similarly, the polynomial interpolation method provides the best relation between performance and CPU utilization, the second best method is the lookup table, and the last, again, is the online method. This last category is interesting, since it enables analyzing which method will leave higher resources, thus providing computing power for other tasks such as data logging, cloud storage, state of charge management, etc.

Table 8. Advantage comparison.

Adaptive Method	Performance–Cost Relation	Performance–CPU Relation
Lookup table	1.93	1.92
Polynomial interpolation	1.97	1.97
Online LQG	1.32	1.08

Finally, this practical analysis confirms that both offline solutions provide an acceptable performance with much lower implementation costs. In particular, the polynomial interpolation method provides the best tradeoff and is thus the best solution for commercial applications.

6. Conclusions

Two adaptive offline methods for adjusting the parameters of a LQG controller and regulating the DC bus of a battery charger/discharger are proposed. The charger/discharger is based on the Sepic/Zeta converter, and the parameters of the controller are determined taking into account the conditions of both the batteries and converter. For the former, conditions may vary between charge, discharge, or standby modes. Moreover, depending on the voltage requirements of the input and output, the converter can be in boost, buck, or equal gain mode. For the LQG controller, a state feedback vector and an observer gain vector must be determined in an adaptive way. The first method consists of calculating the parameters of vectors for several operating points of load and input reference voltage, and those are stored in a lookup table. In the second method, an interpolation is developed among the values of the lookup table. In simulation and experimental results, for several operating conditions, the control signal remains far away from saturation, and the control system is stable, thus providing satisfactory results.

Three indices have been proposed to evaluate the performance of the proposed adaptive methods, which are suitable to evaluate dynamic performance, CPU capacity, and cost. These indices show that the lookup table and polynomial interpolation methods present similar dynamic performance and CPU requirements, but the combined indices show that polynomial interpolation is the best option for practical and commercial application. In addition, compared with the lookup table method, that method exhibits satisfactory performance over the entire operating range, and it does not introduce unexpected transients under any operating conditions. Moreover, polynomial interpolation requires low computational consumption in contrast to the online LQG method, which means considerably lower implementation costs.

The main drawback of both proposed offline solutions is the need for an offline training campaign, which is not needed in the online alternative already reported in the literature. However, with the aim of developing low-cost solutions for deployment in commercial applications, it could be acceptable to spend the time required for training if there is significant reduction of the implementation cost. Finally, for future improvement, evaluating more deployment-oriented platforms for implementation, such as FPGA or ASIC devices, could provide a much better cost/performance ratio.

Author Contributions: Conceptualization, J.A.M.-O., C.A.R.-P., M.L.O.-G., E.F.-M. and S.I.S.-G.; methodology, J.A.M.-O., C.A.R.-P., M.L.O.-G., E.F.-M. and S.I.S.-G.; software, J.A.M.-O., C.A.R.-P., M.L.O.-G., E.F.-M. and S.I.S.-G.; validation, J.A.M.-O., C.A.R.-P., M.L.O.-G., E.F.-M. and S.I.S.-G.; writing—original draft preparation, J.A.M.-O., C.A.R.-P., M.L.O.-G., E.F.-M. and S.I.S.-G.; writing—review and editing, J.A.M.-O., C.A.R.-P., M.L.O.-G., E.F.-M. and S.I.S.-G. All authors have read and agreed to the published version of the manuscript.

Funding: This research was funded by Minciencias, Universidad del Valle, Universidad Nacional de Colombia, and Instituto Tecnológico Metropolitano under the research project "Métodos y Herramientas para el Diseño, Dimensionamiento y Control de Sistemas de Almacenamiento de Energía en sistemas eléctricos basados en fuentes renovables no convencionales, para zonas urbanas y rurales", (Minciencias code 70634, Sicop Code C.I. 21154), which belongs to the research program "Estrategias para el desarrollo de sistemas energéticos sostenibles, confiables, eficientes y accesibles para el futuro de Colombia", (Minciencias code 1150-852-70378, Hermes code 46771, Sicop codes C.I. 21154, 21173, 21174, ITM code RC 80740-178-2021-1).

Institutional Review Board Statement: Not applicable.

Informed Consent Statement: Not applicable.

Data Availability Statement: The data used in this study are reported in the paper figures and tables.

Conflicts of Interest: The authors declare no conflict of interest.

Appendix A

This appendix reports the lookup table data used in the first proposed offline method. The data are distributed in eight tables, one for each parameter of both the controller and observer.

Table A1. Lookup table K_1.

$v_{dc}(V)$ \ $v_b(V)$	10	12	14	16	18	20	22	24	26	28
8	0.02513	0.02462	0.02421	0.02388	0.02359	0.02334	0.02311	0.02289	0.02268	0.02248
10	0.02569	0.02520	0.02480	0.02447	0.02419	0.02394	0.02371	0.02350	0.02330	0.02310
12	0.02600	0.02555	0.02518	0.02487	0.02461	0.02438	0.02417	0.02397	0.02378	0.02359
14	0.02614	0.02573	0.02541	0.02514	0.02490	0.02470	0.02450	0.02432	0.02415	0.02398
16	0.02618	0.02582	0.02554	0.02530	0.02510	0.02492	0.02475	0.02459	0.02443	0.02428
18	0.02616	0.02584	0.02560	0.02540	0.02522	0.02507	0.02492	0.02478	0.02465	0.02452
20	0.02612	0.02582	0.02561	0.02544	0.02529	0.02516	0.02504	0.02493	0.02481	0.02470
22	0.02606	0.02578	0.02559	0.02545	0.02533	0.02522	0.02512	0.02503	0.02493	0.02483
24	0.02600	0.02572	0.02555	0.02543	0.02533	0.02525	0.02517	0.02509	0.02501	0.02493
26	0.02594	0.02566	0.02550	0.02539	0.02532	0.02525	0.02519	0.02513	0.02507	0.02500
28	0.02588	0.02559	0.02544	0.02535	0.02529	0.02524	0.02519	0.02515	0.02510	0.02505

Table A2. Lookup table K_2.

$v_b(V)$ / $v_{dc}(V)$	10	12	14	16	18	20	22	24	26	28
8	0.06072	0.06116	0.06110	0.06075	0.06025	0.05967	0.05905	0.05842	0.05780	0.05720
10	0.05981	0.06010	0.05996	0.05956	0.05903	0.05843	0.05779	0.05716	0.05653	0.05593
12	0.05888	0.05906	0.05886	0.05844	0.05789	0.05728	0.05664	0.05601	0.05538	0.05478
14	0.05796	0.05807	0.05782	0.05738	0.05682	0.05621	0.05558	0.05495	0.05433	0.05373
16	0.05708	0.05712	0.05684	0.05639	0.05583	0.05522	0.05460	0.05397	0.05336	0.05277
18	0.05622	0.05623	0.05593	0.05546	0.05490	0.05430	0.05369	0.05307	0.05247	0.05189
20	0.05541	0.05538	0.05506	0.05459	0.05404	0.05345	0.05284	0.05224	0.05165	0.05107
22	0.05462	0.05458	0.05425	0.05378	0.05323	0.05265	0.05205	0.05146	0.05088	0.05032
24	0.05387	0.05381	0.05348	0.05301	0.05247	0.05190	0.05131	0.05073	0.05016	0.04961
26	0.05314	0.05309	0.05276	0.05229	0.05176	0.05119	0.05062	0.05005	0.04949	0.04895
28	0.05244	0.05240	0.05207	0.05161	0.05109	0.05053	0.04997	0.04941	0.04886	0.04833

Table A3. Lookup table K_3.

$v_b(V)$ / $v_{dc}(V)$	10	12	14	16	18	20	22	24	26	28
8	0.00814	0.00790	0.00768	0.00754	0.00748	0.00748	0.00753	0.00763	0.00775	0.00789
10	0.00845	0.00815	0.00787	0.00766	0.00753	0.00746	0.00746	0.00750	0.00758	0.00768
12	0.00862	0.00831	0.00799	0.00774	0.00756	0.00745	0.00740	0.00740	0.00744	0.00750
14	0.00868	0.00838	0.00806	0.00778	0.00757	0.00743	0.00735	0.00731	0.00732	0.00736
16	0.00865	0.00839	0.00807	0.00779	0.00756	0.00740	0.00730	0.00724	0.00722	0.00724
18	0.00855	0.00834	0.00804	0.00776	0.00754	0.00736	0.00724	0.00717	0.00713	0.00713
20	0.00839	0.00824	0.00798	0.00771	0.00749	0.00731	0.00718	0.00709	0.00704	0.00703
22	0.00819	0.00812	0.00789	0.00764	0.00743	0.00725	0.00711	0.00702	0.00696	0.00693
24	0.00796	0.00797	0.00778	0.00756	0.00735	0.00718	0.00704	0.00694	0.00688	0.00684
26	0.00770	0.00780	0.00765	0.00746	0.00726	0.00710	0.00697	0.00687	0.00680	0.00675
28	0.00742	0.00762	0.00752	0.00734	0.00717	0.00701	0.00689	0.00679	0.00672	0.00667

Table A4. Lookup table K_4.

$v_b(V)$ / $v_{dc}(V)$	10	12	14	16	18	20	22	24	26	28
8	0.04164	0.04310	0.04443	0.04562	0.04666	0.04758	0.04838	0.04910	0.04973	0.05029
10	0.04490	0.04608	0.04717	0.04815	0.04902	0.04979	0.05047	0.05107	0.05160	0.05207
12	0.04762	0.04860	0.04950	0.05032	0.05105	0.05169	0.05227	0.05278	0.05323	0.05364
14	0.04990	0.05073	0.05149	0.05218	0.05280	0.05335	0.05384	0.05428	0.05467	0.05501
16	0.05185	0.05256	0.05321	0.05380	0.05433	0.05480	0.05523	0.05560	0.05593	0.05623
18	0.05351	0.05415	0.05471	0.05522	0.05567	0.05608	0.05645	0.05677	0.05706	0.05732
20	0.05495	0.05553	0.05603	0.05647	0.05687	0.05722	0.05754	0.05782	0.05807	0.05829
22	0.05620	0.05675	0.05719	0.05758	0.05793	0.05824	0.05852	0.05876	0.05898	0.05917
24	0.05730	0.05782	0.05822	0.05857	0.05888	0.05915	0.05940	0.05961	0.05980	0.05997
26	0.05827	0.05878	0.05915	0.05946	0.05974	0.05998	0.06019	0.06038	0.06055	0.06070
28	0.05913	0.05964	0.05998	0.06027	0.06051	0.06073	0.06092	0.06109	0.06123	0.06136

Table A5. Lookup table ℓ_1.

$v_b(V)$ / $v_{dc}(V)$	10	12	14	16	18	20	22	24	26	28
8	7.26×10^3	8.08×10^3	8.88×10^3	9.68×10^3	1.05×10^4	1.13×10^4	1.21×10^4	1.29×10^4	1.37×10^4	1.46×10^4
10	8.11×10^3	8.96×10^3	9.78×10^3	1.06×10^4	1.14×10^4	1.22×10^4	1.30×10^4	1.39×10^4	1.47×10^4	1.55×10^4
12	8.95×10^3	9.82×10^3	1.07×10^4	1.15×10^4	1.23×10^4	1.31×10^4	1.40×10^4	1.48×10^4	1.56×10^4	1.65×10^4
14	9.77×10^3	1.07×10^4	1.15×10^4	1.24×10^4	1.32×10^4	1.41×10^4	1.49×10^4	1.57×10^4	1.66×10^4	1.74×10^4
16	1.06×10^4	1.15×10^4	1.24×10^4	1.33×10^4	1.41×10^4	1.49×10^4	1.58×10^4	1.66×10^4	1.75×10^4	1.83×10^4
18	1.14×10^4	1.24×10^4	1.32×10^4	1.41×10^4	1.50×10^4	1.58×10^4	1.67×10^4	1.75×10^4	1.84×10^4	1.92×10^4
20	1.22×10^4	1.32×10^4	1.41×10^4	1.50×10^4	1.59×10^4	1.67×10^4	1.76×10^4	1.84×10^4	1.93×10^4	2.01×10^4
22	1.30×10^4	1.40×10^4	1.49×10^4	1.58×10^4	1.67×10^4	1.76×10^4	1.85×10^4	1.93×10^4	2.02×10^4	2.10×10^4
24	1.38×10^4	1.49×10^4	1.58×10^4	1.67×10^4	1.76×10^4	1.85×10^4	1.93×10^4	2.02×10^4	2.11×10^4	2.19×10^4
26	1.47×10^4	1.57×10^4	1.66×10^4	1.76×10^4	1.85×10^4	1.93×10^4	2.02×10^4	2.11×10^4	2.20×10^4	2.28×10^4
28	1.55×10^4	1.65×10^4	1.75×10^4	1.84×10^4	1.93×10^4	2.02×10^4	2.11×10^4	2.20×10^4	2.28×10^4	2.37×10^4

Table A6. Lookup table ℓ_2.

$v_{dc}(V)$ \ $v_b(V)$	10	12	14	16	18	20	22	24	26	28
8	5.51×10^3	6.46×10^3	7.38×10^3	8.28×10^3	9.18×10^3	1.01×10^4	1.09×10^4	1.18×10^4	1.27×10^4	1.36×10^4
10	6.19×10^3	7.17×10^3	8.12×10^3	9.05×10^3	9.96×10^3	1.09×10^4	1.18×10^4	1.26×10^4	1.35×10^4	1.44×10^4
12	6.87×10^3	7.89×10^3	8.87×10^3	9.81×10^3	1.07×10^4	1.17×10^4	1.26×10^4	1.35×10^4	1.44×10^4	1.52×10^4
14	7.57×10^3	8.62×10^3	9.61×10^3	1.06×10^4	1.15×10^4	1.24×10^4	1.34×10^4	1.43×10^4	1.52×10^4	1.61×10^4
16	8.27×10^3	9.35×10^3	1.04×10^4	1.13×10^4	1.23×10^4	1.32×10^4	1.42×10^4	1.51×10^4	1.60×10^4	1.69×10^4
18	8.97×10^3	1.01×10^4	1.11×10^4	1.21×10^4	1.31×10^4	1.40×10^4	1.50×10^4	1.59×10^4	1.68×10^4	1.77×10^4
20	9.68×10^3	1.08×10^4	1.19×10^4	1.29×10^4	1.39×10^4	1.48×10^4	1.58×10^4	1.67×10^4	1.76×10^4	1.86×10^4
22	1.04×10^4	1.16×10^4	1.26×10^4	1.37×10^4	1.47×10^4	1.56×10^4	1.66×10^4	1.75×10^4	1.85×10^4	1.94×10^4
24	1.11×10^4	1.23×10^4	1.34×10^4	1.44×10^4	1.55×10^4	1.64×10^4	1.74×10^4	1.83×10^4	1.93×10^4	2.02×10^4
26	1.18×10^4	1.31×10^4	1.42×10^4	1.52×10^4	1.62×10^4	1.72×10^4	1.82×10^4	1.92×10^4	2.01×10^4	2.10×10^4
28	1.26×10^4	1.38×10^4	1.50×10^4	1.60×10^4	1.70×10^4	1.80×10^4	1.90×10^4	2.00×10^4	2.09×10^4	2.19×10^4

Table A7. Lookup table ℓ_3.

$v_{dc}(V)$ \ $v_b(V)$	10	12	14	16	18	20	22	24	26	28
8	-1.06×10^3	-3.46×10^2	3.00×10^2	8.92×10^2	1.44×10^3	1.95×10^3	2.43×10^3	2.89×10^3	3.32×10^3	3.73×10^3
10	-1.92×10^3	-1.17×10^3	-4.95×10^2	1.20×10^2	6.89×10^2	1.22×10^3	1.72×10^3	2.19×10^3	2.63×10^3	3.06×10^3
12	-2.74×10^3	-1.95×10^3	-1.25×10^3	-6.12×10^2	-2.54×10^1	5.20×10^2	1.03×10^3	1.52×10^3	1.97×10^3	2.41×10^3
14	-3.53×10^3	-2.70×10^3	-1.97×10^3	-1.31×10^3	-7.07×10^2	-1.46×10^2	3.79×10^2	8.74×10^2	1.34×10^3	1.79×10^3
16	-4.29×10^3	-3.41×10^3	-2.66×10^3	-1.98×10^3	-1.36×10^3	-7.85×10^2	-2.48×10^2	2.57×10^2	7.37×10^2	1.19×10^3
18	-5.03×10^3	-4.11×10^3	-3.32×10^3	-2.62×10^3	-1.99×10^3	-1.40×10^3	-8.52×10^2	-3.36×10^2	1.52×10^2	6.16×10^2
20	-5.75×10^3	-4.78×10^3	-3.96×10^3	-3.24×10^3	-2.59×10^3	-1.99×10^3	-1.43×10^3	-9.08×10^2	-4.12×10^2	5.99×10^1
22	-6.46×10^3	-5.43×10^3	-4.58×10^3	-3.84×10^3	-3.18×10^3	-2.56×10^3	-2.00×10^3	-1.46×10^3	-9.58×10^2	-4.79×10^2
24	-7.15×10^3	-6.07×10^3	-5.19×10^3	-4.42×10^3	-3.74×10^3	-3.12×10^3	-2.54×10^3	-2.00×10^3	-1.49×10^3	-1.00×10^3
26	-7.84×10^3	-6.69×10^3	-5.77×10^3	-4.99×10^3	-4.29×10^3	-3.66×10^3	-3.07×10^3	-2.52×10^3	-2.00×10^3	-1.51×10^3
28	-8.52×10^3	-7.30×10^3	-6.35×10^3	-5.54×10^3	-4.83×10^3	-4.18×10^3	-3.58×10^3	-3.02×10^3	-2.50×10^3	-2.00×10^3

Table A8. Lookup table ℓ_4.

$v_{dc}(V)$ \ $v_b(V)$	10	12	14	16	18	20	22	24	26	28
8	5.78×10^3	6.26×10^3	6.69×10^3	7.09×10^3	7.46×10^3	7.81×10^3	8.15×10^3	8.47×10^3	8.77×10^3	9.07×10^3
10	6.12×10^3	6.59×10^3	7.02×10^3	7.41×10^3	7.77×10^3	8.11×10^3	8.44×10^3	8.75×10^3	9.06×10^3	9.35×10^3
12	6.45×10^3	6.92×10^3	7.33×10^3	7.71×10^3	8.07×10^3	8.40×10^3	8.72×10^3	9.03×10^3	9.33×10^3	9.61×10^3
14	6.77×10^3	7.23×10^3	7.63×10^3	8.01×10^3	8.36×10^3	8.69×10^3	9.00×10^3	9.30×10^3	9.59×10^3	9.87×10^3
16	7.08×10^3	7.53×10^3	7.93×10^3	8.29×10^3	8.63×10^3	8.96×10^3	9.27×10^3	9.56×10^3	9.85×10^3	1.01×10^4
18	7.37×10^3	7.82×10^3	8.21×10^3	8.57×10^3	8.91×10^3	9.22×10^3	9.53×10^3	9.82×10^3	1.01×10^4	1.04×10^4
20	7.66×10^3	8.10×10^3	8.48×10^3	8.84×10^3	9.17×10^3	9.48×10^3	9.78×10^3	1.01×10^4	1.03×10^4	1.06×10^4
22	7.94×10^3	8.37×10^3	8.75×10^3	9.10×10^3	9.43×10^3	9.73×10^3	1.00×10^4	1.03×10^4	1.06×10^4	1.08×10^4
24	8.21×10^3	8.64×10^3	9.01×10^3	9.36×10^3	9.68×10^3	9.98×10^3	1.03×10^4	1.05×10^4	1.08×10^4	1.11×10^4
26	8.47×10^3	8.90×10^3	9.27×10^3	9.61×10^3	9.92×10^3	1.02×10^4	1.05×10^4	1.08×10^4	1.10×10^4	1.13×10^4
28	8.72×10^3	9.15×10^3	9.52×10^3	9.85×10^3	1.02×10^4	1.05×10^4	1.07×10^4	1.10×10^4	1.13×10^4	1.15×10^4

Similarly, the polynomial interpolation data are also reported. First, the polynomial for the controller is reported in (A1), and the parameters for such an interpolation are given in Table A9. In the same way, the polynomial for the observer is reported in (A2), and the parameters for such an interpolation are given in Table A10.

$$\begin{aligned} Pol_K = (&p00 + p10 \cdot v_{dc} + p01 \cdot v_b + p20 \cdot v_{dc}^2 + p11 \cdot v_{dc} \cdot v_b + p02 \cdot v_b^2 \\ &+ p30 \cdot v_{dc}^3 + p21 \cdot v_{dc}^2 \cdot v_b + p12 \cdot v_{dc} \cdot v_b^2 + p03 \cdot v_b^3 + p31 \cdot v_{dc}^3 \cdot v_b \\ &+ p22 \cdot v_{dc}^2 \cdot v_b^2 + p13 \cdot v_{dc} \cdot v_b^3 + p04 \cdot v_b^4)/1000 \end{aligned} \tag{A1}$$

$$\begin{aligned} Pol_\ell = (&p00 + p10 \cdot v_{dc} + p01 \cdot v_b + p20 \cdot v_{dc}^2 + p11 \cdot v_{dc} \cdot v_b + p02 \cdot v_b^2 \\ &+ p30 \cdot v_{dc}^3 + p21 \cdot v_{dc}^2 \cdot v_b + p12 \cdot v_{dc} \cdot v_b^2 + p03 \cdot v_b^3) \cdot 1000 \end{aligned} \tag{A2}$$

Table A9. Polynomial function coefficients of K.

Coefficient	K_1	K_2	K_3	K_4
$p00$	2.70×10^{1}	4.75×10^{1}	4.70	6.08
$p10$	7.29×10^{-1}	1.29×10^{-1}	7.08×10^{-1}	4.10
$p01$	-9.49×10^{-1}	3.35	3.07×10^{-1}	2.17
$p20$	-4.28×10^{-2}	-1.18×10^{-2}	-4.24×10^{-2}	-1.20×10^{-1}
$p11$	6.81×10^{-3}	-9.29×10^{-2}	-1.15×10^{-2}	-1.46×10^{-1}
$p02$	4.78×10^{-2}	-2.09×10^{-1}	-3.77×10^{-2}	-4.28×10^{-2}
$p30$	7.20×10^{-4}	1.41×10^{-4}	4.65×10^{-4}	1.29×10^{-3}
$p21$	5.49×10^{-4}	1.73×10^{-3}	2.34×10^{-3}	4.15×10^{-3}
$p12$	-3.83×10^{-4}	2.78×10^{-3}	-1.79×10^{-3}	1.70×10^{-3}
$p03$	-1.24×10^{-3}	5.41×10^{-3}	2.12×10^{-3}	4.83×10^{-4}
$p31$	-1.74×10^{-5}	-1.30×10^{-5}	-2.18×10^{-5}	-3.43×10^{-5}
$p22$	9.53×10^{-6}	-2.07×10^{-5}	-1.73×10^{-5}	-3.75×10^{-5}
$p13$	-1.10×10^{-6}	-2.97×10^{-5}	3.54×10^{-5}	1.59×10^{-6}
$p04$	1.36×10^{-5}	-5.28×10^{-5}	-3.40×10^{-5}	-4.72×10^{-6}

Table A10. Polynomial function coefficients of ℓ.

Coefficient	ℓ_1	ℓ_2	ℓ_3	ℓ_4
$p00$	-3.73×10^{-1}	-2.21	-1.82	9.83×10^{-1}
$p10$	3.96×10^{-1}	2.50×10^{-1}	-6.31×10^{-1}	2.32×10^{-1}
$p01$	4.43×10^{-1}	5.70×10^{-1}	6.13×10^{-1}	4.03×10^{-1}
$p20$	-2.29×10^{-3}	6.64×10^{-4}	5.90×10^{-3}	-2.35×10^{-3}
$p11$	8.02×10^{-3}	1.04×10^{-2}	1.39×10^{-2}	-2.81×10^{-3}
$p02$	-4.83×10^{-3}	-8.88×10^{-3}	-1.90×10^{-2}	-8.43×10^{-3}
$p30$	3.73×10^{-5}	6.63×10^{-6}	-5.50×10^{-5}	1.70×10^{-5}
$p21$	-2.58×10^{-5}	-4.04×10^{-5}	-9.80×10^{-6}	2.59×10^{-5}
$p12$	-1.22×10^{-4}	-1.52×10^{-4}	-2.45×10^{-4}	1.19×10^{-5}
$p03$	1.05×10^{-4}	1.58×10^{-4}	2.94×10^{-4}	1.01×10^{-4}

References

1. Payne, J. How to select the Correct Factory Automation Controller for your application: Picking the right hardware is critical, but users also need to make sure they include controller programming software in the evaluation process. *Control Eng.* **2018**, *65*, 22.
2. Jithin, S.; Rajeev, T. A novel stability index-based virtual synchronous machine droop scheme for performance improvement of hybrid AC/DC microgrid under volatile loading conditions. *Electr. Power Syst. Res.* **2023**, *214*, 108901.
3. Babu, N.R.; Saikia, L.C. Optimal location of accurate HVDC and energy storage devices in a deregulated AGC integrated with PWTS considering HPA-ISE as performance index. *Eng. Sci. Technol. Int. J.* **2022**, *33*, 101072. [CrossRef]
4. Mousakazemi, S.M.H. Comparison of the error-integral performance indexes in a GA-tuned PID controlling system of a PWR-type nuclear reactor point-kinetics model. *Prog. Nucl. Energy* **2021**, *132*, 103604. [CrossRef]
5. Balestrino, A.; Landi, A.; Sani, L. Performance indices and controller design for switching converters. In Proceedings of the International Symposium on Power Electronics, Electrical Drives, Automation and Motion, Taormina, Italy, 23–26 May 2006; SPEEDAM 2006; IEEE: Taormina, Italy, 2006; pp. 1316–1321.
6. Blondel, V.D.; Tsitsiklis, J.N. A survey of computational complexity results in systems and control. *Automatica* **2000**, *36*, 1249–1274. [CrossRef]
7. Ribeiro, L.D.; Secchi, A.R. A Methodology to Obtain Analytical Models That Reduce the Computational Complexity Faced in Real Time Implementation of Nmpc Controllers. *Braz. J. Chem. Eng.* **2019**, *36*, 1255–1278. [CrossRef]
8. Yan, S.; Cui, Y.; Li, C.; Gao, X.; Cai, Y. An Improved FCS-MPC Based on Novel Sector Optimization and Capacitor Charge Balance Algorithm for T-Type 3P-3L Converters. *IEEE Trans. Power Electron.* **2023**, *38*, 4559–4571. [CrossRef]
9. Silva, J.J.; Espinoza, J.R.; Rohten, J.A.; Pulido, E.S.; Villarroel, F.A.; Torres, M.A.; Reyes, M.A. MPC Algorithm With Reduced Computational Burden and Fixed Switching Spectrum for a Multilevel Inverter in a Photovoltaic System. *IEEE Access* **2020**, *8*, 77405–77414. [CrossRef]
10. Sands, T. Control of DC Motors to Guide Unmanned Underwater Vehicles. *Appl. Sci.* **2021**, *11*, 2144. [CrossRef]
11. Menezes, J.; Sands, T. Discerning Discretization for Unmanned Underwater Vehicles DC Motor Control. *J. Mar. Sci. Eng.* **2023**, *11*, 436. [CrossRef]
12. Wang, Z.; Sands, T. Artificial Intelligence-Enhanced UUV Actuator Control. *AI* **2023**, *4*, 270–288. [CrossRef]
13. Koo, S.M.; Travis, H.; Sands, T. Impacts of Discretization and Numerical Propagation on the Ability to Follow Challenging Square Wave Commands. *J. Mar. Sci. Eng.* **2022**, *10*, 419. [CrossRef]
14. Shah, R.; Sands, T. Comparing Methods of DC Motor Control for UUVs. *Appl. Sci.* **2021**, *11*, 4972. [CrossRef]
15. Shtessel, Y.; Taleb, M.; Plestan, F. A novel adaptive-gain supertwisting sliding mode controller: Methodology and application. *Automatica* **2012**, *48*, 759–769. [CrossRef]
16. Canciello, G.; Cavallo, A.; Guida, B. Control of Energy Storage Systems for Aeronautic Applications. *J. Control Sci. Eng.* **2017**, *2017*, 1–9. [CrossRef]

17. Russo, A.; Cavallo, A. Supercapacitor stability and control for More Electric Aircraft application. In Proceedings of the 2020 European Control Conference (ECC), Saint Petersburg, Russia, 12–15 May 2020; IEEE: Saint Petersburg, Russia, 2020; pp. 1909–1914.
18. Cavallo, A.; Canciello, G.; Russo, A. Integrated supervised adaptive control for the more Electric Aircraft. *Automatica* **2020**, *117*, 108956. [CrossRef]
19. Canciello, G.; Cavallo, A.; Schiavo, A.L.; Russo, A. Multi-objective adaptive sliding manifold control for More Electric Aircraft. *Isa Trans.* **2020**, *107*, 316–328. [CrossRef] [PubMed]
20. Cavallo, A.; Russo, A.; Canciello, G. Hierarchical control for generator and battery in the more electric aircraft. *Sci. China Inf. Sci.* **2019**, *62*, 192207. [CrossRef]
21. Canciello, G.; Russo, A.; Guida, B.; Cavallo, A. Supervisory Control for Energy Storage System Onboard Aircraft. In Proceedings of the 2018 IEEE International Conference on Environment and Electrical Engineering and 2018 IEEE Industrial and Commercial Power Systems Europe (EEEIC/I&CPS Europe), Palermo, Italy, 12–15 June 2018; IEEE: Palermo, Italy, 2018; pp. 1–6.
22. Cavallo, A.; Canciello, G.; Russo, A. Supervised Energy Management in Advanced Aircraft Applications. In Proceedings of the 2018 European Control Conference (ECC), Limassol, Cyprus, 12–15 June 2018; IEEE: Limassol, Cyprus, 2018; pp. 2769–2774.
23. Cavallo, A.; Canciello, G.; Guida, B. Supervisory control of DC-DC bidirectional converter for advanced aeronautic applications. *Int. J. Robust Nonlinear Control* **2018**, *28*, 1–15. [CrossRef]
24. Kanieski, J.M.; Tambara, R.V.; Pinheiro, H.; Cardoso, R.; Gründling, H.A. Robust Adaptive Controller Combined With a Linear Quadratic Regulator Based on Kalman Filtering. *IEEE Trans. Autom. Control* **2016**, *61*, 1373–1378. [CrossRef]
25. Bimarta, R.; Tran, T.V.; Kim, K.H. Frequency-Adaptive Current Controller Design Based on LQR State Feedback Control for a Grid-Connected Inverter under Distorted Grid. *Energies* **2018**, *11*, 2674. [CrossRef]
26. Kulikowski, K.; Sikorski, A. New DPC Look-Up Table Methods for Three-Level AC/DC Converter. *IEEE Trans. Ind. Electron.* **2016**, *63*, 7930–7938. [CrossRef]
27. Zhang, Y.; Qu, C. Table-Based Direct Power Control for Three-Phase AC/DC Converters Under Unbalanced Grid Voltages. *IEEE Trans. Power Electron.* **2015**, *30*, 7090–7099. [CrossRef]
28. Yan, S.; Yang, Y.; Hui, S.Y.; Blaabjerg, F. A Review on Direct Power Control of Pulsewidth Modulation Converters. *IEEE Trans. Power Electron.* **2021**, *36*, 11984–12007. [CrossRef]
29. Alonso-Martínez, J.; Carrasco, J.E.; Arnaltes, S. Table-Based Direct Power Control: A Critical Review for Microgrid Applications. *IEEE Trans. Power Electron.* **2010**, *25*, 2949–2961. [CrossRef]
30. Ren, X.; Guo, Z.; Wu, Y.; Zhang, Z.; Chen, Q. Adaptive LUT-Based Variable On-Time Control for CRM Boost PFC Converters. *IEEE Trans. Power Electron.* **2018**, *33*, 8123–8136. [CrossRef]
31. Yahagi, S.; Kajiwara, I. Direct Data-Driven Tuning of Look-Up Tables for Feedback Control Systems. *IEEE Control Syst. Lett.* **2022**, *6*, 2966–2971. [CrossRef]
32. Udavalakshmi, J.K.; Sheik, M.S. Comparative Study of Perturb & Observe and Look -Up Table Maximum Power Point Tracking Techniques using MATLABISimulink. In Proceedings of the 2018 International Conference on Current Trends towards Converging Technologies (ICCTCT), Coimbatore, India, 1–3 March 2018; IEEE: Coimbatore, India, 2018; pp. 1–5.
33. Tu, S.H.L.; Yeh, H.W. A PWM controller with table look-up for DC-DC class E buck/boost conversion. In Proceedings of the 2013 IEEE International Conference of Electron Devices and Solid-state Circuits, Hong Kong, China, 3–5 June 2013; IEEE: Hong Kong, China, 2013; pp. 1–2.
34. Chiu, M.M.H.; Tu, S.H.L. A Novel DPWM Based on Fully Table Look-Up for High-Frequency Power Conversion. In Proceedings of the APCCAS 2006 IEEE Asia Pacific Conference on Circuits and Systems, Singapore, 4–7 December 2006; IEEE: Singapore, 2006; pp. 678–681.
35. Zheng, H.; Zou, T.; Hu, J.; Yu, H. An Offline Optimization and Online Table Look-Up Strategy of Two-Layer Model Predictive Control. *IEEE Access* **2018**, *6*, 47433–47441. [CrossRef]
36. Held, M.; Flärdh, O.; Mårtensson, J. Experimental evaluation of a look-ahead controller for a heavy-duty vehicle with varying velocity demands. *Control Eng. Pract.* **2021**, *108*, 104720. [CrossRef]
37. Su, Z.; Shen, X.; Chen, F. Flow Control of Air Compressor in Fuel Cell System Based on Adaptive Look-Up Table Algorithm. In Proceedings of the 2010 International Conference on Electrical and Control Engineering, Wuhan, China, 25–27 June 2010; IEEE: Wuhan, China, 2010; pp. 3761–3766.
38. Montenegro-Oviedo, J.A.; Ramos-Paja, C.A.; Orozco-Gutierrez, M.L.; Franco-Mejía, E.; Serna-Garcés, S.I. Adaptive Controller for Bus Voltage Regulation on a DC Microgrid Using a Sepic/Zeta Battery Charger/Discharger. *Mathematics* **2023**, *11*, 793. [CrossRef]
39. Agrawal, N.; Samanta, S.; Ghosh, S. Modified LQR Technique for Fuel-Cell-Integrated Boost Converter. *IEEE Trans. Ind. Electron.* **2021**, *68*, 5887–5896. [CrossRef]
40. Hwang, C.K.; Huang, K.S.; Lin, K.B.; Lee, B.K. Observer base linear quadratic regulation with estimated state feedback control. In Proceedings of the 2010 International Conference on Machine Learning and Cybernetics, Qingdao, China, 11–14 July 2010; Volume 6, pp. 2802–2805.

41. Brogan, W.L. *Modern Control Theory*, 3rd ed.; Prentice-Hall: London, UK, 1990.
42. *MATLAB*, version 9.12.0.188 (R2022a); The MathWorks Inc.: Natick, MA, USA, 2022.

electronics

MDPI

Article

A New Single-Cell Hybrid Inductor-Capacitor DC-DC Converter for Ultra-High Voltage Gain in Renewable Energy Applications

Ammar Falah Algamluoli * and Xiaohua Wu *

School of Automation, Northwestern Polytechnical University, Xi'an 710060, China
* Correspondence: ammar.algamluoli@coeng.uobaghdad.edu.iq (A.F.A.); wxh@nwpu.edu.cn (X.W.)

Abstract: In this paper, a new single-cell hybrid switched inductor DC-DC converter is proposed to demonstrate the verification of ultra-high voltage gain in renewable energy applications (REA). The modification involves adding a single cell of an inductor with a diode and double capacitor to increase voltage transfer gain. Additionally, this modification helps prevent the input current from becoming zero, pulsating at very low duty cycles. The single cell of the hybrid inductor is interleaved with the main switch to reduce current stress when the capacitor of the single-cell inductor charge becomes zero. Moreover, the addition of a modified hybrid switch inductor with a capacitor, operating in dual boosting mode with a single switch, allows the converter to achieve ultra-high voltage gain. The proposed converter offers several advantages, including ultra-high voltage gain, high efficiency, low voltage stress on power MOSFETs, diodes, inductors, and capacitors, as well as low switching and conduction losses. Furthermore, the proposed converter utilizes transformerless and non-coupled inductors. Mathematical equations have been derived for the discontinuous conduction mode (DCM) and continuous conduction mode (CCM) and implemented using Matlab Simulink software to validate the results. In addition, a dual PI controller is designed for the proposed converter to verify the fixed output voltage. Experimental results have also been obtained for a 200 W prototype, with the input voltage varying between 20 V and 40 V, and an output voltage of 200 V at an efficiency of 96.5%.

Keywords: DC-DC converter; hybrid switch inductor; single cell switched inductor; ZCS

Citation: Algamluoli, A.F.; Wu, X. A New Single-Cell Hybrid Inductor-Capacitor DC-DC Converter for Ultra-High Voltage Gain in Renewable Energy Applications. *Electronics* **2023**, *12*, 3101. https://doi.org/10.3390/electronics12143101

Academic Editors: Diego Bellan and Jelena Loncarski

Received: 16 June 2023
Revised: 12 July 2023
Accepted: 14 July 2023
Published: 17 July 2023

1. Introduction

The world is increasingly interested in utilizing various types of renewable energy sources to generate electrical power, driven by concerns regarding energy security and the environmental impact resulting from carbon dioxide emissions [1,2]. Solar and wind energy, in particular, have gained widespread adoption worldwide [3,4]. For instance, photovoltaic solar panels produce variable low-voltage outputs ranging from 12–40 V, which are unsuitable for applications requiring high DC supply voltage or household appliances [5,6]. Therefore, DC-DC converters are employed to step up very low input voltage to high output DC voltage for various applications such as streetlights, motor drives, microgrid systems, uninterruptible power supplies, fuel cells, and medical equipment [2,4–35]. Different converter topologies, including buck-boost, boost, and Ćuk DC-DC converters, have been introduced to achieve high voltage gain. The choice of converter depends on the specific application requirements. However, these converters encounter challenges when targeting ultra-high voltage gain. Key issues include a lower count of inductors and capacitors, reduced efficiency at high voltage gain with extremely high duty ratios, high voltage stress on power switches and diodes (equivalent to the output voltage), and increased current stress on power devices as the load current rises. In this context, current stress refers to the magnitude of current flowing through the switch during the on state and is influenced by the paths traversed by the current through the MOSFET. Moreover, these converters suffer from elevated switching and conduction losses when the duty cycle exceeds 0.9, as well as low power density [20]. Although boost converter topologies can achieve voltage step-up with high voltage gains of up to 10, their efficiency diminishes at high duty cycles [10,36]. Several researchers have proposed new technologies for DC-DC converters to attain high voltage gain in renewable energy systems (RESs).

A modified DC-DC converter, interleaved with a boost converter based on soft switching between the main and auxiliary power MOSFETs, has been proposed to achieve high voltage gain and solve the inrush current problem [1]. Another modification to the boost converter using a hybrid inductor capacitor has been proposed in [8], and another with a switch capacitor (SC), switch inductor (SL), and voltage multiplier (VM) in [31]. A cascaded conventional SEPIC with a boost converter has been proposed in [24,26], and a modified SEPIC based on an interleaved buck-boost converter in [11]. A modified converter with multi-input has been proposed in [9,25]. These converters, as mentioned, are modified to achieve high voltage gain with high power density. However, they have low voltage gain and a high number of inductors and capacitors. Additionally, they require a high number of power diodes and power MOSFETs to step up the low input to a high output voltage. Furthermore, the power MOSFETs and diodes in these converters experience high voltage stress. Complex control was also required in [8] to achieve high voltage gain. Moreover, these converters achieve high voltage gain with an extremely high duty cycle, resulting in high switching and conduction losses, as well as low performance and efficiency. Other topologies have been proposed, such as a modified DC-DC converter with coupled inductors and a voltage multiplier network (VMN) [3], and a conventional SEPIC with a coupled inductor and VM [23]. These converters have demonstrated high voltage gain, but they feature a high number of passive and active elements, as well as diodes. The large number of components leads to a high parasitic resistance of inductors and capacitors, resulting in reduced efficiency. Additionally, these converters operate at a low switching frequency, which necessitates the use of large values of inductors and capacitors. Moreover, the internal resistance of the power MOSFET (Ron) is high, further diminishing the voltage gain of the system. Furthermore, a major issue with coupled inductor converters is the occurrence of high spike voltages in the off state of the power switch due to the inductance with parasitic capacitance of the power MOSFET switch [36]. To address this problem, a clamped circuit can be added to the power switch to prevent the occurrence of high spike voltages due to the coupled inductor [2,37]. However, adding more components to the circuit increases costs and reduces efficiency due to parasitism. In addition, the system will be heavy and large. Another problem is the pulsating input current at a low duty cycle, which makes these converters unsuitable for RES applications.

Other researchers have developed DC-DC converters to achieve high voltage gain using non-isolated coupled inductors. In [12], a double power switch converter with double switch inductor (SL) was utilized. Additionally, in [13], a modified converter incorporating a hybrid capacitor and inductor, and in [14], a DC-DC converter employing the voltage lift technique, are proposed. Furthermore, in [15,16], a DC-DC converter based on SC with zero voltage switching, and in [10,17], a modified buck-boost converter featuring a single switch and pulsating input current, are described. However, these converters require an extremely high duty ratio to achieve a high voltage gain ratio. This implies high switching and conduction losses, low efficiency, low power density, and high voltage stress on the power switches, diodes, inductors, and capacitors. Additionally, these converters exhibit a high inductor count with low switching frequencies, which results in high parasitic resistance, diminished performance, and efficiency. A modified boost converter with dual power switch was proposed in [18], a modified boost converter with a single switch, multiplier capacitor, and SL in [19,22], and a multi-input converter with multiple capacitors as hybrid energy storage in [21], with the goal of attaining high voltage gain for RESs. A buck-boost converter with SC SL was developed in [27], a pair of cascaded conventional boost converters with dual switches in [28] and with a single switch in [30], and a modified buck interleaved with SEPIC based on (SC) (SL) with two input sources in [29]. These converters achieve high voltage gain; however, they have a large number of power switches and diodes, which has a high impact on system efficiency and performance. In addition, the voltage and current stress on the power switch is high. Furthermore, the gate control circuit is large and complex to implement.

In this paper, a new single-cell hybrid switched inductor-capacitor DC-DC converter is proposed to demonstrate the verification of ultra-high voltage gain in renewable energy applications (REA). The proposed modification involves incorporating a single cell of a hybrid inductor, along with a diode and double capacitors, interleaved with the main switch. This integration enables the proposed converter to attain a high voltage gain while ensuring that the input current does not experience zero pulsations at very low duty cycles. Additionally, the current stress on the main switch will reduce as the duty ratio increases, as shown in Figure 11f when the single-cell capacitor C_1 charge becomes zero. Furthermore, the converter employs a modified hybrid switch inductor in dual boosting mode, working with a single switch. This arrangement allows for the realization of an ultra-high voltage gain. Moreover, one of the passive components, L_2, will be open circuit, and D_1 is working at zero current switching, which means a reduction in the total power loss of the converter.

2. Structure and Operation of the Proposed DC–DC Converter

The DC-DC converter is modified to achieve a high voltage gain by stepping up a low input voltage range of 20–40 V to an output voltage of 200 V. The proposed converter is based on a modified hybrid switched inductor-capacitor configuration to verify high voltage gain. Figure 1a shows the basic connection of a switched inductor, while Figure 1b illustrates the hybrid connection of a switched inductor and a capacitor. Consequently, the proposed converter is designed for Renewable Energy Sources (RES). In Figure 1c, the connection of the proposed converter with PV Panels and Battery is shown for Energy Saving Mode Applications. The structure of the proposed converter includes four inductors, four capacitors, three diodes, and two power switches, as depicted in Figure 1d. The main advantages of the proposed converter are that it utilizes non-coupled inductors and is transformerless. It employs a high switching frequency to reduce the sizes of inductors and capacitors, thereby increasing its efficiency. The proposed converter has a simple structure with a straightforward control circuit. The new single cell of an inductor and capacitor, which is interleaved with the main switch, helps avoid pulsating input current at low duty cycles and minimizes voltage stress on the power switches, diodes, and inductors. Additionally, the current stress on the main switch is reduced when the duty cycle increases. Furthermore, the proposed converter achieves an ultra-high voltage gain compared to previous DC-DC converters. In terms of the power switch PWM generator, it uses a simple design where both MOSFETs turn on and off simultaneously. The proposed converter can operate in DCM under two cases. Firstly, it can function in DCM Case 1 (DCMC1) at a low duty cycle and maximum input voltage, with a duty ratio below 50%. Secondly, the proposed converter can operate in DCM Case 2 (DCMC2) when the input voltage decreases during the day at a high load current, with a duty ratio above 50%. Additionally, the converter can also operate in CCM when the load current increases and the duty ratio exceeds 70%. These scenarios are illustrated in Figure 4a,b, which demonstrates the dynamic performance of the proposed converter.

2.1. *Proposed Converter Operation DCMC1*

The proposed converter can operate in DCMC1 in four states of operation during one cycle when the input voltage is at its maximum, with a low duty cycle, and at light load. The waveform of this mode of operation is shown in Figure 2a. The four states of this operational mode are listed below:

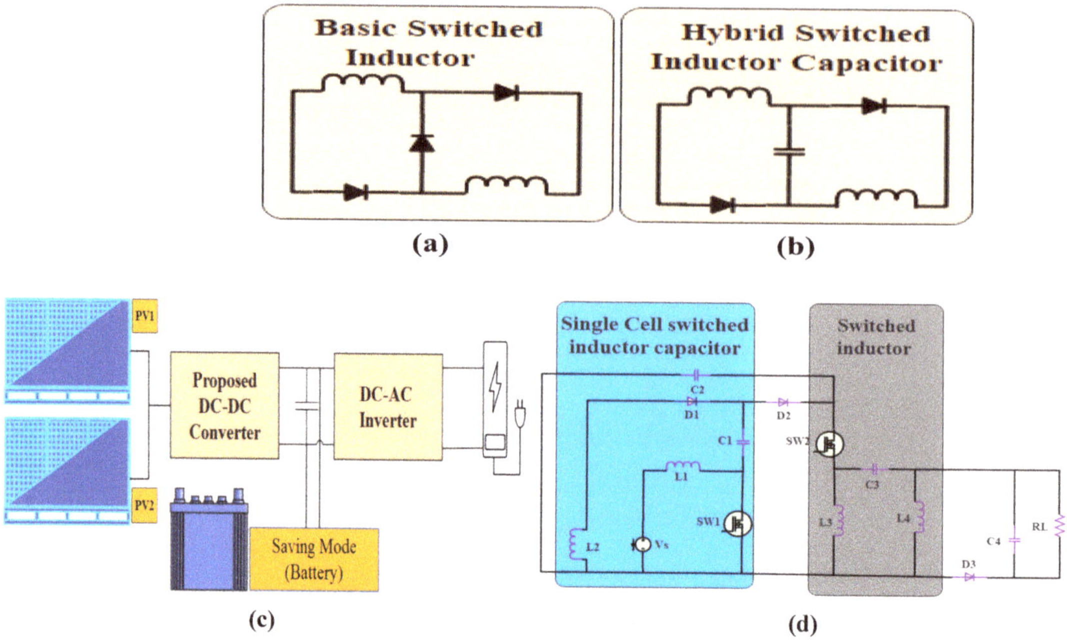

Figure 1. (**a**) Basis circuit of the switched inductor (**b**) Hybrid switched inductor-capacitor connection (**c**) Schematic Diagram: Connection of the proposed converter with PV panels and battery for energy saving mode applications (**d**) The Proposed DC-DC Converter with single-cell switched inductor capacitor interleaved with modified switched inductor.

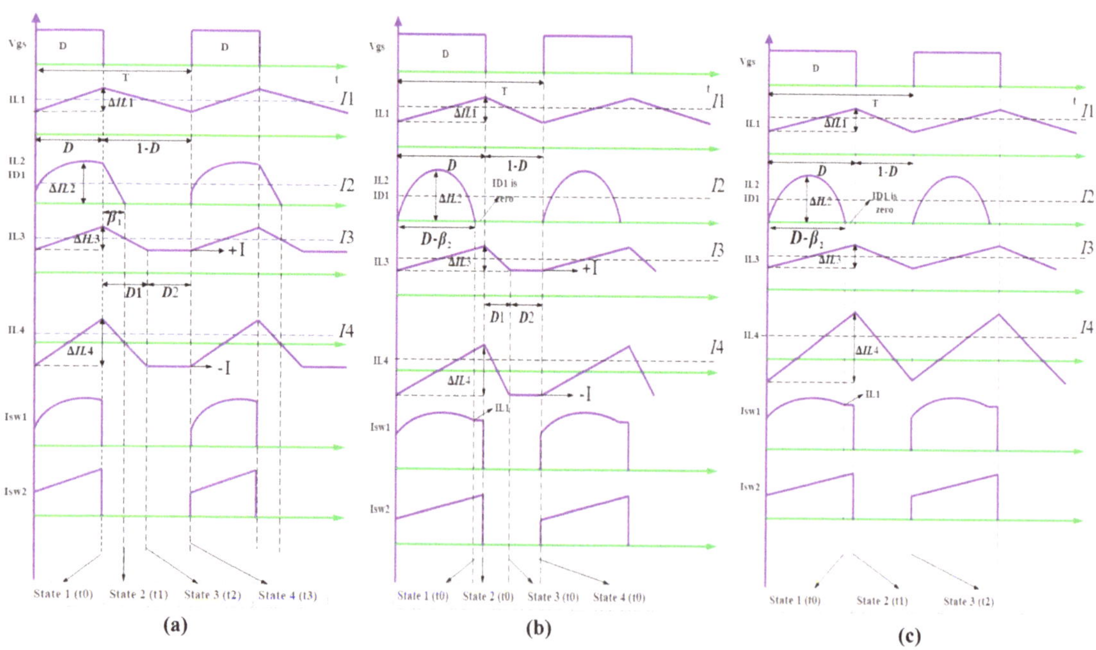

Figure 2. (**a**) Proposed Converter waveforms at DCMC1, (**b**) Proposed Converter waveforms at DCMC2, (**c**) Proposed Converter waveforms at CCM.

State 1: $[0\text{–}t_0]$: In this mode, the two power MOSFETs (Sw_1 and Sw_2) are in the on state, and a PWM generator provides a high source-to-gate voltage to turn both MOSFETs on and keep diodes D_2 and D_3 off. During this mode, L_1 charges with energy from the input source, which is connected in series with it. L_2 begins to charge from C_1, and C_1 discharges the energy stored in L_2 through Sw_1. D_1 is on during this mode, and C_2 stores a significant amount of energy, which is used to charge L_3 through Sw_2. At the same time, L_4 starts charging from C_3. C_4 supplies power to the load and forms the current path for this mode, as shown in Figure 2b.

The voltage equations of this mode for inductors, capacitors, diodes, and MOSFETs are as follows:

$$\left.\begin{array}{l} VL_1 = Vs \\ VL_2 = Vc_1 \\ VL_3 = Vc_2 \\ VL_4 = Vc_2 - Vc_3 \\ Vc_4 = Vo \end{array}\right\} \quad (1)$$

The current equations of this mode for inductors, capacitors, diodes, and MOSFETs are as follows:

$$iL_1 + iL_2 = ISW_1 \quad (2)$$

$$ISW_2 = Ic_2 = iL_3 + iL_4 \quad (3)$$

$$ID_1 = iL_2 = IC_1 \quad (4)$$

$$IC_4 = Io \quad (5)$$

$$\left.\begin{array}{l} iL_1 = \frac{Vs}{L_1} \\ iL_2 = \frac{Vc_1}{L_2} \\ iL_3 = \frac{Vc_2}{L_3} \\ Io = \frac{Vo}{RL} \end{array}\right\} \quad (6)$$

where VL represents the voltage across the inductor, Vc denotes the voltage across the capacitor, Isw represents the current through the power MOSFET switch during the on state, Vo signifies the output voltage, iL represents the current through the inductor, Io represents the output current, RL represents the resistive load, ID is the current through the power diode, and Ic denotes the current through the capacitor.

State 2: $[t_0\text{–}t_1]$: In this mode, the two power MOSFETs are in the off state, and both diodes D_2 and D_3 in the on state. L_1 discharges energy to C_1 and charges C_2. L_2 starts discharging its energy to C_2 through D_1 and remains in the on state. C_2 stores a large amount of energy from L_1 and L_2. L_3 and L_4 start discharging their energy to C_4, which supplies high power to the load and forms the current path of this mode, as shown in Figure 3c.

The voltage equations of this mode for inductors, capacitors, diodes, and MOSFETs are as follows:

$$\left.\begin{array}{l} VL_1 = Vs - Vc_1 - Vc_2 \\ VL_2 = -Vc_2 \\ VL_3 = Vc_3 + VL_4 \\ VL_4 = -Vc_4 \\ Vc_4 = V_o \end{array}\right\} \quad (7)$$

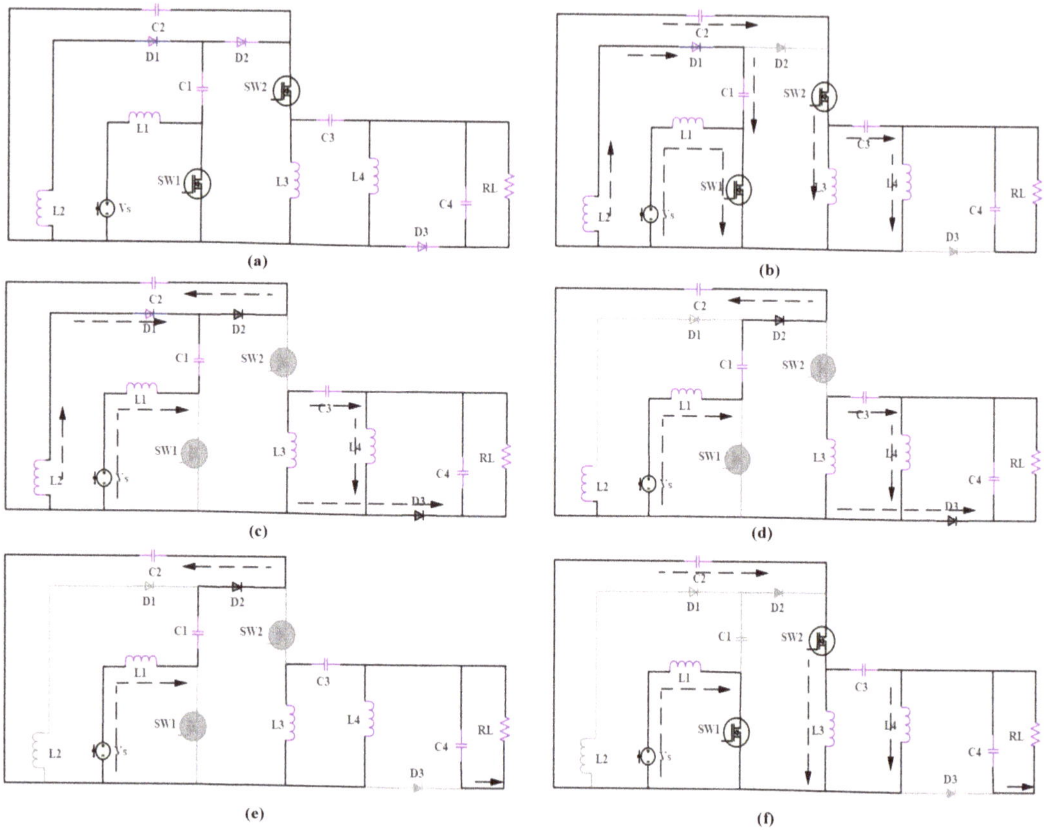

Figure 3. (**a**) proposed DC-DC converter (**b**) state 1 mode DCMC1, state 1 mode DCMC2, state 1 mode CCM (**c**) state 2 mode DCMC1 (**d**) state 3 DCMC1, state 3 DCMC2, state 3 CCM (**e**) state 4 DCMC1 state 4 DCMC2 (**f**) state 2 mode DCMC2, state 2 mode CCM.

The current equations of this mode for inductors, capacitors, diodes, and MOSFETs are as follows:

$$\left.\begin{array}{l} iL_1 = \frac{VS}{L_1} - \frac{Vc1}{L_1} - \frac{Vc2}{L_1} \\ iL_2 = \frac{-Vc_2}{L_2} \\ iL_3 = \frac{Vc_3 + VL_4}{L_3} \\ Io = \frac{Vo}{RL} \end{array}\right\} \quad (8)$$

$$iL_1 + iL_2 = ID_2 \quad (9)$$

$$iL_3 + iL_4 = ID_3 \quad (10)$$

$$ID_1 = iL_2 \quad (11)$$

$$Ic_4 = Io = iL_3 + iL_4 \quad (12)$$

$$iL_1 = Ic_1 = Ii \quad (13)$$

where Ii is the input current, which is equal to iL_1.

State 3: $[t_1–t_2]$: In this mode, the two power MOSFETs are still in the off state, and both diodes D_2 and D_3 are still in the on state. L_1 continues discharging energy to C_1 and charges C_2. L_2 reaches zero charge and iL_2 and ID_1 are zero. C_2 continues receiving energy from only L_1, and D_1 is now in the off state in this mode. L_3 and L_4 continue discharging their energy to C_4, which supplies high power to the load and forms the current path of this mode, as shown in Figure 3d.

The current equations of this mode are as follows:

$$iL_1 = ID_2 \tag{14}$$

$$ID_1 = iL_2 = 0 \tag{15}$$

$$Ic_2 = iL_1 \tag{16}$$

State 4: $[t_2–t_3]$: In this mode, the two power MOSFETs are still off, the PWM generator gives zero gate-to-source voltage to keep them in the off state, and only D_2 is still on. D_3 is now changed to the off state in this mode. L_1 continues discharging energy to charge C_1 and C_2. C_2 stores a large amount of energy for the next pulse to supply it to the load. L_2 has zero charge and iL_2 and ID_1 are zero, and D_1 is still off in this mode. L_3 and L_4 will have the same current values but in opposite directions: $iL_3 = -iL_4$. C_4 supplies high power to the load and forms the current path of this mode, as shown in Figure 3e.

Figure 4a,b illustrate the dynamic performance of the proposed converter when operating in DCM and CCM, respectively. It is evident from the figures that the proposed converter operates in DCM for duty cycles below 70%. However, for duty cycles above 70%, the converter can operate in CCM based on the boundary condition specified in Equation (33).

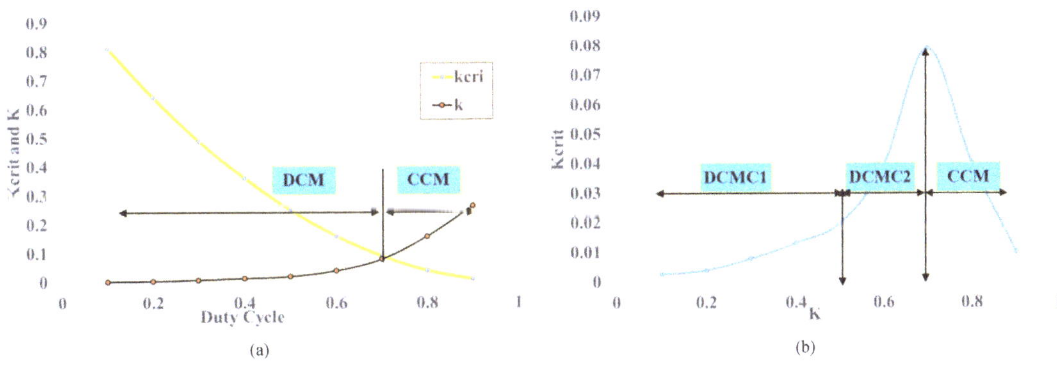

Figure 4. Dynamic performance of the proposed converter (**a**) Loading factor K and Kcrit Vs duty cycle, (**b**) Kcrit Vs K.

2.2. Proposed Converter Operation in DCMC2

The proposed converter can operate in DCMC2 when the input voltage is reduced to the minimum value during the day, while the load current is high. In this mode, the input current is still in CCM, and L_2 will be in resonant mode with C_1, while L_3 and L_4 will continue to operate in DCM. Therefore, this mode has four states of operation, as shown in Figure 2b: the proposed converter waveforms at DCMC2.

State 1: $[0–t_0]$: In this mode, the two MOSFETs Sw_1 and Sw_2 are in the on state, and both diodes D_2 and D_3 in their off states. L_1 starts charging energy from the input source, which is connected in series with it in CCM and never reaches zero. Inductor L_2 starts charging from C_1, and C_1 will discharge energy to L_2 through Sw_1 until $(D - \beta_2)$, the charge across C_1 will be zero, and the current through Sw_1 will come only from IL_1 after this time, as shown in Figure 11f. D_1 is on in this mode, and C_2 stores a large amount of energy, which charges L_3 through Sw_2, and L_4 starts charging from C_3. C_4 supplies power to the load and forms the current path of this mode, as shown in Figure 3b. The current and voltage equations of this mode are the same as for State 1 of DCM Case 1. Only the current through Sw_1 is shown below:

$$\left.\begin{array}{l} Isw_1 = (iL_1 + Ic_1) \text{ from } 0 < \mathrm{t} < (\mathrm{D} - \beta_2) \\ Isw_1 = iL_1 \text{ only from } (\mathrm{D} - \beta_2) < \mathrm{t} < \mathrm{D} \end{array}\right\} \tag{17}$$

$$iL_2 = \frac{Vc_1}{L_2}(D - \beta_2) \tag{18}$$

State 2: [t_0–t_1]: In this mode, the two power Mosfets are still in the on state. The PWM generator continues to provide a high gate-to-source voltage to keep them in the on state, and both diodes D_2 and D_3 remain off. Meanwhile, L_1 continues to charge energy from the input source, while L_2 reaches zero charge due to the charge across C_1 being zero. The current through Sw_1 comes solely from IL_1, reducing the current stress through Sw1. D_1 operates with zero current switching during this mode, meaning that L_2 will be open circuit during the time period (($D - \beta_2$) < t < D). L_3 and L_4 continue to charge from C_2 and C_3, respectively. C4 continues to supply power to the load, and the current path of this mode is shown in Figure 3f. The current equations remain the same as in the previous state 1.

State 3: [t_1–t_2]: in this mode, the two power MOSFETs are off, and both diodes D_2 and D_3 are now on. L_1 now starts discharging energy to C_1 and charges C_2. L_2 does not charge, and iL_2 and ID_1 are zero. L_2 is open circuit during this mode. C_2 receives a large amount of energy from L_1. L_3 and L_4 also start discharging their energy to C_4, which supplies high power to the load and forms the current path of this mode, as shown in Figure 3d.

State 4: [t_2–t_3]: In this mode, the two power MOSFETs are still off, and only D_2 is still on. D_3 is changed to the off state in this mode. L_1 continues discharging energy to charge C_2, and C_2 will have a large amount of energy for the next pulse to supply it to the load. L_2 remains open circuit in this mode. C_2 will continue receiving energy only from L_1. L_3 and L_4 will have the same current but in opposite directions: $iL_3 = -iL_4$. C_4 will supply high power to the load and forms the current path of this mode, as shown in Figure 3e.

When the proposed converter operates in DCMC2, the capacitor C_1 is discharged to zero at ($D - \beta_2$), reducing the current stress across Sw_1. In this case, the current through Sw_1 flows only from L_1. Additionally, the voltage stress across Sw_1 and the power diodes is significantly reduced. Furthermore, the conduction loss of Sw_1 is significantly reduced when the stress current is reduced. In this mode, L_2 will be in an open circuit state, and D_1 will have very low voltage stress. Additionally, as mentioned above, C_1 is discharged to zero. This implies that the components of the proposed converter are reduced during this mode, leading to improved performance and efficiency.

2.3. *Proposed Converter Operation in CCM*

This operation mode occurs when the load current increases to a duty cycle of 70%, as depicted in Figure 4. The input current still operates in CCM, and L_2 enters a resonant mode with C_1, while L_3 and L_4 continue to operate in CCM. Additionally, the voltage gain in this mode will be increased. Thus, this mode consists of three states of operation, as illustrated in Figure 2c, which shows the proposed converter waveforms in CCM.

State 1: [0–t_0]: in this mode, same as state 1 DCMC2.

State 2: [t_0–t_1]: in this mode, same as state 2 DCMC2.

State 3: [t_1–t_2]: In this mode, the two power MOSFETs are off, and both diodes D_2 and D_3 are now on. L_1 will start discharging energy to C_1 and charge C_2, and iL_2 and ID_1 are zero. C_2 receives a large amount of energy from L_1. L_3 and L_4 will also start discharging their energy to C_4, and L_3 and L_4 will have the same current in CCM but in opposite directions: $iL_3 = -iL_4$. C_4 will supply high power to the load and forms the current path of this mode, as shown in Figure 3d.

3. Voltage Gain Calculations of the Proposed Converter

In this section, the voltage gain of the proposed converter is calculated when it operates in DCM and CCM.

3.1. Voltage Gain of the Proposed Converter in DCMC1

In this section, we analyze the voltage gain of the proposed converter while it operates in DCMC1. This particular mode is encountered when the input voltage reaches its maximum value during light load applications, with a duty cycle below 50%. Hence, the equations that describe the voltage gain under the DCMC1 operation of the proposed converter are presented below:

$$\frac{1}{Ts}\left(\int_{0}^{DTs}(Vs+Vc_1)dt+\int_{D}^{Ts}(Vs-Vc_1-Vc_2)dt+\int_{DTs}^{\beta_1 Ts}(-Vc_2)dt\right)=0 \quad (19)$$

$$\frac{1}{Ts}\left(\int_{0}^{DTs}(Vc_2)dt+\int_{D}^{D1Ts}(Vc_2-Vc_3)dt=0\right. \quad (20)$$

$$Vc_2=\frac{Vs}{(1+\beta_1-D)} \quad (21)$$

$$\beta_1=\left(\frac{Vs}{Vc_2}+D-1\right) \quad (22)$$

$$D1=\frac{DVc_2}{Vo} \quad (23)$$

From Equations (1) and (7), applying volt-second balance to L_1 and L_2 results in Equation (19). Solving it results in Equation (21). Applying volt-second balance to L_3 and L_4 from Equations (1) and (7) yields Equation (20). After solving Equation (20), by applying the fact that the average voltage across C_3 is almost zero during the steady state, Equation (23) can be obtained (D1) is the discharging time of L_3 and L_4, and (β_1) represents the discharging time of L_2, which can be found from Equation (22) and is a function of (D, Vs, Vc_2).

$$\left.\begin{array}{l} <I2>=\frac{Vc_1}{2L_2}DTs(D+\beta 1) \\ <I3>=\frac{Vc_2}{2L_3}DTs(D+D1)+I \\ <I4>=\frac{Vc_2}{2L_4}DTs(D+D1)-I \end{array}\right\} \quad (24)$$

$$\left.\begin{array}{l} iL_{2peak}=\frac{VsDTs}{2(1-D)L_2} \\ iL_{3peak}=\frac{VsDTs}{(1-D)L_3} \\ iL_{4peak}=\frac{VsDTs}{(1-D)L_4} \end{array}\right\} \quad (25)$$

$$I3+I4=Io \quad (26)$$

$$Io=\frac{VsDTsD1}{2Le(1+\beta_1-D)} \quad (27)$$

To find the average current through L_2, L_3, and L_4, we can use Equation (24) to determine their respective averages. By adding the average currents I_3 and I_4 in Equation (24), we can calculate the average output current using Equation (27). The peak current of the inductors L_2, L_3, and L_4 can be found using Equation (25).

$$Vc_2=\sqrt{\frac{2LeVo^2}{RLD^2Ts}} \quad (28)$$

$$Le=\frac{L_3L_4}{L_3+L_4} \quad (29)$$

$$Gv(DCMC1) = \frac{VsD^2TsRL}{2LeVo(1+\beta_1-D)^2} \tag{30}$$

$$Gv(DCMC1) = \frac{D}{(1+\beta_1-D)\sqrt{K}} \tag{31}$$

$$k_{crit} = \frac{(1-D)^4}{(1+\beta_1-D)^2} \tag{32}$$

$$k_{crit} = \begin{cases} If \text{ Kcrit} > \text{ K Proposed Converter work in DCM} \\ If \text{ Kcrit} < \text{ K Proposed Converter work in CCM} \end{cases} \tag{33}$$

The voltage across capacitor C_2 can be obtained using Equation (28), which is a function of RL, Vo, D, and Le when the converter operates in DCMC1. Here, Le represents the inductor equivalent of L_3 and L_4, which can be found in Equation (29). Equation (30) provides the voltage gain of the proposed converter at DCMC1. Equation (31) expresses the voltage gain equation of the proposed converter in DCMC1 as a function of the load loss factor (K). The boundary condition of the proposed converter, specifically the critical load loss factor (K_{crit}), can be determined using Equations (32) and (33) by applying the values of (β_1) in Equation (32).

3.2. Voltage Gain of Proposed Converter in DCMC2

In this section, the voltage gain of the proposed converter can be adjusted by varying the input voltage and the load resistance. This mode occurs when the input voltage source reaches its minimum value at a duty cycle above 50%, while the proposed converter provides a high load current. Consequently, the voltage gain of the proposed converter in the DCMC2 equations is presented below.

$$\left.\begin{array}{l} \frac{1}{Ts}\left(\int_0^{DTs}(VS)dt + \int_0^{(D-\beta_2)Ts}(Vc1)dt + \int_{DTs}^{Ts}(Vs - Vc1 - Vc2)dt)\right) = 0 \\ \frac{1}{Ts}\left(\int_0^{DTs}(Vc2)dt + \int_D^{D1Ts}(Vc2 - Vc3)dt = 0\right. \end{array}\right\} \tag{34}$$

$$D(Vin) + (1-D)(Vin - Vc1 - Vc2) + (D-\beta_2)(Vc1) = 0 \tag{35}$$

By applying voltage second balance on L_1, L_2, L_3 and L_4, we obtain Equation (34) as a result in Equation (35). From Equation (36), we determine the value of Vc_1, which is almost equal to the difference between the input voltage and the voltage across capacitor C_2. Vc_2 can be found in Equation (37). The values of (β_2) is adjusted and can be found in Equation (38), depending on the values of C_1 and L_2. After using Equation (34) and concluding with Equation (40), we obtain the voltage gain of the proposed converter when operating in DCMC2.

$$Vc_1 = Vs - Vc_2 \tag{36}$$

$$Vc_2 = \frac{Vs}{(1-D)} \tag{37}$$

$$\beta_2 = D - \frac{\sqrt{C_1 L_2}}{Ts} \tag{38}$$

$$D1 = \frac{DVc_2}{Vo} \tag{39}$$

$$Gv(DCMC2) = \frac{VsD^2TsRL}{2LeVo(1-D)^2} \tag{40}$$

3.3. Voltage Gain of Proposed Converter in CCM

The voltage gain of the proposed converter is derived in CCM as shown in the equations below:

$$\frac{1}{Ts}\left(\int_{0}^{DTs}(Vc_2)dt+\int_{D}^{Ts}(Vc_2-Vc_3)dt=0\right. \quad (41)$$

$$Vo=\frac{Vc2D}{(1-D)} \quad (42)$$

$$Gv=\frac{Vo}{Vs}=\frac{D}{(1-D)^2} \quad (43)$$

By using the same Equations (34) and (35), by applying volt-second balance to L_1 and L_2 and, to L_3 and L_4, we obtain Equation (41) at different times. The result can be found in (42) and (43). The voltage gain of the proposed converter in CCM is given by Equation (43).

The voltage gain equations of the proposed converter in DCM and CCM demonstrate that it has a higher voltage transfer gain compared to previous DC-DC converters. Additionally, two passive components, L_2 and C_1, become open circuits as depicted in Figure 3f. Moreover, D_1 operates at Zero Current Switching and experiences low voltage stress, thereby reducing the total power loss of the converter. This reduction occurs because L_2 reaches zero charging when the charge across C_1 becomes zero. The current flowing through Sw_1 is solely sourced from iL_1.

4. Voltage across Power Diodes, MOSFETs and Capacitor

The proposed converter has two power switches, three power diodes, and four capacitors. Therefore, in this section, the voltage stress across power MOSFETs is calculated. Furthermore, the voltages across the power diodes and capacitors of the proposed converter are also calculated.

$$V_{D_1}=\frac{Vs}{(1-D)} \quad (44)$$

$$V_{D_2}=\frac{Vs}{(1-D)} \quad (45)$$

$$V_{D_3}=\frac{VsD}{(1-D)^2} \quad (46)$$

In order to find the voltage stress across the power diodes in the proposed converter, we can use Equation (44) to determine the voltage stress across D_1, which is a very small value. Equation (45) can be used to obtain the voltage stress across D_2, while Equation (46) provides the means to find the voltage stress across D_3. It can be observed that the voltage across the power diodes is very small and depends on the input voltage, which ranges from 20 V to 40 V.

$$Vsw_1=\frac{VsD}{(1-D)} \quad (47)$$

$$Vsw_2=\frac{Vs}{(1-D)}, average\ voltage \quad (48)$$

To determine the voltage stress across the Power MOSFETs, we can use Equation (47) to find the voltage across MOSFET Sw_1. This voltage is also very small and depends on the input voltage, which varies from 20 V to 40 V. Additionally, Equation (48) allow us to calculate the average voltage stress across Sw_2. During the time period $D < t < D1$, the voltage across Sw_2 is equal to the output voltage. During the time period $D1 < t < Ts$, it is equal to the average voltage across C_2.

$$Vc_1=\frac{Vs}{(1-D)}-\sqrt{\frac{2LeVo^2}{RLD^2Ts}}\ (D<0.5) \quad (49)$$

$$Vc_2 = \frac{Vs}{(1-D)} \tag{50}$$

$$Vc4 = Vo \tag{51}$$

To determine the voltage across capacitors, we can refer to Equation (49) for the voltage across C_1 and Equation (50) for the voltage across C_2. Additionally, Equation (51) provides the means to find the voltage across the filter capacitor, which is equal to the output voltage. By reducing the voltage stress on all power diodes and power MOSFETs, the proposed converter experiences decreased losses, leading to improved efficiency.

5. Features Components of Proposed Converter

In this section, we discuss the main components of 200 W prototype of proposed converter, which include inductors, capacitors, power diodes, power Mosfet, and gate drive circuit. Therefore, these components are crucial for designing and verifying the high voltage gain of the proposed converter. The suggested converter has four small values of inductors, four small values of capacitors. Table 1 provides the parameters of the prototype design for the proposed converter.

Table 1. Prototype parameters design for the proposed converter.

SiC MOSFET	650 V, 40 A. 35 mΩ
SiC Schottcky diode	1200 V 40 A
L_1 L_2 L_3 L_4	100 uh 2.9 mΩ 3 uh 1.5 mΩ 100 uh 2.9 mΩ 15 uh 2.2 mΩ
C_1 C_2 C_3 C_4	1 uF 100 V 200 uF 100 V 2 uF 500 V 100 uF 500 V
Vs	20–40 V
Vo	200 V
Power	200 w
Duty Cycle (D)	0.45
Fs Switching Frequency	150 KHZ
Inductor size ($L_1 = L_3 = L_4$)	(L/2.85 cm·W/2.75 cm·H/2.5 cm)
Inductor size L_2	(L/1.85 cm·W/1.5 cm·H/2 cm)

It can be observed from Table 1 that the values of the inductors become significantly smaller when a high switching frequency is employed. Furthermore, the internal resistance of the inductors is also minimized, resulting in reduced total power losses for the proposed converter. The sizes of the inductors are provided in Table 1. It is evident that by utilizing a high switching frequency, the overall dimensions of the proposed converter are substantially reduced. The inductor specifications used in the proposed converter involve a ferrite core with flat wire to minimize internal resistance. To design the inductors of the proposed converter, L_1 can be designed using Equation (52), C_1 and L_2 can be designed using Equation (53). Inductor L_3 and L_4 can be found using Equation (54). C_2, C_3, and C_4 can be determined using Equations (55), (56), and (57), respectively.

$$L_1 \geq \frac{VsD}{Fs\Delta iL1} \tag{52}$$

$$\left.\begin{aligned} L_2 &= \frac{4\pi^2 Fs^3 RL(1-D)\Delta Vc_1}{DVo} \\ C_1 &= \frac{VoD}{\Delta Vc_1 FsRL(1-D)} \end{aligned}\right\} \tag{53}$$

$$\left.\begin{aligned} L_3 &\geq \frac{VsD}{Fs\Delta iL_3(1-D)} \\ L_4 &\geq \frac{8VsD}{Fs\Delta iL_4(1-D)} \end{aligned}\right\} \tag{54}$$

$$C_2 = \frac{VoD}{\Delta Vc2RLFs(1-D)} \tag{55}$$

$$C_3 = \frac{6VoD}{FsRL} \tag{56}$$

$$C_4 = \frac{VoD}{\Delta Vc_4 FsRL} \tag{57}$$

6. Comparison the Proposed Converter with Previous DC DC Converters

In this section, the proposed converter is compared to previous DC-DC converters. Both the previous works and the proposed converter are simulated in Matlab Simulink under the same conditions. From Figure 5, it can be observed that the voltage gain of the proposed converter is higher than that of the previous converters. A higher gain at a low duty cycle implies lower conduction losses, lower switching losses, higher efficiency, and a reduced number of inductors and capacitors. As shown in Figure 6a, the power MOSFET in the proposed converter experiences lower voltage stress compared to the power MOSFET in the previous converters. However, the voltage stress across the power MOSFET in the proposed converter slightly increases as the voltage gain increases. Regarding the voltage across the diodes, Figure 6b demonstrates that the voltage stress across the diode in the proposed converter is lower than in previous DC-DC converters.

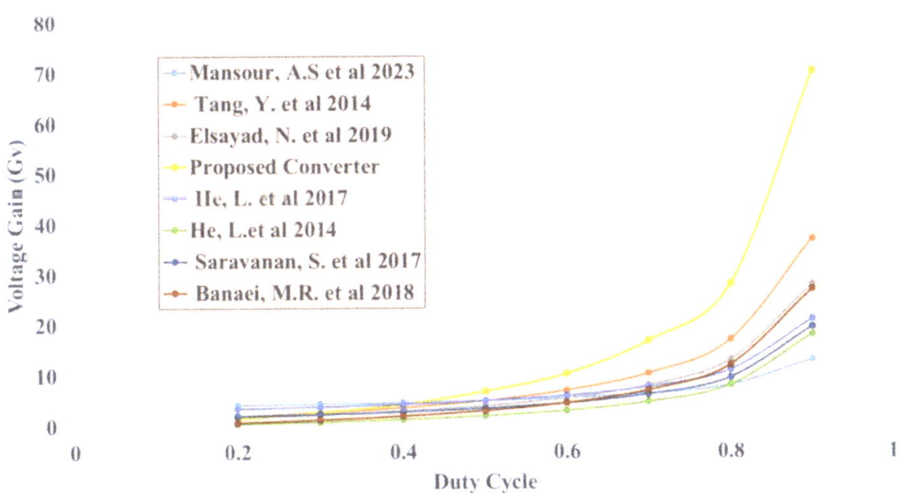

Figure 5. Voltage gain vs. Duty cycle [9,13,15,19,21,31,36].

The proposed DC-DC converter is compared with the previous DC-DC converters in Table 2. It can be seen that the proposed converter operates at a higher switching frequency than previous DC-DC converters. A low switching frequency requires high values of inductors and capacitors with high internal resistance, resulting in high switching and conduction losses, particularly at high duty cycles. Furthermore, high values of passive components result in high weight, high cost, and large size.

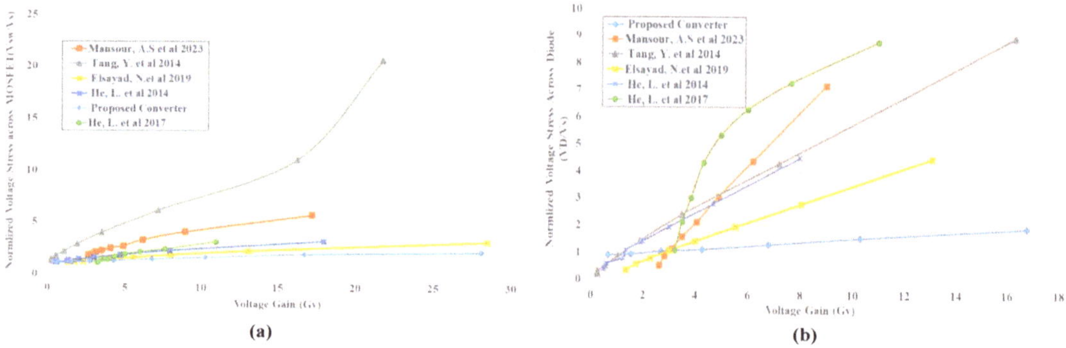

Figure 6. (**a**) Voltage stress across Power MOSFETs Vs. voltage gain, (**b**) Voltage stress across Diode Vs. Voltage Gain [9,13,15,19,36].

Table 2. Comparison between the Proposed Converter and Previous DC DC Converters.

Items	**Proposed Converter**	**Ref [36]**	**Ref [9]**	**Ref [13]**	**Ref [15]**	**Ref [19]**	**Ref [21]**	**Ref [4]**	**Ref [31]**
Switching Frequency	150	5	50	100	50	66	24	45	30
Vs	20–40	24	20–40	50 v	15 v	24	20	20	25 v
Vo	200 v	107 v	200 v	400 v	90 v	180	300 v	Out1 = 155	110 v
Inductors	4	2	4	3	3	3	2	2	4
capacitors	4	3	1	5	4	3	4	8	6
Diodes	3	4	7	4	5	2	4	7	3
switches	2	1	2	1	2	1	1	1	1
Duty cycle	45%	64%	70%	72%	50%	88%	77%	77%	60%
Power (w)	200	52	200	200	200	100	250	250	110
Efficiency	96.5%	91.2%	90%	94.5%	95%	94%	93.5%	90%	94.5%
Input Current	No Pulsating	Pulsating	Pulsating	Pulsating	No Pulsating	Pulsating	No Pulsating	Pulsating	Pulsating
Gain	$\frac{D}{(1-D)^2}$	$\frac{(4-3D)}{(1-D)}$	$\frac{(1+3D)}{(1-D)}$	$\frac{(1+2D)}{(1-D)}$	$\frac{(3-D)}{(1-D)}$	$\frac{(2D)}{(1-D)}$	$\frac{(3+D)}{2(1-D)}$	$\frac{3}{(1-2D)}$	$\frac{(3D)}{(1-D)}$

The proposed converter can step up a low input voltage at a duty cycle of 45% with a load of 200 W, but other converters in Table 2 can step up a low voltage at a high duty ratio. A high diode count leads to high internal resistance and high forward voltage (Vf), limiting the voltage gain and increasing the converter losses. In addition, a high diode count increases the reverse recovery time, which also affects the system's performance. In addition, the proposed converter has zero pulsating input current at low and high duty cycles compared to the input current in the previous DC-DC converter. In terms of the voltage gain equations of the proposed and previous DC-DC converters, as shown in Table 2, the proposed converter can achieve a higher voltage gain than the previous DC-DC converters listed in Table 2. Moreover, this means that the proposed converter is more efficient for applications that require high DC voltage gain at different loads, with more flexibility in the duty ratio. It achieves an efficiency of 96.5%. Additionally, the proposed converter is more suitable for RES.

7. Control Strategy of the Proposed Converter

The suggested controller, as shown in Figure 7, uses dual PI controllers to enable the proposed converter to operate with high performance. The first PI controller, known as the inner loop controller, is designed to control the load current. The second PI controller, called the outer loop controller, is responsible for controlling the output voltage of the proposed converter. The voltage controller takes the difference between the desired voltage and the actual output voltage as its input. It then generates a reference current for the load based on this difference. This reference current is limited intentionally to prevent the converter from using high current. The difference between the reference current and the actual current is then used as input for the current controller. The Proportional and Integral gain parameters of the Pi controller are denoted as Kp and Ki, respectively. The parameters Ki and Kp of the outer loop are set to be ten times faster than the parameters of the inner PI loop controller. The method used for tuning the controller parameters is a trial-and-error approach.

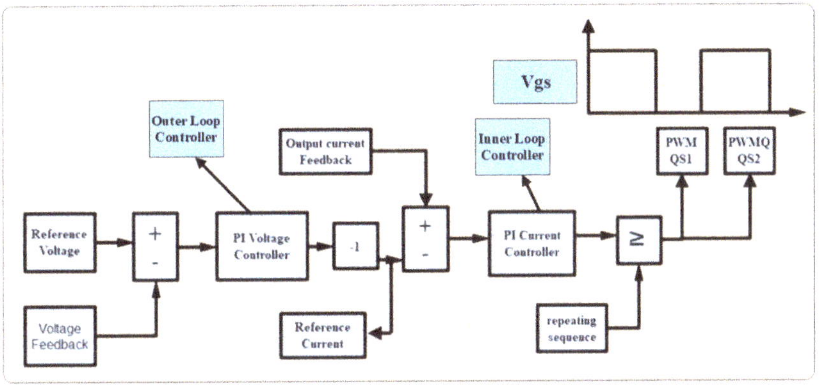

Figure 7. The suggested Controller strategy of the proposed Converter.

After applying the suggested dual PI controller to the proposed converter to verify the fixed output voltage under variable input voltage, Figure 8a shows that the output voltage of the proposed converter stays at 200 V even when the input voltage drops to its lowest values. This means that the suggested controller for the proposed converter is more reliable for applications that need a high consistent output voltage, and it can handle a wide range of duty ratios for higher power density in renewable energy applications. In Figure 8b, it is demonstrated that the converter can provide a variable output voltage ranging from 100 V to 250 V while keeping the input voltage fixed. It also responds quickly to changes, ensuring a swift adjustment in the output voltage.

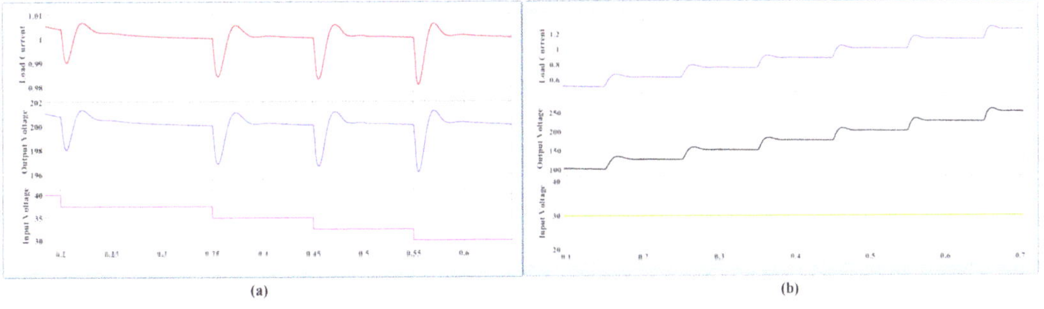

Figure 8. (**a**) output voltage of the proposed converter at variable input voltage, (**b**) variable output voltage at fixed input voltage.

8. Simulation and Experimental Results and Discussion

In this section, a 200 W PCB prototype is designed to validate the experimental results, as shown in Figure 9a. An experimental test is performed in the laboratory for the proposed converter, as shown in Figure 9b. Additionally, MATLAB Simulink and PLECS software are used to verify the experimental results in different cases.

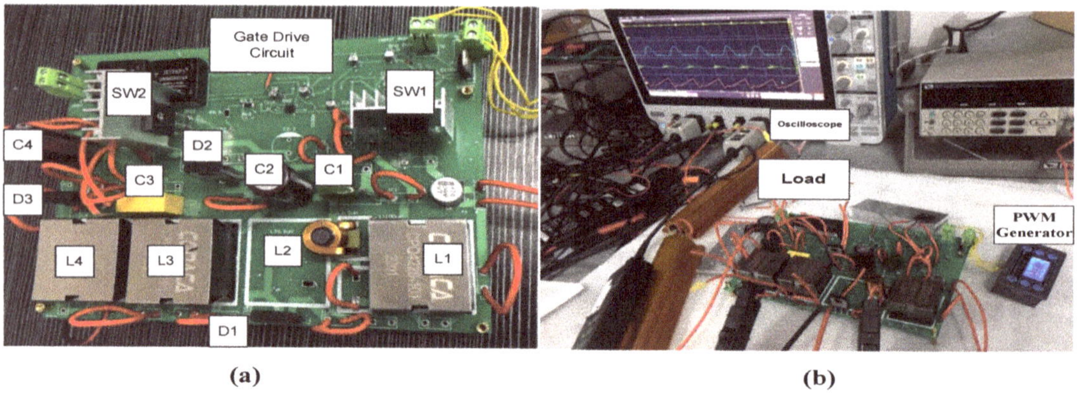

Figure 9. (**a**) PCB prototype of the proposed converter (**b**) experimental test of the proposed converter.

Figure 10a shows the source-to-gate voltage with a duty cycle of 27% and an output voltage of 218 V, with the input voltage at 40 V. It can be seen that the load current is 0.44 A at 95 W. Figure 10b shows the voltage across capacitors C_1, C_2, and C_3. It can be observed that Vc_1 is equal to the difference between Vs and Vc_2, while the voltage across C_2 is 54.6 V. Furthermore, the average voltage across C_3 is zero. In Figure 10c, the current through switching Sw_1 and Sw_2 and the voltage across Sw_1 and Sw_2 are depicted. It can be seen that the current through Sw_1 is equal to iL_1 and iL_2, matching the shape of both inductor currents. Additionally, the voltage across the MOSFET significantly decreases when the converter operates at low duty cycle.

Figure 10d illustrates the current through D_1, D_2, and D_3. The current through D_1 is in series with L_2 to prevent iL_2 from starting in the reverse direction, having the same shape as the current in L_2. This prevents the input current from becoming pulsating at a low duty cycle. The current across D_2 is equal to iL_2 and iL_1 when D_2 is on, while the current through D_3 is equal to iL_3 and iL_4 when D_3 is on. Figure 10e displays the voltage across diodes D_1, D_2, and D_3. It can be observed that the voltage across power diodes significantly decreases when the converter operates at a low duty cycle. Figure 10f shows the voltage across inductors. The voltage across inductor L_1 is substantially reduced in the on state, equal to the input voltage, while in the off state, it is equal to half the input voltage. The voltage across L_2 in the on state has the same shape as Vc_1, and in the off state, VL_2 is equal to Vc_2 for a short period of time. VL_3 and VL_4 are equal to the value of Vc_2, which is very small during the on state and equal to the output voltage during the off state for a very short time as well. This means low voltage stress on the inductors, reducing the total losses of the proposed converter.

Figure 11a indicates that current flows through the inductors during the on-state period. It is evident that the discharging time of L_2 is smaller than that of L_3 and L_4. Additionally, the discharging time of L_2 is denoted as β_1 when the proposed converter operates in DCMC1. Furthermore, the discharging times of L_3 and L_4 are equal to D1, and both discharging times (D1 and β_1) depend on the values of RL, Vo, D. The inductor L_1 exhibits a long discharge time, does not reach zero, and shows no pulsation during the off states. Figure 11b illustrates the current through the inductors when the proposed converter operates in CCM. Figure 11c presents the load current of the proposed converter, which is 1 A at an input voltage of Vs = 20 V and an output voltage of 212 V at 200 W. Figure 11d depicts an input voltage of Vs = 40 V and an output voltage of 208 V at a load current of 1 A at 200 W. It can be observed that the current through L_1 is still in CCM, while the current through L_3 and L_4 is in DCM at 200 W. This approach aims to reduce voltage stress across Sw_2 and inductors L_3 and L_4, thereby enhancing the performance and efficiency of the proposed converter. Figure 11e shows the voltage across D_1, and it can be seen that in the on state, the voltage across D_1 is around 22 V during the period $(D - \beta_2 < t < D)$. D_1 works in ZCS, and inductor L_2 is an open circuit during this time. Therefore, when the input voltage is decreased to the minimum value, the proposed converter can work with high performance with one power diode working in ZCS, reducing the number of passive components. In addition, the voltage stress across D_2 will be reduced when the converter works in DCMC2 when capacitor C_1's charge becomes zero at $(D - \beta_2)$.

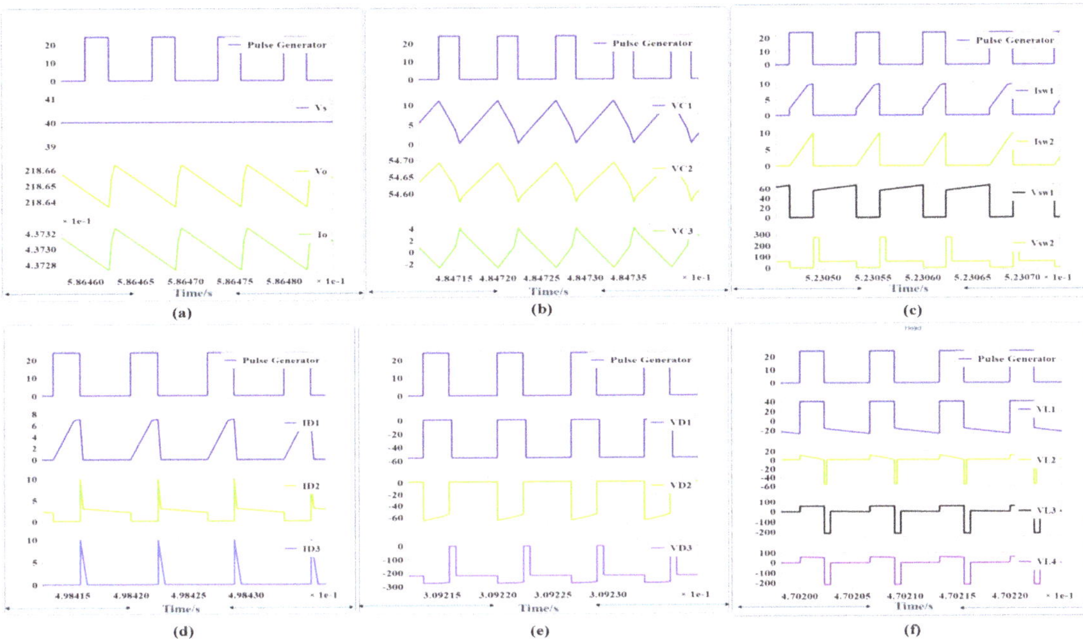

Figure 10. (**a**) Vs, Vo, Io, (**b**) Vc_1, Vc_2, Vc_3, (**c**) Isw_1, Isw_2, Vsw_1, Vsw_2, (**d**) ID_1, ID_2, ID_3, (**e**) VD_1, VD_2, VD_3, (**f**) VL_1, VL_2, VL_3, VL_4.

Figure 11f shows that the current reduction of Sw_1 occurs when capacitor C_1's charge becomes zero at $(D - \beta_2)$. After this period, the current through Sw_1 will come only from L_1. This means that the RMS current of Sw_1 will be significantly reduced when the converter operates at high current. The value of β_2 depends on the values of C_1 and L_2. This means that the proposed converter can supply high load current with high efficiency, especially for battery charging and renewable energy applications.

Figure 12a shows the inductor currents L_1 and L_2 when the converter operates in DCMC1 at D = 0.45. In Figure 12b, the inductor currents L_1 and L_2 can be observed when the proposed converter operates in DCMC2. Figure 12c demonstrates that the current through inductors L_3 and L_4 operates in DCM, with both inductors having the same value but in opposite directions. This method aims to reduce voltage stress across Sw_2 during the very long period of time $(D_1 < t < Ts)$, as previously mentioned.

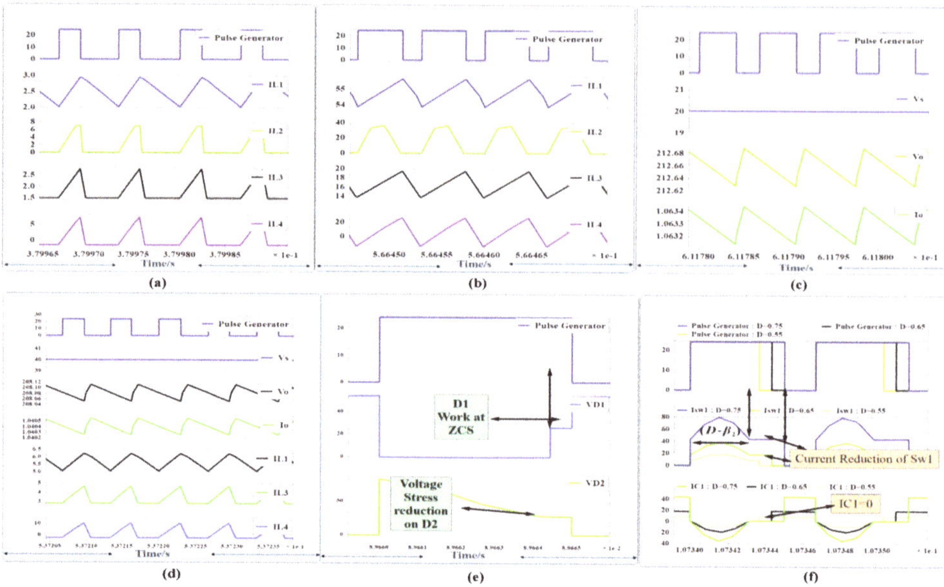

Figure 11. (**a**) iL_1, iL_2, iL_3, iL_4 at low duty cycle, (**b**) iL_1, iL_2, iL_3, iL_4 at CCM, (**c**) Vs, Vo, Io, (**d**) Vs, Vo, Io, iL_1, iL_3, iL_4, (**e**) VD_1, VD_2, (**f**) Isw_1 and IC_1 at different D.

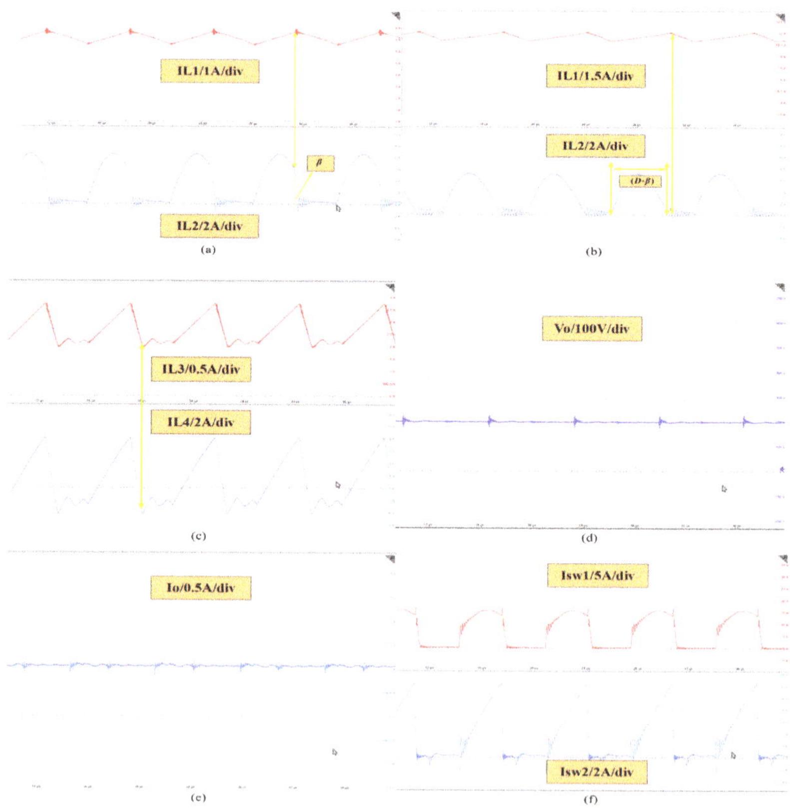

Figure 12. (**a**) iL_1, iL_2, (**b**) iL_1, iL_2, (**c**) iL_3, iL_4, (**d**) Vo = 200 V, (**e**) Load Current 1 A, (**f**) Isw_1, Isw_2.

In Figure 12d, the output voltage of the proposed converter is 200 V at a load current of 1 A, as shown in Figure 12e when the proposed converter operates in DCMC2. The currents through Sw_1 and Sw_2 are depicted in Figure 12f. It can be seen that the voltage across Sw_1 is equal to 54 V, which is a very low voltage as shown in Figure 13a. Figure 13b shows the voltage across Sw_2, which remains at 54 V for a long time period ($D_1 < t < Ts$). Figure 13c illustrates the voltage stress across D_3. Furthermore, Figure 13d shows the voltage across D_1, indicating that during the on state, the voltage across D_1 is around 22 V in the period ($D - \beta_2 < t < D$), with D_1 operating in ZCS, while the inductor L_2 acts as an open circuit during this time. As mentioned earlier, the voltage stress across D_2 will be reduced when the converter works in DCMC2, as capacitor C_1's charge becomes zero at ($D - \beta_2$).

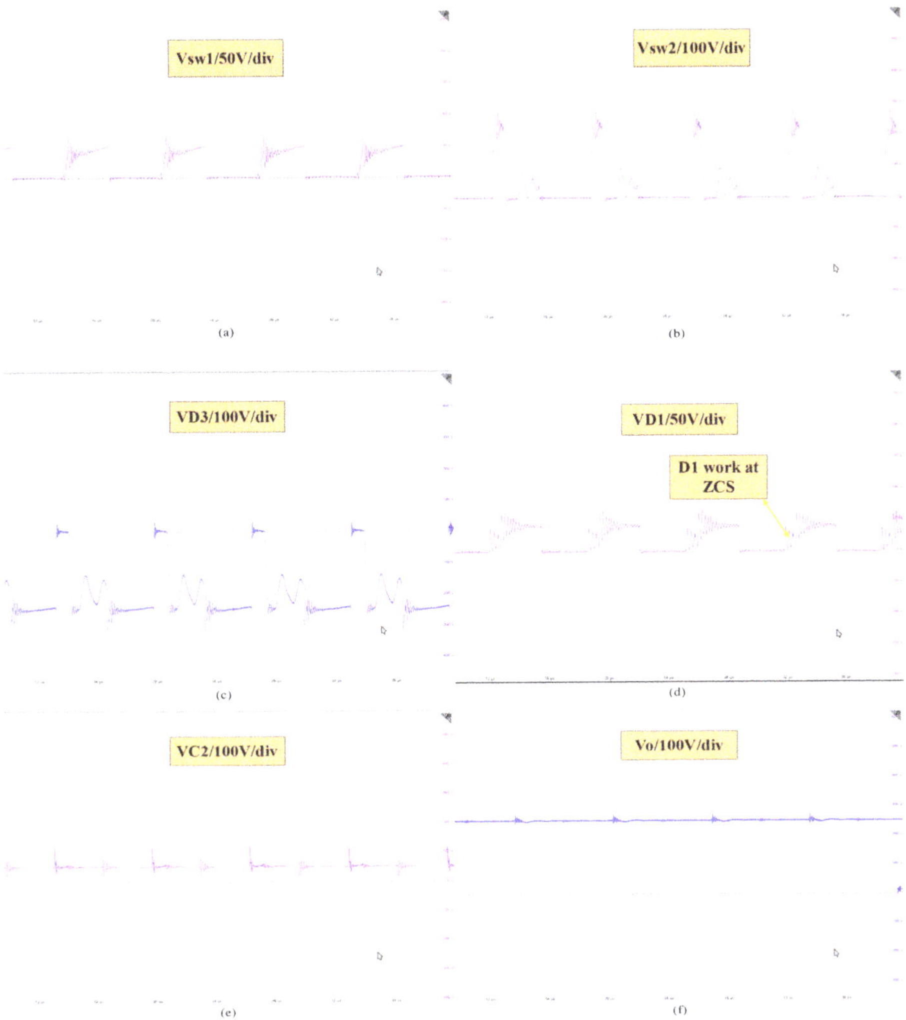

Figure 13. (**a**), Vsw_1, (**b**) Vsw_2, (**c**), VD_3, (**d**) VD_1 at ZCS, (**e**) VC_2, (**f**) Variable output voltage 250 V.

Figure 13e displays the voltage across C_2, which is equal to 52 V at D = 0.45. Additionally, the proposed converter can supply a variable output voltage, as shown in Figure 13f, where the output voltage is 250 V. These qualities make the suggested converter highly efficient and enable it to perform exceptionally well. Moreover, it can supply high currents, even when the duty ratios vary over a wide range.

The proposed converter demonstrates a higher level of performance compared to previous DC-DC converters. Notably, the power MOSFETs in the proposed converter experience significantly reduced voltage stress when operating in Discontinuous Conduction Mode (DCM) for both cases. Additionally, the single-cell switched inductor capacitor operates in resonant mode when the duty cycle exceeds 50%.

Furthermore, one of the passive elements, L_2, becomes an open circuit when the charge on capacitor C_1 reaches zero at $(D - \beta_2)$. This approach effectively reduces the voltage stress across the power MOSFETs and all diodes, while also minimizing the current stress on the main switch. This is achieved by ensuring that the current through the main switch is solely derived from L_1, which is equal to the input current.

The proposed converter also achieves a reduction in power losses, resulting in increased overall efficiency. Specifically, the efficiency of the converter reaches 96.5% at 200 W. Additionally, the input current does not reach zero at low duty cycles, making the proposed converter more efficient and well-suited for Renewable Energy Systems (RESs).

9. The Proposed Converter Efficiency Calculation

The proposed converter consists of four inductors, four capacitors, two power switches, and three diodes. These components, both passive and active, are not ideal. For example, an inductor has internal resistance, which increases as its value increases. The internal resistances of the inductors are denoted as rl_1, rl_2, rl_3, and rl_4. Similarly, the capacitors C_1, C_2, C_3, and C_4 have equivalent series resistances rc_1, rc_2, rc_3, and rc_4. The power diode also incurs power losses due to its internal resistance and forward voltage Vf. Additionally, power losses occur due to conduction and switching losses in the power MOSFET devices. Therefore, it is important to consider all of these losses for the proposed converter. The internal resistances of all active and passive elements are illustrated in Figure 14.

$$Irms = \sqrt{\frac{1}{Ts}\int_{0}^{Ts} (I)^2 dt} \tag{58}$$

$$Isw1rms = \sqrt{\frac{1}{Ts}\left[\int_{0}^{(D-\beta_2)Ts} (iL1 + iL2)^2 dt + \int_{(D-\beta_2)Ts}^{D} (iL1)^2 dt\right]} \tag{59}$$

$$Isw2rms = \sqrt{\frac{1}{Ts}\int_{0}^{DTs} (iL3 + iL4)^2 dt} \tag{60}$$

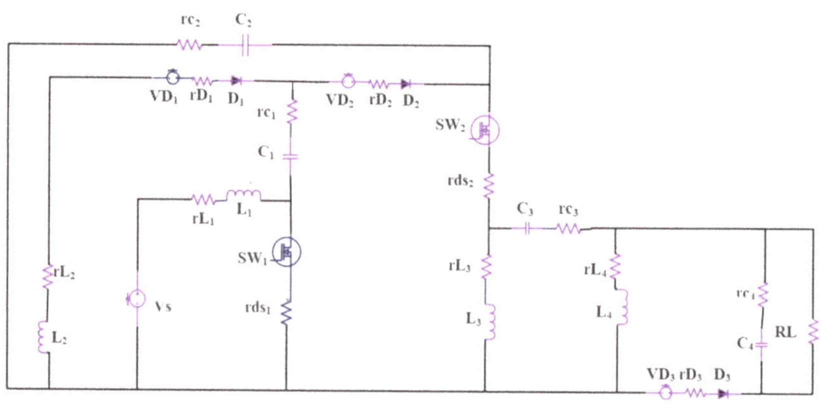

Figure 14. The proposed converter with parasitic components.

To calculate the total power losses of the proposed converter, the rms current is required for calculations related to the inductors, capacitors, power MOSFET, and diodes in both the on and off states. Equation (58) represents the general equation for rms current, and Equations (59) and (60) can be used to obtain the rms current through Sw_1 and Sw_2 during the on state.

$$ID1rms = \sqrt{\frac{1}{Ts}\int_{0}^{(D-\beta_2)Ts}(iL2)^2dt} = iL2rms \tag{61}$$

$$ID2rms = \sqrt{\frac{1}{Ts}\int_{D}^{Ts}(iL1)^2dt} \tag{62}$$

$$ID3rms = \sqrt{\frac{1}{Ts}\int_{D}^{Ts}(iL3+iL4)^2dt} \tag{63}$$

To calculate the rms current through power diodes, Equation (61) provides the rms current through D_1 during the on state, which is equivalent to the rms current through inductor L_2. Equations (62) and (63) describe the rms current across D_2 and D_3, respectively.

$$Ic1rms = \sqrt{\frac{1}{Ts}[\int_{0}^{(D-\beta_2)Ts}(iL2)^2dt + \int_{D}^{Ts}(iL1)^2dt]} \tag{64}$$

$$Ic2rms = \sqrt{\frac{1}{Ts}[\int_{0}^{DTs}(iL3+iL4)^2dt + \int_{D}^{Ts}(iL1)^2dt]} \tag{65}$$

$$Ic3rms = \sqrt{\frac{1}{Ts}[\int_{0}^{DTs}(iL4)^2dt + \int_{D}^{Ts}(iL3)^2dt]} \tag{66}$$

$$Ic4rms = \sqrt{\frac{1}{Ts}[\int_{0}^{DTs}(iL1)^2dt + \int_{D}^{Ts}(iL3+iL4)^2dt]} \tag{67}$$

To calculate the rms current through capacitors C_1, C_2, C_3, and C_4, the values can be determined from Equations (64), (65), (66), and (67), respectively.

$$Isw_1rms = \frac{IoD\sqrt{4D-3\beta_2}}{(1-D)^2} \tag{68}$$

$$Isw_2rms = \frac{Io\sqrt{D}}{(1-D)} \tag{69}$$

After solving the rms values in Equations (59) and (60), the resulting Equations (68) and (69) describe the rms current through Sw_1 and Sw_2, respectively.

$$ID_1rms = \frac{IoD\sqrt{D-\beta_2}}{(1-D)^2} = iL_2rms \tag{70}$$

$$ID_2rms = \frac{IoD}{\sqrt{(1-D)^3}} \tag{71}$$

$$ID_3rms = \frac{Io}{\sqrt{(1-D)}} \tag{72}$$

After solving for the rms values, Equations (70)–(72) provide the rms currents through the power diodes.

$$iL_1rms = \frac{IoD}{(1-D)^2} \tag{73}$$

$$iL_3rms = iL4rms = \frac{Io}{(1-D)} \tag{74}$$

Furthermore, Equations (73) and (74) provide the rms current through inductors L_1, L_3, and L_4. Equations (75)–(78) yield the rms current for capacitors C_1, C_2, C_3, and C_4, respectively.

$$Ic_1rms = \frac{IoD\sqrt{(1-\beta_2)}}{(1-D)^2} \tag{75}$$

$$Ic_2rms = \frac{IoD\sqrt{1+D(1-D)}}{\sqrt{(1-D)^3}} \tag{76}$$

$$Ic_3rms = Io\sqrt{\frac{D}{1-D}} \tag{77}$$

$$Ic_4rms = Io\sqrt{\frac{D}{1-D}} \tag{78}$$

9.1. Conduction Losses Calculation for MOSFET Devices

Conduction losses refer to the power losses that occur when current flows through a device, such as a power MOSFET or any other semiconductor device. These losses are primarily caused by the resistance of the device's conducting path, resulting in power dissipation in the form of heat.

To calculate the conduction losses of the power MOSFET in the proposed converter, the square value of the rms current is multiplied by the internal resistance of the MOSFET, as shown in Equation (79).

$$\begin{aligned} Pcd1 &= \tfrac{PoD^2(4D-3\beta_2)}{RL(1-D)^4} rds1 \\ Pcd2 &= \tfrac{PoD}{RL(1-D)^2} rds2 \end{aligned} \tag{79}$$

From Equation (79), the power conduction losses of power MOSFETs, Pcd1, and Pcd2, can be obtained.

9.2. Switching Losses Calculation for MOSFET Devices

Switching losses, also known as dynamic losses, are the power losses that occur during the switching transitions of a power electronic device, such as a power MOSFET. These losses result from the energy dissipated as the device switches between the on and off states. Switching losses are mainly caused by the charging and discharging of internal capacitances, as well as the voltage and current overlapping during the switching process.

To calculate the switching losses of the power MOSFET in the proposed converter, half of the square of the voltage stress across the MOSFET during the off state is multiplied by the switching frequency and the output capacitor of the power MOSFET (Co).

$$\begin{aligned} Esw &= \tfrac{1}{2}CoVsw^2 \\ P_{SW} &= EswFs \\ P_{SWL1} &= \frac{Vs^2D^2}{2(1-D)^2}FsCo \\ P_{SWL2} &= \frac{Vs^2D^2}{2(1-D)^4}FsCo \end{aligned} \quad (80)$$

where, (Esw) is the energy dissipated during one switching cycle. From Equation (80), the power switching losses of the MOSFETs Sw_1 and Sw_2, denoted as PSWL1,2, can be obtained.

9.3. Total Power Loss in MOSFET Devices

$$PTML1,2 = \frac{PoD^2(4D-3\beta_2)}{RL(1-D)^4}rds1 + \frac{PoD}{RL(1-D)^2}rds2 + \frac{Vs^2D^2}{2(1-D)^2}FsCoFsCo + \frac{Vs^2D^2}{2(1-D)^4}FsCo \quad (81)$$

From Equation (81), PTML1,2 can be obtained as the total power losses of Sw_1 and Sw_2 by adding Equations (79) and (80).

9.4. Losses in Power Diode

Power losses in the diode can be divided into two components: losses due to internal resistance rd and losses due to the forward diode voltage Vf. Therefore, all power losses from the three diodes in the converter must be taken into account. From Equation (82), the average current (IDav) through diodes can be calculated:

$$\left.\begin{aligned} I_{D1av} &= \frac{PoD(D-\beta_2)}{RL(1-D)^2} \\ I_{D2av} &= \frac{IoD}{(1-D)} \\ I_{D3av} &= Io \end{aligned}\right\} \quad (82)$$

In order to find the power losses due to forward voltage (Pvf), Equation (82) is multiplied by Vf, resulting in Equation (83).

$$\left.\begin{aligned} P_{Vf} &= I_{Dav}Vf \\ P_{Vf1} &= Vf_1\frac{PoD(D-\beta_2)}{RL(1-D)^2} \\ P_{Vf2} &= Vf_2\frac{IoD}{(1-D)} \\ P_{Vf3} &= Vf3Io \end{aligned}\right\} \quad (83)$$

To calculate the diode power losses due to internal resistance (rd), the square of the diode's rms current is multiplied by rd, as shown in Equation (84).

$$\left.\begin{aligned} P_{Dr} &= IDrmsrd \\ P_{Dr1} &= \frac{PoD^2(D-\beta_2)}{RL(1-D)^2}rd1 \\ P_{Dr2} &= \frac{PoD^2}{RL(1-D)^3}rd2 \\ P_{Dr3} &= \frac{Po}{RL(1-D)}rd3 \end{aligned}\right\} \quad (84)$$

The total power losses $PDL_{1,2,3}$ across the three diodes can be found in Equation (85), by adding all the losses in the power diodes.

$$P_{DL1,2,3} = P_{Dr1,2,3} + P_{Vf1,2,3} \quad (85)$$

9.5. Power Losses in Inductors and Capacitors

The inductor used in the proposed converter has a very low internal resistance, as shown in Table 1. Additionally, the inductors and capacitors of the proposed converter are designed for high switching frequency, resulting in the converter operating with very high efficiency and performance. Calculation of the power losses due to the internal resistance of the inductors and capacitors are shown below:

$$\left.\begin{array}{l} PL = iLrms^2 rl \\ PL1 = \frac{PoD^2}{RL(1-D)^4} rl1 \\ PL2 = \frac{PoD^2(D-\beta_2)}{RL(1-D)^2} rl2 \\ PL3 = \frac{Po}{RL(1-D)^2} rl3 \\ PL4 = \frac{Po}{RL(1-D)^2} rl4 \end{array}\right\} \quad (86)$$

$$\left.\begin{array}{l} PC = Icrms^2 rc \\ PC_1 = \frac{PoD^2(1-\beta_2)}{RL(1-D)^4} rc_1 \\ PC_2 = \frac{PoD^2(1+D(1-D))}{RL(1-D)^3} rc_2 \\ PC_3 = \frac{PoD}{RL(1-D)} rc_3 \\ PC_4 = \frac{PoD}{RL(1-D)} rc_4 \end{array}\right\} \quad (87)$$

From Equations (86) and (87), the power losses P_L and P_C in the inductors and capacitors can be found, respectively.

9.6. Total Power Losses in Proposed Converter

The proposed converter losses can be divided into MOSFET losses, diode losses, inductor losses, and capacitor losses. The total power loss (T_{PPCL}) of the proposed converter can be found in Equation (88), which involves summing the power losses in the power MOSFETs ($P_{TML1,2}$), the total power losses in the diodes ($P_{DL1,2,3}$), and the losses in the inductors and capacitors ($P_{L1,2,3,4}$) and ($P_{C1,2,3,4}$), respectively. The proposed converter efficiency can be obtained using Equation (89).

$$T_{PPCL} = P_{TML1,2} + P_{DL1,2,3} + P_{L1,2,3,4} + P_{C1,2,3,4} \quad (88)$$

$$\eta = \frac{Po}{Po + T_{PPCL}} 100\% \quad (89)$$

The use of SiC MOSFETs with very low on-state resistance is a better choice to reduce conduction losses. Additionally, using inductors with low values and very low internal resistance can increase the performance and efficiency of the proposed converter.

Figure 15a shows that the proposed converter's output voltage increases as the switching frequency increases. This implies that the design of the proposed converter allows for boosting low voltage to high voltage at a low duty cycle, achieving high power density with an efficiency of 96.5%. Figure 15b provides a comprehensive visual representation of the losses incurred by the proposed converter. It is worth noting that a significant proportion of these losses can be attributed to the power MOSFETs and diodes used in the converter. These losses result from the switching and conduction characteristics exhibited by these components during the converter's operation. Furthermore, a portion of the losses can be attributed to the inherent resistance of the inductors and capacitors present in the system.

Figure 16a,b illustrates the conduction power losses of MOSFETs Sw_1 and Sw_2 at different input voltages. It can be observed that the conduction power losses for both MOSFET switches slightly increased as the input voltage rose from 10 V to 40 V. Additionally, the conduction losses showed a significant decrease as the duty cycle decreased.

This indicates that the SiC MOSFET utilized in the proposed converter exhibits very low power conduction losses, particularly at variable duty cycles. Consequently, the proposed converter achieves higher efficiency compared to previous DC-DC converters. By employing WBG (Wide Bandgap) MOSFETs, both conduction and switching losses can be substantially reduced, leading to a significant increase in the converter's efficiency.

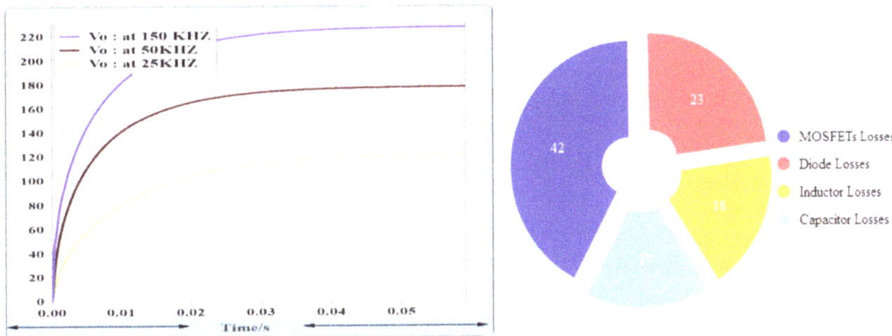

(a) (b)

Figure 15. (**a**) Output voltage of the proposed converter at different switching frequency at D = 45% (**b**) percentage losses of components of proposed converter.

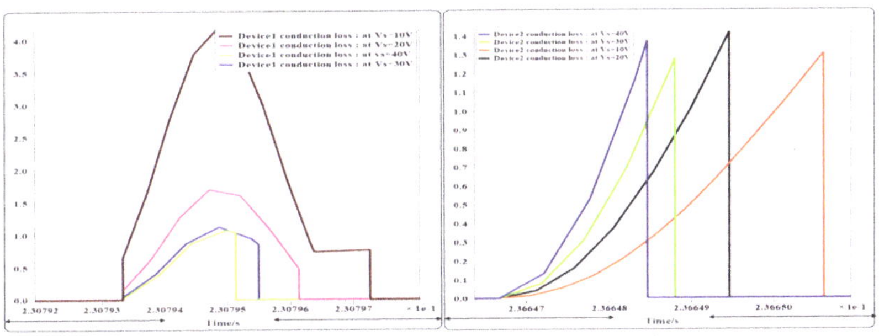

(a) (b)

Figure 16. Conduction losses of power switches at variable input voltage proposed converter (**a**) power Mosfet 1 (Sw_1) (**b**) power Mosfet 2 (Sw_2).

Figure 17a,b displays the efficiency of the proposed converter at various input voltages. The results indicate that at an input voltage of 30 V and a duty cycle of 50%, the converter achieves an efficiency of approximately 97.3% when operating at a power level of 200 W. This demonstrates that the proposed converter is capable of stepping up a low input voltage under light load current conditions, while also providing high load current at the maximum input voltage.

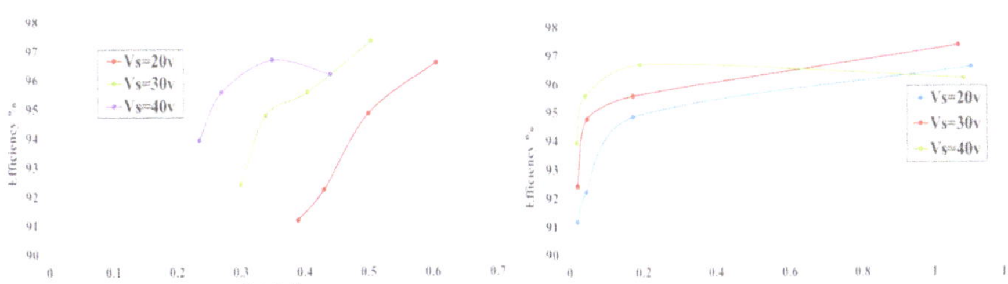

(a) (b)

Figure 17. (**a**) Efficiency of the proposed converter with duty ratio at variable input voltage (**b**) Efficiency of the proposed converter with load current at variable input voltage.

10. Conclusions

As a result, a new single-cell hybrid switched inductor DC-DC converter is proposed to demonstrate the verification of ultra-high voltage gain in photovoltaic applications. The modification in the proposed converter helps prevent the input current from pulsating to zero at very low duty cycles, making it more efficient for renewable energy applications. The single cell of the hybrid inductor is interleaved with the main switch to reduce current stress when the load current increases and the capacitor charge becomes zero. Moreover, the addition of a modified hybrid switch inductor with a capacitor, operating in dual boosting mode with a single switch, allows the converter to achieve ultra-high voltage gain.

The proposed converter offers several advantages, including ultra-high voltage gain, high efficiency, low voltage stress on power MOSFETs, diodes, inductors, and capacitors, as well as low switching and conduction losses. Furthermore, the proposed converter utilizes transformerless and non-coupled inductors. Additionally, the proposed converter's efficiency is around 96.5% when the input voltage is 20 V with a duty cycle of 0.6. The increased flexibility in the duty cycle allows the proposed converter to operate at high power density and convert very low input voltage to high output voltage for renewable energy systems.

In addition, the output voltage of the proposed converter increases when the switching frequency is increased to boost low input to high output voltage at a low duty cycle. The voltage stress on the power devices has been reduced compared to existing DC-DC converters. Moreover, using a high switching frequency reduces the component values and circuit size, resulting in a significant reduction in the weight of the proposed converter. Passive components of the proposed converter are reduced when the converter operates in (DCM) and (CCM), which improves converter efficiency and performance.

Author Contributions: Conceptualization, X.W.; Software, A.F.A.; Formal analysis, A.F.A.; Writing—original draft, A.F.A.; Writing—review & editing, A.F.A.; Supervision, X.W. All authors have read and agreed to the published version of the manuscript.

Funding: This research received no external funding.

Data Availability Statement: The data presented in this study are available on request from the corresponding author.

Conflicts of Interest: The authors declare no conflict of interest.

References

1. Mirzaei, A.; Rezvanyvardom, M.; Najafi, E. A fully soft switched high step-up SEPIC-Boost DC-DC converter with one auxiliary switch. *Int. J. Circuit Theory Appl.* **2019**, *47*, 427–444. [CrossRef]
2. Algamluoli, A.F. Novel Controller for DC-DC Cuk Converter. In Proceedings of the 2019 1st Global Power, Energy and Communication Conference (GPECOM), Nevsehir, Turkey, 12–15 June 2019; pp. 117–121. [CrossRef]
3. Gules, R.; dos Santos, W.M.; dos Reis, F.A.; Romaneli, E.F.R.; Badin, A.A. A Modified SEPIC Converter with High Static Gain for Renewable Applications. *IEEE Trans. Power Electron.* **2013**, *29*, 5860–5871. [CrossRef]
4. Hassani, M.Y.; Maalandish, M.; Hosseini, S.H. A Single-Switch High Step-Up DC–DC Converter with Low Input Current Ripple for Renewable Energy Applications. In Proceedings of the 2022 13th Power Electronics, Drive Systems, and Technologies Conference (PEDSTC), Tehran, Iran, 1–3 February 2022; pp. 186–190. [CrossRef]
5. Górecki, P.; Górecki, K. Methods of Fast Analysis of DC–DC Converters—A Review. *Electronics* **2021**, *10*, 2920. [CrossRef]
6. Gorecki, P.; Wojciechowski, D. Accurate Electrothermal Modeling of High Frequency DC–DC Converters with Discrete IGBTs in PLECS Software. *IEEE Trans. Ind. Electron.* **2023**, *70*, 5739–5746. [CrossRef]
7. Salvador, M.A.; Lazzarin, T.B.; Coelho, R.F. High Step-Up DC–DC Converter with Active Switched-Inductor and Passive Switched-Capacitor Networks. *IEEE Trans. Ind. Electron.* **2018**, *65*, 5644–5654. [CrossRef]
8. Maliat, M.; Rahman, M.S.U. Feasibility Analysis of Modified High Gain Single Ended Primary Inductor Converter (SEPIC) Based DC-DC Converter in PV Applications. In Proceedings of the 2019 2nd International Conference on Innovation in Engineering and Technology (ICIET), Dhaka, Bangladesh, 23–24 December 2019; pp. 1–5.
9. Tang, Y.; Fu, D.; Wang, T.; Xu, Z. Hybrid Switched-Inductor Converters for High Step-Up Conversion. *IEEE Trans. Ind. Electron.* **2014**, *62*, 1480–1490. [CrossRef]
10. Banaei, M.R.; Ardi, H.; Farakhor, A. Analysis and implementation of a new single-switch buck–boost DC/DC converter. *IET Power Electron.* **2014**, *7*, 1906–1914. [CrossRef]

11. Mohamed, A.K.; Makeen, P.; Bastawrous, H.A. Simulation-based Optimized Transformerless Single-Switch Quadratic DC/DC Converter with Continuous Ports Current. In Proceedings of the 2019 IEEE Conference on Power Electronics and Renewable Energy (CPERE), Aswan, Egypt, 23–25 October 2019; pp. 444–447. [CrossRef]
12. Gupta, N.; Bhaskar, M.S.; Almakhles, D.; Sanjeevikumar, P.; Subramaniam, U.; Leonowicz, Z.; Mitolo, M. Novel Non-Isolated Quad-Switched Inductor Double-Switch Converter for DC Microgrid Application. In Proceedings of the 2020 IEEE International Conference on Environment and Electrical Engineering and 2020 IEEE Industrial and Commercial Power Systems Europe (EEEIC/I&CPS Europe), Madrid, Spain, 9–12 June 2020; pp. 1–6.
13. Elsayad, N.; Moradisizkoohi, H.; Mohammed, O.A. A Single-Switch Transformerless DC-DC Converter with Universal Input Voltage for Fuel Cell Vehicles: Analysis and Design. *IEEE Trans. Veh. Technol.* **2019**, *68*, 4537–4549. [CrossRef]
14. Shahir, F.M.; Babaei, E.; Farsadi, M. Voltage-Lift Technique Based Nonisolated Boost DC–DC Converter: Analysis and Design. *IEEE Trans. Power Electron.* **2018**, *33*, 5917–5926. [CrossRef]
15. He, L.; Zheng, Z. High step-up DC–DC converter with switched-capacitor and its zero-voltage switching realisation. *IET Power Electron.* **2017**, *10*, 630–636. [CrossRef]
16. Nguyen, M.-K.; Duong, T.-D.; Lim, Y.-C. Switched-Capacitor-Based Dual-Switch High-Boost DC–DC Converter. *IEEE Trans. Power Electron.* **2018**, *33*, 4181–4189. [CrossRef]
17. Zhang, N.; Zhang, G.; See, K.W.; Zhang, B. A Single-Switch Quadratic Buck–Boost Converter with Continuous Input Port Current and Continuous Output Port Current. *IEEE Trans. Power Electron.* **2018**, *33*, 4157–4166. [CrossRef]
18. Ansari, S.A.; Moghani, J.S. A Novel High Voltage Gain Noncoupled Inductor SEPIC Converter. *IEEE Trans. Ind. Electron.* **2019**, *66*, 7099–7108. [CrossRef]
19. He, L.; Fardoun, A.A.; Zerai, A.A. Step-up/Step-down DC-DC Converter with near Zero Input/Output Current Ripples. *Int. J. Circuit Theory Appl.* **2014**, *42*, 358–375.
20. Kravetz, F.I.; Gules, R. Soft-Switching High Static Gain Modified SEPIC Converter. *IEEE J. Emerg. Sel. Top. Power Electron.* **2021**, *9*, 6739–6747. [CrossRef]
21. Saravanan, S.; Babu, N.R. Design and Development of Single Switch High Step-Up DC–DC Converter. *IEEE J. Emerg. Sel. Top. Power Electron.* **2017**, *6*, 855–863. [CrossRef]
22. Devarapalli, D.; Mehbodniya, A.; Reddy, N.M.; Gopal, A.S.N.R.; Saravanan, S.; Sudha, G.J. High Step-Up Voltage Gain Boost Chopper-Fed DC-DC Converter for Medium Voltage Applications. In Proceedings of the 2022 4th International Conference on Smart Systems and Inventive Technology (ICSSIT), Tirunelveli, India, 20–22 January 2022; pp. 792–798. [CrossRef]
23. Andres, B.; Romitti, L.; Dupont, F.H.; Roggia, L.; Schuch, L. A high step-up isolated DC–DC converter based on voltage multiplier cell. *Int. J. Circuit Theory Appl.* **2023**, *51*, 557–578. [CrossRef]
24. Kumar, S.; Kumar, R.; Singh, N. Performance of closed loop SEPIC converter with DC-DC converter for solar energy system. In Proceedings of the 2017 4th International Conference on Power, Control & Embedded Systems (ICPCES), Allahabad, India, 9–11 March 2017; pp. 1–6. [CrossRef]
25. Muranda, C.; Ozsoy, E.; Padmanaban, S.; Bhaskar, M.S.; Fedak, V.; Ramachandaramurthy, V.K. Modified SEPIC DC-to-DC boost converter with high output-gain configuration for renewable applications. In Proceedings of the 2017 IEEE Conference on Energy Conversion (CENCON), Kuala Lumpur, Malaysia, 30–31 October 2017; pp. 317–322.
26. Yasin, S.A.; Kumar, J.; Kumar, Y.; Athikkal, S.; Peter, J. Analysis of a Dual Input DC-DC Converter Topology Based on SEPIC Configuration. In Proceedings of the 2019 International Conference on Power Electronics Applications and Technology in Present Energy Scenario (PETPES), Mangalore, India, 29–31 August 2019; pp. 1–6. [CrossRef]
27. Banaei, M.R.; Bonab, H.A.F. High-efficiency transformerless buck-boost DC-DC converter. *Int. J. Circuit Theory Appl.* **2017**, *45*, 1129–1150. [CrossRef]
28. Zhang, H.; Park, S.-J. Efficiency Optimization Method for Cascaded Two-Stage Boost Converter. *IEEE Access* **2022**, *10*, 53443–53453. [CrossRef]
29. Eate, V.K.; Veerachary, M. Analysis of Switched Inductor Hybrid Buck-SEPIC two input DC-DC converter. In Proceedings of the 2016 IEEE International Conference on Power Electronics, Drives and Energy Systems (PEDES), Trivandrum, India, 14–17 December 2016; pp. 1–6. [CrossRef]
30. Ramos, J.L.; Ortiz-Lopez, M.G.; Morales-Saldana, J.A.; H., L. Control of a cascade boost converter with a single active switch. In Proceedings of the 2008 IEEE Power Electronics Specialists Conference, Rhodes, Greece, 15–19 June 2008; pp. 2383–2388. [CrossRef]
31. Banaei, M.R.; Ghabeli Sani, S. Analysis and Implementation of a New SEPIC-Based Single-Switch Buck–Boost DC–DC Converter with Continuous Input Current. *IEEE Trans. Power Electron.* **2018**, *33*, 10317–10325. [CrossRef]
32. Wu, Y.-E.; Tai, C.-H. Novel Bidirectional Isolated DC/DC Converter with High Gain Ratio and Wide Input Voltage for Electric Vehicle Storage Systems. *Batteries* **2022**, *8*, 240. [CrossRef]
33. Gorecki, P. Electrothermal Averaged Model of a Diode–IGBT Switch for a Fast Analysis of DC–DC Converters. *IEEE Trans. Power Electron.* **2022**, *37*, 13003–13013. [CrossRef]
34. Yang, L.; Sun, Q.; Zhang, N.; Li, Y. Indirect Multi-Energy Transactions of Energy Internet with Deep Reinforcement Learning Approach. *IEEE Trans. Power Syst.* **2022**, *37*, 4067–4077. [CrossRef]
35. Mitra, J.; Nguyen, N. Grid-Scale Virtual Energy Storage to Advance Renewable Energy Penetration. *IEEE Trans. Ind. Appl.* **2022**, *58*, 7952–7965. [CrossRef]

36. Mansour, A.S.; Zaky, M.S. A new extended single-switch high gain DC–DC boost converter for renewable energy applications. *Sci. Rep.* **2023**, *13*, 264. [CrossRef]
37. Alkhaldi, A.; Elkhateb, A.; Laverty, D. Voltage Lifting Techniques for Non-Isolated DC/DC Converters. *Electronics* **2023**, *12*, 718. [CrossRef]

energies MDPI

Article

Simple Design Scheme for a 300 kW Modular NPC Inverter Using Thermal RC Network Analysis

Jae-Woon Lee [1,2], Ji-Won Kim [1,*], Chee-Woo Lee [2,*], Byoung-Gun Park [1] and Jin-Kyu Lee [1]

1 Electric Machines and Drives Research Center, Korea Electrotechnology Research Institute, Changwon-si 51543, Republic of Korea; lucky_ox@keri.re.kr (J.-W.L.); bgpark@keri.re.kr (B.-G.P.); leejg926@keri.re.kr (J.-K.L.)
2 School of Electrical Engineering, Pusan National University, Busan 46241, Republic of Korea
* Correspondence: jwkim@keri.re.kr (J.-W.K.); cwlee1014@pusan.ac.kr (C.-W.L.); Tel.: +82-55-280-1335 (J.-W.K.); +82-51-510-7377 (C.-W.L.)

Abstract: This paper proposes a simple design scheme for a modular NPC inverter using thermal RC network analysis. The proposed design process is an efficient and straightforward approach to designing the heatsink for a 300 kW modular neutral-point-clamped inverter. The heatsink design plays a crucial role in achieving high power density of a power converter because the weight and size of the heatsink are primarily influenced by its type. The structure and dimensions of the heatsink are mainly determined based on the generated heat by losses of the power semiconductor switches. In this paper, a thermal RC network model was established using parameters from the power switch module and was applied to the simulation of the power converter. The thermal losses of the power semiconductor switches were calculated via this process, and the heatsink was designed according to the calculated thermal losses. The proposed design scheme was analyzed and compared with the thermal fluid dynamic model. To validate the feasibility of the proposed design process, The simulation results were compared with experimental results.

Keywords: thermal model; heat sink design; neutral-point-clamped inverter; linear electrical circuit simulation

Citation: Lee, J.-W.; Kim, J.-W.; Lee, C.-W.; Park, B.-G.; Lee, J.-K. Simple Design Scheme for a 300 kW Modular NPC Inverter Using Thermal RC Network Analysis. *Energies* **2024**, *17*, 393. https://doi.org/10.3390/en17020393

Academic Editors: Diego Bellan and Jelena Loncarski

Received: 20 November 2023
Revised: 3 January 2024
Accepted: 5 January 2024
Published: 12 January 2024

1. Introduction

Recently, electric propulsion systems have replaced mechanical propulsion systems in ships, thereby improving fuel efficiency, reducing emissions, and facilitating the adjustment of ship space [1]. These electric propulsion systems consist of an electric motor, a power converter, and a reducer that is differently used by the characteristic of a ship. Since a ship requires a high-power system, power converters have been widely used by multi-level inverters, which have advantages such as high-power conversion and a reduction in harmonics despite a low switching frequency. Especially, neutral-point-clamped (NPC) inverters have been applied to many different ships because it has relatively simple configurations compared with other multi-level topologies. Moreover, if the NPC inverter consists of modular power stacks, it can easily utilize space and expand the power capacity [2–4]. The loss of the power semiconductor device can be defined as the sum of the conduction loss, which is expressed as the product of the current flowing through the device and the applied voltage, and the switching loss, which is expressed as the product of the current flowing through the switch and the voltage applied during the ON and OFF transition periods. In particular, the switching loss may vary depending on the physical shape of the connected portion of the circuit and the characteristics of the power semiconductor device, such as the switching frequency, which varies according to the control method and the PWM implementation method [5]. In the case of a 3-level NPC inverter, switching and conduction losses are relatively large because the number of switches require a greater number of switching devices compared with a 2-level inverter, whereas the breakdown voltage of the power semiconductor device is halved. A power semiconductor with a low voltage rating is beneficial in miniaturizing the filter size, which in turn can reduce ripples [6].

Generally, the volume and size of the power converter depend on the size of the heat sink considering the topology and basic specification. Moreover, the optimized design, considering the thermal characteristics of the heat sink, can increase the power density of the inverter. The heat sink of a power converter is designed based on the heat generated from losses of a power semiconductor device. Furthermore, a design scheme that reflects the loss and heat information, which is obtained using a circuit simulation analysis program, has been studied in [7,8]. Heat loss analysis, which considers topology, power, and the load of the inverter, can minimize the size of the inverter heat sink with. This analysis can reduce the required development cost and time. As mentioned earlier, the heat generated by the power semiconductor switch is represented by the switching and conduction losses. The foundation for electro-thermal analysis of power electronic systems can be formulated by coupling the thermal model and power loss model with temperature [9,10].

There are existing common methods for thermal analysis such as numerical methods, finite element analysis, and computational fluid dynamics. These analysis methods are solved using 1D, 2D, and 3D heat diffusion equations; however, it is difficult to analyze the switching and conduction losses occurring inside the semiconductor device directly. The estimation of the junction temperature or the overall temperature at several critical locations in the power module is linked to thermal design, thermal management, reliability, and life expectancy. Accordingly, a compact thermal model represented by a thermal RC network is widely used in [11–16].

Foster and Cauer's methods are widely used for thermal RC networks. The Foster method does not thermally ground all capacitors except one in the RC network, whereas in a Cauer network, all capacitors are grounded. Although a non-grounded capacitor does not work as a physical device, it can be easily conducted to the mathematical model, whereas if a grounded capacitor works as a physical device, it is relatively difficult to interpret mathematically. A Foster network can be converted into the equivalent Cauer network, which means that the junction temperature of the Foster network can be analyzed as that of its equivalent Cauer network. The transient junction temperature in the modular inverter can therefore be predicted using the Foster or Cauer network, which can be easily expressed mathematically. The thermal RC network consists of values of the resistor and capacitor and is typically extracted by fitting the transient temperature response of the RC network to the actual temperature response curve, which is determined by the same step input of the power [17,18].

In this paper, the simple design process was proposed by RC network thermal analysis to obtain the heat dissipation of an NPC inverter module for an MW-class high-capacity inverter; the losses of the IGBT module were simply calculated by PLECS, which is an electrical circuit simulation tool, based on the datasheet of IGBT module. The IGBT thermal analysis was analyzed by using RC network thermal analysis. The optimal RC structure and RC value were used; moreover, these were reflected in the temperature saturation process. The proposed design process is an efficient and straightforward approach to designing the heatsink for a 300 kW modular neutral-point-clamped inverter. To verify the feasibility of the heat dissipation design process using circuit simulation, the heat sink design considering the heat loss was performed using the thermal RC network by PLECS. The results of this simulation were compared with the experimental results.

The remainder of this paper is organized as follows. Section 2 describes the 3-level NPC inverter, and Section 3 deals with the inverter heat dissipation design model. Section 4 analyzes the simulation and the experimental results. Finally, Section 5 presents the conclusions.

2. The 3-Level NPC Inverter

During the development of power conversion and motor systems, power conversion systems have recently been regarded as high-capacity and high-pressure systems [19].

Nevertheless, it is difficult to expect low THD and high efficiency in a high-capacity/high-voltage system using a conventional 2-level inverter. The current ripple can be reduced by increasing the switching frequency. Meanwhile, the DC voltage is kept sufficiently high; therefore, the breakdown voltage of the power semiconductor device is required. The 3-level NPC inverter structure was designed to solve this problem and has been recently used in high-voltage/high-capacity power converters [20]. The primary characteristic of the 3-level inverter is that a power semiconductor device with a lower voltage rating than the 2-level inverter can be used while maintaining an excellent ripple performance. Therefore, the filter size of the inverter can be reduced.

The NPC topology comprises four IGBTs in series with the inverter leg and a diode connected to the DC-Link neutral point, as depicted in Figure 1. This neutral point is connected to the output via a clamping diode in each phase. Moreover, the input voltage of the 3-level NPC inverter is a DC voltage, and the output pole voltage has the states VDC/2, 0, and −VDC/2.

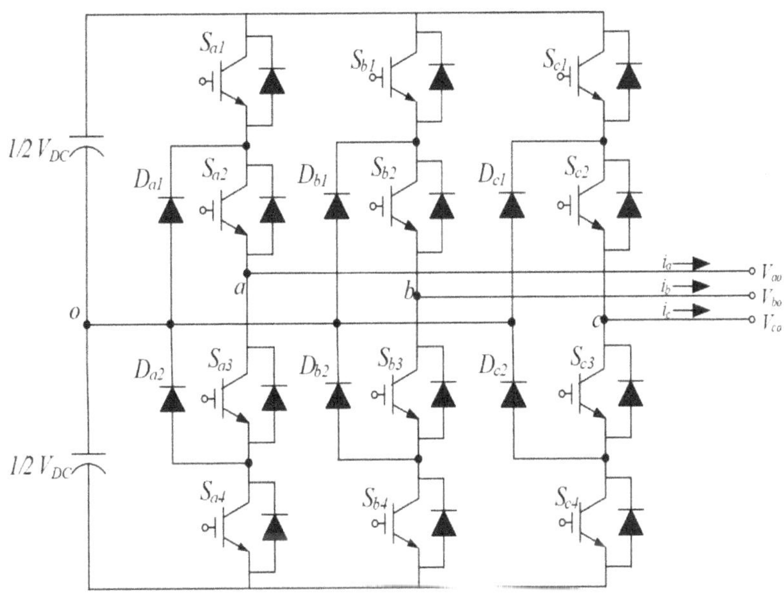

Figure 1. Level NPC type topology.

Compared to a 2-level inverter whose output voltage fluctuates from VDC/2 to −VDC/2, the voltage fluctuation width is small; therefore, the output voltage harmonics can be reduced. As illustrated in Figure 2, when the switches Sx1 and Sx2 are ON and Sx3 and Sx4 are OFF, the switching state is 'P', and the output voltage is VDC/2. Conversely, when Sx1 and Sx2 are turned OFF, and Sx3 and Sx4 are turned ON, the 'N' switching state occurs, and the output voltage is −VDC/2. Finally, when Sx2 and Sx3 are ON and Sx1 and Sx4 are OFF, the switching state is 'O', and the output voltage is 0 V. Therefore, as the state of the switching element changes between 'P-O-N,' conduction and switching losses occur, leading to heat loss.

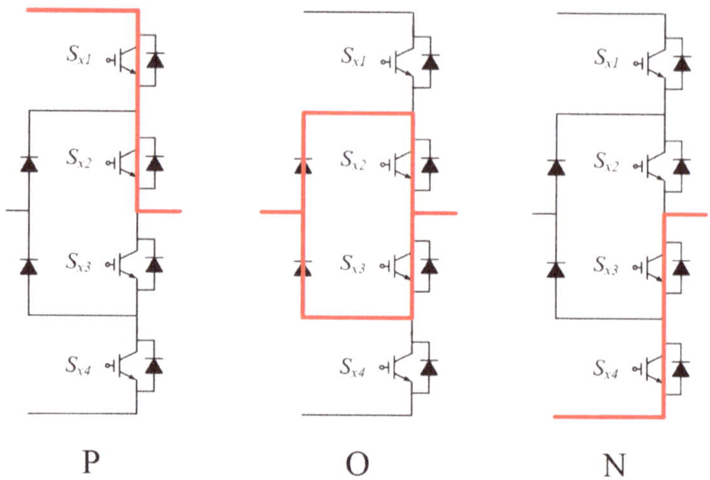

Switching State	Switching Element ($x = a,b,c$)				Output Voltage
	S_{x1}	S_{x2}	S_{x3}	S_{x4}	
P	ON	ON	OFF	OFF	$V_{DC}/2$
O	OFF	ON	ON	OFF	0
N	OFF	OFF	ON	ON	$-VDC/2$

Figure 2. Output voltage and current commutations in 3-level NPC (indicated by red lines representing current flow).

3. Inverter Heat Dissipation Design Model

The power semiconductor device is a vulnerable power electronic component, as shown in Figure 3, and the losses generated in the device directly affect the efficiency of its power conversion system. In addition, heat is generated due to the loss in the semiconductor device, which is the dominant reason behind system failure caused by stress source, as shown in Figure 4 [21–23]. Power semiconductor devices have different maximum junction temperatures, and during operation, they may exceed this temperature, resulting in damage or destruction. Therefore, when designing a power conversion system, conducting a relevant analysis is crucial.

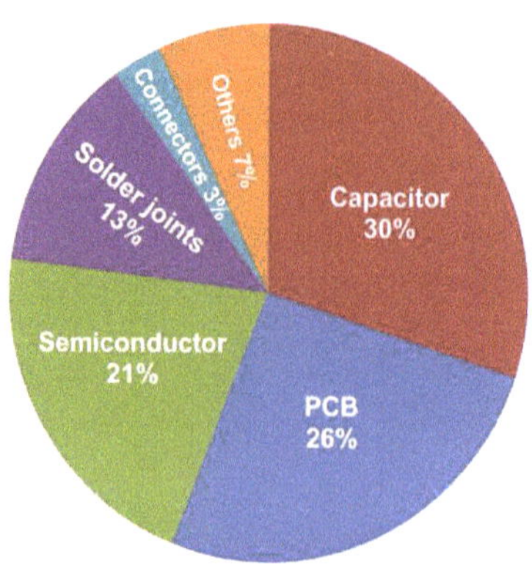

Figure 3. Vulnerable power electronics components [24].

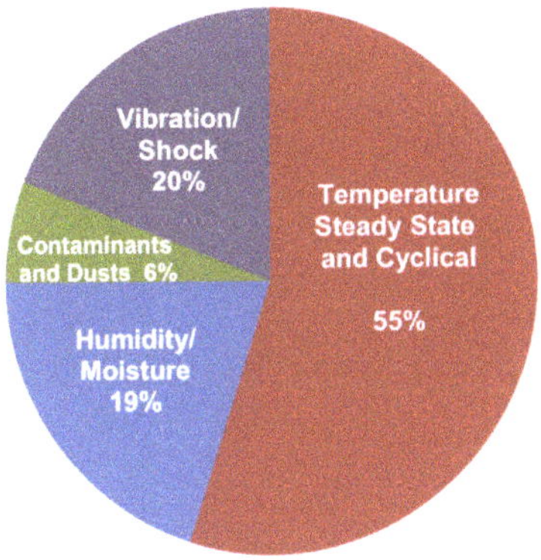

Figure 4. Major failure stress source [24].

The heat dissipation design of the inverter is based on the datasheet of the model. Moreover, the sum of the conduction and switching losses was first calculated to obtain the overall loss, as illustrated in Figure 5. Furthermore, each loss was converted into a thermal model using a simulation tool to track the device temperature in the next process.

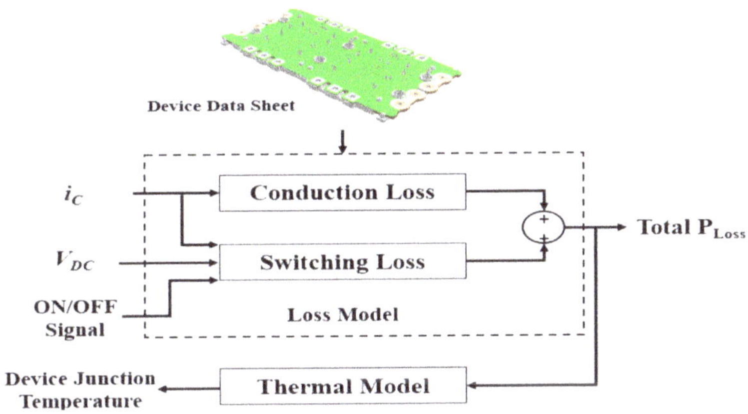

Figure 5. Loss calculation block diagram for heat dissipation design.

The maximum value of each parameter was used to generate the maximum possible power loss at that instant. The temperature distribution tendency of each joint was evaluated based on the analyzed temperature. This was further used as the main parameter of the heat sink design.

3.1. Conduction Loss

Conduction loss occurs after the power semiconductor device becomes conductive, i.e., when it performs in the saturation region. It can be calculated using the following quantities: the current flowing through the power semiconductor device, the voltage applied to it, and the *ON* resistance, as represented by (1) [25].

$$\begin{aligned} P_{cond} &= \frac{1}{T}\int V_{sw}(t)i(t)dt \\ &= \frac{1}{T}\int (V_{ON} + R_{ON}i(t))i(t)dt = V_{ON}I_{avg} + R_{ON}(i_{RMS})^2 \end{aligned} \quad (1)$$

where V_{sw} is the collector–emitter saturation voltage, V_{on} is the *ON* drop voltage, R_{on} is the *ON* resistance, I_{avg} is the average collector current, and i_{RMS} is the *RMS* collector current. In this study, heat loss was calculated by making the device voltage an arbitrary function of device temperature during conduction using PLECS, and this function is displayed as a 2D look-up table in the range of 25–125 °C, as graphed in Figure 6. It is defined as the conduction loss tab of the Thermal Library Browser. The defined function can calculate the actual voltage drop at a given function value using linear interpolation.

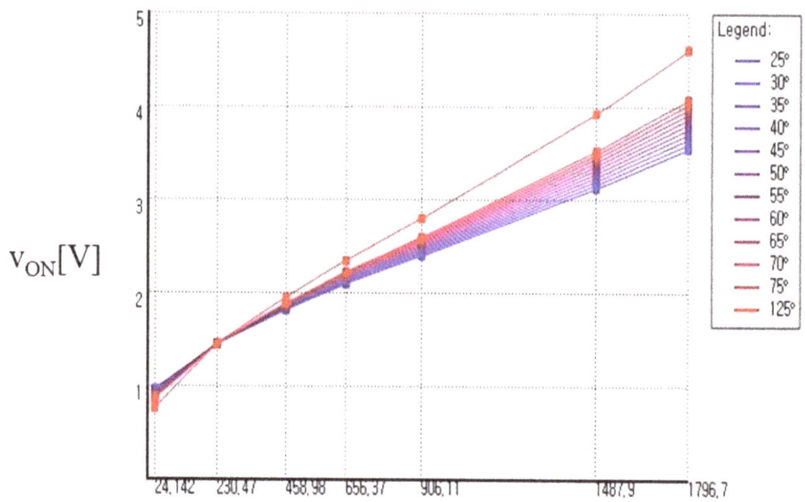

Figure 6. 2D Look-up table of heat loss calculation with arbitrary device voltage function.

3.2. *Switching Loss*

Switching loss occurs as a power loss because the power semiconductor device does not instantly switch from the *ON* state to the *OFF* state or vice versa, and its value is significantly greater than 0, as indicated by the yellow circles in Figure 7. It can be expressed as the relationship between energy loss, switching frequency, and voltage–current when the device is in the *ON* or *OFF* state, as expressed in (2).

$$\begin{aligned} P_{ON,OFF} &= E_{ON,OFF} \times f_{sw} \\ &= \frac{V_{ON,OFF}}{V_{test}} \times \frac{I_{ON,OFF}}{I_{test}} \times E_{ON,OFF} \times f_{sw} \end{aligned} \quad (2)$$

where $E_{ON,OFF}$ is turned *ON* or *OFF* energy loss, f_{sw} is the switching frequency, $V_{ON,OFF}$ is the switch input voltage, $I_{ON,OFF}$ is the switch input current, and V_{test} is the test input voltage on the datasheet. In this study, the switching loss was represented by a 3D look-up table plotted in Figure 8. Each value of the switching energy depends on the current flowing through the device, the temperature, and the blocking voltage. Moreover, the given value calculates the actual loss using linear interpolation and the conduction loss [25].

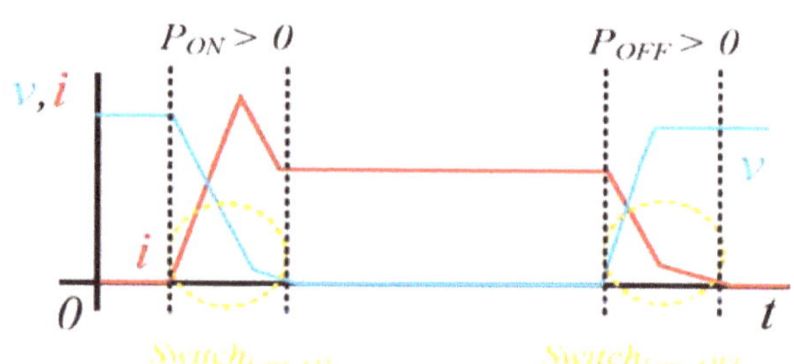

Figure 7. Switching waveforms of voltage and current.

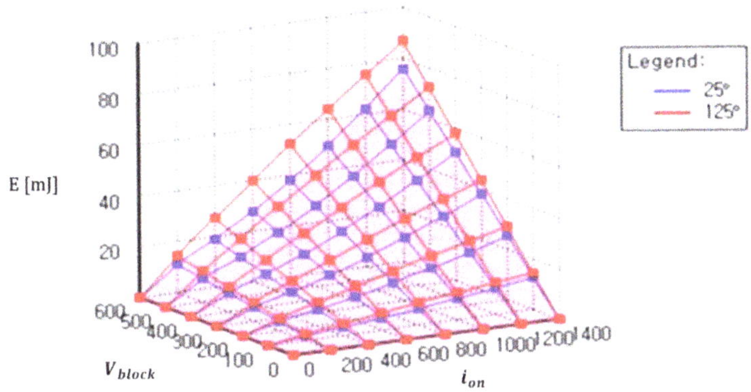

Figure 8. 3D Look-up table of switching loss representation.

3.3. Heat Sink Design

The heat sink absorbs the heat loss of the device, and the core element of the heat loss calculation is based on the conduction and switching losses, leading to heat sink design. The heat transfer is the sum of the thermal resistance and thermal capacitance, as depicted in Figure 9a. This transfer process from the heat sink to the air can be modeled using the thermal impedance, which is based on the equivalence between the electrical and thermal variables, as provided in Table 1 [26].

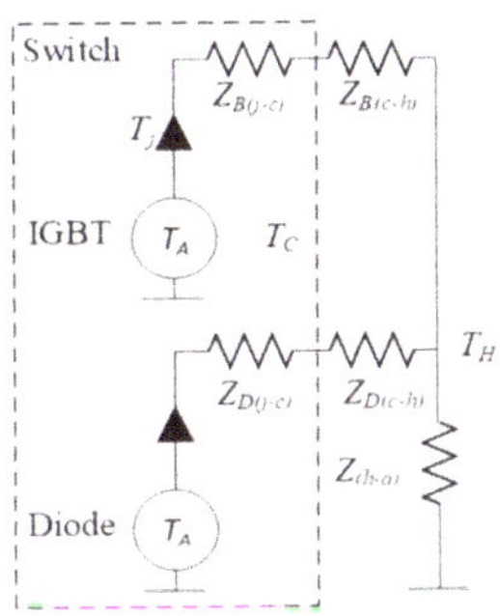

- T_j: junction temperature
- T_C: case temperature
- T_H: heat sink temperature
- T_A: ambient temperature
- $Z_{(j-c)}$: thermal impedance from junction to case
- $Z_{(c-h)}$: thermal impedance from case to heat sink
- $Z_{(h-a)}$: thermal impedance from heat sink to ambient

(**a**)

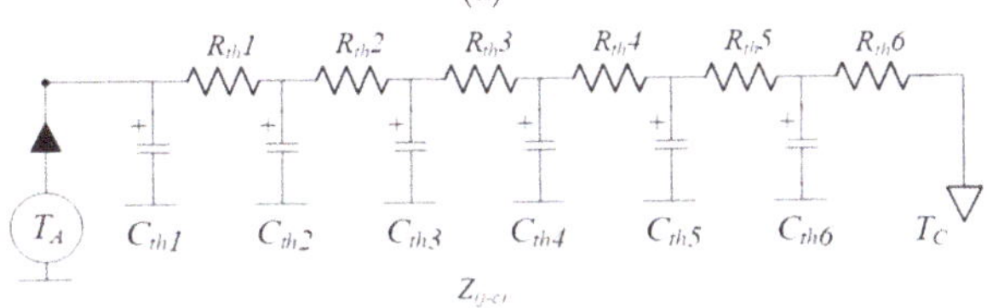

(**b**)

Figure 9. Thermal model for heat sink design. (**a**) Power device. (**b**) Thermal impedance of Cauer.

Table 1. Thermal impedance for heat transfer modeling.

Thermal Parameter	Unit	Electrical Parameter	Unit
Temperature	T in K	Voltage	V
Heat Flux	P in W	Current	A
Thermal Resistance	R_{th} in K/W	Resistance	Ω
Thermal Capacitance	C_{th} in J/K	Capacitance	F

The total loss generated by the switch is calculated using the relationship between the thermal characteristics, as expressed by (3) as follows [25]:

$$\begin{aligned} T_j[^\circ\mathrm{C}] &= P_{Loss}[\mathrm{W}] \times R_{th}\left[\frac{^\circ\mathrm{C}}{\mathrm{W}}\right] + T_A[^\circ\mathrm{C}] \\ R_{th} &= R_{th(j-c)} + R_{th(c-h)} + R_{th(h-a)} \end{aligned} \quad (3)$$

where T_j is junction temperature, P_{Loss} is switch loss power, R_{th} is the total thermal resistance, $R_{th(j-c)}$ is the thermal resistance provided by the junction to the case, $R_{th(c-h)}$ is the thermal resistance provided by the case to the heat sink, $R_{th(h-a)}$ is the thermal resistance of the heat sink to the ambient temperature, and T_A is the ambient temperature.

In this study, using the Cauer model, the system heat transfer function of each RC element is considered as depicted in Figure 9b, and the resultant step response of the thermal impedance can be expressed as follows (4) [27]:

$$Z_{thermal\_transfer}(s) = \frac{1}{s \times C_{th\_1} + \frac{1}{R_{th\_1} + \frac{1}{s \times C_{th\_2} + \ldots + \frac{1}{R_{th\_n}}}}} \quad (4)$$

where $Z_{thermal\_transfer}$ is the thermal transfer function of the Cauer network, and C_{th} and R_{th} are the thermal capacitance and resistance, respectively.

Figure 10 presents the loss model of the PLECS simulation. This loss calculation method incorporates the ideal switch simulation concept combined with measured loss data provided in a lookup table format. The simulation of switch losses is conducted as follows:

- The values of current (I) and voltage (V) applied to the switch are provided to look up the table after the switching event;
- The calculated losses are added to the thermal circuit;
- Temperature is fed back to look up the table for the next event;
- Losses are calculated as functions of current, voltage, and temperature during switching;
- The loss calculation is based on interpolation.

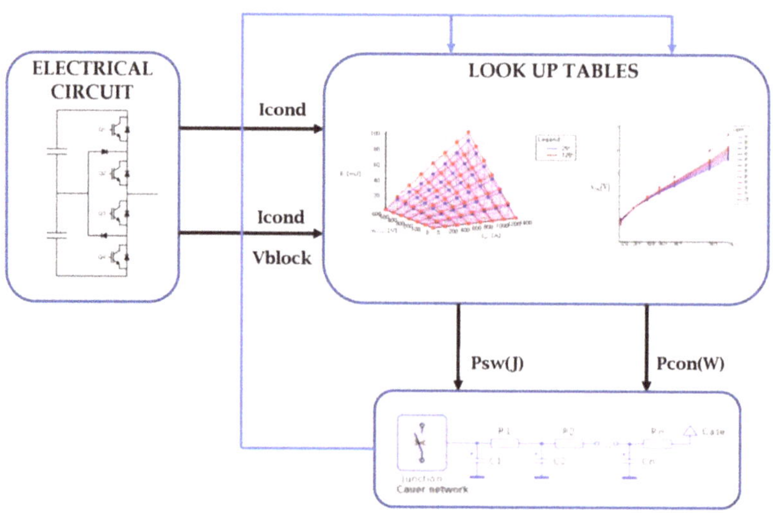

Figure 10. PLECS simulation switch loss model.

In this paper, the approach involves utilizing the switch loss data obtained via PLECS simulation to inform the design of a heat sink using computational fluid dynamics (CFD). The data obtained via simulation comprise both power loss and heat loss of the switch, providing insights into the thermal characteristics. By integrating these data into CFD, it becomes possible to precisely analyze the thermal behavior of the heat sink and design an optimal heat sink.

CFD is a powerful tool used for predicting fluid dynamics and heat transfer, allowing for the calculation of real-world thermal and flow characteristics. It enables the consideration of geometric structure, surface properties, and cooling flow paths in the optimal heat design. Furthermore, CFD allows for the simulation of heat transfer characteristics in real environmental conditions, accounting for the impact of airflow around the heat sink and its exposure to the surrounding environment. This approach facilitates the validation of heat sink design effectiveness by comparing experimental data with simulation results obtained via CFD. This methodology is expected to contribute to the effective design of a heat sink, taking into consideration thermal efficiency and system stability.

4. Simulation and Experiment Results

The heat loss analysis of the proposed 300 kW modular 3-level NPC inverter for the heat dissipation design was simulated by PLECS and is illustrated in Figure 11. The module used was Vincotech70-W424. It consists of the Buck IGBT, diode, Boost IGBT, diode, and Clamp diode, as shown in Figure 12. Each thermal resistance and capacitance are different. The modules and overall simulation parameters are detailed in Tables 2 and 3. As shown in Table 3, since each heat impedance value is different for each heat sink, if all elements are simulated by using the shared heat sink as shown in Figure 11b, the impedance of each element cannot be reflected in the heat sink and only one common impedance is simulated. Therefore, it must be simulated by using the impedance of each element with a separated heat sink, as shown in Figure 11a.

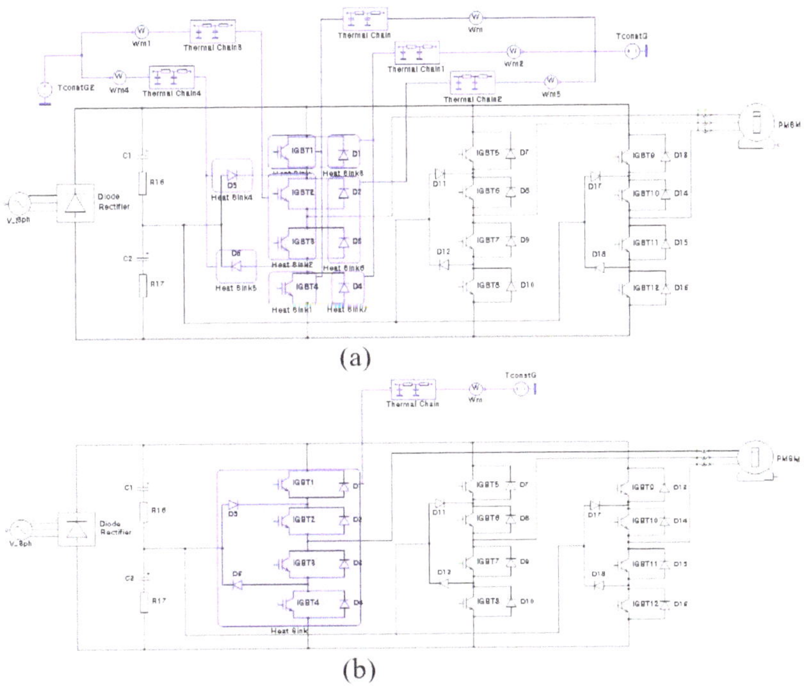

Figure 11. PLECS simulation circuit for 3-level NPC heat loss analysis at 25 °C. (**a**) Separated heat sink; (**b**) shared heat sink.

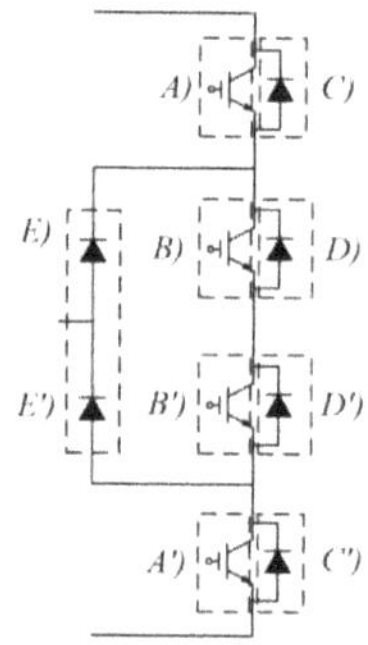

Figure 12. Vincotech 70-W424 module identification.

Table 2. System parameters.

Parameter	Value	Unit
Rated Power	300	kW
Line Frequency	60	Hz
Output Voltage	380	V
Output Current	580	A
DC link Voltage	600	V
DC link Capacitance	3.1	mH
Switching Frequency	1000	Hz

Table 3. The 300 KW thermal resistance (K/W) and thermal capacitance (J/K) simulation parameters.

Thermal Impedance	Z(j−c)					
	Sec1	**Sec2**	**Sec3**	**Sec4**	**Sec5**	**Sec6**
Buck IGBT Resistance	0.003	0.007	0.01	0.008	0.013	0.012
Buck IGBT Capacitance	0.017	1.198	2.022	18.09	44.46	196.2
Boost IGBT Resistance	0.005	0.011	0.011	0.015	0.012	0.002
Boost IGBT Capacitance	1.564	1.604	16.83	40.65	148.7	3687
Buck Diode Resistance	0.009	0.028	0.014	0.013	0.011	0.005
Buck Diode Capacitance	0.3	0.854	5.05	25.6	143.4	1619
Boost Diode Resistance	0.006	0.022	0.016	0.033	0.019	0.005
Boost Diode Capacitance	0.261	0.736	1.935	15.12	81.81	1473
Clamp Diode Resistance	0.005	0.01	0.011	0.015	0.012	0.003
Clamp Diode Capacitance	1.564	1.604	16.83	40.65	148.7	3687

Z(j−c): thermal impedance junction to case; Sec1–6: Thermal resistance and capacitance values for each section of the IGBT module, consisting of six layers (Sec1–6), were obtained from the datasheet.

Switching loss, conduction loss, and thermal impedance data parameters using a thermal editor were input to the IGBT element and the clamping diode, and the resultant values were derived via the set air temperature (25 °C) and heat sink. The thermal impedance values in the datasheet are the Foster network values, which, as mentioned earlier, can be mathematically converted to an equivalent Cauer network. This was executed automatically in the PLECS Thermal Editor to predict the transient junction temperature in the inverter. The average inverter loss value generated by the device was compared with the heat flow of the thermal circuit, and it was observed that the two values converged in the steady state, as plotted in Figures 13 and 14. This means that thermal and electrical variables correspond with each other. Figure 13 is the result of using a separated heat sink, and Figure 14 is the result of using a shared heat sink.

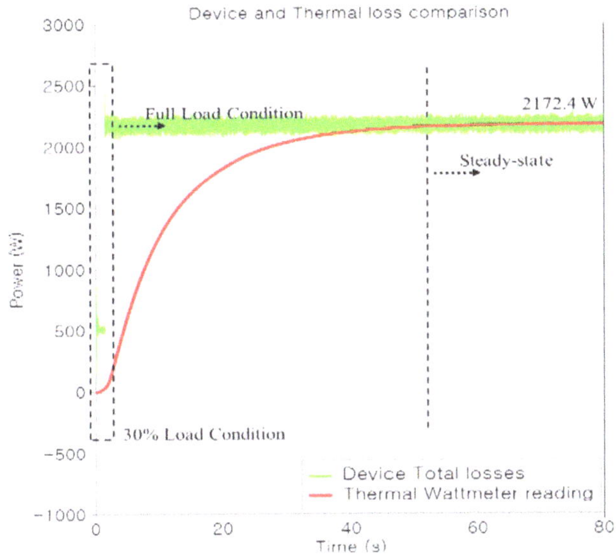

Figure 13. Comparison between the variation in the total inverter loss by the device and the thermal wattmeter reading with separated heat sink.

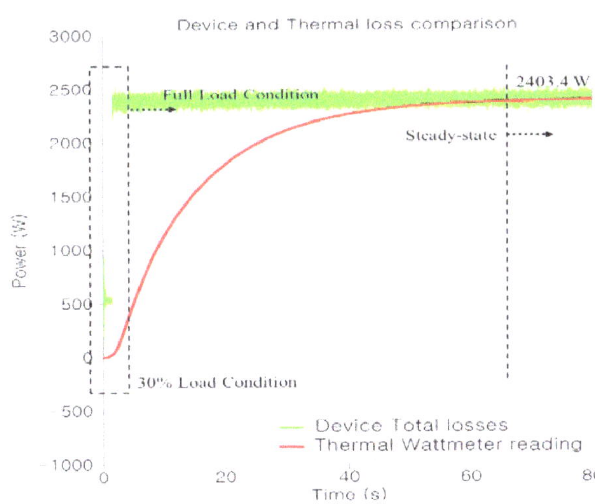

Figure 14. Comparison between the variation in the total inverter loss by the device and the thermal wattmeter reading with shared heat sink.

As mentioned earlier, as a result of using the shared heat sink, the impedance of each element was the same. It leads to a larger impedance applied to a relatively small impedance. Hence, it causes more loss. The conduction loss observed in Boost IGBT was higher than that of Buck IGBT. This was due to the long duration of IGBT conductivity, as depicted in Figure 15. The switching losses in Buck IGBT were higher than those observed in the case of Boost IGBT because the former had more voltage and current crossing points when switching, as illustrated in Figure 16. As a result of this, the derived results showed that the inverter loss occurring at the device was approximately 2.1 kW, specified in Table 4.

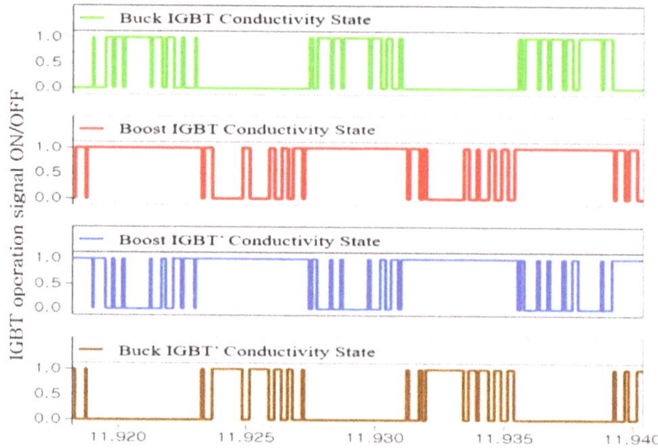

Figure 15. IGBT ON/OFF conductivity state.

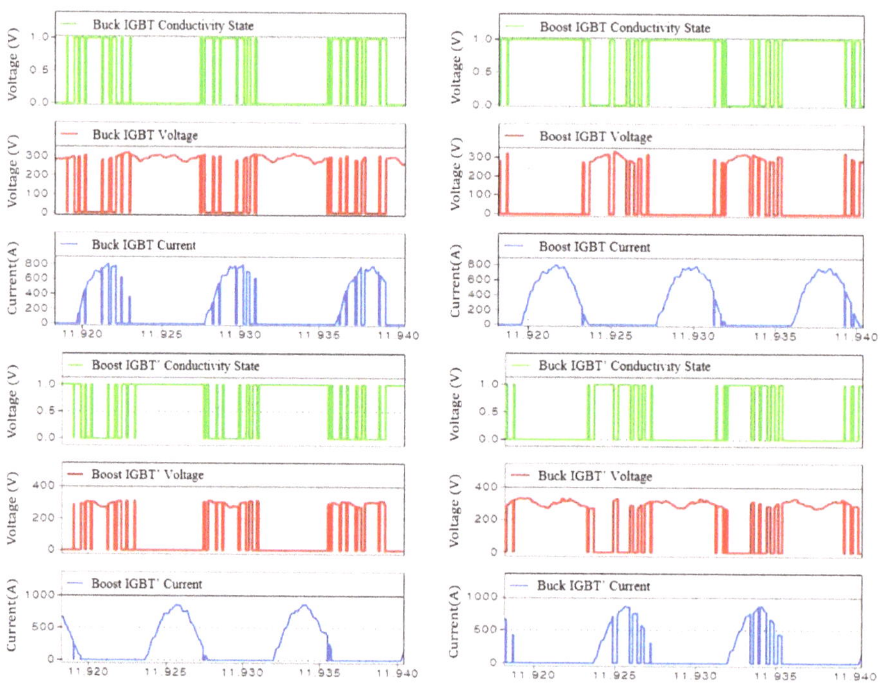

Figure 16. Switching voltage and current waveforms.

The overall 3-phase system underwent an analysis for a total heat loss of 6.3 kW, yielding an inverter efficiency of 97.9%. The highest device temperature recorded was approximately 101 °C, with the peak temperature of the heat sink reaching 78 °C. To validate the reliability of the PLECS simulation, design, and product manufacturing, thermal impedance and device loss data were incorporated via CFD analysis, as presented in Figure 17. The summarized results in Table 5 indicate a maximum temperature difference of about 1.6 °C for the heat sink. Figure 18 illustrates the heat sink designed via simulation and CFD analysis.

Table 4. Device loss and temperature simulation results using separated heat sink (1-phase).

	Buck IGBT	Boost IGBT	Boost IGBT′	Buck IGBT′	Total
SW_{Loss} (W)	11	2.8	2.7	10.7	27.2
$Cond_{Loss}$ (W)	388	490	491	390	1759
Junction °C	89.7	105.1	105.4	93.6	-
Heatsink °C	72	78	78	72	-
	Boost Diode	**Buck Diode**	**Buck Diode′**	**Boost Diode′**	
SW_{Loss} (W)	0.45	0.002	0.0005	0.47	0.9225
$Cond_{Loss}$ (W)	11.6	11.6	12.03	12.01	47.24
Junction °C	35	34	34	35	-
Heatsink °C	28.7	27.4	27.2	28.5	-
	Clamp Diode		**Clamp Diode**		
SW_{Loss} (W)	4.2		3.97		8.17
$Cond_{Loss}$ (W)	165.1		164.8		329.9
Junction °C	65.3		65.6		-
Heatsink °C	55		55		-
					2172.4

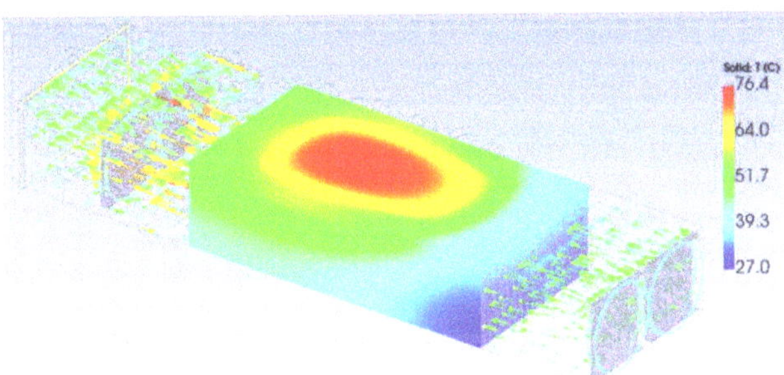

Figure 17. Heat sink design using CFD.

Table 5. Heat sink analysis using CFD.

Flow Rate	Air Temperature	Thermal Impedance/Device Loss	Max Temperature	Pressure Drop
4 Fans	25 °C	datasheet	76.4 °C	75 Pa

The complete 3-level NPC modular experimental system is depicted in Figure 19. It consists of a control module, a three-phase power module, a capacitor bank, and a rectifier module, and its size can be adjusted in a rack with easy attachment and detachment. In the experiment, the temperature sensor could not be attached directly to the semiconductor element inside the NPC module; therefore, the temperature generated at four points around the NPC module was measured, as illustrated in Figure 20. In Figure 21, Channel 1 represents the motor's angular positions utilizing encoder signals, while Channel 4 illustrates the output phase current. Channels 2 and 3 depict the voltage across the upper and lower capacitors of the DC link, respectively. In the context of the neutral-point-clamped (NPC) inverter, maintaining the voltage balance between the capacitors in the DC link is crucial, as any voltage disparity among them can have a direct impact on the output.

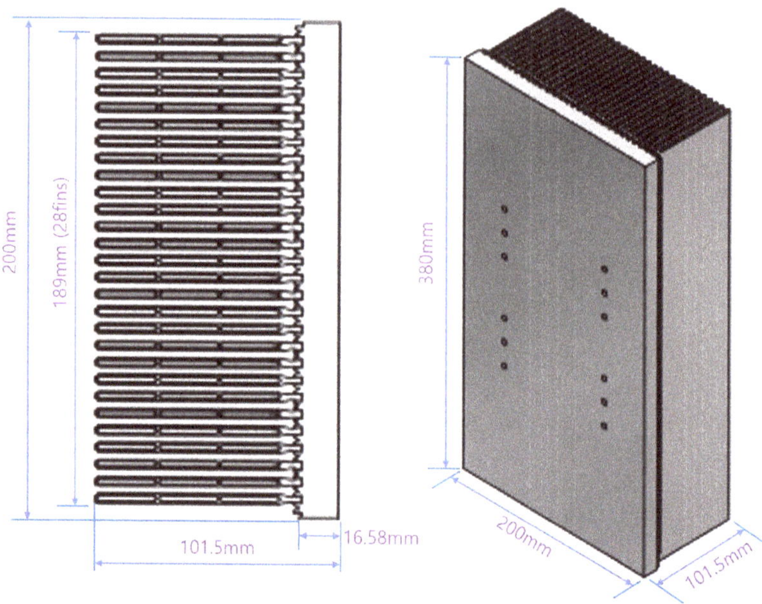

Figure 18. Actual heatsink design via CFD analysis.

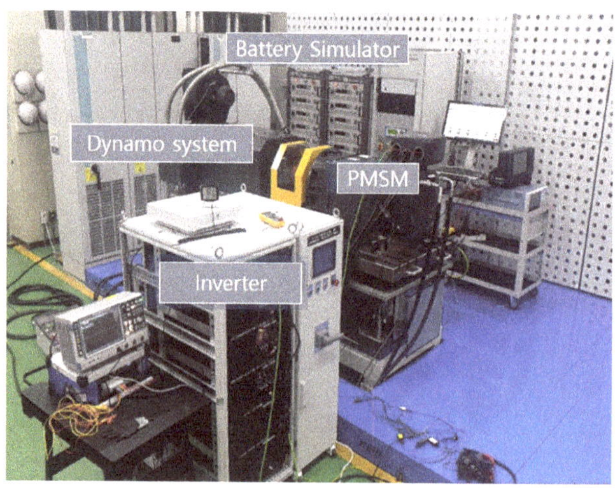

Figure 19. Module type 3-level NPC inverter and complete system.

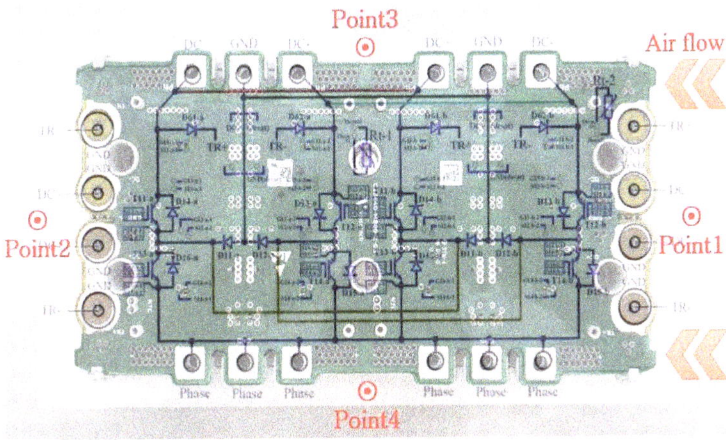

Figure 20. Vincotech 3L-NPC module temperature points 1–4.

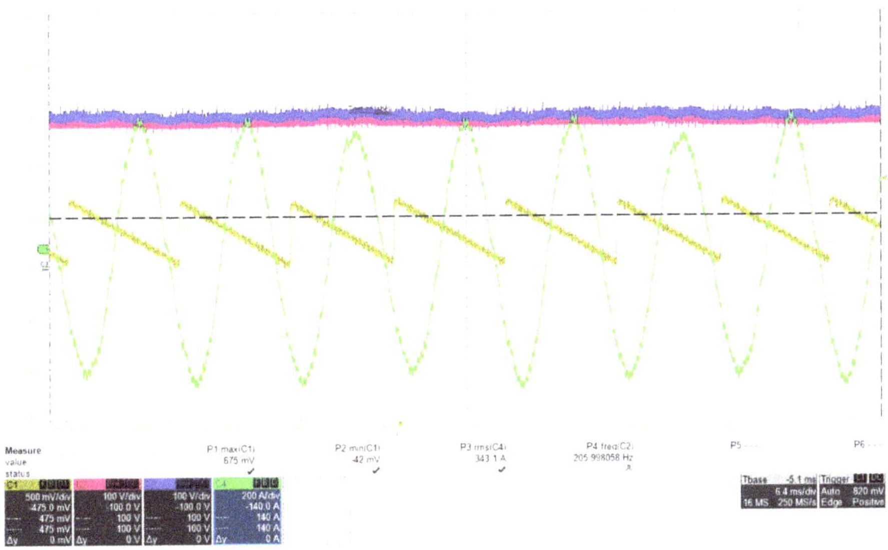

Figure 21. Upper and lower capacitors voltage of the DC link, the output phase current, and the motor's angular position signal.

The voltage difference between the DC link capacitors in the NPC inverter is significant, as it plays a pivotal role in influencing the overall output. Therefore, ensuring voltage balancing is imperative for optimal performance and stability in the system. The motor used in this experiment was a permanent magnet synchronous motor from Siemens, and the output power of a 300 kW modular NPC inverter was confirmed, as demonstrated in Figure 22. Inverter input and output and motor output were measured using a WT5000 power analyzer from Yokogawa. The measured torque value and command value were compared using the HBM eDrive recorder.

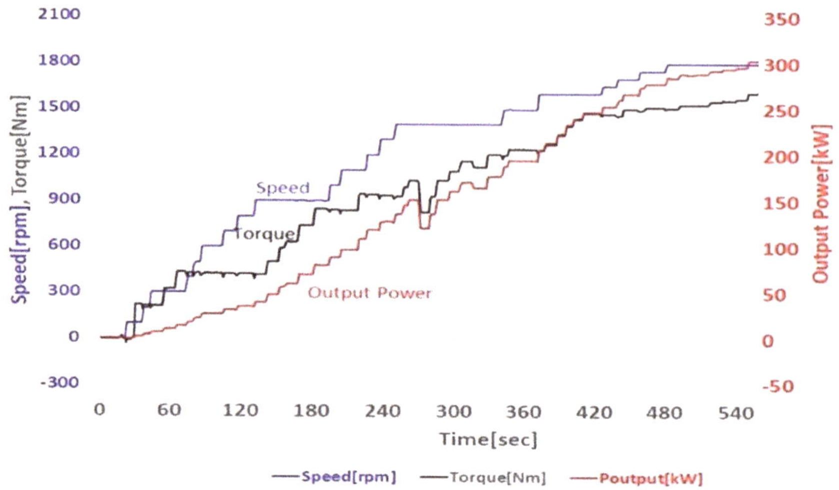

Figure 22. The 300 kW dynamo test results.

For confirmation, the actual temperature measurement results of the four points of the module and inverter efficiency are provided in Table 6. It can be seen that the overall experimental results were similar to the simulation results. The error generated between the values of heat sink temperature obtained by experiment and simulation was thought to be a result of the difference in the actual atmospheric temperature of the experimental environment and the heat flow, such as the flow separation phenomenon and swirl flow that occurs when cooling using a fan in the heat sink.

Table 6. Simulation and experiment results of the 3-Level NPC module temperature and inverter efficiency (3-phase).

	Point1	Point2	Point3	Point4
Saturated Heat Sink Temperature in experiment	71.3 °C	74.6 °C	60.2 °C	65.5 °C
Inverter Efficiency	Simulation		Experiment	
	97.9%		97.1%	
Max Temperature	PLECS	CFD		Experiment
	78 °C	76.4 °C		74 °C

Inverter efficiency and heat sink temperature were measured using the WT5000 equipment. For in-depth measurements, temperature sensors were strategically installed on the heat sink. The equipment allowed us to gather precise data on inverter input voltage, current, output voltage, and current. This comprehensive approach ensured accurate efficiency assessment and temperature monitoring, enhancing the reliability of the experimental results.

5. Discussion

5.1. Differentiating Factors from Existing Heat Dissipation Design Studies

The primary distinction of this research lies in the development of a comprehensive switching loss analysis model tailored to various NPC or inverter switching patterns. This model sets our study apart from conventional heat dissipation design papers, emphasizing the importance of understanding switching patterns for optimized control strategies.

5.2. Unique Contribution of Conduction Loss Analysis Model

Unlike conventional studies, our research focuses on a sophisticated conduction loss analysis model, specifically considering ANPC NPC or inverter switching patterns. This unique approach allows for accurate predictions of conduction losses under real-world operating conditions, contributing to the advancement of inverter efficiency.

5.3. Integration of Heat Dissipation Design with Drive Control System

A distinctive aspect of our study is the integrated consideration of heat dissipation design with the drive control system. By establishing a strong connection between these two elements, we aim to enhance the overall stability and performance of the system, setting our work apart from traditional heat dissipation design studies.

5.4. Anticipated Societal and Economic Impact

Our research envisions a significant societal and economic impact by enhancing the efficiency of large-capacity high-voltage inverters. This impact extends to strengthening industrial competitiveness, fostering innovation, and contributing to energy savings and environmentally friendly technological advancements.

5.5. Innovation in Various Applications of Large-Capacity Inverters

In contrast to existing studies, our work emphasizes the potential for innovation in various applications of large-capacity inverters. We anticipate driving technological advancements across diverse industries, leading to the creation of new business models and innovative solutions.

6. Conclusions

6.1. Validation of Heat Dissipation Design

- The study extensively analyzes heat dissipation for the design of a 300 kW modular neutral-point-clamped (NPC) inverter, with a specific focus on the heat sink design of the inverter module;
- Both the simulation model and practical implementation outcomes exhibit a close alignment, providing strong validation for the accuracy of the conducted heat loss analysis.

6.2. Future Application of the Loss Analysis Model

- The developed loss analysis model holds substantial potential for application in the systematic development of inverter systems across diverse operating conditions;
- Leveraging this model allows for the achievement of an optimized heat dissipation design, contingent upon the selected switching techniques.

6.3. Prospects for Cost Reduction and Accelerated Development

- The proposed approach offers promise for cost reduction and a shortened development timeline in practical system design;
- By integrating the loss analysis model, developers can efficiently design inverter systems with enhanced heat dissipation, addressing critical challenges in a more streamlined manner.

6.4. Foundational Insights for Megawatt-Class Inverters

- The comprehensive insights derived from the 300 kW NPC modular inverter design process, including production and performance evaluation results, serve as foundational data;
- Intended to advance Megawatt (MW)-class large-capacity inverters, these contributions offer a robust foundation for reference and further exploration in the design and implementation of high-capacity inverter systems.

Author Contributions: Conceptualization, J.-W.L. and B.-G.P.; Formal analysis, J.-W.K.; Funding acquisition, J.-W.K.; Investigation, J.-W.L.; Methodology, J.-W.L.; Project administration, B.-G.P.; Resources, J.-W.L. and B.-G.P.; Software, J.-W.K. and J.-K.L.; Supervision, C.-W.L.; Validation, C.-W.L. and B.-G.P.; Visualization, J.-K.L.; Writing—original draft, J.-W.L.; Writing—review and editing, J.-W.L. All authors have read and agreed to the published version of the manuscript.

Funding: This research was supported by the Korea Electrotechnology Research Institute (KERI) primary research program through the National Research Council of Science & Technology (NST), funded by the Ministry of Science and ICT (MSIT) (No. 24A01057).

Data Availability Statement: Data are contained within the article.

Conflicts of Interest: The authors declare no conflicts of interest.

References

1. Kim, S.-G.; Dujic, D. Extending Protection Selectivity in DC Shipboard Power Systems by Means of Additional Bus Capacitance. *IEEE Trans. Ind. Electron.* **2020**, *67*, 3673–3683. [CrossRef]
2. Kouro, S.; Malinowski, M.; Gopakumar, K.; Pou, J.; Franquelo, L.G.; Wu, B.; Rodriguez, J.; A Pérez, M.; Leon, I.J. Recent Advances and Industrial Applications of Multilevel Converters. *IEEE Trans. Ind. Electron.* **2010**, *57*, 2553–2580. [CrossRef]
3. Javaid, U.; Freijedo, F.D.; Dujic, D.; van der Merwe, W. Dynamic Assessment of Source–Load Interactions in Marine MVDC Distribution. *IEEE Trans. Ind. Electron.* **2017**, *64*, 4372–4381. [CrossRef]
4. Ghimire, P.; Park, D.; Zadeh, M.K.; Thorstensen, J.; Pedersen, E. Shipboard Electric Power Conversion: System Architecture, Applications, Control, and Challenges. *IEEE Electrif. Mag.* **2019**, *7*, 6–20. [CrossRef]
5. Sadigh, A.K.; Dargahi, V.; Corzine, K.A. Analytical Determination of Conduction and Switching Power Losses in Flying-Capacitor-Based Active Neutral-Point-Clamped Multilevel Converter. *IEEE Trans. Power Electron.* **2016**, *31*, 5473–5494. [CrossRef]
6. Barros, J.D. Optimal Predictive Control of Three-Phase NPC Multi level Converter for Power Quality Applications. *IEEE Trans. Ind. Electron.* **2008**, *55*, 3670–3681. [CrossRef]
7. Raciti, A.; Cristaldi, D.; Greco, G.; Vinci, G.; Bazzano, G. Electrothermal PSpice Modeling and Simulation of Power Modules. *IEEE Trans. Ind. Electron.* **2015**, *62*, 6260–6271. [CrossRef]
8. Gilson, G.M.; Pickering, S.J.; Hann, D.B.; Gerada, C. Piezoelectric Fan Cooling: A Novel High Reliability Electric Machine Thermal Management Solution. *IEEE Trans. Ind. Electron.* **2013**, *60*, 4841–4851. [CrossRef]
9. Ma, M.; Guo, W.; Yan, X.; Yang, S.; Zhang, X.; Chen, W.; Cai, G. A Three-Dimensional Boundary-Dependent Compact Thermal Network Model for IGBT Modules in New Energy Vehicles. *IEEE Trans. Ind. Electron.* **2021**, *68*, 5248–5258. [CrossRef]
10. Yang, X.; Heng, K.; Wu, X. Distributed Thermal Modeling for Power Devices and Modules With Equivalent Heat Flow Path Extraction. *IEEE J. Emerg. Sel. Top. Power Electron.* **2023**, *11*, 5863–5876. [CrossRef]
11. Chen, H.; Yang, J.; Xu, S. Electrothermal-Based Junction Temperature Estimation Model for Converter of Switched Reluctance Motor Drive System. *IEEE Trans. Ind. Electron.* **2020**, *67*, 874–883. [CrossRef]
12. Sodan, V.; Stoffels, S.; Oprins, H.; Decoutere, S.; Altmann, F.; Baelmans, M.; De Wolf, I. Fast and Distributed Thermal Model for Thermal Modeling of GaN Power Devices. *IEEE Trans. Compon.* **2018**, *8*, 1747–1755. [CrossRef]
13. Shahjalal, M.; Ahmed, R.; Lu, H.; Bailey, C.; Forsyth, A.J. An Analysis of the Thermal Interaction Between Components in Power Converter Applications. *IEEE Trans. Power Electron.* **2020**, *35*, 9082–9094. [CrossRef]
14. Boglietti, A.; Carpaneto, E.; Cossale, M.; Vaschetto, S. Stator-Winding Thermal Models for Short-Time Thermal Transients: Definition and Validation. *IEEE Trans. Ind. Electron.* **2016**, *63*, 2713–2721. [CrossRef]
15. Deepak, K.; Tran, M.-T.; Tran, D.-D.; El Baghdadi, M.; Hegazy, O. Design Investigation of Liquid Cooled Heat Sink for GaN FET Dual-Three Phase Inverter. *Int. Symp. Power Electron.* **2022**, 149–154. [CrossRef]
16. Qiao, G.; Cao, W.; Hu, Y.; Li, J.; Sun, L.; Hu, C. Surrogate Model-Based Heat Sink Design for Energy Storage Converters. *Energies* **2023**, *16*, 1075. [CrossRef]
17. Yu, Y.; Lee, T.-Y.T.; Chiriac, V.A. Compact Thermal Resistor-Capacitor-Network Approach to Predicting Transient Junction Temperatures of a Power Amplifier Module. *IEEE Trans. Compon.* **2012**, *2*, 1172–1181. [CrossRef]
18. An, T.; Zhou, R.; Qin, F.; Dai, Y.; Gong, Y.; Chen, P. Comparative Study of the Parameter Acquisition Methods for the Cauer Thermal Network Model of an IGBT Module. *Electronics* **2023**, *12*, 1650. [CrossRef]
19. Muller, N.; Kouro, S.; Malinowski, M.; Rojas, C.A.; Jasinski, M.; Estay, G. Medium-Voltage Power Converter Interface for Multigenerator Marine Energy Conversion Systems. *IEEE Trans. Ind. Electron.* **2017**, *64*, 1061–1070. [CrossRef]
20. Lee, J.-S.; Yoo, S.; Lee, K.-B. Novel Discontinuous PWM Method of a Three-Level Inverter for Neutral-Point Voltage Ripple Reduction. *IEEE Trans. Ind. Electron.* **2016**, *63*, 3344–3354. [CrossRef]
21. Choi, U.-M.; Blaabjerg, F.; Lee, K.-B. Study and Handling Methods of Power IGBT Module Failures in Power Electronic Converter Systems. *IEEE Trans. Power Electron.* **2015**, *30*, 2517–2533. [CrossRef]
22. Oh, H.; Han, B.; McCluskey, P.; Han, C.; Youn, B.D. Physics-of-Failure, Condition Monitoring, and Prognostics of Insulated Gate Bipolar Transistor Modules: A Review. *IEEE Trans. Power Electron.* **2015**, *30*, 2413–2426. [CrossRef]

23. Falck, J.; Felgemacher, C.; Rojko, A.; Liserre, M.; Zacharias, P. Reliability of Power Electronic Systems: An Industry Perspective. *IEEE Ind. Electron. Mag.* **2018**, *12*, 24–35. [CrossRef]
24. Wang, H.; Zhou, D.; Blaabjerg, F. A reliability-oriented design method for power electronic converters. *IEEE Appl. Power Electron. Conf. Expo. (APEC)* **2013**, 2921–2928. [CrossRef]
25. Sadigh, A.K.; Dargahi, V.; Corzine, K.A. Investigation of Conduction and Switching Power Losses in Modified Stacked Multicell Converters. *IEEE Trans. Ind. Electron.* **2016**, *63*, 7780–7791. [CrossRef]
26. Andresen, M.; Ma, K.; Buticchi, G.; Falck, J.; Blaabjerg, F.; Liserre, M. Junction Temperature Control for More Reliable Power Electronics. *IEEE Trans. Power Electron.* **2018**, *33*, 765–776. [CrossRef]
27. Codecasa, L. Canonical forms of one-port Passive Distributed thermal networks. *IEEE Trans. Compon. Packag. Technol.* **2005**, *28*, 5–13. [CrossRef]

electronics

Article

A Multiple-Sensor Fault-Tolerant Control of a Single-Phase Pulse-Width Modulated Rectifier Based on MRAS and GPI Observers

M. Dardouri [1], M. Salman [2], S. Khojet El Khil [1,*], C. Boccaletti [2] and K. Jelassi [1]

1 LR11ES15 Laboratoire des Systèmes Electriques, Ecole Nationale d'Ingénieurs de Tunis, Université de Tunis El Manar, Tunis 1002, Tunisia; meriem.dardouri@enit.utm.tn (M.D.); khaled.jelassi@enit.utm.tn (K.J.)
2 Sapienza Electric Machines and Power Electronics Research Group (SEMPER), Department of Astronautical, Electrical and Energy Engineering, University of Rome 'La Sapienza', 00185 Roma, Italy; muhammad.salman@uniroma1.it (M.S.); chiara.boccaletti@uniroma1.it (C.B.)
* Correspondence: sejir.kek@edu.isetcom.tn

Abstract: Due to their advantages in ensuring low harmonic distortion and high power factors, single-phase Pulse-Width Modulated (PWM) rectifiers are widely employed in several industrial applications. Generally, the conventional control loop of a single-phase PWM rectifier uses both voltage and current sensors. Hence, in case of sensor fault, the performance and the availability of the converter can be seriously compromised. Therefore, diagnosis approaches and fault-tolerant control (FTC) strategies are mandatory to monitor these systems. Accordingly, this paper introduces a novel multiple-sensor FTC scheme for a single-phase PWM rectifier. The proposed fault diagnosis approach relies on joining several Generalized Proportional Integral (GPI) and Model Reference Adaptive System (MRAS) observers with a residual generation technique to detect and isolate sensor faults in a simple and reliable manner. While conventional sensor FTC methods dedicated to PWM rectifiers can only deal with single faults, the suggested approach guarantees a very good effectiveness level of sensor fault detection, isolation (FDI) and FTC of multiple-sensor fault occurrence scenarios. Consequently, the single-phase PWM rectifier can work with only the survivable single sensor with the guarantee of very good performance as in healthy operation mode. The effectiveness of the proposed sensor FDI approach and its control reconfiguration performance are demonstrated through both extensive simulation and experimental results.

Keywords: single-phase pulse-width modulated rectifier; sensor fault detection; fault-tolerant control; observers; GPIO; MRAS

Citation: Dardouri, M.; Salman, M.; Khojet El Khil, S.; Boccaletti, C.; Jelassi, K. A Multiple-Sensor Fault-Tolerant Control of a Single-Phase Pulse-Width Modulated Rectifier Based on MRAS and GPI Observers. *Electronics* **2024**, *13*, 502. https://doi.org/10.3390/electronics13030502

Academic Editor: Mohamed Benbouzid

Received: 12 December 2023
Revised: 16 January 2024
Accepted: 18 January 2024
Published: 25 January 2024

1. Introduction

In recent years, Pulse-Width Modulation (PWM) Rectifiers have arisen as one of the most important components in various industrial applications such as renewable energy [1–3], EVs [4,5], electric traction [6–8], and so on. They ensure several advantages such as sinusoidal input currents, stable DC-Link voltage and unit power factor [9,10].

For such applications, downtimes caused by different types of failures have a direct impact on the productivity of the system as well as on production costs. Hence, ensuring reliability is becoming an issue of utmost importance. The overall components of the system should be monitored to avoid any possibility of performance degradation due to faults. According to this understanding, fault-tolerant control policies are currently needed more than ever to ensure the system functioning even under fault occurrence.

During the last years, fault diagnosis and fault tolerance of PWM rectifiers have been important research topics that still present challenges, as addressed by several recent publications [11–33]. Apart from failures occurring in the power semiconductors [11–14], a grid-connected voltage source converter is also sensitive to failures in the sensors that provide vital information used by the main control system. In this context, the continuity of service of the system depends intimately on the availability of reliable measurements.

Regarding sensor FDI and FTC issues in single-phase grid-connected converters, several studies have been conducted. State observers have been investigated for a robust sensor fault-tolerant control of single-phase PWM rectifiers [15–19]. In [15], a Luenberger observer (LO)-based FDI is presented to allow DC-link and grid current sensors fault tolerance. After fault detection and isolation, the fault reconfiguration procedure is based on an open-loop state estimator. The stability and robustness of the observer during parameter variation are verified. Poon et al. [16] addressed a model-based state estimator general FDI approach that can be used to diagnose and identify faults in components and sensors in switching power converters. In [17], an open-loop state estimator is developed for the detection of grid current and DC-link voltage sensors for a single-phase grid-connected PWM rectifier. Scaling and offset and drift fault detection have been investigated. However, the proposed approach may be sensitive to the system's parameters' mismatch. Unknown input observers for sensor fault detection and localization are presented in [18]. A reduced-order observer-based resilient control for single- or multiple-sensor faults for cascaded H-bridge multilevel converters is provided in [19]. Grid current and DC-link voltage sensor faults are considered, and five types of faults for each sensor are investigated.

Xia et al. [20] proposed a sliding-mode observer (SMO)-based algorithm for the fault detection and fault reconfiguration of catenary current and DC-link voltage sensor faults of a single-phase PWM rectifier for electric traction applications considering unipolar and bipolar modulation methods. A similar approach is discussed in [21], where the sliding-mode observer design is improved by considering the DC load current as an unknown input of the observer. In [22], an incipient voltage sensor fault isolation method is developed for a single-phase three-level rectifier. A bank of sliding-mode and adaptive estimators was designed in order to isolate different DC-link voltage sensor fault modes. However, only FDI is discussed, and the FTC is not presented. An interval estimator and a sliding-mode estimator have been designed in [23]. The proposed estimator was used for an incipient sensor fault detection method for a class of nonlinear control systems with observer-unmatched uncertainties. The voltage and current sensor fault-tolerant control of the LC branch in the active power decoupling single-phase PWM converter through SMO are discussed in [24]. In [25], a robust control strategy for a three-phase grid-tied inverter with an LCL filter under grid voltage and grid current sensor faults is presented. A quadrature filter of a dual second-order generalized integrator (DSOGI) architecture is used for the grid current estimation. An internal model-based estimator is proposed for grid voltage estimation. A fault-tolerant space-vector hysteresis current control for a three-phase grid-connected converter is discussed in [26]. The analytical redundancy introduced by the use of three measurements allows the continuous work of the system. The parity space equations and virtual flux estimation have been discussed to ensure grid voltage sensor fault tolerance in [28], whereas a hybrid model-based and data-driven method is proposed for the grid current sensor fault-tolerant control [29].

In addition to residual generation-based FDI approaches, fault estimation (FE)-based approaches have gained more and more interest during the last years [30–32]. A state observer fault estimation to compensate for DC-link voltage and catenary current sensor faults for a single-phase two-level PWM rectifier in a high-speed railway electric traction system has been proposed in [30]. In [31], a grid current sensor fault estimation and compensation for a grid voltage sensorless single-phase grid-connected converter using proportional integral observers is described. A robust grid voltage sensor FE based on an unknown input observer with disturbances decoupling is addressed in [32].

The state-of-the-art review has shown that the research field of sensor FTC in electrical power converters is still very attractive with many challenging issues. Recently, more attention has been paid to the use of data-driven approaches for fault diagnosis purposes. In [33], a just-in-time learning (JITL) and modular Bayesian network (MBN) is proposed to diagnose sensor faults in a high-speed electric railway traction application, whereas the time series feature patterns of a single-phase PWM rectifier are employed to identify several types of faults such as the following: traction transformer faults, traction converter faults, main grid poor contact or inrush current [34]. Zhu et al. [35] presented an automatic crack damage diagnosis in hydraulic structures by combining computer vision and artificial intelligence. Long short-term memory networks and phase space reconstruction are used for the prediction of a dam deformation [36]. Data-driven-based methods are of great interest for fault diagnosis or predictive maintenance purposes. Nevertheless, they are still not able to handle the fault tolerance step, and the use of estimators/observers is very useful for the continuous operation of the system under sensor faults. In this way, model-based and signal-based methods are still very welcomed to ensure reliable and effective sensor fault tolerance [37].

Moreover, multiple-sensor fault-tolerant control is becoming an attractive research topic but is still limited to three-phase induction motor drives [38] or PMSM drives [39]. The single-phase PWM rectifier's control requires mainly three sensors [15]: one grid voltage sensor, one grid current sensor and a DC-link voltage sensor, as depicted in Figure 1. The state-of-the-art review demonstrates that the previously proposed sensor FTC methods are focused on single-sensor faults only, in particular on the grid current sensor and the DC-link voltage sensor. Indeed, the design of an FTC approach that takes into account the faults in the three sensors has not received much research attention yet. To the best knowledge of the authors, very few contributions have discussed multiple-sensor fault tolerance of a single-phase PWM rectifier [19]. On the other hand, the proposed FTC methods provided by the literature need accurate knowledge of the load current [15,20,30]. Hence, a load current sensor [15] or a load current observer [24] is mandatory, which may increase the system's cost and the FTC algorithm complexity. In this context, a robust sensor FDI and FTC approach focused on the three sensors of the PWM rectifier and able to tolerate multiple-sensor faults is of great significance and may considerably contribute to improving the availability and safety operation of the studied system.

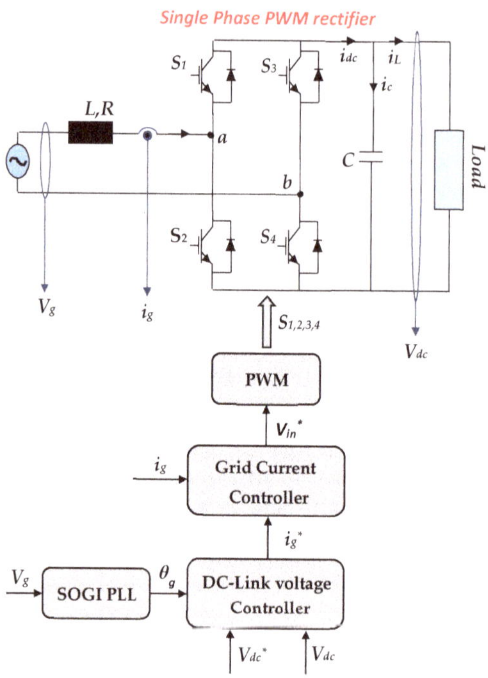

Figure 1. Single-phase PWM rectifier: Converter Topology and its multiloop control scheme.

To address the problem of multiple-sensor fault-tolerant control for single-phase PWM rectifiers, this paper proposes a new structure that aims to ensure a simple, reliable and robust single- and multiple-sensor fault detection and fault-tolerant control of a single-phase PWM rectifier. The main features and contributions of this work are:

(1) The proposed method is capable of tolerating single- or multiple-sensor faults, and the rectifier's control loop can be driven by the single survivable sensor. For this purpose, two multiple-sensor fault-tolerant controls are investigated. In the first scenario, a DC-link voltage sensor fault followed by a grid voltage sensor fault is discussed. In the second scenario, a DC-link voltage sensor fault followed by a grid current sensor fault is investigated.

(2) For this purpose, a residual generation-based FDI approach is presented. Hence, a bank of two Generalized Proportional Integral Observers (GPIO) and one Model Reference Adaptive System (MRAS) observer are developed in order to accurately estimate the grid voltage, the DC-link voltage and the grid current, respectively. Once the residual exceeds its corresponding threshold, a fault flag switches from 0 to 1 and the faulty sensor is identified. Thereafter, the faulty measurement is substituted by the virtual sensor into the rectifier's control loop, to guarantee the continuous work of the rectifier.

(3) The proposed FTC approach does not need any additional sensors, in particular load current sensors. Hence, in this work, the load current is considered as an unknown disturbance. It should not have any impact on the performance of the proposed FTC approach, neither in healthy operation nor during sensor post-fault operation of the system. Moreover, the system's parameters' mismatch, including grid-side filter inductor and DC-link capacitance, does not affect the robustness of the proposed method.

(4) During sensor post-fault operation, the proposed FTC method allows the start-up of the process with the single survivable sensor with good accuracy and the same performance of the closed-loop control as in healthy operation mode.

(5) The effectiveness of the proposed method is demonstrated through extensive simulation and experimental studies. The proposed FDI/FTC approach can be deployed on the commonly used digital controllers, improving the cost and the performance of real-time control of single-phase PWM converters in several industrial applications.

The paper is organized as follows: First, in Section 2, the single-phase PWM rectifier model and its control strategy are presented. Section 3 is devoted to analyzing the design of the different GPI and MRAS observers. FDI and FTC techniques are discussed in Section 4. Simulation results are carried out in Section 5. The experimental validation of the proposed FDI and FTC algorithm is discussed in Section 6. Finally, the paper is summed up in a conclusion in Section 7.

2. Mathematical Model of the Single-Phase PWM Rectifier and Control Strategy

The system under study and its control strategy are presented in Figure 1. The single-phase PWM rectifier can be modelled using the following equations:

$$\begin{cases} V_g = Ri_g + L\frac{di_g}{dt} + V_{ab} \\ i_c = i_{dc} + i_L = C\frac{dV_{dc}}{dt} \\ V_{ab} = (S_1 - S_3)V_{dc} \\ i_{dc} = (S_1 - S_3)i_g \end{cases} \tag{1}$$

where i_g is the grid current, V_g the grid voltage, V_{ab} is the rectifier input voltage, V_{dc} is the DC-link voltage, i_{dc} is the DC-side output current, i_c is the current in the DC-side capacitor, i_L is the load current and S_1 and S_3 are the binary switching control signal. x^* corresponds to the reference of the variable x. L denotes a filter used for the connection with the grid, R is its internal resistance, and C is the DC-link capacitor. The single-phase PWM rectifier provides a high-energy quality: unity power factor, energy bi-directionality, a sinusoidal form of the grid current and a continuous and constant DC-link voltage. The loop control of the single-phase PWM rectifier is composed of:

- DC-Link Voltage Control Loop: The main purpose of the DC-link voltage loop is to maintain this voltage at a constant reference value. The voltage control loop is based on a PI controller due to its simplicity, easy implementation and good regulation results of DC quantities.
- PR-Based Grid Current Control: The grid current controller allows i_g to track its reference $i_g{}^*$ with good accuracy. For this purpose, a proportional resonant (PR) controller is employed [15].
- SOGI-PLL-Based Grid Synchronization: *To* ensure grid-side unity power factor, the phase angle of the grid current i_g should be synchronized with the grid voltage V_g. Different synchronization techniques are presented in the literature [40]. Among these, the second-order generalized integrator-based phase-locked loop (SOGI-PLL) is an effective approach due to its fast dynamic response and high filtering capacity [15].

3. Design of GPI and MRAS Observers

In this paper, two generalized proportional integral observers (GPIO) are designed to estimate the grid voltage V_g and the DC-link voltage V_{dc}, respectively. In the same way, a model reference adaptive system (MRAS) observer is employed to estimate the grid current i_g.

From (1), the dynamics of the single-phase PWM rectifier are described by:

$$\begin{cases} \frac{d}{dt}\begin{pmatrix} i_g \\ V_{dc} \end{pmatrix} = \begin{pmatrix} -\frac{R}{L} & -\frac{S_{13}}{L} \\ \frac{S_{13}}{C} & 0 \end{pmatrix}\begin{pmatrix} i_g \\ V_{dc} \end{pmatrix} + \begin{pmatrix} \frac{1}{L} \\ 0 \end{pmatrix} V_g + \begin{pmatrix} 0 \\ \frac{-1}{C} \end{pmatrix} i_L \\ y = \begin{pmatrix} 1 & 0 \\ 0 & 1 \end{pmatrix}\begin{pmatrix} i_g \\ V_{dc} \end{pmatrix} \\ S_{13} = S_1 - S_3 \end{cases} \tag{2}$$

It can be seen from (2) that the state matrix contains the discrete switching control signal S_{13}. For unipolar sinusoidal PWM, S_{13} has three possible states during one switching period: ± 1 and 0. So, changes in S_{13} may affect the performance and the convergence of the observer [19,21]. In order to overcome this issue, the switching control signals S_1 and S_3 may be replaced by their respective duty cycles d_1 and d_3. During one switching period, the voltage V_{ab} may be expressed as follows:

$$V_{ab}(t) = (d_1 - d_3)V_{dc}(t) = d_{13}V_{dc}(t), kT_s < t \leq (k+1)T_s \tag{3}$$

Since the switching frequency (10 kHz) is much larger than the grid voltage frequency (50 Hz), then the duty cycle d_{13} is considered as a constant in one switching period T_s. Finally, d_{13} is a continuously varying parameter instead of a discrete parameter S_{13}. Furthermore, neglecting high-frequency components, d_{13} may be considered the fundamental component of S_{13}, and V_{ab} can be expressed as:

$$V_{ab}(t) = m_{ab}V_{dc}\sin(\theta_g + \varphi_{ab}) \tag{4}$$

where m_{ab} corresponds to the modulation index, and φ_{ab} is the phase difference with respect to the grid voltage. Finally, a corresponding equivalent circuit of the grid-side loop of Figure 1 is presented in Figure 2.

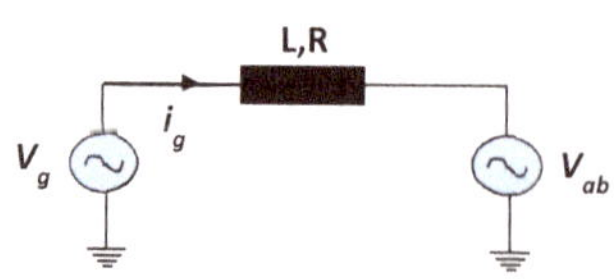

Figure 2. Equivalent Structure of the single-phase PWM rectifier.

The grid current state model is described by:

$$\frac{di_g}{dt} = Ai_g + B(V_g - V_{ab}) \tag{5}$$

where $A = -R/L$ and $B = 1/L$.

3.1. Grid Voltage GPI Observer Design

In this section, a first GPI observer is designed for observing the grid voltage V_g. Generalized proportional integral observers use not only the estimation error information but also the integral of the estimation error, which makes them useful for the estimation of system disturbances and unknown inputs [41,42].

The structure of the GPIO applied to the grid current state Equation (5) is depicted in Figure 3. The variable f_a corresponds to the disturbance and/or unknown input that the GPIO aims to estimate.

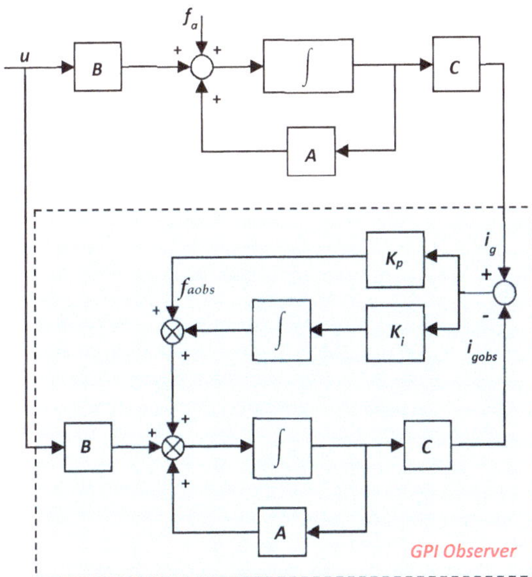

Figure 3. Structure of the GPI Observer.

In order to analyze the system dynamics, the output of the integrator is extended to a new state variable f_a, which denotes the system unknown input and is described by:

$$f_a = \frac{V_g}{L} \tag{6}$$

The structure of the GPIO applied to the grid current state Equation (5) is depicted in Figure 3, where the dynamics of the observer are described by:

$$\frac{d}{dt}\begin{pmatrix} i_{gobs} \\ f_{aobs} \end{pmatrix} = \begin{pmatrix} A & 1 \\ 0 & 0 \end{pmatrix}\begin{pmatrix} i_{gobs} \\ f_{aobs} \end{pmatrix} + \begin{pmatrix} B \\ 0 \end{pmatrix}u + \begin{pmatrix} K_{p1} \\ K_{i1} \end{pmatrix}(i_g - i_{gobs}) \tag{7}$$

where K_{p1} and K_{i1} are the observer's proportional and integral gains, respectively. The variable f_{aobs} is the estimated value of the unknown input f_a, and it corresponds to the integral of the estimation error $(i_g - i_{gobs})$. According to (6) and (7), if the grid voltage is estimated by the GPI observer, $u = -V_{ab}$ and then:

$$f_{aobs} = \frac{V_{gobs}}{L} \tag{8}$$

3.2. Observer Stability Analysis

The error model of the proposed observer is obtained for stability analysis, and the estimation errors of the state variables defined in (7) are given by:

$$e_i = i_g - i_{gobs} \quad and \quad e_f = f_a - f_{aobs} \tag{9}$$

Consequently, the state error equation is given by the following [43]:

$$\frac{d}{dt}\begin{pmatrix} e_i \\ e_f \end{pmatrix} = \begin{pmatrix} A - K_{p1} & 1 \\ -K_{i1} & 0 \end{pmatrix}\begin{pmatrix} e_i \\ e_f \end{pmatrix} + \begin{pmatrix} 0 \\ \frac{df_a}{dt} \end{pmatrix} \tag{10}$$

From (10), it is observed that the system will be Hurwitz stable if both of the characteristic polynomial's roots in the matrix D_c, i.e., are in the left half-plane [43].

$$F_1(s) = s^2 + \left(\frac{R}{L} + K_{p1}\right)s + K_{i1} \tag{11}$$

In order to simplify the design process, we assume that both observer's poles are located at $-w_0$ and are expressed by:

$$F_1(s) = s^2 + \left(\frac{R}{L} + k_{p1}\right)s + k_{i1} = (s + w_0)^2 \tag{12}$$

Therefore:

$$\left(\frac{R}{L} + K_{p1}\right) = 2w_0 \quad \text{and} \quad k_{i1} = w_0^2$$

From (12), it is shown that the design of $\frac{R}{L} + K_{p1}$ and K_{i1} is related to the GPI observer's bandwidth w_0. Hence, the key process is the appropriate bandwidth value w_0 selection. The observer's bandwidth is generally selected to be 5–15 times lower than the switching frequency by considering the tradeoff between the fast observation performance and the noise sensitivity immunity [44]. In this paper, the observer bandwidth is selected around 1 kHz, which is 10 times lower than the switching frequency.

In Figure 4, the root locus of the PI observer with grid inductance L as a variable value is shown. It can be seen that the root locus remains on the left side of the s-plan, and the stability of the observer is guaranteed.

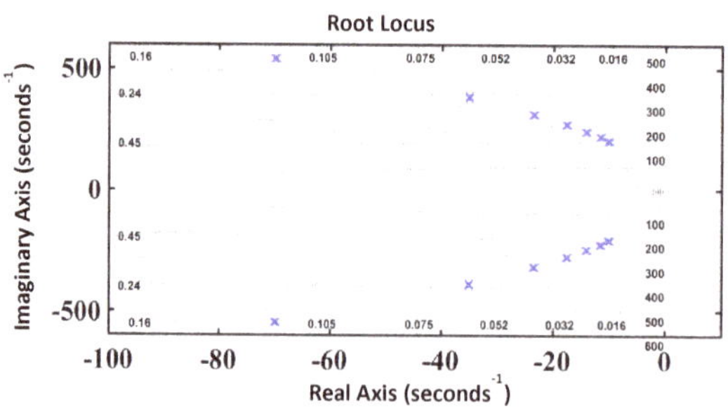

Figure 4. Root locus plot of the grid voltage GPIO with L as a variable value.

3.3. DC-Link Voltage Estimation Based on SOGI-GPIO

Similarly, a second GPIO is used to estimate the DC-link voltage V_{dc}. The DC-link voltage estimation is achieved in two steps. First, the fundamental component of the rectifier's input voltage V_{ab} is estimated through the GPIO. The proposed GPIO structure is similar to the observer used to estimate the grid voltage. Consequently, V_{ab} is considered as an unknown input of the system described by (5) and, therefore, $f_a = -V_{ab}/L$. According to (7), if V_{ab} is estimated by the GPI observer, $u = V_g$, and the estimated unknown input is expressed as:

$$f_{aobs} = \frac{V_{abobs}}{L} \quad (13)$$

Since the dynamics of the GPIO of V_{ab} are similar to the observer of the grid voltage V_g, then the stability of the observer is guaranteed by the appropriate choice of its proportional and integral parameters K_{p2} and K_{i2}, respectively.

Once the input rectifier voltage is estimated, a SOGI block is used to generate its estimated orthogonal components $V_{ab\alpha\text{-}obs}$ and $V_{ab\beta\text{-}obs}$ of V_{abobs}, then the estimated value of the DC-link voltage V_{dcobs} is obtained, as presented in Figure 5.

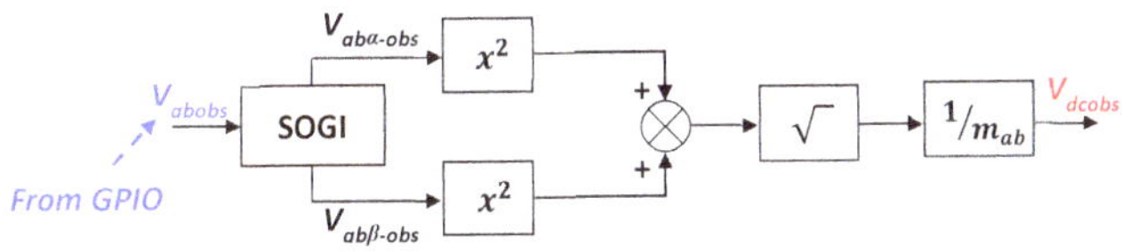

Figure 5. Structure of the SOGI-based DC-Link voltage estimator.

3.4. Virtual Flux MRAS-Based Grid Current Estimation

The MRAS observer main design concept is based on the use of a reference model and an adaptive model. The idea is to use a correction/adaptation mechanism so that the output of the adaptive model is adjusted to the reference model [45].

Recently, the MRAS observer has been successfully used for grid-connected power converters applications to estimate AC voltages [46] or grid-side impedance parameters [47].

The grid current estimation is based on a grid virtual-flux-based MRAS observer. Its structure is depicted in Figure 6a.

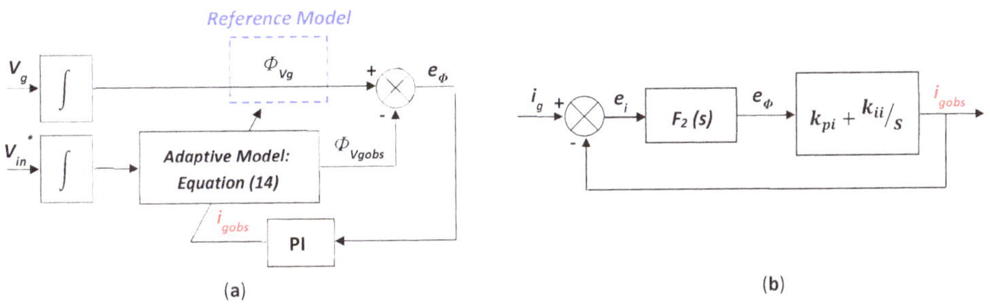

Figure 6. MRAS-based estimator: (**a**) MRAS-based grid current estimator's structure; (**b**) Equivalent block diagram of the MRAS estimator.

The reference model of the MRAS observer is the virtual flux Φ_{Vg}, which can be obtained from the grid voltage measurement. The adaptive model aims to generate the estimated virtual flux value Φ_{Vgobs}, expressed as follows:

$$\begin{cases} \phi_{V_g} = \int V_g dt = \int V_{ab} dt + R\int i_g dt + L i_g \\ \phi_{V_{gobs}} = \int V_{in}^* dt + R\int i_g dt + L i_g \end{cases} \quad (14)$$

The estimation errors are given by:

$$e_\phi = \phi_{V_g} - \phi_{V_{gobs}} = \left(\frac{R}{s} + L\right)\left(i_g - i_{gobs}\right) = \left(\frac{R}{s} + L\right)e_i = F_2(s)e_i \quad (15)$$

The virtual flux estimation error is used to drive a suitable adaptation mechanism which generates the estimated grid current for the adaptive model.

Therefore, a PI corrector is sufficient, and i_{gobs} is given by:

$$i_{gobs} = \left(k_p + \frac{k_i}{s}\right)e_\phi \quad (16)$$

Figure 7 shows the root locus of the closed-loop transfer function, presented in Figure 6b, of the MRAS estimator for the grid current with the grid inductance L filter variations. One can clearly deduce the estimator's stability, since, for all inductance values, the root locus remains on the left side of the s-plan.

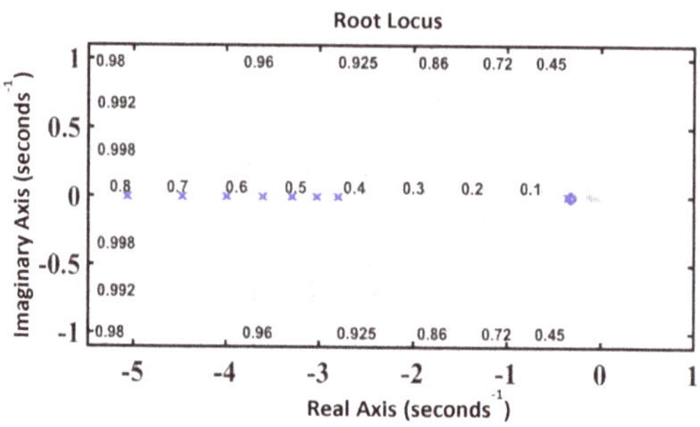

Figure 7. Root locus plot of the grid current MRAS estimator with L as a variable value.

4. Proposed Sensor FDI and FTC Strategy

The control strategy of the single-phase PWM rectifier requires accurate and exact measurements of the grid voltage V_g, the grid current i_g and the DC-link voltage V_{dc}. In case of erroneous measurements, the converter's performance would be seriously perturbed [15]. The overall structure of the proposed multiple-sensor FDI and FTC algorithm is illustrated in Figure 8, while the flowchart of the proposed multiple-sensor fault-tolerant control is depicted in Figure 9. As illustrated by Figure 8a, once the estimated state information is obtained through the GPIOs and MRAS observers, they are compared to the measured information in order to generate the desired residuals. To reduce the impact of the load charge variation and/or system's parameters' mismatch on the fault detection process, all residuals are defined as normalized quantities and expressed as:

$$R_{V_g} = \frac{|V_g - V_{gobs}|}{V_{gnom}}, \quad R_{V_{dc}} = \frac{|V_{dc} - V_{dcobs}|}{V_{dc}^*}, \quad R_{i_g} = \frac{|i_g - i_{gobs}|}{i_{gnom}} \quad (17)$$

where V_{gnom} and i_{gnom} represent the rated values of V_g and i_g, respectively. The detection of the sensor fault is achieved by comparing each residual R_x $\{x = i_g, V_{dc}, V_g\}$ to a corresponding threshold, T_x $\{x = i_g, V_{dc}, V_g\}$. The robustness of the FDI algorithm depends precisely on the residuals' sensitivity to the system's parameters and operating point variations. In healthy mode, all residuals should maintain low values (close to zero) below their corresponding thresholds. However, in faulty operating mode, only the residual corresponding to the faulty sensor exceeds its threshold, while all residuals related to healthy sensors remain below their specified thresholds. Hence, the choice of the thresholds should be carefully made taking into account all system variations to minimize false alarms and increase the FDI algorithm robustness while maintaining a fast detection time. Thus, the fault is detected, and the faulty sensor is isolated when the corresponding fault flag $F_x\{x = i_g, V_{dc}, V_g\}$ is generated:

$$F_x = \begin{cases} 1, & \text{if } R_x > T_x \\ 0, & \text{if } R_x < T_x \end{cases} \quad (18)$$

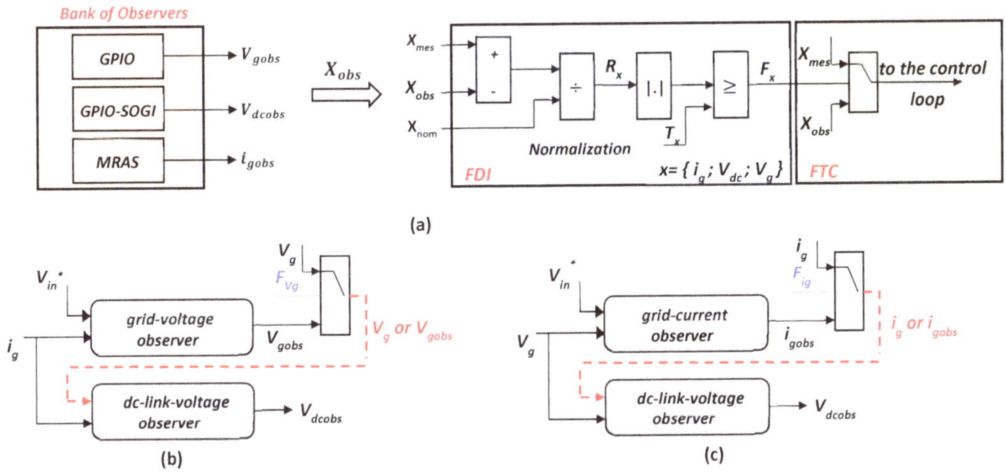

Figure 8. Structure of the proposed multiple-sensor faults FDI and FTC approach: (**a**) FDI general structure; (**b**) FTC for case 1: DC-link voltage and grid voltage sensors failure; (**c**) FTC for case 2: DC-link voltage and grid current sensor failure.

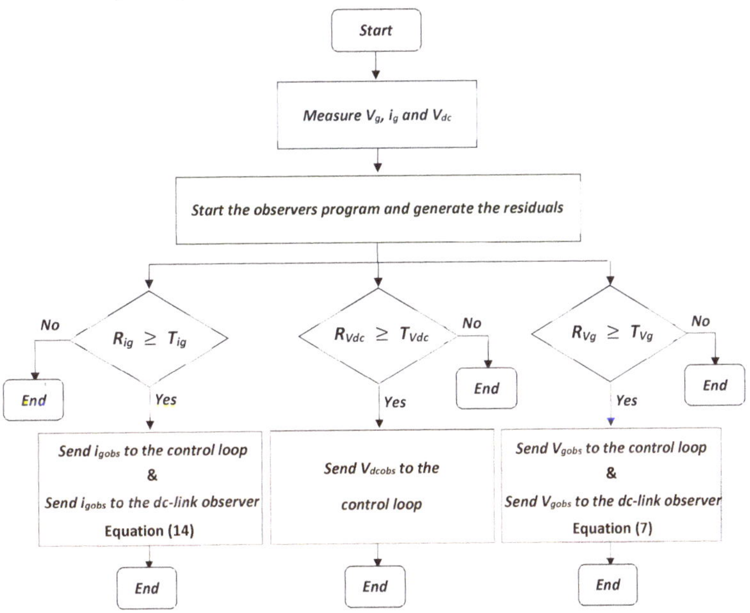

Figure 9. Flowchart of the proposed single- and multiple-sensor fault detection and FTC.

Once the faulty sensor is identified, the fault-tolerant process is initiated. The faulty measurement is substituted by the estimated one in the feedback control loop. It should be noticed that only two multiple-sensor fault scenarios can be treated: DC-link voltage and grid voltage sensors failure and DC-link voltage and grid current sensors failure. The simultaneous failure of the grid voltage sensor and the grid current sensor cannot be handled by the proposed approach. Figure 8a describes the FDI scheme by the residual generation and their evaluation in order to trigger the FTC process due to the switching value of the faulty sensor flag. Figure 8b presents the fault-tolerant control algorithm in case of V_g and V_{dc} multiple-sensor faults, whereas Figure 8c describes the FTC algorithm in case of i_g and V_{dc} multiple-sensor faults.

5. Simulation Results

In order to investigate the performance of the proposed FDI and FTC strategy, the single-phase PWM rectifier model is built in the Matlab-Simulink software R2016b environment. The system parameters used for simulation are presented in Table 1.

Table 1. Parameter values for simulation.

Description	Symbol	Value
RMS voltage supply	V_g	230 V
DC-bus voltage	V_{dc}	400 VDC
Sampling time	T_s	100 µs
Switching frequency	f_{sw}	10 kHz
Line impedance	L, R	20 mH, 0.2 Ω
DC-Link capacitor	C	1100 µF
Rated Load resistance	R_L	100 Ω

5.1. Multiple-Sensor Fault-Tolerant Control

The simulation results reported in Figure 10 illustrate the system performance before and after the occurrence of a 100% gain sensor fault of the DC-link voltage sensor at $t = 2$ s followed by an 80% offset fault of the grid voltage sensor which appears at $t = 3$ s. In healthy operating mode, i.e., for $t < 2$ s, all residuals are below the pre-defined fixed thresholds. At the occurrence of the DC-link voltage sensor fault (for $t = 2$ s), the residual R_{vdc} exceeds its fixed threshold T_{Vdc}, and the fault flag signal F_{Vdc} switches from 0 to 1, as presented in Figure 10c,d. This indicates the presence of a DC-link voltage sensor fault. At the same time, both residuals R_{Vg} and R_{ig} remain below their respective thresholds T_{Vg} and T_{ig}. Once the DC-link voltage sensor fault is correctly located, the faulty measurement is replaced by its estimated value V_{dcobs} in the converter's control loop. As presented in Figure 10a,b, V_{dcobs} follows its reference value, and the grid current i_g also follows its reference, in phase with the grid voltage. Similarly, at the occurrence of the grid voltage fault (at time $t = 3$ s), the residual R_{Vg} exceeds its fixed threshold T_{Vg}, as shown in Figure 10c,d, and the fault signal F_{Vg} switches from 0 to 1. This indicates the presence of a grid voltage sensor fault. Immediately after, the input of the SOGI-PLL block switches from the measured value of the grid voltage to the estimated value V_{gobs}. After the control reconfiguration, the grid current i_g keeps its sinusoidal shape, in phase with the grid voltage V_g, ensuring a unit power factor.

The simulation results depicted in Figure 11 illustrate the system performance before and after the occurrence of a 100% gain fault of the DC-link voltage sensor at $t = 2$ s followed by a 100% gain fault of the grid current sensor at $t = 3$ s. At the occurrence of a fault (for $t \geq 2$ s), the residual R_{Vdc} exceeds its fixed threshold T_{Vdc} (Figure 11c), and the fault signal F_{Vdc} passes from 0 to 1 (Figure 11d). This indicates the presence of a DC-link voltage sensor fault. At the same time, both residuals R_{vg} and R_{ig} remain below their thresholds T_{ig} and T_{Vg}, the fault signals F_{Vg} and F_{ig} remain zero and the faulty measurement is replaced by the estimated value of V_{dc}. At $t = 3$ s, residual R_{ig} exceeds its fixed threshold T_{ig}, and the fault signal F_{ig} passes from 0 to 1 (Figure 11d). This indicates the presence of a grid current sensor fault. Once the grid current sensor is correctly located, the faulty measurement is replaced by the estimated value of i_g. One can clearly notice that the grid current i_g keeps its sinusoidal shape and is in phase with the grid voltage V_g, ensuring a unit power factor (Figure 11b), and the system stability is not affected.

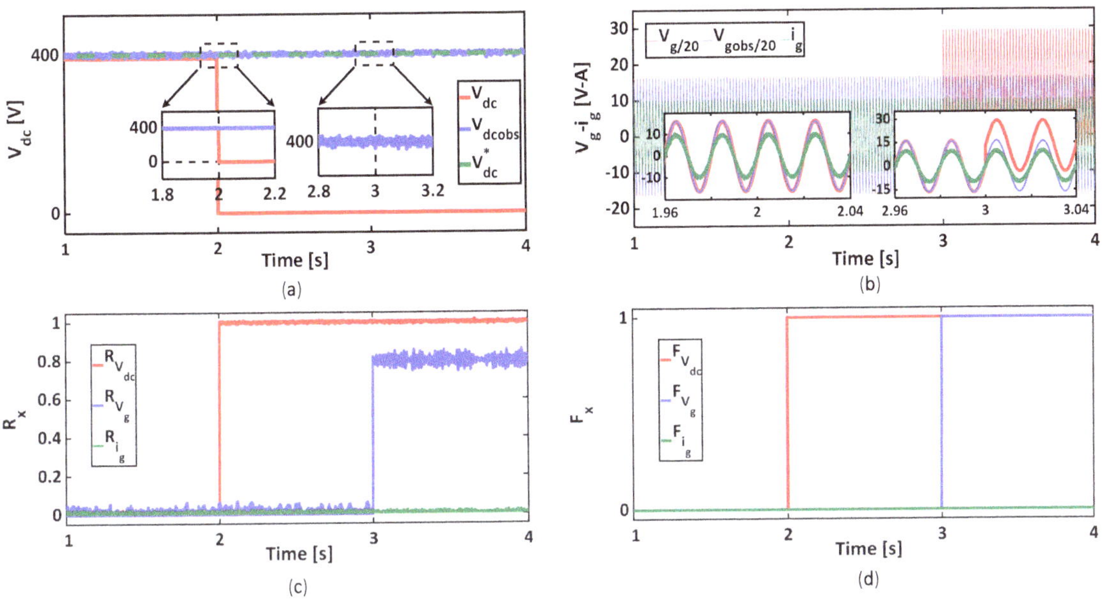

Figure 10. Simulation results for control reconfiguration in case of 100% gain fault of the DC−link voltage sensor and 80% offset fault of the grid voltage sensor, (**a**) V_{dc} fault tolerant control, (**b**) V_g fault tolerant control, (**c**) Residuals evolution, (**d**) Fault flags variations.

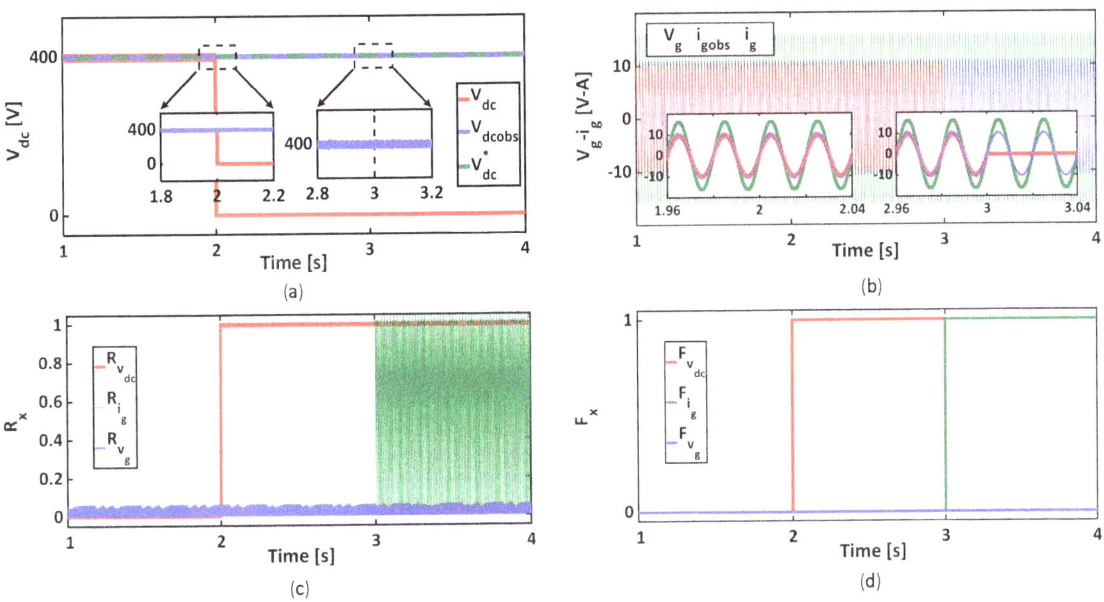

Figure 11. Simulation results for control reconfiguration in case of 100% gain fault of the DC−link voltage sensor and 100% gain fault of the grid current sensor. (**a**) Vdc fault tolerant control, (**b**) ig fault tolerant control, (**c**) Residuals evolution, (**d**) Fault flags variations.

5.2. Post-Fault System Restart with the Single Survivable Sensor

The effectiveness of the proposed FTC approach has been investigated in the previous section by means of simulation. Nevertheless, and since the converter is not continuously working (the converter may be switched off by the user), it is important to investigate the capability of the proposed FTC algorithm to start up the converter using only the single survivable sensor.

Figure 12a–c present the simulation results regarding the start-up of the converter without DC-Link voltage and grid voltage sensors under 50% and 120% variation of nominal values of *L* and *C* as well as the operating point variation. As can be seen, all of the observed variables are accurately estimated under different operating conditions. More precisely, when the converter controller is activated, the DC-link estimated voltage reaches its reference value and is equal to the actual DC-link voltage. Moreover, it takes about 40 ms for the estimated grid voltage to reach the actual value. In addition, the resultant grid current presents a highly sinusoidal waveform with a unity power factor operation even with 100% variation of the output load resistance introduced at time $t = 1$ s (see Figure 12c).

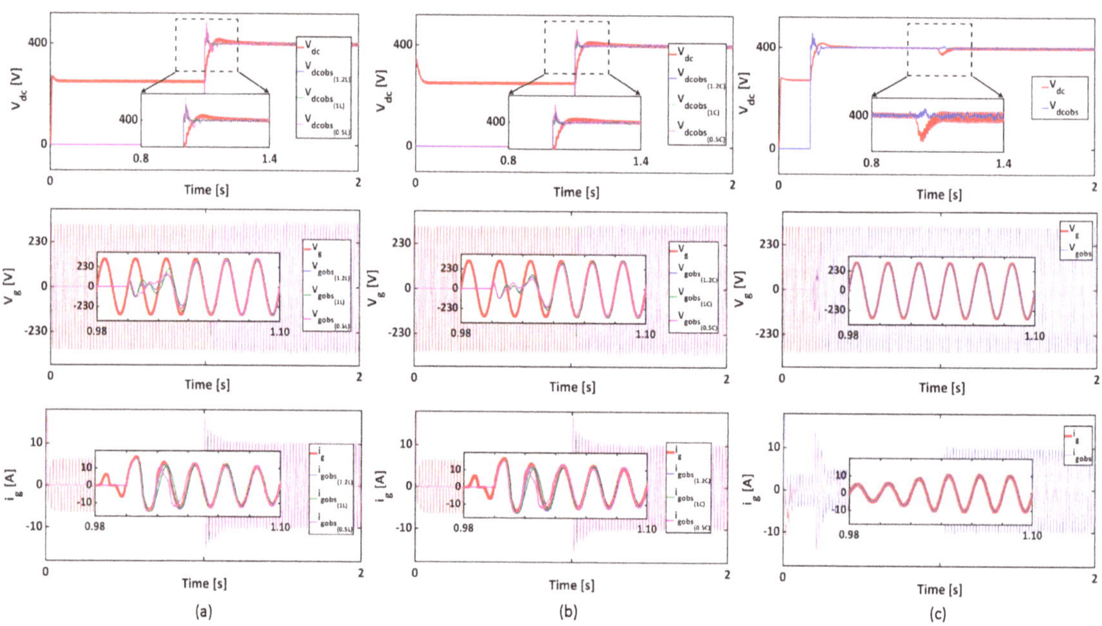

Figure 12. Simulation results for converter's start-up without DC-link voltage and grid voltage sensors considering parameter mismatches: (**a**) grid inductance mismatches; (**b**) DC-link capacitance mismatches; (**c**) DC load variation.

A second test for the start-up of the converter closed-loop operation without DC-link voltage and grid current sensors was performed (Figure 13a–c). It can be observed that the estimated DC-Link voltage follows its reference and actual values. Similarly, the estimated grid current perfectly follows the actual current under *L*, *C* and DC load variation. Moreover, the estimated grid voltage follows the actual grid voltage with accuracy even under extreme conditions and parameter uncertainties.

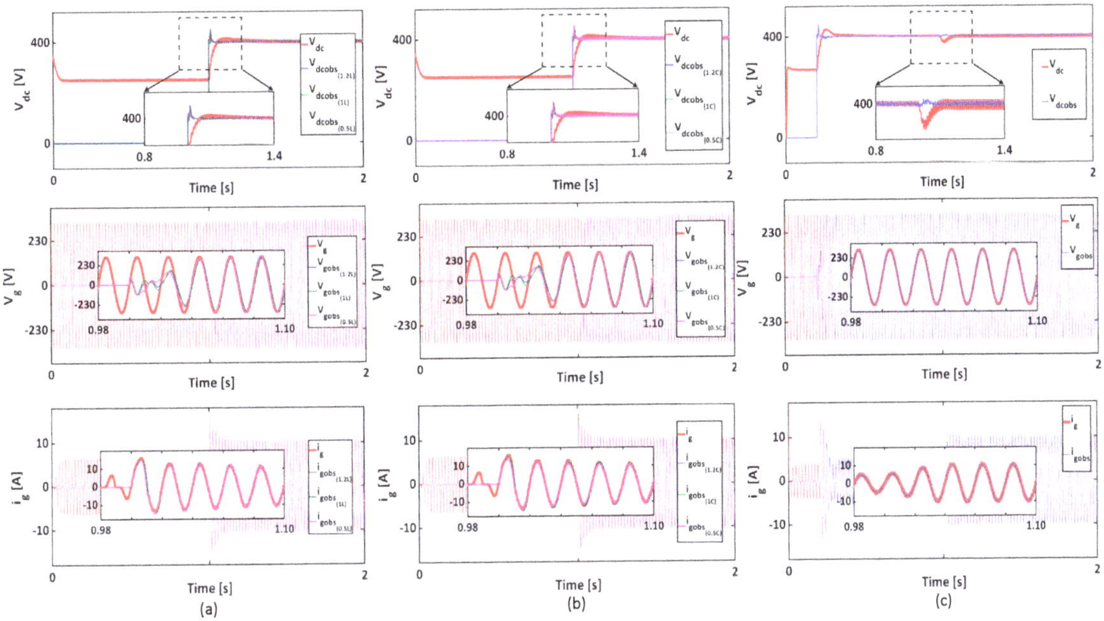

Figure 13. Simulation results for converter's start-up without DC-link voltage and current voltage sensors considering parameter mismatches: (**a**) grid inductance mismatches; (**b**) DC-link capacitance mismatches; (**c**) DC load variation.

These different case studies prove the robustness of the proposed observation approach considering different operation conditions.

6. Experimental Results

To demonstrate the feasibility and effectiveness of the proposed FTC method, experimental results, based on a reduced-scale experimental setup, are carried out. The structure of the experimental setup is presented in Figure 14. It comprises one Semikron SKiiP voltage source converter used as a single PWM rectifier, an input filter (20 mH, 0.2 Ω), a DC-bus capacitor bank of 1100 µF and a variable resistive load. The rectifier's PWM frequency is set to 10 kHz. A current sensor (LEM PR 30) is used to measure the grid current i_g. The DC-link voltage and the grid voltage are measured by two Tektronix p5200 voltage sensors. A 48 V/230 V single-phase transformer is used for the converter's grid connection. The FTC algorithm is implemented through the Matlab-Simulink environment into the dSPACE DS1104 digital controller board using a sampling time of 100 µs. All experimental setup parameters are reported in Table 2. The values of the different proportional and integral observers' gains K_p and K_i used in simulations and experiments are listed in Table 3.

Table 2. Laboratory prototype system parameters.

Description	Symbol	Value
RMS voltage supply	V_g	230/48 V
DC-Link voltage	V_{dc}	80 VDC
Sampling time	T_s	100 µs
Switching frequency	f_{sw}	10 kHz
Line impedance	L, R	20 mH, 0.2 Ω
DC-Link capacitor	C	1100 µF
Rated Load resistance	R_L	36 Ω

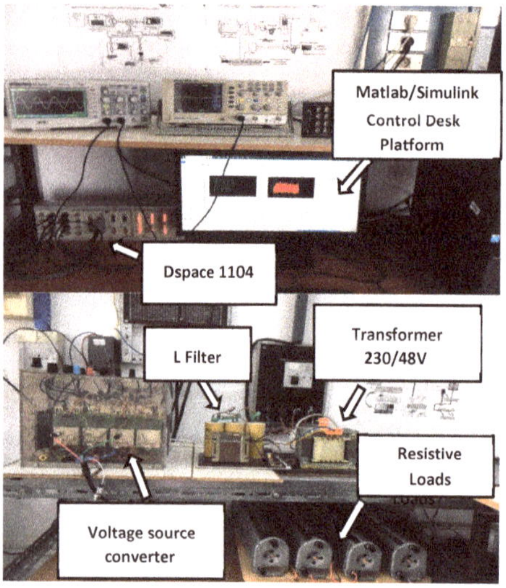

(a)

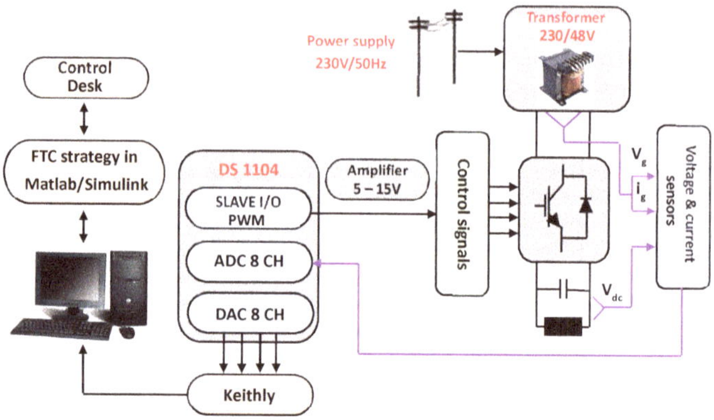

(b)

Figure 14. Experimental implementation of the proposed FDI and FTC: (**a**) experimental test bench; (**b**) descriptive diagram of the experimental implementation.

Table 3. GPI and MRAS observers' gains.

Estimator	Simulations		Experiments	
	k_p	k_i	k_p	k_i
V_g estimator	2.5×10^3	8×10^6	100	1×10^5
V_{dc} estimator	10	1×10^4	500	5×10^5
i_g estimator	0.7×10^3	5×10^2	31	10

6.1. Residual Robustness Analysis and Threshold Selection

The robustness of the FDI algorithm under healthy operation depends on the residuals' sensitivity to the system parameters' variations and to the operating point variations. In Figure 15a, the experimental results regarding the variations of the three residuals under load variations are reported. The load resistance is shifted from 60 to 40 Ω at time t = 1.70 s and again from 40 to 60 Ω at time t = 3.10 s. It can be seen that during resistance load variations, all residuals maintain low values below 0.08. In Figure 15b, the residuals' evolution under a DC-link voltage variation from 80 V to 100 V is detailed. All residuals present small variations, always below 0.08. Finally, the sensitivity of the proposed fault diagnosis approach is tested under a system parameter variation, namely the L filter inductance variation. The filter inductance was first changed from 30 mH to 20 mH and then to 14 mH. As can be seen, slight variations of the three residuals occurred; however, such variations are still low, around 0.05.

Therefore, the behavior of the three residuals following these different tests showed that their oscillations are still very small, close to zero. The thresholds should be carefully selected in order to guarantee the proposed method high immunity under healthy operation, as well as an accurate/fast detection and localization of the faulty sensor [15]. As presented in Figure 15, the defined fault detection residuals variations under healthy operation are small and always below 0.08. Under faulty operation mode, the variations of the residuals correspond to the measurement error. According to this, thresholds T_{Vdc}, T_{Vg} and T_{ig} may have the same value, fixed at 0.1.

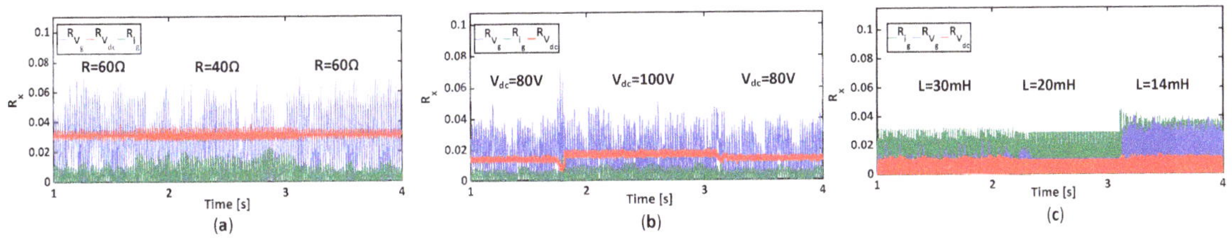

Figure 15. Experimental results regarding residuals evolution under: (**a**) DC load variation, (**b**) DC-link voltage variations, (**c**) grid filter inductance variations.

6.2. Multiple-Sensor Fault-Tolerant Control

The experimental results shown in Figure 16 present the performance of the proposed FDI and FTC in the case of DC-link voltage and grid voltage sensor faults. At time t = 3.90 s, the DC-link voltage fault is introduced (Figure 16a). In healthy operation mode, V_{dc} = 80 V; thus, the output of the DC-link voltage sensor value is set to 0. While grid voltage and current residuals remain at small values below their respective thresholds, the DC-link voltage R_{vdc} exceeds its threshold T_{vdc}. The fault flag F_{vdc} is set to 1, which indicates that the V_{dc} sensor is faulty. Immediately, the fault tolerance process is triggered, and the virtual sensor V_{dcobs} is sent to the control loop of the system. Then, an 80% offset fault of the grid voltage sensor is applied at time t = 10.05 s, as illustrated in Figure 16b. As a result, the grid voltage residual exceeds its threshold, and its corresponding fault flag is then set to 1 (Figure 16d,e). Furthermore, when the grid voltage sensor's detection is achieved, the input of the SOGI-PLL block switches from the measured grid voltage to the estimated grid voltage obtained from the GPIO. Now, the converter's control loop is employing only the single survivable grid current sensor. The DC−link voltage control loop, as well as the grid current reference generation, are based on the estimated voltages values V_{dcobs} and V_{gobs}, respectively. The single-phase PWM rectifier can still operate normally after the DC-link and grid voltage sensors malfunction. It can be seen in the zoomed parts of Figure 16a–c that the DC-link voltage is well regulated at 80 V, the grid current is sinusoidal with low harmonic distortion and the grid power factor is equal to 1.

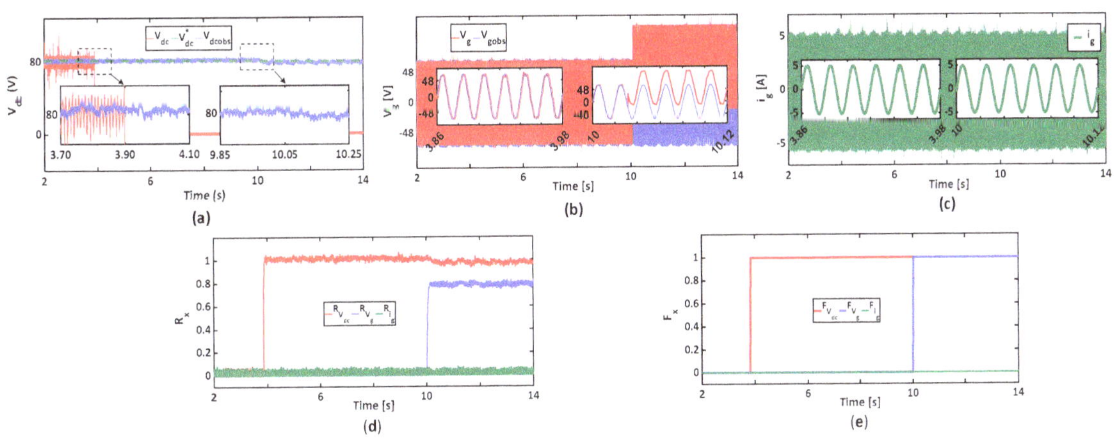

Figure 16. Experimental results for the fault-tolerant control in case of 100% gain fault of the DC−link voltage sensor and 80% offset fault of the grid voltage sensor. (**a**) V_{dc} fault tolerant control, (**b**) V_g fault tolerant control, (**c**) grid current i_g, (**d**) Residuals evolution, (**e**) Fault flags variations.

The experimental results concerning the control reconfiguration for the single-phase converter in case of 100% gain fault of the grid current sensor fault and 100% gain fault of the DC−link voltage sensor are detailed in Figure 17. When the DC−link voltage sensor fault is introduced at time $t = 5.32$ s, the residual R_{vdc} exceeds its fixed threshold T_{vdc}, and therefore, the fault flag F_{vdc} is equal to 1. Consequently, V_{dcest} substitutes the erroneous measurement in the loop control. It can be seen from Figure 17a–c that the DC−link voltage sensor fault tolerance is successfully achieved, and the system maintains its performance. At time $t = 10.193$ s, the output of the grid current sensor is forced to 0. As a consequence, as indicated in Figure 17d, the grid current residual exceeds its threshold, and the fault flag F_{ig} is set to 1. Once the grid current sensor is correctly located, the faulty measurement is replaced by the estimated value of i_{gobs}. After the control reconfiguration of the grid current sensor fault, only the grid voltage sensor survives. The zooms of Figure 17a–c clearly show that V_{dcobs} and i_{gobs} are following their respective reference values, ensuring a unit power factor, and that the converter stability is not affected.

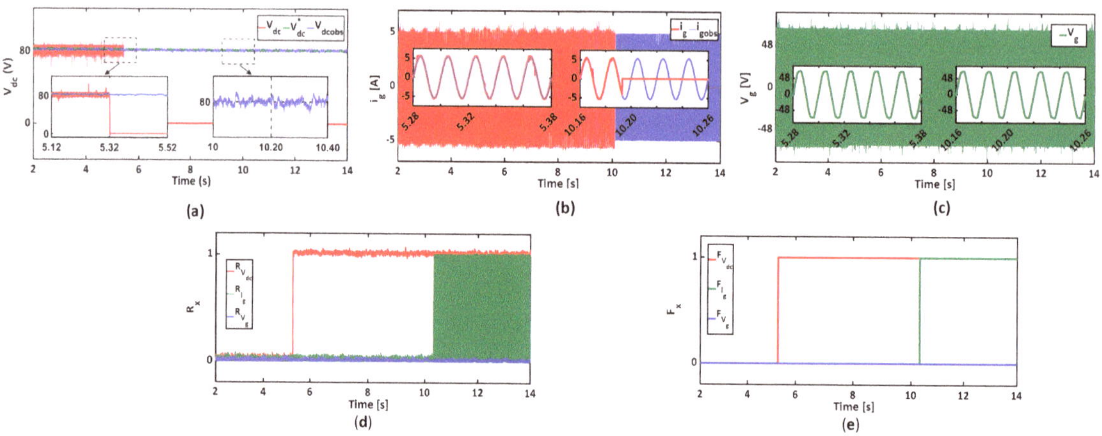

Figure 17. Experimental results for the fault-tolerant control in case of 100% gain fault of the DC−link voltage sensor and 100% gain fault of the grid current sensor. (**a**) V_{dc} fault tolerant control, (**b**) i_g fault tolerant control, (**c**) grid voltage V_g, (**d**) Residuals evolution, (**e**) Fault flags variations.

6.3. Post-Fault System Restart with the Single Survivable Sensor

Figure 18 presents the experimental results in the case of DC−link and grid voltages sensorless operation of the converter. Hence, the rectifier's control loop is only driven by the grid current sensor. Initially, the converter is operating as a diode full bridge. At time $t = 1.88$ s, the feedback control is activated. Figure 18a shows the variation of the observed quantity V_{dcobs} and of the actual voltage V_{dc}. It can be seen that the observed DC−link voltage, as well as the actual DC−link voltage, are correctly reaching their reference value. Figure 18b,c show the variations of the estimated grid voltage V_{gobs}, the actual grid voltage V_g and the grid current i_g. The estimated grid voltage follows the measured one, and the grid current i_g has a sinusoidal waveform in phase with the grid voltage. Moreover, at time $t = 8.25$ s, the load is changed from 100 Ω to 36 Ω. It can be observed that the proposed control strategy keeps its good performance. The estimated values V_{dcobs} and V_{gobs} are following their respective actual values while ensuring a unit power factor, and the converter stability is not affected.

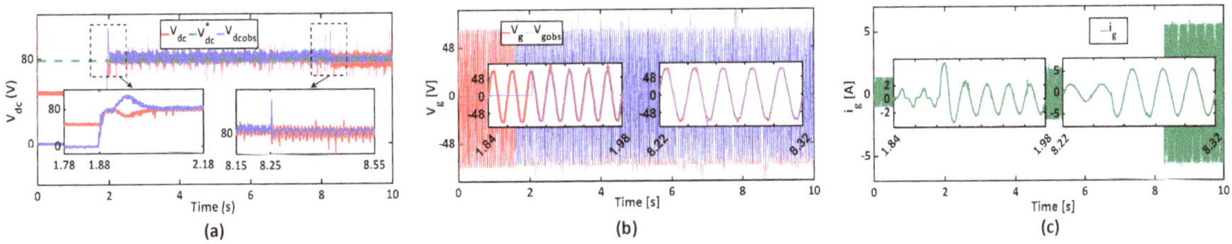

Figure 18. Experimental results for the system's start−up without DC−link voltage and grid voltage sensors. (**a**) dc-link voltage, (**b**) Grid voltage, (**c**) Grid current.

The start-up of the PWM rectifier operation without DC−link voltage and grid current sensors is illustrated in Figure 19. The closed-loop control is initiated at time t = 4.25 s. Similarly to the simulation results, the observed voltage V_{dcobs} and the actual voltage V_{dc} are correctly reaching their reference value. As shown in Figure 19c, the estimated grid current i_{gobs} is following the actual grid current i_g.

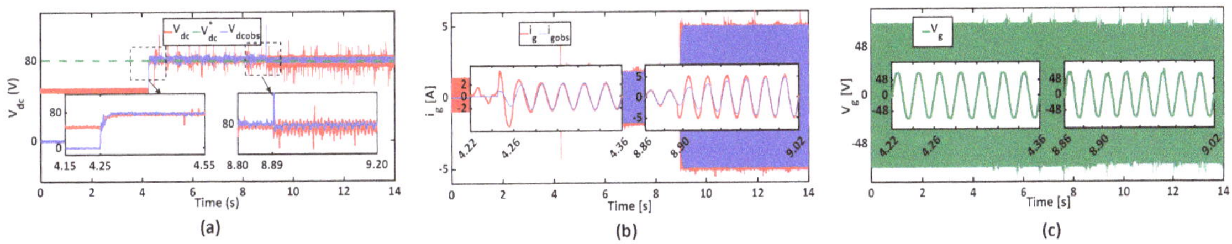

Figure 19. Experimental results for the system's start−up without DC−link voltage and grid current sensors. (**a**) dc-link voltage, (**b**) Grid current, (**c**) Grid voltage.

Moreover, the grid unity power factor is guaranteed according to Figure 19b,c, where the grid current is in phase with the grid voltage. At t = 8.89 s, the load is changed from 100 Ω to 40 Ω. It can be observed from Figure 19a that the proposed control strategy keeps its good performance. The estimated values V_{dcobs} and i_{gobs} are following their respective reference values while ensuring a unit power factor, and the converter stability is not affected.

6.4. Comparative Study with Previously Proposed Methods

To better highlight the benefits of the proposed multiple-sensor fault diagnosis algorithm, its performance is compared with some previously reported approaches, as summarized in Table 4. It can be seen that the proposed approach has similar sensor fault detection and sensor FTC performance to those discussed by [15,17,21,25,30]. Nevertheless, the abovementioned methods cannot handle multiple-sensor failures. Moreover, they need the use of an additional DC load current sensor [15,25,30], which increases the system's cost.

The approach presented by [27] ensures a multiple-sensor fault-tolerant control for a three-phase grid-connected inverter fed by PV arrays. It has focused on the three grid-side AC current sensor faults and DC-link voltage sensor faults. The proposed sensor fault detection and isolation decision logic can guarantee the detection of a single- or multiple-AC-sensor fault; however, if the DC-link voltage sensor is faulty, it violates one or more analytical redundancy relations on the AC side. In order to overcome this issue, when both AC and DC sensor faults occur, the isolation high priority is given to the DC sensor, while the AC sensors are given low-priority fault isolation signals.

Table 4. Fault detection and FTC performance comparison for different methods.

Fault Diagnosis Method	Fault Modes	Single- and Multiple-Sensor Fault Tolerance	Detection Time	Need of Additional Sensor	Algorithm Accuracy	Algorithm Complexity
Luenberger observer [15]	Current and DC-link voltage sensor faults	Single-sensor FTC	<0.02 T_f	Yes	Robust against variations	Medium
Open-loop estimator [17]	Current and DC-link voltage sensor faults	Single-sensor FTC	<0.2 T_f	No	Affected by parameters' variations	Medium
Sliding-mode observer [21]	Current and DC-link voltage sensor faults	Single-sensor FTC	≈0.05 T_f	No	Robust against variations	Medium
Fault estimation [30]	Current and DC-link voltage sensor faults	Single-sensor FTC	---	Yes	Robust against variations	Medium
DSOGI estimator [25]	Current and AC voltage sensor faults	Single-sensor FTC	---	Yes	Robust against variations	Medium
State estimator [27]	Current and DC-link voltage sensor faults	Single- and multiple-sensor FTCs	≈0.05 T_f	Yes	Robust against variations	Medium
Reduced-order observer [19]	Current and DC-link voltage sensor faults	Single- and multiple-sensor FTCs	<0.015 T_f	Yes	Robust against variations	Medium
Proposed approach	Current, DC-link voltage and AC voltage sensor faults	Single- and multiple-sensor FTCs	≈0.05 T_f	No	Robust against variations	Medium

In the proposed FTC approach, it has been shown that an AC sensor fault can be discriminated from a DC sensor fault without any constraints. When DC and AC sensor faults occur, the proposed FTC method can easily handle both of them.

The method discussed by [19] ensures multiple-sensor fault-tolerant control in cascaded H-bridge multilevel converters and can identify the fault type (open-circuit fault, stuck fault, gain fault, offset fault, and abnormal noise). However, only one multiple-sensor FTC scenario has been discussed, which involves the grid current sensor and the DC-link voltage sensor. In our work, two scenarios have been presented: in the first scenario, a DC-link voltage sensor fault followed by a grid voltage sensor and, in the second scenario, a DC-link voltage sensor fault followed by a grid current sensor fault. It is important to highlight that for the multiple-grid current sensor and DC-link voltage sensor faults, the extreme case with 100% gain faults for both sensors has been discussed.

Both work in [19,27] and our work address multiple-sensor faults, but the proposed work emphasizes a novel approach that simplifies the system by not requiring additional sensors, which represents an improvement in terms of system complexity and cost.

Finally, none of these methods has proposed a unified sensor fault detection approach for the main three sensors used in the closed-loop control of the single-phase PWM rectifier without any extra hardware. The extensive experimental results demonstrated that the proposed method can be easily implemented into a real-time digital controller, which may improve the cost and the performance of a single-phase PWM converter in several industrial applications such as EVs, electric traction systems, and so on.

Likewise, the proposed method has some limitations. It cannot identify the sensor fault type (scaling, offset, open-circuit, and so on). Additional features should be extracted in order to enhance its performance in that sense. Moreover, open-circuit faults would change the rectifier's input voltage, which results in high estimation errors and may lead to a sensor false alarm. Hence, in future work, an effective algorithm that can diagnose and discriminate sensor faults from power semiconductor faults in a single-phase PWM converter will be considered.

7. Conclusions

This paper proposes a novel, simple, and fast unified sensor fault detection and isolation without extra hardware and sensor fault-tolerant control of a single-phase pulse PWM rectifier based on residual generation through a robust bank of GPI and MRAS observers. Compared with the previous published methods, the proposed approach offers some benefits:

(1) The proposed approach is a unified sensor fault detection method for the main three sensors used in the closed-loop control of the single-phase PWM rectifier without any extra hardware.

(2) In addition to single-sensor fault detection, the proposed approach is able to diagnose multiple-sensor faults and ensure a post-fault operation of the PWM rectifier with the single survivable sensor.
(3) It has the capability to guarantee the start-up of the single-phase PWM rectifier with a single survivable sensor and with the same performance as in healthy operation, even under DC-link load variations or system's parameters' mismatch.

In particular, if compared with other methods addressing multiple-sensor faults, the proposed method relies on a novel approach that does not require additional sensors, with a moderate mathematical complexity, based on a clear mechanism for fault detection and isolation. Additionally, it is characterized by robustness and practical viability, which are important features in real-world applications.

Several simulations and experimental results have been discussed to demonstrate the robustness and effectiveness of the proposed approach under distinct simultaneous faulty scenarios and to show the very good performance of the post-fault operation of the single-phase PWM rectifier in terms of DC-link voltage regulation and grid-side power quality.

In future work, some improvements to the proposed method should be introduced in order to identify the sensor fault type and to discriminate sensor faults from power semiconductor faults. In this way, the combination of model-based approaches with data-driven approaches is of great benefit.

Author Contributions: Conceptualization, M.D. and S.K.E.K.; methodology, M.D. and S.K.E.K.; software, M.D. and M.S.; validation, M.D., M.S. and S.K.E.K.; formal analysis, S.K.E.K., C.B. and K.J.; investigation, M.D. and S.K.E.K.; resources, M.D. and M.S.; writing—original draft preparation, M.D., M.S. and S.K.E.K.; writing—review and editing, S.K.E.K., C.B. and K.J.; supervision, S.K.E.K., C.B. and K.J.; project administration, S.K.E.K., C.B. and K.J. All authors have read and agreed to the published version of the manuscript.

Funding: This work was supported by the Tunisian Ministry of Higher Education and Scientific Research (MHESR) and the department of Astronautics, Electrical and Energetic Engineering, Sapienza University of Rome, Italy.

Data Availability Statement: Data are contained within the article.

Conflicts of Interest: The authors declare no conflicts of interest.

Nomenclature

ADC	Analogue to Digital Converter
C	DC-link capacitor
d_1, d_3	Duty Cycle
DAC	Digital to Analogue Converter
DSOGI	Dual Second-Order Generalized Integrator
EVs	Electric Vehicles
f_a	Disturbance and/or unknown input
f_{aobs}	Observed value of the disturbance and/or unknown input
FDI	Fault Detection and Isolation
FE	Fault Estimation
FTC	Fault-Tolerant Control
f_{sw}	Switching frequency
F_x	Fault Flag, $x = \{i_g, V_g, V_{dc}\}$
GPIO	Generalized Proportional Integral Observer
i_c	Current in the DC-side capacitor
i_{dc}	DC-side output current
i_g	Grid current
i_{gobs}	Observed value of the grid current
$i_g{}^*$	Reference value of the grid current
i_{gnom}	Nominal value of the grid current
i_L	Load current

T_f	Grid current fundamental period
K_i	Integral Gain
K_p	Proportional Gain
L	Line inductance
LO	Luenberger observer
m_{ab}	Modulation index
MRAS	Model Reference Adaptive System
PI	Proportional Integral
PR	Proportional Resonant
PWM	Pulse-Width Modulated
R	Line resistance
R_L	Rated Load resistance
R_x	Residual, $x = \{i_g, V_g, V_{dc}\}$
SMO	Sliding-Mode Observer
SOGI	Second-Order Generalized Integrator
SOGI-PLL	Second-Order Generalized Integrator Phase-Locked Loop
S_1, S_3	Binary Switching Control Signal
T_s	Sampling Time
T_x	Threshold, $x = \{i_g, V_g, V_{dc}\}$
V_{ab}	Rectifier input voltage
V_{abobs}	Estimated value of the rectifier input voltage
$V_{abobs\alpha}, V_{abobs\beta}$	Estimated orthogonal components of the rectifier input voltage
V_{dc}	DC-link voltage
V_{dc}^*	Reference value of the DC-link voltage
V_{dcobs}	Observed value of the DC-link voltage
V_{in}^*	Reference value of the rectifier voltage
V_g	Grid voltage
V_{gnom}	Nominal value of the grid voltage
V_{gobs}	Observed value of the grid voltage
θ_g	Phase angle
ϕ_g	Grid virtual flux
ϕ_{gobs}	Observed value of the grid virtual flux
φ_{ab}	Phase difference

References

1. Ndabarushimana, E.; Qin, N.; Ma, L. Disturbance Decoupling for a Single-Phase Pulse Width Modulation Rectifier Based on an Extended H-Infinity Filter. *Electronics* **2023**, *12*, 2765. [CrossRef]
2. Zhu, Y.; Wang, Z.; Wang, C.; Zhu, Y.; Cao, X. A Novel Improved Coordinate Rotated Algorithm for PWM Rectifier THD Reduction. *Electronics* **2022**, *11*, 1435. [CrossRef]
3. Kang, L.; Zhang, J.; Zhou, H.; Zhao, Z.; Duan, X. Model predictive current control with fixed switching frequency and dead-time compensation for single-phase PWM rectifier. *Electronics* **2021**, *10*, 426. [CrossRef]
4. Bi, Y.; Wu, C.; Zhao, T.; Li, H.; Xu, J.; Shu, G.; Wang, Y. Modified deadbeat predictive current control method for single-phase AC–DC PFC converter in EV charging system. *IEEE Trans. Ind. Electron.* **2023**, *70*, 286–297. [CrossRef]
5. Ramos, L.A.; Van Kan, R.F.; Mezaroba, M.; Batschauer, A.L. A Control Strategy to Smooth Power Ripple of a Single-Stage Bidirectional and Isolated AC-DC Converter for Electric Vehicles Chargers. *Electronics* **2022**, *11*, 650. [CrossRef]
6. Yang, C.; Wang, J.; Wang, C.; You, X.; Yu, S.; Su, P. Tuning method of resonant current controller with DC elimination for PWM rectifiers in electric multiple units. *IEEE Trans. Transp. Electrif.* **2020**, *6*, 740–751. [CrossRef]
7. Tasiu, I.A.; Liu, Z.; Wu, S.; Yu, W.; Al-Barashi, M.; Ojo, J.O. Review of recent control strategies for the traction converters in high-speed train. *IEEE Trans. Transp. Electrif.* **2022**, *8*, 2311–2333. [CrossRef]
8. Liu, H.; Ma, L.; Peng, L.; Song, W.; Cheng, S. Robust current control scheme for single-phase PWM rectifiers based on improved μ-synthesis in electric locomotive. *IET Power Electron.* **2020**, *13*, 4068–4078. [CrossRef]
9. Peng, L.; Ma, L.; Song, W.; Liu, H. A simple model predictive instantaneous current control for single-phase PWM converters in stationary reference frame. *IEEE Trans. Power Electron.* **2022**, *37*, 7629–7639. [CrossRef]
10. Lee, G.-Y.; Park, H.-C.; Ji, M.-W.; Kim, R.-Y. Digitalized Control Algorithm of Bridgeless Totem-Pole PFC with a Simple Control Structure Based on the Phase Angle. *Electronics* **2023**, *12*, 4449. [CrossRef]
11. Deng, Q.; Gou, B.; Ge, X.; Lin, C.; Xie, D.; Feng, X. A High-accuracy-light-AI Data-driven Diagnosis Method for Open-circuit Faults in Single-Phase PWM Rectifiers. *IEEE Trans. Transp. Electrif.* **2023**, *9*, 4352–4365. [CrossRef]

12. Qin, N.; Wang, T.; Huang, D.; You, Y.; Zhang, Y. VWM-DCRNN: A Method of Combining Signal Processing With Deep Learning for Fault Diagnosis in Single-Phase PWM Rectifier. *IEEE Trans. Power Electron.* **2023**, *38*, 8894–8906. [CrossRef]
13. Zhang, K.; Gou, B.; Feng, X. Online Fault Diagnosis for Single-Phase PWM Rectifier Using Data-Driven Method. *CPSS Trans. Power Electron. Appl.* **2022**, *7*, 49–57. [CrossRef]
14. Li, Z.; Wang, B.; Ren, Y.; Wang, J.; Bai, Z.; Ma, H. *L*- and *LCL*-Filtered Grid-Tied Single-Phase Inverter Transistor Open-Circuit Fault Diagnosis Based on Post-Fault Reconfiguration Algorithms. *IEEE Trans Power Electron.* **2019**, *34*, 10180–10192. [CrossRef]
15. Youssef, A.B.; El Khil, S.K.; Slama-Belkhodja, I. State Observer-Based Sensor Fault Detection and Isolation, and Fault Tolerant Control of a Single-Phase PWM Rectifier for Electric Railway Traction. *IEEE Trans Power Electron.* **2013**, *28*, 5842–5853. [CrossRef]
16. Poon, J.; Jain, P.; Konstantakopoulos, I.C.; Spanos, C.; Panda, S.K.; Sanders, S.R. Model-Based Fault Detection and Identification for Switching Power Converters. *IEEE Trans Power Electron.* **2017**, *32*, 1419–1430. [CrossRef]
17. Yu, Y.; Song, Y.; Tao, H.; Hu, J. Voltage and Current Sensor Fault Diagnosis Method for Traction Rectifier in High-Speed Trains. *Electronics* **2024**, *13*, 197. [CrossRef]
18. Ndrabarushimana, E.; Ma, L. Robust Sensor Fault Detection for a Single-Phase Pulse Width Modulation Rectifier. *Electronics* **2023**, *12*, 2366. [CrossRef]
19. Xie, D.; Lin, C.; Zhang, Y.; Sangwongwanich, A.; Ge, X.; Feng, X.; Wang, H. Diagnosis and Resilient Control for Multiple Sensor Faults in Cascaded H-Bridge Multilevel Converters. *IEEE Trans. Power Electron.* **2023**, *38*, 11435–11450. [CrossRef]
20. Xia, J.; Guo, Y.; Dai, B.; Zhang, X. Sensor Fault Diagnosis and System Reconfiguration Approach for an Electric Traction PWM Rectifier Based on Sliding Mode Observer. *IEEE Trans Ind. Appl.* **2017**, *53*, 4768–4778. [CrossRef]
21. Xia, J.; Li, Z.; Gao, X.; Guo, Y.; Zhang, X. Real-time sensor fault identification and remediation for single-phase grid-connected converters using hybrid observers with unknown input adaptation. *IEEE Trans. Ind. Electron.* **2023**, *70*, 2407–2418. [CrossRef]
22. Zhang, K.; Jiang, B.; Yan, X.-G.; Mao, Z. Incipient Voltage Sensor Fault Isolation for Rectifier in Railway Electrical Traction Systems. *IEEE Trans Ind. Electron.* **2017**, *64*, 6763–6774. [CrossRef]
23. Zhang, K.; Jiang, B.; Yan, X.-G.; Shen, J. Interval Sliding Mode Observer Based Incipient Sensor Fault Detection with Application to a Traction Device in China Railway High-Speed. *IEEE Trans Veh. Technol.* **2019**, *68*, 2585–2597. [CrossRef]
24. Xiong, J.; Zhang, J.; Xu, Z.; Din, Z.; Zheng, Y. Active power decoupling control for PWM converter considering sensor failures. *IEEE J. Emerg. Sel. Top. Power Electron.* **2023**, *11*, 2236–2245. [CrossRef]
25. Chowdhury, V.R.; Kimball, J.W. Robust Control Scheme for a Three Phase Grid-Tied Inverter with LCL Filter During Sensor Failures. *IEEE Trans Ind. Electron.* **2021**, *68*, 8253–8264. [CrossRef]
26. Merai, M.; Naouar, M.W.; Slama-Belkhodja, I.; Monmasson, E. FPGA-Based Fault-Tolerant Space Vector-Hysteresis Current Control for Three-Phase Grid-Connected Converter. *IEEE Trans Ind. Electron.* **2016**, *63*, 7008–7017. [CrossRef]
27. Mehmood, F.; Papadopoulos, P.M.; Hadjidemetriou, L.; Charalambous, A.; Polycarpou, M.M. Model-based fault diagnosis scheme for current and voltage sensors in grid side converters. *IEEE Trans. Power Electron.* **2023**, *38*, 5360–5375. [CrossRef]
28. Ben Youssef, A.; El Khil, S.K.; Belkhodja, I.S. Open-circuit fault diagnosis and voltage sensor fault tolerant control of a single phase pulsed width modulated rectifier. *Math. Comput. Simul.* **2017**, *131*, 234–252. [CrossRef]
29. Xia, Y.; Gou, B.; Xu, Y. Current sensor fault diagnosis and fault tolerant control for single-phase PWM rectifier based on a hybrid model-based and data driven method. *IET Power Electron.* **2020**, *13*, 4150–4157. [CrossRef]
30. Gong, Z.; Huang, D.; Jadoon, H.U.K.; Ma, L.; Song, W. Sensor-Fault-Estimation-Based Tolerant Control for Single-Phase Two-Level PWM Rectifier in Electric Traction System. *IEEE Trans Power Electron.* **2020**, *35*, 12274–12284. [CrossRef]
31. Dardouri, M.; El Khil, S.K.; Jelassi, K. Current Sensor Fault Reconstruction and Compensation of an AC-Voltage Sensorless Controlled Single Phase Grid Connected Converter using PI-Observers. *Int. J. Renew. Energy Res. (IJRER)* **2022**, *12*, 248–258.
32. Gong, Z.; Huang, D.; Ma, L.; Qin, N.; Jadoon, H.U.K. Grid voltage sensor fault-tolerant control for single-phase two-level PWM rectifier. *IET Electr. Power Appl.* **2021**, *16*, 776–788. [CrossRef]
33. Chen, Z.; Chen, W.; Fan, X.; Peng, T.; Yang, C. JITL-MBN: A realtime causality representation learning for sensor fault diagnosis of traction drive system in high-speed trains. *IEEE Trans. Neural Netw. Learn. Syst.* **2022**. early access. [CrossRef] [PubMed]
34. Ni, Q.; Zhan, Z.; Li, X.; Zhao, Z.; Lai, L.L. A Real-time Fault Diagnosis Method for Grid-Side Overcurrent in Train Traction System using Signal Time Series Feature Pattern Recognition. *IEEE Trans. Ind. Electron.* **2023**, *71*, 4210–4218. [CrossRef]
35. Zhu, Y.; Tang, H. Automatic damage detection and diagnosis for hydraulic structures using drones and artificial intelligence techniques. *Remote Sens.* **2023**, *15*, 615. [CrossRef]
36. Zhu, Y.; Xie, M.; Zhang, K.; Li, Z. A Dam Deformation Residual Correction Method for High Arch Dams Using Phase Space Reconstruction and an Optimized Long Short-Term Memory Network. *Mathematics* **2023**, *11*, 2010. [CrossRef]
37. Gao, Z.; Cecati, C.; Ding, S.X. A survey of fault diagnosis and fault-tolerant techniques—Part I: Fault diagnosis with model-based and signal-based approaches. *IEEE Trans. Ind. Electron.* **2015**, *62*, 3757–3767. [CrossRef]
38. Chakraborty, C.; Verma, V. Speed and Current Sensor Fault Detection and Isolation Technique for Induction Motor Drive Using Axes Transformation. *IEEE Trans. Ind. Electron.* **2015**, *62*, 1943–1954. [CrossRef]
39. Kommuri, S.K.; Bin Lee, S.; Veluvolu, K.C. Robust Sensors-Fault-Tolerance with Sliding Mode Estimation and Control for PMSM Drives. *IEEE/ASME Trans. Mechatron.* **2018**, *23*, 17–28. [CrossRef]
40. Han, Y.; Luo, M.; Zhao, X.; Guerrero, J.M.; Xu, L. Comparative Performance Evaluation of Orthogonal-Signal-Generators-Based Single-Phase PLL Algorithms—A Survey. *IEEE Trans. Power Electron.* **2016**, *31*, 3932–3944. [CrossRef]

41. Xu, J.; Mi, C.C.; Cao, B.; Deng, J.; Chen, Z.; Li, S. The State of Charge Estimation of Lithium-Ion Batteries Based on a Proportional-Integral Observer. *IEEE Tans Veh. Technol.* **2014**, *63*, 1614–1621.
42. Lu, J.; Savaghebi, M.; Ghias, A.M.Y.M.; Hou, X.; Guerrero, J.M. A reduced-order generalized proportional integral observer-based resonant super-twisting sliding mode control for grid-connected power converters. *IEEE Trans. Ind. Electron.* **2020**, *68*, 5897–5908. [CrossRef]
43. Söffker, D.; Yu, T.-J.; Müller, C. State estimation of dynamical systems with nonlinearities by using proportional-integral observer. *Int. J. Syst. Sci.* **1995**, *26*, 1571–1582. [CrossRef]
44. Qi, Y.; Zafarani, M.; Gurusamy, V.; Akin, B. Advanced severity monitoring of interturn short circuit faults in PMSMs. *IEEE Trans. Transp. Electrif.* **2019**, *5*, 395–404. [CrossRef]
45. Mukherjee, S.; Chowdhury, V.R.; Shamsi, P.; Ferdowsi, M. Grid Voltage Estimation and Current Control of a Single-Phase Grid-Connected Converter Without Grid Voltage Sensor. *IEEE Trans Power Electron.* **2018**, *33*, 4407–4418. [CrossRef]
46. Dardouri, M.; El Khil, S.K.; Jelassi, K. Sensorless Active Damping of LCL_Filter based on MRAS Estimator for Single-Phase Grid-Connected Inverter. In Proceedings of the 2019 10th International Renewable Energy Congress (IREC), Sousse, Tunisia, 26–28 March 2019; pp. 1–5.
47. Mehreganfar, M.; Saeedinia, M.H.; Davari, S.A.; Garcia, C.; Rodriguez, J. Sensorless Predictive Control of AFE Rectifier with Robust Adaptive Inductance Estimation. *IEEE Trans Ind. Inform.* **2019**, *15*, 3420–3431. [CrossRef]

World Electric Vehicle Journal

MDPI

Article

Diagnosis of Multiple Open-Circuit Faults in Three-Phase Induction Machine Drive Systems Based on Bidirectional Long Short-Term Memory Algorithm

Badii Gmati [1], Amine Ben Rhouma [1,2], Houda Meddeb [1] and Sejir Khojet El Khil [1,*]

1 LR11ES15 Laboratoire des Systèmes Electriques, Ecole Nationale d'Ingénieurs de Tunis, Université de Tunis El Manar, 1002 Tunis, Tunisia; badii.gmati@enit.utm.tn (B.G.); amine.benrhouma@enit.rnu.tn (A.B.R.); meddebhouda@gmail.com (H.M.)

2 Department of Electrical Engineering, ENSIT, Université de Tunis, 1008 Tunis, Tunisia

* Correspondence: sejir.kek@edu.isetcom.tn

Abstract: Availability and continuous operation under critical conditions are very important in electric machine drive systems. Such systems may suffer from several types of failures that affect the electric machine or the associated voltage source inverter. Therefore, fault diagnosis and fault tolerance are highly required. This paper presents a new robust deep learning-based approach to diagnose multiple open-circuit faults in three-phase, two-level voltage source inverters for induction-motor drive applications. The proposed approach uses fault-diagnosis variables obtained from the sigmoid transformation of the motor stator currents. The open-circuit fault-diagnosis variables are then introduced to a bidirectional long short-term memory algorithm to detect the faulty switch(es). Several simulation and experimental results are presented to show the proposed fault-diagnosis algorithm's effectiveness and robustness.

Keywords: open-circuit fault; induction motor; sigmoid function; bidirectional long short-term memory; deep learning

Citation: Gmati, B.; Ben Rhouma, A.; Meddeb, H.; Khojet El Khil, S. Diagnosis of Multiple Open-Circuit Faults in Three-Phase Induction Machine Drive Systems Based on Bidirectional Long Short-Term Memory Algorithm. *World Electr. Veh. J.* **2024**, *15*, 53. https://doi.org/10.3390/wevj15020053

Academic Editors: Diego Bellan and Jelena Loncarski

Received: 13 November 2023
Revised: 21 January 2024
Accepted: 30 January 2024
Published: 5 February 2024

1. Introduction

Three-phase pulse-width modulation (PWM) voltage source inverters (VSIs) have been widely used for grid-connected converters or AC machines' variable-speed drive applications. For most of these applications, high reliability and availability are of utmost importance. Consequently, the condition monitoring, fault detection, and fault tolerance of three-phase PWM VSIs are widely requested functions that should be added to the drive system's controller [1,2].

An open-circuit (OC) fault is one of the most relevant faults that may affect induction-motor variable-speed drives [3]. This type of fault may affect one or more power semiconductors. As discussed in [4,5], OC faults introduce severe perturbations: a DC current injection, an overcurrent, and pulsating electromagnetic torque. In some cases, the electric drive must be shut down. Consequently, fault diagnosis and fault tolerant control [5,6] are of utmost importance. Several studies have focused on OC faults, and different algorithms have been proposed, as summarized by many survey articles [1–3,7,8]. OC fault-diagnostics methods are mainly classified as model-based approaches [9–21], signal-based approaches [22–30], and data-driven approaches [31–41].

Model-based fault-diagnosis approaches use the mathematical model of the electric machine and/or the model of the voltage source inverter. The main idea is to compare estimated quantities obtained from the system's model to the measured quantities. In most cases, electric machine currents [9–14] and output voltages of the voltage source inverter [15–20] are the estimated quantities.

In a well-functioning operation, the measured and estimated quantities are almost the same. Hence, the estimation error converges to zero. When an open-circuit fault is affecting one or more power switch(es), the deviation of the estimation error is used to detect the fault and to identify the faulty switch(es).

In [9,10], Jlassi et al. use the current form factor (CFF) to detect open-circuit faults in permanent magnet synchronous machine (PMSM) drives for both generator and motor applications. The main idea is to compare the CFF obtained from the measured currents to the CFF obtained from estimated currents based on the state Luenberger observer. A similar approach is discussed in [11] to detect current sensors and open-circuit faults in grid-connected three-level NPC inverters. In this study, the currents are estimated using a sliding-mode observer.

The current residual approach is proposed to diagnose open-circuit faults in PMSM, and induction-motor drive systems are discussed in [12,13], respectively.

Voltage-based OC fault detection has been discussed in [15–20]. To avoid the use of additional voltage sensors, which may increase the system's cost, the output voltage of the VSI should be estimated. In [15], an OC fault-diagnosis scheme is proposed based on voltage estimation for an induction machine drive system. A state estimator is used to estimate the phase voltage based on the induction machine model. The estimated voltage is compared to the voltage obtained from the DC-link voltage and the inverter switching signals. The residual of the estimation error is used to detect the faulty switch(es). Nevertheless, this approach needs the use of low-pass filters that require high tuning and introduce delays in the detection time. Freire et al. present a similar approach applied to PMSM drive systems [16]. In [17], the calculated common-mode voltage behavior is used to detect and locate OC faults in three-phase, two-level inverters for induction machine drive systems.

Recently, with the increase in the use of model predictive control (MPC) in variable-speed drive systems, some studies propose the use of motor-current prediction errors to detect OC faults for PMSM drive systems [22–25]. Thanks to the robustness of MPC, the motor-current prediction errors are very low in well-functioning (healthy) operation modes. It has been shown that in the case of OC faults, the motor-current prediction errors increase, allowing for the detection and identification of the fault switch [22–25].

Signal-based methods typically involve motor-current signature analysis [26–30]. In [26], a single current sensor is employed to detect OS faults in the PMSM drive system. A DC-link current sensor is used to reconstruct the motor stator currents. Then, the normalized average value of the reconstructed current is used to detect the faulty switch. Sejir et al. [27] present a current analysis-based algorithm to detect OS faults in PMSM drive systems. The fault-detection variables use the interaction between two stator currents, which allows for the detection of 27 types of OC faults and current-sensor faults. Reference current errors are adopted to detect OC faults in voltage source inverters [28]. However, this approach can only be used for closed-loop controlled electric drives.

More recently, data-driven fault-diagnosis approaches have become more attractive for fault-diagnosis and fault-classification purposes [31–41]. Indeed, data-driven approaches are only based on recorded data obtained from measured quantities, instead of specific complex mathematical models.

In [31,32], a Fast Fourier Transform (FFT) algorithm is used to extract open-circuit fault features from motor currents [31] or the inverter's output line-to-line voltage [32]. Then, a fast-learning technology is applied to diagnose the faulty switch(es). Xia et al. [33] present a transferrable data-driven algorithm for open-circuit switch-fault diagnosis in three-phase inverters. A deep learning-based approach for the open switch-fault diagnosis of three-phase PWM converters is discussed in [34]. Current behaviors in healthy and faulty operation modes are analyzed for fault feature extraction. Hang et al. [36] propose an OC fault-diagnosis algorithm for PMSM drives using a wavelet convolutional neural network (WCNN). The normalized current vector trajectory graph obtained by the Clark transform is sent to the WCNN model to detect and localize the faulty power semi-conductor. However, this method requires high-computation execution time and needs preprocessing for noise robustness. In [37], the online data-driven fault diagnosis of dual three-phase PMSM drives is discussed.

Long short-term memory networks (LSTMs) have been investigated for fault-diagnosis issues due to their high accuracy compared to other fault-diagnosis techniques [38–45]. The effectiveness of the method is confirmed in [38], where it is used to detect multiple open-circuit switch faults of the back-to-back converter in doubly fed induction generator (DFIG)-based wind turbine systems. A similar approach has been used to diagnose OC faults in multilevel converters [39–41]. The LSTM approach is applied for diagnosing faults in electric vehicles [42,43] and for motor electrical faults [44] and mechanical fault diagnosis [45].

A state-of-the-art review has shown that model-based OC fault-detection approaches are effective. However, these approaches need a good knowledge of the studied system's model. Furthermore, they are sensitive to the variations in the system's parameters. Moreover, they need a robust fault-detection threshold, which makes them complicated to elaborate. Signal-based OC fault-detection methods present a good alternative to model-based ones. They are attractive especially since they do not require either highly computational resources or extra hardware. Data-driven approaches have the advantage of being only dependable on recorded data obtained from measured quantities instead of specific complex mathematical models. The main concerns of such algorithms are their complexity and the need for large-scale data for feature extraction, training, and validation. These requirements make real-time implementation difficult.

Therefore, by combining signal-based methods with those based on data, this scheme may achieve high-accuracy fault detection and fault localization results. Consequently, this paper proposes a bidirectional long short-term memory (BiLSTM)-based algorithm to diagnose open-circuit power semi-conductor faults in a three-phase PWM voltage source inverter for an induction-motor drive system. The main contribution of the proposed approach can be summarized as follows:

- √ The proposed method can achieve an accurate diagnosis of single and multiple open-circuit faults without any extra hardware requirements. Only already measured induction-motor stator currents are used.
- √ A new, robust current-normalization approach is developed to keep the motor currents free from load-torque and motor-speed transient variations.
- √ The normalized currents are then combined in order to generate three OC fault indicators. Then, the fault-detection variables are introduced to a BiLSTM network to identify the faulty switch(es). The BiLSTM network does not need to set any fault-detection threshold, which increases the accuracy and the effectiveness of the proposed approach.

The rest of the paper is organized as follows: Section 2 presents the fault-detection variables and analysis of the fault features. Section 3 describes the BiLSTM network and the fault-diagnosis algorithm. The performance of the proposed OC fault-diagnosis algorithm is analyzed through simulations in Section 4 and evaluated through experiments in Section 5. Finally, conclusions are drawn in Section 6.

2. Fault Features Analysis

The structure of the three-phase induction-motor drive system is depicted in Figure 1. The two-level VSI is composed of six IGBTs ($T_1 \rightarrow T_6$) and their anti-parallel diodes. Under healthy operation conditions, the induction-motor stator currents are expressed as:

$$\begin{cases} i_a(t) = I_m \sin(\omega t) \\ i_b(t) = I_m \sin(\omega t - \frac{2\pi}{3}) \\ i_c(t) = I_m \sin(\omega t - \frac{4\pi}{3}) \end{cases} \quad (1)$$

where I_m is the induction-motor stator current amplitude and ω is the synchronous electrical pulsation.

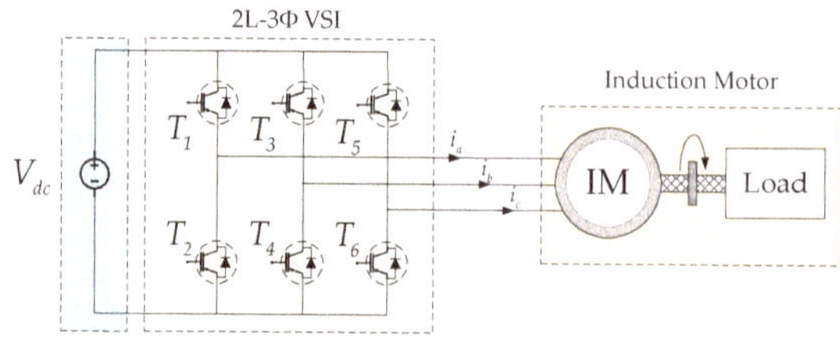

Figure 1. Structure of 2L-3Φ VSI-fed IM system.

When an open-circuit fault occurs, it produces a distortion of the stator currents with an increase in their amplitude, electromagnetic torque oscillations, and excess heat, which can lead to motor failures [27].

The motor currents can provide accurate signatures to ensure an effective diagnosis of OC faults. However, motor currents are sensitive to the load torque and motor-speed variations. To reduce the dependency of the OC fault signatures on load-torque and/or motor-speed variations, it is necessary to normalize the motor currents. In this way, several approaches discussed in the literature have been proposed as normalization tools: Park's vector modulus $\left\| i_{dq} \right\| = \sqrt{i_d^2 + i_q^2}$ in [23], the maximum absolute value of the motor-phase currents $\max\{|i_a|, |i_b|, |i_c|\}$ in [27], and the average absolute values of the motor-phase currents $\langle |i_n| \rangle_{n=\{a,b,c\}}$ in [28]. Although these methods provide good results, they also require additional computational effort and prior knowledge of the motor parameters to ensure a real-time normalization of diagnosis variables, which increases computation time and decreases the performance of the OC fault methods.

In this work, a new motor current-normalization approach is introduced. The main idea consists of applying the sigmoid function to the motor currents for the normalization process. The sigmoid function of the real variable x, $F(x)$, is defined as:

$$F(x) = \frac{2}{1 + e^{-\lambda x}} - 1 \tag{2}$$

where λ is a positive real. In Figure 2, the sigmoid function of the sinusoidal variable x is presented. It can be seen that $F(x)$ varies by ± 1 and has the same period as the variable x, apart from its maximum value x_m. The impact of the positive real λ on the dynamic of $F(x)$ is depicted in Figure 2. A greater λ value is higher and more $F(x)$ variation is faster.

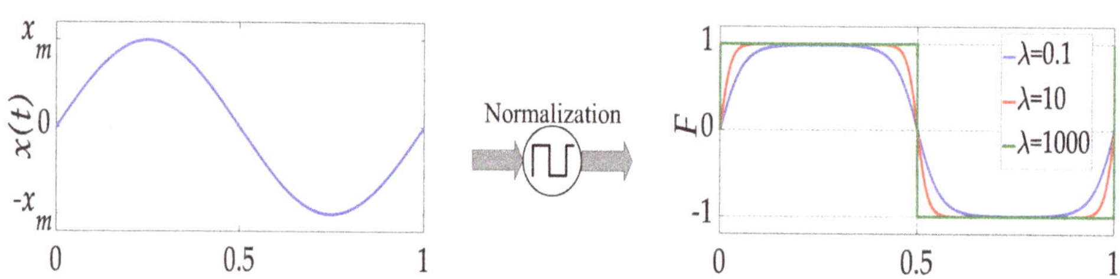

Figure 2. Sigmoid function for different coefficient values λ.

However, the motor currents do not have a pure sine waveform due to the impact of sampling, PWM, and measurement noises. As a result, the chattering problem appears in the normalized function and can be seen during the zero crossing of the current signal, as shown in Figure 3a.

To analyze this feature, the positive-current half-period is taken as an example. The current can be divided into a chattering zone, where the current is small, and a non-chattering area, due to an adequate current drop. The oscillations of the sigmoid function due to the chattering zone reduce the performance of the motor current-normalization approach. To fix this issue and to reduce the chattering effect of the sigmoid function, the normalization function of the induction-motor stator current is modified; this is expressed below:

$$F(x) = \begin{cases} 0 & if \quad x \in [-x_h, x_h] \\ \frac{2}{1+e^{-\lambda x}} - 1 & else \end{cases} \quad (3)$$

where x_h is the minimum value of signal $x(t)$ that avoids the chattering problem. Figure 3b shows the output of the modified normalization function, where it is forced to be zero during the zero-crossing of the current signal. It should be noticed that, when the fault occurs in the non-chattering zone $[t_3; t_4]$, it can be immediately detected. However, if it occurs when $|x(t)| \leq x_h$, $[t_1; t_2]$, it cannot be detected, and we should wait for the next cycle.

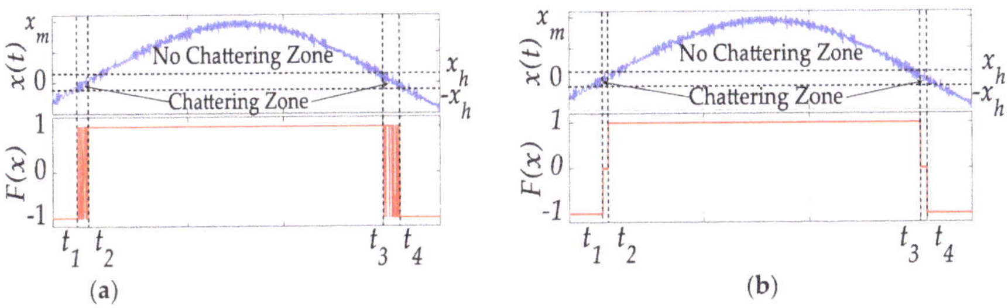

Figure 3. Current-normalization method: (**a**) with chattering problem, (**b**) without chattering problem.

By applying Equation (3) to the induction-motor stator currents, we obtain the variations of the sigmoid variables F_a, F_b and F_c of the motor currents i_a, i_b and i_c, respectively, as presented in Figure 4a. It can be seen that F_a, F_b and F_c have the same period as their corresponding motor currents and vary by ± 1, which makes them free from load-torque variations and motor-speed transient variations. Thereafter, three fault-diagnosis indictors—χ_a, χ_b and χ_c—are defined as follows:

$$\begin{cases} \chi_a = 2F_a - F_b - F_c \\ \chi_b = 2F_b - F_c - F_a \\ \chi_c = 2F_c - F_a - F_b \end{cases} \quad (4)$$

The variations in the fault-diagnosis indicators for one period of motor currents are depicted in Figure 4b. During the healthy mode of the induction-motor drive, the fault-detection variables exhibit similar behavior to the motor stator currents with the same period, and there is a phase shift between them equal to $2\pi/3$. Moreover, they vary by ± 4. It should be noted that the variations in the fault-detection variables are also independent of motor speed or load torque.

When an open-switch fault occurs, it affects the stator currents waveforms, as well as the F_a, F_b and F_c and χ_a, χ_b and χ_c waveforms. Figure 5a describes the stator currents and variations in the detection variables χ_a, χ_b and χ_c in the case of an open-switch fault of IGBT T_1 applied at time t = 0.6 s. Instantly, the behavior of χ_a, χ_b and χ_c is not the same as in the healthy operation mode. Indeed, χ_a loses its positive sequence, whereas χ_b and χ_c vary between 4 and -3.

A similar analysis was conducted for an open-switch fault of T_1 and T_2 applied at time t = 0.6 s, (open phase fault), as presented in Figure 5b. χ_a becomes equal to 0, and χ_b and χ_c are opposite and vary between 3 and -3. A third fault case is analyzed considering an open-switch fault of T_1 and T_4, as shown in Figure 5c. In this case, χ_a loses its positive sequence, χ_b loses its negative sequence, and χ_c varies between 3 and -3. The analysis of these three faults scenarios has shown for each open-switch fault that the detection variables $\chi_{a,b,c}$ have a specific feature. This feature makes them suitable to be used to achieve a robust and reliable open-switch fault-detection algorithm for the studied electric drive system.

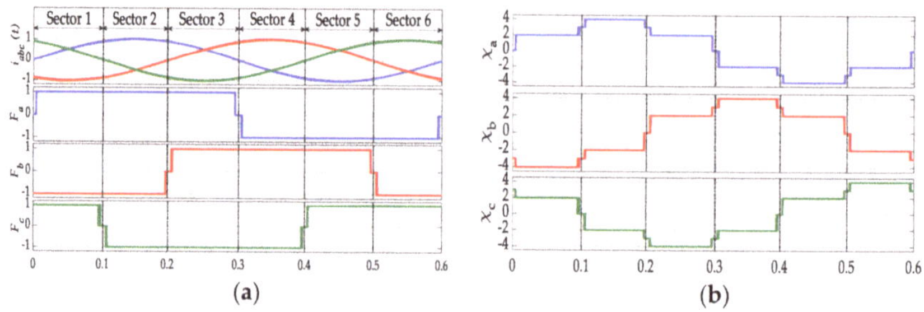

Figure 4. Time-domain waveforms of: (**a**) three-phase currents and sigmoid functions, (**b**) detection fault variables in healthy mode.

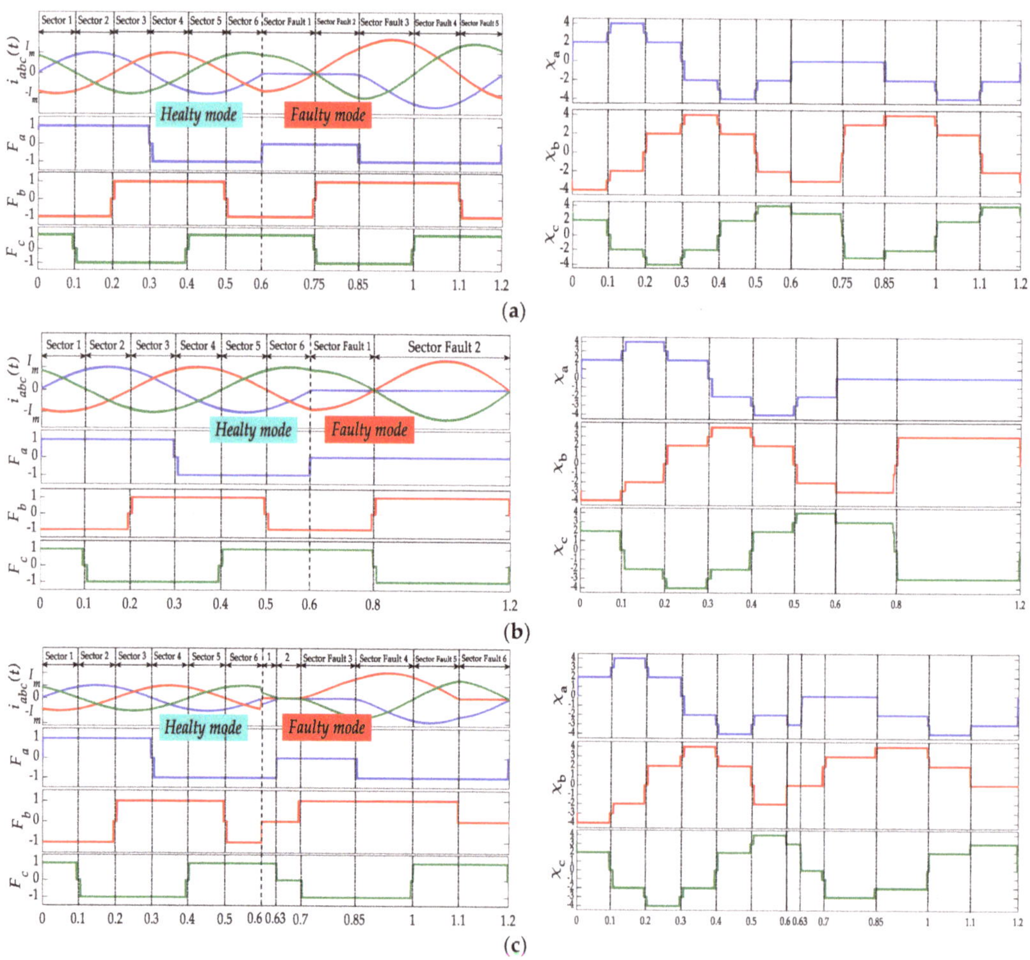

Figure 5. Time-domain waveforms of three-phase currents, sigmoid functions and detection fault variables for: (**a**) Fault in T_1, (**b**) Fault in T_1 and T_2, (**c**) Fault in T_1 and T_4.

3. LSTM Approach for Fault Diagnosis

3.1. LSTM Structure

LSTM is an improved version of a recurrent neural network (RNN), which has achieved satisfactory performance in sequence learning and temporal modeling [38].

LSTM has a special structure, which allows for solving the challenge of gradient vanishing or explosion in a simple RNN since it replaces the iterative transformation with addition in the calculation of hidden state [45]. The structure of an LSTM unit is illustrated in Figure 6.

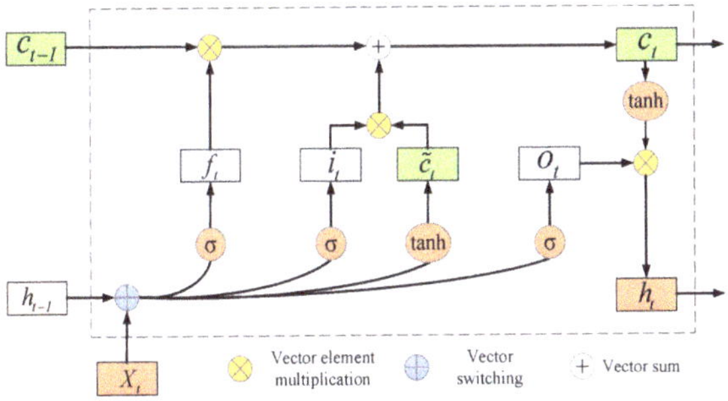

Figure 6. LSTM Network structure.

LSTM mainly consists of three gates: a forget gate f_t, an input gate i_t, and an output gate o_t. The forget-gate layer f_t determines which information should be forgotten. The equation of the forget-gate layer can be expressed as:

$$f_t = \sigma(w_f.[h_{t-1}, X_t] + b_f) \tag{5}$$

where X_t and h_{t-1} represent, respectively, the input at the current time and the output at the previous time of the LSTM network. W_f and b_f represent the weight and the bias of the forget-gate layer. σ is the sigmoid function and [ ] represents the concatenate operation.

The input-gate layer i_t updates the cell state based on the input at the current time X_t and the output at the previous time h_{t-1}. The equations of the input gate can be described as:

$$i_t = \sigma(w_i.[h_{t-1}, X_t] + b_i) \tag{6}$$

$$\tilde{c}_t = \tanh(w_c.[h_{t-1}, X_t] + b_c) \tag{7}$$

where X_t and h_{t-1} represent, respectively, the input at the current time and the output at a previous time. W_i and b_i are the weight and the bias of the sigmoid function in the input-gate layer. tanh is a hyperbolic tangent function, W_c and b_c are the weight and the bias of the tanh function in the input-gate layer.

The output-gate layer o_t selects information that should be the next output, which depends on the cell state c_t. The equation of the output-gate layer can be expressed as Equation (8); w_o and b_o are the weight and the bias of the output-gate layer.

$$o_t = \sigma(w_o.[h_{t-1}, X_t] + b_o) \tag{8}$$

The cell state and the output (hidden) state at time step t are described by the following equations:

$$c_t = f_t \times c_{t-1} + i_t \times \tilde{c}_t \tag{9}$$

$$h_t = \tanh(c_t) \times o_t \tag{10}$$

3.2. Stacked LSTM (MLSTM)

A stacked LSTM is a deep LSTM that consists of multiple LSTM layers, where each layer contains multiple memory cells. The inputs of the first LSTM layer are the sequence data, and the input of other LSTM layers is the hidden state of the previous LSTM layer. Therefore, the stacked LSTM hidden layers make the model deeper and more accurate. This type of network becomes a powerful method for challenging sequence prediction problems. The structure of a stacked LSTM with n hidden layers is shown in Figure 7.

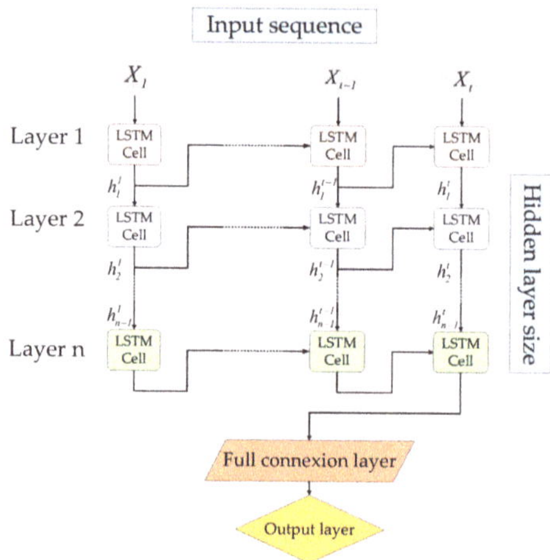

Figure 7. Stacked LSTM structure.

3.3. Bidirectional LSTMs (BiLSTM)

BILSTM is an emerging approach to address fault-diagnosis problems [46–50]. This technique has been shown to provide more accurate results than other classical methods and LSTM [50,51]. In this study, the BiLSTM network is used to identify and locate multiple open-circuit faults in a three-phase, two-level voltage source inverter for an induction-motor drive system. Compared with LSTM, BiLSTM can procure information from both earlier and later segments in sequence [51]. The BiLSTM structure consists of a forward LSTM layer and a backward LSTM layer, which reverses the direction of the input sequence flow [51,52]. Applying the LSTM twice makes the prediction results more integrated and leads to improving the accuracy of the model. Furthermore, it should be mentioned that BiLSTM is a much slower model and requires more time for training. The BiLSTM architecture is presented in Figure 8.

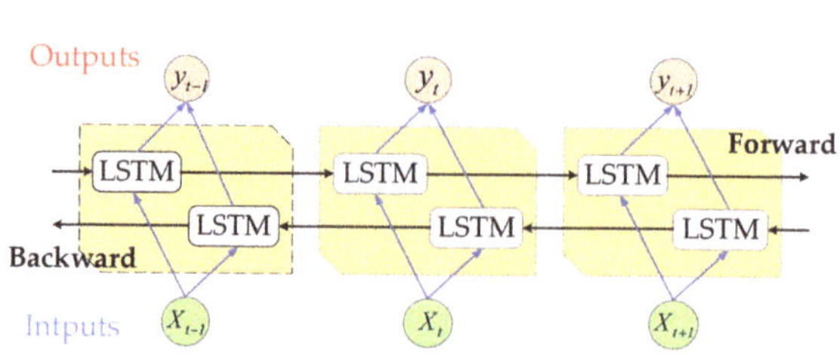

Figure 8. BiLSTM structure.

3.4. Evaluation Metrics

After building prediction models, several metrics can be used to evaluate the performance of models and compare them.

3.4.1. Root-Mean-Square Error (RMSE)

The root-mean-square error (RMSE) is the most commonly used performance measure for prediction tasks. The RMSE can be calculated by using (11).

$$RMSE = \sqrt{\frac{1}{n}\sum_{i=1}^{n}(y_i - \hat{y}_i)^2} \quad (11)$$

where $\hat{y}_i$ and y_i are the prediction and the real output value, and n is the number of data.

3.4.2. Mean Absolute Error (MAE)

The MAE is the other criterion used to evaluate the model performance. The MAE is expressed as (12).

$$MAE = \frac{1}{n}\sum_{i=1}^{n}|y_i - \hat{y}_i| \quad (12)$$

3.4.3. Mean Absolute Percentage Error (MAPE)

The MAPE is one of the most common metrics used to measure the prediction accuracy of a model and it is described as (13).

$$MPAE(\%) = \frac{1}{n}\sum_{i=1}^{n}\frac{|y_i - \hat{y}_i|}{y_i} \times 100 \quad (13)$$

The summation ignores observations where $y_i = 0$. In general, the lower the MAPE value is, the more accurate the model is.

The last metric used in this paper to evaluate the performance of the proposed algorithm is the accuracy of the prediction model, which is expressed as:

$$Accuracy(\%) = 100 - MPAE \quad (14)$$

3.5. Diagnostic Network Implementation and Validation

The proposed diagnosis method consists of the BiLSTM network with one hidden layer. The BiLSTM-based network is adopted for the fault diagnosis, using the normalized diagnosis variables χ_a, χ_b and χ_c of the current-sensor signals as input. Firstly, two-phase current signals are collected by the sensors installed between the inverter and the motor, respectively. Then, a normalization process is applied to extract suitable features. Finally, the normalized input sensor data $[\chi_a\ \chi_b\ \chi_c]$ are put into the BiLSTM network.

The output of the BiLSTM network is a concatenation of the forward and backward hidden states. The final output of the network contains six flags $[T_1\ T_2\ T_3\ T_4\ T_5\ T_6]$, which represent the healthy or faulty state of each power switch of the inverter that controls the IM. Each flag can take either the value 1 to indicate that the desired power switch is infected, or the value 0 to indicate that the desired switch is healthy. The proposed BILSTM enables us to predict and identify 21 open-circuit faulty states and 1 normal state. The structure of the fault-diagnosis method based on the BiLSTM network is illustrated in Figure 9.

To train the BiLSTM network to work efficiently, the system requires normal and faulty feature data. For this, we generated all possible single and multiple open-circuit fault scenarios. Thirty thousand samples are used as training data for each class of default, with a sampling time of 10^{-4} s. The cost function used in the training process is the root-mean-square error (RMSE).

After many trials ranging from 10 to 100, as shown in Figure 10, the size of the hidden layer of BiLSTM network is set as 100 units to make a tradeoff between accuracy and computation time.

The maximum training epoch is set as 1000 to ensure that the error after training is small enough. The BiLSTM model is trained using the LSTM MATLAB toolbox vR2020b.

The parameters of the BiLSTM models are summarized in Table 1.

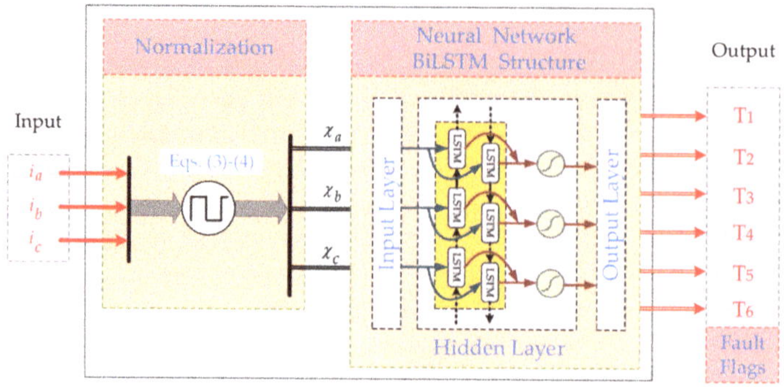

Figure 9. Structure of the fault-diagnosis module based on BiLSTM network.

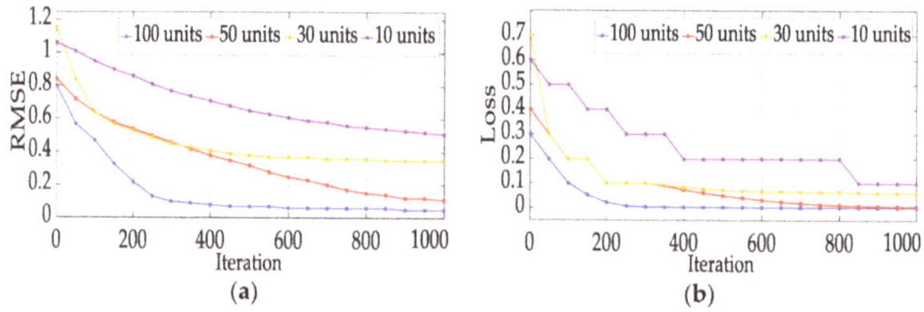

Figure 10. (**a**) Evolution of the RMSE as a function of hidden layer size in training progress of BiLSTM Network, (**b**) Evolution of the Loss as a function of hidden layer size.

Table 1. BiLSTM parameter.

Parameter	Value
Max training epochs	1000
Loss function optimizer (solver)	Adam
Initial learning rate	0.001
Loss function	RMSE
Gradient threshold	0.001
Hidden units	100

The principle of the proposed BiLSTM-based fault-diagnosis method considering the data detailed above is shown in Algorithm 1.

Algorithm 1: BiLSTM-based Fault-Diagnosis Algorithm

Step 1: Data set collection.

– Fault-detection variables $[\chi_a\ \chi_b\ \chi_c]$.
– Fault flags $[T_1\ T_2\ T_3\ T_4\ T_5\ T_6]$.

Step 2: Parameter initialization.
Step 3: Set the BiLSTM model.

– Forward hidden layer output h_t^f: $h_t^f = \sigma(w_x^f X_t + w_x^f h_{t-1}^f + b_h^f)$.
– Backward hidden layer output h_t^b: $h_t^b = \sigma(w_x^b X_t + w_h^b h_{t-1}^b + b_h^b)$.
– The model output y_t: $y_t = w_y^f h_t^f + w_y^b h_t^b + b_y$.

Step 4: Train the BiLSTM model.
Step 5: Return the network model (w, b).
Step 6: Test model. If the evaluation metrics are not satisfactory, then adjust network parameters and go to Step 3.

After training, the performance and efficiency of our network were tested in a situation in which open-circuit switch faults occur simultaneously in T_5 and T_6 at t = 1 s. The simulation results are shown in Figure 11a,b.

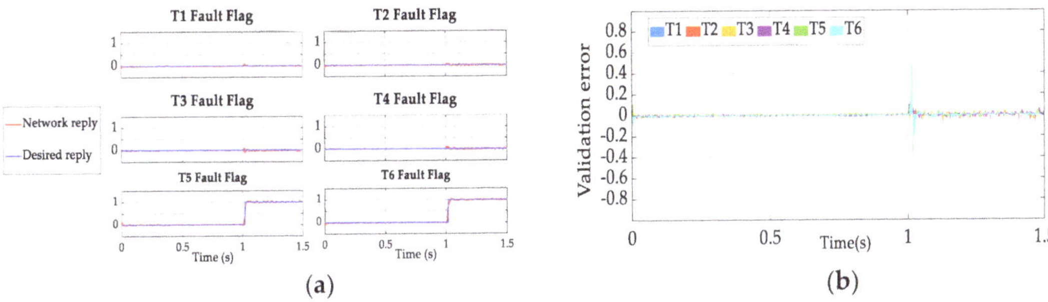

Figure 11. Simulation results: (**a**) Test of BiLSTM Network in the presence of T_5 and T_6 fault, (**b**) Validation error of BiLSTM model.

It can be seen from Figure 11b that the validation error value varies between −0.05 and 0.05, with a little increase when the fault begins to occur (t = 1 s); this result demonstrates the efficiency of the proposed model in the prediction of default class and the time at which the fault occurs.

4. Simulation Results

The performance of the proposed open-switch fault-detection approach is first analyzed through simulations using MATLAB/Simulink Software vR2020b for a variable-speed three-phase induction-motor drive. In this section, the robustness of the proposed method under speed/load-torque variations is studied first. Then, the effectiveness of the proposed approach in detecting single and multiple open-switch faults is presented.

4.1. Robustness under Operating Point Variations

The performances of the proposed fault-diagnosis approach under motor-speed and load-torque variations are presented in Figures 12 and 13, respectively. Figure 12 presents variations in stator current, motor speed, detection variables, and the output of the BiLSTM network ($T_1 \rightarrow T_6$ fault flags) when the motor speed varies from 1200 to 1500 rpm at time t = 1 s with no load. The detection variables maintain the same behavior during steady-state operation as well as during a transient state. Finally, all fault flags remain at low levels, equal to 0. Figure 13 describes the same variables in the case of torque-load change from 0 N·m to 3 N·m at time t = 1 s with motor speed equal to 1000 rpm. Here, again, all fault flags remain at a value of 0, and no false alarm is triggered.

4.2. Open-Switch Fault Detection

In Figure 14, the stator currents for an OC fault in the upper switch T_1 of phase a, with 4 N·m as the load torque and 800 rpm as the rotor speed, are presented. When the open-switch fault of IGBT T_1 is applied at t = 1 s, the positive half-cycle of the current i_a is deleted. Therefore, the lost variable χ_a is a positive sequence and varies between the values −4 and 0, whereas both the other diagnostic variables, χ_b and χ_c, vary between values −3 and 4. Hence, the T_1 fault flag increases instantly to 1 after 4 ms of fault appearance, relevant to 10% of the stator current's fundamental period.

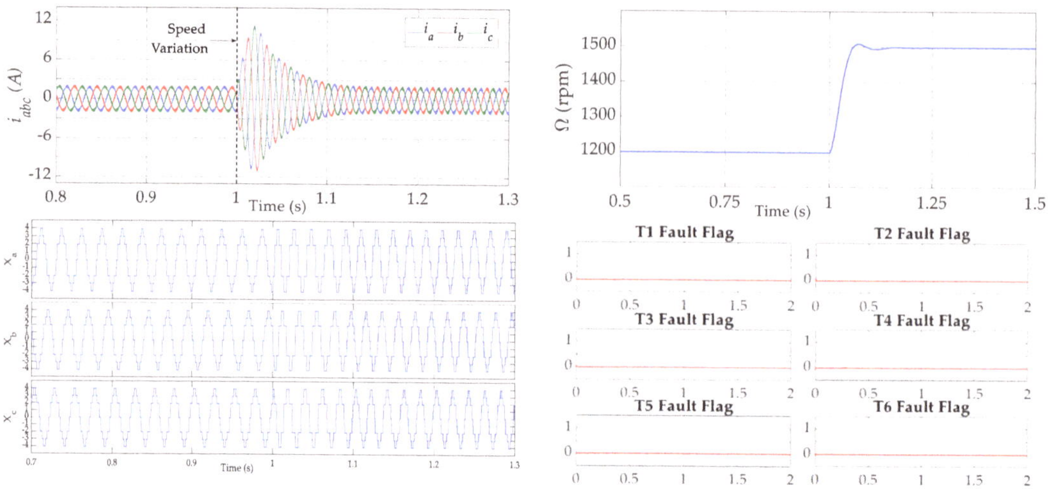

Figure 12. Simulation results for speed variation from 1200 to 1500 rpm without load.

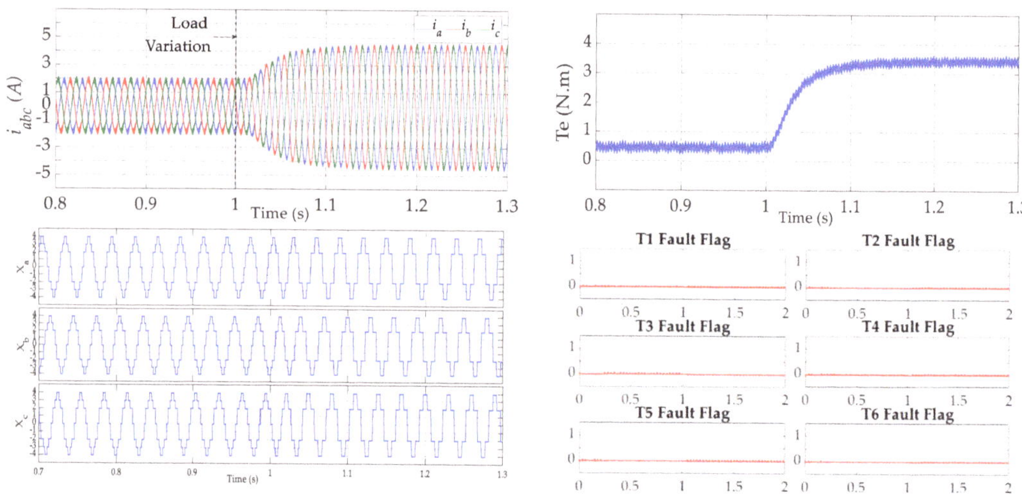

Figure 13. Simulation results for load variation from 0 to 3 N·m under a speed of 1000 rpm.

The performance of the proposed approach relating to the diagnosis of an open-phase fault is shown in Figure 15. Primarily, a single OC fault in power switch T_1 occurred at time $t = 1$ s, where the motor is running at the speed of 1400 rpm and the load torque is equal to 3 N·m. The faulty power switch is identified when the T_1 fault flag increases to 1 at time $t = 1.003$ s, equal to 14% of the motor current's fundamental period. At $t = 1.5$ s, the fault in IGBT T_2 is added, and the behavior of the fault-diagnosis variables automatically changes. Regarding all diagnosis variables in this failure situation (open-phase fault), the diagnosis variable χ_a becomes equal to 0, but the other variables, χ_b and χ_c, take the values of -3 and 3. Then, the open-phase fault is distinguished when both T_1 and T_2 fault flags take the value 1, 4 ms after fault occurrence, equal to 19% of the motor current's fundamental period.

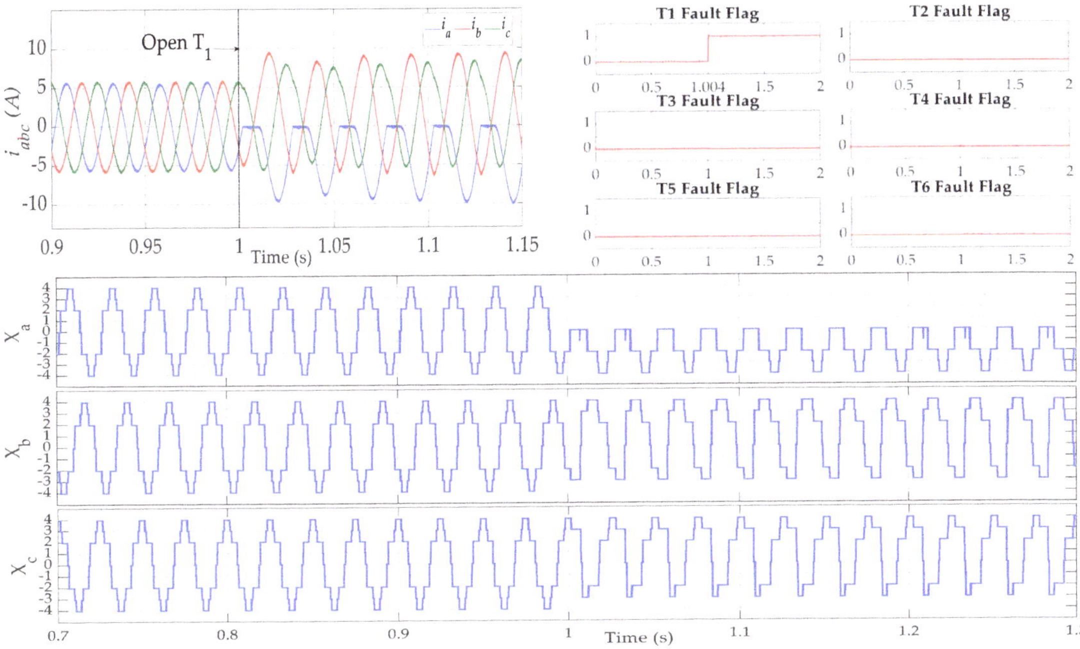

Figure 14. Simulation results for a fault in T_1 under a speed of 800 rpm and a load torque of 4 N·m.

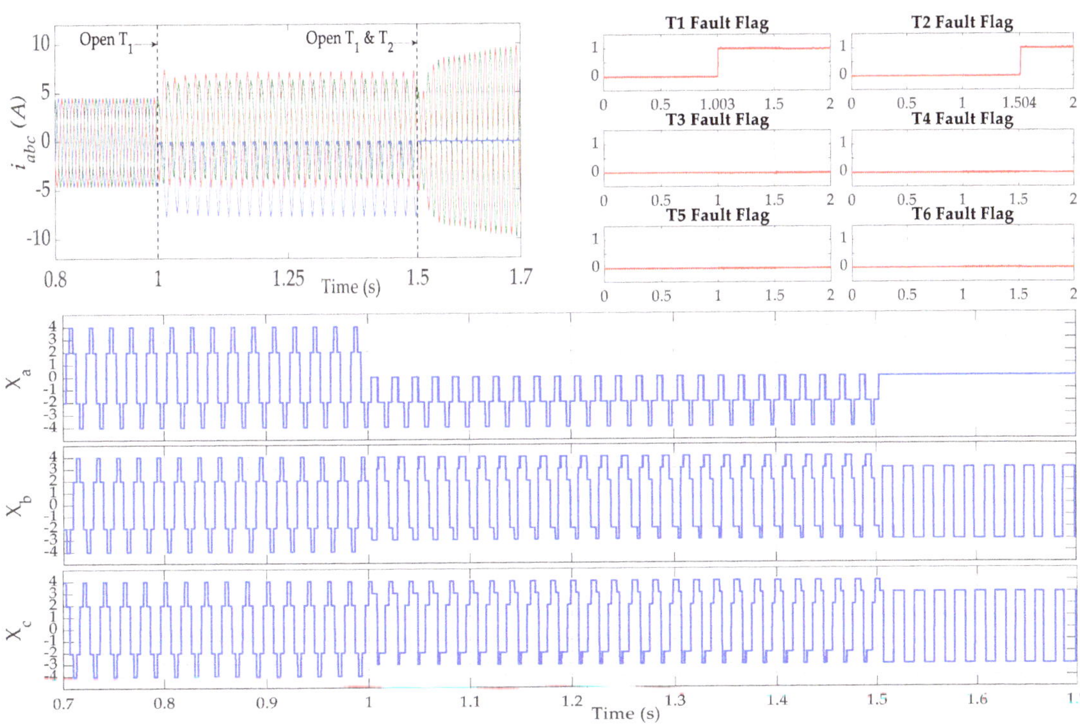

Figure 15. Simulation results for a fault in T_1 and T_2 under a speed of 1400 rpm and a load torque of 3 N·m.

The steady-state performance of the proposed AI approach for both open-switch faults in the VSI is presented in Figure 16. Firstly, an OC fault in IGBT T_1 appears at $t = 1$ s, under the operating speed of 1200 rpm and with 3 N·m as the load torque. The infected IGBT is localized by switching the T_1 fault-flag value to 1 at $t = 1.004$ s, with 10% of the fundamental period as a delay. At $t = 1.5$ s, the second fault in T_4 is added, and the first diagnostic variable χ_a maintains its negative sequence. However, the second diagnostic variable χ_b changes its behavior by losing its negative sequence, varying between 0 and 4, whereas χ_c varies between -3 and 3. For this state, the T_4 fault flag needs 3.5 ms as a time delay to take the value 1, which would denote that an open-circuit fault has occurred in power switch T_4. The detection time of this fault is equal to 8.75% of the current's fundamental period.

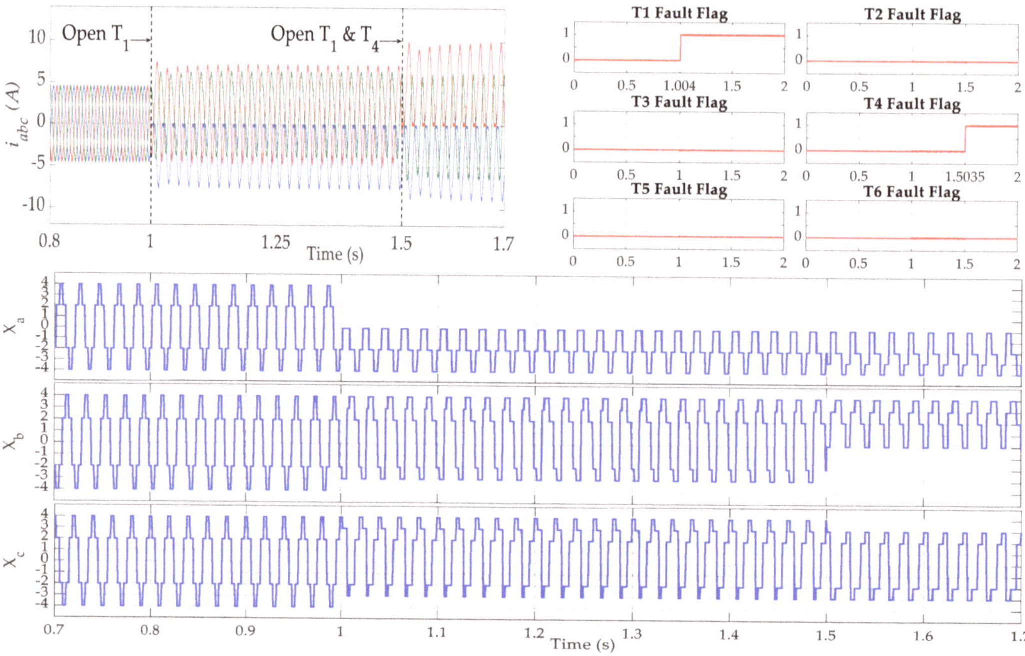

Figure 16. Simulation results for a fault in T_1 and T_4 under a speed of 1200 rpm and a load torque of 3 N·m.

5. Experimental Results

The performances of the proposed open-circuit fault-diagnosis method were examined using a 3 kW induction-motor drive prototype. The experimental test bench, shown in Figure 17, comprises a three-phase voltage source inverter (SEMIKRON), fed by a 3 kW PV array, and an induction motor coupled with a permanent magnet synchronous generator (PMSG). A digital signal processor (DSP-Dspace1104) is used for the motor drive system control. The SEMIKRON inverter is composed of three IGBT modules (SKM50GB12T4) and a 1100µF DC capacitor supporting 750VDC. The switching frequency was set to 5 kHz, and the sampling period was 100 µs. Two Hall-effect current sensors (LEM LA55P) were used for sensing motor-phase currents. A Keithley (DT9834) data acquisition module with a 16-bit resolution analog input and a sample rate of 500 kS/s throughput was used for recording the test results. A fault-gate generator box was used to generate an IGBT opening fault. The idea is to switch the PWM signal input of the gate driver to zero in the fault case. The mechanical load could be established with the help of the PMSG coupled to the PD3 rectifier and a variable resistive load. The OC fault was introduced by removing the gate command signal of the considered faulty IGBT. The parameters of the IM are shown in Table 2.

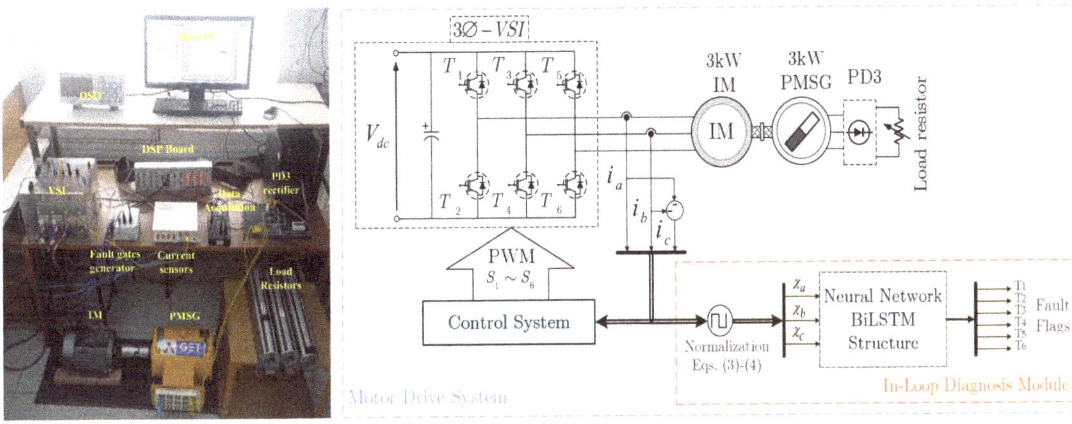

Figure 17. Experimental test bench with the architecture of the complete system.

Table 2. IM parameters.

Parameter	Symbol	Value	Parameter	Symbol	Value
Rated Power	P_a	3000 W	Stator Resistance	R_s	2.26 Ω
Rated Voltage	V	380 V	Rotor Resistance	R_r	1.45 Ω
Rated Current	I_n	6.2 A	Stator Inductance	L_s	0.249 H
Rated Frequency	F	50 Hz	Rotor Inductance	L_r	0.249 H
Rated Speed	Ω	1430 rpm	Mutual Inductance	L_m	0.237 H
Rated Torque	T_e	20 N·m	Moment of Inertia	J	6.84×10^{-3} Kg·m^2
Pair of Poles	P	2	Friction Coefficient	f	3.745×10^{-4} N·m·s/rad

5.1. Robustness under Operating Point Variations

The experimental results, reported in Figure 18, show the time-domain waveforms of the phase currents and the IGBT fault flags used for open-circuit fault diagnosis in the VSI. In this evaluation, a fast transient process was conducted using a speed step from 700 to 1000 rpm under no load for the first test and a rated load torque under 60% for the second test. Figure 19 provides the experimental results when the 3Φ-IM operates at a mechanical rotor speed equal to 1000 rpm. The transient states consist of applying two-step transitions of the load torque to the IM: the rated load torque from 0 to 60% and then the rated load torque from 60% to 0%. For the diagnostic variables, even though transient states are observed, they present the same behavior, which corresponds to a healthy operation mode of the VSI. Regarding the outputs of the BiLSTM network, all fault flags remain at their 0 values. The obtained results confirm the high performance and robustness of the proposed FDI approach under load and speed change.

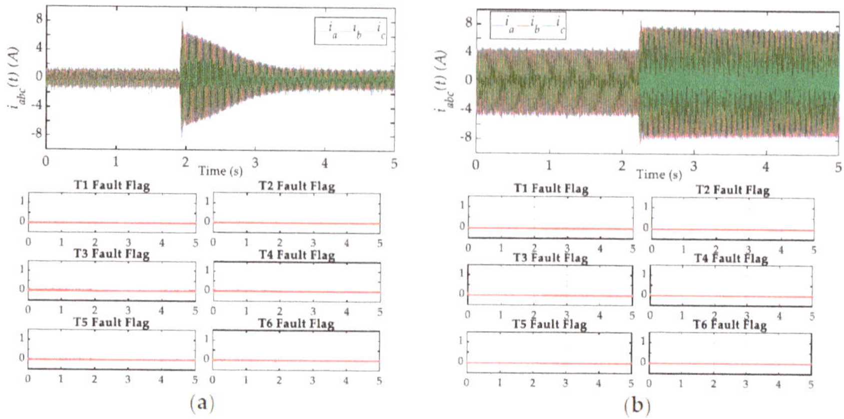

Figure 18. Experimental results for speed variation from 700 to 1000 rpm with: (**a**) no load, (**b**) 60% of rated torque.

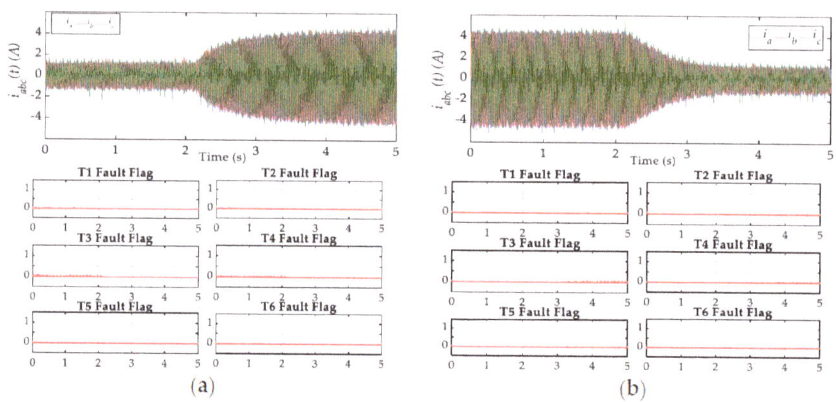

Figure 19. Experimental results for load variation under a speed of 1000 rpm: (**a**) from 0 to 60% of rated torque, (**b**) from 60% to 0% of rated torque.

5.2. Open-Switch Fault Detection

To further examine the practicality of the proposed method, several tests were performed under fault conditions. The first test was conducted when the fault occurred in phase b, resulting from an open-switch fault of IGBT T_4. In this test, the operating point of the motor was fixed, respectively, at $\Omega = 780$ rpm and $T_{em} = 50\%$ of rated load torque. The time-domain waveforms of three-phase stator currents; the diagnostic variables χ_{abc}; and the $T_{1...6}$ fault-flag behavior are reported in Figure 20. First, before the introduction of the OC fault, the behavior of the fault-detection variables χ_a, χ_b and χ_c corresponds to the healthy operation of the VSI. At $t = 3.605$ s, an open-circuit fault occurs in IGBT T_4 of the second inverter leg by fixing its switching signal in the « 0 » state. As a result, the negative half-cycle of the current i_b is cut and is now limited to only flowing in the positive direction, while other currents (i_a and i_c) undergo a light deformation and flow in negative and positive directions. Consequently, the behavior of three diagnostic variables ($\chi_{a,b,c}$) is not yet the same as in the healthy operation mode. Indeed, χ_b loses its negative sequence, varying between 0 and 4, whereas χ_a and χ_c vary between -4 and 3. As a result, the fault flag corresponding to the faulty IGBT T_4 immediately takes the value of 1 at $t = 3.609$ s, and there is a time delay equal to 12% of the current's fundamental period.

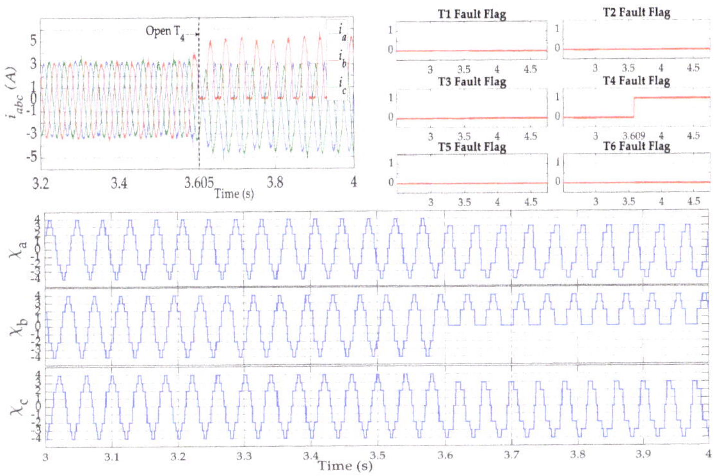

Figure 20. Experimental results for a fault in T_3 under a speed of 780 rpm and 50% of rated torque.

The second test corresponds to an open-phase fault involving two IGBTs in the same inverter leg (T_1 and T_2). Figure 21 presents the experimental results of the output-inverter currents together with the diagnostic variables used for FDI in the VSI and the outputs of the BiLSTM network. However, the motor speed and the load torque are, respectively, fixed at Ω = 990 rpm and T_{em} = 50% of the rated load torque. The fault is introduced into the first inverter leg at t = 1.378 s by keeping the switching signals of both IGBTs simultaneously in the « OFF » state. In this case, current i_a becomes equal to zero over the whole current cycle, while the other currents maintain their sinusoidal shapes but oscillate in phase opposition between −4 A and 4 A. As soon as the fault occurs, the diagnostic variable χ_a takes the value of zero, and the other diagnostic variables χ_b and χ_c are opposite and vary between 3 and −3. As learned in offline conditions, the diagnostic algorithm based on the BiLSTM network decides that the fault is an open phase involving the first inverter leg. As a result, open-phase fault identification is achieved at t = 1.387 s when both T_1 and T_2 fault flags switch from 0 to 1, and there is a time delay equal to 30% of the current's fundamental period.

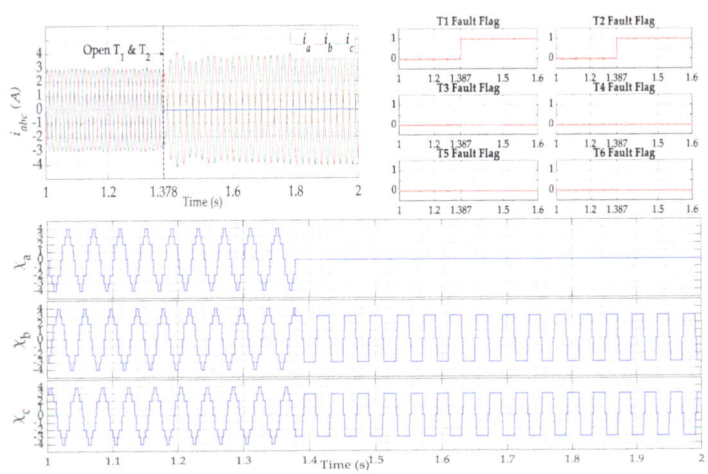

Figure 21. Experimental results for a double fault in T_1 and T_2 under a speed of 990 rpm and 50% of load.

The last test, which presents the performance of the proposed algorithm regarding the diagnosis of a double fault in the power switches T_2 and $T_{4,}$ is depicted in Figure 22. In this case of faulty condition, the 3Φ-IM runs under a speed of 1000 rpm and no load. The fault is applied between t = 2.54 s and t = 6.4 s. During this time, current i_a loses its negative half-cycle and current i_b loses its positive half-cycle, whereas current i_c is slightly deformed but maintains its sinusoidal form. Immediately, diagnostic variables χ_a and χ_b lose their negative and positive sequences, respectively, while variable χ_c varies between −3 and 3. As a consequence, the FDI approach replies immediately—the T_2 and T_3 fault flags take the value of 1 between times 2.544 s and 6.405 s. The detection time of the fault condition corresponds to 16% of the motor current's fundamental period.

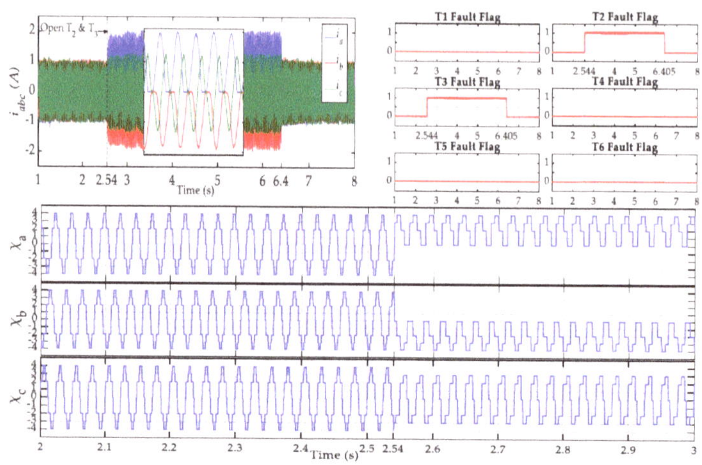

Figure 22. Experimental results for a double fault in T_2 and T_3 under a speed of 1000 rpm and no load.

5.3. Performance Evaluation and Comparison

The proposed fault-diagnosis method is applied to a 3Φ-induction-motor system for certain faults: T_1, T_2, T_3, T_4, T_5, T_6, T_1 and T_2, T_3 and T_4 and T_5 and T_6. About 1500 testing samples of each fault were used to validate the performance of the proposed BiLSTM-based method and other comparison methods, including standard LSTM, stacked LSTM (MLSTM), and feed-forward neural network (FFNN). These network structures are engaged as comparison methods to prove the efficiency and robustness of the proposed method. The FFNN and LSTM are trained with 100 units in 1 hidden layer, while the MLSTM models are trained with 50 units in each hidden layer. The training parameters of the BiLSTM model were considered for the other network structures. The different methods were tested in a situation where an OC fault occurs at t = 0.5 s. Table 3 presents the RMSE, MAE, time detection, and accuracy of each model.

Figure 23 shows MAE evolution in a situation when an open-circuit switch fault happens in T_1 and T_2 at t = 0.5 s. This metric is illustrated for the BiLSTM and the other comparison methods.

Figure 24 shows the RMSE evolution for each fault (T_1, T_2, T_3, T_4, T_5, T_6) in a situation when an open-circuit fault happens in T_1 and T_2 at t = 0.5 s. The RMSE is illustrated for the BiLSTM and the other comparison methods.

According to Table 3, the RMSE and MAE values of the BiLSTM method converge towards 0, which explains its high prediction accuracy in fault detection and identification (98.07%). This percentage is better than the accuracy reported by the FFNN and classic LSTM networks. Moreover, the BiLSTM's detection time of fault conditions varies between 2.5 ms and 6 ms, which corresponds to 12–30% of the motor current's fundamental period. Consequently, the BiLSTM approach can identify and predict various fault scenarios accurately and quickly.

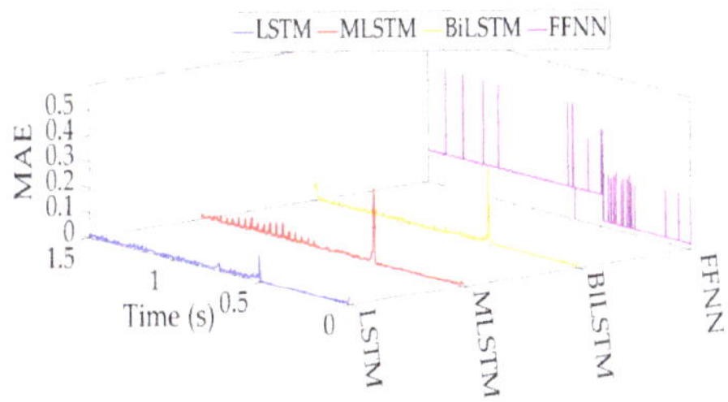

Figure 23. MAE evolution when open-circuit switch fault happens in T_1 and T_2 for the BiLSTM and the other comparison methods.

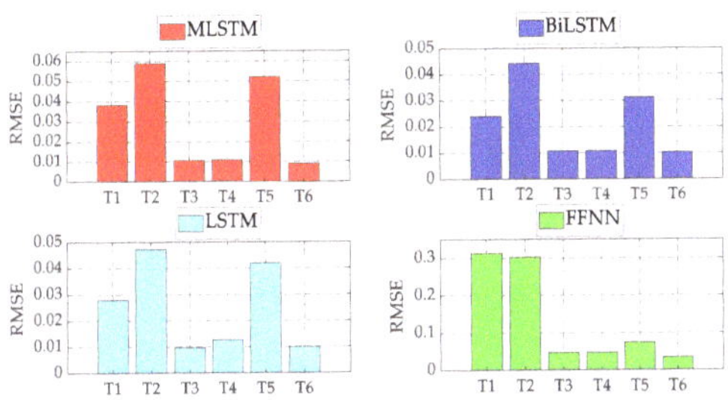

Figure 24. Performance evolution: RMSE evolution for each fault (T_1, T_2, T_3, T_4, T_5, T_6) when an open-circuit switch fault happens in T_1 and T_2 for the BiLSTM and the other comparison methods.

Moreover, appending additional layers for a classic LSTM network does not greatly improve the performance of the detection model since the LSTM and the MLSTM have the same accuracy—97%. However, it increases the calculation time. In addition, there is not a big difference in the fault-detection time, which can reach up to 24 ms.

On the other hand, the FFNN is the fastest for fault detection because the detection time is around 1 ms, but it has the lowest accuracy—48%. To prove the robustness of the BiLSTM network, the proposed method is tested on several faults at different operating points by varying the motor speed and the rated torque. The results are shown in Figures 18–22. In all cases, the method based on BiLSTM ensures the best performance. This again demonstrates the effectiveness and the robustness of the proposed method.

Consequently, the BiLSTM model can be used for effective open-circuit fault detection in a 3Φ-induction-motor-based drive system.

Table 4 presents a comparative analysis of the proposed fault-diagnosis technique, specifically for OC faults, against techniques previously used for IGBT faults, with a particular emphasis on detection time, accuracy, and implementation effort. The detection time is evaluated in relation to the current's fundamental period. The data in Table 4 clearly show that most of the techniques based on signal or/and model approaches present a low detection time that does not exceed the fundamental period for diagnosis, with little implementation effort. On the other hand, the fault-diagnosis techniques based on the artificial intelligence approach present a higher detection time and implementation effort but an accuracy near 100%. Compared to the two types of OC fault-diagnosis techniques, the proposed method based on the BiLSTM neural network demonstrates strong performance across the evaluated parameters. Considering various fault modes, including multiple IGBT open-circuit faults, the tests show that the proposed method balances the diagnostic speed and accuracy well: the fault-detection time is lower than 30% of one current's fundamental period; the accuracy value is equal to 98.08%; and there is a low implementation effort, which outperforms state-of-the-art learning algorithms.

Table 3. Performance evaluation for each method.

Faulty Switch	FFNN				LSTM				MLSTM (3 Layers)				BiLSTM			
	RMSE	MAE	T_d (ms)	Accuracy (%)	RMSE	MAE	T_d (ms)	Accuracy (%)	RMSE	MAE	T_d (ms)	Accuracy (%)	RMSE	MAE	T_d (ms)	Accuracy (%)
T_1	0.2073	0.1268	**1**	32.9996	0.0142	**0.0093**	11.5	97.8281	0.0150	0.0101	11	97.5032	**0.0135**	0.0095	3.45	**98.2613**
T_2	0.2002	0.1241	**1.1**	34.4586	0.0209	**0.0128**	21	96.5231	0.0207	0.0137	23	95.7970	**0.0192**	**0.0128**	5.2	**96.5596**
T_3	0.1915	0.1222	**1**	34.8000	0.0186	0.0116	23	96.5823	0.0174	**0.0094**	24	96.8548	**0.0178**	0.0116	6	**97.3394**
T_4	0.1977	0.1241	**1**	33.5121	0.0160	0.0105	16	97.3179	0.0157	0.0102	13.5	97.5014	**0.0146**	**0.0099**	4.88	**97.9273**
T_5	0.1983	0.1235	**1**	33.8192	0.0167	0.0107	21	97.3560	0.0192	0.0121	24	95.1774	**0.0158**	**0.0098**	2.5	**98.1525**
T_6	0.2061	0.1242	**1**	33.6194	0.0162	0.0102	12	97.4939	0.0149	0.0092	13	97.0993	**0.0132**	**0.0089**	4.02	**97.8641**
T_1 & T_2	0.1369	0.0935	**1**	74.1148	0.0250	0.0109	14	98.5637	0.0300	0.0132	17	98.1124	**0.0220**	**0.0105**	2.57	**98.8193**
T_3 & T_4	0.1242	0.0775	**1**	75.8453	**0.0244**	**0.0112**	24	98.2274	0.0270	0.0126	19	96.8244	0.0253	**0.0109**	4.66	**98.3066**
T_5 & T_6	0.1304	0.0796	**1**	76.2560	0.0302	**0.0092**	10	99.3372	**0.0278**	0.0101	16	99.2203	0.0302	0.0101	3.75	**99.4112**
Mean	0.1769	0.1108	**1.0111**	47.7139	0.0202	0.0107	16.94	97.6922	0.0208	0.0111	17.83	97.1211	**0.019**	**0.0104**	4.11	**98.07**

The bold values represent optimum evaluation metrics.

Table 4. Comparative study with other methods.

Method	Research Plant	Faulty Modes	Detection Parameter	Detection Time *	Accuracy	Implementation Effort
System model-based Sliding-Mode Observer (SMO) [13]	PWM VSI-fed sensorless IM drive	IGBT open-switch fault	Current signals, speed signals and IM model	20%	--	Medium
Output line voltage residuals [17]	IM drive	IGBT open-switch fault	Three-phase currents	5–83%	--	Low
Predictive current errors and Fuzzy Logic approach [23]	PMSM drive	IGBT open-circuit fault	Predictive current errors	12–75%	--	Medium
Online data-driven Random vector functional link (RVFL) [31]	PWM VSI-fed IM drive	IGBT open-circuit fault and current-sensor faults	Three-phase currents and speed signals	110%	98.83%	High
Machine learning-based transferrable data-driven method [33]	Three-phase inverter	IGBT open-circuit fault	Three-phase currents	100%	96.76%	Medium
Wavelet Convolutional Neural Network (WCNN) [36]	PMSM drive	IGBT open-circuit fault	Three-phase currents	1000%	100%	High
Classification of open-circuit faults based on Wavelet Packet and LSTM network [40]	Five-level nested NPP converter	IGBT open- and short-circuit switch faults	Current flying capacitor and voltages switch	120%	99.58%	High
Proposed approach (Prediction of open-circuit faults based on BiLSTM network)	PWM VSI-fed IM drive	IGBT open-circuit fault	Three-phase currents	12–30%	98.08%	Low

*: % of the current's fundamental period.

6. Conclusions

This paper proposes a new BiLSTM-based approach that aims to detect and localize open-circuit faults in three-phase, two-level VSIs for induction-motor drive systems. The proposed method uses only the measured induction-motor stator currents. To keep the proposed fault-diagnosis approach free from load-torque and/or motor-speed variations, an innovative approach is presented for motor-stator current normalization through the sigmoid function. Then, three detection variables—χ_a, χ_b and χ_c—are defined and sent to the BiLSTM network to localize the faulty switch(es). The performances of the proposed algorithm have been analyzed through simulations and experiments and have shown:

i. The robustness of the proposed fault-diagnosis algorithm to load-torque and motor-speed variations, and all switches' fault flags remain at their respective low levels.
ii. The accuracy and capability of the proposed algorithm to diagnose single and multiple open-circuit power-switch faults. Moreover, the detection time is acceptable since it is less than the stator current's period.

The proposed work provides evidence of practical viability through extensive simulations and experimental results, demonstrating real-world applicability and robustness. Therefore, the proposed work represents a more promising and effective solution for the fault diagnosis of single and multiple open-circuit faults without extra hardware for three-phase electric speed drives.

Our team's next step is to investigate the ability of the proposed method to detect current-sensor faults and to discriminate them from open-circuit faults.

Author Contributions: Conceptualization, B.G.; methodology, S.K.E.K.; formal analysis, A.B.R. and S.K.E.K.; investigation, H.M.; resources, S.K.E.K.; writing—original draft preparation, B.G. and H.M.; writing—review and editing, A.B.R. and S.K.E.K.; supervision, S.K.E.K.; project administration, A.B.R. and S.K.E.K. All authors have read and agreed to the published version of the manuscript.

Funding: This work was supported by the Tunisian Ministry of Higher Education and Research under Grant LSE–ENIT-LR 11 ES15.

Data Availability Statement: Data are contained within the article.

Acknowledgments: The authors are grateful for the support from the Tunisian Ministry of Higher Education and Research.

Conflicts of Interest: The authors declare no conflicts of interest.

References

1. Yepes, A.-G.; Lopez, O.; Gonzalez-Pietro, I.; Duran, M.-J.; Doval-Gandoy, J. A Comprehensive Survey on Fault Tolerance in Multiphase AC Drives, Part 1: General Overview Considering Multiple Fault Types. *Machines* **2022**, *10*, 208. [CrossRef]
2. Yepes, A.-G.; Lopez, O.; Gonzalez-Pietro, I.; Duran, M.-J.; Doval-Gandoy, J. A Comprehensive Survey on Fault Tolerance in Multiphase AC Drives, Part 2: Phase and Switch Open-Circuit Faults. *Machines* **2022**, *10*, 221. [CrossRef]
3. Lu, B.; Sharma, S. A literature review of IGBT fault diagnostic and protection methods for power inverters. *IEEE Trans. Ind. Appl.* **2009**, *45*, 1770–1777.
4. Tang, H.; Weili, L.; Zhigang, W. Influence of Inverter Open Circuit Fault on Multiple Physical Quantities in the PMSM. *IEEE Trans. Power Electron.* **2022**, *38*, 901–916. [CrossRef]
5. Zhang, Z.; Hu, Y.; Luo, G.; Gong, C.; Liu, X.; Chen, S. An Embedded Fault-Tolerant Control Method for Single Open-Switch Faults in Standard PMSM Drives. *IEEE Trans. Power Electron.* **2022**, *37*, 8476–8487. [CrossRef]
6. Pires, V.F.; Cordeiro, A.; Foito, D.; Pires, A.J. Fault-Tolerant Multilevel Converter to Feed a Switched Reluctance Machine. *Machines* **2022**, *10*, 35. [CrossRef]
7. Choi, U.T.; Blaabjerg, F.; Lee, K.B. Study and handling methods of power IGBT module failures in power electronic converter systems. *IEEE Trans. Power Electron.* **2015**, *30*, 2517–2533. [CrossRef]
8. Orlowska-Kowalska, T.; Wolkiewicz, M.; Pietrzak, P.; Skowron, M.; Ewert, P.; Tarchała, G.; Krzysztofiak, M.; Kowalski, C.T. Fault diagnosis and fault-tolerant control of PMSM drives–state of the art and future challenges. *IEEE Access* **2022**, *10*, 59979–60024. [CrossRef]
9. Jlassi, I.; Estima, J.O.; El Khil, S.K.; Bellaaj, N.B.; Cardoso, A.J.M. Multiple open-circuit faults diagnosis in back-to-back converters of PMSG drives for wind turbine systems. *IEEE Trans. Power Electron.* **2015**, *30*, 2689–2702. [CrossRef]

10. Jlassi, I.; Estima, J.O.; El Khil, S.K.; Bellaaj, N.B.; Cardoso, A.J.M. A Robust observer-based method for IGBTs and current sensors fault diagnosis in voltage-source inverters of PMSM drives. *IEEE Trans. Ind. Appl.* **2017**, *53*, 2894–2905. [CrossRef]
11. Xu, S.; Huang, W.; Wang, H.; Zheng, W.; Wang, J.; Chai, Y.; Ma, M. A Simultaneous Diagnosis Method for Power Switch and Current Sensor Faults in Grid-Connected Three-Level NPC Inverters. *IEEE Trans. Power Electron.* **2023**, *38*, 1104–1118. [CrossRef]
12. AN, Q.-T.; Sun, L.; Sun, L.-Z. Current residual vector-based open-switch fault diagnosis of inverters in PMSM drive systems. *IEEE Trans. Power Electron.* **2015**, *30*, 2814–2827. [CrossRef]
13. Maamouri, R.; Trabelsi, M.; Boussek, M.; M'Sahli, F. Mixed model-based and signal-based approach for open-switches fault diagnostic in sensorless speed vector-controlled induction motor drive using sliding mode observer. *IET Power Electron.* **2019**, *12*, 1149–1159. [CrossRef]
14. Zhou, X.; Sun, J.; Cui, P.; Lu, Y.; Lu, M.; Yu, Y. A Fast and Robust Open-Switch Fault Diagnosis Method for Variable-Speed PMSM System. *IEEE Trans. Power Electron.* **2020**, *36*, 2598–2610. [CrossRef]
15. Hu, K.; Liu, Z.; Tasiu, I.; Chen, T. Fault Diagnosis and Tolerance with Low Torque Ripple for Open-Switch Fault of IM Drives. *IEEE Trans. Transp. Electrif.* **2020**, *15*, 133–146. [CrossRef]
16. Freire, N.M.A.; Estima, J.O.; Cardoso, A.J.M. A voltage-based approach without extra hardware for open-circuit fault diagnosis in close-loop PWM AC regenerative drives. *IEEE Trans. Ind. Electron.* **2014**, *61*, 4960–4970. [CrossRef]
17. Cheng, Y.; Sun, Y.; Li, X.; Hanbing, D.; Lin, J.; Su, M. Active Common-Mode Voltage-Based Open-Switch Fault Diagnosis of Inverters in IM-Drive Systems. *IEEE Trans. Ind. Electron.* **2020**, *68*, 103–115. [CrossRef]
18. Li, Z.; Ma, H.; Bai, Z.; Wang, Y.; Wang, B. Fast transistor open circuit faults diagnosis in grid tied three phase VSIs based on average bridge arm pole to pole voltages and error adaptive thresholds. *IEEE Trans. Power Electron.* **2018**, *33*, 8040–8051. [CrossRef]
19. Li, Z.; Wheeler, P.; Watson, A.J.; Costabeber, A.; Wang, B.; Ren, Y.; Bai, Z.; Ma, H. A Fast Diagnosis Method for Both IGBT Faults and Current Sensor Faults in Grid-Tied Three-Phase Inverters with Two Current Sensors. *IEEE Trans. Power Electron.* **2020**, *35*, 5267–5278. [CrossRef]
20. Manikandan, R.; Selvaraj, R.; Singh, R.R. Voltage Signature based Open Circuit Switch Fault Diagnosis Strategy for IM Drives with MPC. *IEEE Trans. Ind. Appl.* **2023**, *59*, 6780–6791. [CrossRef]
21. Diao, N.; Zhang, Y.; Sun, X.; Song, C.; Wang, W.; Zhang, H. A Real-Time Open-Circuit Fault Diagnosis Method Based on Hybrid Model Flux Observer for Voltage-Source-Inverter Fed Sensorless Vector Controlled Drives. *IEEE Trans. Power Electron.* **2023**, *38*, 2539–2551. [CrossRef]
22. Hang, J.; Wu, H.; Zhang, J.; Ding, S.; Huang, Y.; Hua, W. Cost Function-based Open-Phase Fault Diagnosis for PMSM Drive System with Model Predictive Current Control. *IEEE Trans. Power Electron.* **2021**, *36*, 2574–2583. [CrossRef]
23. Gmati, B.; Jlassi, I.; Khojet El Khil, S.; Cardoso, A.J.M. Open-switch fault diagnosis in voltage source inverters of PMSM drives using predictive current errors and fuzzy logic approach. *IET Power Electron.* **2021**, *14*, 1059–1072. [CrossRef]
24. Zhang, Y.; Mao, Y.; Wang, X.; Wang, Z.; Xiao, D.; Gaoliang, F. Current Prediction Based Fast Diagnosis of Electrical Faults in PMSM Drives. *IEEE Trans. Transp. Electrif.* **2022**, *8*, 4622–4632. [CrossRef]
25. Huang, W.; Du, J.; Hua, W.; Lu, W.; Bi, K.; Zhu, Y.; Fan, Q. Current-based open-circuit fault diagnosis for PMSM drives with model predictive control. *IEEE Trans. Power Electron.* **2021**, *36*, 10695–10704. [CrossRef]
26. Yan, H.; Xu, Y.; Zou, J.; Fang, Y.; Cai, F. A novel open-circuit fault diagnosis method for voltage source inverters with a single current sensor. *IEEE Trans. Power Electron.* **2018**, *33*, 8775–8786. [CrossRef]
27. Khojet El Khil, S.; Jlassi, I.; Cardoso, A.J.M.; Estima, J.O.; Bellaaj, N.B. Diagnosis of open-switch and current sensor faults in PMSM drives through stator current analysis. *IEEE Trans. Ind. Appl.* **2019**, *55*, 5925–5937. [CrossRef]
28. Estima, J.O.; Cardoso, A.J.M. A new algorithm for real-time multiple open-circuit fault diagnosis in voltage-fed PWM motor drives by the reference current errors. *IEEE Trans. Ind. Electron.* **2013**, *60*, 3496–3505. [CrossRef]
29. Cui, R.; Yu, S.; Li, S. Open-Switch Fault Detection Based on Open-Winding Five-Phase Fault-Tolerant Permanent-Magnet Motor Drives. *Machines* **2022**, *10*, 829. [CrossRef]
30. Abdelkader, R.; Cherif, B.D.E.; Bendiababdella, A.; Kaddour, A. An Open-Circuit Faults Diagnosis Approach for Three-Phase Inverters Based on an Improved Variational Mode Decomposition, Correlation Coefficients, and Statistical Indicators. *IEEE Trans. Instrum. Meas.* **2022**, *71*, 1–9. [CrossRef]
31. Gou, B.; Xu, Y.; Xia, Y.; Qingli, D.; Ge, X. An Online Data-driven Method for Simultaneous Diagnosis of IGBT and Current Sensor Fault of 3-Phase PWM Inverter in Induction Motor Drives. *IEEE Trans. Power Electron.* **2020**, *35*, 13281–13294. [CrossRef]
32. Cai, B.; Zhao, Y.; Liu, H.; Xie, M. A data-driven fault diagnosis methodology in three-phase inverters for PMSM drive systems. *IEEE Trans. Power Electron.* **2017**, *32*, 5590–5600. [CrossRef]
33. Xia, Y.; Xu, Y. A transferrable data-driven method for IGBT open-circuit fault diagnosis in three-phase inverters. *IEEE Trans. Power Electron.* **2021**, *36*, 13478–13488. [CrossRef]
34. Li, Z.; Gao, Y.; Zhang, X.; Wang, B.; Ma, H. A Model-Data-Hybrid-Driven Diagnosis Method for Open-Switch Faults in Power Converters. *IEEE Trans. Power Electron.* **2021**, *36*, 4965–4970. [CrossRef]
35. Gonzalez-Jimenez, D.; del-Olmo, J.; Poza, J.; Garramiola, F.; Sarasola, I. Machine learning-based fault detection and diagnosis of faulty power connections of induction machines. *Energies* **2021**, *14*, 4886. [CrossRef]
36. Hang, J.; Shu, X.; Ding, S.; Huang, Y. Robust Open-Circuit Fault Diagnosis for PMSM Drives Using Wavelet Convolutional Neural Network with Small Samples of Normalized Current Vector Trajectory Graph. *IEEE Trans. Ind. Electron.* **2023**, *70*, 7653–7663. [CrossRef]

37. Jin, L.; Mao, Y.; Wang, X.; Lu, L.; Wang, Z. Online Data-Driven Fault Diagnosis of Dual Three-Phase PMSM Drives Considering Limited Labeled Samples. *IEEE Trans. Ind. Electron.* **2023**. [CrossRef]
38. Xue, Z.Y.; Xiahou, K.S.; Li, M.S.; Ji, T.Y.; Wu, Q.H. Diagnosis of Multiple Open-Circuit Switch Faults Based on Long Short-Term Memory Network for DFIG-Based Wind Turbine Systems. *IEEE Trans. Emerg. Sel. Top. Power Electron.* **2020**, *8*, 2600–2610. [CrossRef]
39. Han, Y.; Qi, W.; Ding, N.; Geng, Z. Short-time wavelet entropy integrating improved LSTM for fault diagnosis of modular multilevel converter. *IEEE Trans. Cybern.* **2021**, *52*, 7504–7512. [CrossRef]
40. Ye, S.; Jiang, J.; Junjie, L.; Liu, Y.; Zhou, Z.; Liu, C. Fault Diagnosis and Tolerance Control of Five-Level Nested NPP Converter Using Wavelet Packet and LSTM. *IEEE Trans. Power Electron.* **2020**, *35*, 1907–1921. [CrossRef]
41. Wang, Q.; Yu, Y.; Ahmed, H.O.; Darwish, M.; Nandi, A.K. Open-Circuit Fault Detection and Classification of Modular Multilevel Converters in High Voltage Direct Current Systems (MMC-HVDC) with Long Short-Term Memory (LSTM) Method. *Sensors* **2021**, *21*, 4159. [CrossRef]
42. Kaplan, H.; Tehrani, K.; Jamshidi, M. A Fault Diagnosis Design Based on Deep Learning Approach for Electric Vehicle Applications. *Energies* **2021**, *14*, 6599. [CrossRef]
43. Li, D.; Zhang, Z.; Liu, P.; Wang, Z.; Zhang, L. Battery Fault Diagnosis for Electric Vehicles Based on Voltage Abnormality by Combining the Long Short-Term Memory Neural Network and the Equivalent Circuit Model. *IEEE Trans. Power Electron.* **2021**, *36*, 1303–1315. [CrossRef]
44. Husari, F.; Seshadrinath, J. Stator Turn Fault Diagnosis and Severity Assessment in Converter Fed Induction Motor Using Flat Diagnosis Structure Based on Deep Learning Approach. *IEEE J. Emerg. Sel. Top. Power Electron.* **2023**, *11*, 5649–5657. [CrossRef]
45. Guo, J.; Lao, Z.; Hou, M.; Li, C.; Zhang, S. Mechanical fault time series prediction by using EFMSAE-LSTM neural network. *Measurement* **2021**, *173*, 108566. [CrossRef]
46. Saadi, M.N.; Bouteraa, N.; Redjati, A.; Boughazi, M. A novel method for bearing fault diagnosis based on BiLSTM neural networks. *Int. J. Adv. Manuf. Technol.* **2023**, *125*, 1477–1492.
47. Guo, Y.; Mao, J.; Zhao, M. Rolling Bearing Fault Diagnosis Method Based on Attention CNN and BiLSTM Network. *Neural Process. Lett.* **2023**, *55*, 3377–3410. [CrossRef]
48. Wang, J.; Guo, J.; Wang, L.; Yang, Y.; Wang, Z.; Wang, R. A hybrid intelligent rolling bearing fault diagnosis method combining WKN-BiLSTM and attention mechanism. *Meas. Sci. Technol.* **2023**, *34*, 085106. [CrossRef]
49. Zhang, Y.; Zhang, S.; Jia, X.; Zhang, X.; Tian, W. A novel integrated fault diagnosis method of chemical processes based on deep learning and information propagation hysteresis analysis. *J. Taiwan Inst. Chem. Eng.* **2023**, *142*, 104676. [CrossRef]
50. Xu, Z.; Mei, X.; Wang, X.; Yue, M.; Jin, J.; Yang, Y.; Li, C. Fault diagnosis of wind turbine bearing using a multi-scale convolutional neural network with bidirectional long short term memory and weighted majority voting for multi-sensors. *Renew. Energy* **2022**, *182*, 615–626. [CrossRef]
51. Yan, X.; Guan, T.; Fan, K.; Sun, Q. Novel double layer BiLSTM minor soft fault detection for sensors in air conditioning system with KPCA reducing dimensions. *J. Build. Eng.* **2021**, *44*, 102950. [CrossRef]
52. Zheng, X.; Chen, W. An attention-based Bi-LSTM method for visual object classification via EEG. *Biomed. Signal Process Contr.* **2021**, *63*, 102174. [CrossRef]

electricity

MDPI

Article

Single-Stage LLC Resonant Converter for Induction Heating System with Improved Power Quality

Anand Kumar [1], Anik Goswami [1,*], Pradip Kumar Sadhu [2] and Jerzy R. Szymanski [3]

1 School of Electrical Engineering, Vellore Institute of Technology, Chennai 600127, India; anand.kumar@vit.ac.in
2 Department of Electrical Engineering, Indian Institute of Technology (ISM) Dhanbad, Dhanbad 826004, India; pradip@iitism.ac.in
3 Faculty of Transport, Electrical Engineering and Computer Sciences, Department of Electrical Drives and Industrial Electronics, Casimir Pulaski University of Radom, Malczewskiego 29, 26-600 Radom, Poland; j.szymanski@uthrad.pl
* Correspondence: anik91_go@rediffmail.com; Tel.: +91-8777-383-221

Abstract: This paper proposes a single-stage direct AC to high-frequency (HF) AC resonant converter based on LLC configuration for induction heating (IH) systems or HF applications. Unlike conventional converters for IH systems, the proposed topology converts the utility frequency to HF AC in a single stage without using a DC link inductor and capacitors and takes the advantages of LLC configuration. Additionally, it improves the power factor to 0.9–1, lowers the THD (3.2% experimentally), and protects against the high-frequency components. An embedded control scheme was designed to keep the HF current oscillating at a resonant frequency, ensuring zero-voltage switching. The operating principle of the proposed topology was investigated using mathematical equations and equivalent circuits. Finally, it was verified using computer simulation, and an experimental prototype of 1.1 kW was developed to demonstrate the proposed topology's uniqueness.

Keywords: induction heating (IH) system; resonant converter; embedded control; passive filter

Citation: Kumar, A.; Goswami, A.; Sadhu, P.K.; Szymanski, J.R. Single-Stage LLC Resonant Converter for Induction Heating System with Improved Power Quality. *Electricity* **2024**, *5*, 211–226. https://doi.org/10.3390/electricity5020011

Academic Editor: Andreas Sumper

Received: 20 February 2024
Revised: 20 April 2024
Accepted: 24 April 2024
Published: 26 April 2024

1. Introduction

Induction heating (IH) is an efficient technique to generate very high temperatures for a wide range of applications, such as industrial heating (brazing and melting of steel) and domestic cooking applications [1,2]. Every application has a unique operating frequency that is determined by the shape of the workpiece and the required skin depth [3,4]. Typically, a high-frequency (HF) current supply is needed for the IH technique in order to produce the HF eddy current in the workpiece that causes the heating effect [5,6]. The skin depth (δ) can be stated as:

$$\delta = \sqrt{\frac{\rho}{\pi f_s \mu}} \tag{1}$$

where ρ is the electrical resistivity, f_s is the switching frequency of the resonant inverter, and μ is the magnetic permeability of the workpiece.

Generally, a series resonant inverter is employed to produce HF alternating current (AC) in IH technology [7,8]. A wide variety of inverter topologies have been developed for the same purpose. Voltage and current resonant inverters are among the most commonly used types [9,10]. Out of these, the voltage-fed resonant inverter is widely used because it has numerous control possibilities [11,12]. Typically, quasi, class D/E, and half-bridge resonant inverters are mostly employed for low-power applications [13,14]. In contrast, full-bridge series resonant topology (FB-SRI) is used for high-power applications [15,16]. Along with these converters, some popular output power control techniques have been developed, including (a) phase shifting (PS), (b) pulse frequency modulation (PFM), (c) asymmetrical voltage cancellation (AVC), (d) pulse density modulation (PDM), and

(d) asymmetrical duty cycling (ADC) [17–20]. Each of the proposed converter topologies and power control techniques has its own merits and demerits, depending on the applications.

The merits and demerits of some conventional resonant converters, along with their control techniques, are given in Table 1.

Table 1. Merits and demerits of conventional resonant converters with their control algorithms [1,3].

Converter Topology	Control Algorithm	Merits	Demerits
Two-output series resonant inverter with common capacitor	Pulse width modulation control	✓ Reduced the component counts, size, and cost ✓ Common capacitor is used to create resonating capacitor	✓ Can't achieve independent power control ✓ High switching losses
Full bridge converter	Pulse density modulation, frequency modulation, phase shift modulation	✓ Higher input power factor (≈0.95) can be achieved	✓ More power losses across the switches
Single-stage boost full bridge resonant inverter	Phase shift PWM technique	✓ High efficiency has been achieved ✓ Less switching loss	✓ Higher harmonic injection
Multi-modulated converters using full bridge topology	Deadband current control technique	✓ Phase-locked loop is used to vary the switching frequency	✓ More switches are used ✓ More switching losses ✓ Complex control technique is used
Multi-frequency resonant converter	Deadband current control technique	✓ Can achieve high eddy current ✓ Centre-tapped transformer is used	✓ Calculation of resonant frequency is difficult ✓ High switching loss at the load side

In the last few years, several configurations of IH converters have been developed, comprising rectification and inversion modes. The input power factor (PF) degradation and the generation of an HF component due to the high switching frequency operation in the inversion mode are the two main problems with the IH system. Recently, boost power factor circuits (BPFCs) have been used as front-end converters to mitigate these problems [21,22], although their extra switches and energy storage elements makes them costly and bulky. Additionally, multiple power conversion stages might lead to various severe issues including poor system reliability, efficiency deterioration, and increases in power loss and electro-magnetic interference (EMI) [23,24]. A direct AC–HF AC converter could be a great way to solve these issues because it not only uses fewer components but also enhances PF and blocks the HF component by adding a passive filter to the input side.

Additionally, it has been observed that the series resonant configuration of the IH load is used in the majority of IH applications. However, previous research works have shown that LLC configuration provides better performance as compared to series resonant configuration (RLC) in terms of short circuit immunity and low current stress in the switches (that are used in the inverter) [25–27]. Additionally, with only a small series inductance, the LLC resonant configuration enables exceptional performance with a high quality factor (Q) [28–30]. Owing to this, the requirement of an output transformer can be eliminated.

A direct AC to HF AC LLC resonant converter has been proposed in light of the aforementioned power factor issues, high-frequency component generation, multiple power conversion stages, and the desire to benefit from an LLC load resonant configuration. In addition to enhancing the PF at the input side, this converter will enable us to convert grid frequency AC mains to HF AC in a single conversion stage. Additionally, a passive filter comprising an input inductor and a capacitor has been used to prevent the flow of HF components from the load side to the source side.

This paper is classified as follows. In Section II, the proposed converter's circuit design and operation are described. Circuit analysis, which includes calculation of frequency, output power, voltage, and current gain, is described in Section III. Section IV contains discussion of simulation and experimental findings. In Section V, the conclusion is offered.

2. Proposed Direct AC–HFAC LLC Resonant Converter

2.1. Circuit Description

The proposed topology for the IH system is shown in Figure 1. It consists of four bi-directional switches, and each bi-directional switch comprises two insulated gate bipolar transistors (IGBTs) and two diodes. A bi-directional switch simultaneously conducts the current and blocks the voltage of both polarities based on the control signal.

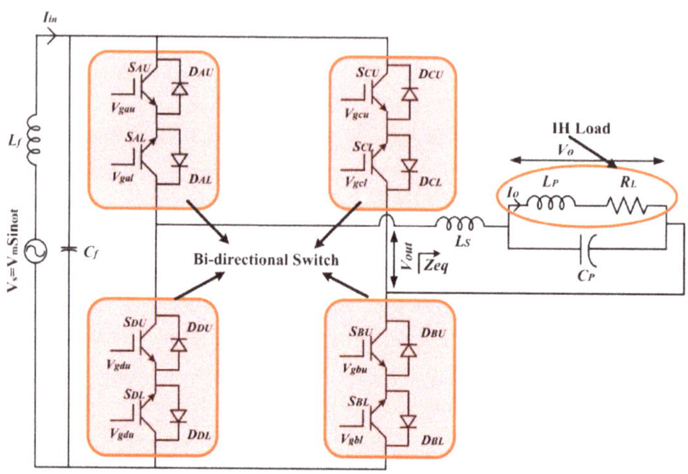

Figure 1. Proposed configuration of the direct AC–HFAC LLC resonant converter.

A series inductor (L_S) is connected to a parallel combination of a resonating capacitor (C_P) and an IH load in order to create the LLC configuration. A typical induction heating load consists of a workpiece (that is to be heated) and a heating coil or a litz wire-based coil in the case of a domestic IH system. Typically, the IH coil and workpiece (IH load) are considered the transformer's primary and secondary. Thus, the IH load is referred to as the transformer's primary, and the obtained equivalent impedance after transferring is represented as a series R–L circuit.

Consequently, to know the electrical behaviour of the proposed topology, the IH coil and its workpiece are modelled as series R_L and L_P, where R_L and L_P are the referred equivalent resistance and inductance of the heating coil, respectively.

Thereafter, an LC circuit or passive filter is designed which ensures the elimination of the HF component at the input side, as shown in Figure 2. HF components cannot flow back to the input side because the designed passive filter has a high input equivalent impedance ($Z_{eq(in)}$) at a frequency above 4 kHz. Consequently, power quality at the grid side improves. In Figures 1 and 2, V_s, I_{in}, V_{out}, V_o, and I_o represent the input voltage, input current, output voltage of the converter, output voltage across the IH load, and output current flowing through the load, respectively.

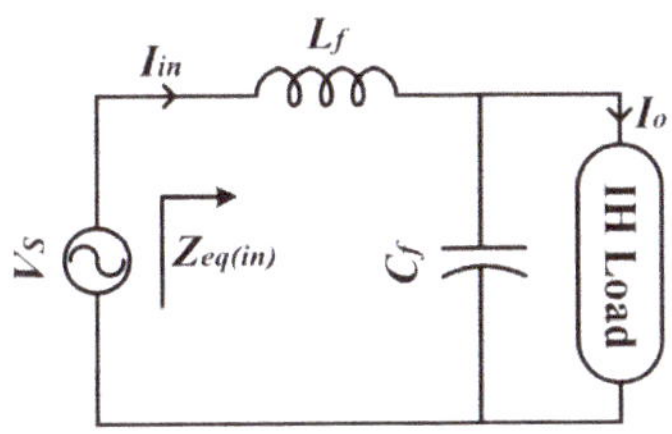

Figure 2. Passive filter.

2.2. Modes of Operation

In accordance with the polarity of the input voltage, the proposed direct AC–HF AC LLC resonant converter has four distinct operating modes. For the positive input half cycle, modes 1 and 2 are defined, and for the negative input half cycle, modes 2 and 3 are defined. The four modes of operation are explained with the assumption of a resistive load because the proposed converter has been made to operate as a resonant converter. Also, the dead time between the pulses and the effect of the input filter is neglected. However, in the prototype implementation, a very small dead time has been considered between the gate signals. Figure 3 shows the circuit topology for each mode of operation and corresponding waveforms.

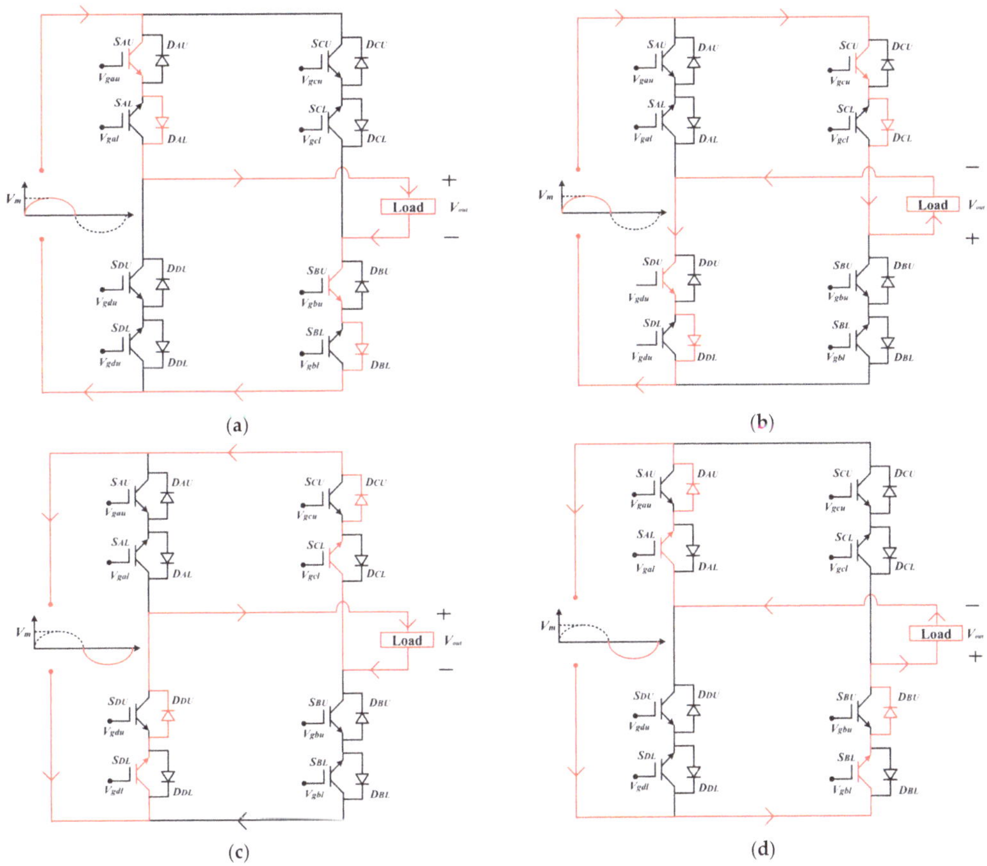

Figure 3. Operational modes of the proposed direct AC–HF AC converter: (**a**) Mode 1; (**b**) Mode 2; (**c**) Mode 3; (**d**) Mode 4.

Positive input half cycle, $V_s > 0$

Mode 1 (t_0–t_1): In the positive input half cycle, the upper IGBTs (S_{AU}, S_{BU}, S_{CU}, and S_{DU}) and lower diodes (D_{AL}, D_{BL}, D_{CL}, and D_{DL}) of bi-directional switches are forward-biased (FB), respectively. Among these FB IGBTs, S_{AU} and S_{BU} receive the gate signal at t_0 to t_1, as shown in Figure 4. Therefore, the direction of the current at t_0 to t_1 will be: $S_{AU} \rightarrow D_{AL} \rightarrow$ load $\rightarrow S_{BU} \rightarrow D_{BL}$.

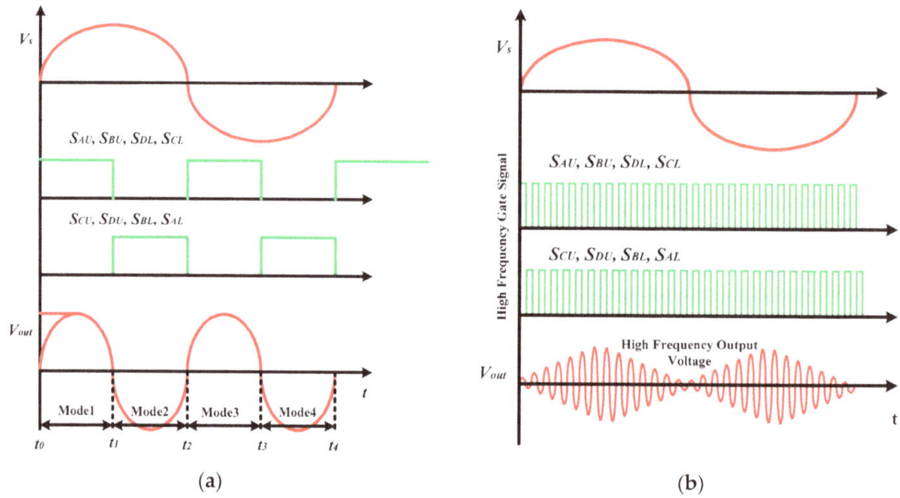

Figure 4. Frequency synthesization of the output voltage using the proposed topology: (**a**) 100 Hz-output voltage synthesization; (**b**) High-frequency output voltage synthesization.

Mode 2 (t_1–t_2): This mode is also for the positive input half cycle. In this mode, S_{AU} and S_{BU} are turned off, as these IGBTs do not receive the gate signal between t_1 and t_2, as shown in Figure 4. S_{CU} and S_{DU} do receive the gate signal between t_1 to t_2. Therefore, the direction of the load current reverses in the positive input half cycle: $S_{CU} \rightarrow D_{CL} \rightarrow$ load $\rightarrow S_{DU} \rightarrow D_{DL}$.

Negative input half cycle, $V_s < 0$

Mode 3 (t_2–t_3): In the negative input half cycle, the lower IGBTs (S_{AL}, S_{BL}, S_{CL}, and S_{DL}) and upper diodes (D_{AU}, D_{BU}, D_{CU}, and D_{DU}) of bi-directional switches are FB, respectively. Among these FB IGBTs, S_{DL} and S_{CL} receive the gate signal at t_2 to t_3, as shown in Figure 4. Therefore, the direction of the current at t_2 to t_3 is: $S_{DL} \rightarrow D_{DU} \rightarrow$ load $\rightarrow S_{CL} \rightarrow D_{CU}$.

Mode 4 (t_3–t_4): This mode is also for the negative input half cycle. In this mode, S_{DL} and S_{CL} turn off, as these IGBTs do not receive the gate signal between t_3 and t_4, as shown in Figure 4. S_{BL} and S_{AL} do receive the gate signal between t_3 and t_4. Therefore, the direction of the load current reverses in the negative input half cycle: $S_{BL} \rightarrow D_{BU} \rightarrow$ load $\rightarrow S_{AL} \rightarrow D_{AU}$.

The above modes of operation are explained for 100 Hz-output voltage synthetization, and it can be concluded that the direction of the load current in each input half cycle could be changed by allowing the time period of the conducting IGBTs to be changed in a single stage, as shown in Figure 4a. Consequently, this frequency changer operation of this topology could be deployed in IH technology that requires an HF AC as shown in Figure 4b. The detailed possible switching strategies are given in Table 2.

Table 2. Switching Strategies.

Input Voltage (V_s)	Forward-Biased Switches	Modes	IGBTs/Diodes Status	Time Interval	Output Voltage (V_{out})
$V_s > 0$	(S_{AU}, S_{BU}, S_{CU}, S_{DU}) and (D_{AL}, D_{BL}, D_{CL}, D_{DL})	Mode 1	($S_{AU}/D_{AL}/S_{BU}/D_{BL}$) ON ($S_{CU}/D_{CL}/S_{DU}/D_{DL}$) OFF	t_0–t_1	$V_{out} > 0$
		Mode 2	($S_{AU}/D_{AL}/S_{BU}/D_{BL}$) OFF ($S_{CU}/D_{CL}/S_{DU}/D_{DL}$) ON	t_1–t_2	$V_{out} < 0$
$V_s < 0$	(S_{AL}, S_{BL}, S_{CL}, S_{DL}) and (D_{AU}, D_{BU}, D_{CU}, D_{DU})	Mode 3	($S_{DL}/D_{DU}/S_{CL}/D_{CU}$) ON ($S_{BL}/D_{BU}/S_{AL}/D_{AU}$) OFF	t_2–t_3	$V_{out} > 0$
		Mode 4	($S_{DL}/D_{DU}/S_{CL}/D_{CU}$) OFF ($S_{BL}/D_{BU}/S_{AL}/D_{AU}$) ON	t_3–t_4	$V_{out} < 0$

3. Mathematical Analysis of Proposed Topology

The mathematical analysis of the proposed direct AC–HF AC LLC resonant converter is based on the following assumptions:

(a) All components in the circuit are ideal.
(b) The AC input voltage (V_s) is purely sinusoidal.
(c) The effects of parasitic capacitance are neglected.
(d) The load current is purely HF sinusoidal.

3.1. Calculation of Switching Frequency

The equivalent circuit of the proposed topology based on the LLC tank is shown in Figure 5. In the figure, V_{out} is the HF output voltage, L_s is the series inductance, and C_P is the resonating capacitor connected parallel to the IH load comprised of L_P and R_L. The proposed direct AC–HF AC LLC resonant converter is made to work at resonant frequency. Thus, in order to obtain resonant frequency, first the equivalent impedance of the LLC resonant tank is determined.

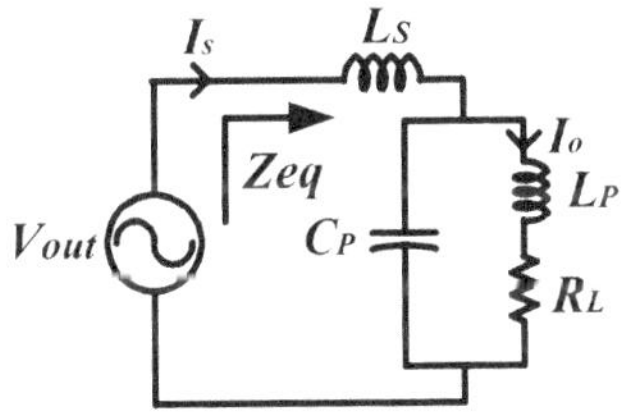

Figure 5. Equivalent circuit of the proposed topology.

The equivalent impedance of the LLC resonant tank can be determined as:

$$Z(j\omega) = j\omega L_S + \frac{(1/j\omega C_P) \times (R_L + j\omega L_P)}{1/j\omega C_P + (R_L + j\omega L_P)} \tag{2}$$

On solving Equation (2):

$$Z_{eq}(j\omega) = j\omega L_S + \frac{(R_L + j\omega L_P)}{1 + j\omega C_P(R_L + j\omega L_P)} \tag{3}$$

On complete rationalization of Equation (3):

$$Z_{eq} = \frac{R_L(1 - \omega^2 L_P C_P) + \omega^2 R_L C_P L_P}{K} + \frac{j\omega\left[L_S + L_P(1 - \omega^2 L_P C_P) - R_L^2 C_P\right]}{K} \tag{4}$$

where $K = \left(1 - \omega^2 L_P C_P\right)^2 + (R_L \omega C_P)^2$.

To have the active power, the imaginary part of Equation (4) should be equal to zero. Also, $R_L << L_P \sim 0$.

Therefore,

$$L_S + \frac{L_P(1-\omega^2 L_P C_P)}{(1-\omega^2 L_P C_P)^2} = 0 \quad (5)$$

$$L_S + \frac{L_P}{(1-\omega^2 L_P C_P)} = 0$$
$$\omega = \sqrt{\frac{L_P + L_S}{L_S L_P C_P}} \quad (6)$$

where ω is the angular resonant frequency.

$$f = \frac{1}{2\pi}\sqrt{\frac{L_P + L_S}{L_S L_P C_P}} \quad (7)$$

In Equation (7), f is the resonant frequency of the LLC tank, and at this frequency, the maximum power is transferred to the IH load. However, to enable the zero-voltage switching (ZVS), the switching frequency (f_s) is selected higher than the calculated resonant frequency (f), ensuring less switching and fewer power losses.

3.2. Quality Factor

At a higher quality factor, (Q), the resonant inverter normally operates close to the resonant frequency (ω_o). In contrast to the induction coil current (I_o), as shown in Figure 6, the Q factor has a negligible effect on the resonant frequency, i.e., the peak value of I_o occurs at the same frequency, regardless of the Q factor. The peak value of I_o is related to the RLC parallel end, as shown in Figure 5, where L_S does not play a role in the frequency response of I_o. The quality factor (Q) for the second order system can be defined as:

$$Q = \frac{\omega_o}{2\alpha} \quad (8)$$

where ω_o and α are the resonant frequency and damping coefficient of the second order system, respectively.

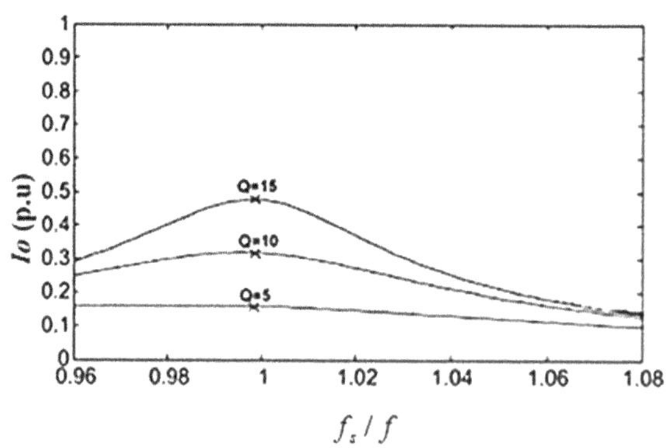

Figure 6. Output current (I_o) in p.u at different Q factors.

Let $\lambda = -\alpha \pm j\sqrt{\omega_0^2 - \alpha^2}$ be the solution of the characteristic polynomial of the system. Now, based on Figure 5, the parallel circuit impedance is:

$$Z_P(S) = \frac{L_P S + R_L}{C_P L_P S^2 + R_L C_P S + 1} \quad (9)$$

The characteristic polynomial of the above equation is given as:

$$C_P L_P \lambda^2 + R_L C_P \lambda + 1 = 0 \quad (10)$$

Taking λ into account, the damping coefficient is:

$$\alpha_P = \frac{R_L}{2L_P} \quad (11)$$

The resonant frequency is:

$$\omega_{oP} = \frac{1}{\sqrt{L_P C_P}} \quad (12)$$

Therefore, using Equation (8), the quality factor of the parallel circuit is:

$$Q_P = \frac{1}{R_L}\sqrt{\frac{L_P}{C_P}} \quad (13)$$

In this work, the values of L_P and R_L for the IH coil were measured through an LCR meter and the switching frequency was kept constant and was usually higher than the resonant frequency (f) to create zero-voltage switching. Thereafter, the values of L_S and C_P were calculated using Equations (7) and (13).

3.3. Current Gain

As shown in Figure 5, I_s is the current that flows through the IGBTs/switches, and it must be the lowest magnitude possible to obtain low current stress in the switches. Alongside it, I_o is the IH load current and must be very intense. Thanks to the LLC circuit, it is possible to have low current stress in the switches and to provide great power dissipation in the IH load. This merit of the LLC tank makes it preferable to the SRI (where all the current flows to the IH load through the switches) for IH applications. Therefore, the maximum ratio of I_o and I_s defined as current gain is very much needed and occurs at resonant frequency (f), as calculated in Equation (7).

Current gain is calculated as:

$$G_I(\omega) = \frac{I_o(\omega)}{I_s(\omega)} \quad (14)$$

$$G_I(\omega) = \frac{1/j\omega C_P}{1/j\omega C_P + j\omega L_P + R_L} = \frac{1}{1 - L_P C_P \omega^2 + j\omega R_L C_P} \quad (15)$$

Thus, modules of G_I become:

$$|G_I(\omega)|^2 = \frac{1}{(1 - L_P C_P \omega^2)^2 + (R_L C_P \omega)^2} \quad (16)$$

The current gain at resonant frequency, ($\omega_o = \sqrt{(L_P + L_S)/(L_P L_S C_P)}$), is given as:

$$|G_I(\omega_o)| = \frac{L_S}{L_P} \times \frac{1}{\sqrt{\frac{C_P L_S R^2 (L_S + L_P)}{L_P^3} + 1}} \quad (17)$$

A high current gain allows us to obtain a high heating effect with small I_s.

3.4. Voltage Gain

Voltage gain can be defined as the ratio of voltage of the parallel resonant capacitor (C_P) to the first harmonic amplitude of the output voltage of the converter, V_1. This ratio allows us to evaluate the voltage stress in the parallel resonating capacitor (C_P). The voltage gain (G_v) is given as:

$$G_v(\omega) = \frac{V_{C_P}(\omega)}{V_1} \tag{18}$$

Using the voltage division rule in Figure 5:

$$\frac{V_{C_P}(\omega)}{V_1} = \frac{1/j\omega C_P//(R_L + j\omega L_P)}{j\omega L_S + (1/j\omega C_P//(R_L + j\omega L_P))} \tag{19}$$

At resonant frequency (ω_o), capacitor voltage (V_{CP}) is given as:

$$\frac{V_{C_P}(\omega_o)}{V_1} = \left(-\frac{L_P}{L_S} - j\frac{L_P^2}{R_L L_S}\sqrt{\frac{L_P + L_S}{L_P L_S C_P}}\right) \tag{20}$$

The magnitude of voltage gain ($G_V(\omega)$) can be calculated as:

$$|G_V(\omega_o)| = \left|\frac{V_{C_P}(\omega_o)}{V_1}\right| = \sqrt{\left(\frac{L_P}{L_S}\right)^2 + \frac{L_P^2}{R_L^2 L_S^2}\left(\frac{L_P + L_S}{L_P L_S C_P}\right)} \tag{21}$$

On solving Equation (21), we obtain:

$$|G_V(\omega_o)| = \frac{L_P}{L_S}\sqrt{1 + Q^2\left(\frac{L_P}{L_S} + 1\right)}, \text{ where } Q = \frac{1}{R_L}\sqrt{\frac{L_P}{C_P}} \tag{22}$$

3.5. Calculation of Output Power

Theoretically, the resonant frequency (f) to impart the maximum power to the load is given in Equation (7). Thus, at this frequency, the approximated Z_{eq} is equal to $R_L(^{L_S}/_{L_P})^2$. Therefore, the maximum average output power ($P_{o,ave(m)}$) transferred to R_L at the resonant frequency (f) can be determined as:

$$P_{o,ave(m)} = \frac{(V_{O1})_{rms}^2}{R_L(^{L_S}/_{L_P})^2} \tag{23}$$

However, the proposed topology is designed in order to operate at switching frequency (f_s), which is kept higher than the resonant frequency (f) to enable ZVS operation.

V_o can be evaluated by using the Fourier series, which is as follows:

$$V_o(t) = \begin{cases} \sum_{n=1}^{\infty} \frac{4V_m(t)}{n\pi}\left(cos\left(\frac{n\pi(1-D)}{2}\right) - cos\left(\frac{n\pi}{2}\right)\right)sin\ (n\omega_o t), & \text{for odd n} \\ 0, & \text{for even n} \end{cases} \tag{24}$$

The fundamental output voltage (V_{o1}) can be determined as:

$$V_{01} \cong 4\frac{V_m(t)}{\pi}\left(\cos\left(\frac{\pi(1-D)}{2}\right)\right)\sin(\omega_o t) \tag{25}$$

Therefore, the rms value of V_{o1} can be written as:

$$V_{01} \cong 4\frac{V_m/\sqrt{2}}{\pi}\left(\cos\left(\frac{\pi(1-D)}{2}\right)\right) = \frac{2\sqrt{2}}{\pi}V_m\left(\cos\left(\frac{\pi(1-D)}{2}\right)\right) \tag{26}$$

On assuming maximum average output power ($P_{o,ave(m)}$) for the above equation, we obtain:

$$P_{o,ave(m)} = \frac{8V_m}{\pi}\left(\cos\frac{(\pi(1-D))}{2}\right)\frac{1}{R_L\left(\frac{L_S}{L_P}\right)^2} \tag{27}$$

As can be seen from Equation (27), it is obvious that maximum average output power ($P_{o,ave(m)}$) can be controlled by the duty cycle (D) or switching frequency (f_s).

4. Results

Based on the above theoretical analysis, a computer simulation using MATLAB 2015a was performed, and then an experimental prototype with 1100 W was built and tested, validating the feasibility of the proposed topology. The circuit parameters and obtained operational parameters for the developed simulation model and experimental prototype are given in Table 3.

Table 3. Circuit and operational parameters.

Circuit Parameters		Operational Parameters	
Parameters	**Value**	**Parameters**	**Value**
Input voltage (V_s)	110 $V_{r\cdot m\cdot s}$	Output voltage (V_{out})	88 V_p
Series inductance (L_S)	56 µH	Output current (I_{out})	28 V_p
Induction coil (L_P)	52.7 µH	Average output power (P_{out})	1100 W
Workpiece resistance (R_L)	1 Ω	Switching frequency (f_s)	30 kHz
Resonant frequency (f)	24 kHz	Input power factor	0.92

Figure 7 depicts a block diagram of a prototype implementation for the suggested topology. In this block diagram, the proposed converter receives 1-∅, 230 V utility frequency AC (UFAC) and creates HF AC directly without intermediate stages. The HF AC then flows to the IH load via the series inductance L_S, reducing switch current stress (I_s). An embedded controller has been designed to generate gate signals for the switches, consisting mostly of a zero-crossing detector (ZCD), an Arduino (Atmega 2560), and a driving circuit.

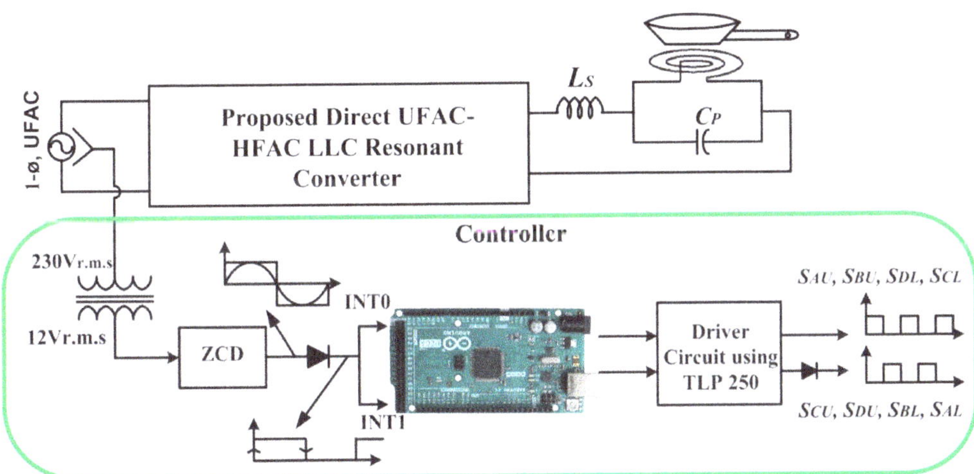

Figure 7. Block diagram of the prototype implementation.

To ensure zero-crossings of input mains, a step-down transformer is used to lower 1-∅, 230 V to 12 V. Then, 12 V AC is delivered to the ZCD, as illustrated in Figure 7. Following that, a diode is employed to remove the negative half of the rectangular pulse. The resulting pulse is perfectly synchronized with the input mains. Subsequently, a synchronized pulse is sent to the Arduino's interrupt pins (INT0 and INT1). Furthermore, interrupt pins INT0 and INT1 detect the rising and falling edges of the synchronized pulse and create pulses (V_{g1} and V_{g2}) at the required frequencies according to the programming.

Finally, V_{g1} and V_{g2} are given to the IGBT switches S_{AU}, S_{BU}, S_{DL}, S_{CL} and S_{CU}, S_{DU}, S_{BL}, S_{AL} of the proposed converter via driver circuits, respectively. The prototype implementation of the proposed IH power supply system is depicted in Figure 8. Each bidirectional switch in this setup is made up of two diodes and two IGBTs. The different voltage and current waveforms are recorded using a digital storage oscilloscope (DSO) and a current sensor probe. Figure 9 presents the modelling and experimental findings concerning the input voltage and current, implying that the input current is devoid of the HF component created during high-switching frequency operation. Furthermore, the obtained input PF (experiment) is 0.92. Thus, it enhances the power quality of the input mains while also protecting against HF components, owing to the passive filter shown in Figure 2.

Figure 8. Experimental setup of the proposed IH power supply system.

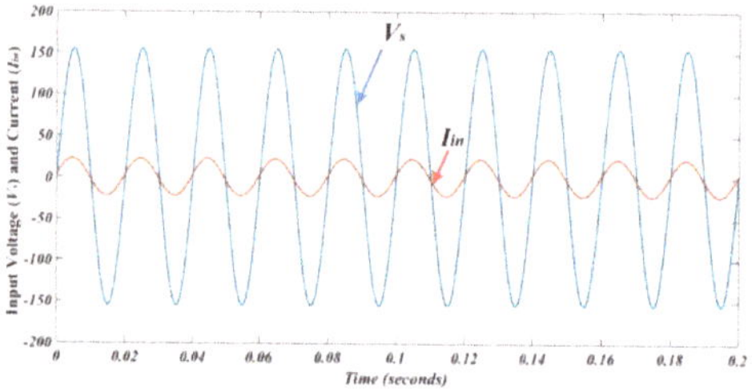

(**a**)

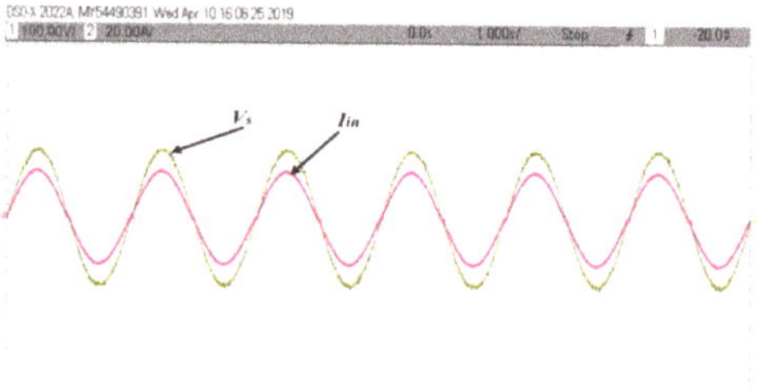

(**b**)

Figure 9. Input voltage and current: (**a**) Simulation result; (**b**) Experimental result (V_s: 100 V/div; I_{in}: 20 A/div).

A digital oscilloscope and current sensor captured the voltage and current waveforms. Figure 9 showcases the simulation and experimental results, demonstrating the input current's absence of high-frequency (HF) components typically generated by high switching frequencies. This translates to a measured power factor (PF) of 0.92, indicating efficient power utilization and reduced harmonic distortion. This improvement in power quality and HF component protection stems from the passive filter depicted in Figure 2.

Further analysis using a fast Fourier transform (FFT) on the input current (Figure 10) reveals a total harmonic distortion (THD) of 2.03% in simulations and 3.2% experimentally. Both values fall within acceptable ranges for IH applications. Additionally, the recorded root mean square (RMS) values for the input voltage and current are 110 V and 15.87 A, respectively.

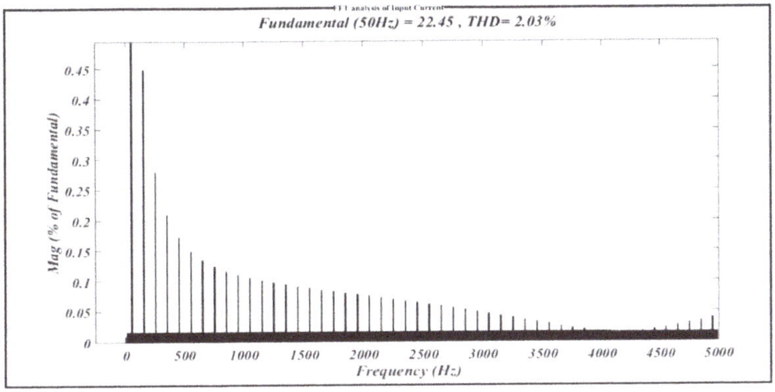

Figure 10. FFT spectrum of the input current.

The switching frequency of the gate signals, generated from the controller, is kept at 30 kHz, which is greater than the calculated resonant frequency using Equation (7), i.e., 24 kHz, to ensure ZVS condition. ZVS condition not only reduces the switching and power losses but increases the overall efficiency of the converter. The simulation and experimental results of synchronizing gate signals with the input mains through ZCD are shown in Figure 11.

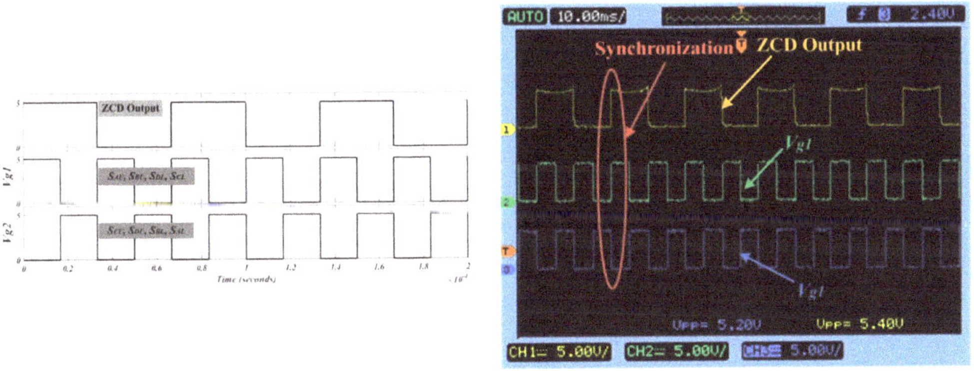

(a) (b)

Figure 11. Gate signals: (**a**) Simulation result; (**b**) Experimental result (V_{g1} and V_{g2}: 5 V/div).

In this work, the IH coil and its load is modelled as L_P ($\approx$52.7 µH) and R_L ($\approx$1 Ω) to study the electrical behaviour of the proposed converter. As is well known, when the converter functions as a resonant inverter, the voltage and current are always in the same phase. Consequently, it can be inferred from Figure 12 that the output voltage and current waveforms are in the same phase, both experimentally and through simulation. Figure 12 shows the simulation and experimental results of the output voltage and current across the IH load, having R.M.S values of 62.22 V and 19.79 A, respectively. The experimentally obtained PF between the output voltage and current was found to be 0.97 (almost unity), which shows the resonant property of the proposed converter. Therefore, the experimental average output power ($P_{o,(ave)}$) can be calculated as $P = VIcos\phi = 62.22 \times 19.79 \times 0.97 = 1194$ W (approximately). The average output power ($P_{o,(ave)}$) is shown in Figure 13a.

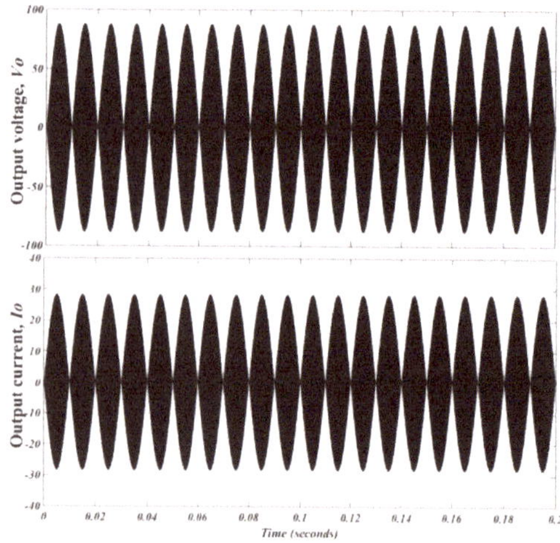

(**a**)

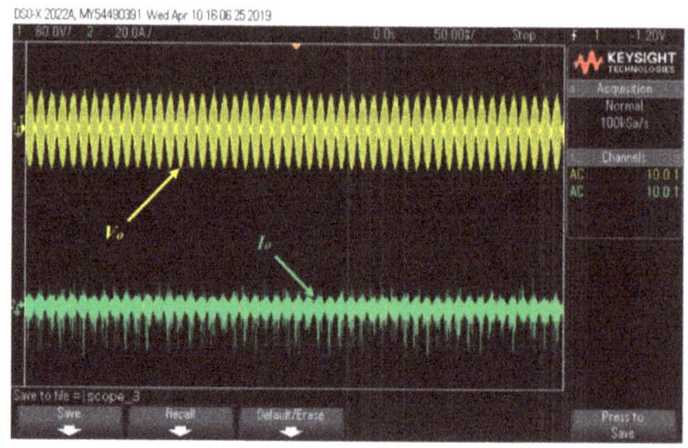

(**b**)

Figure 12. Input voltage and current results: (**a**) Simulation result; (**b**) Experimental result (V_o: 80 V/div; I_o: 20 A/div).

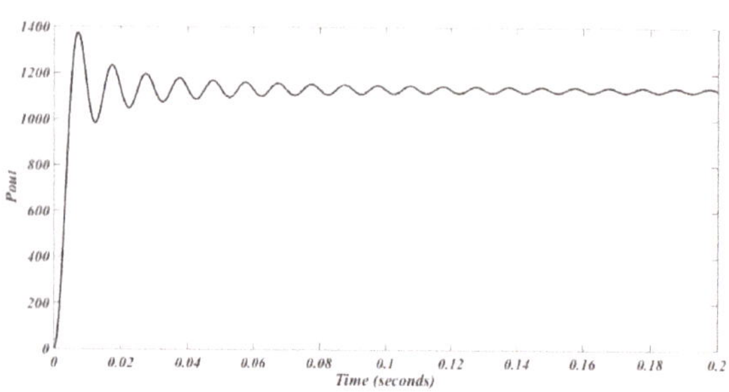

(a)

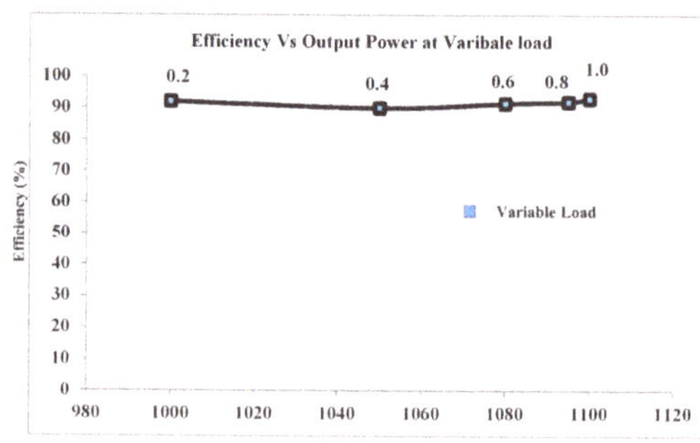

(**b**)

Figure 13. (**a**) Average output power (P_{out}) and (**b**) efficiency analysis.

Furthermore, an extensive efficiency analysis was carried out. As previously indicated, the inductance and resistance of the coil and its load (pot, vessel, etc.) are represented by L_P and R_L in a series connection, which represents the IH coil and its load. In order to perform an efficiency analysis, load resistance (R_L) is varied between 0.1 and 1 Ω, while L_P is maintained at a constant value. The power and efficiency of the input/output are calculated for each value of load resistance. Figure 13b displays the resulting output power and efficiency graph at a variable load. On average, 93% efficiency is attained.

As can be seen from the obtained simulation and experimental results, a comparison has been made between the proposed topology and the conventional topology for the IH system in terms of PF, THD, input filter design, controller complexity, and design costs and is shown in Table 4. As can be observed from this table, the proposed topology also improves the power quality in terms of THD and PF. Consequently, the proposed single-stage direct AC to HF AC LLC resonant converter can be effectively deployed in IH systems. This proposed converter's single drawback is its high switch count, which increases switching losses.

Table 4. Comparison table.

Comparison Criteria	Conventional Topologies for IH Systems [1,21,31]	Proposed Topology for IH Systems
Input PF	Low ~(0.7–0.8)	High ~(0.9–1)
THD	High	Low
Input filter design	Complex	Moderate
Switch stress	High	Low
Control complexity	Moderate	Easy
Design cost	High	Moderate

5. Conclusions

In this paper, an 1100 W single-stage direct AC–HF AC LLC resonant converter for IH systems has been proposed; it generates 30 kHz current directly from the utility 50 Hz power supply. An embedded control scheme (using Arduino 2560) is used to generate gate signals. The proposed converter is implemented with four bi-directional switches. The associated circuit operation and mathematical model has been explained to analyze the converter operation. Further, the potency of the proposed topology is validated through MATLAB simulation and experimental results. From the obtained results, the presented work has the following peculiar merits:

1. The presented topology converts utility grid-frequency AC to HF AC in a single stage and also maintains the input power factor close to unity simultaneously.
2. The implemented control scheme is based on an embedded system having a simple configuration and is easy to implement.
3. As LLC configuration is used, it reduces the current stress across the switches and also enables an intense current magnitude across the IH load simultaneously.
4. Finally, the proposed topology not only lowers the THD of the input current but also blocks the HF component that receives back flows towards the utility side. Thus, the power quality of the input mains improves.

Therefore, it can be concluded that the presented topology can be effectively applied in the IH application and can also be deployed where HF applications are required.

Author Contributions: Conceptualization, A.K. and A.G.; methodology, A.K.; software, A.K.; validation, P.K.S. and J.R.S.; formal analysis, A.K.; investigation, A.G.; resources, A.G.; data curation, J.R.S.; writing—original draft preparation, A.K.; writing—review and editing, A.G.; visualization, A.K.; supervision, P.K.S. and J.R.S.; project administration, P.K.S. and J.R.S.; funding acquisition, P.K.S. and J.R.S. All authors have read and agreed to the published version of the manuscript.

Funding: This research received no external funding.

Data Availability Statement: The data are contained within the article.

Acknowledgments: Authors acknowledge VIT Chennai and IIT (ISM) Dhanbad for the lab facilities.

Conflicts of Interest: The authors declare no conflicts of interest.

References

1. Lucía, O.; Maussion, P.; Dede, E.J.; Burdío, J.M. Induction heating technology and its applications: Past developments, current technology, and future challenges. *IEEE Trans. Ind. Electron.* **2013**, *61*, 2509–2520. [CrossRef]
2. Patil, M.; Choubey, R.K.; Jain, P.K. Influence of coil shapes on temperature distribution in induction heating process. *Mater. Today: Proc.* **2023**, *72*, 3029–3035. [CrossRef]
3. Acero, J.B.L.B.J.; Alonso, R.; Burdio, J.M.; Barragán, L.A.; Puyal, D. Analytical equivalent impedance for a planar circular induction heating system. *IEEE Trans. Magn.* **2005**, *42*, 84–86. [CrossRef]
4. Chaboudez, C.; Clain, S.; Glardon, R.; Rappaz, J.; Swierkosz, M.; Touzani, R. Numerical modelling of induction heating of long workpieces. *IEEE Trans. Magn.* **1994**, *30*, 5028–5037. [CrossRef]

5. Ngoc, H.P.; Fujita, H.; Ozaki, K.; Uchida, N. Phase angle control of high-frequency resonant currents in a multiple inverter system for zone-control induction heating. *IEEE Trans. Power Electron.* **2011**, *26*, 3357–3366. [CrossRef]
6. Musii, R.; Lis, M.; Pukach, P.; Chaban, A.; Szafraniec, A.; Vovk, M.; Melnyk, N. Analysis of Varying Temperature Regimes in a Conductive Strip during Induction Heating under a Quasi-Steady Electromagnetic Field. *Energies* **2024**, *17*, 366. [CrossRef]
7. Millán, I.; Burdío, J.M.; Acero, J.; Lucía, O.; Llorente, S. Series resonant inverter with selective harmonic operation applied to all-metal domestic induction heating. *IET Power Electron.* **2011**, *4*, 587–592. [CrossRef]
8. Dimitrov, B.; Hayatleh, K.; Barker, S.; Collier, G. Design, Analysis and Experimental Verification of the Self-Resonant Inverter for Induction Heating Crucible Melting Furnace Based on IGBTs Connected in Parallel. *Electricity* **2021**, *2*, 439–458. [CrossRef]
9. Neogi, K.; Sadhu, M.; Das, N.; Sadhu, P.K.; Chakraborty, A.; Ganguly, A.; Banerjee, A. A new approach for the stability analysis of high-frequency series resonant inverter-fitted induction heater. *Ain Shams Eng. J.* **2019**, *10*, 185–194. [CrossRef]
10. Aunsborg, T.S.; Duun, S.B.; Munk-Nielsen, S.; Uhrenfeldt, C. Development of a current source resonant inverter for high current MHz induction heating. *IET Power Electron.* **2022**, *15*, 1–10. [CrossRef]
11. Yu, A.; Zeng, X.; Xiong, D.; Tian, M.; Li, J. An Improved Autonomous Current-Fed Push-Pull Parallel-Resonant Inverter for Inductive Power Transfer System. *Energies* **2018**, *11*, 2653. [CrossRef]
12. Kim, N.G.; Jo, S.W.; Han, B.; Choi, H.H.; Kim, M. Highly efficient bidirectional current-fed resonant converter over a wide voltage gain range. *IEEE Trans. Ind. Electron.* **2020**, *68*, 10913–10927. [CrossRef]
13. Miyamae, M.; Ito, T.; Matsuse, K.; Tsukahara, M. Performance of a high frequency quasi-resonant inverter with variable-frequency output for induction heating. In Proceedings of the 7th International Power Electronics and Motion Control Conference, Harbin, China, 2–5 June 2012; pp. 2877–2882.
14. Mendonça, L.S.; Naidon, T.C.; Raposo, R.F.; Bisogno, F.E. An Unit-less Mathematical Model for Analysis and Design of Class-E Resonant Converters. In Proceedings of the 2019 IEEE 15th Brazilian Power Electronics Conference and 5th IEEE Southern Power Electronics Conference (COBEP/SPEC), Santos, Brazil, 1–4 December 2019; IEEE: Piscataway, NJ, USA, 2019; pp. 1–6.
15. Gomes, R.; Vitorino, M.A.; Acevedo-Bueno, D.; Correa, M. Multi-Phase Resonant Inverter with Coupled Coils for AC/AC Induction Heating Application. *IEEE Trans. Ind. Appl.* **2019**, *56*, 551–560. [CrossRef]
16. Devara, V.B.; Neti, V.; Maity, T.; Shunmugam, P. Capacitor-sharing two-output series-resonant inverter for induction cooking application. *IET Power Electron.* **2016**, *9*, 2240–2248. [CrossRef]
17. Esteve, V.; Jordán, J.; Dede, E.J.; Bellido, J.L. Enhanced asymmetrical modulation for half-bridge series resonant inverters in induction heating applications. *IET Power Electron.* **2023**, *16*, 2482–2491. [CrossRef]
18. Carretero, C.; Lucía, O.; Acero, J.; Burdío, J.M. Phase-shift control of dual half-bridge inverter feeding coupled loads for induction heating purposes. *Electron. Lett.* **2011**, *47*, 670–671. [CrossRef]
19. Shen, J.; Ma, H.; Yan, W.; Hui, J.; Wu, L. PDM and PSM Hybrid Power Control of a Series-Resonant Inverter for Induction Heating Applications. In Proceedings of the 2006 1ST IEEE Conference on Industrial Electronics and Applications, Singapore, 24–26 May 2006; pp. 1–6.
20. Burdio, J.M.; Barragan, L.A.; Monterde, F.; Navarro, D.; Acero, J. Asymmetrical voltage-cancellation control for full-bridge series resonant inverters. *IEEE Trans. Power Electron.* **2004**, *19*, 461–469. [CrossRef]
21. Meziane, B.; Zeroug, H. Comprehensive Power Control Performance Investigations of Resonant Inverter for Induction Metal Surface Hardening. *IEEE Trans. Ind. Electron.* **2016**, *63*, 6086–6096. [CrossRef]
22. Franco de Souza, A.; Ribeiro, E.R.; Vicente, E.M.; Tofoli, F.L. Experimental evaluation of active power factor correction techniques in a single-phase AC-DC boost converter. *Int. J. Circuit Theory Appl.* **2019**, *47*, 1529–1553. [CrossRef]
23. Gonçalves, J.T.; Valtchev, S.; Melicio, R.; Gonçalves, A.; Blaabjerg, F. Hybrid three-phase rectifiers with active power factor correction: A systematic review. *Electronics* **2021**, *10*, 1520. [CrossRef]
24. İnci, M.; Büyük, M.; Demir, M.H.; İlbey, G. A review and research on fuel cell electric vehicles: Topologies, power electronic converters, energy management methods, technical challenges, marketing and future aspects. *Renew. Sustain. Energy Rev.* **2021**, *137*, 110648. [CrossRef]
25. Yang, Y.; Wang, H.; Sangwongwanich, A.; Blaabjerg, F. Design for reliability of power electronic systems. In *Power Electronics Handbook*; Butterworth-Heinemann: Oxford, UK, 2018; pp. 1423–1440.
26. Pérez-Tarragona, M.; Sarnago, H.; Lucía, Ó.; Burdío, J.M. Design and Experimental Analysis of PFC Rectifiers for Domestic Induction Heating Applications. *IEEE Trans. Power Electron.* **2018**, *33*, 6582–6594. [CrossRef]
27. Espi-Huerta, J.M.; Santamaria, E.J.D.G.; Gil, R.G.; Castello-Moreno, J. Design of the L-LC Resonant Inverter for Induction Heating Based on Its Equivalent SRI. *IEEE Trans. Ind. Electron.* **2007**, *54*, 3178–3187. [CrossRef]
28. Zgraja, J.; Lisowski, G.; Kucharski, J. Autonomous Energy Matching Control in an LLC Induction Heating Generator. *Energies* **2020**, *13*, 1860. [CrossRef]
29. Chudjuarjeen, S.; Sangswang, A.; Koompai, C. LLC resonant inverter for induction heating with asymmetrical voltage-cancellation control. In Proceedings of the 2009 IEEE International Symposium on Circuits and Systems, Taipei, Taiwan, 24–27 May 2009; pp. 2874–2877.

30. Kranprakon, P.; Sangswang, A.; Naetiladdanon, S. Model predictive control of LLC resonant inverter for induction furnace. In Proceedings of the 2017 International Electrical Engineering Congress (iEECON), Pattaya, Thailand, 8–10 March 2017; IEEE: Piscataway, NJ, USA, 2017; pp. 1–4.
31. Vishnuram, P.; Ramachandiran, G.; Sudhakar Babu, T.; Nastasi, B. Induction heating in domestic cooking and industrial melting applications: A systematic review on modelling, converter topologies and control schemes. *Energies* **2021**, *14*, 6634. [CrossRef]

MDPI AG
Grosspeteranlage 5
4052 Basel
Switzerland
Tel.: +41 61 683 77 34

MDPI Books Editorial Office
E-mail: books@mdpi.com
www.mdpi.com/books

www.ingramcontent.com/pod-product-compliance
Lightning Source LLC
Chambersburg PA
CBHW061500220926
51850CB00004B/139
* 9 7 8 3 7 2 5 8 1 6 1 1 8 *